The Tube Amp Book

by Aspen Pittman

DELUXE REVISED EDITION

The Tube Amp Book
DELUXE REVISED EDITION

by **Aspen Pittman**

Edited by **Dave Hunter**

Technical consultant **Myles Rose**

A BACKBEAT BOOK
First edition 2003
Published by Backbeat Books
600 Harrison Street,
San Francisco, CA 94107, US
www.backbeatbooks.com

Copyright 1986, 1988, 1991, 1993, 1995, 2003 Aspen Pittman, Except chapters: The Wide World of Preamp Tubes, Rectifiers, Power Modelling chapters copyright 2003 Myles Rose; Thoughts on 12AX7 Type Tubes, Where It All Began copyright 2003 Mark Baier; What Not To Do, Maintenance Checklist, copyright 2003 Tom Mitchell; The Amplifier Signal Circuits copyright 2003 Jack Darr; The Last Word on Class A, Amp Terms Glossary copyright 2003 Randall Aiken; Ampeg Cathode Bias and other Ampeg, Marshall, and Fender mods copyright 2003 Ken Fischer ; Maximizing Silverface Fender Amps copyright 2003 Brinsley Schwarz; Vox AC30 Check-Up copyright 2003 David Petersen; Replacing Output Transformers copyright 2003 Doug Conley; Speakers & Speaker Cabs and selected Amp Companies contributions copyright 2003 Dave Hunter; Small Amps Vs. Big Amps in the Studio copyright 2003 Huw Price

Art Director: **Nigel Osborne**
Design: **Paul Cooper Design**
Editorial Director: **Tony Bacon**
Production: **Phil Richardson**

Origination by Hong Kong Scanner Arts
Print by Colorprint Offset

03 04 05 06 07 5 4 3 2 1

Contents

Introduction 4	**Section Two:**	Early 50W Bias/Standby Mod
	TECHNICAL ARTICLES 101	Hum Reduction Tip
Section One:		
TUBE AMP COMPANIES 5	General safety warning and disclaimer	**MODS FOR VOX** 145
	Behind The Tube Mystique	Vox AC30 Check-Up
Alessandro	Why Change Tubes?	
Ampeg	A Word About Changing Tubes	**THE LAST WORD ON CLASS A** 147
Ashdown	Tube Reference Guide: "What's the best tube	Replacing Output Transformers
Bad Cat	for my amp?"	Speakers & Speaker Cabs
Bogner	Report On Existing Tube Factories	Small Amps vs. Big Amps for Recording
Bruno	The Wide World of Preamp Tubes	Power Modelling: Suiting Amp Size to Venue
Budda	Thoughts on 12AX7 Tube Types	
Clark	Vacuum Tube Rectifiers	
Cornell	Our Philosophy on Amp Modifications	**Section Three:**
Cornford	A Word About Biasing Your Amp	**REFERENCE** 155
Danelectro & Silvertone	Biasing An Amp With Fixed Bias	Tube/Amp Replacement Guide
Dr Z	The Bias Probe	Tube Specs Sheets and Info
Dumble	Boogie Bias Mod	Tube Cross-Reference Guide
Engl	GT Survival Tips for Tube Amps	European Tube Nomenclature
Epiphone	What Not To Do	Directory of Tube Amp Companies
Fender	Maintenance Check List	Amp Terms Glossary
Garnet	The Amplifier Signal Circuits	
Gibson		
Gretsch	**MODS FOR AMPEGS** 132	**About the author** 191
Groove Tubes	Conversion Of 12DW7 To 12AX7	
Hammond & Leslie	Conversion Of Ampeg Amps from 7027a To	
Hawaiian Steel Guitar Amps	EL34	**Acknowledgements** 192
Hiwatt	Converting An SVT from 6164B/8289A To 6550	
Hughes & Kettner		**Section Four:**
Kelley	**MODS FOR FENDERS** 133	**SCHEMATIC DIAGRAMS**
K&M (Two Rock)	Post-Tweed Amps Noise Reduction	Reading Schematic Diagrams
Laney	Preamp Mod for Warmer, Cleaner Sound	Introduction to GT Schematics File
McIntosh	Conversion to EL43s	A-Z Indez of Printed Schematics
Magnatone	Maximizing Silverface Fenders: "Blackfacing"	
Marshall	and modification	**On The CD**
Park	Where It All Began: Essential Ingredients of	More than 800 schematic diagrams in
Matchless	the Tweed Tone	printable format, with active index
Mesa/Boogie		
Music Man	**MODS FOR HIWATTS** 144	
Orange	Hiwatt Bias Doubling Circuit	
Peavey		
Premier	**MODS FOR MARSHALLS** 144	
Rivera	100W to 50W	
Selmer	Converting 6550s to EL34s	
Soldano	Converting EL34s to 6550s	
Sound City	Marshall Master Volume Mod	
Supro	Transient Suppression (Tube Saver) Mod	
THD		
TopHat		
Trainwreck & Komet		
Traynor		
Valco		
Victoria		
Vox		
Watkins/WEM		
Wizzard		
TOP TEN ALL-TIME 93		
CLASSICS		
MODERN CLASSICS 99		

An Introductory thought

Guitar amps get no respect, or so it seemed to me after 25 years in the music business. Guitars get the girls, and the press. Amps got the shuffle, to the rear of the van. Change your strings weekly, change your tubes whenever the amp "blows," or... just get a new amp man.

My fellow tube amp fans argue that the amp is at least half of rock guitar. I disagree; to me it's more like 100 per cent. The electric guitar was around from the early '30s, but there couldn't *be* an electric guitar until the amp was invented! Unfortunately, as the new tool in the box the amp was treated just like an accessory... sometimes they even built them into the guitar cases to save space. But nobody needed (or wanted) to play that loud for Hawaiian music. The swing era also had electric guitar, but those guys played chords at a low volume in time with the drums to back up the horn solos. Nope, we didn't have rock until some crusty bluester cranked up (yes Leo, all the way to 12) that old Fender over the horns (and the drums)... then we had magic. The amp was the last missing piece of a 50-year-old, ever-evolving puzzle of music styles – and the electronic revolution that came out of radio – that finally came together in the '50s to give us rock'n'roll. It was then that the really incredible sounding tube amps began to emerge and become the headliners of the back lines of bands.

I was really intrigued to learn more about the beast, but the details about tube amps were hard to come by. There were far too many books on guitars – enough already! The TUBE AMP made it happen... where is their story told? I used to think I was alone in this revelation, but now – with 80,000 copies in print of five progressive editions dating back 20 years, the development of which has advanced a little 32-page publication into an 800-plus-page 4.1th edition and, finally, into this incredible *Deluxe Revised Edition* of the large-format full color Encyclopedia Tubemania you are holding in your hands – I guess I've got some company out there!

This *Tube Amp Book* is my personal obsession that keeps on growing. A day doesn't go by that I don't get a letter or call about it. Generally, the comments are kind and reassuring, but we get the occasional English major who rips us on everything from spelling to our homemade typesetting. I am embarrassed to say that for the first four versions I was my own editor, spellchecker, layout "artist" and photographer – but this time I think I am finally on to something. I have now partnered with Backbeat Books from London, who coincidentally have an all-English staff of professionals, with the exception of the transplanted all-American Editor Dave Hunter to cover my back, so they'd better get it right!

I guess the most rewarding result of writing *The Tube Amp Book* is the change it seems to have brought to the "respect level" of both vintage and modern amps. I have heard from dozens of young amp builders working for the big companies or out on their own who cite the TAB for their first inspiration to start building and designing their own amps. It was a sky hook for a generation of talented and passionate amp builders, and the players are also getting more out of their music than ever. When I started, the amp was the door stop, but now – almost 24 years after I founded Groove Tubes – they are treated with the respect they deserve, and are as important a signature to a player's tone and style as his guitar, maybe more!

Of course, there is a downside to this current popularity for vintage and/or boutique custom shop amps. I can no longer acquire a mint condition Fender Deluxe "not working" for $100 at a guitar store in LA, and a truly well-built modern classic from Bogner or Fender will cost as much as a Paul Reed Smith 10 Top with birds... but that is as it should be, in my humble opinion.

Thank you for your support and enjoy the book,

God bless, **Aspen Pittman**

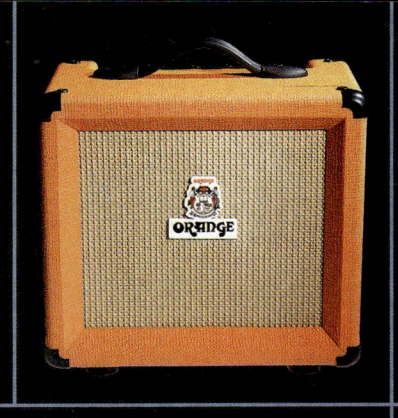

Tube Amp Companies

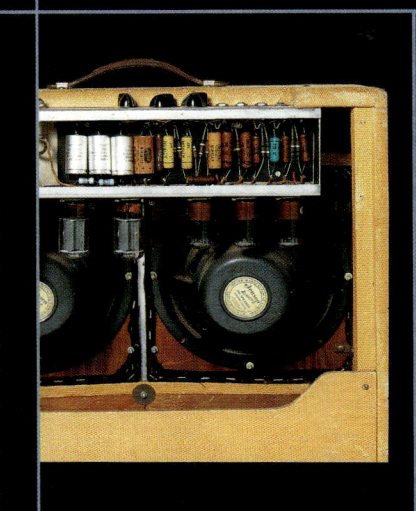

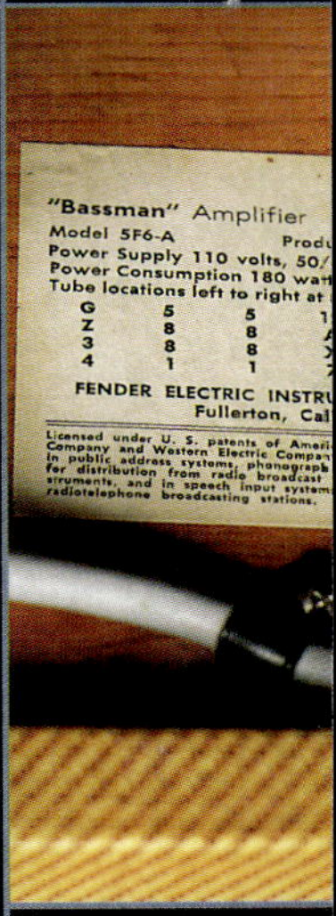

The Amp Companies

HUNDREDS of makes and models of tube amp have been sold over the last 60 years. I couldn't begin to list all the ones I've come across, but I do want to go into some detail about those that I think are the most interesting and those that have had a definite influence on our music. On addition, a lot of amps created by dedicated smaller builders – and some mass-manufacture brands – deserve a quick look, too.

Alessandro

George Alessandro's amps differ from many of those in what have come to be called the 'boutique' market, partly because his designs are less directly based on the usual classic vintage tube amps (tweed Fenders, Marshall JTMs, Vox AC30s…) than most others, and partly because he has so closely aligned his work with the esthetics of the high-end tube audio world. Alessandro amps, built in Pennsylvania, are meticulously crafted, all hand-wired using high-grade components and built in exotic hardwood cabinets. They are also interesting for offering a somewhat extreme list of optional upgrades: solid pure silver chassis, anyone? It'll only cost you an extra $4,000. Or if your taste runs more to gold, he'll build your amp in a solid gold chassis for a mere $50,000 premium. I'm not joking.

On a less cosmetic note, Alessandro takes his parts choices seriously, and components like high-end filter and signal capacitors, high-purity grain-oriented copper and silver conductors, and selected NOS tubes are standard for him. He also makes some interesting choices in tube type, using the octal-style 6SL7 dual triode in the preamps for his Beagle, Plott Hound and Black'n Tan amps, and the big, tight-sounding 6550 output tubes in some of his higher-powered amps such as the Bloodhound. Like that of others building tube amps to this standard, Alessandro's work is expensive, but a number of players have clearly decided it's worth it. For guitarists without a bottomless pit full of cash, he has also recently introduced his Working Dog series of relatively more affordable amps.

*These Alessandro tops and cabs – on show at the **1999** winter NAMM – display the builder's skill with highly figured hardwood cabinets.*

Ampeg

The Ampeg company was easily the most popular of the east coast amplifier companies. They gave Fender a run for their money during the 1950s and 1960s with a full line of self-contained amps, as well as some more powerful piggyback units brought out during the 1970s.

My personal favorites of the older models are the 1-12" Jet and Rocket amps. These amps use two 7591 power tubes and put out about 30W. The 7591 tube is similar to the 6V6 but has a cleaner tone and produces a better sound for jazz or country music, and was also very popular in tube hi-fi systems at this time. These amps later became available with reverb, and I must say that these old Ampegs' reverb sound is the best I've ever heard. These amps are called the Jet-12R and the Reverb-o-Rocket. I've found these amps usually sell for between $100 and $200 and make an excellent value in a small club amp or a practice amp.

Not quite rock'n'roll, but the "amplified peg" for double bass gave Ampeg its name, and the company has remained primarily a bass and bass-amp specialist ever since.

This is one of the oldest Ampeg amps we've seen. Simply called the 'Bassamp,' it has 35W with 1-15" speaker. Courtesy Andy Brauer

6 THE TUBE AMP BOOK

Ampeg

The Ampeg B-15S was the last evolution of the tube-powered Portaflex bass amps – with 60W and 1-15", it records great. Courtesy of Leeds Rentals.

The Ampeg SVT puts out 300W from a sextet of 6550s and is the undeniable king of bass amps (with full 8x10" cab middle photo above, head alone below it). Amps courtesy of Leeds Rentals

Ampeg's B-15N Portaflex – featured in the catalog page pictured right – was a popular choice for both club-sized gigs and studio work. Before giants like the SVT came along, it was also one of the better-designed bass amps of its day, and still packs good punch and a great tone where lower volume is required.

Another favorite series of amps from Ampeg were the Portaflex amps made in the 1960s and 1970s (and again after that in a made-in-Japan copy of the originals). The basic idea behind the design of the Portaflex amps was that they could be conveniently packed up for travel by flipping the amp top over into the speaker cabinet, which gives the circuitry extra protection within the cabinet. When you were ready to play, you just flipped the top over and clipped it to the top of the cabinet.

Perhaps the most famous of these Portaflex amps are those made for bass guitar. These are the B-15, B-15N, B-15S, and B-18. They were built with 15" bass speakers and put out between 40W and 60W, depending on the model. The B-15S was the final production of the bunch and had a pair of 7027 power tubes that could be upgraded to 6550s. The B-15S is also the most powerful of the range and is definitely collectable. These amps were the standard for recording bass in the studio for years, and are still used today to get that special round tone that only they can get.

Perhaps the most famous bass amp Ampeg ever made for rock'n'roll was the SVT. It has a whopping 300W and can drive two cabinets with 8-10" specially designed speakers in a sealed cabinet. At the time of its introduction the AVT was a radical departure from anything bass players had seen before. This bass amp has punch, to say the least, and became popular when the Rolling Stones toured worldwide with the amp in the early 1970s.

The SVT originally used six

This early '60s Ampeg Reverb-o-Rocket had 1-12" Jensen and 35W. They have about the best reverb I have ever heard, and were popular with jazz guitarists. Amp and mid '50s Ricky Combo guitar from the GT Collection, photo by Jeff Veitch.

THE TUBE AMP BOOK 7

Ampeg

6146b power tubes but these were a hard-to-find variety, and Ampeg soon changed to the newer 6550a, made by GE. If you have an SVT with 6146s, we highly recommend that you convert to the 6550 (see our Amp Servicing section for complete details on this modification).

The SVT gets the Groove Tubes award for the 'most tubes used in any amp' – there are 14 tubes in total. One of the preamp tubes used in the SVT was the 12DW7 tube. The reason I say 'was' is that the 12DW7 tube has been discontinued by the only company which was still making it.

The 12DW7 is actually half a 12AX7 and half a 12AU7; that is to say, it is built so that the two plates inside the dual-triode tube have different specifications from each other.

We have included a simple modification which involves changing three resistors so the amp will run with 12AX7s instead (see Amp Servicing and Modification section. Also see Top Ten All Time Classics for more information on the SVT).

St Louis Music has the rights to the Ampeg name now, and has reissued a lot of the great models of the past – though none of these is built strictly to vintage specifications. Even so, the name should be welcomed back to the tube amp world, and the range includes some good-sounding models for both bass and guitar.

The Super Rocket Reverb, above, is a newer guitar combo from St Louis Music's revived Ampeg brand, with 100W from a quartet of 6L6s, channel switching, reverb and vibrato. Photo courtesy of The Guitar Magazine.

Ashdown

Britain's Ashdown Engineering has been known for its powerful, punchy bass amps for some time, but entered the tube guitar amp market with a big fanfare in 2001 with the release of their hand-wired Peacemaker 50 and 100 stacks and combos. The former yields 50W from a 6550 duet, while the latter gets 100W from a quartet of EL34s. Both use the same fairly straightforward, switchable dual-gain preamp with shared EQ. One groovy feature – previously seen on Ashdown bass amps, too – is a front-panel VU meter that reads the output level in standard mode, but can be manually switched to read the bias level of the corresponding output valve for easy re-biasing.

Ashdown quickly followed up the big Peacemakers with a range of smaller, more affordable PCB-based combos, the Peacemaker 20 and 40, built around an EL84 duet and quartet respectively. Then, in 2003, they offered the Fallen Angel range of mid-priced high-gain tube amps aimed at alt-rockers and nu-metal players.

Ashdown's Peacemaker 40 combo, right, is packed with modern features, and brings a taste of their high-end, hand-wired Peacemaker amp heads to lower budgets. It features a quartet of EL84s, and is built around a printed circuit board (PCB). Photo courtesy of The Guitar Magazine.

Bad Cat

Although they never mention the word themselves, Bad Cat sprang up in the wake of Matchless's temporary closure in 1999. Although it has moved on in some of its own directions, its roots certainly begin with many designs familiar to Matchless fans. This is obviously more than coincidence – but it's easily justified, too. While conducting his early research to set up the company, founder/president James Heidrich began contacting engineers in the amp industry and interviewing artists, and, as he puts it himself, "All roads seemed to lead to Mark [Sampson]." Heidrich hired Mark Sampson – co-founder of Matchless in 1989 with Rick Perrotta – as a design consultant; soon after, he brought Perrotta on board to oversee production. By the time Bad Cat amps hit the stores in 2001, Matchless was up and running again, but its founders were firmly entrenched in another company.

"In my own search for 'the tone,' I found that a class A circuit using the arduous and expensive point-to-point wiring technique produced the richest, most sonically pleasing results of any amplification design out there," says Heidrich. "Not every class A design is a good one, however. It takes the technical prowess, ears and yes, imagination of a master to come up with a circuit that can truly be called a classic." While early versions of Bad Cat models such as the Black Cat and the Cub were almost indistinguishable from their inspirations – the D/C30 and Lightning respectively – barring the logo badges (one early Cub 15R inspected was even kitted out with Matchless-branded signal caps!), the designs of these had already begun to evolve at the time of printing, as well as being joined by original new models like the high-gain Hot Cat.

Whatever its roots, Bad Cat is looking like an amp company to be reckoned with, and is fast garnering rave reviews and new players alike.

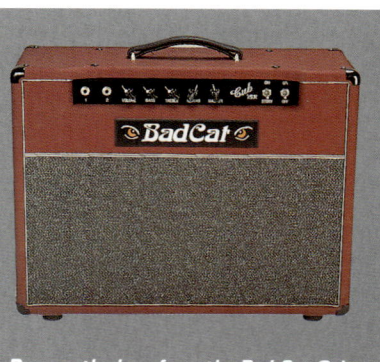

Remove the logo from the Bad Cat Cub 15R, above, and it looks a lot like a 2x10" Matchless Lightning. In fact, both are great-sounding, point-to-point, 15W class A amps based around the same designed by Mark Sampson.

8 THE TUBE AMP BOOK

Bogner

After packing up his basket of high-gain rock amp tricks in 1989 and relocating from Germany to Los Angeles, Reinhold Bogner quickly gained a notable and varied client list – including a wide range of major players like Steve Vai, Allan Holdsworth, Steve Stevens, Dan Huff and others – who came to him for repairs, mods, and custom building of original amps. His reputation really soared, though, after he 'revitalized' Edward Van Halen's favorite Marshall Plexi; off the back of that success, Bogner Amplification was born.

Today, Bogner is a major name in high-end, mega-fierce tube rockers. Notable models include the Ecstasy 100W head or 2x12" combo, with either 4x6L6s or 4xEL34s, and three different vintage-voiced (but definitely max-gain) channels; the two-channel Shiva; the simpler, EL84-based 15W or 30W single-channel Metropolis; or the ultra-heavy 120W 4xEL34 Uberschall.

Bogner amps are in no way shy, retiring creatures, to say the least; and Reinhold's promotional prose is just as in-your-face as his styling and preamp designs. Of the latter amp, he says: "Uberschall, the German word for 'Super Sonic,' is really Armageddon in a box! Everything this amp touches is Chernobylized! … Imagine a Ferrari stuck in sixth gear, going 212 mph, and slamming into a nuclear reactor. This is how the Uberschall sounds even before you turn it on." A warning to apartment dwellers: probably not neighbor-friendly, then.

Bruno

New Yorker Tony Bruno bases his models broadly on the classics from Vox, Fender and Marshall, though he approaches each with the epitome of hand-craftsmanship and personal care.

He further distinguishes models by mixing-and-matching some 'vintage' style features for more original results – as seen in his ultra-expensive 10th Anniversary Underground 30: essentially an upgraded AC30-type head, with a three-knob Fender-type tube reverb in front of the preamp stage, plus tremolo. Pretty cool.

Bruno uses only high-grade premium and NOS components, and wires, builds and finishes ever amp himself. More unusual than this, he also goes the extra mile to play each amp personally, by itself and with a full band, to ensure satisfactory results. And get this: he even uses a 'mixed-audience' test to – as he puts it – 'woman proof' every amp because of a woman's more sensitive high hearing!

Other Bruno models include the standard Underground 30, the Tweedy Pie 18 & 35 (based on the tweed Deluxe and Super respectively), Cow Tipper 22, 35, 45 & 90 (blackface Deluxe Reverb, Pro, Super, and Twin Reverb-style circuits), Pony 50 and 100 (Plexis), and others, plus more original custom designs.

Budda

While most channel-switching amps let you jump between preamp stages with different gain settings, Budda's more unusual Dual Stage 30 feeds a 3x12AX7 preamp into two different selectable tube output stages. Photo courtesy of The Guitar Magazine.

Budda Amplification is headed by Scott Sier, Jeff Bober and Dan van Riesen of Zen Engineering (in a past incarnation, Riesen designed the Triaxis preamp and V-Twin pedal for fellow Californians Mesa/Boogie). Behind their distinctive purple panels, many Budda amps mark a practical compromise between the single-channel so-called 'boutique' amps and the modern channel-switching do-it-alls that many players feel they require today. To achieve this, Budda mixes point-to-point wiring, hand-wired circuit boards and some PCB construction to make its amps.

The Budda Dual Stage 30, designed by Bober, offers an unusual spin on most channel-switching amps, with a fairly straightforward (though versatile) 3x12AX7 preamp section feeding into a switchable dual-stage output section. You select its four EL84s for compressed crunch or creamy, British-flavored leads, or two 6L6s for tighter American sounds.

Both are in cathode-biased class A, and available at the stomp of a footswitch – a real twist on the usual footswitchable clean/high-gain preamps. The 18W Twinmaster, on the other hand, is a straightforward 2x10" combo based around an EL84 duet output, with only volume, treble and bass controls, plus low and high gain inputs.

The latter cleverly adds a tube stage to overdrive the normal input's more standard preamp stage – fairly unusual. A fun look, interesting designs, and good sounding tube amps.

Clark

Michael Clark's impressive, accurate tweed-era Fender replicas have earned him a lot of respect in the few years since his work has become more widely known. The deserved kudos are the result of his hand-building every amp himself, using top quality, often military grade or NOS parts, as well as constructing and hand finishing his own finger-jointed cabs from solid #1 grade southern yellow pine. Time-consuming work, but it helped earn this builder from West Columbia, South Carolina, a coveted 'Editor's Pick Award' from Guitar Player magazine for his Tyger, an homage to the 35W 3x10" tweed Bandmaster.

Clark models cover most of the later 1950s tweed Fenders, among them the Piedmont, a 45W 5F6-A-style 4x10" Bassman; the Beaufort, an 18W 5E3 Deluxe-style 1x12"; and the 80W 2x12" Twin-style Low Country. He has also ventured into more original designs, like the Anthony Wilson Model, first built for Diana Krall's guitarist as a custom-order and now offered as a standard model. The 35W combo uses two 7581A/6L6GC in fixed bias, one 12AX7, one 12AY7, and has solid state rectification. The main features are simple: just normal and bright channel volumes with a shared tone control, plus tube reverb – all belted out through an EV12L for increased punch and clean headroom. And like the rest of Mike's amps, it's covered by an impressive 10-year warranty, which is transferable to a subsequent owner should you sell on the amp. As if you would.

Cornell

Based near Southend-on-Sea in Essex, England, Denis Cornell has been building great-sounding custom-order and small-run production amps since the early '90s under the banner of DC Developments, and designed and built well known amps for larger British firms for years before that. His work has been much-loved by British players, without Denis himself ever quite getting the recognition he deserves.

After doing an apprenticeship with a division of the Pye/Phillips group, Denis went to work for Dallas Musical Instruments (part of Dallas-Arbiter) in 1969. This eventually found him working on the design and development of a number of early '70s Sound City amps – a brand which DMI owned – such as the Sound City 50 Plus, 120W and 200W heads, and the 50W Concord amplifier. When Dallas-Arbiter acquired the Vox name in the mid '70s, Denis worked with Tom Jennings on the relaunch of Vox. In addition, he has worked throughout his career as a technical consultant to Fender Europe and Fender's UK distributor, Arbiter.

Cornell's standard range of tube amps – all of them hand-wired using quality components – runs from the 10W single-ended 6L6-based Romany combo (born out of a tube amp kit project, called the Stinger, originally developed in 1999 for readers of The Guitar Magazine in the UK), through the 15W Rambler and 35W Journeyman, to the 40W or 80W Voyager (the latter using a pair of big KT88 output tubes). All are classically Brit-flavored, but with plenty of original twists, such as optional on-board power soaks, or the occasional use of the less-seen EF86 pentode preamp tube. In addition to these, Cornell has recently put his 'Eric Amp' on the market, a production version of two 80W tweed Fender Twin-style reproductions which he custom-built at Eric Clapton's request when the guitar legend's original vintage Twins "couldn't cut it" any longer. Clapton first used the Cornell reproductions in the Queen's 'Golden Jubilee' concerts, and has taken them out on the road since then.

Denis Cornell's 'Eric Amp,' based on a pair of custom order combos originally built for Eric Clapton himself, follows the template of a narrow-panel tweed Fender Twin, with a few clever twists.

This Cornell Journeyman, left, from the mid '90s, is a sturdy and toneful hand-wired combo for the gigging guitarist, based around the popular output stage of a quartet of EL84s.

Cornford

Cornford Amplifiers was founded by British guitarist and amp fanatic Paul Cornford, with help from fellow designer Martin Kidd, after years of self-acknowledged "messing around with amps trying to find 'the tone.'" In its relatively brief existence, it has become one of the highest regarded small-run, hand-made amp builders in the UK in its relatively brief existence.

While its bigger amps, such as the 50W (optional 100W) MK50H or 35W Hell Cat, look like upmarket Marshall or Vox-types at first glance, Cornford has taken a number of classic designs and run in very much its own direction. The result is a vibe that combines US and UK tones, and often bridges the two. Their aforementioned flagship MK50H is a 'single channel' but essentially dual dual-voiced class AB 50W head with four 12AX7s in the preamp and a dual 5881 (6L6GC) output section and footswitchable overdrive. It's therefore very different from the EL34-based designs that most Brit makers still use for their larger amps, and manages a smooth run from biting, shimmering Fender clean tones to smooth, warm Marshall-type overdrive (virtues that have made it a favorite amp of blues-rocker Gary Moore in recent years).

Other models include the 6W 1xEL84 single-ended class A Harlequin, and the 20W 2xEL84 class A/B Hurricane combos.

Cornford's 20W 1x12" Hurricane combo features hand-wiring and a solid pine cab. Amp courtesy of Chandler Guitars, photo by Miki Slingsby.

Danelectro & Silvertone

Nathan I Daniels' Danelectro company made amps and guitars in Red Bank, New Jersey, from 1953 to 1958, and then in Neptune, New Jersey, from 1958 to 1969, when the original company went under. It also produced OEM products for chain stores like Sears (Silvertone) and Montgomery Ward. (Note that the reissue Danelectro amps built in the far east are currently all solid state models, and nothing like the originals.)

Our oldest Dano amp is an unusual early 1950s Twin Twelve Series CA (pictured left), with twin Jensens, twin 6L6 duets and even twin output transformers! But to me, the most amazing thing about this amp is the 3/4" solid paper board material its cabinet is made out of. Still sounds good, but I'd avoid outdoor concerts during the rainy season. The other unique feature is the location, or rather the angle, of the control panel/chassis, which is mounted diagonally while the controls read horizontally. I've never seen another amp laid out like it.

Another of our Danelectros, the Model 42, had 1-12" speaker, 25W and vibrato. The

This old Danelectro Twin Twelve is one of the more unusual amp designs you will see. Cool styling, but the diagonal chassis – with control legends that remain horizontal – can't have been the most practical format from a manufacturing perspective.

Here's a couple of my favorites: a 1950s Danelectro Convertable guitar that could also be played acoustically coupled with a neat example of the Danelectro amplifier line, the Model 42. Both from the GT Collection, photo by Jeff Veitch.

THE TUBE AMP BOOK 11

Danelectro & Silvertone

This mid 1960s Silvertone model 1484 Twin Twelve is another cool and unusual design from the Danelectro camp, and this time a little more practical than the 'diagonal' older Dano model.

This amp has had a resurgence of popularity with newer grunge and alt-rock bands, too. Photo by Aspen Pittman.

nicest thing about it, however, is the two-tone 1950s styling.

The Silvertone amps were kinda like the poor man's Fender. Available from the Sears & Roebuck department stores for very reasonable prices, they could also be bought through the Sears mail order catalog. There are many Silvertone amps that I could have shown, but the mid 1960s model 1484 Twin Twelve pictured here is my favorite. It has twin 12" blue-frame alnico Jensens, a duet of 6L6s, reverb (kinda cheesy sounding, though), and best of all, the head fits into the lower rear of the speaker cabinet for compact traveling. If you've ever cranked one of these up, I don't have to tell you what a great blues tone it gets. I'm sure many a rising star began careers playing through Silvertone amps, and if you're looking for that affordable first vintage amp, check one out.

Dr Z

Ever since he made his first original design in 1988, Dr Z's amps have been gaining respect around the world and a growing roster of professional players who swear by them. "Call it good fortune, or call it destiny, my father was a radio/TV repair man in the mid 1950s and early 1960s, and this is what drew me into the world of glowing glass," he says. The good Dr – known as Mike Zaite to the non-amp world – began his training in all things tubular as a teenager by performing "unauthorized and un-requested" modifications on his bandmates' amps that were stored in his basement, and eventually earned a college degree in electronics. After years of working on more modern devices, he was inevitably drawn back to the world of the tube amp. A few tweaks down the road, that first original amp of 1988 ended up on stage with Joe Walsh of The Eagles for the band's 'Hell Freezes Over' tour. Not a bad start in the biz, and he has never looked back.

At the time of going to print the Dr Z line includes 11 different amp models and four unique speaker enclosures. Though they doubtless derive elements from a few classic designs, the majority of these are more original than the amps of many other great, new hand-built makes. From the affordable 18W Carmen Ghia to the unique and high-end 32W Route 66, a minimalist look remains a major Dr Z trademark. With only volume and tone controls, the 2xEL84 Carmen Ghia makes a great-sounding, hand-wired tube amp available to the gigging guitarist at less than $1,000 for the amp top version; offering the unique coupling of an EF86 preamp into a KT-66 duet power stage, the Route 66 still features just volume, bass and treble controls, which were nevertheless enough to earn it the first ever 'Editor's Pick Award' from Guitar Player magazine.

Don't let the minimalism fool you into thinking these designs are run of the mill, however: the simple-seeming Carmen Ghia features a unique circuit for its single tone control, designed to make finding the sweet spot for any guitar a snap, along with a conjunctive filter and fixed DC biased phase inverter – elements you don't see much on other guitar amps.

Dr Z thanks his father and Ken Fisher for his success – oh, and Edition II of The Tube Amp Book, which he "read till the pages literally disintegrated." This edition should hold together longer, Doc!

12 THE TUBE AMP BOOK

Dumble

The Dumble Overdrive Special is one of the most famous – and expensive – 'boutique' amps of all time. Photo by Aspen Pittman

These are perhaps the rarest and most expensive boutique amps on the market, if you could even classify them as 'on the market' at all. You will never see an ad for one. To my knowledge there has never been a Dumble 'dealer.' Most all are are sold directly by Dumble to famous end users at extremely high prices.

The most known model is the Overdrive Special, which has a normal channel and a switchable high gain channel and uses 4x6L6s (early models) or 4xEL34s (later models) in a common push/pull circuit design to produce somewhere between 100W and 150W. The only other well-known model is the Steel String Singer, which is a more straightforward but higher-powered design using multiple 6550 tubes with lots of clean headroom, much like a Marshall Major.

There is little consistency between the Dumble amps I have seen. Some have been really great sounding amps, while others left me flat. Repairing one is a real nightmare as the reclusive tech guru guards his 'secrets' by pouring tubes of black silicone over the finished circuit board to discourage 'counterfeiters' (as he calls them).

Dumble amps are hand-made by a reclusive amp tech by name of Howard Alexander Dumble; there is no actual production line as such. Howard started modifying blackface or tweed style Fender amps, and later developed into making a few random Fender amp 'reproductions' in the late 1960s, adding some extra tricks like fuzz tone or an extra gain stage. This was similar to early Boogie amps and any number of 'modded' Fender amps that many techs were producing at that time.

But Dumble's amps have gained some fame over the years through being used by some pretty well-known musicians. Right place, right time in the career of a Larry Carlton or a Robin Ford and – *presto* – the legendary Dumble amp was born. However, unlike Boogie and the others, Howard never had a desire to become a businessman, so he has remained the reclusive 'amp tech to the stars,' and is reportedly very difficult to contact.

His production is very small. Informed estimates are that there have been fewer than 250 amps, of all models, produced to date. His reclusive lifestyle and elitist attitude about whom he will accept orders from further reinforce this limited availability. In fact you must send him tapes of your playing if you are not a known artist, and you'd better be a good guitarist!

On top of this, his standard sales policy of payment in advance and long delivery timetables have driven up prices for his amps to spectacular levels. A used Dumble Overdrive Special for example can today sell for $10,000+ for a head, and $14,000 for a combo!

Sadly, because of this many 'Dumble Copy' amps have sprung up recently from makers who actually advertise themselves as copies of the hard-to-find amp. These companies are making reasonable sounding 'Dumble' amps for much less money than a true Dumble, and much more money than a nice reissue Twin Reverb… it's a crazy world!

In the final analysis, these are basically nice sounding amps based on modified Fender circuit designs. The build quality of those I have examined is average. Having seen and repaired quite a few Dumbles over the years, I can tell you the components are, in most cases, off-shelf transformers with surplus NOS 'Orange Drop' capacitors and NOS tubes, and they are packaged in rather simple Tolex covered plywood boxes with off-shelf speakers. But then again, so were the early tweed Fenders and/or early Plexi JTM45 Marshalls. I love this business… and to think my Mom wanted me to play the clarinet and become a lawyer.

Engl

For years this German maker has been selling amps that are much loved by players within their homeland and have occasionally earned a devoted follower from over the border, too.

Ritchie Blackmore, for one, has called Engl amps "The best I have ever used" – and Engl have built him a Signature Model to express its thanks.

On the whole, Engl amps lean heavily toward the rock side of the market: high-gain channel switchers, with lots of knobs. In addition to the Blackmore Signature, Engl's chief designer, Horst Langer, has produced the Savage head, a 120W, 4-channel amp based on a 6550 output stage, and the Classic Tube 50, producing 50W from a 5881 duet.

The latter is relatively simple for Engl, in terms of its features, including only two footswitchable gain levels and a mere nine knobs with which to control them.

Beyond the guitar amps, Engl also offers a wide range of tube-powered pedals and rack devices.

Epiphone

Far rarer than their guitars, Epiphone amps have nevertheless had a place in the spotlight over the years. The Electar Zephyr was a fixture on many bandstands in the swing era of the 1940s (and certainly looks the part, with its stylized, scroll 'E' on wooden fretwork in front of the grille). In the 1960s, amps like the EA-32RVT Comet and EA-33RVT Galaxie – both made by Gibson in Kalamazoo, Michigan – found a few fans, too.

Of all these, though, the Epiphone Professional was certainly the most innovative for its time. It illustrates the epitome of the 'inseparability' of the guitar and amp: all of the amp's controls for volume, tone, reverb and tremolo are located on the pickguard of the guitar, and connected to the amp by a large multipin cable. There's even a five-section 'Tonexpressor' switch! This was a fairly rare package, even during the early 1960s when it was made.

In this old ad, session guitarist Al Caiola displays one of the archtop jazz boxes that Epiphone was still best known for in the day, with an Epiphone Electar amp getting second billing below it.

This Epiphone Professional combination is really advanced and innovative for its time. All of the amp's controls for volume, tone, reverb and tremolo are on the pickguard of the guitar and connected by a large multipin cable to the amp. There's even a five-section 'Tonexpressor' switch panel on the pickguard. The only control on the amp is the on/off switch! This was a fairly rare package, even during the early 1960s when this unit was made. I bought this package from my pal Dan at Norm's Rare Guitars, Reseda, California. Photo by Jeff Veitch.

14 THE TUBE AMP BOOK

Fender

The 1945 K&F amp was painted gray wrinkle, had a leather strap handle, coarse metal mesh grill, two inputs, an 8" Jensen speaker and no controls at all. Sounding surprisingly good, this amp was a gift to the GT Collection from Red Rhodes.

The first amp to bear Leo Fender's name – the Model 26 – is shown in two variations above. It was available with a 6", 10", or 15" Jensen speaker. The cabinet and matching handle were natural finished hardwoods like mahogany and walnut. These beauties are owned by John Sprung of American Guitar Center/Parts Is Parts (specialist in vintage instruments) who also took the picture.

The first production amps from Leo Fender were made in Fullerton, California, some time around late 1945 or early 1946. He had, however, made some guitars and amps on a custom basis for musicians during World War II, from his radio shop. Exact details are sketchy, but we do know that these early amps, and the matching lap steel guitars, were designed and manufactured in collaboration with Doc Kauffman, a friend and fellow engineer.

Accordingly, these early products were made under the name of the K&F Manufacturing Corporation, using a beautiful logo of a K&F surrounding a lightning bolt with a bass and treble clef in the center of a red circle. This was very intricate for the 1940s. The K&F amps had two unlabeled inputs and no volume or tone controls. This was because the amps were packaged with the matching lap steel guitar and the controls were on the guitar, which was typical of the way electric guitars and amps were viewed at this time – as a total package, and not as separates.

Leo Fender began making amps under the Fender name from 1947. The early amp designs were nothing new or unusual, but rather were straight out of the RCA tube application manuals that were widely available at the time. Most of my early Fender amps carry the words "Licensed under U.S. Patents of American Telephone and Telegraph Company and Western Electric Company". The first actual Fender amp series were the Model 26s. They

A rare couple, the 1946 Fender Model 26 with matching lap steel on the right, with a mid 1950s White amp and matching steel on the left. The Model 26 was the first amp and guitar set that carried Fender's name. The White is an obscure Fender combo, named in respect for Forrest White, a principal Fender engineer and designer responsible for many of the innovative Fender designs. From the GT Collection, photo by Jeff Veitch.

THE TUBE AMP BOOK 15

Fender

This 1949 Champion 600 has a two tone vinyl covering, a single volume control, two inputs and 6" speaker. GT Collection

came in hardwood cabinets with a natural wood finish, and even had a custom made wood handle! Red, Blue or Gold colored velvet cloth with three thin chromed steel strips were used to cover the speaker baffle board.

The Model 26 was available with 8", 10" and 15" Jensen speakers, but the most popular was probably the 10" version. This amp section used two 6V6 output tubes and a 5Y3 rectifier. The preamp section employed a 6SC7 (an octal fore-runner of the 12AX7) and a 6N7 driver. The Jensen speakers used on these early Model 26s were the type with the field output transformer mounted to the frame of the speaker.

The amp has a very warm, sweet tone, and the Model 26 I have is easily one of my best-sounding amps. This design is the fore-runner of the tweed Deluxe amp of the 1950s, which many famous guitarists have used for recording.

Fender soon began to fill out his line of amps, and added smaller models called Champions, later shortened to Champ. The Champion 400 and 600 models had just one 6V6 output tube, and were usually sold as beginner amps for students. The other interesting feature about the Champion amps was that, for the first time, Fender used vinyl covering in a two-tone color scheme of a dark brown leatherette on the sides and a white leatherette for the center panel of the amp.

The grille cloth was a matching dark brown cloth. The control panel was dark brown with cream colored legends – and just one volume knob, but it goes to 12! The overall look reminds me of a pair of 1950s saddle shoes.

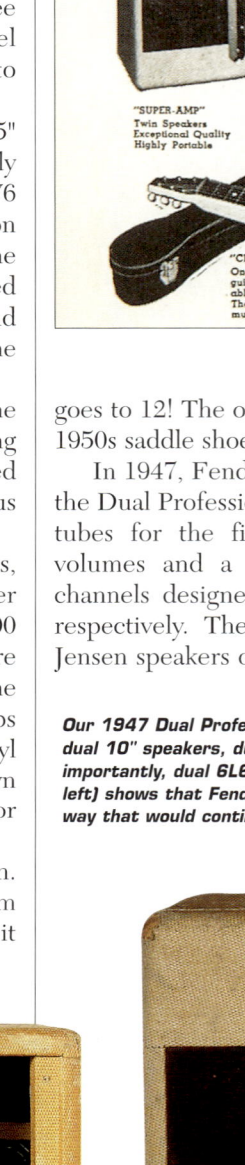

In 1947, Fender brought out a larger amp called the Dual Professional, and used a pair of 6L6 output tubes for the first time. Controls included two volumes and a shared tone, with two separate channels designed for a microphone and a guitar respectively. The Dual Professional had two 10" Jensen speakers on what were actually two separate

Our 1947 Dual Professional amp was the first Fender to offer dual 10" speakers, dual channels for guitar and voice and, most importantly, dual 6L6s. A look inside the back of the amp (below-left) shows that Fender was already wiring their circuits in a way that would continue throughout the 1950s. GT Collection.

16 THE TUBE AMP BOOK

Fender

TWIN AMP

The Fender Twin Amp is an Amplifier featuring the latest in electronic advances plus offering physical advantages over any single speaker amplifier.

It is capable of tremendous distortionless power and wide range tone characteristics which make it the favorite of topflight musicians.

The Twin Amp is housed in a beautiful three-quarter inch solid stock, lock jointed cabinet, covered with the highest grade airplane luggage linen. It employs two heavy duty 12" Jensen speakers which greatly improve the overall sound distribution as compared to the usual single speaker amplifier.

It is recommended where extremely high fidelity sound is required at high volume, and is truly one of the finest musical instrument amplifiers made.

Features of this fine unit include a top mounted chrome plated chassis on which are mounted the following controls: ground switch, "On" and "Off" switch, stand-by switch, bass, treble and presence tone controls, two separate volume controls and four input jacks. It also features a panel mounted pilot light and a top mounted extractor type fuse holder.

Tubes used are one 12AY7, two 12AX7, four 5881, one GZ34.

This amplifier is recommended for any high quality amplified musical instrument purpose.

Size: 20½" high, 24" wide, 10½" deep.

BASSMAN AMP

The Bassman Amplifier is believed by most players to be the finest amplifier produced for the reproduction of low bass frequencies. And today it has been brought up to date with the incorporation of the latest advancements in amplifier design. While its characteristics have been designed to accommodate string bass, at the same time, it makes an excellent amplifier for use with other musical instruments. It is truly a most flexible unit.

It has been engineered to handle the low powerful bass notes without distortion and to reproduce the bass tones in their true form. It is ruggedly constructed to withstand the tremendous pounding it receives in its usage.

The Bassman Amplifier incorporates another Fender First — the mid-range control which provides brilliance to the mid-frequencies and is used in conjunction with the treble and bass boost controls. Every player will readily notice the fullness and balance of tone through the audible frequency range with the use of this control.

This amplifier features the Fender solid stock lock jointed cabinet, covered with brown and white striped airplane luggage linen. It has the famous top mounted chrome plated chassis featuring the following controls: ground switch, "On" and "Off" switch; stand-by switch; bass; mid-range, treble and presence tone controls; and four input jacks. It employs four heavy duty 10" Jensen PM speakers.

Tubes used: two 5881, one GZ34, two 12AX7, one 12AY7. This amplifier will produce tremendous high fidelity audio power and is truly one of the most versatile musical instrument amplifiers on the market.

Size: 23" high, 22½" wide, 10½" deep.

Fender

The tweed Fender Twin, left, is seen even less than an original Bassman these days – much less in great condition like this one. It was Fender's most powerful amp at the time of its release, with first 50W and then 80W from a duet and quartet of 6L6s respectively, and 2x12" Jensen speakers. Thanks to its massive, high quality output transformers, the Twin also had more clean headroom and a fuller low-end response than any other Fender amp of the 1950s, yet developed a smooth, bluesy, monstrous overdrive sound when cranked up, too.

baffles mounted side-by-side, and a split-angled front grille.

Incidentally, it was common for these early amps to use output transformers located on the speaker frame and not on the amp chassis. These early Fenders introduced the now famous tweed covering that was to trademark Fender amps all through the 1950s and into the 1960s.

THE TWEED AMPS

The Fender amps made in the 1950s and early 1960s were to become known as the 'tweed' Fenders, and have great collector value and interest. Most of them sound great for rock'n'roll, probably because the music form was invented on them. Most of the rock'n'roll biggies used Fender amps, not to mention the popular country and western stars of the time.

The amp line ranged from the student models like the Champ and Harvard – with, respectively, 4W driving a 6" or 8" speaker or 10W driving a 10" speaker – up to the Twin model with a whopping 80W driving two big Jensen 12" speakers.

Arguably the best amp of all time, the 1959 Bassman has been reissued by the Fender company and copied by dozens of 'repro' builders. Crisp and clean at low volumes, and real crunchy when cranked up, it's the favorite of many top guitarists. The 1963 Strat is my personal guitar, amp from the GT Collection, photo by Jeff Veitch.

Fender's tweed amps evolved through three main cabinet and decorative styles from the late 1940s until 1960, when most were phased out for Tolex. The three Deluxe amps above illustrate the change from TV-front, to wide-panel, to narrow-panel cab. The diminutive Champ, below, remained available in tweed with a top-mounted control panel until 1964.

18 THE TUBE AMP BOOK

Fender

This ad for the Fender Bassman, right, points out the many quality design and build features of the larger tweed Fenders, and gives a great back view of the sweet 10" Jensen blue-frame alnico speakers. Fender was boasting about nearly every aspect of its amp construction at this point – and certainly had a right to!

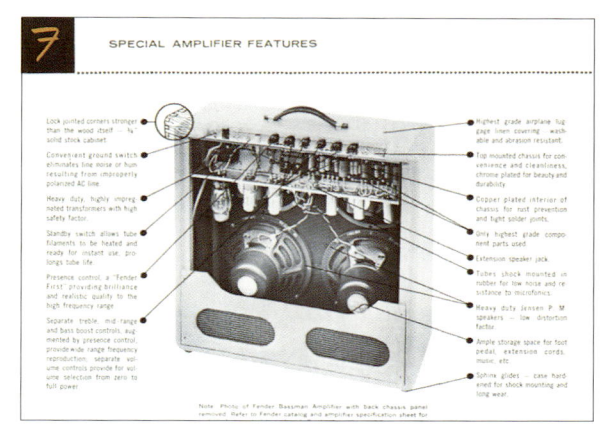

The power sections used 6L6 output tubes in all models over 30 watts and 6V6 power tubes for models under that. The preamp sections first used the larger 8-pin style tubes like 6SC7 and 6SL7.

In the early 1950s, Fender preamp sections began using 9-pin miniature tubes like the 12AY7 and the more popular 12AX7. This change coincided with the introduction of more sophisticated controls on the larger models, with features like bass, treble and middle instead of just the overall tone control which came on student model amps. The quality of the new 9-pin miniature tubes was superior for audio purposes because they had less adverse microphonics and noise. Another advance for Fender, the tweed models were the first to have 'vibrato,' pulsating change of pitch, and 'tremolo,' pulsating change of volume.

Fender made many different models in the 'Tweed Years' but stuck to just a few basic amp circuits. For instance, in the mid-1950s the circuit which used two 6L6 output tubes and three 12AX7 preamp tubes was found in similar form in the Pro Amp, Bandmaster, Bassman, Tremolux, Vibrolux, and Super Amp, though the use of different output transformers in certain models accounted for some tonal differences and disparities in output power. In the late-1950s, however, varying changes to the driver/phase inverter stages on different amps and wider alterations in circuitry – including the addition

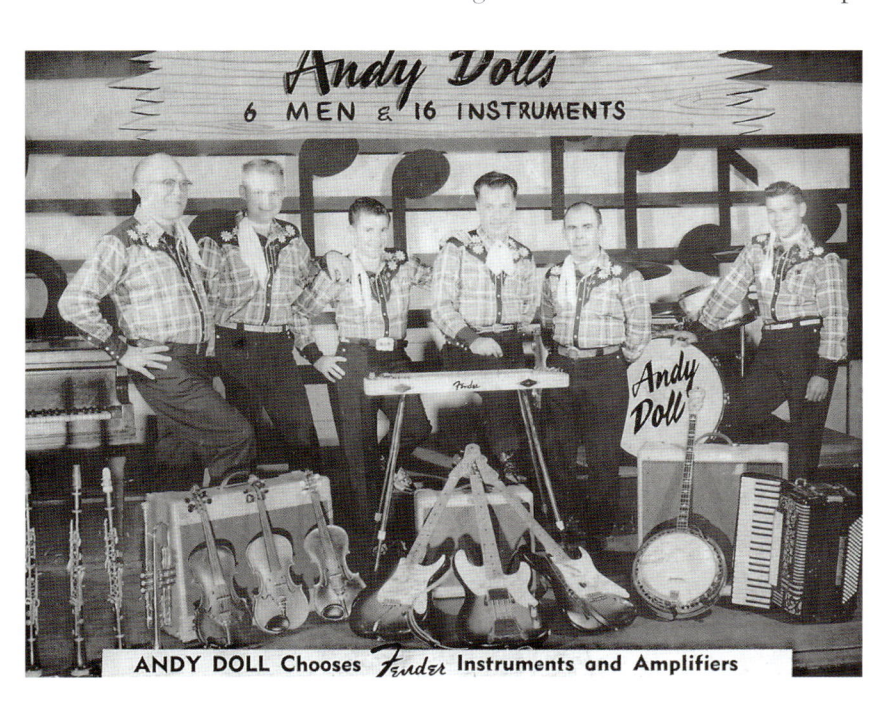

This much-gigged Tremolux is an interesting transitional piece. Made in May, 1960, it's one of the last medium-to-large Fenders you will find in tweed, and its circuit design includes three significant advances: it is one of few pre-Tolex Fenders to have its power supply filter caps mounted under a 'cap can' on the back of the chassis (notice their absence from the internal photo, above); it uses the same preamp as the Deluxe, but is fixed rather than cathode biased, and it's the only 2x6V6 Fender of its era to use the advanced 'long-tailed pair' phase inverter as seen on the bigger amps. On top of that, it has a great new feature – tremolo! Amp and 'Tele-ized' 1957 Esquire courtesy of Dave Hunter, photo by Miki Slingsby.

THE TUBE AMP BOOK 19

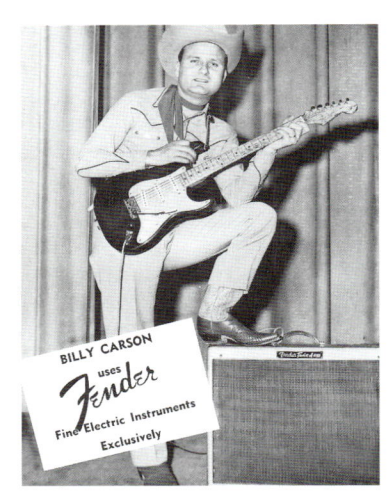

Fender

of a middle control to the Bassman, for example – made for more distinct sounding models. Naturally, all these amps also sound different because a single 15" speaker is different from four 10" speakers.

My favorite sounding amps from this period are still the Fenders with 10" speakers, namely the 4x10" Bassman and the 2x10" Super. These amps used the dark-blue Jensen 'Alnico 5' speakers (as did most Fender amps during the 1950s) in an open-back cabinet design.

These Jensen have a great clean tone and sound terrific when overdriven for a lead sound. Hundreds of your favorite guitar tracks have been created on this combination of Fender amp and early Jensen speakers. Incidentally, the sound of these early Jensens was similar to the early alnico-magnet Celestions that Marshall and Vox would soon be using. When Jim Marshall began to make his amps in England during the early 1960s, he copied the circuit of the Fender 4x10" Bassman, possibly because it was the best sounding amp around for either guitar or bass. The different cab designs used – the open-back 4x10" of the Bassman and the closed-back 4x12" of the Marshall JTM45 – probably account for the biggest tonal differences between the two.

It's interesting to note that the most desirable and collectable Fenders and Marshalls are amps that have this circuit design in common. If you are lucky

This 1958 3x10" Bandmaster is a much rarer find than the famous Bassman. The two are similar, but differ somewhat in circuit design, cab size and, of course, speaker format. The Bandmaster sounds slightly different, too, with a little more middle response and a quicker onset of distortion. It was the amp of Doobie Brother Pat Simmons, and one of the very first amps I had the pleasure of 'Groove Tubing.' The extremely rare 1960 Bandmaster, above, has the transitional tan Tolex and 'backward' control panel: Bass, Treble, Volume. Amps GT Collection, gold sparkle Strat courtesy of Norm Harris. Photos by Jeff Veitch.

Here's my favorite color of Fenders, the rough white Tolex with oxblood grille. This 1960 Twin-Amp has 80W from a quartet of 5881s, 2x12" Jensens, and goes great with the matching Fender reverb unit. Photo Jeff Veitch.

20 THE TUBE AMP BOOK

Fender

FENDER QUALITY...... INSIDE AND OUT!

These are a few of the latest developments and design features of Fender Professional Amps. They combine with those features described on the opposite page to provide you with the tops in amplification qualities, plus complete satisfaction of ownership.

DUAL CHANNEL CIRCUITS

Showman 15"
Showman 12"
Vibrasonic
Twin
Bandmaster
Pro
Super
Concert

— NORMAL / VIBRATO

Bassman — BASS / NORMAL

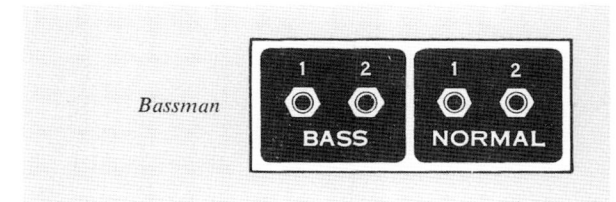

Tremolux
Vibrolux
Deluxe — BRIGHT / NORMAL

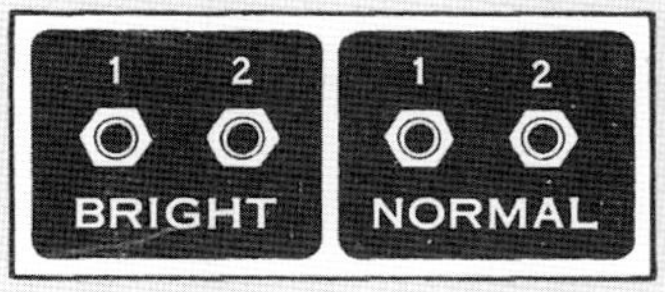

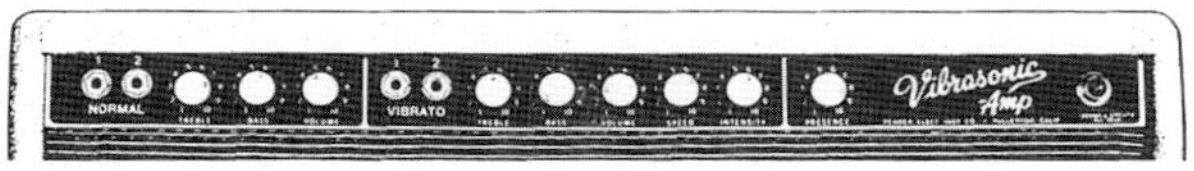

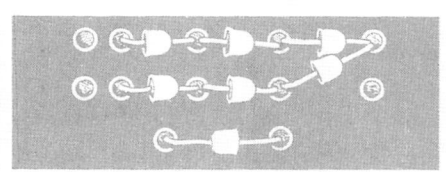

Modern silicon rectifiers are used rather than glass tubes. This feature reduces chassis heating and reduces servicing problems.

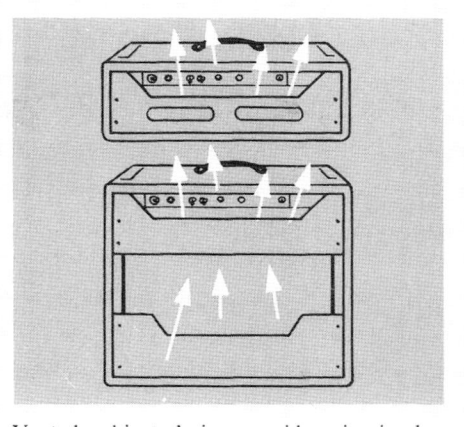

Vented cabinet design provides air circulation around chassis. Components operate at cooler temperatures thus prolonging amplifier life.

FRONT PANEL CONTROL. More convenient to use; control settings easier to read.

Attractive "Tolex" vinyl covering material resists abrasions and scuffs and is unaffected by climate variations.

Solid wood ¾ inch cabinet stock is lock-jointed at corners providing rugged construction and adequate chassis support.

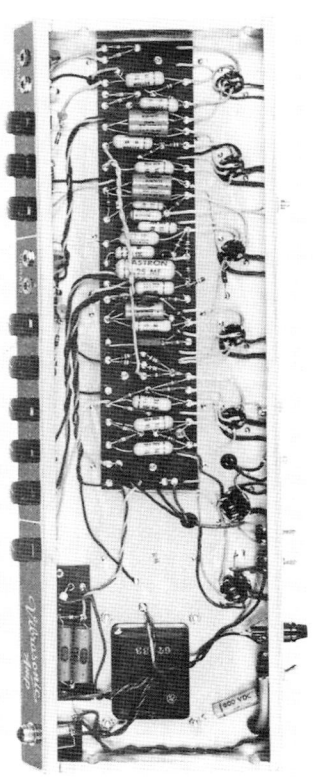

Component layout and chassis design eliminate circuit interaction, facilitate servicing. Note unit parts panel . . . all small parts are securely soldered eliminating vibrations and rattles.

Modern cabinet styling is striking in appearance and is enhanced by the textured Tolex covering material.

17

Fender

enough to get an old Fender tweed amp, restore it and never let it go, for this is clearly the sound that gave birth to rock'n'roll.

THE TOLEX YEARS

The early 1960s gave way to an entirely new design for Fender amps, which displayed the controls on a front-facing panel. A new feature introduced at this time was reverb, just in time for surf music!

Perhaps the most immediately notable change in these amps was the covering, a plastic-coated cloth called Tolex, made by General Tire. Fender would use three basic colors of Tolex during the next 25 years; they were white, brown, and black. The first color Fender used for this new amp line was a tan or light brown Tolex with a rough texture and a slightly pinkish hue. The grille cloth on these earliest Tolex amps was similar to the tweed amps, though amps with this combination of material are quite rare. I've been lucky enough to find just one 3 x 10" Bandmaster made this way.

Fender quickly developed this early Tolex color scheme into the more familiar off-white or cream colored Tolex, also with a rough texture. These early white Tolex amps were introduced in 1961 and had a dark brown-ish maroon colored grille cloth. The controls featured a matching round white knob with an indicator line engraved on the top. Fender logos on these early white amps were flat with matching enameled paint in the engraved section of the nameplate that underlined the Fender name. The first Fender Reverb units had this color scheme and are the most rare and sought after of all the vintage Reverb units.

The next evolution of the Tolex years was a switch to a smoother brown Tolex with more of a chocolate shade. These amps sported a flashy wheat-yellow grille cloth and brown control knobs with an engraved indicator line, but retained the same flat Fender logo.

This 1961 Super with oxblood colored grille cloth was the last of the rough textured tan era amps. However, Fender color schemes frequently overlapped so many combinations of the Tolex and grille colors can be found. The amp is from the GT Collection. The candy apple red Jaguar courtesy of Norm's Rare Guitars, photo by Jeff Veitch.

Here's a transitional 1960 Super-Amp. The Super has always been a great rock'n'roll amp, and with 35W into 2x10" speakers it really cooks.

The 1963 Vibroverb was based on the Super, but with the major addition of reverb – making it the first Fender amp to carry the effect. It lasted only about a year in this format, and is one of the most desirable Fenders of all. This one belongs to David Swartz of California Vintage Guitar. Photo by Aspen Pittman & Ed Ouellette.

22 THE TUBE AMP BOOK

Fender

Fender's next change was to a smooth white Tolex, ivory grille cloth with gold highlights, and black control panel that used a skirted, numbered knob for the first time instead of numbering the panel itself, as with all of their previous amps. It also switched to the raised 'Fender' script logo that is still used today (though it has undergone some evolution and variations along the way).

The really confusing thing about all these evolutions of early Tolex Fenders is that many of the catalogs published between 1960 and 1965 showed several different amps on the same page. For instance, one catalog I've got has a tweed Bassman, a white Showman and a brown Reverb unit all on the same page, set up like a bandstand. This indicates that there was probably a lot of spill-over from one era to another.

This second brown era of Fender amps introduced the first amp with built-in reverb, the Vibroverb. The brown Vibroverb amp had 2-10" gold-frame Jensen speakers, and was the last model introduced before Fender finally settled on black Tolex for its entire line of amps. The Vibroverb's big contribution was the built-in reverb that was to become Fender's biggest selling attraction over the years. The tube complement of this amp was three 7025s and three 12AX7s in the preamp section with 2-6L6s for the power section and a 5AR4 (aka: GZ34) rectifier.

This 1964 brown Vibroverb was the first Fender amp to feature built-in reverb and the last amp made in the brown Tolex. The 2-10" gold frame Jensens give a warm, crisp tone. This is perhaps one of the rarest and most desirable vintage amps.

Fender did finally convert all its Tolex amps to black in 1964, introducing models that would be produced cosmetically unchanged for more than a decade, barring a few minor details. The most famous of these were the Princeton Reverb, Deluxe Reverb, Super Reverb, Dual Showman, and Twin Reverb. The early black Tolex amps had Silver grille

With the introduction of their now-legendary 'blackface' amps, Fender once again set the standard for modern guitar amplification. The amps above are four of the most popular of all time: the Deluxe Reverb, Vibrolux Reverb, Pro Reverb, and Twin Reverb. All from the GT Collection, photos by Jeff Veitch. Right, the James Cotton Blues Band, with both blonde and black Tolex amps.

This example of the smooth brown Tolex 1962 Deluxe, which generally followed the rough white Tolex era, came with a single 12" Jensen, two channels and tremolo. The Gibson Melody Maker is from the same era, and together these were a likely combination for the intermediate guitar player of that time. Ted Nugent has one of these Deluxes that I've re-Groove Tubed a couple of times, and it's the amp he recorded 'Cat Scratch Fever' with. This one's in the GT Collection. The Gibson is courtesy of Norm's Rare Guitars. Photo by Jeff Veitch. In the endorsement photo of The Esquires, above, notice how they have blonde, tan and black amps on the bandstand – not uncommon in that era.

THE TUBE AMP BOOK 23

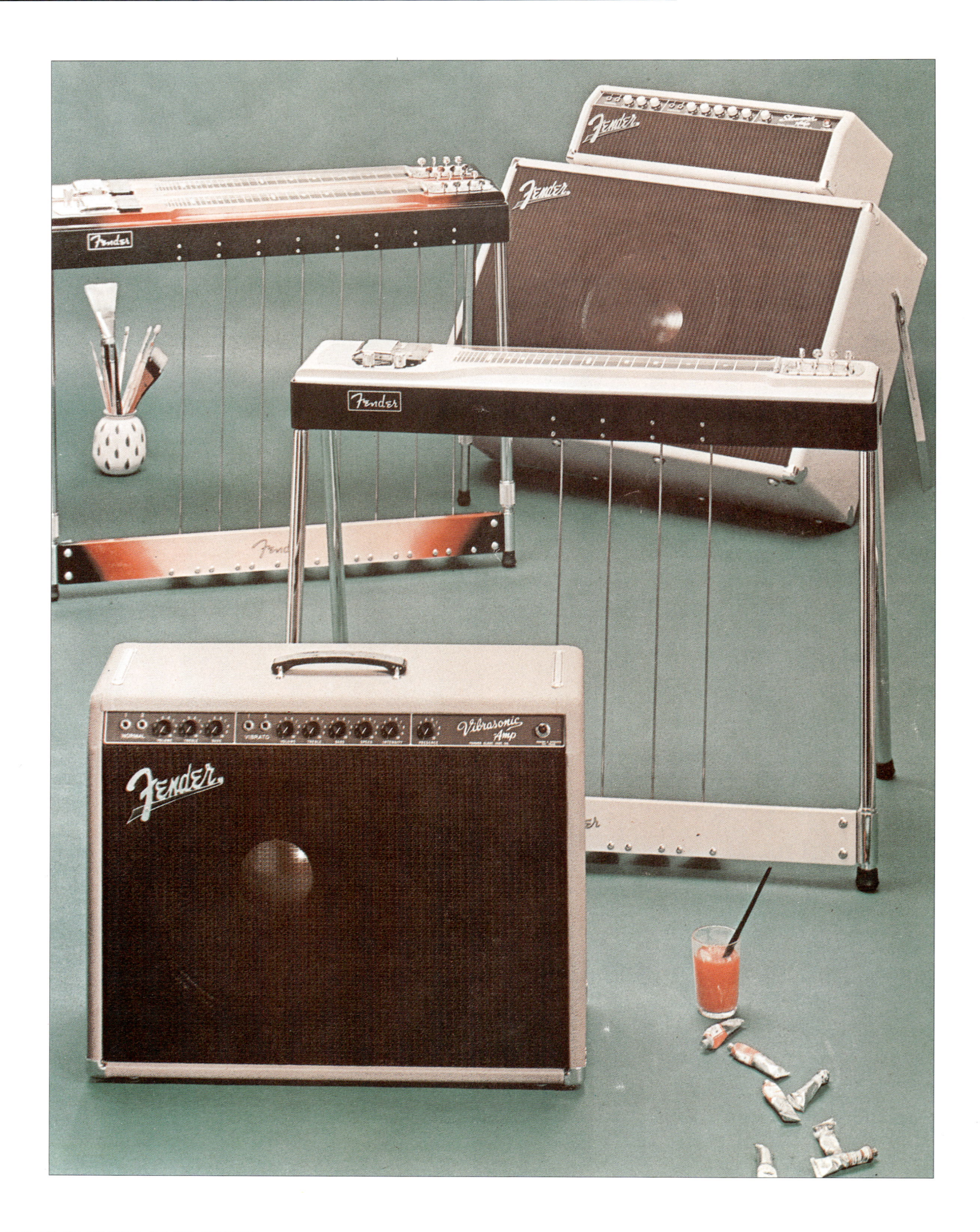

Fender

24 THE TUBE AMP BOOK

Fender

The great James Burton, above right, shows of his Fender Telecaster and Twin Reverb, while Ricky Nelson strums a somewhat less classic Fender acoustic guitar.

Blackface Fenders of 1967 gave way to silverface amps in 1968. Though little was changed internally on most models between these transitional years, the former remain much more highly regarded. Note the aluminum trim around the speaker baffle on the 1968 Twin Reverb below. This is a feature that lasted only one year.

cloth and black face panels with black control knobs numbered 1 to 10. These are often referred to by collectors as the 'blackface' Fenders. Many of the blackface Fenders were made before Columbia Records bought the company from Leo Fender in 1965. Thus they are also referred to as the 'pre-CBS' amps. There was plenty of overlap for the next two years, however, and many of the blackface amps produced by the early CBS company are identical to the pre-CBS versions in specs and performance. Many of the silverface amps that followed from 1968 onward, on the other hand, suffered from engineering changes that made them less desirable for players compared to the previous generation of amps. But with early silverface amps in particular, many of these changes can easily be reversed and the amp returned to blackface standards by a knowledgeable repairman.

When Leo sold the company to CBS, changes in personnel caused changes in the product. The engineering department, for example, implemented some 'reliability' changes to the basic Fender circuits of the Twin Reverb and many other amps.

These changes added several resistors and capacitors at various points in the circuit to give more 'stable' operation and reduce 'distortion.' Leo Fender had resisted these changes over the years because he felt they adversely altered the sound that made his amps famous. Leo knew musicians loved the sound of his amps and basically believed that if it wasn't broke, don't fix it! Musicians seemed to agree with Leo on this because sales of the 'new, improved' Fender amps fell off immediately.

The most striking cosmetic alteration instituted at this time was the change of the control panel from the traditional black to a 'modern' looking silver color, while the grille cloth was changed to add blue highlights to the traditional silver thread.

Despite Fender's wide success with tube amps, they tried several times to put out transistor versions of their traditional amp models. Again, the musicians snubbed the new models, and kept their old blackface amps. After a year or so of music dealers complaining of poor sound quality from the new silverface amps, Fender changed many models back to their original circuit designs of the pre-CBS era. The damage was already done, however, and the stigma of the CBS Fenders had set in. And more dramatic changes were on the way besides – including master volume controls, pull-knob overdrive, and more.

To say that the blackface Fenders are universally superior to the later silverface versions is probably an unfair statement, because there were relatively few amps made during this transition period of Fender, and many models such as the Deluxe Reverb were never altered in any way.

The 'King of Klean' – my 1965 Fender blackface Super Reverb has really got the rhythm sound of doom. Four early alnico Jensens (with the horseshoe magnets) powered by a duet of 6L6s, tube rectification, tilt-back legs and a Middle control make this the evolutionary son of the 4x10" Bassman. The 1965 ES-335 makes a nice mate to the Super Reverb, with enough power from those early humbuckers to overdrive the amp nicely. Both are from the GT Collection, photo by Jeff Veitch.

THE TUBE AMP BOOK 25

Fender

However, like most American companies that survived from the 1960s to the 1980s, the overall quality changed for the worse as suppliers were replaced for budget reasons and production methods rationalized. It is my opinion, however, that almost any Fender amp made in the 1960s or the 1970s can be made to sound great if reconditioned and fitted with quality tubes. In the interest of restoring all those nice old Fenders out there to their original condition and optimum performance, we have published detailed schematics and layout diagrams for the most common Fender amps. We have also included servicing and reconditioning information. (See: Tube Amp Service Tips and Modification.)

Recently the new Fender company has brought out new models with channel switching, gain boost, and active EQ. You can see these models at any of the Fender dealers in your town and they can explain all the wondrous new features. The new Fender amps are some of the best amps out there today for the money, but I'm personally fond of the old ones. In my humble opinion, the amp designers today go too far past the basic idea that an amp should just amplify the signal with the best possible tone and the most consistent, reliable performance possible. I could make a good argument with anybody that the tone and quality of most of the Fender amps made prior to 1968 is at least the equivalent of any amps made today, probably better. When it comes to guitar amps, in my opinion, oldies are goodies!

Since the early 1990s Fender has reissued many of its most popular vintage amps such as the 1959 4-10 tweed Bassman; the 1963 brown Vibroverb; the blackface Twin Reverb, Deluxe Reverb, and Super Reverb, and others. They have also brought out some new but 'vintage-styled' hand-wired Custom Shop models like the Vibro-King and Tone-Master, which many players have raved about in recent years.

I must admit that, in the case of the reissues, many of these are very close to the originals in both appearance and performance. They have changed many of the constructional details as compared to the 1950s and 1960s versions, but this is consistent with the way Fender has always built things. That is to say they take advantage of the modern materials and construction methods to produce an affordable product.

For example, the reissue Bassman uses a printed circuit board, as opposed to the hand-stuffed oil board the original amps used, with all components wired by hand. The printed circuit board integrates all the components on one PC board and thus saves time (and cost) in construction. The real purist will argue that the point-to-point, hand-wired method is superior. The fact is if the low cost PCB was available to Leo in 1959, he would have used it.

Since the primary components are so important to the sound and feel of the amp, it is gratifying to see that the new Fender company has

A look under the hood of two popular Fenders of then and now – an original 1951 TV-front Deluxe, top, and a 2001 Blues Junior, below it – shows how Fender's construction techniques have changed over the years, evolving from hand-wired eyelet board circuits to printed circuit boards (PCBs). Both approaches, however, aimed at producing a relatively affordable amp for the novice player, and succeeded. Photo by Miki Slingsby.

The powerful 2x10" Prosonic combo was an all-tube Fender of the late 1990s, with channel switching between high-gain modern and vintage-voiced channels. Amp and 1996 Fender Custom Shop Anniversary Strat courtesy of Chandler Guitars, Kew, London; photo by Miki Slingsby.

Fender

Fender's general amp lines and Custom Shop alike have continued to be busy through the 1990s and into the new millennium, releasing both vintage reissues like the 1964 Super Reverb Reissue, top, and vintage-inspired new models like the Vibro-King. Amps from both camps have generally had a very warm welcome. Photos courtesy of The Guitar Magazine.

striven to duplicate these. The old Jensen Alnico 5 speakers used in the original Bassmans were great units and a big part of the sound. The reissue Fender have used an Alnico magnetic element, first in speakers manufactured by the Eminence company in Kentucky, and more recently by a revitalized Jensen company which now builds its speakers in Italy.

I have tested the reissue Bassman against my several original examples in the GT Collection and have the following observations. All of the original Bassmans I have sound a little different from each other, but all of them sound great. That's probably because most the resistors, capacitors and other parts have 20 per cent tolerances and that variable effects the tone accordingly. Having said that, I have to say that any of my original Bassmans sounded and played sweeter and nicer than the reissue. Not by all that much, but noticeably better.

Although the new Bassman was a great sounding amp, I thought it sounded a little harsher and brighter that my old ones. This could be explained mostly by the differences in the speakers, tubes, transformers and the age difference between all of these. I might get a sweeter sound by fooling around with some different tube sets, but I don't think I could ever get the reissue to sound exactly like the original. However, the reissue gets closer to it than most other amps around today, with the exception of the THD Bassman amps out of Seattle and a few other good tweed Fender reproduction builders, such as Victoria and Clark.

The very latest contributions from Fender's amp Custom Shop finds them getting even closer to the specs of their vintage originals, while taking some clever and popular mods onboard, too. Just as this *Revised Edition* goes to press, the new 1964 Vibroverb Custom is on its way to the guitar stores. This amp marries a faithful reproduction of the very rare 1964 1x15" blackface Fender Vibroverb – with hand-wired circuit on fiber eyelet board, solid finger-jointed pine cabinet, and other vintage specs – with a number of popular modifications proposed by the late, great guitarist and amp tech Cesar Diaz.

Before he passed away in April, 2002, Cesar approached Fender with the idea of just such a Custom model, particularly including mods he had added himself to Stevie Ray Vaughan's own original Vibroverb. These included beefed up power supply filtering for tighter lows, a Rectifier switch to access diode rectification for a higher voltage supply, and a Modified switch to tap an alternate preamp tube bias supply for more gain from the front end.

The amp promises to be a winner, great for hot Texas-style blues and plenty of other styles. The user can also change from the 12AT7 phase inverter standard with the model to a 12AX7 for even more gain at that stage, and the factory stock power tubes are now the new US-made 6L6GE Groove Tubes.

Fender has also updated its popular Vibro-King Custom for 2003, giving it the transitional blackface dress of black Tolex, black control panel and white knobs, and adding Groove Tubes 6L6GEs here as well. With so many other hand-wired amps making a big splash in the market, it's great to see that Fender still knows how to do it right themselves, too.

The Fender Custom Shop's latest release is the 1964 Vibroverb Custom, which blends the classic 50W 1x15" Vibroverb of 1964 with a selection of popular mods suggested by the late amp tech Cesar Diaz.

THE TUBE AMP BOOK 27

Fender

Overview

HOW OLD IS THAT FENDER IN THE WINDOW?
By Richard Smith (Reprinted from *Guitar Player* magazine)

Leo Fender did not invent the guitar amp. His amps used basic circuits that could be found in the Radiotron Designer's Handbook, most of which were invented and patented by the Western Electric scientists working at Bell Labs. (Vacuum-tube circuits were first used in audio amplifiers and radios rather than in musical instrument amplifiers.) By the time Leo entered the amp field, Rickenbacker, Gibson, National, and other manufacturers had done much of the pioneering work. Moreover, one of Leo's repairmen at Fender Radio Service, Ray Massie, probably helped with Leo's earliest designs.

Fender started making musical instrument amplifiers at his radio shop in Fullerton, California, during World War II. Pieced together with surplus parts, the first few were custom-built for professional musicians. Commercial production began shortly after the war, when he and Doc Kauffman started building K&F lap steel guitars. At that time, many people, including Leo, saw the electric guitar and the amplifier as a single instrument. As a result, he approached guitar and amp design in an interconnected, hand-in-glove way.

Leo had an immeasurable impact on the development of the electric guitar, but what did he contribute to the amp industry? Little at first, because most of Leo's early effort went into his guitar designs. After several years, however, Leo concluded that the limitations of the speakers and alnico-magnet pickups available to him kept him from attaining the perfect electric guitar sound. After he developed the Telecaster, Stratocaster, and Jazzmaster pickups – each somewhat deficient in Leo's mind – he focused on improving the tone circuits in his amps to make up for inherent pickup and speaker problems.

Tone controls were Leo's foremost contribution to the field of amp design. When Fender amps started sporting treble, bass, middle, and presence controls in 1957, the modern guitar amp had more or less arrived. And it arrived in one piece; Fender amps were rugged and roadworthy. Starting in 1948, Leo built virtually indestructible cabinets with ultra-strong lock joints. To test their durability, Leo sent his first amps on the road with bands like Bob Wills and his Texas Playboys. These Fenders were knocked around in buses over thousands of miles of highways and endured countless nights in smoky dance halls and bars. Leo saw the abuse his products got from working musicians, and he built his amps to take it.

But Leo had little control over the manufacture of the component that seemed to wear out the fastest, the speaker. As guitarists started playing louder, the speaker emerged as the weak link in the sound chain. Leo knew that the only solution was tougher speakers with better frequency response. As the Fender company became a major speaker customer, Leo's influence over speaker design grew. Working with manufacturers such as Jensen and JBL, Leo probably forced more improvements in loudspeaker construction than anyone else in the music industry. His efforts paid off. By the mid 1950s, many guitarists considered Fender amps the best made, best sounding, and best priced units on the market. Fender, in effect, spurred other manufacturers to build better amps. Some copied the Fender details down to the tweed (luggage linen) covering, while others came up with their own innovations. In the 1950s and 1960s, Gibson, National, Ampeg, and other amps improved, but Fender still set the pace and the standards. Leo Fender was the father of the modern guitar amplifier. His ideas are still readily apparent in most of today's amps.

COLLECTING AND DATING EARLY FENDER AMPS:
AN ILLUSTRATED SUMMARY

1. K & F amps (late 1945 through mid-1946): There were three K & F models. Here's the small one with a 8" speaker. Originals have a stenciled K & F name-plate. (GT Collection)

2. Wooden Cabinets (mid 1946 through mid 1948): The first Fender amps had hardwood cabinets fastened with angle iron and finished in a variety of natural wood grains. Leo built three models in this manner: the Princeton (with an 8" speaker), the Deluxe (Model 26, with a 10") and the Professional (with a 15"). (Amp and photo courtesy John Sprung)

3. The Dual Professional: Produced in 1946 and early 1947, this amp had two 10"s and beveled cabinet covered in tweed, like later Super-Amps. (GT Collection)

4. TV-Front Tweed (mid 1948 through mid/late 1952): Fender introduced tweed-covered Princetons, Deluxes, and Pros in the summer of 1948 – dubbed 'TV' because they looked like a television set. Occasionally, these amps had a piece of masking tape signed and dated at the factory and stuck inside the chassis. If there is no date on the chassis, check the speaker codes. Pictured here is a 1952 Deluxe. (GT Collection)

5. Two Tone Leatherette (1949 and 1950): This 1949 Champion 600 is covered in two-tone leatherette. (GT Collection)

28 THE TUBE AMP BOOK

Fender

Overview

For many reasons – but mainly because they sound so good – Fender amps made between 1945 and 1965 have become collector's items. Certain models, such as tweed 4x10" Bassmans, Twins, and 2x10" brown Vibroverbs, are especially revered. Fender did not have 'model years' in the strict sense, but several periods were characterized by distinct cabinet styles and coverings. Any semblance of a model year revolved around the summer trade shows where Fender usually introduced its new products. Moreover, changes of cabinet styles and coverings usually overlapped for different models. For example, Fender produced both tweed Bassmans and the brown Tolex amps in 1960. Dating old Fender amps does not necessarily involve minutia such as knob types, nameplates, or handles (although mind-numbing knowledge of these details is important to serious amp collectors). What follows is an illustrated summary of Fender amp styles from 1945 to the mid 1960s and some suggestions for dating them. These guidelines assume that you have an original Fender amp with its original speakers – highly modified amps present different problems and are considerably less interesting to collectors.

DATE CODES ON SPEAKERS

Nearly all the speakers used in early Fender amps had date codes applied by the speaker's manufacturer. These date codes tell the speaker's date of manufacture, a clue that can help approximate an amp's date of manufacture. From a late 1940s through the early 1960s, most Fender amps featured Jensen Concert Series Alnico 5s, which collectors call 'blue-cap' speakers.

Starting in the early 1960s, Fender usually used Oxford and JBL speakers, but later reintroduced Jensens, as well. In time, Fender's amp production increased to the point where the company had to rely on additional companies. Still, most pre-CBS Fender amps (that is, amps made before 1965) had either Jensen or Oxford speakers. Jensen and Oxford stamped their manufacturer and date codes in ink onto the back of each speaker's frame. Jensen's company code number was 220, and Oxford's code was 465. These codes were followed by a three-digit date code. The Jensen pictured here has the code 220122, where the fourth digit (1) stands for the year of the manufacture (1951), and the fifth and sixth digits (22) stand for the 22nd week of 1951. Similarly, the Oxford speaker pictured here has the code 465-244 standing for the 44th week of 1962.

There is a slight problem with using these codes because both companies repeated the numbering every decade. The '1' in our first example could have meant either 1951 or 1961. A '2' could have meant 1952 or 1962, a '9,' 1949 or 1959, and so on. Nevertheless, the speaker's origin and appearance should suggest from what decade it came. (There's still no substitute for a keen eye and experience.) The clue that this Jensen speaker came from 1951 rather than 1961 is an older-style blue label out of use by the late 1950s.

How close are these speaker date codes to the date Fender completed the amplifier? If a speaker in an amp is original, the speaker date will always be earlier than the amp's completion date. Fender bought speakers in large quantities, but since they built so many amps, speakers had a short stay at the factory. Unless a speaker was hidden away by accident, it sat around the Fullerton plant for several months at most. Speaker codes, then, should suggest the amp's date of manufacture within a range of four or five months, and probably less.

FENDER DATE CODES ON TUBE CHARTS, SPEAKERS AND CHASSIS

Fender amps made after mid 1953 usually had a date code stamped onto their tube charts. The company glued the chart inside the cabinet on the self-contained units and inside the amp head on the piggy-back models. If your

6. Wide-panel tweed (mid/late 1952 through mid/late 1954): Fender introduced the wide-panel Twin Amp during the summer of 1952. Soon the whole line, including the 1953 Deluxe shown here, adopted this style. To date all Fender amps made after mid 1953, check the tube chart. (GT Collection)

7. Narrow-panel tweed (mid/late 1954 through mid/late 1959): Shortly after the 1954 summer trade shows, Fender amp cabinets began their transition to the narrow-panel tweed style. Fender used these cabinets until the transition to Tolex-covered amps. Here's a 1959 Bassman. (GT Collection)

8. Brown Tolex (mid/late 1959 through 1963: the Fender Vibrasonic Amp (essentially a Pro with a stock 15" JBL), was the first brown Tolex amp; others soon followed. The earliest brown Tolex – as on this 1960 Super – had a rough texture with a pinkish hue, compared to the later smooth brown Tolex more commonly seen. (GT Collection)

9. Blonde Tolex (1961 through *64): The first versions of piggy-back amps such as the Showman, Bandmaster, and Bassman had blonde Tolex coverings. Like the brown Tolex, the earlier blonde material had a coarse texture, compared to the smooth blonde material used in late 1963 and 1964. Smooth blonde Tolex amps appeared simultaneously with black Tolex models. Pictured are a Twin and reverb unit. (GT Collection)

10. Black Tolex with black panel (mid/late 1963 through 1968): Fender introduced the first blackface, black Tolex amps in mid 1963. By 1965 the entire line had made the transition to this style. Pictured here, a 1964 Twin Reverb. (GT Collection)

THE TUBE AMP BOOK 29

Fender

Overview

First Letter Year		Second Letter Month
A 1951	A	JANUARY
B 1952	B	FEBRUARY
C 1953	C	MARCH
D 1954	D	APRIL
E 1955	E	MAY
F 1956	F	JUNE
G 1957	H	AUGUST
H 1958	I	SEPTEMBER
I 1959	J	OCTOBER
J 1960	K	NOVEMBER
K 1961	L	DECEMBER
L 1962		
M 1963		
N 1964		
O 1965		

*Note: There is info on dating Marshall amps on page no. 67.

By the 1960s there were several ways to date Fender amps. But be wary of some dates; Fender Service Center employees often stamped an actual, uncoded date on units returned to the factory. The date indicates when the amp was serviced. This chart shows the letter date correlations in the code stamped onto Fender tube charts after mid 1953.

DATING AMERICAN TUBE AMPLIFIERS BY POTENTIOMETERS

There is a manufacture/date code stamped on the pots in your American made tube amplifier. Keep in mind that for this system to be accurate, the pots must be the original pots and also that some smaller manufactures buy a year's supply (or more) of pots at one time. Therefore assuming the pots haven't been changed, the code will indicate a date that the amp could not have been made before, and was likely made soon after.

The pot codes will have six or seven digits. The first three digits indicate the manufacturer. If there are six digits total, the 4th digit indicates the last number of the year the pot was made. If there are seven digits total, the 4th and 5th digit indicate the last two numbers of the year the pot was made. The last two digits in the code indicate the week of the year (01 to 52) the pot was made. Some codes will have all the numbers together, while some separate them with a hyphen or period. Also, many manufacturers may also use their own part numbers, so if the last two digits are larger than 52, it is not the date code number.

The common manufacturer numbers are 137 (CTS), 304 (Stackpole), 140 (Clarostat), 134 (Centralab), 381 (Bourns), and 106 (Allen-Bradley). Examples: pot #304-6443 (Stackpole 43rd week of 1964; pot #1377528 (CTS 28th week of 1975); pot #137731 (CTS 31st week of 1947 or 1957).

amp still has a tube chart (most do), look for two small letters in rubber-stamped ink. Don't confuse the date code with the production number, the model number, or the serial number. The first letter of the tube chart date code stands for the year of manufacture, the second stands for the month. The code is based on January 1951, even though it was introduced sometime in 1953. 1951 is 'A,' 1952 is 'B,' 1953 is 'C,' etc. 'A' represents January, 'B' represents February, and so forth. The 'CH' date code found in the wide-panel Deluxe in Fig. 6 tells us that Fender made the unit in August 1953. The tube chart pictured here shows the code 'IH' from August 1959. Sometimes the factory stamped the tube chart code onto the amp chassis, as well. Fender used still another date code system for early 1960s speakers and amp chassis. Shown here is a rubber-stamp code, AB 0763 (7th week of 1963), applied to an Oxford speaker at the Fender factory.

The date codes on the two Jensen speakers, far left, show us the top one was made in the 25th week of 1957, and the bottom one in the 2nd week of 1960. The Bassman's tube chart, near left, shows the date stamp 'IK' for Novermber, 1959, while the Tremolux's stamp 'JE' dates it to May, 1960.

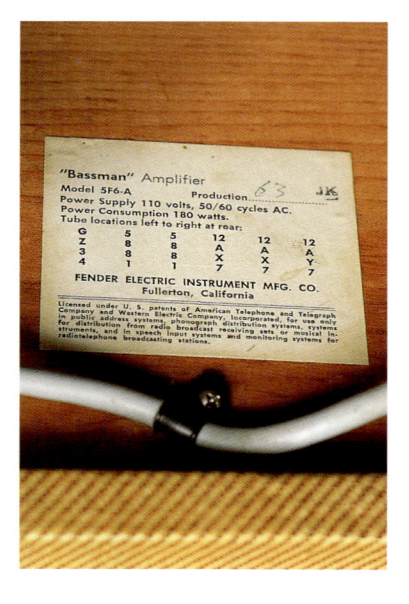

Garnet

'Gar' Gillies founded Garnet amps in the mid 1960s after a career of repairing radios in the day and playing trumpet on jobs at night. He made his first amp for a PA so he could amplify his voice and muted trumpet. As the rock 'craze' flashed through Canada, Gillies was quick to see a trend for amps to fuel the music, so he began working with the newer generation to build amps stable enough for their music: rock amps. He formed the new company out of his older radio repair business (sounds a bit like Leo!) together with his two sons Russel and Garnet (whose name was used for the amps).

Garnet's first PAs were incredibly well built, using heavy duty transformers (they weigh a ton!) and high-grade components through out; of course they were also hand-wired and meticulously well made. The earliest production went to a local Winnipeg group then called Chad Allen and the Expressions (soon to change their name to the Guess Who?). At that time Gar was manager and roadie for the struggling band. As they played around, Gar's amps quickly found favor a with other groups on the bills and orders soon followed.

I remember that it was in an LA swap meet that I found my first Garnet amp, which was a PA. I had already started Groove Tubes and knew a lot about old McIntosh hi-fi amps, RCA recording consoles, and tube guitar amps from Marshall and Fender... but this Garnet PA looked like a blend of all three! It had cool RCA knobs and old-time build quality, transformers and KT88 output tubes that reminded one of a McIntosh 75, and passive tone controls like a guitar amp – making for a loud and very musical amp. A few years later I was blessed to find a BTO Special with a cool distortion and tone section called the Sound Fountain – and it all lights up when activated. It has black Plexi panels and those classic RCA control knobs. These amps have muscle, like an early Hiwatt, but also possess their own character. Most of all, however, the high build quality and choice of components puts these amps at the top of the list for vintage-quality tube amps.

The company made relatively low numbers so there aren't scads of them around today, but some of the more popular models were the Pro, Rebel, and BTO. I have never seen most of the models they made, as few traveled far out of Canada. I do have a BTO. It is an amazing amp, and ahead of its time, with the Stinger fuzz tone built in! Amps in these three series all had two guitar heads and a suitably powered PA system.

The BTO is perhaps the most sought-after of Garnet amplifiers. These brutes have the power to cut through the loudest bands, and features that are not only cool but were way ahead of other amps of the period. The Stinger fuzz circuit can drive its dedicated tube so hard you'd swear it was begging for mercy. The built-in trem has beautiful depth and a wide speed range that will turn you off transistor trems forever. Later production models were inverted, ie, the chassis was mounted to the bottom of the cabinet with the tubes pointing up for better cooling of this very hot amp. Contrary to popular opinion (and the TAB 4th edition) these amps were not named after the 1970s rock band BTO (Bachman Turner Overdrive), which did use Garnet amps, as production of the amp began prior to inception of the band. The name stood for Big Time Operator. Later another variation on this design was made for Randy Bachman, which was named the Herzog. The Stinger fuzz tone was the sound of that incredible sustaining lead solo in The Guess Who?'s hit *American Woman*.

To follow the success of these lines, the popular PRO 200, PRO 400, and PRO 600 all-tube heads were introduced, followed by three Deputy models, produced as combo amps and heads.

One-piece combo amps would include the Banshee, Gnome, L'il Rock, Mach 5, Revolutions I, II, and III, Enforcer, and Sessionman. The Herzog, H-zog, and two stand-alone reverb units were all-tube effects devices designed by Gar in the late 1960s and early 1970s.

In 1989 the Garnet Amplifier Company Ltd officially closed its doors. Since then, Gar has been doing what he likes best, working one-on-one with

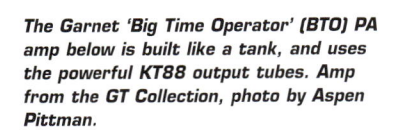

The Garnet 'Big Time Operator' (BTO) PA amp below is built like a tank, and uses the powerful KT88 output tubes. Amp from the GT Collection, photo by Aspen Pittman.

Garnet

the musician, repairing, upgrading and designing custom amps, and always working towards that special sound.

Thanks to Russell Gillies for fact-checking and updates.

Randy Bachman on Developing the Herzog Amp

"Gar and I developed this unit together back in '65-'66. Growing up playing violin, I loved the sustain, especially of a viola or cello. So early on in mid 1960s, I found out that by plugging a small amplifier into a bigger amplifier I could get this sound. Now I was taking the power out (which normally would go to the speaker) and plug it into the input of another amp. The result, for a few short minutes was a cool, great new sustained sound.

"Gar Gillies had a TV/radio repair shop, was a cool musician, and I wasn't embarrassed to take in my amps, which were literally burned by the power misuse – fried, to say the least. Gar asked what the heck I was doing, and when I told him, he said, 'You're insane to do this, it's very dangerous!' He offered to help me do it a safer and less destructive way. So Gar proceeded to build me a tube preamp which, when put into another amp, got me the desired sound. But not really; the sound was a little weird in a Fender amp, which was all that was around, so Gar decided to get parts from Heathkit and build an amplifier to go with the unit.

"We were looking for a name, and at the time I was reading [Saul Bellow's novel] with Herzog written across the cover. Hence the name, so we could stop referring to the unit as the 'noise thing.' I used to go to his shop on Osborne Ave late at night after gigs, and stay till the wee hours of the morning, making the most incredible moose and ox bellowing sounds, distortion/blotto screeching sounds that would make us laugh. But once it was smoothed out, it was smooth. So that became my lead guitar solo sound. It was first featured on The Guess Who? CBC weekly show, Let's Go and Music Hop in 1967-68, but really came to the forefront as the 'sound' of 'No Time,' 'American Woman' and many other sounds on The Guess Who? albums, *Wheatfield Soul*, *Canned Wheat*, and *American Woman*. I continued using it later on *Brave Belt I* and *II*, and the on Bachman Turner Overdrive albums.

That's about the story. There is still no unit that sounds like it today. One that comes close is the SansAmp rackmount, which has an actual setting called 'American Woman.' It's close.

John Johnson of Johnson Amps, Digitech, and DOD pedals and amps told me that, as a kid, he tried to make his own pedal to get the American Woman guitar sound and couldn't get it. However, the pedals he did make in trying to get that sound have gone on to sell hundreds of thousands for Digitech and DOD, but none got the 'American Woman' sound."

CHEERS, RANDY BACHMAN

This Garnet Pro amp features the famous Stinger fuzz section – seen to the left of the control panel in the top photo – followed by a 'Sound Fountain' tone stack, topped off with tremolo... and sounds great! GT Collection, photo by Aspen Pittman.

Gibson

The Gibson company is primarily known for its fine guitars, but the Gibson amp predates anything made by Fender or Marshall. Gibson amps go way back, and there have been some really interesting models. There must be 70 different Gibson amps listed in our official service manual, but my favorite ones are the stereo models, in particular the GA-79RVT. This amp was the first stereo amp I know of, and it gets a great tone using four 6BQ5 power tubes (two for each side). The 6BQ5 is the equivalent of the European EL84 power tube that was used in the famous Vox AC15 and AC30 amps. Each side of the GA-79RVT amp produces 20W and drives a 10" inch Jensen alnico speaker.

The front of the amp is a V-shaped split grille which spreads the two Jensens and gives a really big sound when played in stereo or a nice separation if two player share the amp. Also when playing in stereo, only one channel has Reverb, which creates an interesting opportunity for mixing a dry channel with the Reverb channel. The GA-79RVT has a stereo/mono switch on the top panel and so can be set up to be a 40W mono amp too. I had sold a really nice example of this amp to Joe Walsh (through Fred at Westwood Music) a few years ago and decided I really wanted one back. It took me nearly three years to finally locate one, and it's definitely not for sale! Gibson made several variations on the stereo amp over the years. These were made under the Bell name as well as Gibson, and you can find a sample of the Bell Stereo in one of the GT amp collection shots in this book. I believe the Bell line was marketed

Gibson was into the stereo craze in a big way in the 1960s, as seen in the set advertised here and in other designs.

32 THE TUBE AMP BOOK

Gibson

LES PAUL

★ With built-in variable tremolo

Endorsed by leading artists for its thrilling tremolo and brilliant treble. Outstanding in its price range for power and distortion-free sound reproduction, for flexibility and ease of handling.

Features

Handsome case with top-mounted control panel . . . woven Saran grille cover . . . jewelled pilot light. Tremolo with separate intensity and frequency controls . . . on-off switch in foot pedal. 12" Jensen speaker . . . 7 tubes. Two channels . . . separate volume controls for each . . . bass-treble voicing control . . . two inputs for each. Polarity switch eliminates interference. 16 watts output.

22" wide, 20" high, 10½" deep.

No. 3053. GA-40T .. .. 90 gns.

LES PAUL Control Panel

INVADER

★ With Gibson reverberation ★ dual speakers

Handles three instruments and microphone with ease, and gives you Concert Hall sound—all the extra clarity and quality of Gibson reverberation. A big-voiced professional amp with two speakers, ample-power, and a wide range of tonal qualities.

Features

Modern styling with woven Saran grille cover . . . jewelled pilot light. 12" and 8" best-quality speakers. 6 tubes . . . 16 watts output. Combination treble and bass voicing control plus tone expanders. 4 input circuits: two for instruments or microphones, two for instruments only . . . separate volume controls for each. Combination polarity switch with off-standby-on positions.

22" wide, 20" high, 10⅛" deep.

No. 4011. GA-30RV .. 103 gns.

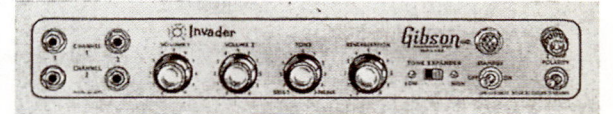

INVADER Control Panel

EXPLORER

★ With built-in tremolo

Small in size, but powerful of voice . . . with a tremolo that regulates from extremely fast to a slow wobble. It adds fine tone quality and wide sound range to make it second to none for its size and low price.

Features

Smart new slant grille, lock-corner case. Wide swing tremolo and powerful sound. 10" heavy duty concert speaker. 5 tubes—2 dual purpose. 3 input jacks . . . 14 watts output. Quick-change controls for volume, tone, depth, and frequency of tremolo. Foot control switch for tremolo.

20" wide, 16½" high, 9" deep.

No. 2031. GA-18T .. 64 gns.

EXPLORER Control Panel

Gibson

Gibson was already the biggest name in archtop acoustic guitars in the '30s when the quest for more volume brought amplification into the picture. The catalog pages pictured left show the many and varied homes their early amps found with players of 'electric Spanish' and Hawaiian (lap steel) guitars.

especially for the electric accordion but they also work well for guitar. Remember Gibson was one of the few guitar companies with stereo guitars at this time.

The earliest Gibson amp I have is an EH-150 that dates back to the 1940s. It has an interesting cabinet design that closes up like a suitcase for safe traveling.

It also has an 'echo speaker' output which sounds very innovative for the 1940s but is actually just their name for an extension speaker. I guess you place the extension speaker in the corner and therefore it 'echoes'.

Most of the Gibson amps made in the 1940s used the octal type preamp tubes and 6V6 power tubes. Later in the 1950s and 1960s they used the modern 9-pin preamp tubes and the common 6L6 power tubes. One of my favorite examples of these later amps is the Les Paul amp. Gibson made several Les Paul signature amps, including the Les Paul Junior, Les Paul TV, and, largest of all, the GA-40 Les Paul with a whopping 40W driving a sweet old Jensen Alnico 5 12" speaker.

I guess my favorite of all the Les Paul amps is the GA-40 because it

The 1963 GA-79RVT Gibson Stereo amp uses two independent amps that can be switched to mono or stereo. It has two 10" Jensens in a split-front cabinet, each driven by a separate amp using a pair of 6BQ5 (EL84) output tubes, with reverb and tremolo. This amp is great for recording or just sitting 'n picking with a friend in the stereo mode. Amp from the GT Collection, ES-355 stereo guitar courtesy of my pal and GT family member Paul Patronete; photo by Jeff Veitch.

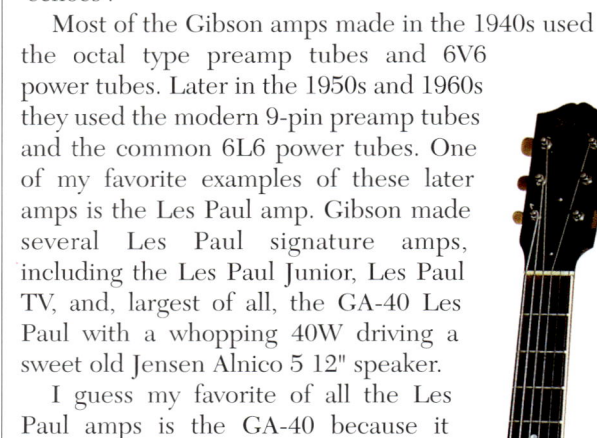

An early Gibson BR-1 amp with a 12" speaker from the mid 1940s shown here with a '37 Gibson ES-150 featuring a Charlie Christian pickup. The amp is from the GT Collection, the guitar courtesy of Guitar Gallery of Pasadena, California, photo by Jeff Veitch.

34 THE TUBE AMP BOOK

Gibson

Pictured in the photo to the right are a late 1940s Gibson EH150 'suitcase' amp with 1x12" speaker and a latching cabinet back coupled with a 7-string Gibson lap steel that has the only 7-string Charlie Christian pickup I've ever seen. To its right is a Richenbacker (the early spelling) 8-string and matching 1x12" amp combo from the same era. From the GT Collection, photo by Jeff Veitch

The Gibson Les Paul amp is a favorite of mine. With 40W and a Blue Jensen 12", it sounds really big and fat (early 1950s model main photo below, with a rear shot of a later 1950s model inset above it). The nice Les Paul TV model was supplied by Norman Harris of Norm's Rare Guitars, photo by Jeff Veitch.

sounds so sweet and looks really great. It has a two-tone brown vinyl covering, and a real 1950s checkered grill cloth with a plastic grill overlay that spells out 'LP' in large script. If you've got an old Les Paul guitar, you should really try to get one of these to go with it, just because.

Perhaps the most famous tube product made by Gibson isn't an amp at all, but the early tube model Echoplex. These were a tape echo system that could give a player a great deep, true echo effect, or on the later models a sound-on-sound effect so you could play along with yourself – although I could never get just the right number of bars recorded to make it work for me. Jimmy Page used on of these Echoplexes on his famous recording of 'Stairway to Heaven'.

Through much of the 1970s, 1980s and 1990s Gibson found themselves in the wilderness as far as amps were concerned, and certainly you could

THE TUBE AMP BOOK 35

Gibson

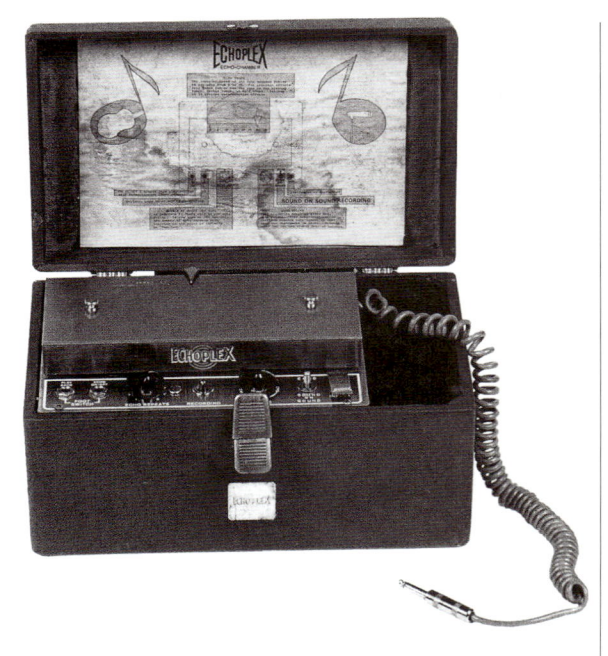

An early Gibson Echoplex with 'Sound on Sound,' above, although the straight echo effect of this all-tube device is what makes it so desirable. GT Collection

This Maestro SG System model 115 amp with reverb and built-in phase shifter – an effect pioneered in different forms under the Maestro brand name – was built for the Gibson-partner company by CMI Electronics of El Monte, California. It features a solid-state front end for preamp and effects, into a tube output stage. Output is either to an external guitar speaker cab or via a line out to the PA rig or recording desk.

forgive young players of the day for thinking Gibson never built an amp. They were, for a time, allied with Lab, a maker of occasionally successful solid state amps that was also owned by Norlin; but as for tube amps, nothing to speak of came out with a Gibson brand. Then at the close of the 1990s, Gibson's purchase of British bass and guitar amp builder

The '59 GA-80 Vari-Tone amp below-left puts about 35W into a 15" Jensen and has an unusual pushbutton tone control section – and makes a great partner for the '63 ES-345 guitar. The early 1960s Maestro RTV amp to its right – with the fun Harmony H-75 guitar – features an unusual complement of four 8" speakers. Both sets courtesy of Charlie Chandler, photos by Miki Slingsby.

36 THE TUBE AMP BOOK

Gretsch

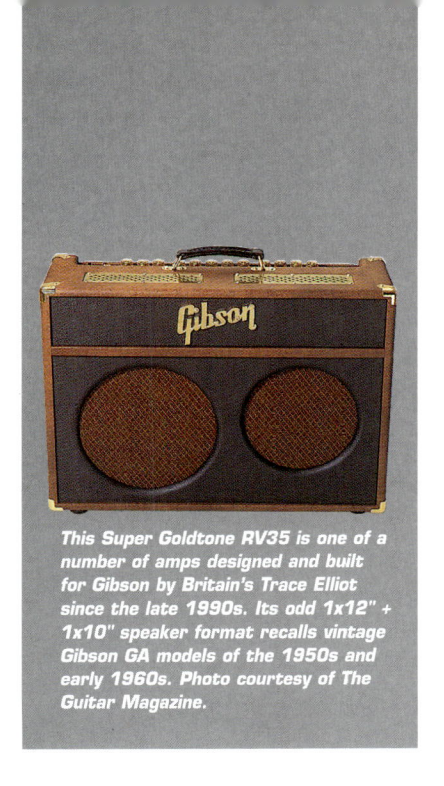

This Super Goldtone RV35 is one of a number of amps designed and built for Gibson by Britain's Trace Elliot since the late 1990s. Its odd 1x12" + 1x10" speaker format recalls vintage Gibson GA models of the 1950s and early 1960s. Photo courtesy of The Guitar Magazine.

Trace Elliot launched a new range of tube amps in the revitalized Gold Tone series. Built in the British factory, with some clever designs by Trace's Paul Stevens, these resulted in some good-sounding and relatively affordable Gibson tube amps hitting the guitar stores once again.

The first of these to arrive was the petite Gibson GA-15, a simple 15W 1x10" built around 2xEL84s in class A, and essentially a rebadging of what had been Trace Elliot's own short-lived Velocette combo. It has only volume and tone controls plus a bright switch, in true vintage tradition.

The dual channel, 30W Super Gold Tone GA30RV combo has more contemporary clean/overdrive switching, but stylistically is a throwback to some Gibson amps of earlier years like the GA30 of the late 1950s, with 12" and 10" Celestions housed in the same cab, and 4xEL84s in the classic Vox-like class A output stage. Then there's the big Super Gold Tone GA60RV, a footswitchable two-channel, 2x12" combo which develops 60W in class A/B from a pair of EL34s. It has a nifty external biasing facility that allows easy swapping for 6L6s, KT88s, 6550s and so forth.

The range is built to good standards for modern mass-produced amps, using thick glass epoxy circuit boards and star grounding for low noise and hum, tube reverb and effects loop circuits, and chassis mounted tube sockets throughout.

However you slice it, it really seems there *should* be a playable Gibson tube amp on the market, so you have to welcome these newcomers.

Gretsch

The unusual Gretsch 'Anniversary Stereo' set in the ad below featured a splitter box to send the stereo guitar's signal to two different amps.

Vintage Gretsch amps are probably among the best known of the many amps manufactured by Valco for various suppliers from the 1940s to the 1960s, including National, Oahu, Supro, Airline, Danelectro and others. On top of that, the Gretsches are undoubtedly some of the coolest looking of all these, even if the control features – and the circuits beneath them – are largely the same as those found in models offered by a number of other brands.

The Gretsch Amps evolved through about half a dozen different cosmetic styles from the early 1950s to the late 1960s, when Valco went out of business. The first models appeared in a light tweed cloth, which was changed for a gray/black cloth with silver

Here's a 'Great Gretsch' combo; the amp's from the early 1950s while the guitar is mid 1960s. Gretsch made interesting amp/guitar combinations, especially for the Country & Western markets. Amp and guitar courtesy of Norm's Rare Guitars, Reseda, California. Photo by Jeff Veitch

THE TUBE AMP BOOK 37

Gretsch

highlights by the mid 1950s. By the end of the decade, many of the Gretsch Electromatic amps took on a groovy wraparound speaker baffle and grille – another cool and distinctive Gretsch look that further distinguished them from the crowd.

Throughout the mid 1950s some of these amps were also given the Cowboy treatment, decked out in full western trim of ivory leatherette, tooled and studded brown leather belts, and rodeo-style graphics to match the Chet Atkins signature model guitars of the period (a cornpone style that Chet objected to from the start, and eventually convinced Gretsch to abolish).

By the early 1960s, Gretsch amps began to take on more standard shapes and lost their wraparound grilles, and came to look a little more like the square amps sold by other brands. Styles and models evolved a little further up until Baldwin purchased Gretsch in 1967, after which their amp sales declined pretty rapidly (and indeed this marked the beginning of a period of confused directions and declining sales for many great American guitar and amp companies).

Most of these early Gretsch tube amps are well built, feature great blue-frame alnico Jensen speakers (sometimes in unusual configurations) and other quality components, with tremolo and reverb on many, and sound pretty good indeed. Even early on, many kicked out 30W or more, thanks to a pair of 5881 output tubes, and bigger models certainly were often designed for a clean, rich sound – that Chet influence again. Like many good American-made amps of the 1950s, they are getting more and more expensive and harder to find all the time, but most are still more affordable than Fender amps of the same era – and there can be few better choices to complete the Rockabilly vibe for the guitarist searching for ultimate twang!

This Rockabilly Duet features a late 1950s Gretsch model 6161 amp with a rather outrageous '59 6120 Chet Atkins Hollow Body. The amp has a wild two-way speaker system and a wrap around grille. Courtesy of Guitar Gallery, Pasadena, California. Photography by Jeff Veitch

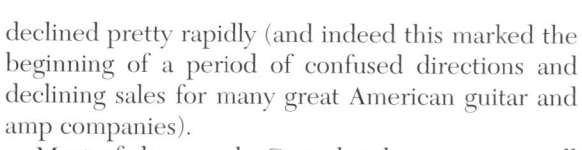

This unusual Gretsch Chet Atkins model amp has a built-in tuner and remote control capacity for many of its features. Two 12" blue Jensens and 40W produce a sweet, clean tone, great for those Chet licks. The lefty Country Gentleman is courtesy of Scott, owner of Guitar Gallery of Pasadena, California, the amp from the GT Collection. Photo by Jeff Veitch

Groove Tubes

Through my years of collecting and researching vacuum tube musical equipment, I developed the urge to build some classic gear for myself. I wasn't interested in 'me too' products or just knock-off copies of Fender and Marshall amps, but rather products that might break new ground, if that were still possible.

Of course, there is a lot to be learned from these early classic amp builders. Those lessons coupled with today's superior materials and components inspired our company's first motto: "Modern Designs, Classic Tone." We believe the secret to building good-sounding tube gear is to keep it simple and use the best components available. Therefore, component engineering is as important as the design itself. More importantly, we never forget that the primary goal is to get more tone and touch!

Our very first product, the STP-G (Studio Tube Preamp for Guitar), would blend the classic tone and response of a tube guitar amp (borrowed heavily from the blackface-era Fender Deluxe Reverb power stage design) with the modern idea of our Speaker Emulator, which was an invention for recording guitar direct to the mixer without speaker or microphone. We concentrated on controlling the quality of the most critical components, such as tubes (that was relatively easy for us), the transformers (this took quite a bit of research), and of course the speakers we wanted to 'emulate.'

The very best of the vintage guitars and amps all had great sounding components, either by design or by divine accident. Fender used a type of light ash called swamp ash... and presto! the resonant, lightweight Tele was born which today, 50-plus years later, is a rare and sought after vintage guitar. For amps, Fender's use of the GE 6L6, Schumacher transformers and Jensen Alnico 5 guitar speakers all contribute to the famous tweed Bassman's classic sound. In Europe, this was similar to the Mullard tubes' and the Celestion speakers' impact on Vox and Marshall amps.

Although the STP-G was marketed as a direct recording preamp, its authentic tube distortion and dynamic response comes from its power tubes driving the patented Speaker Emulator loading. The STP-G is actually a complete tube amp that can drive a speaker with about 25W RMS. Its built-in Speaker Emulator circuitry converts that power tube signal down to a more manageable low-level preamp signal, compatible with any recording console. The STP-G output tube complement could use either 6V6s or EL34s. It was used on many early rock recordings, most notably for the haunting lead sound on Blue Oyster Cult's 'Don't Fear the Reaper' as played by Buck Darma, one of our first customers.

Groove Tubes' development of elaborate processes for testing and matching tubes single-handedly opened up a brave new world of tone shaping to the guitarist. Above, GT's new generation of packaging bids farewell to the previously familiar plastic cylinders.

GT Electronics' first guitar amp product, the STP-G, is a studio preamp that doubles as a 25W all-tube amp top.

The Speaker Emulator (our first US patent # 4,937,874) was developed first as an integral part of our STP-G, which was released in 1986. Our Speaker Emulator stand-alone product came a few years later and signaled a new and innovative approach to recording a classic tube amp. At the heart of this product is a reactive load design that parallels the load of a particularly good sounding speaker, the early Celestion G12 15W speaker that was in early Marshall and Vox amps. The Speaker Emulator combines this patented reactive load with a sophisticated pole filtering system (a type of EQ) which creates resonance to emulate various cabinet designs. This EQ section can also be defeated for driving another guitar amp rather than a full-range PA or studio monitor.

Our first SE could handle the output power of most any tube amp, up to 150 watts, and had many additional controls which allowed fine tuning of the filter section to dial in the perfect 'cabinet' resonance sound the player was looking for. However, it was strictly for the studio, and naturally many players wanted one made for touring after using it in the recording process.

The SE II, the touring version, was released in June 1993 and added speaker return outputs so it could be used 'in line' and reduce speaker return power by 50 per cent or 75 per cent. Also, the SE II added a parallel effects loop (with mix control) and separate outputs for the loop or the dry amp signal.

No amount of fancy overdrive circuits common to the newer amps can come close to the classic rock'n'roll sound of your favorite smelly 'ol tube amp cranked up into a Speaker Emulator. Furthermore, once the SE has converted that powerful output signal to a preamp signal, you can add all the effects processing you want with maximum effect because your distortion is before the effects! The SE can drive the PA board, another stage amp, or a recording console, turning your classic old tube amp into the ultimate tube preamp.

We subsequently licensed several well known

Groove Tubes

amp companies like Marshall, who built their own version of the SE. All these are off the market today, but we still produce the Speaker Emulator II, proving the case that nothing sounds or feels like a classic old tube amp cranked up. Our SE IIs were first used on tour with Eddie Van Halen, ZZ Top and many others back in the 1980s and 1990s but are still popular with many recording and touring bands. We recently outfitted Joe Satriani with multiple channels for his road rig.

Our second project was a three channel preamp for recording and live stage work which used three very different designs for each of its three channels. We dubbed this preamp the Trio, and its three 'voices' – Clean, Mean and Scream – were designed to offer distinctive classic tube amp tones. We kept the circuitry very simple, using passive tone controls like those of the early Fender, Vox, and Marshall amps. Each channel has Gain, Bass, Middle, Treble and Volume controls, but also each channel has its own unique sound and touch. Later, we made a mod for Mike Rutherford of Genesis and Mike and the Mechanics allowing the #3 Scream channel to be mixed with the #1 Clean channel to gave a two-amp tone thing like Keith Richards got playing through two amps simultaneously. The Trio's channels can be switched in four ways via the front panel, a footswitch, by rear panel rack switching, and the optional MIDI switching.

The Trio also was our first product to feature our Parallel Effects Loop , which allows the insertion of effect devices with zero loss of tone! This circuit was used on all subsequent GT amps and preamps. Perhaps the most famous of the Trio users were Jerry Garcia and Bob Weir of the Grateful Dead, who were using both the Trio and our D75 in their rack setup for many years and right up to Jerry's death. The Trio preamp is still in production today, basically unchanged except for a move to our new silver/black color scheme.

After we finished with the Trio – which was the world's first three-channel-switching all-tube preamp – we needed the amp to go with it, so we developed another 'GT World's First': our Dual 75 rack mount guitar amp. This was the world's first selectable tube stereo amplifier to allow the player to customize his sound by selecting basically any type of common power tube for either of the two channels. Furthermore, the player can then switch between two different sounding channels (A/B), or combine them (A+B) for a true Keith Richards-style two-amp tone. Either side of the Dual 75 can use any of four very different basic types of power tubes, such as:

The 6L6 – à la Fender amps, this tube gives a hard, clean tone at lower levels with a fat middle and smooth clipping as it's pushed into distortion. The overdriven tone stays pretty smooth, very controllable and round.

The EL34 – à la Marshall amps, this tube has more powerful, and produces a clean tone with strong lows and highs but a slight midrange dip or hollow spot as compared to the 6L6. However, when this tube is pushed into distortion, the EL34 has a beautiful compression and a raspy edge that is pure rock; often called 'crunch tone.'

The 6550 – a more powerful tube than the EL34 or 6L6, with a even, tight sound that accurately reproduces all types of input signals. It's a tube that's late to overdrive, distorting only at very high volume levels, then producing a brash, metallic tone.

The KT88 – capable of even more power in certain amps than the 6550, also with a full range tone but a smooth and more controllable transition into the distortion mode, a bit softer and rounder tone than the 6550.

The Dual 75 is all about customizing, not modifying your amp set-up. It allows a player to mix or match any of these power tubes and drive any combination of speaker cabinets. Use a duet of 6L6s to drive a small 1x12" cab, together with a duet of EL34s to drive a big 4x12" cab... the best of both tonal worlds! The D75's footswitch can alternate between these two channels (A/B), or combine them (Both). I prefer the 'Both' setting, I guess because I never could settle on just one amp. The D75 also has separate Volume and Presence controls for each channel, as well as a separate Parallel Effects Loop for each channel.

We made a few higher-powered versions of the D75, the Dual 120, by special order. The D120 is capable of producing in excess of 120W RMS per channel with 6550s – great for the stage, but too much for the club musician, so we discontinued making it... just too loud!

The STP-B was our first product for the bass player, and it doesn't look like any of our other products because it was designed and manufactured by our good friends at SWR, the bass amp guys. Steve W. Rabe (SWR) and I worked together at Acoustic for many years. We were both frustrated guitar players trapped in an amp company famous for its solid state bass amps. I left to start Groove Tubes, and Steve left to form SWR, building amps that used a tube in the preamp driving a solid state power amp. As I was just introducing my STP-G we both decided it would make sense if Groove Tubes marketed just the tube preamp section out of his bass amp under Groove Tube Electronics. We shook hands (no

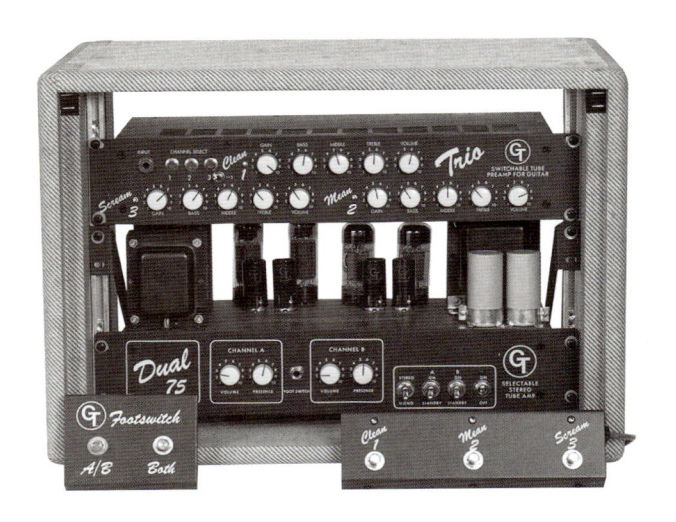

The Trio preamp and Dual 75 power amp, housed in a tweed rack case above, completes the GT 'Touch-Tone' system as represented in the earlier days of GT amplifier production.

GT's Soul-o 75 captures the Clean and Scream channels from their Trio preamp in a rack-able unit that can be bolted into any number of combo GT cabs, with a spare slot for an effects unit to boot.

Groove Tubes

Above, one of the laboriously hand-painted 'graffiti grilles' on an early Soul-o amp.

On this rear photo of the GT Soul-o amp with koa cab, note the raw, bark-covered edge of the timber used for the back panel.

The photo to the right displays a couple hunks of very tasty timber together in one place, in the guise of a GT Soul-o combo with highly polished koa cabinet, and koa bodied Brian Moore guitar.

contracts!), and we started selling Bass Preamps. This product was later discontinued as SWR began to market its own tube preamp for bass.

Our next adventure was the Soul-o Series self-contained guitar amps, introduced in January of 1993. We wanted a traditional format amp line and so combined two preamp channels of the Trio (the #1 Clean and #3 Scream channels to be exact) with one channel of the selectable D75 power amp. The basic rack mount Soul-o amp could be installed into a rack system, or also be packaged into a rubber suspension combo cabinet with an extra space for an effects unit. The Soul-o's parallel effects loop and channel switching (Clean, Scream or Both) were foot switchable.

Perhaps the most innovative new feature of the Soul-o Series, however, was its ability to alter its basic power tube stage design to emulate the three classic designs vintage amps had used over the years: Class A/B (à la blackface Fender and Marshall); Class A Normal (early Fender tweed era); and Class A Gnarly (à la Vox AC30). The class A designs produce about one-third less power than to the class A/B; however, they offer a soft touch, shimmering distortion and a warm, round clean tone. These three optional output designs, and the many optional types of power tubes possible with the Soul-o's selectable output stage, gave the Soul-o Series the widest variety of tone and feel ever offered in a guitar amplifier up to that time.

The Soul-o 45 amp was next, our 'pedal friendly' amp, specifically designed to use footpedal FX devices driving the amp's first input stage. This design evolved because I noticed that while all guitar amps were designed using a typical single or double-coil guitar pick up as the 'signal' source, most players were not plugging their guitar into the amp directly at all. Instead, most players we knew were going through pedal boards first… which changed the dynamics of the signal path tremendously.

A guitar pickup is a high-impedance, low-output device, but after exiting the typical foot pedal the guitar's signal becomes a low impedance, relatively high output device. So this kinda' overloads the amp's input stage, which mushes up the sound and masks the dynamics of the pedal and the amp. We decided to use the pedal output as the signal reference when designing the S45 preamp stage. With this new design, the pedals come through with more dynamics and they do not compress the first stage of the amp. Additionally, we didn't give the S45 a high-gain distortion channel because we reasoned the various pedals would provide the overdrive tones, and this worked out great. The only 'feature' we added was a built-in tube-driven Reverb because no pedal can achieve that tube reverb tone. We still make the S45 in both a top and a combo format.

A COUPLE OF EARLY SOUL-O TRIVIA NOTES

1) The first 100 amps we made were labeled Solo. Then Hartley Peavey called to say he had already used (and therefore trademarked) that name on a discontinued club PA amp and demanded I change the name. Not wanting to piss off one of the biggest gorillas in the jungle, I decided to simply change the name to Soul-o, which I felt was actually a better description of the amp anyway… make tone, not war, is my motto!

THE TUBE AMP BOOK 41

Groove Tubes

2) The Soul-o cabs were available with an optional 1950s style painted swirl 'graffiti' design. I had made a custom amp for the NAMM show using some original 1950s grille cloth I got from an old cabinet shop; these grilles were also on some more interesting 1950s amps made by Danelectro. Well, I only had enough for a few amps and naturally we got dozens of orders... so I eventually figured out how to hand-paint these designs onto each grille (man, was I a mess after making a few dozen of those!). Thankfully, we discontinued graffiti grilles several years ago, so I have more time to work on amp and mic designs these days.

As times changed, our customers were more interested in simpler, lower-powered amps for smaller clubs and recording. So we discontinued the S75 (last ones were made in 2002) in favor of producing smaller, simpler class A and class A/B amps like the long-time GT standard amp the S45 (35W to 60W depending on tube selection, a class A/B amp design), the new S30 (dedicated class A amp using the 6L6 family of tubes), and the new Single (7-15W, depending on tube selection, which is self-biasing and a single-ended class A design). We still produce the Trio and the D75 stereo amp too, as well as the Speaker Emulator II.

The next step was our entry into the tube-based studio product range, making tube mics and signal processors.

The Groove Tubes SFX Chapter (US patent #6219426-B1)

You go through 40 years of playing and listening to guitar amps and you think you've seen and heard it all; that was me before SFX came into my life. Drew Daniels, an old friend who makes a living as a sound designer and audio engineer, came to me with an interesting idea to make a stereo instrument amp that had just a single cabinet, and yet produced realistic stereo sound everywhere in the room. He asked me to partner in the development. I jumped at the chance, and it was one of my better decisions. The idea was simple, and had sound audio physics behind it, so I took over the project of developing an idea we named Center Point Stereo.

Of course musicians record in stereo all the time, but rarely can play the same parts live because the audience and bandmates are all politically incorrect – that is, they are either too far to the right or left because in traditional stereo systems there is only one sweet spot, and it's pretty small. So while we have all tenaciously tried time and time again to make stereo 'happen' in a club, we always sound better in mono because at least it sounds the same everywhere... but the sound has no dimension. So, you can imagine I was very motivated to make this idea work.

After only a few weeks of hard work on the electronics and some time at my favorite cabinet maker, we had our first prototype. I used a GT Trio preamp, an old Alesis Quadraverb to generate the left/right effects, then ran it into the SFX electronics and onto the GT Dual 75 amp. The SFX speaker had a 12" Celestion facing the club and another two 12" speakers in a facing dipole arrangement that pointed left and right in a 360-degree open sided bottom cab. It looked pretty weird – and you shoulda' seen the cabinet maker's expression when I showed him the drawings... It's hard to get an old dog excited – but this brought quite a smile to his face.

Surprisingly, the first time we switched it on it worked perfectly. No matter where you stood you heard stereo, it was simply stunning! We had stumbled onto something bigger than I'd bargained for, and I decided to enlist a Big Brother company to partner in the production and distribution. So I called my old pal Richie Fleigler, who was then the head of the Fender Amp Division. After a quick demo for Richie and the 'Boss' Bill Shultz, we had a handshake deal. That handshake would produce an award winning amp, the Fender Acoustisonic with SFX, which won 'Product of the Year' at that year's NAMM show. Fender trademarked the new name of SFX Technology (stood for Stereo Field Expansion), and we later received a US patent for this technology.

Fender is currently developing other SFX amps, and Groove Tubes makes some custom SFX systems that adapt to any tube amp, preamp and effects systems. We also make several custom shop SFX amps for keyboard and an all-tube stereo amp and cab for guitar.

GT Audio

Our last (well maybe not last) Big Adventure into Tubesville was into the wonderful world of vacuum tube recording gear. I had spent years in and out of studios recording guitar amps and naturally got well into the mics and methods of recording tube guitar

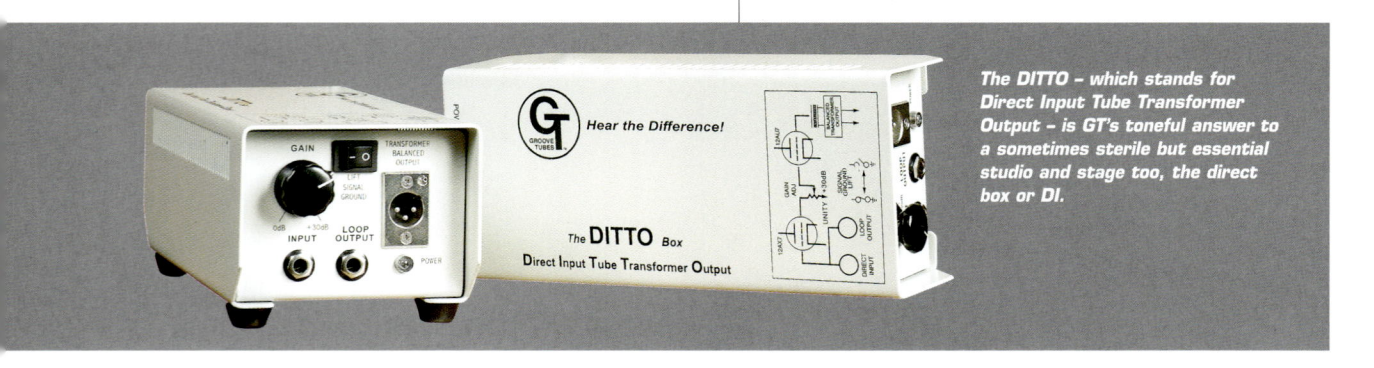

The DITTO – which stands for Direct Input Tube Transformer Output – is GT's tuneful answer to a sometimes sterile but essential studio and stage too, the direct box or DI.

Groove Tubes

amps (ie, Speaker Emulator). The pursuit of great tone in a tube guitar amp or a tube mic is basically the same process: a close study of great sounding vintage tube gear as your starting point, following through with creative engineering and component development. The challenge was just too tempting to resist!

Once again, it would seem that all the great classic recordings of Elvis, the Beatles, Ray Charles, you name it, were made on vacuum tube audio equipment. In these early days, tube-based condenser mics, tube mic preamps, tube EQs, and tube compressors were all mixed down on studio playback tube amplifiers and the cutting heads of the recorders were driven by tube amps! Even the headphone amps that the musicians tracked with were powered by tube amps.

Then came the transistor revolution of the late 1950s and early 1960s and all the studio products companies who made the mics and signal processing brought out 'new and improved' models replacing tubes with these newfangled transistors. They all thought tubes were over, and solid state was the future. Well, they were partly right. It is interesting to note that most, if not all, of these companies have returned to offer 'reissues' of their earlier tube designs and dozens of new companies have sprouted up making similar tube-based recording products.

This is most likely because many producers and engineers (the ones with ears) never quit using that old tube gear. Many top artists like Phil Collins, Linda Ronstadt, and Mick Jagger are still using old tube condenser mics and outboard gear to get the 'real deal' tone. A studio's reputation is often bolstered by its selection of classic tube mics and signal processing gear. Of course now that everybody has caught on, this classic recording gear has become very expensive, not to mention the expense in restoration and maintenance for new tubes, caps, and possibly a capsule rebuild. For example, an early Neumann U47 tube mic in mint condition can fetch up to $10,000, and single replacement tube for this mic, the VF14M, sells for $1,000 a piece, if you can find one for sale!

We saw this as our golden opportunity to develop a whole new line of tube recording equipment combining that Classic Tone with Modern Designs. Our first product, the Model One vacuum tube condenser microphone (the MD1) released in 1992, was followed closely by the Direct Tube One direct box for instrument recording (the DT1). The MD1 offered every musician, engineer and producer an opportunity to experience that great tube mic tone and warmth, coupled with modern performance specifications for a fraction of the normal price, because we used current sources for tubes and transformers while developing our own capsule designs and manufacturing capacity.

We went on to develop the MD3 with a very large diameter capsule, and were the first to use 3 micron diaphragm film with evaporated 24kt gold as the conductor. We went on to develop more tube and FET based mics over the next 10 years, and currently offer eight different mic models, which sell very well throughout the world.

We have produced over 15,000 microphones over the last 12 years and our mics have received many awards, great reviews, and are used in studios both large and small throughout the world today. In fact, by popular demand we just recently reissued the original Model One; you know you're old when you reissue you own products!

Perhaps our most ambitious 'tube amp' achievement was the development of the ViPRE mic preamp which was introduced in 2001. It was the world's first all-tube Variable Impedance mic PREamp (hence the name ViPRE), and again we brought a Modern approach to get that Classic tone. The ViPRE has five features never before combined on a single product:

1) a variable transformer input that allows the mic to load five different ways and so change the behavior of the mic as needed for the recording situation. Loading ranges between 300 Ohms, 600 Ohms, 1200 Ohms, and 2400 Ohms, or can go directly to the first preamp tube.
2) It has a variable rise time, so the electronics 'speed' can be slowed down to emulate the older mic preamp. Five optional setting range from Slow to Fast. This really changes the dynamics of the recording, and can roll off some unwanted highs and vocal sibilance in bright female voices.
3) ViPRE has a fully differential signal path (all balanced) which means it has two parallel signal paths throughout the box (a design which doubles the component costs), in contrast with 99 per cent of the other mic pres that use an unbalanced approach. Balancing the signal in this way extends Vipre frequency response to over 100,000 cycles and greatly reduces hum and distortion.
4) Precision gain controls means that ViPRE has no pots! Instead all levels are controlled with multi-decked ceramic switches that change the gain by 1dB and 5 dB increments. These very expensive switches are required because only a switch can maintain a 180 degree balanced signal path
5) Accurate VU metering with five range options; this is a real VU like those used on the old recording consoles and it is accurate to 1 per cent.

All this is expensive to make, as we had to develop or recreate most of the ViPRE components from scratch. It also requires more assembly time than a high-priced guitar, but the results are worth it as ViPRE received a Technical Achievement Nomination from the MIX foundation, which showcases the industry's best products at the annual Audio Engineering Society convention. In the end, we found the same attention to detail and component engineering made the difference with Groove Tubes Studio products, as it did for our guitar series electronics. A simple new song, but sung well, will always find favor!

GT's Model Two tube microphone further developed the company's exploration into the wide world of professional tube recording gear.

Hammond & Leslie

Perhaps the most famous rock'n'roll tone other than an electric guitar is the legendary tone of the Hammond organ connected to a Leslie speaker system. As you may know, this is a tube audio system. The tone generators in the Hammond are driven by 6AU6 preamp tubes, and the sound of solid-state organs has never equaled them. The Leslie speakers had a pair of 6550s driven by a 12AU7 phase inverter, which was in part responsible for that incredible tone. The harder a Leslie was driven by the organist, the more distortion it would create, so it had a special 'touch' that just can't be matched, even though there are many new organs that 'sound like' the original.

It's the same as all the transistor amps that supposedly have that great 'tube sound' but feel cold as ice. The variable range of distortion that follows the amount of signal strength is what gives tube amps a playable response, just like a drum or an acoustic piano. In the final analysis, it's the 'touch' that gives a tube amp its special advantage for making good music.

Harmony

While Harmony was, for a time, the largest manufacturer of guitars in America, their amps never amounted to much more than a few cheap student models. But any time from the 1940s until they went out of business in the mid 1970s, there were always a few Harmony amp models on offer. They could be had pretty cheap then, and they're not worth much now. With simple controls and very basic circuits, they were never great sounding amps, either, but floor one to the max and, like any little tube amp, it might just wail out some useful sounds for recording or messing around.

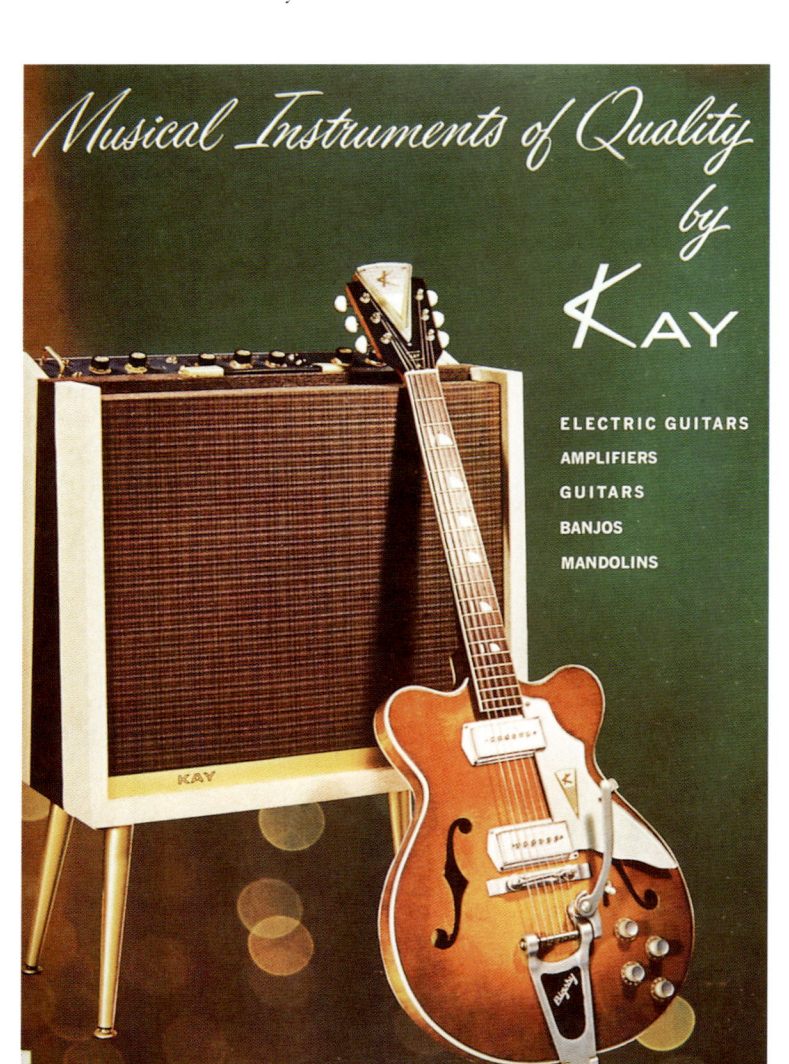

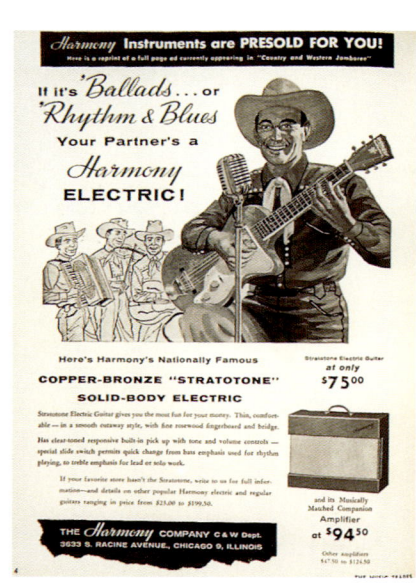

Harmony guitars were always considered budget options, but their amps afforded even less respect. The company never offered a combo deserving even of the limited admiration given to some of their cool but affordable semi or archtop guitars, such as that in the ad below. In truth, Calvin probably plugged it into a Fender or Gibson amp when he wanted to 'sing.' Kay, far left, was another company offering some interesting but underwhelming amps.

Hawaiian Steel Guitar

It's hard to imagine today, but at one time the most popular commercial music in the world was Hawaiian music, and the early electric lap steel and amp combo were a big part of that sound and music.

Back in the 1940s, everybody wanted to learn how to play the 'new' electric guitar, not for rock, but for making that sensational Hawaiian music that had taken the entire world by storm. To be sure, most of the Hawaiian band was acoustic, but the lead tone playing the melody and solos was this incredible newfangled electric lap steel guitar, which was plugged into an amplifier.

The lap steel guitar was the natural evolution of the acoustic, lap steel, open-tuned 'Weissenborn' style guitar. The high number sold to beginner students is obvious when you notice how many of these guitars and amps have survived over the years – many more than the first Telecasters.

The lap steel guitars were mostly the six-string variety, tuned to an open G and played through their companion amps. These combined to give the music that eerie, dreamy tone cutting through with an Island melody over the pounding drums and percussion and acoustic rhythm instruments. Most of these sets are from the 1940s and 1950s, and many were made and sold with a matching motif made up of gray, green or blue 'mother of toiletseat' plastic that was applied over both the guitar and amp's wood bodies. The more elaborate combos had Hawaiian scenes 'flocked' onto the grille cloth, such as an island boy climbing a coconut palm in a tropical bay.

The amps were very simple. Many of the early ones had no controls except an on/off switch, so the volume and tone were entirely controlled from the guitar (not a bad idea when you think about it… no tune-in trips to the instrument panel between every song!). Every one I've had has been a class A, singled-ended (one power tube) design using mostly the 6V6 output tube and early octal preamp tubes such as the 6SC7 or 6SL7. Of course they also used tube rectifiers, usually a 5Y3.

The speaker was usually a 6" or 8" design, generally made by either Jensen or CTS. These amps often had the output transformer attached to the speaker frame, instead of the amp chassis. This speaker style is called a "field core output transformer." This system made for a more compact amp, which suited the limited space of the early bandstands, and made the amp very portable. When restoring one these amps that has a bad speaker (very likely after 50+ years of abuse), the recone is not always an easy one. But it may be easier than finding a replacement speaker and transferring the output transformer to the new frame. You can also find that the transformer has gone bad too, and finding a small enough output transformer to drive a modern speaker can also be a bear.

It is amazing to me, however, how many of these old sweethearts need nothing more than a few new caps, a squirt of lubricant on the single pot (usually the off/on switch too) and new tubes to really bring them to life. Very seldom have I had to go the full route of replacing the transformers or reconing the speaker… which is a tribute to the way our grandfathers built things in those days!

Restoring these early amps is usually just a simple job of replacing a few caps to reduce hum, finding either an NOS octal-style preamp tube (BTW, if you can't find a NOS tube, then GT makes a 'Substitube' adapter that allows substitution with a modern 12AX7 style tube), a new 6V6 output tube and maybe a 5Y3 rectifier (not usually). Invariably, when I am finished performing such restorations, the amps sound incredible – especially when used with higher-output pickups. These early lap steel guitar pickups did not spare the magnet nor the wire… so they usually had good output. These old amps can be quickly pushed into clipping with a Gibson P90 or humbucker, but are slightly less aggressive (cleaner) with the single-coil Fender type guitars.

These three beauties were typical of the amps sold to students of the Hawaiian steel guitar. These amps are commonly referred to as 'Mother of Toiletseat' amps (because of the fake abalone covering, as seen on toilet seats of the day) and were mostly made in California by OEM factories like the Gourley company and the Factor's Manufacturing Company for student programs and chain stores' music departments. From the GT Collection, photo by Jeff Veitch

Hawaiian Steel Guitar

But remember, they were never intended to be played into distortion, but rather for the accent to those gently swaying hula hips… heaven forbid we should use these for rock'n'roll. Actually, when I started collecting Hawaiian amps back in the '70s, it was just for the look. But later as I started to restore them (and remember most of them out there are 50+ years old so they need some attention) they proved real usable recording amps. These early amps were the inspiration for several single-ended class A designs on the market today, including our own GT Soul-o Single that is slightly larger, with a 10" Jensen, and has more gain for rock and blues recording.

Our modern GT Single still has a single output tube and is self-biasing, so the tube can be 'hot swapped' between dozens of tube options to produce a pretty versatile sounding recording amp, still with that same sweet, long sustaining class A tone found on many a rock recording today – but a tone that originated long, long ago on an island far, far away…

The rear view of this Oahu Hawaiian steel guitar amp – labeled 'The Master' on its back-facing control panel – reveals the simplicity of such designs, and shows off its sweet-sounding silver frame 10" Jensen alnico speaker.

Feel the balmy Pacific breezes blowing yet? At least the stack of Lockola, Leilani, Oahu, Supro and other Hawaiian amps pictured left beats any wall of Marshalls for looks, if not power.

Hiwatt

The Hiwatt company was founded by Dave Reeves in the early 1960s as an 'up market' alternative to Marshall. The product was of the highest quality and had a reputation for being louder and cleaner than the other British amps of the day.

Dave's first big customer was Pete Townshend of The Who. Dave built several amps for Pete, who used nothing else for many years. The heads were available in 50W and 100W models; later some combos were added. The best of these is the Bulldog, a 50W 1-12" combo.

Hiwatt amps were built in the south of England, near London, and remained virtually unchanged until the death of Dave Reeves in 1981. Dave left control of the company to several of his employees. It then passed to Eric Dixon, and the company started to make amps a little different in design and quality from the original Hiwatts made by the Reeves factory.

One design change that caused problems was the mounting of both the preamp and power tube sockets directly on the circuit board. This type of design saves labor and costs in the building of the amp, but the vibrations of a circuit board can cause adverse microphonics in high gain stages and the heat generated by power tubes can cause cracking or warping of a low quality circuit board. The net result of these labor and cost reductions in the second series of Hiwatts was to cause some problems in the field and adversely affect the Hiwatt reputation for quality. This naturally has made the original amps more desirable.

Why do I bother to mention these facts? For two reasons: first, the player who's looking for a Hiwatt should know the difference between the two eras of Hiwatt (similar to the pre- and post-CBS eras of Fender); second, because recently the Hiwatt company has changed hands again and the new owners have made a commitment to reinstate the

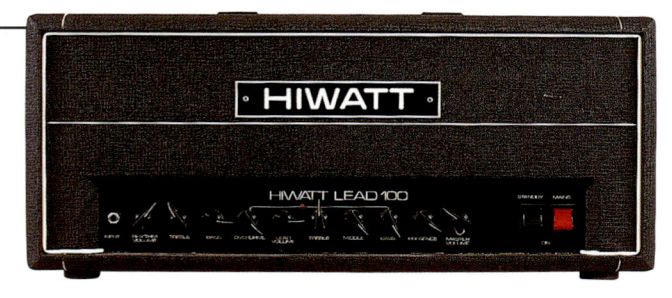

This early 1980s Hiwatt 100 – from the final era of the original company, shortly after Dave Reeves' death – has footswitchable rhythm and lead channels, with a quartet of EL34s for that 'crunch of doom' roar. Amp courtesy of Chandler Guitars in Kew, London, photo by Miki Slingsby.

46 THE TUBE AMP BOOK

Hughes & Kettner

original quality level of Hiwatt in the image of the amps made under Dave Reeves. Players should look for these newest amps to be greatly improved and much like the originals in quality and performance.

The new owners have redesigned the L100R, LB100, and the LC50R combo, using construction techniques that have better reliability and superior performance. They have also reissued a series of signature custom models that have been out of production for many years. They are hand building the original DR103, DR504, SA112 (formally called the Bulldog, a 1-12" combo from a later period of HIWATT) the SC412 cabinet with the original Fane speakers. In addition, they have also introduced a very impressive Studio-Stage combo with two heavy duty Fane 12" speakers and switchable 20W/40W output from four EL84s (switched to two on low power), which captures much of the big Hiwatt sound in a smaller studio or club-sized package. Open up new Hiwatts like this Studio-Stage and the bigger reissues, and they are hand-wired as neatly as the Dave Reeves Hiwatts of old – a real joy to behold! We've had occasion to retube a few of the DR103 and DR504 amps here at Groove Tubes and can tell you they are very close to the mark, and great sounding amps. Naturally, whenever you try to make a reissue, the old materials may not be the same. Even modern environmental concerns prohibit reproducing exact duplicate finishes and/or components. Considering these pitfalls, I wish the new owners the best of success, as Hiwatts have always been a special favorite of mine.

The original 1960s Custom 100 half-stack above epitomizes the big sounds and meticulous build quality of the golden age of Hiwatt, as heard in the playing of Pete Townshend, David Gilmour, and other great British rockers.

This new Custom Studio/Stage 40W 2x12" combo – manufactured by the current English owners of the Hiwatt name – marks a return to the hand-wired, built-like-a-tank Hiwatt chassis of old... and sounds fantastic. Photo courtesy of The Guitar Magazine.

You think those 100W Hiwatts weren't loud? Why do you think Roger Daltrey spent all that time leaping around the stage shouting his head off?

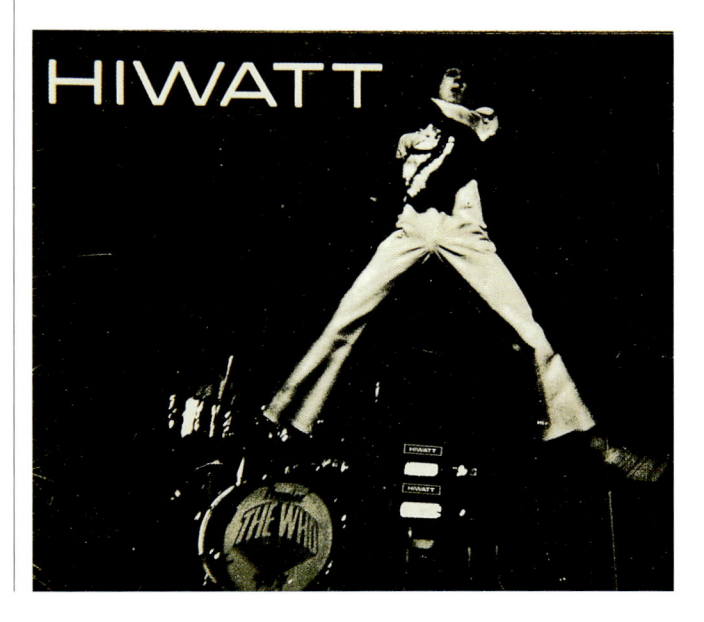

Hughes and Kettner

We tend to think of German rock bands as heavy to the point of industrial, but while H&K – long the big boy of German amp builders – certainly plays up the hard-rock image, it has been catering to a wide and varied range of European guitarist for many years. Despite the brand's longevity, however, it has sometimes had difficulty shaking its slightly stolid, Teutonic image outside continental Europe. That is part of the reason it is as well known in the USA today for its tube loaded pedals (the Rotosphere Leslie-cab simulator, for one) as for its amps.

These days H&K mostly follows the line of mid-priced modern tube amp building, with designs based around PC boards, and often including PCB-mounted tube sockets and controls, too. This is no different than plenty of other modern makers, and is seen by most big companies as the only way to bring a mass-produced amp in at an affordable price.

Like other big names in this market, H&K has concentrated as much, or more, on solid state and modeling guitar amps in recent years, but it still offers some respectable-sounding tube jobs. The Duotone 100W top and 50W combo offer straightforward, Marshall-esque, two-channel rock performance from four or two EL34s respectfully, but give the formula a sleek, modern twist with a backlit, transparent Plexiglas front panel, extra switchable boost, dual switchable master volumes

Hughes & Kettner

and other features. The 100W TriAmp MK2, on the other hand, goes as madly do-it-all as the most excitable of its Californian rivals: three "authentic dual-channel tube amps" in one chassis, six footswitchable channels total, all linked to MIDI-controllable loop system designed to take you from classic clean to hyper-gain overdrive, with a designated stage board to help you handle it all.

Harking back to simpler times, the one-channel Puretone uses two EL34s in a class A output stage to generate 25W, but still features the modern, fluorescent-style H&K cosmetics.

Hughes & Kettner's Duotone offers clean/overdrive channel switching and 100W output from a quartet of EL34s – all packed behind a see-through Plexiglas front panel.

Kelley

I have always loved the Jim Kelley amps manufactured during the early 1980s. They were real simple amps with the emphasis on quality construction and good materials (they were the 1st company to use Groove Tubes in production). They had a built in power attenuator and used the output tubes to create distortion. The later models had a unique biasing circuit with LEDs to indicate the correct bias point, real cool.

But then Jim tired of the business world, and just closed his doors one day to take up a career as a college professor.

K&M (Two-Rock)

The high-end, well crafted K&M Two-Rock and Custom amps, made in Cotati, California, are inspired by the ultra-rare works of a certain legendary Californian amp builder (rhymes with 'humble'). Like many great amps some relatively straightforward designs are at the heart of many models – matched with careful build techniques and attention to detail – but blended with a range of original adaptations and clever switching arrays designed by co-owners Joe Mloganoski and Bill Krinard. All in all, the result is a range of dynamic, playable, musical tube amps. K&Ms are all hand-wired using top grade components. Main models include the 50W or 100W Two-Rock Topaz, the 50W Two-Rock Opal, the 50W or 100W K&M Custom, and the 35W Two-Rock Ruby. The latter is based around a pair of Svetlana EL34s in the output stage, while the others feature duets or quartets of NOS Sylvania 7581s (a military grade 6L6GC). Interestingly, K&M favors the older version Chinese 12AX7 preamp tube, because they say they have found remaining stocks of NOS American 12AX7s to be fairly microphonic. Between the hard labor and the high-end components, these are expensive amps, but they have quickly gained a lot of rave reviews among players.

Komet see Trainwreck

48 THE TUBE AMP BOOK

Laney

The Laney company was started by Lyndon Laney and Bob Thomas in the late 1960s in the city of Birmingham, England. When the late 1970s saw a demand for a more sophisticated tube amp with tightly controlled overdrive and sustain, Laney's popularity faded and tube amp production was suspended in 1980.

More recently, Laney has released a new line of amps named A.O.R. (Advanced Overdrive Response) and has had some success being marketed as an 'affordable' tube amp from the UK.

The overall quality and performance is somewhat less than other, better-known British amps like Marshall and Hiwatt.

However, they are a British tube amp for a player on a budget. With Black Sabbath guitarist Tony Iommi's resurgent popularity among young metal players, Laney – always his amp of choice – has issued an all-valve signature model in his honor.

For recent Black Sabbath concerts, Tony has only been using eight of the 100W amp heads into eight 4x12" cabinets.

McIntosh

A tube amp guide that failed to mention the granddaddy of all tube amps would fall short of being complete.

Anybody who has ever heard a hi-fi system powered by a classic McIntosh tube amp needs no further introduction to the subject. For those of you who haven't, let's just say your education is not yet complete.

Tubes for hi-fi are making a big comeback. It seems that there are types of distortion that impair good audio imaging and which occur in transistor amps but not in tube designs. McIntosh amps have long been the favorite of many audio engineers – although, until recently, they couldn't explain exactly why they sounded better.

One of the unique features of McIntosh amps is their transformer design, which they refer to as 'unity coupled.' This special design allows for incredible accuracy and high reliability. These amps were the standard for the laboratory as well as the recording studio.

For a real treat, try hooking up an old Mac 30W amp to a high efficiency horn system like the Altec 604E co-axial speaker or a pair of Klipsh horns and stand back. The old horn systems work better with tubes because tubes have a natural compression effect on unwanted high frequency peaks.

Transistor amps reproduce everything so clinically that these peaks come off sounding harsh through horn systems. This resulted in more speaker and less horn as transistor amps got more popular. This applied to PA systems as well as hi-fi systems. I guess the Bose speakers are the best example of an 'all speaker' approach.

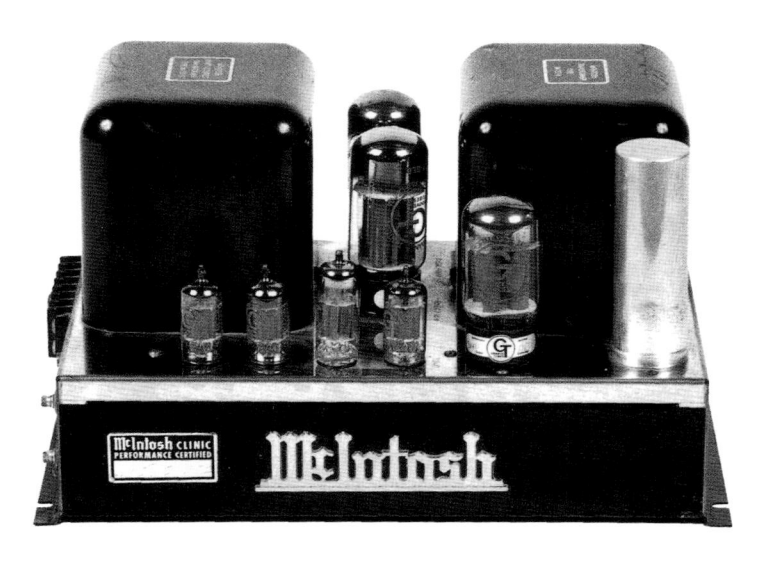

The McINTOSH 30 (pictured above) uses a duet of 6L6s and the unique 'unity coupled' transformer design to produce a very linear output. GT Collection.

While systems like this make transistor amps less harsh, they are much more inefficient, so it takes loads of amplifier power to achieve the same volume as a little old tube amp one-tenth the power driving a horn system.

I wouldn't be surprised if tube amps make a comeback for PA, as they have in the hi-fi business, because tube amps make horns sounds smooth and natural, and horns are a lot less weight and take up less room on stage.

McINTOSH made mono amps from 30W (MC-30) up to 75W (MC-75). They also had stereo models ranging from 30W a side (MC-230) up to 75W a side (MC-275).

Magnatone

Ah, the Magnatone 280-A Stereo-Vibrato twin-12 combo amp. This was the Cadillac of the Magnatone line, made by Magna Electronics of Inglewood, California. Magna Electronics, better known for student lap steel/amp combos, also made real professional amps with unique designs and revolutionary features, such as stereo vibrato. The 280 had, without a doubt, the most advanced guitar amp circuit design of its time. Consider this 12-tube design had four tubes devoted just to the stereo vibrato effect, which actually altered pitch as true vibrato should. Fender and the others often labeled their amps 'Vibrato' when the amp really only had tremolo, a pulsating volume change which was much easier (and cheaper) to build.

A true vibrato amp like the 280 was more complex and expensive to make. Remember, in 1957 a tweed Fender Bassman had just six tubes, and its latest technological advance was the midrange knob. The 280's isolated dual output stages, which used two duets of the rather unusual 6973 (6CZ5) pentode, completed the vibrato effect, as each power section alternately pulsed in time with the vibrato drive circuitry. It was also possible to further enhance the stereo-vibrato effect by plugging in stereo remote speakers in place of, or in addition too, the two stock internal 12" Alnico Jensens. The sound of this amp through two stereo cabs separated by 10-15 feet is indescribable, to say the least. This weird and wonderful amp was the forerunner to the Japanese-made Roland stereo chorus amps that would become popular in the 1980s. The 260 and 280 series were made from 1957 to 1959.

Magnatone had great tone and features, but were underpowered, too heavy, and overpriced compared to the market leaders, Fender and Gibson. The company was bought by Estey Organ Company around 1960, and the Estey company redesigned many of the existing amps, and also brought out new models. For example, they added reverb to the 280 and created the new 480, a great amp if you manage to find one.

Estey expanded the Magnatone amp line to offer lower priced models covered in standard black vinyl with silver grilles, trying to compete with Fender. These economy amps sounded fine, although I prefer the look of the earlier brown and cream motif of the 1950s amps. Perhaps the most extraordinary

This Custom 280 from the GT Collection featured the first (and last) all-tube stereo vibrato to employ two 12" speakers independently to create an incredibly deep vibrato effect. The Estey company bought Magnatone and the later amps bear the Estey name as well. They continued for several years to pioneer exotic guitar amps like the M-20. The Gretsch Nashville is courtesy of Norman Harris, Norm's Rare Guitars, Reseda, California. Photo by Jeff Veitch

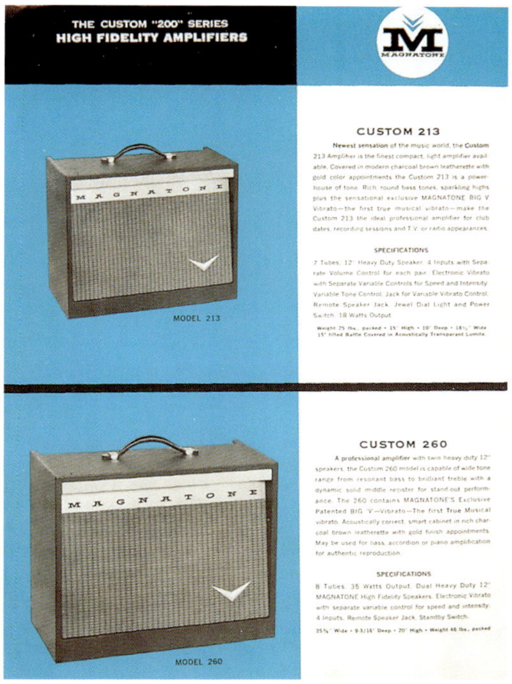

Magnatone

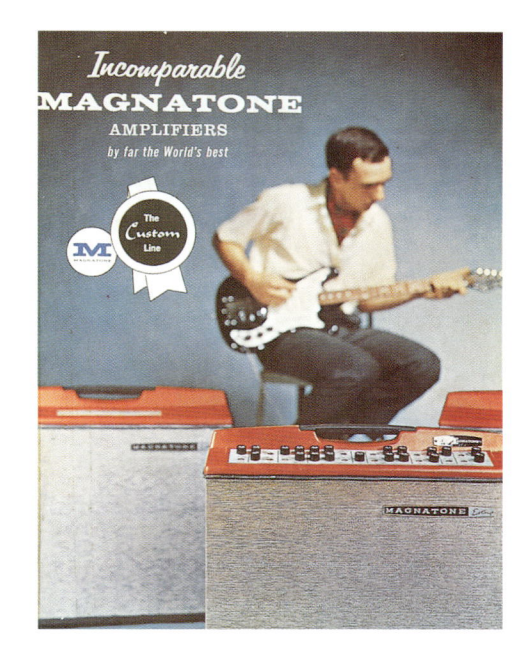

This Magnatone Custom 480 is a slightly later evolution of the magical 280. Just plug in the great later-era Gibson Flying V and get ready to do your best Lonnie Mack impersonation (though the vibrola tailpiece will have to stand in for Lonnie's Bigsby). Amp and guitar courtesy of David Swartz at California Vintage Guitar, photo by Aspen Pittman & Ed Ouellette.

Magnatone model to come from this Estey era was the M-20 stereo amp. Estey bought a quantity of molded plastic TV chassis and produced the 'M' series amps, beginning around 1962. The M-20 used 6CA7s (EL34) and was the most powerful Magnatone amp ever made, sporting about 40W RMS per channel. See the ad for the M-20 amp reprinted in this book, which lists the many features of this spaceship-looking amp. Red Rhodes, our late GT amp guru, hated it when I would bring one of these amps in for service; they are a rats nest of wires, with the preamp located under the control panel and the power amp on the bottom of the cabinet. If the Stereo-Vibrato isn't functioning, you're in for a wild ride tracing all the various interactive components involved, from the preamp, to the pulsators, to the speed regulators, through the varistors which do the pitch shifting and finally to the power section which completes the vibrato effect… lots of work!

The Estey-era Magnatone amps were made from 1960 to about 1968 and are truly unique and desirable guitar amps, with a sound all their own. The original Magnatone amps manufactured by Magna Electronics are perhaps more collectable, and better sounding if you've got a true vintage ear. Lonnie Mack is one of my favorite players who has always used these early Magnatone amps as part of his unique tonal signature. The fabulous Stereo-Vibrato effect is still the best vibrato sound I've ever heard, and for my money is unequaled by anything made today.

Many thanks to Mr. G.A. Rhoads, Ph.D, of Westboro, Mass., for contributing many of the Magna and Estey schematics as well as interesting Magnatone facts used in this item.

THE TUBE AMP BOOK 51

Marshall

Marshall

Hard to believe, but at the time this edition of *The Tube Amp Book* goes to press Jim Marshall is celebrating his 40th anniversary of building amplifiers. This would be an impressive feat for anyone in any field, let alone in a wacky and unpredictable industry that supplies musicians with the tools of their trade. Consider the fact that he has been at the top of the heap for nearly all of those 40 years, however, and this accomplishment seems truly amazing.

The incredible success story of Jim Marshall, and the company bearing his name, is a model for entrepreneurs. As with most such stories, Jim's amazing success is due in part to hard work, a commitment to excellence, and good timing. But perhaps it was Jim's intuitive ability to adapt and change with the times that made him such a success story. I often wonder which came first, the Marshall amp or the metal music for which it became famous. They probably helped shape each other!

In the beginning, there was tone, and it was good, but it was not loud enough.

Jim made his first amp back in 1962, in an effort to keep pace with changing demands on rock music amplification. Simply, Jim's clients were asking for amps that could keep up with the pounding of the emerging rock drummers, who were playing a more aggressive – and *louder* – form of that 1950s rock'n'roll. The music was changing fast in those days, but players also wanted amps they could afford. Jim, a musician himself (well, a drummer anyway), recognized an opportunity to kill two birds with one stone: he would make his amps louder and more affordable than the imported American amps by doing it himself. And why not?

Fender was the leading brand of amps which Jim sold pretty well in his retail store, in Hanwell, England, in a corner of northwest London. Jim has commented that he thought Fender made the best sounding amps at that time. Jim's favorite model was the 4x10" Bassman, which was a tweed-covered combo amp (amp and speakers in the same cabinet). Of course for an English musician, a Fender amp was really expensive, imported all the way from California. Most of the local musicians simply could not afford them. These obvious market conditions certainly helped influence Jim to start building and selling domestically made amps that could rival the imports in sound but at affordable prices. It was just good business sense. However, as a musician, Jim also had the good business sense to make amps for the new rock'n'roll customer, who was playing louder and more aggressively than ever before. His customers asked for more, he gave it to them.

Jim relied heavily on Ken Bran, his technician in the store, to carry out the design work and oversee the production

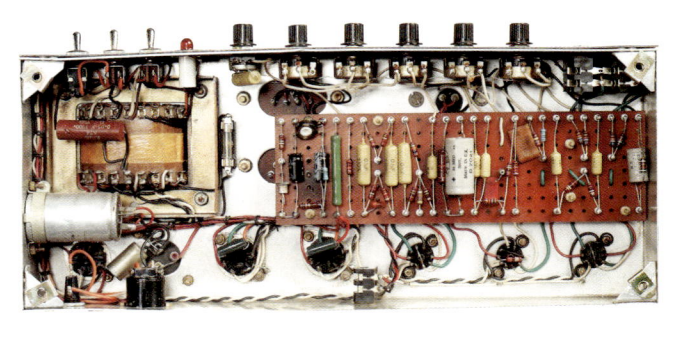

The JTM45 pictured below with matching cabinet, and above in the under-chassis shot, is one of the rarest and most precious of all Marshalls in existence. Supposedly only two 'offset' panel JTM45s were ever built; this one, however, carries serial number 009, but it might be that previous examples were prototypes... or they just didn't begin at 001! (The serial number is shown later in this chapter in the 'Ten Most Desirable Marshalls' section.) Notice, also, Jim Marshall's signature on the front baffle of the amp top, below. Amp owned by Bill Bass, cab GT Collection, '58 Les Paul courtesy of David Swartz of California Vintage Guitar. Photo by Aspen Pittman & Ed Ouellette.

52 THE TUBE AMP BOOK

Marshall

panel was bare aluminum and had just one speaker output along with the detachable power cord socket. No fuse, no voltage selector (this was selectable on the top of the transformer), and no speaker impedance selector (the output impedance of these early models was always 16 ohms to match with his 16-ohm 4x12 cabinets).

The sound was not identical, however, because Marshall was using domestic British components from the many post-war electronic surplus houses. Jim's first power tubes were surplus original 5881 power tubes, a USA military version of the RCA 6L6 design. Soon after, Jim switched to using the UK-produced KT66, which was interchangeable with a 6L6/5881, but produced yet another tonal difference from the Bassman sound.

These first Marshalls had a softer, smoother sound – not at all the bright, crunchy tone that later Marshalls were to have (remember, they were originally designed for bass players). The amps were called JTM45s and had hand-formed aluminum chassis. The JTM stood for Jim and Terry Marshall (Terry was Jim's son).

The first 100 JTM45s made had enameled metal name plates with Marshall in block letters. These amps are now quite rare and valuable. Later, the JTM45s came with large gold Plexiglas nameplates, still with the block lettering. The front panels were also changed to a more appealing gold Plexiglas, with the familiar gold-capped brown control knobs. The back panels were now upgraded to white plastic with gold silk-screened letters to identify the various functions.

Jim's amps now started coming with two speaker outputs and an external fuse holder. It was at about this time that Jim started using a stronger steel chassis, which made the amps more rugged for road tours. They had also by now started using the more common European EL34 power tube, and the GZ34 rectifier was dropped in favor of solid-state rectification.

Naturally, the power supply transformers and speaker output transformers were designed around the EL34 tube – and so Marshalls moved farther away from the sound and performance of the early Fender amps upon which they were patterned. This change from the 6L6 to the EL34 in particular

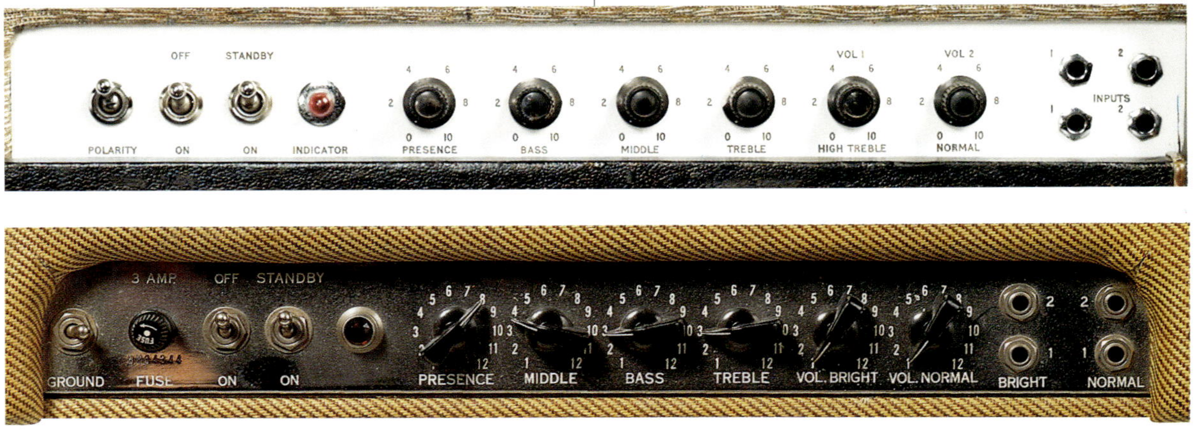

This rare example of the early (post-offset) 1963 Marshall JTM45 amp had a hand-bent aluminum chassis, KT66 output tubes, a tube rectifier, and the small metal logo – all limited to these early Marshalls. About 100 were made, according to Jim Marshall. This very clean example - complete with its original GEC KT66s – is owned by Rick Batey.

of his amps. As Ken also preferred the sound of the early Bassman, they agreed that the Marshall amp should start from there. In fact, the first Marshall amp was virtually identical in design. The controls were identical in function and layout to those of the tweed Fender Bassman – complete with four inputs, two separate volumes, treble, bass, middle, and presence. The front panel also had a power-on switch, standby switch and pilot light. The back

Here's an example of an early Marshall control panel (top-right) as compared to the '59 Fender Bassman control panel. The early Marshalls evolved directly out of the tweed Bassman amp. The early Marshalls also used 6L6 power tubes – or the KT66 direct replacements – and had a tube rectifier. The sound of the early Marshall was more of a blues tone like the Fenders.

THE TUBE AMP BOOK 53

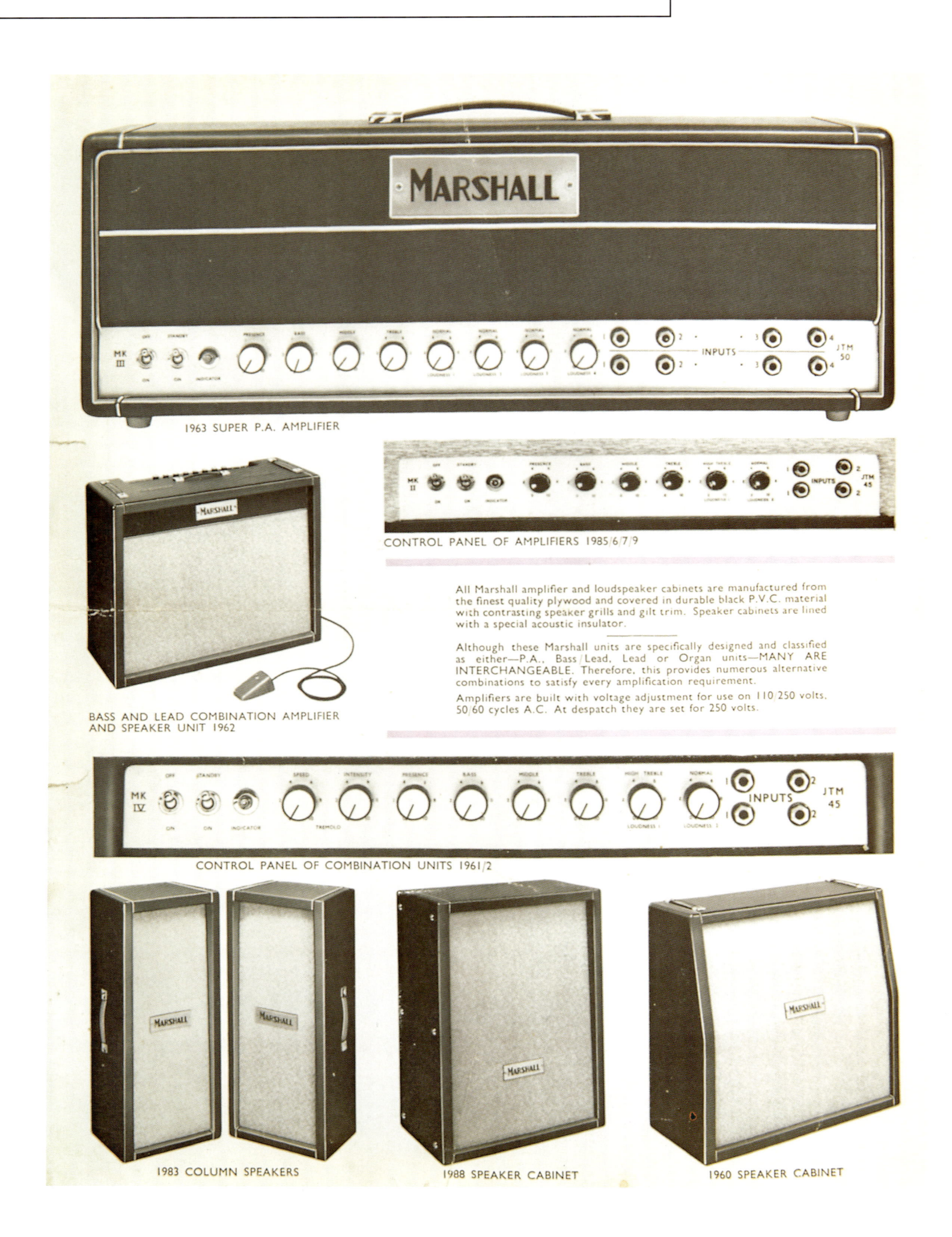

Marshall

1963 SUPER P.A. AMPLIFIER

BASS AND LEAD COMBINATION AMPLIFIER AND SPEAKER UNIT 1962

CONTROL PANEL OF AMPLIFIERS 1985/6/7/9

All Marshall amplifier and loudspeaker cabinets are manufactured from the finest quality plywood and covered in durable black P.V.C. material with contrasting speaker grills and gilt trim. Speaker cabinets are lined with a special acoustic insulator.

Although these Marshall units are specifically designed and classified as either—P.A., Bass/Lead, Lead or Organ units—MANY ARE INTERCHANGEABLE. Therefore, this provides numerous alternative combinations to satisfy every amplification requirement.

Amplifiers are built with voltage adjustment for use on 110/250 volts, 50/60 cycles A.C. At despatch they are set for 250 volts.

CONTROL PANEL OF COMBINATION UNITS 1961/2

1983 COLUMN SPEAKERS

1988 SPEAKER CABINET

1960 SPEAKER CABINET

54 THE TUBE AMP BOOK

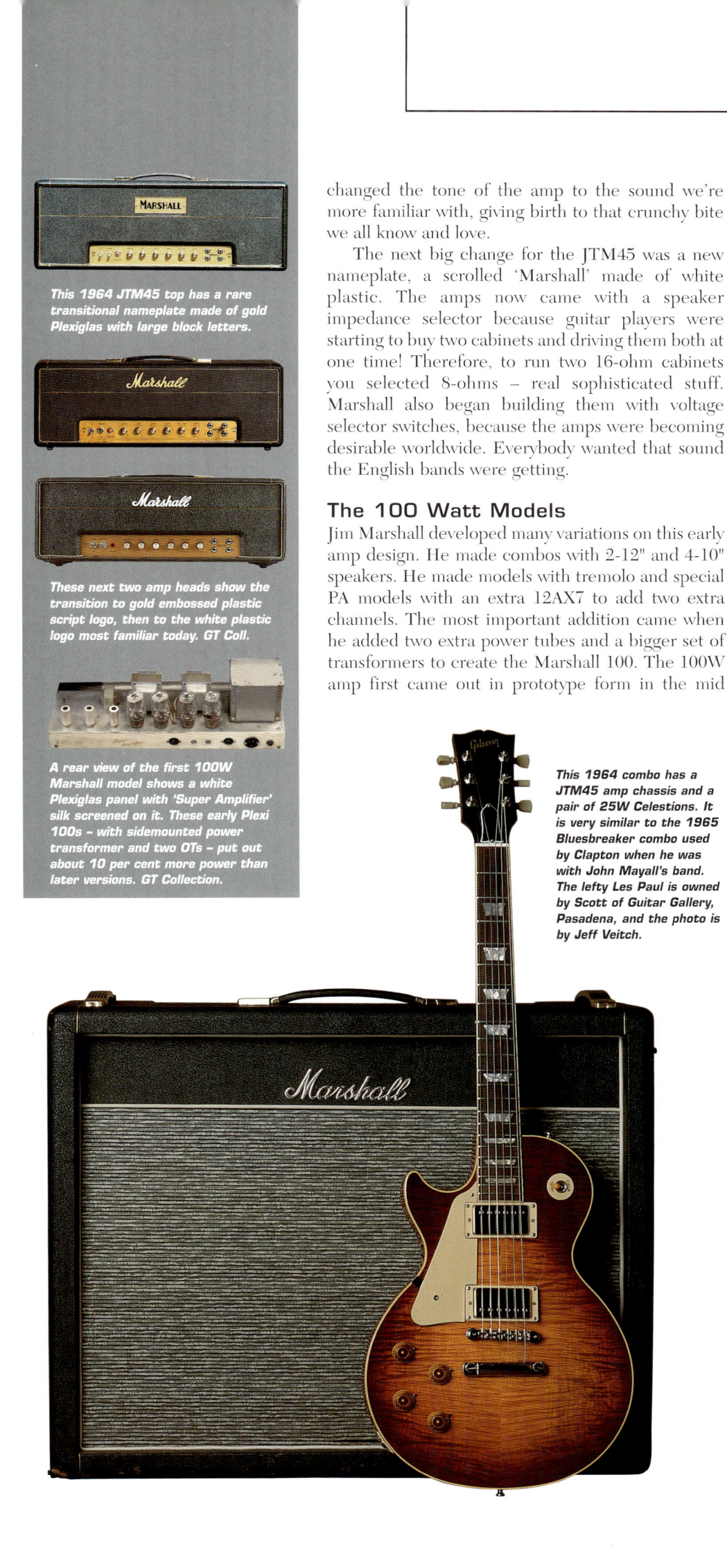

Marshall

This 1964 JTM45 top has a rare transitional nameplate made of gold Plexiglas with large block letters.

These next two amp heads show the transition to gold embossed plastic script logo, then to the white plastic logo most familiar today. GT Coll.

A rear view of the first 100W Marshall model shows a white Plexiglas panel with 'Super Amplifier' silk screened on it. These early Plexi 100s – with sidemounted power transformer and two OTs – put out about 10 per cent more power than later versions. GT Collection.

changed the tone of the amp to the sound we're more familiar with, giving birth to that crunchy bite we all know and love.

The next big change for the JTM45 was a new nameplate, a scrolled 'Marshall' made of white plastic. The amps now came with a speaker impedance selector because guitar players were starting to buy two cabinets and driving them both at one time! Therefore, to run two 16-ohm cabinets you selected 8-ohms – real sophisticated stuff. Marshall also began building them with voltage selector switches, because the amps were becoming desirable worldwide. Everybody wanted that sound the English bands were getting.

The 100 Watt Models

Jim Marshall developed many variations on this early amp design. He made combos with 2-12" and 4-10" speakers. He made models with tremolo and special PA models with an extra 12AX7 to add two extra channels. The most important addition came when he added two extra power tubes and a bigger set of transformers to create the Marshall 100. The 100W amp first came out in prototype form in the mid

This 1964 combo has a JTM45 amp chassis and a pair of 25W Celestions. It is very similar to the 1965 Bluesbreaker combo used by Clapton when he was with John Mayall's band. The lefty Les Paul is owned by Scott of Guitar Gallery, Pasadena, and the photo is by Jeff Veitch.

1960s, with the first ones keeping the JTM45 front panel, and again with a chassis fashioned from aluminum. The back panel came with the name 'Super Amplifier' silk-screened in gold ink. These early prototype 100W amps were larger, and so required larger wood cabinets.

After a few years, Marshall standardized the size of its wood cabinets and the 50W began coming in the larger cabinet along with the 100W. Cabinets for the 100W amps were vented for the additional heat of four power tubes. Marshall completely switched over to the larger box design around 1970 and, as a result, the older 'small' top 50W amps made in the late 1960s are more collectible.

Many players say the small-top 50W amps sound better but I think these are isolated comparisons. My own experience with Marshall amps is that after they have been properly reconditioned and set up with quality tubes, they all can perform about the same – outrageous! I know lots of guys will disagree with this because they've had amps that sounded better or worse than a friend's amp. This is likely due to the fact that a lot of Marshalls out there have been repaired with the wrong parts, or have had really bad tubes and/or filter caps installed.

I know that there are slight variations in all transformers, and that other parts Marshall has used over the years have had 10 per cent tolerances – so that no two Marshalls, in theory, can sound 'exactly' the same (which holds true for vintage Fender and Vox amps, too). Some Marshall power supply transformers can vary as much as 20 per cent and certain years had a higher voltage supply from the transformers, which can make for really powerful 100-watters (I've had some putting out 150 watts). However, I can't tell you how many times I've taken a friend's Marshall that was sounding bad and

THE TUBE AMP BOOK 55

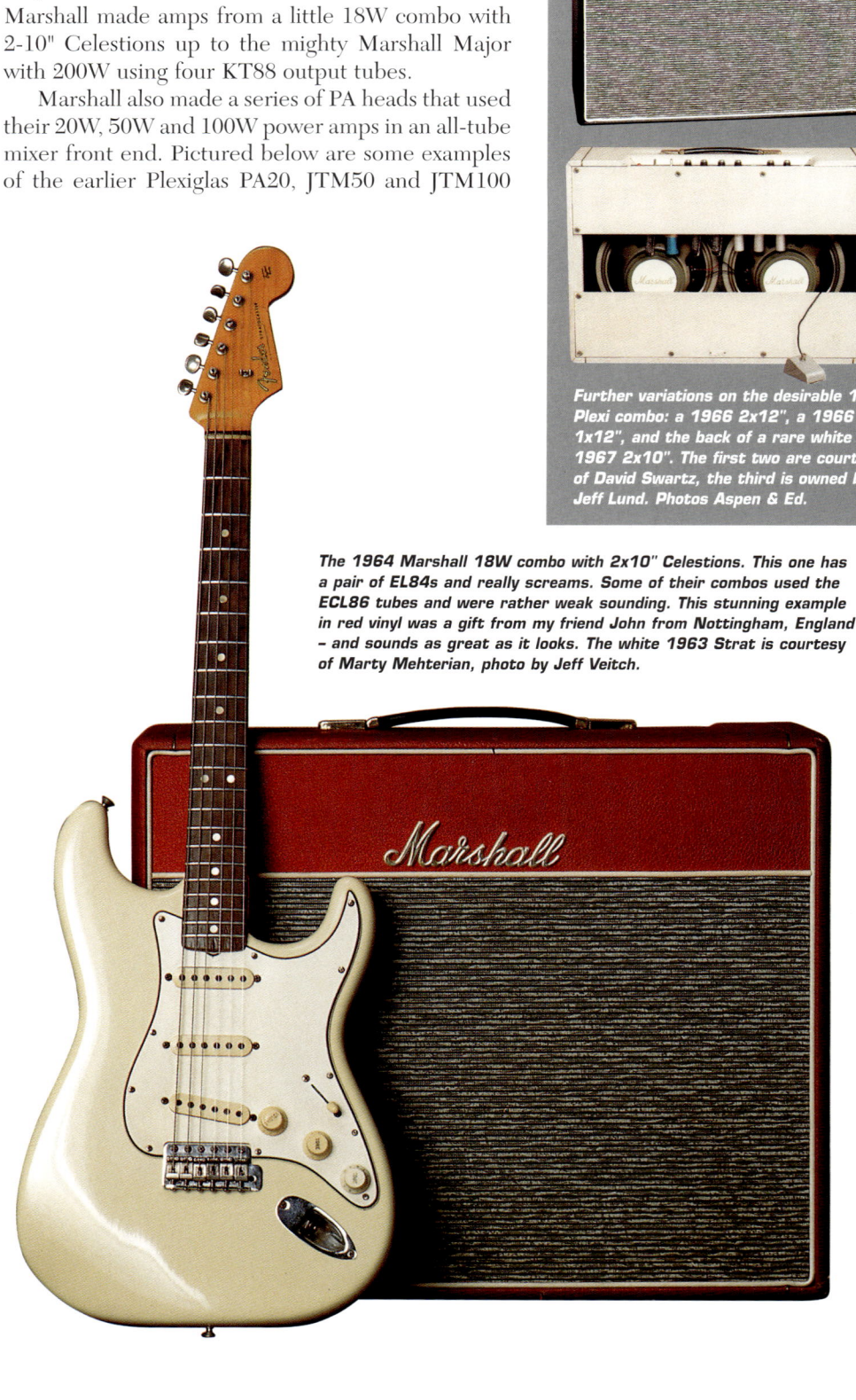

Marshall

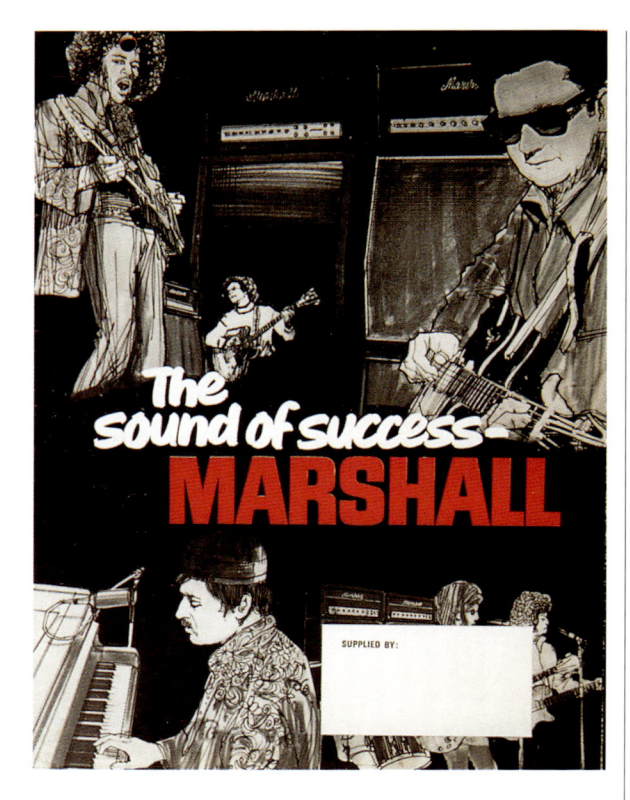

performed simple filter cap replacement, or removed somebody's 'custom modification' while installing our tubes, and had the amp come alive!

I have yet to find a standard Marshall 50 or 100 that can't be made to sound just like the ones on the classic albums. Naturally, the tone and response character of any Marshall can be changed dramatically depending on which type of tubes you use (more on this later), and also which performance rating you choose (there are 10 in the Groove Tubes range alone, for example). I will say, however, that the small tops do look real cool and I've got several in my personal collection.

The Master Volume Model
The Master Volume models came about in 1975, the result of a demand for players to get distortion at lower levels for recording and playing small clubs.

The type of master volume Marshall used was the type that limits the output drive to the power tubes, while increasing the gain in the preamp section. This creates a buzzy overdrive sound many players liked, and it did have many practical advantages for the all-around club player. However, it wasn't the same sounding distortion guitarists had come to know from the earlier non-master volume Marshalls, which came from the power tubes being driven to their limit.

Power tubes have many shades of distortion that change as they go from a soft volume to their loudest volume. The power tubes react with the speaker in a type of electronic feedback that changes as you play harder. Therefore, the player can change the overtones of the distortion simply by hitting the string differently. This is why power tube distortion is generally considered warmer and more flexible.

Many players prefer this kind of distortion. As a result, the standard non-master volume Marshalls are still very popular with players.

I personally like the sound of the non-master volume Marshalls and feel they get a sound the master volume models just can't reproduce, although it's probably not a giant difference. After all, any Marshall is just fine!

The Other Marshall amps
There were many interesting Marshall amps made, other than the straight 50W and 100W guitar amps. Most of these amps were never imported into the US, so I call them the 'Unknown Marshalls.' Marshall made amps from a little 18W combo with 2-10" Celestions up to the mighty Marshall Major with 200W using four KT88 output tubes.

Marshall also made a series of PA heads that used their 20W, 50W and 100W power amps in an all-tube mixer front end. Pictured below are some examples of the earlier Plexiglas PA20, JTM50 and JTM100

Further variations on the desirable 18W Plexi combo: a 1966 2x12", a 1966 1x12", and the back of a rare white 1967 2x10". The first two are courtesy of David Swartz, the third is owned by Jeff Lund. Photos Aspen & Ed.

The 1964 Marshall 18W combo with 2x10" Celestions. This one has a pair of EL84s and really screams. Some of their combos used the ECL86 tubes and were rather weak sounding. This stunning example in red vinyl was a gift from my friend John from Nottingham, England – and sounds as great as it looks. The white 1963 Strat is courtesy of Marty Mehterian, photo by Jeff Veitch.

56 THE TUBE AMP BOOK

MARSHALL — LEAD, BASS & ORGAN SPEAKER CABINETS

MODEL 1960 4 x 12" SPEAKER UNIT

MODEL 1960B SPEAKER UNIT

This illustration also serves to show the cabinet design of all the 4 x 12" speaker cabinets detailed below.

MODEL 1990 8 x 10" LEAD SPEAKER UNIT

ALL UNITS LISTED ON THIS PAGE ARE FITTED WITH CASTORS AND RECESSED CARRYING HANDLES

8 x 10" LEAD SPEAKER UNIT
MODEL 1990. Containing eight 10" (25·40 cms) Celestion loudspeakers with a power output capacity of 80 watts. Because of its particularly bright tonal response this makes an exceptionally good Lead unit.
Cabinet dimensions: height 46" (116·80 cms), width 26" (66·08 cms), maximum depth 13" (33·02 cms).

4 x 12" SPEAKER UNITS
Models 1982 and 1982B below each contain four 12" (30·48 cms) high power Celestion loudspeakers giving a power output capacity of 100 watts, either one being suitable for use with 100-watt amplifiers 1959 or 1992.

One model 1982 and one model 1982B, mounted one on top of the other being necessary when used with a 200-watt Lead or Bass amplifier.

MODEL 1982 Angled front. Dimensions: height 29¼" (74·30 cms), width 30" (76·24 cms), maximum depth 14" (35·56 cms).
MODEL 1982B Straight front. Dimensions: height 29¼" (74·30 cms), width 30" (76·24 cms), depth 14" (35·56 cms).

The 1960 and 1960B models below each contain four 12" (30·48 cms) heavy duty Celestion loudspeakers giving a power output capacity of 75 watts, either unit being suitable for use with 50-watt amplifiers *1986, 1987 or 1989.
One each of these units, mounted one on top of the other make a very impressive set-up when used with a 100-watt Lead or *Bass amplifier.

MODEL 1960 Angled front. Dimensions: height 29¼" (74·30 cms), width 30" (76·24 cms), maximum depth 14" (35·56 cms).
MODEL 1960B Straight front. Dimensions: height 29¼" (74·30 cms), width 30" (76·24 cms), depth 14" (35·56 cms).
*See special 4 x 12" Bass speaker cabinets below.

SPECIAL 4 x 12" BASS SPEAKER UNITS
Each with a power output capacity of 75 watts the following units contain four specially designed 12" (30·48 cms) Goodman loudspeakers capable of handling exceptionally low Bass frequencies. These units are particularly recommended for use with 50-watt and 100-watt Bass amplifiers.
MODEL 1935 Angled front. Dimensions: height 29¼" (74·30 cms), width 30" (76·24 cms), maximum depth 14" (35·56 cms).
MODEL 1935B Straight front. Dimensions: height 29¼" (74·30 cms), width 30" (76·24 cms), depth 14" (35·56 cms).

MODEL 1972 2 x 12" OR MODEL 1988 1 x 18" SPEAKER UNIT

All the speaker units listed on this page are supplied complete with a 5' (152·40 cms) lead and waterproof cover.

2 x 12" LEAD & ORGAN SPEAKER UNIT
MODEL 1972. Containing two 12" (30·48 cms) heavy duty Celestion loudspeakers with a power output capacity of 50 watts. Cabinet dimensions: height 30" (76·24 cms), width 26" (66·08 cms), depth 12" (30·48 cms).

1 x 18" BASS SPEAKER UNIT
MODEL 1988. Containing one 18" (45·72 cms) Celestion loudspeaker with a power output capacity of 50 watts. Cabinet dimensions: height 30" (76·24 cms), width 26" (66·08 cms), depth 12" (30·48 cms).

Marshall

MARSHALL — LEAD, BASS & ORGAN COMBINATION AMPLIFIERS

CABINET & CONTROL PANEL OF 50-WATT MODELS 1961 & 1962

CABINET & CONTROL PANEL OF 18-WATT MODELS 1958 & 1973

50 WATTS OUTPUT

LEAD MODEL 1961. Fitted with four 10" (25·40 cms) Celestion loudspeakers. Four inputs: separate volume controls for high treble inputs (channel 1) and normal inputs (channel 2). Presence, Bass, Middle and Treble tone controls common to both channels. Tremolo unit with separate Speed and Intensity controls; detachable Tremolo footswitch control. Cabinet dimensions: height 23" (58·40 cms), depth 10½" (26·65 cms), width 32" (81·32 cms). Fitted with castors.

BASS & LEAD MODEL 1962. Amplifier and cabinet specification as model 1961 above, but fitted with two 12" (30·48 cms) Celestion loudspeakers.

All the models listed on this page have high impedance inputs and interchangeable output impedance. Voltage adjustable for use on 110/250 volts, 50/60 cycles A.C. At despatch they are set for 250 volts. Complete with mains lead and waterproof cover.

18 WATTS OUTPUT

LEAD MODEL 1958. Providing a distortion-free volume level and quality of reproduction far above that suggested by the modest 18 watts output rating. Containing two 10" (25·40 cms) Celestion loudspeakers. Two-channel amplifier, each channel with two inputs, separate volume and tone controls: ON/OFF switch and Standby switch: Tremolo Speed and Intensity controls: detachable Tremolo footswitch control. Cabinet dimensions: height 20" (50·80 cms), width 24" (61·00 cms), depth 9" (22·86 cms).

LEAD, BASS & ORGAN MODEL 1973. Amplifier specification and controls (including Tremolo) as model 1958 above but fitted with two 12" (30·48 cms) Celestion loudspeakers. Cabinet dimensions: height 20" (50·80 cms), width 28" (71·16 cms), depth 9" (22·86 cms).

REVERBERATION. 18-watt models 1958 and 1973 are available fitted with Reverberation unit at small extra cost.

Marshall SupaFuzz

MODEL 1975. Three transistor, battery-operated unit contained in smart gravity-moulded metal case. Two controls—volume and filter; jacksocket input and output; foot control pushbutton ON/OFF switch. Extra long sustain (15 seconds). Battery automatically cuts out when unit is disconnected from Amplifier. Durable stove-enamelled finish in a range of attractive colours.

58 THE TUBE AMP BOOK

Marshall

Here's the front of Jeff Lund's gorgeous 1967 18W 2x10", this time with a lovely '57 LP Jr. courtesy of Norman Harris of Norm's Rare Guitars. Photo Aspen & Ed.

Marshall LEAD & ORGAN SET-UPS

Lead
A UNIT 1. 50 watt. Comprising 1987 Amplifier and 2045 speaker cabinet
B UNIT 2. 50 watt. Comprising 1987 Amplifier and 2053 speaker cabinet
C UNIT 3. 100 watt. Comprising 1959 Amplifier and 1960/1960B speaker cabinets (stacked)
D UNIT 4. 100 watt. Comprising 1959 Amplifier and 2052 speaker cabinet
E UNIT 5. 100 watt. Comprising 1959 Amplifier and 2054 speaker cabinet
F UNIT 6. 200 watt. Comprising 1967 Amplifier and 2056 speaker cabinet
G UNIT 7. 200 watt. Comprising 1967 Amplifier and 2-2052 speaker cabinets linked together.

Organ
A UNIT 8. 50 watt. Comprising 1989 Amplifier and 2045 speaker cabinet
B UNIT 9. 100 watt. Comprising 1959 Amplifier and 2054 speaker cabinet
C UNIT 10. 100 watt. Comprising 1959 Amplifier and 1980 speaker cabinet
D UNIT 11. 100 watt. Comprising 1959 Amplifier and 2053/2053B speaker cabinets (stacked)
E UNIT 12. 200 watt. Comprising 1967 Amplifier and 2056 speaker cabinet
F UNIT 13. 200 watt. Comprising 1967 Amplifier and 2054/2054B speaker cabinets (stacked)

Marshall LEAD/ORGAN & BASS SET-UPS

Lead/Organ
A UNIT 14. 50 watt. Comprising 1987 Amplifier (for organ use 1989)and 2053 or 2054 speaker cabinet
B UNIT 15. 50 watt. Comprising 1987 Amplifier and 1960 speaker cabinet
C UNIT 16. 100 watt. Comprising 1959 Amplifier and 2053 or 2054 speaker cabinet
D UNIT 17. 100 watt. Comprising 1959 Amplifier and 1982 speaker cabinet
E UNIT 18. 100 watt. Comprising 1959 Amplifier and 2052 speaker cabinet
F UNIT 19. 200 watt. Comprising 1967 Amplifier and 1979/1979B speaker cabinets (stacked)

Bass
A UNIT 20. 50 watt. Comprising 1986 Amplifier and 2045 speaker cabinet
B UNIT 21. 100 watt. Comprising 1992 Amplifier and 1980 speaker cabinet
C UNIT 22. 100 watt. Comprising 1992 Amplifier and 1935/1935B speaker cabinets (stacked)
D UNIT 23. 200 watt. Comprising 1978 Amplifier and 1979/1979B speaker cabinets (stacked)

These set-ups are recommended as being ideal combinations of Marshall equipment. There are many combinations not shown but bearing in mind price, power, and the type of sound needed, these set-ups cover most requirements.

Here's another match made in tone heaven: a 1966 Bluesbreaker paired with an all-gold '55 Gibson Les Paul – and both of them outrageously clean. Can you hear the thunder? Amp courtesy of Jeff Lund, guitar owned by Norman Harris of Norm's Rare Guitars. Photo by Aspen Pittman & Ed Ouellette.

Marshall PA tops. These actually are fine for guitar with minor changes to the front end, or they make great keyboard amps for medium-sized clubs.

Marshall also built amps under different names, such as the Park line, and I have a little section devoted to Park amps later on in the book. Perhaps the rarest and least known of the Marshall varieties were the 'NARB' amps, made for just a few months during a kind of crazy period for Marshall. Until recently, I only heard rumors about the NARB amps – and you know about music business rumors. Anyway, Jim Marshall, for some crazy reason (probably profits), decided to make Marshall amps under another name and chose to use his friend and designer Ken Bran's name spelled backwards, hence 'NARB.' Shortly after Jim decided to make these he stopped, and so as far as I know there were probably less than 100 made, mostly 100W tops from the early 1970s

Without doubt, the most colorful products Marshall ever put out were the custom-order guitar stacks they made between 1970 and 1974. Marshall dealers and the endorsement bands had a field day with Marshall amps available in white, red, purple, orange, tan, green, yellow and blue.

I've been lucky enough to acquire several, and many are shown in the 'Ten Most Desirable Marshalls' section that follows. My favorite is an orange Tremolo 50 with a Plexiglas panel, which is quite rare because almost all of the colored Marshalls came after the company switched to brushed gold metal control panels. I also have a purple Super Lead 100 stack and a red Park 50 made in late 1969. (Park amps are another interesting

THE TUBE AMP BOOK

Marshall

Marshall amp story; see the section on Park amps for more details.)

Marshall put out many interesting combos, like the early JTM45 1962 model (that is the model number, rather than the year) with two 12" Celestion 20W speakers. This was the famous amp that Eric Clapton is pictured with while recording John Mayall's *Blues Breakers* album, some of the best-sounding lead guitar of all time. See the 'Ten Most Desirable Marshalls' section for more on this great amp.

Marshall has historically offered combo versions of its 50W and 100W amps, and the later ones have even had built-in reverb, the first of which was the Super Artist, built in the mid 1970s. The amp was designed for organ and electric piano so it doesn't sound dirty enough for many guitarists. This is probably the main reason the amp didn't catch on and was discontinued after a few years.

The Reissue Models

Leave it to the young guys at Marshall to bring back the best of Marshalls past. I mean, of course, the famous Bluesbreaker 2x12 combo and the matching JTM45 half stack. These amps are great reproductions in every detail, from the covering and grille materials right down to the tubes. They sound great and feel great, too. I've played them both and have to say I like them a lot, even more than the Fender-reissued Bassman, which I also like. I won't say they're right for everybody, especially players who need effects loops, lots of knobs, and 5,000 MIDI presets controllable with your feet.

No they don't have all the fancy gadgets. What they do have, that is lacking in so many of the new amps and preamps, is big tone along with real touch. The only catch is you have to get that tone with your fingers, not your feet. If that sounds like a challenge you'd enjoy, and you don't have a small fortune to shell out for the original article, these new/old amps are for you.

THE TEN MOST DESIRABLE MARSHALLS

1) CIRCA 1962: FIRST 'BLOCK' LOGO JTM45 HEAD AND 4X12" CAB

Ken and fellow worker Dudley Craven hand-built the first all-aluminum-chassis amp in 1962, right in Jim's shop, and started taking orders. Jim had the foresight to separate the amplifier from the speakers, and this became an early Marshall trademark; these amps were fairly powerful and he needed more speakers to handle them. Hence, out of simple necessity, was born perhaps the most significant product Marshall ever introduced, the 4x12" cab with it's slanted top baffle board. The idea for this slanted baffle board occurred to Jim because he thought it would be a more balanced look when the relatively thinner amp top (or 'head') was sitting on it. Another component feature that further moved the amp from the Bassman predecessor was the Celestion 12" speaker. These early Celestions had alnico magnets and are, in my opinion, among the best sounding guitar speakers ever made.

These first Marshall 'pre-model-name' 45W tops have a soft, creamy tone, which is full of harmonics and perfect for the blues player of any era. But they were also just a bit louder than anything else in the store when they arrived, and had more gain, which suited them for rock players of the day. The simple cathode follower tone stack borrowed from the Bassman allows the tone to be shaped by the style of the player: play soft for a smoother bass tone, play harder and the tone gets brighter with more edge. You can hear the personality of the player – it has 'touch.'

Of all components, tubes have perhaps the most influence on the tone of an amp. I have already

This 'offset' panel JTM45 carries serial number 009, though it may well be earlier than the ninth production unit built. A back view of the cabinet shows how positioning the chassis to the far right of the cab (as viewed from the front) puts the heavy transformers nearer the center, for a better balance when carried by the handle. The photo also reveals Jim Marshall's signature in two different places on the rear panel of the chassis. Amp courtesy of Bill Bass, photo by Aspen Pittman & Ed Ouellette.

I walked into Music Exchange in Birmingham, England and saw this 1963 Marshall JTM45 PA in its original plastic. I just threw £20 notes at Dave and Gary until they gave it up. I've never seen another so clean. Jim told me he made 100 early metal Marshall nameplates with the red letters. Well, here are three of them! The cabs have two Celestion 12" speakers each and sound rather nice. GT Collection, photo by Jeff Veitch.

60 THE TUBE AMP BOOK

Marshall

mentioned the KT66 power tubes, but the preamp tubes they chose also greatly influenced that signature Marshall tone. Fender and Marshall both used the 12AX7/ECC83 tube for the first stages, but by about the mid 1960s Fender switched to the 12AT7/ECC81 driver/phase-inverter tube to satisfy its country players; though the Bassman upon which the JTM45 was based, along with other tweed and early tan Tolex Fender amps, had used a 12AX7 in this position. The driver tube 'drives,' or pushes, the power tubes. Marshall stayed with the 12AX7/ECC83, which gave its amps a bit more drive and distortion and made them more suited for the rock sound.

A fun and easy tonal experiment can be made by simply changing the driver tube (that's the tube closest to the power tubes) of any Marshall to a 12AT7/ECC81. Or, you can also 'mod' the blackface and early silverface Fender amps by replacing the 12AT7/ECC81 with a 12AX7/ECC83. These tubes are safely interchangeable, and the difference in performance is amazing.

This early 45W top with the metal block logo would evolve to become the Plexi block logo JTM45 over the next few years. All of the early amps from this era sounded slightly different because of the wide tolerance of the resistors and caps and ongoing changes of components. A tolerance of 10 per cent or more was common for all amp components of this era and explains the slight differences between otherwise identical amps.

Remember, this was not rocket science, but Rock-It science. Precision and expensive components were seldom used by amp manufacturers. One exception was the tubes, which were more consistent due to the many competing factories in those days. These NOS ('new old stock') tubes from RCA, GE, Mullard and Gold Lion are coveted today, selling for as much as $300 each! (For your information, Groove Tubes was able to reverse-engineer the Gold Lion KT66 tube, and we currently produce an exact replica, indistinguishable from the original, which sell for about $75 for a matched pair.)

2) CIRCA 1965: THE 'BLUESBREAKER' MODEL 1962 2X12" COMBO AMP

It was a natural evolution to package Jim's early 45W top into a 2x12" combo. The earliest version of the model 1962 had a larger cabinet with a slightly upward-slanted baffle board holding a pair of Celestion speakers. These 20W 12" Celestions were barely capable of handling the powerful dual KT66 output section, but they sounded great. The first model 1962 was produced in late 1964 or early 1965, with 'Snakeskin' grilles and the early script Marshall logos (remember: rather confusingly, Marshall model numbers have nothing to do with years of manufacture, even though they look that way!). The second version, which had evolved into a noticeably thinner cabinet, appeared mid 1965.

Part of the growing popularity of the Marshall

The Bluesbreaker is arguably the sound that launched British blues-rock in the mid 1960s – at the hands of Eric Clapton and a PAF-loaded Les Paul. Here's a great one from 1966, with a back view that reveals its Celestion G12M Greenback speakers. Amp courtesy of Jeff Lund, photo by Aspen Pittman & Ed Ouellette.

amp was thanks to the fact that it was more affordable. Jim excelled at making good amps at a competitive price. A Vox AC30 combo sold for one third more, and a Fender combo for about twice the price of Jim's 1962 model combo amp. This amp would have been a great seller based on value alone, but it was destined to become perhaps the most important Marshall amps ever. Now the stuff of legend, most Brit-rock fans know it as the amp used by a young Eric Clapton when he was still a relatively unknown British guitarist playing on John Mayall's 1965 hit *Blues Breakers* album. Eric set new standards for electric guitar tone, and Marshall was at the heart of that tone… and the 1962 combo therefore became known as the 'Bluesbreaker'.

Several important design and component factors contribute to the incredible tone of this Bluesbreaker amp:

1) The KT66 power tubes yield a round, bell like tone with soft distortion character.
2) The 12AX7/ECC83 phase inverter produces more compression and distortion.
3) The cabinet is open backed and made of dense Baltic birch plywood. The second version with the

Marshall

thinner cabinet has its own special sonic signature.

4) The two 20W 12" Celestion speakers placed side by side sound different than the 4x12"s in a closed-back cab, and better suited to recording.

5) The cathode follower tone stage gives it that reactive 'touch' for more expressive range. Marshall maintained this tone circuit from the tweed Fender days after Fender's own amps had moved on to a different design. The major differences of the resistor and capacitor values of the passive tone circuit used by Marshall allowed much more gain, and shifted the frequency response curve up and out – all factors that rock players loved.

6) The relatively high plate voltages (as compared to Fender) gave the Bluesbreaker amp that aggressive, full out tone, which combined with Eric's early Les Paul (with hot alnico humbucker pickups!) to make Rock history.

7) This point may sound a bit technical, but needs to be mentioned to further show the evolution of the Marshall amps from their early Bassman roots. Ken Bran wisely chose to increase the negative feedback in his amp designs (as compared to the Bassman his design was based on), and this greatly contributed to the even harmonic characteristics of the Marshall amps.

It was truly a match made in heaven, partly planned and partly by divine accident. This amp and the records it made have been an inspiration to guitarists and amp builders ever since.

3) CIRCA 1965: THE MODEL 1958 2X10" 18W COMBO

Introduced in 1965, this little tone machine was a natural product for Marshall to offer. Its lower price competed with the many amps on the market from Vox, Watkins and others. It used three 12AX7/ECC83 preamp tubes like the 'big boys' in the line, a pair of EL84 power tubes and a tube rectifier (EZ81). The tube rectifier slightly sagged when the amp was pushed, which many players liken to using a compressor. Coupled with the lower-powered early ceramic-magnet Celestion speakers in a smaller, lighter cab with snakeskin grille and a Plexi control panel, these amps looked and sounded like the little brother to the Bluesbreaker amps.

Smaller amps would cost less (so Jim could sell more), which was the prime motivation for making these amps, I am sure. But they also were better suited to the smaller UK clubs (and the recording studio), and sold very well in Europe. Unfortunately, few of these were exported to the US and so they are not easy to find in the colonies today. I have photographed a red vinyl-covered model 1958 here, which is extremely rare.

They also had added tremolo as a standard feature (footswitch included). Reverb was available on special request, adding an extra EF86 for the drive and recovery stage. But perhaps the component that most influenced tonal departure in this series was the use of its lower-powered EL84 output tubes.

The EL84 tubes have their own distinct and sweet tone, are very gain (touch) sensitive, and quite articulate. Then, as they are pushed into full power, they have a brassy and slightly brittle distortion character that was more typical of the Vox amps The Beatles were using. As they use a 9-pin miniature socket like an ECC83 preamp tube, they cost much less than octal sockets and tubes, so these were – and are – a popular power tube choice for the lower-powered amps from many companies both then and today.

Marshall's beloved 18W combo came in a broad range of formats, and was a budget-priced item for the British amp maker when it was introduced. Left, a 1965 2x10" model. Amp courtesy of David Swartz of California Vintage Guitar, photo by Aspen Pittman & Ed Ouellette.

The 18W Marshall pictured left is a 2x12" combo from 1966 – and sounds as sweet as your dreams partnered with this slab-board '59 Stratocaster with gold hardware. Amp and guitar courtesy of David Swartz of California Vintage Guitar, photo by Aspen & Ed.

Marshall

As these early models had the period Marshall preamp section, however, they were more aggressive, having much higher gain, and made incredible recording amps.

The amp was priced to sell, at just £55 (about $80)! The same amp chassis was fitted into a 1x12" cab (model 1974, also at £55) and a 2x12" cab (model 1973 £70). In 1967 the amp evolved into a 20W design using just two ECC83/12AX7 tubes, no reverb option, and Marshall switched to a solid state rectifier… fewer tubes, more profit. They sold better and so are fairly common today. The new version kept the same model names, and still sounded pretty cool, but the design changes reduced much of the raw tone and feel from the first 18 W model.

4) CIRCA 1967: JTM50 'PLEXI TOP' MODEL 1987 (THE FIRST EL34 MODEL)

This amp's release brought the final key ingredient of the Marshall sonic recipe: the change to the EL34 power tube. The KT66 was being discontinued, and they were expensive. But this popular European power pentode was much less costly and very easily available. These EL34s were made by Mullard, a British tube company, and had a bigger bass, crunchy highs, and a scooped out midrange. But more importantly, they delivered more power, about 10 per cent more than the previous 6L6/KT66 type.

Marshall also phased out the tube rectifier, replacing it with a modern solid-state rectifier design that increased the amp's power. The amp's rectifier stage converts AC to DC for the signal path tubes. The tube rectifier that Marshall had been using, specifically a GZ34/5AR4, was the highest rated tube rectifier commonly available at that time. But as more and higher power tubes were being added to the Marshall amps the GZ34 began to prove a weak link.

These two tube changes both significantly changed the tone and increased the power of the new 50W models; both were desirable for the evolving sound of rock.

Cosmetically, the amps were evolving too. These new amps had a plastic silkscreened control panel, that used a gold-colored background for the black silkscreened lettering. The earlier JTM model name, which stood for Jim and Terry Marshall (Terry was Jim's son), was changed to JMP, which stood for Jim Marshall Products. These earlier amps were also labeled MKII, indicating the evolution of the beast, but later amps dropped this off the name plate, which was a good idea because if every slight change in design and appearance had a number, we'd likely be on MK101 by now, and that doesn't have the right marketing ring to it, now does it?

Another marketing dimension for Marshall was to change the electronics slightly and use the same basic amp for other purposes, hence the 1986 model for bass and the model 1985 for PA. The Bass head actually just added a capacitor to one of the two channels to soften the highs (and make it more 'bassy'), and the PA amp added an extra preamp tube (was three, now four) for two more channels, totalling four channels with independent volume controls for each. These amps are not often listed in the Top Ten by collectors, so they usually sell for much less than the 'lead' amps like the 1987. However, for smart players on a budget – and who can use a wire clipper – these bass models can easily be converted to great lead amps simply by removing the small cap at the input jacks. And the PA amps usually need just a slight resistor change to become a guitar amp again – same tubes, transformers and tone you'd get from the lead models, and for half the price!

5) CIRCA 1969: MODEL 1987 SECOND-VERSION 50W TOP WITH METAL FACE PLATE

Okay, so why do I include this amp if it's got the same model number as my #3 Plexi 50W model? Simple, the same reason I'd list a 1959 Fender Strat and a 1963 Fender Strat in my top 10 all-time guitars: they are different. While many of these evolving Marshall amps had the same model number, and were still the same in many respects, the evolution from one

This rare Orange Super Tremolo 50 has a Plexiglas panel and was made for Canada, most likely for a Guess Who tour, in late 1968. This has to be one of the earliest examples of a colored top. The 1960 Tele Custom and amp are from the GT Collection, photo by Jeff Veitch.

THE TUBE AMP BOOK 63

Marshall

transformer to another, or one tube to another made a significant change to the power, sound, and feel of the amp.

It is this evolving and perfecting of the beast that had interested me the most when I first started collecting and exploring the Marshall amp many years ago. I found that the Plexi era amps had a slightly cleaner, softer tone, while the amps that evolved a few years later had more output, more gain, were slightly brighter, and were way more aggressive. This evolution was following (or leading?) the developing styles of the rock bands. British bands especially were evolving from the early R&B and Delta Blues tonal roots into the British heavy metal brand of rock. Sometimes I wonder which came first, the chicken or the egg? In other words, was the music evolving the Marshall amp, or was it the other way around? In the end, I guess it would be a bit of both, but probably Jim was just providing what his customers were asking for, as any good businessman would.

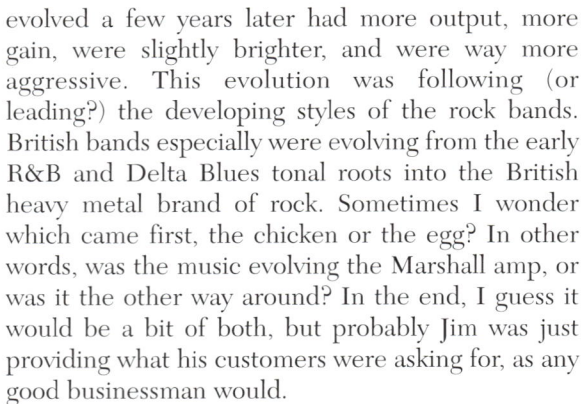

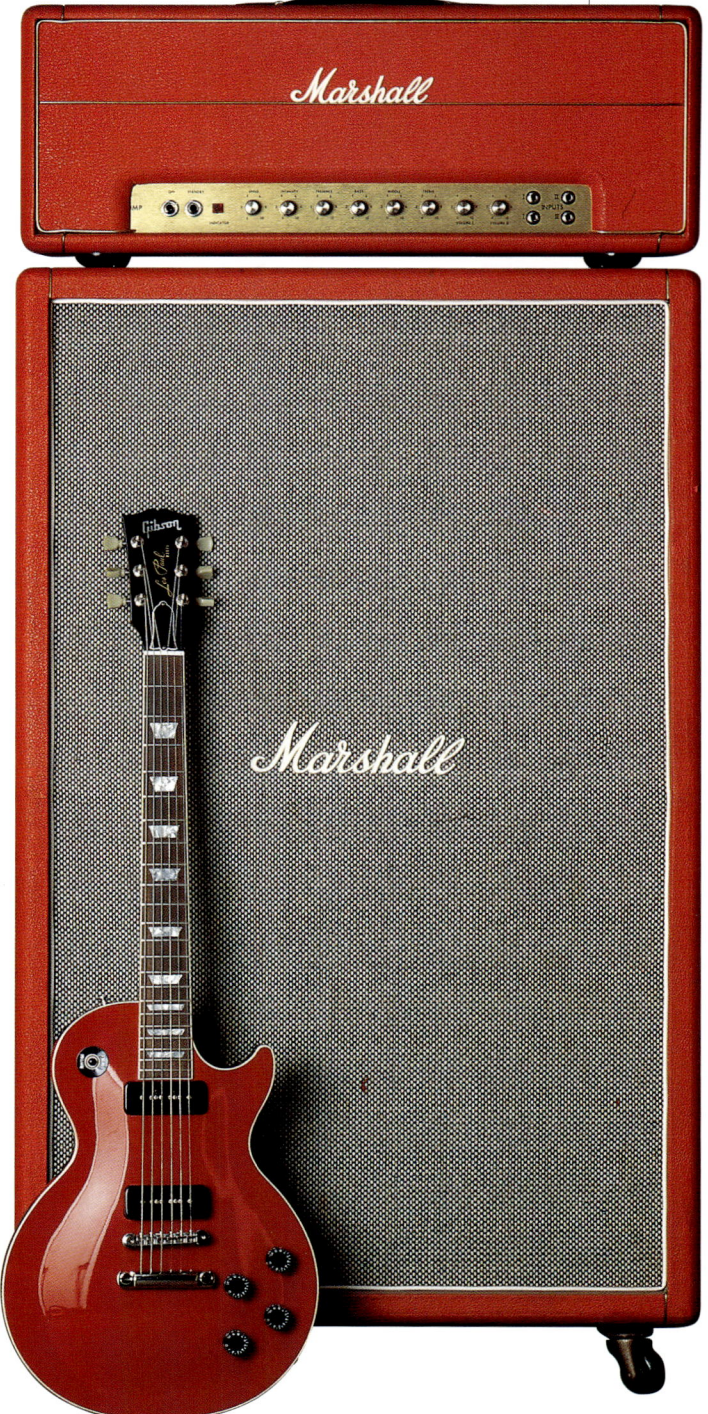

So, while this model had basically the same circuit design as its predecessor, it added 'new and improved' support components, especially the transformer sets, which changed this amp into an even higher gain and more aggressive guitar amp.

When Jim converted to the EL34, the increased voltage handling capacity of this tube meant he could up the power transformer supply voltage a bit from the older 6L6 designs. As he increased these plate voltages, the tubes got more juice and produced a little more power and gain; they got more juicy and touch-sensitive, which players loved!

These newer amps also had some minor cosmetic changes, including scripted Marshall logs (instead of the rectangular box logo on a Plexi) and aluminum control panels replaced the earlier Plexiglas panels. Collectors commonly call amps from this era 'metal panel' Marshalls. They were first built into smaller cabs, but as the Model 1987 evolved these were housed in the larger, standard cab. While the earlier Plexi panel 50W tops are prized and bring real serious collector bucks, players in the know are real happy with this later-era amp as they cost half as much as amps made just a year earlier.

This looks like a red 8x10 50W, but it's an enlarged 4x12 cabinet made on special order during the early 1970s. Jim Marshall's signature is on two of the inside cabinet panels and he explained to me that he probably covered this one as an example, likewise insisting each employee sign their work. The matching red Les Paul is owned by Norman Harris of Norm's Rare Guitars, Reseda, California. The photo is by Jeff Veitcfh.

6) CIRCA 1966/67: THE MODEL 1959 JTM 100W – PLEXI TOP

Pete Townshend of The Who had asked Jim for a louder 50W top… but twice as loud. So Jim's design team of Ken Bran and assistant Dudley Craven started experimenting with a double-50 prototype that could deliver more power and drive a second 4x12" cab. Their earliest model 1959 used four KT66 tubes pushing two 50W output transformers built on a single chassis with a massive power transformer. Perched on a pair of enlarged 4x12" cabs, the first Plexi-paneled 100W lead

Marshall

In its day, the 100W Marshall Plexi must have seemed like the final evolution of the rock guitar amp... until the 200W Marshal Major came along. For huge crunch, scorching leads, and mucho volume, however, the 100W Plexi still ruled. It was the rig of Pete Townshend until his move to Hiwatt, as well as the amp that Jimi Hendrix was best known for playing through. The full stack shown right displays the 'Hendrix sound chain.' from white Strat to Vox Wah to Fuzz Face to Marshall; the solo amp top reveals enough detail to show that the amp had changed little – outwardly – from the 50W version.

emerged (early ones used 'JTM45' panels!) onto Pete Townshend's backline. The bottom cabs were made straight so as to fully support the angled cab – or, as it would now forever be known, the 'top' cab. This seven-foot stack towered above the Twin Reverb or Vox AC30. It didn't take long for the big touring bands to completely switch over to these tone monsters, just for the look alone.

Later, Marshall began to offer Celestion G12H-30 speakers in their 4x12" 'Bass' cabinets. The 'H' stood for 'heavy magnet,' and they handled more power (Hendrix preferred these for both tone and durability). Eric Clapton was now in Cream, Jimi Hendrix was touring, Jimmy Page was using them in The Yardbirds and later would use them in Led Zeppelin, and of course Pete Townshend was playing to packed arenas with The Who, and it was the new Marshall 'stack' behind each band. Rock was growing in power and volume, responding to the growing audiences. The PA systems of the day couldn't keep up, so the stage backline provided the volume; we knew the words anyway. Many of these bands had several 100W stacks 'daisy chained' together for an even more overpowering effect. Ears were likely damaged forever… but nobody was sober enough to notice this at the time.

One thing for sure, though: you went to concerts to hear it *loud*. In those days it was all about getting as close to the stage as humanly possible, and you weren't close enough until the music started vibrating your skull and you couldn't hear yourself scream!

I remember seeing Jimi Hendrix in Santa Barbara, California, in about 1967. I couldn't afford a ticket, so I stood outside until I found a chance to hop the fence and get inside. It was the loudest thing I had ever heard. You felt it more than heard it – and only then you had 'Been Experienced.'

Some cosmetic changes were necessitated by the 100W models; the cabinets increased in size, and cooling vents were added to the top. And when you picked up one of these tone monsters, you knew it. They had an extra output transformer and a larger chassis and cabinet – so you knew right away this amp was made of, and for, heavy metal.

7) CIRCA 1969: MODEL 1959 100W TOP – METAL PANEL

By the time the metal panel amps were introduced in 1969, production was ramping up, consistency was steadily improving. A new single output transformer and enlarged power transformer producing higher plate voltages was designed for Marshall by Dagnall, a new supplier (although Drake continued to make the 50W transformers). Celestion was also improving its speakers to handle more power. The amps made from 1969 through 1975 were more consistent than any made before them, and became the classic design many have followed, but few have equaled!

These new 100W amps had EL34 power tubes

THE TUBE AMP BOOK 65

Marshall

Ah... the GT collection's most famous amp: the 1972 Purple Marshall 100 as used by George Lynch for his last few albums. George has had scads of modern tube equipment (including some of our own GT stuff) but always falls back to this amp for that classic tone, and of course his technique helps a lot. The 1968 Les Paul courtesy of Marty Mehterian, photo by Jeff Veitch.

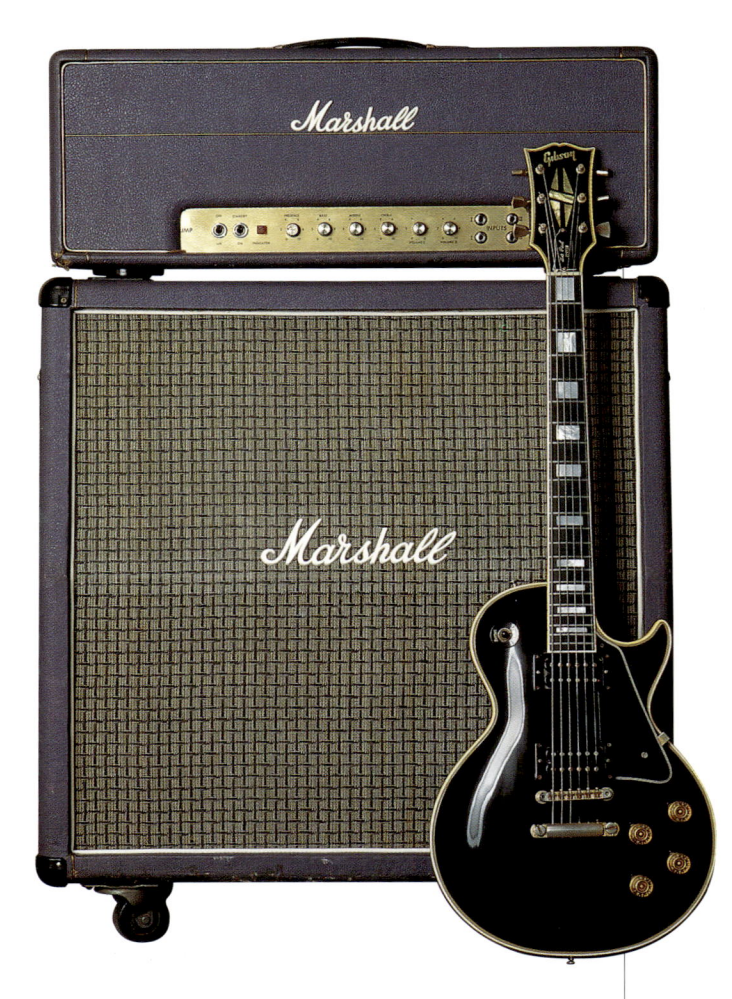

(replacing the KT66), which were the final ingredient of a special recipe that would last for decades. Little would change with regard to these basic power component building blocks for years to come, except that after 1976 they lowered the voltages a bit for improved reliability. Since Marshall was now selling thousands of amps – many exported to the US – reliability was becoming an issue, and it's this that lead to an infamous 'tube chapter' in the Marshall saga.

The EL34 tubes were physically rattling apart during shipments to the US stores. Many 'out of box' failures led to the use of a more rugged 6550 tube for the US, Japan, and Canada for many years. But while the 6550 was more reliable, its sound was much stiffer and not as harmonically rich as the EL34. Today, most of these imported amps have gradually been refitted for the EL34s (we at GT have been recommending this easy conversion since the late 1970s). Eventually Marshall returned to the EL34 worldwide by popular demand. The lesson here is that everything about the recipe makes a difference; make just one little change like this and a major tone controversy erupts!

This was a colorful era in music, and for Marshall in particular, as they started taking special orders for wildly colored amps covered in purple, red, and orange vinyl. These amps cost more, so they are very rare today. Most had cabs with the grey checkered basketweave material and were made between 1969 and 1973. I had lent a purple 1974 model 1959 to George Lynch several times while he was in the studio (it's pictured here). He just mentioned this fact once in an interview and now I get more letters and emails about that purple amp and his tone on those Dokken records than I do about any amp I've even owned. Purple power!

Perhaps the most significant addition to the line was the 1960B straight-front cabs, which were made to be placed underneath the 1960A (angled) cabinets. Marshall now also offered 8x10" and 2x15" cabs as well. However, the Marshall 4x12" from this era was as good as a guitar cab ever got. The company still makes it today, sticking close to the original design and components. Why fix it if it ain't broke?

8)) CIRCA 1983-90: MODEL 2210 – THE JCM 800 100W TOP

For those who have played one of these later Marshall master volume amps, this choice will come as no surprise. Many great guitarists achieved that signature 1980s rock tone with this amp on stage and in the studio from its inception in 1981 through 1990, when it evolved into similar models. Bands like AC/DC and Bad Company recorded with them, and Jeff Beck toured with them. At the time, it was enormously successful, outselling the 'non' master model 1987 lead amps about 30 to 1!

This new series was a departure from earlier designs, adding several new and usable features: dual footswitchable channels that could be set for lead and rhythm; master volume for each channel; an FX loop; and reverb… something Marshall 'stack' players had wanted for a long time. The master volume can also be set full up to use the gain control just like a non-master design, so you could also get a classic Marshall sonic signature from these amps if you wanted – pretty cool.

When these first came out, the 'purists' (myself included) found them somewhat flat tonally as compared to the earlier amps. When discussing this commonly held 'bad rap' with Marshall historian Mike Doyle, author of *The History of Marshall*, he

Don't give up on the mid 1980s JCM800 too soon. It can be a great sounding amp for the right applications, especially with the right speakers.

66 THE TUBE AMP BOOK

Marshall & Park

said he also was unimpressed initially with the JCM800, but had changed his mind several years later. Mike changed his opinion quite by accident one night when he was looking around for an amp to play a gig. The only one available was a JCM800 model 2210 top, but there was no matching cab. He found an older production 4x12" cab, so off he went to the gig. When set up the rig, he discovered it sounded really amazing! After some research, Mike discovered the introduction of the JCM800 included a change over to a new Celestion speaker model, the G12M-70, a departure from earlier G12T-65 which was in the older cab he had used. So the fault was in the speaker, and not in the amp. Apparently Marshall agreed, and the M-70 speaker was soon replaced with the G12T-75, and the tonal crisis was over.

Many of these 'most desirable' amps I have talked about here are out of reach for mere humans with day jobs and kids. But here's an amp that can be found for the price of coal and polished into a diamond with just some new tubes. Remember: all tube amps rely heavily on the quality of the tubes you install, so save a lot on the amp and spend a little more on a matched set of quality valves and you'll be rockin' for years to come. By the way, Mike Doyle now rates the model 2210 among his very favorites, and still has his kept in the original box.

9) CIRCA 1968: THE MODEL 1967 200W MARSHALL MAJOR

This unusual Marshall was made from to 1967 through 1974, and in relatively small numbers. Pro drivers talk about the 'need for speed'; heavy metal guitarist were talking about the need for volume! Bands were playing extremely loud by using several Marshall stacks daisy chained together, and they wanted more. Jim obliged, and began to develop the 200W Marshall Major amp. The first ones had Plexi panels, a larger cab, and were simply labeled 'Marshall 200.' They were billed as the "World's most powerful distortion-free amplifier".

The first Plexi-paneled amps were a design departure from the 100W series, featuring 'active' tone controls. In my opinion, 'active' tone controls are rather unmusical because each EQ band is separated and controlled by a preamp stage – and is sucking tone like little FX loops. The passive approach is more direct and player-interactive. But if you wanted loud and dynamic, this was the amp.

The next version that came out in late 1968 had gone back to passive tone controls (hooray!), and 'Marshall Major' appeared on the rear panel. The signature tone came from the larger KT88 power tubes and new, more powerful transformer sets built to match. The tone was incredibly tight and loud, yet still had the Marshall 'touch.' It weighed and cost

The Marshall Major, right, was the ultimate in 'stadium loud' – and in many ways still is, though it's a rare amp today. They were Ritchie Blackmore's amp of choice, Mick Ronson injected his stripped LP Custom through one, and even Stevie Ray Vaughan dabbled with a Major for a time. Amp from the GT Collection, photo Aspen & Ed.

Marshall serial numbers
by FRANK LEVI

Located on Rear Panel

ST/A_ _ _ _C = 1971, 100W Super Tremolo
  ①   ②   ③

1. Type & Output Power **2.** serial no **3.** date code

Through Dec. 1983 (Note: serical no's of early units are often meaningless)

Serial No Section 1
Key for Type and Output Power codes

A/ = All 200W models	/A = 200W Models (Plexi)
SL/A = 100W Lead	SL/ = 100W Lead Models (Plexi)
SB/A = 100W Bass	SB/ = 100W Bass Models (Plexi)
SP/ = 100W PA	SP/ = 100W PA (Plexi)
ST/ = 100W Tremolo	ST/ = 100W Tremolo Models (Plexi)
S/A = 50W Amp	S/ = 50W 'JTM' (Plexi)
T/A = 50W Tremolo	T/ = 50W 'JTM' -Tremolo (Plexi)

Serial No Section 3

Key for date codes 1969-89
NO LETTER = 30 June 1969 and earlier.
A = 1 July, 1969 through 31 Dec, 1970
C = 1971   D = 1972   E = 1973
F = 1974   G = 1975   H = 1976
J = 1977   K = 1978   L = 1979
M = 1980   N = 1981   P = 1982
R = 1983   *S = 1984   T = 1985
U = 1986   V = 1987   W = 1988
X = 1989   O = Silver Jubilee Amp

* On post-1983 Amps the date code and serial number sections are reversed.

NB: Many Canadian & European Marshalls do not use the prefix 1)

Marshall & Park

more, however, and could really blow some speakers if not properly loaded, so this was not the amp for everyone. Only real volume freaks used them – like Ritchie Blackmore of Deep Purple. Actually, as Mike Doyle points out in his excellent *The History of Marshall*, Ritchie's amps were specially modified to cascade the two channels and thereby produce more preamp distortion, becoming the first unofficial 'master volume' Marshall amp.

Unfortunately the KT88, the most powerful pentode of the time, was discontinued. The US-designed 6550 couldn't stand the plate voltages (and many amps were labeled with this warning), so when the KT88 supply dried up, so did the Major. It was discontinued in 1974.

The good news for collectors who have this rare Marshall is that several tube suppliers today offer a KT88 model, including our own special high power version, the KT88SV, with heat sinks added to the plates to keep the tube more stable in higher power applications. We designed this tube and have it made in the old Tesla factory, now renamed JJ, in Slovakia. There are also Russian and Chinese versions, and a similar KT90 equivalent is made in Serbia. The KT88 has re emerged mostly for the hi-fi market, where there is quite a demand for these powerful pentodes. The famous McIntosh 275 hi-fi amp uses them, and there are many others. Who knows: if rock turns back to loud again, maybe Jim will reissue the Major... but don't count on it.

10) CIRCA 1965-82: THE PARK AMP – EARLY 45W HEAD AND LATER 2X12" COMBO.

Perhaps the most interesting and desirable series of Marshall amps were actually not known as Marshalls at all, but were made under the 'Park' label. This came about because Marshall had entered into a long-term distribution agreement with the Rose-Morris company that conflicted with a prior regional exclusive distribution agreement with a northern retailer by name of Johnny Jones, whose shop Jones and Crossland was in the Birmingham area.

As the now 'former' exclusive distributor for Marshall in the north, you might imagine Johnny wasn't happy to lose a good seller. Since Johnny was a pal of Jim's they soon worked out a nice compromise to build amps under a different name, as Jim's contract didn't exclude this possibility. Johnny preferred his wife's maiden name, Park, or these amps might forever have been labeled 'Jones' – which wouldn't quite have the same ring.

My favorite Park amp is the early red 45W top pictured here; it has tone for days and gain like a TS-9 tube screamer! It was made in the early days, most likely around 1965 or so, and has many Marshall components, such a Plexiglas panel and logo, but with cool white space-aged knobs that look right out of a Flash Gordon instrument panel. In fact, Park amp models closely followed the Marshall line as they used similar components, but Park gave Jim the opportunity to try different variations – like the 75W park combo with two KT88 power tubes, which was well suited for a keyboard combo. I found this later era 2x12" park Combo (pictured here with the white corners) and it had an interesting input option that allowed using either channel independently or a middle input jack that paralleled the channels (1+2) for more gain and power. This was probably because some Marshall users had discovered this trick using a short jumper cord between their dual channel 4 input models; it really works great! This combo model also has reverb, a pair of the G12M 25W, speakers and a presence control, labeled 'Edge'.

Sadly, this alternative Marshall amp line with the interesting twists in design was to be discontinued in 1982 as Marshall had his full distribution rights back

Celestion date codes

1956-1962

YEAR	CODE
1956	A
1957	B
1958	C
1959	D
1960	E
1961	F
1962	G

MONTH	CODE
JANUARY	A
FEBRUARY	B
MARCH	C
APRIL	D
MAY	E
JUNE	F
JULY	G
AUGUST	H
SEPTEMBER	I
OCTOBER	J
NOVEMBER	K
DECEMBER	L

the date codes from 1956 to 1962 are written in the form: day, month, year. for example:
15DE = 15TH APRIL 1960

1963-1967

YEAR	CODE
1963	H
1964	J
1965	K
1966	L
1967	M

MONTH	CODE
JANUARY	A
FEBRUARY	B
MARCH	C
APRIL	D
MAY	E
JUNE	F
JULY	G
AUGUST	H
SEPTEMBER	J
OCTOBER	K
NOVEMBER	L
DECEMBER	M

note the loss of the 'i' from the month codes. the date codes from 1963 to 1967 are written in the form: day, month, year. for example:
19MK = 19TH DECEMBER 1965

1968-1991

YEAR	CODE
1968	A
1969	B
1970	C
1971	D
1972	E
1973	F
1974	G
1975	H
1976	J
1977	K
1978	L
1979	M
1980	N
1981	P
1982	Q
1983	R
1984	S
1985	T
1986	U
1987	V
1988	W
1989	X
1990	Y
1991	Z/A

MONTH	CODE
JANUARY	A
FEBRUARY	B
MARCH	C
APRIL	D
MAY	E
JUNE	F
JULY	G
AUGUST	H
SEPTEMBER	J
OCTOBER	K
NOVEMBER	L
DECEMBER	M

the date codes from 1968 to 1991 are written in the form: month, year, day. for example:
KH7 = 7TH OCTOBER 1975

1992 – 2000

YEAR	CODE
1992	B
1993	C
1994	D
1995	E
1996	F
1997	G
1998	H
1999	J
2000	K

MONTH	CODE
JANUARY	A
FEBRUARY	B
MARCH	C
APRIL	D
MAY	E
JUNE	F
JULY	G
AUGUST	H
SEPTEMBER	J
OCTOBER	K
NOVEMBER	L
DECEMBER	M

the date codes from 1992 to 2000 are written in the form: day, month, year. for example:
29EJ = 29TH MAY 1999

Ever since 1956, Celestion speakers have been stamped with a date code consisting of two numbers and two letters, which indicate the date of manufacture. These codes are added on the assembly line, stamped on one spoke of the speaker's frame (chassis) or on the edge of the magnet. Of course the letter code recycled as the end of the alphabet was reached, and the order of the coding was sometimes changed, too. The following list will help you precisely date any post-1956 Celestion and, therefore, can help in dating an amp or cabinet that contains such a speaker, providing it is original equipment.

for the Rose-Morris contracts and wanted to focus on a single brand name. The Park name would later reappear on a line of solid state amps manufactured by Marshall in the Far East. I doubt Jim will ever reissue my red Park with the Flash Gordon knobs, but who knows… if there were a strong customer demand, I am sure Jim, ever the gentlemen's shopkeeper of rock'n'roll amps, would oblige.

Park

This old red Park is my favorite sounding 'Marshall' out of the whole GT Collection. The split volume controls and the taller cabinet were the only significant difference from Marshalls of the era. I don't know why this one sounds so good. GT Collection.

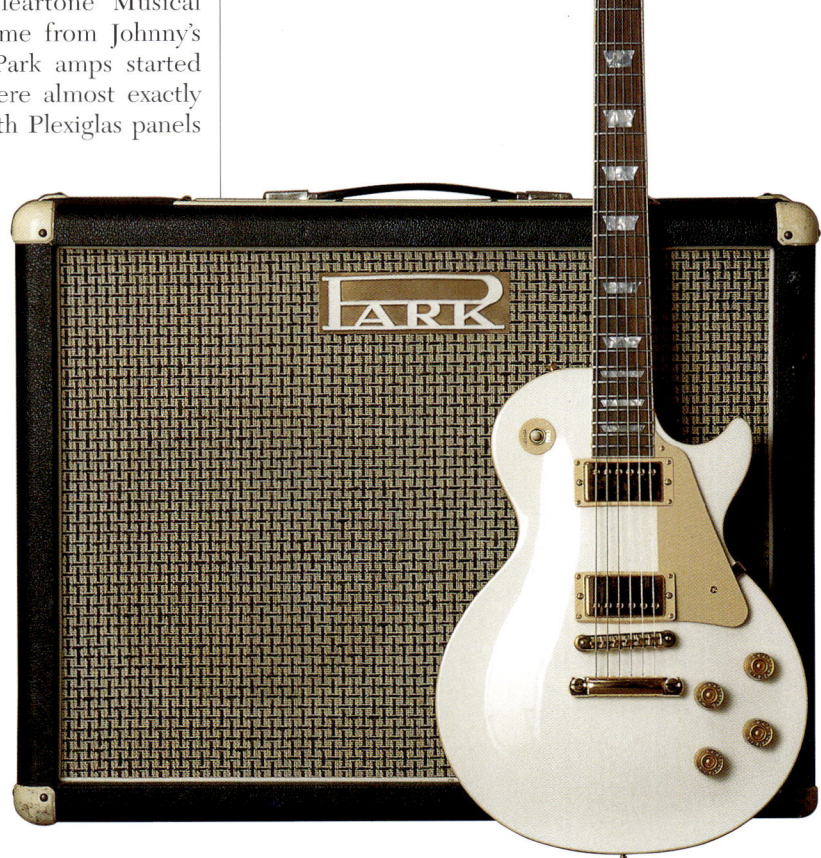

A real pretty Park amp from the last days of Park in 1978 (far right), this combo has 100W and 2x12" Celestions, and is part of the GT Collection. The white Les Paul is courtesy of Norm's Rare Guitars, Reseda, California, photo by Jeff Veitch

The Park amps were made by Jim Marshall in the same factory as the amps that bore his name. The reason he started a second amp line under another name was partly economic and partly political.

Jim Marshall had a good friend by the name of Johnny Jones, who also had a music store, located in Birmingham, in the Midlands of England. Johnny distributed Marshall amps from the beginning for the north of England. But then Jim Marshall made a deal to sell the marketing and distribution rights for Marshall amps, for a period of 15 years, to the Rose-Morris company, a large distributor in England. Naturally, Johnny was upset, because under this new agreement he lost the right to distribute the Marshall amp line in the north. Although Jim got the money he needed to expand his manufacturing operation, this agreement would ultimately put him in a poor position financially because he was limited to the amount of money he could make, since all his production of Marshalls went to Rose-Morris.

Jim Marshall decided to kill two birds with one stone and agreed to make amps for Johnny Jones under another name. Together they formed a distribution company called Cleartone Musical Instruments. The Park name came from Johnny's wife's maiden name. The first Park amps started showing up around 1965 and were almost exactly like the Marshalls of that era, with Plexiglas panels (they were black instead of gold), and the same standard control features. The exception was that the Park's two volume controls were separated by two of the input jacks, rather than having two volume controls side-by-side followed by four inputs, as on the Marshalls. Park amps used the same transformers and came with the same Celestion speakers as the Marshalls, so they sounded much the same. The shape of the cabinet was changed slightly and Parks carried silver-trimmed piping, panels and handles instead of the gold color scheme used on the Marshall amps.

Over the following years, Jim made changes and put out Park amps that were different designs from the Marshalls. A good example of this was the Park 75 that used two KT88 output tubes rather than the stock EL34 tube common to the earlier Parks and Marshalls.

There is some speculation that Jim actually made Park amps a little hotter in the gain stages to better

The 1978 Park combo on its own: note how Park amps returned to more 'vintage looking' cosmetics than contemporary Marshalls.

THE TUBE AMP BOOK 69

Park

compete with the Rose-Morris product, but this would be tough to prove.

I will say that the old red Park 50W that I have is the best-sounding 'Marshall' I've got in my whole collection, so perhaps there is some truth to the theory.

In my own travels in England, I found that the Park and Marshall amps hold a similar resale value in the UK, and are known to be Marshalls by most musicians much the same way we think of Ford and Mercury cars built today as the same basic car. Park amps were discontinued by Jim in the late 1970s when his 15-year agreement with Rose-Morris expired and he regained all rights to his name. The collector value of these Park amps is quite good, since they are relatively scarce even in the UK. They are very rare over here and collectors I know have paid very high prices for the early Plexiglas Park amps.

This late 1970s Park Lead 50 combo has master volume, low and high-gain inputs (with a footswitch to switch between the two), reverb, a single 12" Celestion, and a crunch sound from two EL34s reminiscent of many great Marshalls. It's another amp from the later-era Park range, and uses the grille cloth of the combo from the previous page, with a more modern control panel. Amp owned by Alex Theodossi, G&L Commanche guitar courtesy of Chandler Guitars, photo by Miki Slingsby.

Matchless

Perhaps one of the most successful modern classics would be the Matchless amps. Matchless was founded by Mark Sampson and Rick Perrotta in 1989, with their first designs roughly based on the Vox AC30 class A circuit. Matchless amps have the big, wet class A tone that instantly feels right when you plug into one. Many variations of the design followed, such as smaller 15W combo amps like the Lightning and Spitfire, ranging up to higher-powered 100W amps with four EL34s – but still using that spongy class A circuitry.

These amps are great for recording, and low volume stage work; if you want to play loud and clean, however, or need versatility like switchable channels with parallel effects loops… look elsewhere (although I should mention their highly regarded Super Chief, which Mark designed with the "loud and clean" player in mind). Some friends of mine have commented that Matchless amps are kind of a

Here's an early example of a Matchless Lightning '15 in the 1x12" format in eye-catching red vinyl. Also available as a 2x10" combo – with or without reverb in either – it's a gorgeous-sounding point-to-point, 15W class A combo for recording or club gigs. Photo courtesy of The Guitar Magazine.

Mesa/Boogie

This 1998 T/C30 is a variation on Matchless' flagship amp, the D/C30, which carries two 10" Celestion G10L-35 speakers in place of the usual 12"s. It has one 'Top Boost' style channel and another channel driven by a hot EF86 pentode, all through 4xEL84s.

The preamp in this mid 1990s Clubman 35 carries a 12AX7 in its first gain stage, then employs a 6SH7 pentode before hitting the P/I and a class A output stage with two EL34s.

Matchless's Chieftain 2x12" combo (also available as 1x12", 4x10", and head) was designed for a somewhat rockier sound than the D/C30. It pumps 40W from a duet of EL34s, with controls for Volume, Bass, Middle, Treble, Brilliance, Master, and Reverb.

'one trick pony,' but I point out that their concept is just that, and also that their 'one trick' is done really well.

Matchless reached its peak of production around 1997, then was hit hard by a collision of unfortunate circumstances, including the crash of the Yen (Japan was a big customer at that time) and other factors. It was forced to close temporarily in 1999.

It opened again about a year later, by which time Mark Sampson and Rick Perrotta had gone elsewhere (see the earlier entry for Bad Cat). Phil Jamison, however, remained at the helm as production manager, a position he had held at Matchless since 1991.

Since then, new examples of Matchless' flagship amps have begun appearing in stores and on stages once again, built – according to all reports – to the same standards as before the closure.

Mesa/Boogie

Above, Carlos Santana – the man who named the Boogie – rests his tush on a cream-Tolex covered Mark I. Right, a hardwood-cab Mark IIB from around 1981. The MkII signaled the first major evolution in the design; it had footswitchable clean/lead channels with master volumes for each, with both linked in when lead mode is selected.

Mesa/Boogie is often credited with ushering in the modern era of tube amps, and the company was one of the first to have sizeable production numbers in what has often become known as the 'boutique' amp market, though it is too big for that category today.

The Boogie amp developed out of the 'hot rod' Fender amp phenomenon that took off during the early 1970s. It seemed that stock Fender amps didn't have enough poop (power) needed to play the 'new music,' so amp techs everywhere started fooling around with increased-gain mods, beefing up power stages, and adding master volume controls to Fender amps.

Randall Smith was one such technician, and part owner of Prune Music in the San Francisco Bay Area during the late 1960s. Its customers read like a Who's Who of San Francisco-area musicians, including the Dead, the Airplane, Moby Grape, the Doobies, Quicksilver and Santana. The first 'Boogie' amp was built for Bay Area guitarist Barry Melton as a prank. Smith took a 12W Fender Princeton, stripped the chassis bare and rebuilt it, using a souped-up 60W 4x10 tweed Bassman circuit and carefully squeezin in a 12" JBL D-120, thereby keeping the amp and cabinet looking totally stock.

The first finished unit was shown to Carlos Santana, who remarked after playing it, "Shit man, that little thing really boogies!" The name was born. These early Boogie/Fender amps marked the transition from the classic vintage era into the modern high-gain amp era. Remember that at this time Fender was putting out wimpy sounding tube amps (the dreaded CBS era) and several evolutions

THE TUBE AMP BOOK 71

Mesa/Boogie

A look in the back of this export model Mark IIC from around 1983-4 reveals the 'Class A – Simul Class' switch (far right on the control panel) that marked one of Mesa/Boogie's major selling points at that time. Amp courtesy of Chandler Guitars, photo by Miki Slingsby.

of some really bad sounding transistor amps with names like Scorpio and Libra. Then along comes some hippy from Mill Valley whose souped up Princetons blow away the current Fender product. Small wonder the Boogie company took off the way it did!

These early Boogies had two input channels: Input 2 offered traditional 'blackface' performance for rhythm playing, and Input 1 activated the extra high-gain performance for the new lead voice. This lead channel had more gain than any production amp had ever offered, allowing the note to sustain seemingly forever. Santana's *Abraxis* is the first example on record, and perhaps the best, of how good these early Boogies sound.

Early production, using Fender Princeton amps, was somewhere between 150 and 200 units. Demand grew for Randall's hot-rodded Fenders, but the supply of old Princetons was dwindling. So Smith formed MESA Engineering and started building his own amps from his converted dog kennel. These early production Boogie amps – numbering around 3,000 – were called Mark Is (well, after the Mark IIs came along anyway) and added an extra tube for even more gain and overdrive. These new amps were offered in various coverings like cream Tolex and snakeskin, and also in polished hardwood cabs, something new and exciting for guitar amps. My favorite Boogies are these early MKIs made out of figured koa or maple: tube amps with sex appeal! Tube amps were finally getting some respect.

The Boogie model names reflect the evolution of the design. The Mark II introduced footswitching between the independent lead and rhythm channels. The Mark IIB series debuted in the early 1980s, and during its run – in the mid 1980s – a new patented power tube configuration called Simul-Class was introduced. Finally, the Mark IIC and C+ added an improved effects loop, and quieter footswitching to become the fully evolved Mark II amp, and very collectable.

Simul-class is basically two different classes of amplifier working together simultaneously. To understand it, first picture a normal 100W, four-power-tube amplifier: it's really two identical 50W two-tube power sections wired in parallel. But with Simul-Class, the two amplifiers wired in parallel are not identical. One is the standard Class A/B pair of 6L6s providing the bulk of the horsepower. The other pair of tubes (usually EL34s) is coupled through a special output transformer and biased for class A triode operation. This approach blends these two different-sounding tubes and output designs together to produce a unique sounding amplifier.

Inevitably two channels were no longer enough in the 1980s guitarist's quest for sonic diversity at the tap of a toe. Welcome, then, the third 'crunch' channel, as found on this 1985 Mark III combo. Amp courtesy of Chandler Guitars in Kew, London, photo by Miki Slingsby

Mesa,s Heartbreaker seeks to capture more vintage-like tones, but in an equally versatile package. It has two channels ' Love and Lust ' each with multiple voices, and comes powered by a quartet of 6L6s, but the bias can be switched for EL34s or 6V6s. Amp courtesy of Chandler Guitars, photo by Miki Slingsby.

If you're looking to win the prize for 'most tubes on the block,' this Triple Rectifier head stands a good chance of doing it for you. In addition to the sextet of 6L6s it carries to generate a claimed 150W, it uses three 'Coke bottle' rectifier tubes to keep the power comin'.

72 THE TUBE AMP BOOK

Music Man

A Simul-Class/Class A switch allows the user to turn off the center pair of 6L6s and run just the EL34s, at about 25W before clipping. Players seeking more warmth and less bite from these amplifiers can install 6L6s in these two end sockets instead.

For the Mark III, Boogie added a third, preset 'crunch' sound to the lead/rhythm configuration, and the Mark IV sported three fully independent channels – and probably more knobs and options than ever seen on an amp before that time. Since then, new models have diversified into so many directions it can be difficult to keep up. Current production Mark IV Boogies are shipped with a quartet of Russian 6L6s.

A peculiarity of the Boogie Mark series amps is that they do not have bias adjustment controls, even though virtually all other quality guitar amplifiers do. Boogie sets the bias at the factory using fixed resistors. It felt that an adjustable bias control (a pot instead of a fixed resistor) can lead to future problems if this control is improperly set, and offered its own matched replacement tubes to suit the factory preset bias level. When a bias change is needed on a Boogie, it is not too much trouble to make bias voltage changes by soldering in a small fixed resistor. If a player wishes to experiment with new and different tube types, however, it would be a good idea to install a bias pot. This way a tech can more easily adjust the bias voltage so that the tubes will wear normally and sound their best.

The simple addition of a bias pot in place of the fixed resistor expands your tube options and improves reliability. (For more info on biasing, and our recommended procedure of installing a bias in a Boogie see the Servicing and Modification section of this book.)

The preamp section of Boogies is especially demanding of quality tubes since Boogies get most of their distortion from overdriving their multi-stage 'cascading gain' preamp stages. When the first tube gain stage is re-amplified throughout the circuit, adverse microphonics or high noise gets louder. It is wise, therefore, to purchase preamp tubes tested at high gain levels so they work well in all stages of a Boogie.

MESA/Boogie, a true American rags-to-riches success story, now offers a wide variety of tube amps, ranging from retro-minded models like the Blue Angel and Maverick, to the versatile power of the Road King and Heartbreaker, to the hyper-gain Dual and Triple Rectifier Solo Heads, along with Mark IV and MKI reissues.

Music Man

He started with the classic tones of a Marshall Bluesbreaker and has now come almost full-circle, by playing through as-near-as-dammit the same circuit in his tweed Fender Twins (and reproductions thereof). But Eric Clapton took the seemingly odd sidestep of endorsing Music Man amps, with their solid state preamps, in his wayward late 1970s period. Gotta love the way the Explorer's headstock keeps your ciggy burning at just the right angle, though.

Music Man was started in the mid 1970s by some ex-Fender employees, including Leo Fender himself in a consulting role, although he had nothing to do with the amp line. These were some of the very first hybrid amps, using a solid-state preamp to drive a tube output section. Music Man amps were great for country, jazz or for clean rhythm sounds – much the way Fender amps were leaning at about the same time – but not too good for the crunchy overdrive rock'n'roll sounds made by Boogies and Marshalls. Surprisingly, perhaps, in light of this, Music Man counted Eric Clapton among its endorsees during his mid '70s Explorer-playing phase, along with country pickers like Albert Lee, James Burton, and Speedy West.

I think Music Man's best sounding amps were the very first models, which employed a 7025 tube driver to push the output tubes. It later changed this driver section to a solid-state device and – I feel – lost some of the warmth and touch the early ones had. The early models to look for are those with 'Sixty-five' and 'One Thirty' written out in script on the badge in the lower right-hand corner of the cabinet. You can also just look behind at the chassis for the one little preamp tube in its shield to check whether it's one of the good ones.

It is generally agreed that the early amps sound warmer and have a little more natural tube distortion. The distortion on this amp is produced by the master volume effect, and of course with a transistor preamp this results in a harsher type of distortion. Later, Music Man released the popular RD50, which uses a 7025 in an overdrive circuit, and its distortion was generally much better received by players.

The Music Man company was sold in the early 1980s to the Ernie Ball company. Amplifier production was suspended at that time.

Music Man 210-HD with a solid-state preamp coupled with a duet of 6L6s to produce a hybrid 150W tube amp. Courtesy of Buster McNeil

Orange

The Orange amps were designed by Matthew Mathias and manufactured by Cliff Cooper in Huddesfield, Yorkshire, England. Cliff Cooper also had a retail store in London by the name of Orange. Cliff marketed his amps under this name in London and the South while selling them under the name of Matamp in the Midlands and the North. This was a common practice at the time, because the north and south of England were so different in style and taste – as are America's east and west coast, though in somewhat different ways. This is similar to the arrangement with Marshall and Park amps: both made in the same factory while marketed as Park in the North.

The Orange amps were considered an 'up-market' Marshall copy and sold for more money. The design of the amp is pretty standard, but the graphics were unique and unusual, with little pictures above the controls to denote the various functions. I especially like the echo in/out jacks identified by twin-peaked mountains complete with snow on the top. The covering is an orange vinyl and the chassis are all painted in hard enamel orange paint. A small number of Orange amps were made with black vinyl covering, but these are few and far between. Orange sold mostly the two basic 80W and 120W tops, along with a stackable 4-12" speaker cabinet. They also had many other models, like their 2-12" combo, a nice tube reverb unit, and my favorite – the Hustler, a 1-12" combo with master volume, 40W, and built like a tank.

The last amp introduced in the first era or Orange was the Overdrive 120. This was the most desirable of all their tops because it had an extra gain stage along with the master volume, and predates any of the British amps with the feature. I guess the thing that impresses me most about Orange amps is the overall high quality of the materials and the workmanship. Premium speakers were usually stock equipment, not extra. The power tubes were always EL34, while the pre-amp section used the ECC83 (7025, 12AX7). These amps are still held in high regard in Europe and sell for as much as a second-hand Marshall. Very few were imported into the US, so they are scarce over here. I have found that when they are sold in the States, they sell for considerably less than a used Marshall, and so make an excellent value for the low-budget rock'n'roller.

Orange amps went quietly out of business in the late 1970s when Cliff Cooper simply closed the doors. My best reports say he just lost interest and closed the company to pursue other interests. In recent years, however, Cooper has begun building Orange amps again, under the banner of his OMEC distributorship in London. The big amps are back boasting similar specs, but some of the best reviews have come in response to his new smaller models, such as the 15W 1-12" AD15/12 combo, a classic Brit 'class A' style amp powered by EL84s and sold at a reasonable price (thanks in part to PCB circuit construction), and the hand-built 5W AD5 combo – like a cute little old Vox AC4, only orange… and Orange.

Cliff Cooper revived Orange to rave reviews in the late 1990s, and this new TC30 captures a lot of the company's original vibe, with added features like channel switching.

This single-ended AD5 from the revived Orange company squeezes 5W from a single EL84 output tube, and has a reissue Jensen P10R speaker. While most of the non-custom shop Oranges built today are PCB-based, this little cutie has a hand-wired circuit – pictured below-left. Photos courtesy of The Guitar Magazine.

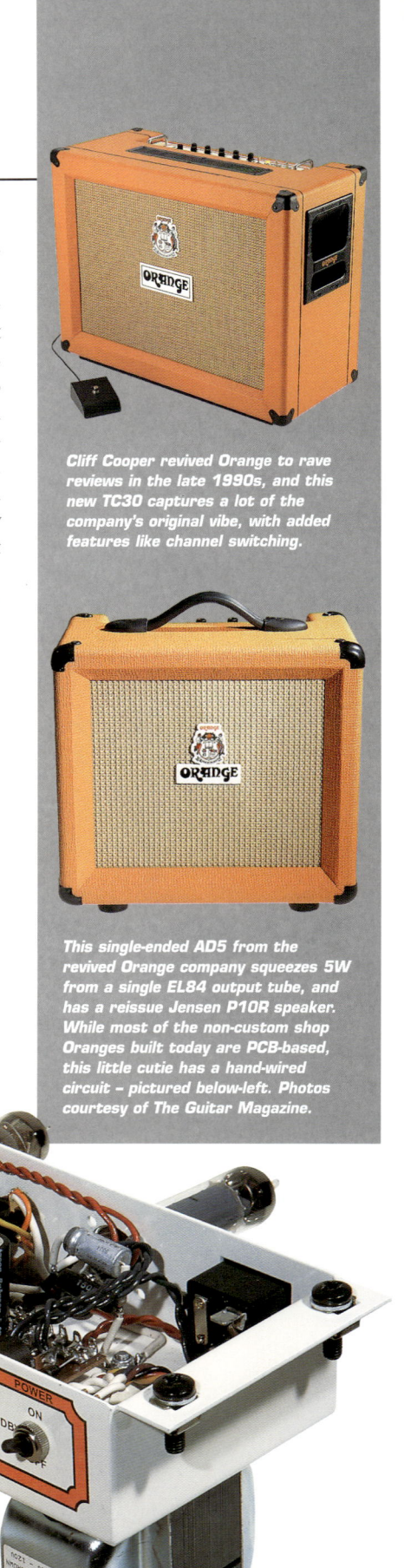

This is an original Orange 120 designed by Matthew Mathias and built by Cliff Cooper in London during the first era of the company. The accompanying reverb unit has EQ and a blend switch. Amps courtesy of Paul Patronete, photo by Jeff Veitch.

74 THE TUBE AMP BOOK

Peavey

Hartley Peavey's company gained prominence as an affordable rival to Fender in the '70s, and has forged a successful line in the tube-amp market mainly by sticking to that formula ever since. The tweed-wearing Classic series has long been a budget choice – and a popular base with DIY amp buffs for a variety of modifications. In what we might call 'the modern era,' Peavey has positioned itself as a builder of relatively wallet-friendly guitar stacks for hard rockers, thanks mainly to a major artist endorsement program – topped by the signing of one Edward Van Halen himself.

The Van Halen signature 5150 and 5150 MkII head/stack and combo have become the amp of choice of many an aspiring young heavy rock guitar hero. The original 5150 generates a rated 120W from a 6L6 quartet, with footswitchable rhythm and lead channels and shared EQ. The MkII edition adds bright/crunch switches and independent channel EQs. Aimed more at the nu-metal player, the 120W Triple XXX model has footswitchable clean, crunch and 'ultra' channels, a three-way resonance switch, and a switchable power stage which accepts either 4x6L6s or 4xEL34s.

Even with these big heads on board, however, Peavey is moving further and further away from its primarily tube-amp roots, selling vast numbers of solid state and modelling amps, guitars, PAs, microphones, keyboards and even drum kits. A massive musical force, without a doubt.

Peavey's 5150 and 5150 II heads and cabs give flash-fingered rock heroes a taste of the Van Halen sound in a (relatively) affordable package; this 2x12" 5150 combo, far right, performs the same tricks in a more portable package.

Premier

Another vintage American make whose models look as good as they sound (and sometimes better), Premier amps were made in New York City from 1940 and were popular on the East Coast music scene during the 1940s and early 1950s. They were built by Multivox, who also made products for JC Penny. Inside the chassis they weren't quite as well built as Fender amps of the era, but feature circuits with some similarities (as do most American amps of the period), and they can be great for smooth blues overdrive when cranked up. Their cabinets weren't nearly as well made as Fender's, either, though they certainly do look cool: the metal 'cross hairs' grille protector of the early Model 120, the overhanging upper deck of the later 110 and 120 cabs, the suitcase-model 88 and 76, and the colored knobs and two-tone coverings on most all add up to a pretty groovy line in tube combos.

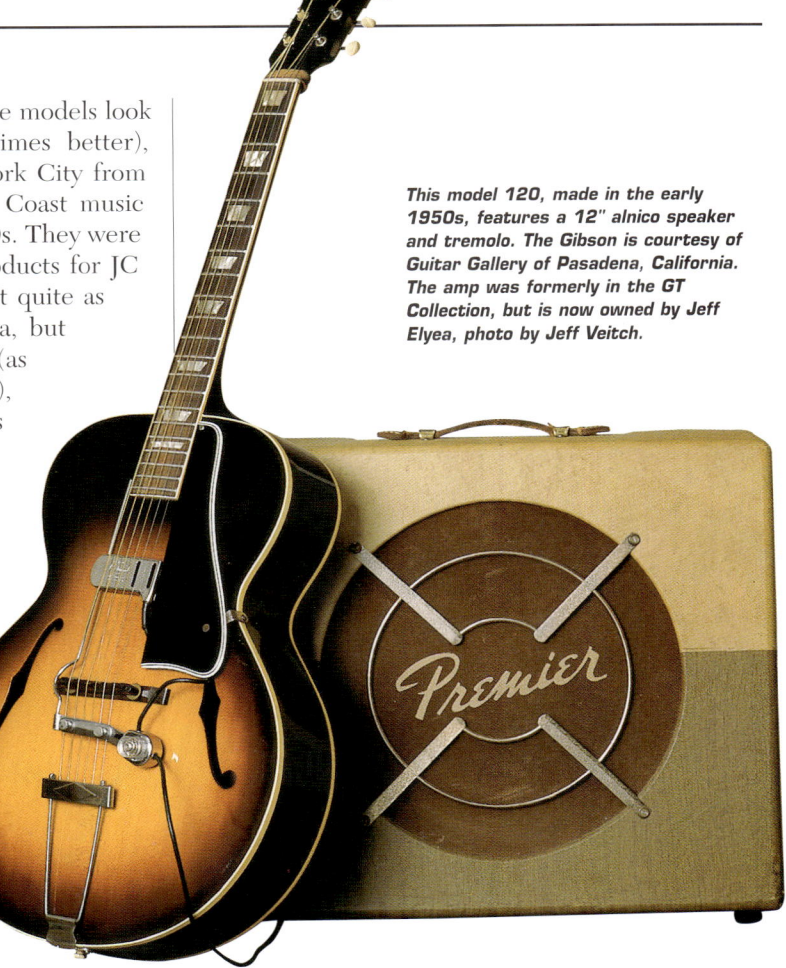

This model 120, made in the early 1950s, features a 12" alnico speaker and tremolo. The Gibson is courtesy of Guitar Gallery of Pasadena, California. The amp was formerly in the GT Collection, but is now owned by Jeff Elyea, photo by Jeff Veitch.

Rickenbacker

Rickenbacker

Rickenbacker was never even a fraction as well known for its amplifiers as for its guitars (should have bagged that Beatles endorsement for the amps too, huh!). Still, it offered a few interesting pieces from the 1930s to the early 1960s, and as one of the early major names in the Hawaiian lap steel boom their early amps are particularly interesting (see also the 'Richenbacker' tweed suitcase style amp pictured in the Gibson section).

Rickenbacker guitars were there at the dawn of the electrification of the instrument, and like everyone else at the time, they built amps to accompany their 'Spanish' and 'Hawaiian' (lap steel) instruments.

Rivera

The California-based Rivera company has come to be known as a force in high-end, modern-design, channel-switching tube amplifiers, and few builders today have a pedigree to match that of founder and owner Paul Rivera.

Paul's early career consisted of repairing, modifying and custom building effects racks, amplifiers and pedal boards for session guys in Hollywood, New York and Nashville, such as Larry Carlton, Robben Ford, Lee Ritenour, Chet Atkins, Jerry Reed, Steve Gibson, Steve Kahn and Reggie Lucas. In the late 1970s – having founded Rivera R&D in 1976 in Tujunga, California – he went to work for Bill Shultz and Roger Balmer of Yamaha to design the Mark 2 series of amps, which became the Japanese company's best selling guitar amps of all time.

When Shultz and Balmer were hired by CBS in 1980 to revitalize Fender, they brought in Rivera to "rebuild Fender back into an amp company." He did an impressive job of taking them there and, along the way, inaugurated what we could call the first modern era of Fender tube amps. Rivera began the job at the start of 1981, and by the NAMM show in January 1982 he was able to present his new Concert, Princeton Reverb 2, Super Champ, Deluxe Reverb 2, Twin Reverb 2 and a few more models. Tube amps built with a range of modern features, including footswitchable overdrive, these would usher Fender into the age of the high-gain rock amp at last. The following January saw the Rivera-led amp department issue a large range of solid state amps to accompany the tube models, and by 1984 the new range had taken Fender from 15,000 amp sales a year to an impressive 125,000 – a serious leap in market share.

Come 1985, however, CBS had Fender up on the

Two great Californians – a 2002 Rivera Quiana and a Tom Anderson T-style guitar. The Quiana was Rivera's first amp to use 6L6 output tubes as standard. Amp and guitar courtesy of Chandler Guitars in Kew, London. Photo by Miki Slingsby

76 THE TUBE AMP BOOK

Selmer

The Quiana Studio is a more compact version of the full 2x12" combo intended for recording use, but it packs more than enough punch and volume for use on stage.

sales block once again, and Rivera went to work on designing his own range of amps, a path he has followed through to this day.

"Looking back over the 25 years of all the work with the different companies," says Paul, "my work with Fender, where I was involved in over 75 products, was quite an adventure and a learning experience. Creating your own products, however, and enduring all of the challenges of owning and managing your own company – especially manufacturing – is one of the great times of my life. But it all comes back to working with musicians and designing equipment that pleases them. Thirty years from now I want to see my equipment on stage and still working. It is quite an incentive to build the best. My family name is on the front, and there is no corporate shield or name to hide behind."

The Rivera approach to tube amp design has been based largely around a dual-voice philosophy, with most models featuring much the same two-channel preamp – one Brit voice, one American, each with individual footswitchable boost – routed to different output configurations and combo or head formats.

The R Series has remained Rivera's bread'n'butter line, thanks to the combos' compact, portable designs and relative affordability. The range features the standard footswitchable channels, and includes 30W, 55W and 100W models with EL34-based output stages.

In fact, the English-toned EL34 has been Paul's tube of choice for the most part, and was used exclusively on Rivera amps until the Quiana in the late 1990s, which produces 55W or 100W from two or four 6L6s.

One of their more up-market amps, the Quiana includes deluxe features such as a Focus control to alter the speaker's response characteristics (akin to the 'resonance' controls found on some other makes), and a built-in Warm pull-switch which fattens and sweetens up the lower frequencies, as well as triode switching to offer a lower-powered output.

Other notable models include the Knucklehead series of rock warriors, the compact 40W 1x12" Chubster combo – sounding something like a JTM45 half-stack in a box – the LosLobottomy guitar sub-woofer cab, and the first all-tube combo for acoustic guitar, the Sedona, developed in conjunction with fingerpicking supremo Doyle Dykes. Mucho cool.

Selmer

Selmer, a distributor in the UK, made an excellent valve amp that was quite advanced for the time. The most famous model is the Thunderbird Twin Thirty, which has twin 12" Celestion speakers and 30W from a pair of EL34s. It has twin channels, with a five-position push-button tone selector, reverb, and tremolo. Perhaps my favorite features are its cosmetic details: it sports silver alligator-pattern vinyl, black-and-gold sparkle grille cloth, and a lighted green eye in the center that pulsates in time with the tremolo. That aside, these Selmers were well made amps, and sounded great.

Major English beat combos like The Beatles, The Shadows and The Animals played through Selmer amps before being persuaded to endorse Vox, and they often played second fiddle to the big Jennings brand, before Marshall started making big waves,

This Selmer Thunderbird Twin Thirty (the larger amp) has a five push-button tone section, reverb, and vibrato that has a pulsating green center eye, along with 2-12" Celestions and mock-snakeskin covering. The Little Giant model is from their beginners' line, with about 4W and an 8" speaker. Both amps and Hagström Futurama guitar GT Collection, photo Jeff Veitch.

Selmer

too. A Selmer model – though I'm not sure which – is responsible for the sound of the main riff in The Animals' hit 'The House Of The Rising Sun.'

Other popular Selmers included the Treble'n'Bass (another great 'budget alternative' vintage Brit amp), the Tremolo 50, the early Selector Tone, and the Zodiac. Although I say 'budget,' these amps have become more and more collectable in recent years and the prices are rising fast, though they are still nowhere near vintage Vox or Marshall levels.

Perhaps this slow start on the collector's market is caused in part by the hampered image in rock'n'roll circles of a company that built all the wind instruments; though of course a guitar amp is a wind instrument, if you crank it up high enough.

Soldano

Since the release of his Super Lead Overdrive amp in 1987, Michael J Soldano's designs have been much-emulated in rock amp circles. The two-channel, footswitchable, 100W SLO 100 was one of the first big rivals to Mesa/Boogie's high-gain crown, and was taken up by an impressive range of major guitar players, including Eric Clapton, Joe Satriani, Mark Knopfler, Gary Moore and others. Mike Soldano has always approached his amp building with a blend of classic principles and modern production techniques. Printed circuit boards are used for consistency, for instance, but these utilize extra thick boards and wide traces, with careful, logical circuit runs, and sturdy hand-wired flying lead connections to all components. Tube sockets are all chassis mounted, too, while cabinetry is robust and ready to take the knocks – overall, built for the abuses of the rock road trip. More crucially than this, they have managed to make purple a truly hard-rockin' color!

Soldano amps are often referred to as having a 'hotrod Marshall' sound, though Mike Soldano's tubes of choice have generally been 6L6 and 5881 types – the big Fender tube – rather than EL34s, and his designs don't follow those of Marshall religiously by any means. Considering how Fender and Marshall dominated practically everything in the 50W-plus range of classic amps, however, and Marshall has

Left, a 2000 Soldana Decatone head and cab. With 100W and channel switching, the more recently developed Decatone is intended to offer additional tonal flexibility to that of the flagship SLO model. Amp and PRS Single Cut guitar courtesy of Chandler Guitars, photo by Miki Slingsby.

The two internal shots of the same Decatone, below, show the build quality of the current Soldano amps. The top shot reveals hefty transformers mounted at opposite ends of the chassis, while the bottom photo shows that a PCB-based amp can still be carefully and sturdily constructed.

78 THE TUBE AMP BOOK

always had the rockier image – perhaps it is difficult not to look that way for comparisons. Then again, the Mesa/Boogie MkI that set new standards for high-gain tube amps in the 1970s was based on a modified Fender, so these things are often beyond precise comparison. Either way, Soldano has moved quickly from being the new kid on the block to securing status as a leading elder statesman of the rock amp world.

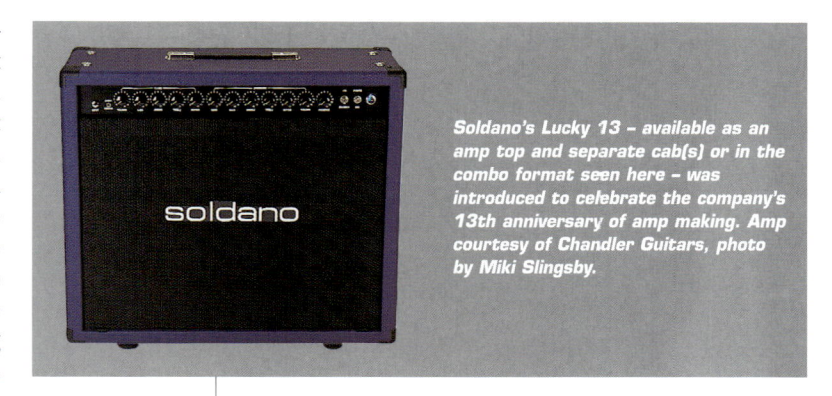

Soldano's Lucky 13 – available as an amp top and separate cab(s) or in the combo format seen here – was introduced to celebrate the company's 13th anniversary of amp making. Amp courtesy of Chandler Guitars, photo by Miki Slingsby.

Other significant models include the relatively diminutive 20W Astroverb, the two-channel 50W or 100W Lucky 13 head or combo (released in 2000 to celebrate Soldano's 13th anniversary), and the three-channel (clean, crunch, overdrive) 100W Decatone head and cab. There are far more rivals now than when Soldano came into being in the late 1980s, but its amps still appear to be holding on as one of the first choices for high-gain stadium rock, short of heavy thrash or nu-metal.

Sound City

In Britain in the late 1960s and early 1970s, Sound City amps – built by Dallas-Arbiter – were probably about fourth or fifth on the list for rock guitarist seeking big sounds. They had Marshall, Hiwatt, Orange, and Laney to compete with, but they were well-built amps, hand-wired on sturdy turret strips, with good Partridge transformers in many, and employed some clever and unusual preamp circuits, too. Models like the Sound City 50 Plus, the 100 head, and 120 head use two, four and six EL34s respectively, and can be good value in a vintage British valve amp.

Supro

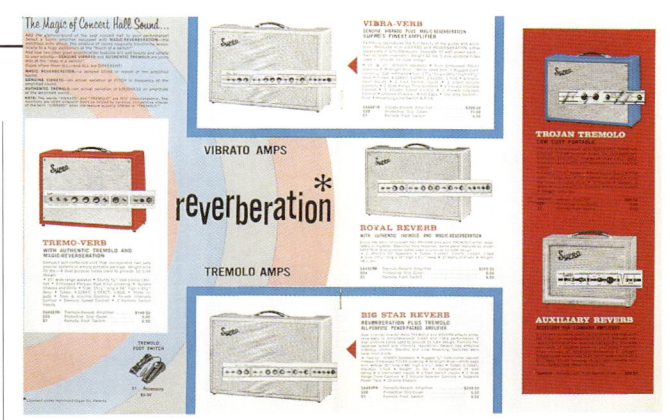

Although they were primarily a budget brand of the National Dobro company, first, and Valco from 1943-onward, Supro always had a lot more 'cool' going for them than plenty of other cheap amps of the 1950s and 1960s. Like most of these makes, their early amps were designed for Hawaiian lap steel

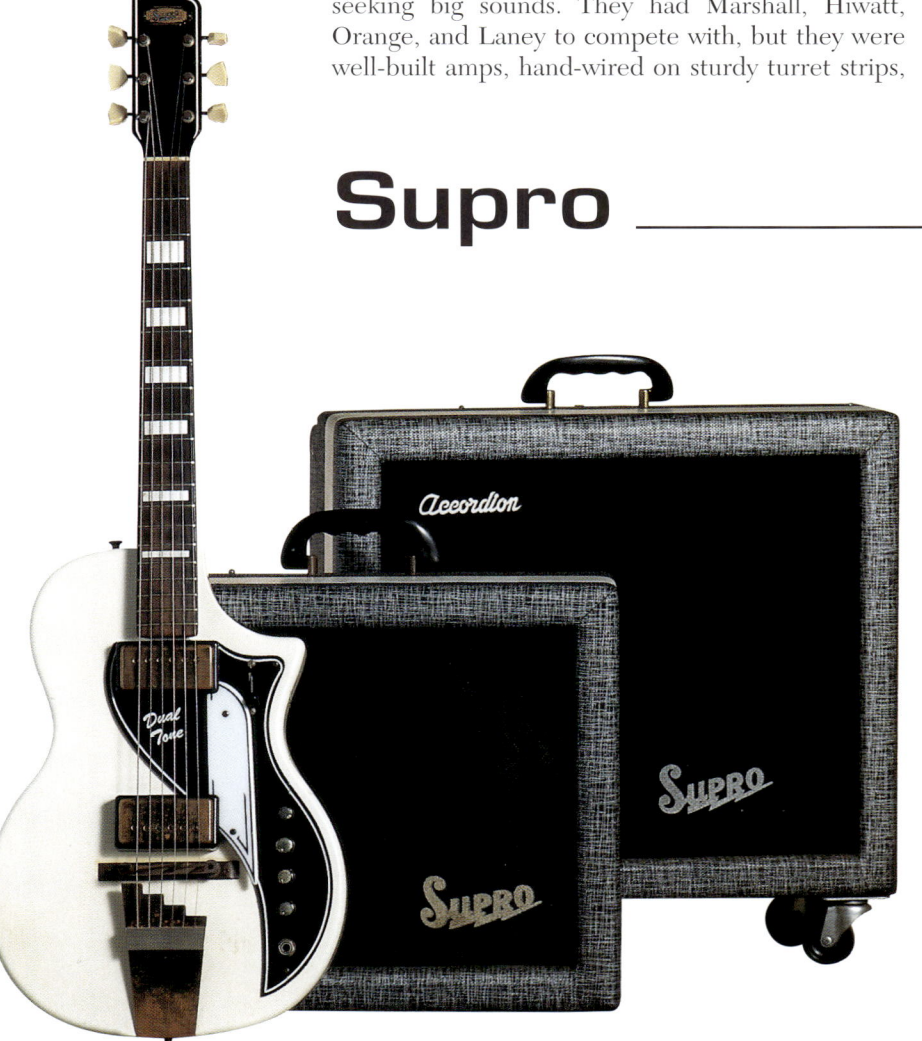

Supro amps were very popular during the 1940s and 1950s and sound great. First made for the accordion and lap steel market, they later had many real flashy rock'n'roll models. Link Wray is pictured with one of these during the period when he recorded Rumble in 1958. Amps are from the GT Collection, Supro Dual Tone guitar courtesy of Erhard Bochen. Photo by Jeff Veitch

THE TUBE AMP BOOK 79

Supro

guitar and even accordion first, but they make great little guitar amps when kept in good shape, with soft, juicy overdrive when cranked up. Link Wray is known to have used these amps in the early days, and Jimmy Page is reported to have recorded most of his guitar tracks for the first two Led Zeppelin albums on a small Supro, too… using his '58 Fender Telecaster, as it happens (sounds like a Les Paul and a Marshall stack, right!). See the entry for Valco for more information on amps of this type.

THD

Another one of my personal favorites is the THD amp, handbuilt by Andy Marshall of Seattle, Washington. Andy was the first to reissue a reproduction 4-10 Fender tweed Bassman, long before Fender saw the light. Andy has upgraded the whole construction approach of the early tweed era amps to improve reliability and consistency of performance. He built his first amp at nine years old, which started his lifelong love affair with tube amps. While studying in Vienna to get his degree in German, he engineered many recording sessions and began to develop ideas for improving the consistency of guitar amps. He built his first THD amp in April of 1987, using double thick PC boards with copper traces 4 times heavier than normal, Teflon wire and hi-temp silver solder. THD output tubes are rubber isolated to reduce interaction with the combo speakers… and did I forget to mention that THD amps has always used Groove Tubes as stock valves?

THD developed its own vintage replacement speaker, then went on to offer a 50W bass amp top developed from its popular tweed Bassman amp. In recent years THD has expanded beyond the 'tweed clones' to offer original designs like the single-ended Uni-Valve recording and small club amp head, the larger Bi-Valve, and a range of tube amp-related products, such as the popular Hot Plate power attenuators.

This 2002 THD Bi-Valve is intended as a step up power-wise from their single-ended Uni-Valve, and offers two channels, built-in Hot Plate power attenuator and other modern features. Most fun of all, the output tube types can be swapped without rebiasing, for a wide range of tonal changes at the output stage. Pictured here with a THD 2x12" cab and a gorgeous original '52 Les Paul Goldtop, all courtesy of Chandler Guitars, photo by Miki Slingsby.

TopHat

Brian Gerhard of TopHat is another more recently established maker of high-quality, hand-built tube amps who adds a number of clever and very original twists to some classic designs. He started the company with his wife, Joan, in 1994 – having developed an understanding of tube amplification, like so many, by building Heathkit stereo kits as a kid, studying high school and college electronics, and dabbling with guitar amp mods as a player. By 1995 they had wound up other business efforts to focus on TopHat full time. At first they made an assortment of custom-built models, based on tweed Bassmans, Marshall circuits (predecessors to their own Emplexador model), and of course Top Boost Vox circuits (the inspiration for TopHat's King Royale). At the start, all amps were built to order by Brian. With the good reception the amps soon got from players and critics, however, manufacture eventually expanded into a range of flagship production models.

It wasn't long before Brian also started what would become the Club Series, with half-power versions of his larger amps. The Club amps were released at the 1997 NAMM show and became TopHat's best selling series, mostly due to the more affordable price of an amp that still retained the careful hand building, quality components, and intelligent design of the bigger amps. The first year production of the Club Deluxe was a three-knob version – sort of a mating of a brown-era Fender Deluxe preamp and a tweed Deluxe power section.

The Club Royale was TopHat's elaborated Top Boost Vox preamp ("We added a mid control, full-time master and, later, the fat-off-bright pre-EQ switch," says Brian). By 1998, they added a 2x12" combo version of the Club Series and, in 2002, a new 6L6 version, the Super Deluxe.

One early model, the Embassy (a 4x6V6, non-reverb amp) was eliminated by the end of 1997, in favor of the newer Ambassador model (a blackface-based design with reverb). TopHat later added to this a 35W class A version (or cathode biased, no negative feedback for the purist), and 50W and 100W class AB versions with 6L6s.

Over the years they have adapted most of their standard models to add or alter features to better allow them to cover a wider range of applications – to play cleaner, brighter, fatter – and whatever players have commonly requested.

The effort has paid off, and has seen TopHat grow to become one of the most respected smaller builders around today.

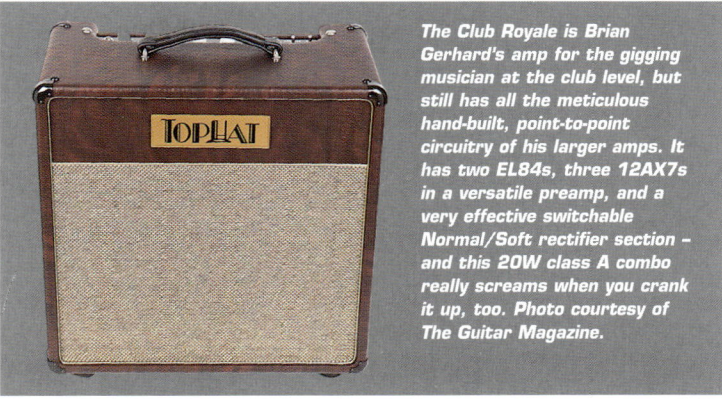

The Club Royale is Brian Gerhard's amp for the gigging musician at the club level, but still has all the meticulous hand-built, point-to-point circuitry of his larger amps. It has two EL84s, three 12AX7s in a versatile preamp, and a very effective switchable Normal/Soft rectifier section – and this 20W class A combo really screams when you crank it up, too. Photo courtesy of The Guitar Magazine.

Trainwreck

Ken Fischer was an early Groove Tubes friend and customer; we became acquainted back in 1981 as I recall. He was one of the more notable repair techs in our small world of tube amps and was located on the East coast, and we referred many customers to him for amp service. They always called back and thanked us, so I guess Ken knew what he was doing. Ken had begun an interest in electronics at an early age, and later learned more in the Navy technical service departments. After that, he worked at Ampeg as a tech and finally was advanced to the engineering department. Ken left Ampeg the day the company was sold, and never looked back. There was a shortage of qualified repair techs out there, especially for tube amps, so Ken opened his shop doing repairs and modifications for musicians in his area.

Ideas began to develop in Ken's mind to make his own amp and, finally, repeated requests from loyal customers prompted him to start building interesting custom amps that used variations of the 4xEL84 format in the power stages. His first amps were called the Liverpool 30. Many companies have closely emulated the AC30 design – some of these turned out better than others – but my personal favorite of these is Ken's Trainwreck Liverpool amp. The Trainwreck name came from an old biker handle he had earned with his riding group… I guess he drove kinda' crazy.

His amps are all laboriously hand made – to high manufacturing standards, using good materials and sturdy construction techniques – so production has been low to be sure, but he's been making them for about 20 years and they have been found on stages and studios all over the world. Ken is very open and opinionated, and we at Groove Tubes have always listened. Ken knows his stuff. One peculiarity of Trainwreck is Ken's refusal to make combo amps, preferring to make separate top/cab systems for more options and to reduce speaker/electronics interaction.

The three basic Trainwreck models are the

This Trainwreck Liverpool 30 built by Ken Fischer illustrates perfectly how the best-sounding amp you can imagine doesn't necessarily have to be loaded with deluxe features, channel switching, or controls for ever possible tonal parameter. Amp courtesy of David Swartz, photo by Aspen Pittman & Ed Ouellette.

THE TUBE AMP BOOK 81

Trainwreck

Liverpool and the Rocket (both of which use EL84s in the power stage), and the Express, which uses octal-base tubes such as EL34s, 6L6s and 6550s which produce the extra power. The preamp stages designs are simple and classic sounding, using only passive tone controls. The cabinets are nicely finished hardwood, and the look is completely handmade and cool. Like so many of these hand-built amps, the price reflects the hours spent in assembly, so they are not cheap.

One interesting quirk of Trainwreck amps is that they were never serial numbered, but instead were each named after women – so even though there are just a few models they are all really individuals! Any of these original Trainwreck amps is pretty rare these days, and very collectable.

Despite their high cost, Trainwreck amps gained fame among the most elite players on the east coast thanks to the way they delivered that great tone, so demand quickly outgrew the supply. Soon Ken's back-order list surpassed his ability to build them, as his amps didn't lend themselves to high volume manufacturing. Enter Komet amps...

In June 1998, Holger Notzel and Michael Kennedy had opened Riverfront Music in downtown Baton Rouge, Louisiana. They specialized in vintage guitar and amplifier repair and sales. After many years of collecting and servicing classic vintage amps, they had only come across one type of amplifier that surpassed the tonal qualities of the best of those Golden Age amps: Ken Fischer's Trainwreck designs.

Holger had been on the Trainwreck order list for years. It became obvious with Ken's declining health and the huge back-order list that delivery of an actual Trainwreck amp was not to be expected any time soon.

Holger approached Ken about the possibility of building a Trainwreck-inspired amp himself. Ken was gracious enough to donate a circuit design, which was quickly turned into a prototype. That prototype sounded very special and clearly had a lot of the qualities that players admired in an actual Ken Fischer Trainwreck.

Everyone who played the amp fell in love with the tone and they received many requests to make the amp available to the public. Holger approached Ken with an idea to build these for their customers and Ken agreed. Ken tweaked the circuit and designed proprietary transformers for the Komet amps, but he left the amps cosmetics and layout up to Holger and Michael.

A licensing agreement between Komet Amplification and Trainwreck Circuits was reached and the Komet 60W head was born, and remains in production unchanged.

The Komet 60 uses two EL34 output tubes in a class A/B design, three 12AX7 preamp tubes and a 5AR4 rectifier. It can drive into 4/8/16 Ohms with an adjustment to the impedance selector, and can also be biased with a simple DVM by using the convenient rear panel access jacks.

Additionally, the potentiometer used for bias adjustment is externally mounted for ease of access. It also incorporates a repeatable 1-100 dual-rotating knob that allows you to dial in exact bias points. They can then be exactly duplicated when changing to different tubes: very cool.

The amps also have a unique feature/option in their 'touch response' switch (on the rear panel) that allows for a change in the dynamic interaction and change in the players touch to produce a faster or more gradual transition into the distortion modes as the amp is cranked up.

Quality is the hallmark of this special Trainwreck amp, and if you can't wait to find an original Ken Fischer Trainwreck, this might well be the ideal amp for you.

Basically, all of Ken's designs are simple, not relying on a lot of switching or EQ, just simple straightforward and well-built amps that had a lot of component engineering go into them to perfect the tone. Maybe that's why for many years Trainwreck amps only used and recommended Groove Tubes, tested and performance matched components he found consistent and reliable.

These kinds of limited production high quality tube amps are my personal favorites, as they have the dynamics to make the player's personality come shining through and do not need a lot of channel switching or EQ tweaking.

They always sound good, no matter what level you are playing, and get their cool and responsive distortion from turning the amp up and from the player's touch factor, and not just from overdriving a preamp tube early in the gain chain.

These amps are not for beginners. You have to know how to play well to get the most out of them, but for players who are blessed to own one they get a lot of use!

Traynor

Jack Long, who was an owner of Long and McQuade retail music shops in Canada, joined forces with a young Canadian bass player by name of Pete Traynor and founded Yorkville Sound in 1963. Traynor was the brand name they chose for the PAs and amps they would make, and if you were alive and into music in Canada from the early 1960s on you certainly knew about Traynor amps and PAs.

Traynor amps were not so popular in California where I grew up, but we had a few stores that sold or

Traynor

This Traynor YGM-3 combo uses EL34 output tubes and has a great, ballsy crunch tone when cranked up – thanks in part, no doubt, to the solid lows offered by its closed-back cabinet. GT Coll, photo Aspen & Ed.

This 2000 Custom Valve 40 marks Traynor's return to the tube guitar amp market. It is a good-sounding and affordable combo. Photo courtesy of The Guitar Magazine.

This early 1970s YGL3-A Mark 3 has a quartet of EL34s, master volume, and a boost switch on each channel, and was accompanied by a cab with four Celestion G12H speakers. GT Collection, photo by Aspen Pittman & Ed Ouellette.

rented Traynor amps and PAs. Through Long and McQuade's retail experiences, Traynor pioneered a unique rental system here in the USA that they ran through local music stores. This system is still in existence today in the Long and McQuade retail chain throughout Canada. Jack's brilliant system allowed poor musicians to have great gear at low rental rates for their weekend gigs. Also, it introduced the players to Traynor amps they could rent, and later apply the rental charges toward the eventual purchase of the gear. Both Jack and Pete were working musicians, so this was a great idea born out of their understanding of the music business.

It should be noted that Jack Long is still active in his company today, although he has the help of several children he brought up in the business. Jack is truly one of the nicest, most gracious men I have ever met in this business (or any other!). He befriended me (and many others) on so many occasions when I was up and coming in this business, and he is one of my inspirations for how to conduct business and treat customers. We could all learn a lesson in humanity from this fine man, who is still at the helm of this successful company bearing his name.

The early Traynor amps and PAs were simple but extremely well-built tube amp designs for the day, and reliability was their strongest feature. And because they used construction techniques, designs and components of the highest quality, they also sounded really good: as good, in my opinion, as any of their counterparts of the day. I know the few Traynors I've got in my collection are amazing sounding amps, with loads of tone and a terrific reverb that rivals the legendary Ampeg reverb.

According to the official Traynor history (found at www.yorkville.com), the company was the first to introduce master volume in a production amp back in 1969, which was years before Marshall had a Master Volume amp. This amp was the YSR-2 'Signature Reverb' combo with four 10" speakers, and was an incredibly good-sounding amp reminiscent of the Fender Super Reverb, another favorite of mine. Since I can not find an earlier example of a master volume amp from my research, I believe this claim to be true.

Additionally Traynor made loud/clean tube amps in the tradition of a military production unit and had put out an amp in 1968 that produced 250W (measured before distortion @ 2 Ohms!) using four 6KG6A TV vertical hold tubes for the output. It was

THE TUBE AMP BOOK 83

Traynor

called the YBA-3A 'Super Custom Special,' had a fan cooling unit built in, and powered two 8x10" cabs model YC-810, nicknamed the 'Bib B's'.

The most popular, perhaps, were their 130W tops like the YBA-3, first released in 1967, which used an 8x10" speaker cab to make what was at that time the most powerful bass amp in North America. The guitar version with built-in Reverb used a 4x12" cabinet, and they were also quite loud and clean. These amps used the UK EL34 or USA made 6CA7 output tubes and had massive output transformers that kept them particularly clean, loud, and heavy! I have a slightly later version in my collection (YGL3-A Mark 3) with a black Plexiglas control panel, master volume, and a boost switch on either channel. It came to me with what looks like original Mullard EL34s and Celestion G12H speakers (the ones Hendrix liked best). This incredibly well-built amp sounds like a cross between an early Marshall 100W top and an Ampeg V4 with that great long-spring reverb… but if you want distortion, you used a pedal!

Perhaps my absolute favorite Traynor amps, however, were their lower-powered small combo amps that mostly used EL84/6BQ5 outputs, such as the YGM-3 (the earliest version was called the YGM-1).

These appeared in 1969 and also featured a great sounding built-in reverb. I still have one of these little tone monsters in my collection, and it just amazes my friends with its crunch tone and fat reverb each time I crank it up. And this was an amp born in 1967, the height of the Golden Years of the tube amp!

Yorkville has recently brought the Traynor name back to grace a new range of affordable but good sounding tube amps. The Traynor Custom Valve 40, for example, has received rave reviews from guitar magazines and players alike, and is hailed by many as a 'Fender beater' in its price range.

Valco

While Valco is a name that all vintage amp buffs recognize, there aren't a whole lot of examples of 'Valco' tube amps out there, because the majority of their offerings appeared under other names. Up until 1969 when they closed shop, Valco provided amps sold as Supro, Gretsch, Oahu, National, Airline and others. In fact, in the 1940s, 1950s and 1960s, if you manufactured an electric or lap steel guitar, you pretty much had to market an amplifier, too, and if you didn't have a workshop to build your own, you probably got them from Valco.

The circuits behind most of these brand names' amps in any particular era were, therefore, largely similar, and the differences were mainly cosmetic. In this regard, some were certainly more impressive than others – see the Gretsch entry for some stunning examples, and Supro for others.

Though they were mainly a mass-manufactured OEM supplier, Valco had a long and colorful history in the musical instrument business. Louis Dopyera of Dobro fame moved his National Dobro company to Chicago in 1936, and in 1943 changed the name to Valco, after the initials of the three owners at that time (Victor Smith, Al Frost and Louis Dopyera). After working on wartime necessities in the early 1940s, like most manufacturers, Valco returned to instrument building in the mid 1940s, just in time to cash in on the Hawaiian guitar craze. Whatever the brand logo on the front, most any of these vintage Valco tube amps can be great little blues and rock'n'roll amps after a little tune-up and a fresh set of tubes (see the Hawaiian Steel Guitar Amps entry for more information on this.) Valco became part of the Kay company in the late 1960s, and the pair went out of business soon after, in 1969.

Two cool Valco-built amps from the early 1960s: top, a Wards Airline and, below, a Supro Thunder Bolt (reputedly the model Jimmy Page recorded with). The rear-view shows how they are identical in everything but cosmetics, right down to the ceramic-magnet Jensen speaker. Both courtesy of David Swartz.

Victoria

Victoria's founder and owner Mark Baier started his amplifier company in 1994 from the simple inspiration of a player frustrated by his efforts to find a 'new' amp with all the tonal integrity of the vintage tweed Fenders. With a career in the securities business, a wife, and two children, Mark had returned to playing the guitar again and decided to treat himself to a new amp; after sampling Fender's (then recently) reissued Bassman and the respected repro models built by other makers and coming away somewhat disappointed each time, he plunged headfirst into the challenge of dissecting and recreating a narrow-panel Fender amp in all its glory, and the obsession escalated from there.

"I figured that I couldn't be the only poor slob out there that wanted an old tweed Fender that wasn't beat up and priceless," says Baier, "so I decided to make a couple. Easier dreamt than done. I soon found out… the little details were the most important to get correct. My goal was (and still is) to reproduce that old Fender amp as exactly as possible."

Baier's quest found him contracting Raytheon to make fresh carbon comp resistors, locating the original cloth-covered wire manufacturer and convincing it to gear up for short-run production again and, ultimately, tracking down the retired Triad technician responsible for designing Fender's transformers in the 1950s. It paid off. Victoria Amplifiers, based in Naperville, Illinois, quickly became recognized as some of the best tweed-repro amps available at any price.

Models like Victoria's 35210-T, 35310-T, 45410-T and others (named according to their wattage rating – first two digits – and speaker configuration – following digits) have been come firm favorites with players like Buddy Guy, Junior Watson, Debbie Davies, Kid Ramos and plenty of others. Newer models like the Victorilux and Victori-Ette have pushed the boundaries a little by adding reverb and tremolo to the tweed brew. None of 'em is cheap, but Victoria represents another hand-builder convinced that if things were done so right by Leo in the 1950s, why do them differently?

Victoria's line is inspired by classic tweed Fenders like the Deluxe, Bassman, Twin and Bandmaster, but has expanded to include some original and British-flavored designs, too. All are totally hand built and beautifully put together, and they sound great, too.

Vox

The Vox company was founded by Tom Jennings, who had run a music store under the Jennings name in Dartford, England, and had been retailing and marketing musical instruments in the UK since the 1930s. He had formed the Jennings Organ Company in the late 1940s, supplying Hammond-style organs to the UK and Europe at affordable prices.

One of his smaller organs was a single keyboard called a Univox, with a matching amp and speaker unit, that was designed to simulate everything from vocal parts to a piano. This was the first amp with the Vox amp name.

As rock'n'roll took the British Isles by storm during the 1950s, Jennings formed a lasting relationship with a clever electrical engineer he had come to know, by the name of Dick Denney. Tom was impressed by an amp Dick had built for a guitarist friend and was anxious to cash in on the insatiable demand for rock'n'roll amps. Together they launched the Vox line of guitar amps in 1956, beginning with the 15W AC15 model, which was Dick Denney's original two-channel amp and included two new Celestion G12 speakers in a combo design. This model was shortly followed by the smaller AC4 and AC10 for the beginner market – a 4W 1-8" and a 10W 2-10" respectively.

It's an interesting point that this AC15 was the product of an original amp design meant for the guitar. Most, if not all, guitar amps marketed up to this point were circuit designs converted from

Vox

The AC15 was originally the Vox flagship amp from Jennings Musical Instruments, and claims the distinction of being the first tube amp circuit designed exclusively with guitar in mind. The example far left dates from 1958-59 – note the pre-Celestion Goodmans Audiom 60 speaker in the rear view – and the one near left from 1956-57; it's the earliest Vox amp I own, or have ever seen for that matter. Amps owned by Jim Elyea.

accordion amps or simple designs from the RCA Tube Application manual. These new AC15s used a cathode-biasing circuit employing a pair of EL84 output tubes in a class A output design. Pretty unusual for those times. This type of circuit biases the output tubes at maximum dissipation, but at a relatively lower power supply voltage to sacrifice linearity for a particularly desirable distortion characteristic. In other words: less power, sweeter distortion. Not ideal where maximum wattage is the goal, but really great for the guitar player who wants toneful overdrive at all output levels. This design produces lots of second and third-order harmonics that noticeably change as the volume is turned up. In other words, these amps are really playable – and so became instantly very popular with many of the British bands like Hank Marvin and the Shadows.

The Vox sound was quickly established, but the type of rock'n'roll being played was getting louder all the time. By the end of the 1950s bands still wanted that Vox sound, but at higher levels. This led to the development of the AC30, AC50 and AC100 amps during the early 1960s.

Of these, the most notable by far is the legendary AC30, an amp that actually was originally just a doubling of the AC15, louder and just as sweet sounding. Initially it was available only as a combo, fitted with Vox Bulldog speakers – alnico-magnet Celestion G12s labeled for Vox use. Later it would be offered as a two-piece head-and-cab unit, with a chrome stand available. The AC30 was immediately

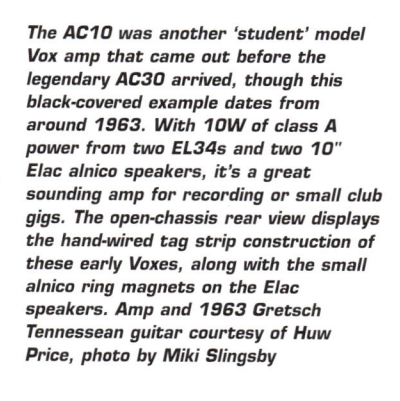

The AC10 was another 'student' model Vox amp that came out before the legendary AC30 arrived, though this black-covered example dates from around 1963. With 10W of class A power from two EL34s and two 10" Elac alnico speakers, it's a great sounding amp for recording or small club gigs. The open-chassis rear view displays the hand-wired tag strip construction of these early Voxes, along with the small alnico ring magnets on the Elac speakers. Amp and 1963 Gretsch Tennessean guitar courtesy of Huw Price, photo by Miki Slingsby

86 THE TUBE AMP BOOK

THE NATURAL CHOICE OF LEADING ARTIST'S

Some of the stars that feature VOX AMPLIFICATION

THE BEATLES, THE SHADOWS, JET HARRIS, THE TORNADOS, CLIFF RICHARD, JOHN BARRY SEVEN, LONNIE DONEGAN, ADAM FAITH & GROUP, GENE VINCENT, DAVE GOLDBERG, MALCOLM MITCHELL, THE ALLISONS, JOHNNY & THE HURRICANES, THE JAYWALKERS, SHIRLEY DOUGLAS & CHAS. McDEVITT, THE EAGLES, JOE BROWN, THE DALLAS BOYS, SHANE FENTON, THE ECHOES, THE KING BROTHERS, ERIC DELANEY, KARL DENVER TRIO, HELEN SHAPIRO & RED PRICE, SOUNDS INCORPORATED, EMILE FORD & THE CHECKMATES, FREDDIE & THE DREAMERS, THE DAKOTAS

FREDDIE

ADAM FAITH — BILLY J. KRAMER — CLIFF RICHARD — JET HARRIS

VOX AMPLIFICATION

A.C.4
A fine little amplifier for the soloist. Two inputs and separate tone and volume controls. Fitted with 8" loudspeaker and foot operated Vibrato on/off control. Attractive standard presentation, stowage space for accessories. Dimensions—16" × 7½" × 16½".

A.C. 10
The A.C.10 is the popular size for the average group, fitted with 10" loudspeaker this unit also employs four inputs, two normal and two vibrato, with separate volume and tone controls and foot operated on/off Vibravox/Tremolo switch. Presentation is the same as the larger models. Dimensions—18" × 18" × 9".

A.C. 10 TWIN
This model has two special 10" loudspeakers for more efficient distribution of the 10 watts undistorted output. With 4 inputs and Vibravox. Foot controlled on/off switch. Dimensions 24" × 18" × 9".

A.C.10 REVERB TWIN
A new Vox high performance amplifier at a moderate price. The standard Reverb Twin features twin loudspeakers with internal reverberation and separate pressurised tone cabinet to provide maximum output and drive with Vibravox. On/off foot control. Dimensions 24" × 18" × 9" and 24" × 9" × 9".

A.C. 15
Employing one 20 watt heavy duty Vox loudspeaker (15/20 watts output). Four inputs: 2 Vibravox and 2 normal channels. Foot controlled on/off Vibravox. Handsome rugged carrying case. Dimensions 20" × 21" × 10". Bass Model available 5 gns. extra.

A.C. 15 TWIN
Fitted with two 12" heavy duty Vox loudspeakers, for improved sound distribution, incorporating exclusive Vibravox. 15/20 watts undistorted output. 4 inputs. Special Bass Frequency model available 5 gns. extra. Dimensions 27" × 21" × 10".

A.C. 30 SUPER-TWIN
This model has 6 inputs providing 30 watts output. There are two cabinets; Amplifier and controls in one and two heavy-duty 12" Vox loudspeakers in a separate pressurised cabinet. Incorporates the exclusive Vibravox. Dimensions 27" × 21" × 10" and 27" × 10" × 9".

A.C. 30 SUPER REVERB TWIN I
As above with built-in Reverb.

A.C. 30 SUPER REVERB TWIN II
As above with built-in Reverb and two separate pressurised loudspeaker cabinets.

A.C. 30 TWIN (The Shadows Model)
Fitted with two 12" heavy duty Vox Loudspeakers. Output 30 watts undistorted; 6 inputs in 3 channels—2 normal, 2 Vibravox, 2 Brilliance—separate controls per channel. Attractive rugged carrying case. Special Bass Frequency model available 5 gns. extra. Dimensions 27" × 21" × 10".

VOX REVERB UNIT (Cliff Richard Model)
Jennings pioneered the introduction of echo and reverberation in the United Kingdom. This new sound—the third dimension—is an essential addition to modern amplification. Unquestionably the finest Reverberation equipment available. Variable control. Separate volume and tone controls. Independent channel input sockets. Operable on any high impedance amplifier. Non-mechanical. For Public Address—Guitar—Organ, etc.

VOX ECHO (The Shadows Model)
The original in the United Kingdom and still the finest. Portable and compact. Three inputs. Push/Pull controls for variety of echoes, multi-echoes and stunt incidental tones. Separate on/off control. Brings all the exhilarating effects of a large auditorium in even the smallest space. Presented in sturdy attractive carrying case with clear operating instructions and stowage compartment for cables and accessories.

LINE SOURCE SPEAKER
This new scientifically developed model has been created with the same technical perfection that is associated with all Vox equipment. Having the advantages of giving a full spread of sound and minimising microphone feedback, this model is not only ideal for Public Address installations but is now fast becoming popular with the modern groups. Fitted with an adjustable Stand which enables the Speaker to be set at any angle. Waterproof cover supplied. Incorporates four 10" Dia. Speakers, 40 Watts output per Cabinet. 15 ohms impedance. Dimensions—45½" high × 13" across × 7½" deep.

THE BEATLES

Vox

The jewel of the GT Collection, this is a very early Vox AC30 in the tan covering. The first AC30s were covered in this tan vinyl and had a reddish metal control panel, reddish-brown checked grille cloth, and of course Bulldog 12s. The early amps were available in either black or tan, although later Vox discontinued the tan. The amp is from the GT Collection; the '67 Candy Apple Red Esquire was once mine by now belongs to my pal Marty Mehterian, however, he lets me borrow it on weekends. Photo by Jeff Veitch.

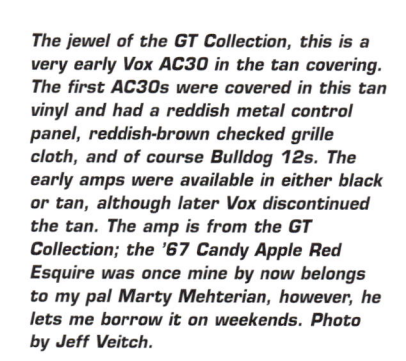

Here are two more early 1960s Voxes, a 15W 1x12" AC15 and a 4W 1x8" AC4, both in the highly collectable tan vinyl with reddish-brown grille cloth. GT Collection, Photo by Jeff Veitch.

popular with all the great British musicians, most notably The Beatles, whose early recordings used the AC30 for both the rhythm and lead tracks.

The Vox amps made and sold in the US are not the same amp as their British counterparts. In fact, the US-made amps were designed and manufactured by the Thomas Organ company under a poorly advised agreement Tom Jennings entered into that gave up his rights to distribute products with the Vox name in the US and Canada. He thought the Thomas company would be buying his amps to distribute, but instead they set up a factory in Southern California and made their own Vox amps. Their amps were basically transistor organ amps with lots of 'bells and whistles' that developed a reputation for unreliability and sounded nothing like the Vox amps used by the Beatles in England. I was working in a retail music store in Hollywood around this time, called the 'The Vox Guitar Center' (later shortened to just 'The Guitar Center'). We sold hundreds of these solid-state Vox amps and took half of them back for repair within the first month. I didn't know at this time that there really were great sounding Vox amps out there, I was just in the wrong country.

Incredibly, all the advertising put out by the US Vox company showed the Beatles using the real tube Vox amps. Kids bought the transistor amps and couldn't figure out why they couldn't get those authentic English rock'n'roll tones. In fact, these amps sounded pretty awful.

Meanwhile, back in England, the Jennings company was growing, even without the US distribution deal. The early amps were covered in a tan vinyl with a distinctive reddish brown grill cloth in an 'X' (or 'diamond') pattern. This would later be changed to a black vinyl with a primarily black grill cloth of the same design. Tom added chrome stands as an accessory that allowed the speaker cabinet to be tilted to better project the sound. The later US-made amps also offered this feature.

Most of the Vox amps we find in America are of the black vinyl variety, but maybe you've seen Tom Petty using one of these early tan AC30s on most of his tours and recording sessions.

Vox also made a very few custom-colored amps in red, white and green. Although I don't have exact facts, employees I've talked to remember making these amps for certain English dealers in the late 1960s. I estimate that less than a few

This rare JMI Vox model 710 amp from the mid 1960s (top photo) is similar to an AC10, with 2xEL84s power tubes, a 9-pin EZ81 tube rectifier, and a 2x10" speaker cabinet. The rare 1965-66 Jennings combo (bottom photo) is almost identical to a Vox AC15 – though it looks entirely different – and even shares that model name. Both courtesy of David Swartz, photos Aspen Pittman & Ed Ouellette.

Vox

The early to mid 1960s copper-panel AC30 shown above is stamped 'BASS' on the lower-right edge of the chassis, but there were few significant differences between the AC30s built for bass and for guitar. The silver-frame Celestion alnico speakers are almost identical to the famed 'Blue Bulldogs,' though the color – as much as anything – makes them slightly less collectable. Amp owned by Steve Harris, photo courtesy of The Guitar Magazine.

This is an ultra rare example of a colored Vox amp. During a short period in Vox history, JMI (Jennings Musical Instruments) took some special orders for colored AC30 amps from a few good dealers in England. I spoke to an employee who supervised the covering in this era and he said they made fewer than 10 in red, but also made a few green and blue amps as well. The amp was formerly in the GT Collection, but is now owned by Jim Elyea. The 1968 Tele Thinline is courtesy of Erhard Bochen, photo by Jeff Veitch

dozen of these custom colored amps exist, and so are quite rare. I have been lucky enough to acquire one example through my friends at Music Ground in Doncaster. It's a very early red AC30 with the reddish-brown control panel from around 1966.

The AC30 was also available with reverb or extra gain in the tone section – they were labeled either AC30 Rev or AC30 Top Boost. The Top Boost model is considered the best-sounding – and therefore most desirable – of the bunch because it has two extra ECC83 preamp tubes and a complex interactive tone-boost section that gives it extra punch compared to the standard AC30. The Top Boost models are easily recognized by the additional Treble and Bass controls that use a common Cut control. The three controls make up the tone section of the AC30 Top Boost, whereas the standard AC30 has just a single Tone control. As the Top Boost models became more and more popular, many early non-TB AC30s were retrofitted with the additional tone circuit; on these the extra controls are generally found on a plate mounted on the upper rear panel of the amp.

The AC50 was another great Vox, though somewhat undersung compared to the AC15 and AC30, and maybe not as sweet sounding overall. Developed by Dick Denney specifically to suit The Beatles' demands for more power for larger arenas, the AC50 used two EL34s to generate an output that, I have heard it reported, can reach as high as 70W on well serviced models. It's a big, bold, clean amp to be sure, only easing into crunch when you really get it cranked up. Early 'small box' models with a tube rectifier and just one channel are relatively rare, after which they added a second channel and, soon after, converted to a solid state rectifier to handle the power draw, which was pushing the GZ34 tube rectifiers to their design limits. Either of these could be bought relatively cheaply until pretty recently, but have lately become very collectable. When their most important customer (yep, The Beatles again) needed still more power, Vox developed the AC100 – but these are very rare indeed.

The team of Tom Jennings and Dick Denney was a successful one that produced some of the world's best sounding amps, but Tom wasn't such a great businessman. This fact was evident because, despite a successful product line of guitars and amps, the company folded several times in the course of the late 1960s and 1970s. This was the result of many management mistakes and bad deals (like the US distribution deal he made with Thomas Organ mentioned earlier). Tom Jennings and his Jennings Musical Instrument company went in and out of business frequently during the 1960s and 1970s, and wound up broke at the end. A sad fate for the man

THE TUBE AMP BOOK 89

Vox

Compare these AC30 top panels. The one on top is the standard model. The one on the bottom is the Top Boost model with the additional controls in the tone section. The Top Boost model also had more gain and thus is more desirable for lead playing. GT Collection

who designed and built perhaps the best-sounding amp of the day, arguably ever, and certainly one of the most emulated.

The Vox name changed hands several times over the following years, finally being acquired by the Rose-Morris company from the Dallas Arbiter company in the late 1970s after Rose-Morris lost the Marshall distribution for England. (Dallas Arbiter had manufactured the short lived Sound City amps and, sadly, they didn't have Vox for long either.)

Korg now owns the Vox name, and is manufacturing a new AC30 in the Marshall factory. This is a version I have been very impressed with. These days, when so many newly formed amp companies are offering close copies of this amp for small fortunes, it's refreshing to see a truly accurate reissue amp come along that gets it right (although these amps are not cheap either, and of course are built around PCBs rather than hand-wired tag strips like the originals – though keep a lookout for rare examples of the 2002 Ltd Edition Hand Wired Vox AC30). Since the enormous success of the first JMI Vox tube amps, and subsequent demise of JMI, the AC30 has been 'reissued' many times by companies called Vox Sound Ltd, A Vox Product, Vox Limited, Foxx Electronics (US),

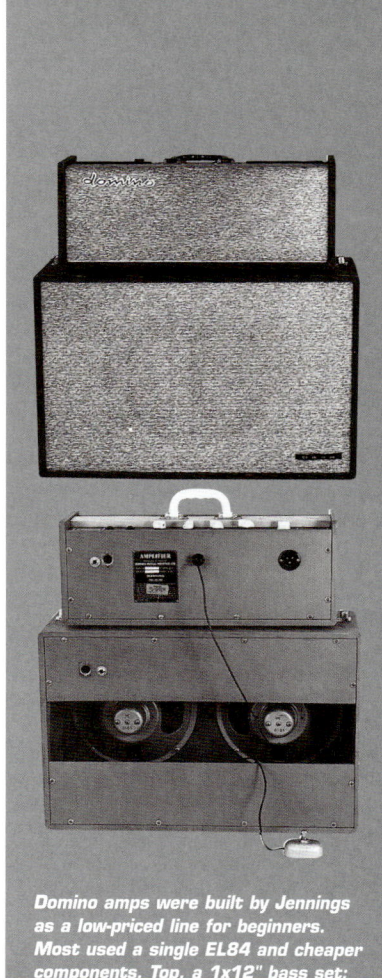

Domino amps were built by Jennings as a low-priced line for beginners. Most used a single EL84 and cheaper components. Top, a 1x12" bass set; below it, back of a 2x10" guitar amp.

When The Beatles and other big groups in England in the mid 1960s needed even more power, Vox designed the AC50 to fit the bill (pictured below left). With a duet of EL34s pumping into a heavyweight output transformer, it stays crisp and clean right up near max, with plenty of bite, and its power also made it a popular choice for bass in the 1960s. When still greater power was required, Vox 'doubled' the AC50 to produce the AC100 (below). The well-gigged AC50 and Gibson Les Paul Jr courtesy of Al Duncan, photo by Miki Slingsby.

90 THE TUBE AMP BOOK

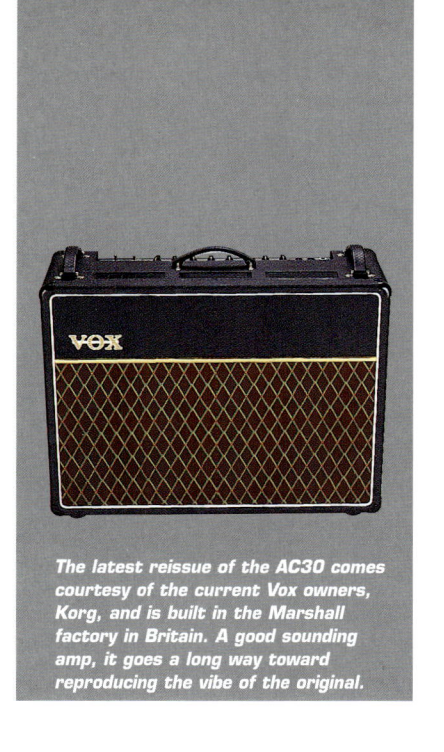

The latest reissue of the AC30 comes courtesy of the current Vox owners, Korg, and is built in the Marshall factory in Britain. A good sounding amp, it goes a long way toward reproducing the vibe of the original.

and most recently before now, by Rose-Morris; many of these versions of the amp have been pretty disappointing, too. There have also been many companies like Trainwreck, Matchless, and others which have closely emulated the AC30 design. Some of these turned out better than others (my personal favorite of these is the Trainwreck Liverpool amp – see their entry for details).

The new Vox reissue amp is offered with either the newly reissued Celestion Blue Bulldog Alnico 12", or a less costly but great sounding Celestion G12M Greenback. While these are still relatively expensive amps, they are less money than the copycat amps, they're built better than many of them, all the parts are correct to the originals, and in my opinion, they sound better! I haven't been too impressed with many of the reissues as compared to the originals, but this amp gets a lot closer in spirit and sound.

The current-issue AC15, on the other hand, is not as highly regarded as the AC30 reissue. Not strictly a reissue, it has features like master volume, etc.

By far, the most collectable Vox amps are the early ones made by the Jennings Musical Instrument Company, labeled 'JMI,' usually on the control panel. Good luck finding a clean AC30 Top Boost made by JMI at a reasonable price. They just started drying up a few years ago and a good one today can cost you several thousand. A few years ago they could be had for several hundred.

Watkins/WEM

Charlie Watkins never acquired the fame of Jim Marshall or Tom Jennings, but he was the first successful manufacturer of amps in Britain. He sold a broad range of equipment from 1954 right into the early 1980s – and has even returned in the new millennium to reissue his Copicat tube tape echo and V-front Dominator amp.

After serving in the Merchant Navy in World War II, Charlie returned to his native south London, where he played his beloved accordion in bands for a few years to earn some money, and eventually opened a record shop in 1949. In those days, record shops were where you went to buy your musical instruments, too. As the English 'skiffle' craze took off, Charlie saw a booming market ahead and started distributing – then making – his own gear. After first importing guitars from Germany, he realized he needed amps to go with them; Arthur

The 15W, split-front 2x10" Watkins Dominator was extremely popular in England in the late 1950s, and you can't get much cooler styling than that! With two EL84s in the output, it sounds not unlike a Vox AC10 or AC15 – and really disperses some sound with that V-front cab. Far right, the promo brochure for the reissue model. Amp from the GT Collection, photo by Aspen Pittman.

O'Brien at Premier (UK) came to his aid, and designed what would be the first proper Watkins amp, the EL84-based 10W Westminster, released in 1955, complete with stylish two-tone covering and gold string trim.

It was a near-instant success, and was quickly followed by the extremely cool V-front Dominator – belting out 15W from a Vox-like 2xEL84 output stage – and the cute little single-ended 5W Clubman. On top of that, Charlie soon offered what was arguably an even more revolutionary piece of equipment: his Copicat tube-powered tape echo unit, which preceded models by Meazzi, Binson and Vox. Inspired by singers using two massive Revox tape machines to create a tape loop, Charlie and engineer Bill Purkis compacted the idea down to a portable 12"x8" unit – and sold out their entire debut run of 100 Copicats on the first day it appeared in the record shop!

"The Copicat, Tom Jennings' AC30, and the Fender Strat were really the three main elements of the 1960s sound [in Britain]," says Charlie. "The mystique of the Copicat really comes from a bit of

Watkins/WEM

The tube-powered Copicat tape echo was one of Watkins's most popular products, and its sales helped the company step up to bigger premises and become a major player on the British amp scene throughout the 1960s. Charlie Watkins tells us the unit's great tone was partly due to a 'flaw' in its design – but he ain't telling us what he did wrong! Early 1960s Copicat courtesy of Huw Price, photo by Miki Slingsby.

bad engineering that went into it… but I won't tell you what it is! Every firm that's tried to produce their own version would do it faithfully until they came to that part, and couldn't bring themselves to do it as badly." In fact, Watkins and WEM amps were never the best built in the world, but they were sturdy enough, and certainly affordable.

In 1961 Watkins Electric Music started building its own solidbody electric guitars, too, which became a popular choice for young musicians in Britain. A couple years later, inspired by the impact and success of the striking, three-letter VOX logo, Charlie started badging all of his amps with a new, bright-red WEM logo. Other than the natural evolution of circuits and designs, nothing radical changed inside, but the new name marks the end of the 'vintage' era of Watkins amps for many collectors. Nevertheless, amps that followed this period are some of the best-sounding, and best-value, Watkins/WEM amps made, including more powerful square-cab versions of the Dominator, and the 30W HR30, WEM's rival to the AC30.

It would be remiss not to mention that Charlie Watkins also deserves credit as the father of the modern PA system in Britain, and was the first manufacturer in Europe to take on the revolutionary idea of slaving together numerous high-powered amps to produce a live vocal system.

His early PAs supported The Byrds' British tour in 1966, and by 1967 bands such as Fleetwood Mac, The Who, Pink Floyd and Ten Years After were playing in front of the biggest PAs ever seen – a whopping 600W – and all WEMs. By this time, however, the technology of choice had turned to solid state output stages.

Now in his eighties, Charlie has gone back into business to reissue limited numbers of his Copicat tape echo, along with an updated version of the V-front Dominator. The latter now has footswitchable clean and overdrive channels, and develops 30W from four EL84s with four 12AX7s in the preamp. It now carries a pair of Celestion Vintage 10s in place of the original's 10" Elac 10N alnico speakers. Stand back a few feet, and that angle-front cab really disperses some sound!

Thanks to David Petersen for supplementary information and the interview with Charlie Watkins.

Wizzard

If you like the style, quality, and tone of the old Hiwatt amps, check out the Wizzard amps being built in Canada by Rick St. Pierre. AC/DC has toured and recording with Wizzard amps, and they have had a lot of respect from other players, too. Rick's amps are loud, clear and fat, just like an old Dave Reeves original but with a bit more muscle.

After going out of production for a while, Wizzard amps are now being built in Canada again. Current models include the Vintage Classic and Modern Classic in top or 2x12" combo format, both in 50W or 100W versions and with options to run on either EL34s or 6L6s. Wizzard also offers a Joe Perry Signature Model, and a range of other variations on their classic formats. All models are hand-wired on military type turret boards, with high-grad capacitors, resistors and pots, all housed within rugged 14 gauge nickel and brass plated steel chassis – nearly bulletproof. Sales are factory direct only and with what appears to be a pretty long waiting list, but if you're after a modern take on that muscular, clean Hiwatt-style sound, a Wizzard might be worth waiting for.

The top ten all time classics
Reprinted from Guitar Player

Imagine having to choose your absolute all-time favorite song – or even your top 50, for that matter. That would be child's play compared to deciding which vintage tube amps to feature for this article. Give me 50 choices, and I'll make everybody happy. But give me 10, and I just know I'll ruffle some feathers. Given the space constraints of this article, however, that's all I can do justice to. I apologize in advance for leaving out somebody's sweetheart, but I don't think I'll get many arguments on the 10 terrific tone machines I've chosen.

Fender Tweed Champ

The 1958 Fender Tweed Champ

Power	3 1/2W RMS into 4-ohm load
Tubes	One 6V6 output, one 12AX7 preamp, and one 5Y3 rectifier
Speaker	Usually a 3.2-ohm, 8" Jensen (though some used a 6" Jensen)
Special features	One knob, the on/off switch and volume control.

This model – covered in luggage-style tweed – has to be the cornerstone of all tube guitar amps. Designed in 1953 for beginners in music schools, most of these tweed beauties were originally purchased by lap steel guitarists for either country or Hawaiian music. Nevertheless, they were to have a big role in rock. First off, they were inexpensive, a priority for young rock'n'rollers. Hundreds of great solos were recorded through these little giants during the early days of rock'n'roll, most likely because Champs distort so nicely when cranked up to 12. That's two more than 10, so eat your heart out, Spinal Tap. But the Champ also provided that great distortion at a relatively low volume level, and most studios and engineers preferred working at lower volume levels in the 1950s and early 1960s (and in many cases they still do). It simply suited their mikes and studios better.

Technically speaking, the Champ couldn't be simpler. One 12AX7 dual-triode tube acts as the first gain stage and also as the output driver to a single 6V6 output tube, which puts out a mighty 3 1/2 watts RMS into a 3.2-ohm load. It's the simplest example of how to use a tube from the RCA tube applications manual, a reference book supplied to amp manufacturers in the 1940s and 1950s. To keep your Champ in top form, change your 6V6 output tube every few years if you use the amp regularly. The 12AX7 shouldn't need regular replacement if it's a good-quality one.

If your amp still doesn't sound right, check to see that the speaker is truly a 3.2-ohm model (or a close-enough 4-ohm unit); blown Champ speakers are often replaced with the more common 8-ohm variety, resulting in a big loss of power and tone. If you're looking for a little more power and a tighter sound, try replacing the 5Y3 with a solid-state rectifier. Stay with the tube rectifier for a sweeter, softer distortion tone, however.

The 1959 Fender 4x10" Tweed Bassman

Power	40W into a 2-ohm load
Tubes	Duet of 6L6s output, one 12AY7 and two 12AX7 preamp, and one 5AR4 rectifier
Speakers	Four 8-ohm 10" Special Design Jensens
Special features	Four inputs, two channels with one tone section featuring bass, treble, middle, and presence.

Among Fender fanatics, this is undoubtedly the most sought-after vintage amp. It may also be the best-sounding guitar amp Fender ever made. The Bassman series was introduced in 1951 along with the Fender Precision Bass.

Initially equipped with a 15" Jensen in a 'TV-type' cabinet, the Bassman pumped out about 40W. In 1955 Fender changed the configuration to four 10" Jensen Special Design speakers wired in parallel at a two-ohm impedance load. A year later a middle (midrange) control was added and inputs were increased to four (circuit 5F5), followed soon after by some minor circuit changes to the model 5F6-A, which marked the final evolution of the tweed Bassman. It was manufactured in that form until 1960. In my collection, it is clearly the best-sounding amp of the era.

The change from a single 15" to four 10" speakers definitely contributed to the Bassman's famous tone. Fender had trouble with the open-back 15" 'flapping out' on the bass's low E string; switching to four 10"s rolled off the fundamental and produced a tighter bass response. Keep in mind that Fender was still trying to produce a bass amp, and these evolutionary

Fender Tweed Bassman

THE TUBE AMP BOOK 93

The top ten all time classics

changes were in that interest. Little did the company know the amp would become so popular with early blues guitar greats, and later rock stars, for its piercing tone.

Another great Fender feature during this era of tweed amps was the presence control, seen only on the higher-priced models. The control is an interesting circuit that taps the negative energy from the speaker – called the negative feedback loop – and reintroduces this signal to the output driver of the power tubes. The result of winding up the presence control is a distinct increase in highs, much different to that provided by a treble control. It was this later-version Bassman that was emulated by Jim Marshall in 1963 in an attempt to make an affordable amp available to bassists in the United Kingdom, thus laying the foundation for the Marshall amp empire.

If you're lucky enough to own a 4x10" Bassman, you should always keep fresh, matched sets of power tubes in the output stage. Though many tube charts call for 5881 output tubes, I've found that the 6L6 is a satisfactory substitution. Most of the Bassmans I've re-tubed sound best with General Electric 6L6s. They are most like those used by Fender during the 1950s and 1960s and provide smoother overdrive and a sweeter, rounder tone in all the Tweed-series amps' clean mode. The later-style 6L6s – made by Sylvania and used by Fender, Peavey, and MESA/Boogie (until these tubes also started drying up) – produce a harder, crunchier sound. It's also important to have a serviceman check the amp's many electrolytic capacitors, which filter 60 Hz hum out of the power supply. As the amp ages, these caps dry up, producing a mushy bass response and a low, following harmonic tone that plays out of tune as you push the amp for solos.

The 1962 Gibson Stereo 79RVT

Power	15W per channel (it's stereo!) into an 8-ohm load
Tubes	Quartet of 6BQ5 (EL84 in the UK) output, three 6EU7, one 12AU7, one 7199 preamp
Speakers	Two 10" 8-ohm Jensen Special Designs in an angled cabinet
Special features	Five inputs (one Stereo and two for each channel), reverb, tremolo, stereo/mono switch, and separate bass and treble for each channel.

I've included this little-known amp for a couple of reasons. Its stereo design and great reverb circuit make it arguably the best example of a long line of really neat and innovative amps built by Gibson in the 1950s and 1960s. And, as one of the first amps I ever collected, it's also a sentimental favorite. I was talked into selling mine to one of my favorite players, Joe Walsh, because I thought I could find another right away.

As it turned out, it took six years of hard looking to find another 79RVT, and I'm keeping this one, no matter what.

The 79RVT can be used as a 30W mono amp with two 10" Jensen Special Designs, or it can be split into two 15W amps, with each driving one speaker. Although this feature was originally intended for stereo accordions or for club players who wanted to sing through one side and play guitar through the other, I've found a much better application. Two players can plug into this amp at once (although only one channel gets the reverb), which provides a great natural mix. Because the sound comes from one point, it's easy to compensate for each player's volume by ear.

But the really great thing about this amp is the overdrive tone when you crank it up. Each side of the output stage employs a pair of 6BQ5 power tubes in a classic class A/B design. These tubes have a smooth, natural compression and are the same output tubes used in Vox AC30s and early Marshall combos. They are called EL84s in Europe, and it's interesting to note that Gibson was using them in great-sounding amps long before Vox and Marshall. The reverb is also unusual for amps of this era. Because it functions as a completely separate amplifier stage, it produces reverb without an increase in the channel volume. This means you can get various strengths of reverb by balancing the channel volume and the reverb volume, including 100 per cent reverb by turning the channel volume completely off. We're talking about real deep reverb in this mode.

The 79RVT came covered either in a tweed similar to Fender's tweed with a dark brown grille cloth or, later, in a chocolate-brown Tolex with a tan-and-red checkered grille cloth. It also came with a solid-mahogany footswitch, which looks suspiciously like the leftover piece from a Gibson Firebird guitar's cutaway.

The 1964 'Blackface' Fender Twin Reverb

Power	80W into 4 ohms
Tubes	One 6L6 quartet output, four 12AX7s, two 12AT7 preamps
Special features	Chrome tilt-back legs to project the sound upward, two channels with bass, middle, treble, and bright switches, and one channel with reverb and vibrato

Commonly referred to as the workhorse of the industry, the Twin Reverb has been favored by jazz, country, and rock players since its introduction. During its heyday, the Twin was the ultimate club amp because of its power and compactness. Its open-back cabinet provides a broad dispersion pattern, perfect for filling up a club from a small stage.

Much has been made about the 'pre-CBS' era of the Twin Reverb (and other Fender products), and

Gibson 79RVT Stereo Amp

Blackface Fender Twin Reverb

The top ten all time classics

it's my opinion that there is definitely something to it. When Columbia Records (a division of Columbia Broadcasting System) purchased Fender Musical Instruments in 1965, Leo Fender was removed from the management loop and a team of previously frustrated engineers quickly made several circuit changes to the amp. These changes, they said, "would make the amp more stable and extend tube life while eliminating most of that nasty distortion." The changes involved adding resistors and capacitors at several points in the circuit. These post-CBS Fender models sound constipated and tight, with none of the life and feel of the earlier ones.

When amp sales fell drastically, the sales force demanded a return to the pre-CBS design, which was quickly agreed to by Fender management. Though these lame Fenders were produced for just eight months, it took much longer for retailers to sell them off, and the damage was done; pre-CBS amps were almost immediately more desirable to musicians in the know. Because Fender replaced its traditional black-faced control panel with a shiny silver plastic panel after 1967 – by which time most of the changes had come in – it's easy to distinguish between pre- and post-CBS amps.

The silver-faced amps built between the CBS takeover and the later ones with a master volume and a pull-distortion knob are very good-sounding amps, and a bargain for players who can't pay the inflated vintage price of an old blackface Twin.

Blackface Fender Deluxe Reverb

The 1965 Fender 'Blackface' Deluxe Reverb

Power	22W into 8 ohms
Tubes	One 6V6 duet output, four 12AX7, two 12AT7 preamps, one 5U4GB rectifier
Speaker	One 12" 8-ohm Jensen Special Design
Special features	Dual sensitivity inputs on dual channels that have bass and treble controls; the second channel has reverb and vibrato that can be activated with a foot switch.

This is the amp I would choose if I were stranded on a desert island and could only have one. It's small and simple to operate, sounds terrific for any kind of music, and gets a great overdrive distortion when it's cranked up. The pure tone of a single 12" Jensen shows why it's really one of the greatest guitar speakers ever made. Fender used them because they were cheap and reliable. Perhaps the primary secret to the great sound of this amp, however, is the complement of 6V6 power tubes. Essentially, the 6V6 is a smaller version of the 6L6 with a similar impedance and sound; it puts out about 60 per cent of the 6L6's power. This means it can be a clean amp for a small country club gig, or provide just the right level for smooth overdriven leads in a small rock club band. It's great to use a Super Reverb for rhythm and switch to a Deluxe Reverb for leads. The two amps are a perfect match, and both can be bought – in silverface guise, at least – for less than any of the fancy new channel-switching combos around today.

The 1965 Vox AC30 Top Boost

Power	30W into 16 ohms
Tubes	One EL84 quartet output, seven 12AX7s, one 12AU7 preamp, one GZ34 rectifier
Speakers	Two 12" 8-ohm Vox Bulldogs (made by Celestion)
Special features	Three channels, a master tone section with treble, bass, and a cut control, vibrato

Anybody's list of top 10 amps would have to include this English legend. Vox amps were, after all, what the Beatles used, and the AC30 was the one they most often toured and recorded with. The AC30 TV (for Top Boost) had two extra tubes and an expanded tone control circuit that gave it more gain and versatility. Both AC30s were very popular with most 1960s and 1970s British bands and appeared on many classic recordings. The AC30 usually came in an open-back combo version with two 12" Vox Bulldog speakers, although it was also offered as a closed-back piggy-back unit with a chrome stand that tilted the cabinet for better projection.

The Bulldog speakers were an important part of the AC30's distinctive sound. Built by Celestion, they were virtually identical to the Celestion G12s made for the early Marshalls. Although the Bulldogs were rated for just 15W, I know of people who have used them in their Twins for years to get more crunch; that's at least 40W into each speaker.

The AC30's sound is distinct and very different from the Fender amps of the time. Listen to the guitar sound on the Beatles' Taxman [Revolver, Capitol, SW-2576]; that's an AC30 Top Boost. Contributing to the amp's smooth tone in both the clean and distorted modes is its very unusual class A circuit designed by Dick Denney. While a class A design produces less power from the same number of output tubes (four EL84s in this case), the tone is very smooth and the distortion is rich in overtones. To my knowledge, Vox AC30s were the only commercial amps at that time that utilized a pure class A circuit – other than small, single-ended amps like the Fender Champ and the Vox AC4, which are always class A.

When the Beatles began to break as world sensations, Vox founder Tom Jennings struck an ill-fated distribution deal with the Thomas Organ Company for Vox marketing rights in the United States and Canada. Instead of buying tube amps from the British manufacturer, however, Thomas Organ set up a manufacturing line in its Sepulveda, California, organ factory and started cranking out transistorized Voxes with names like Super Beatle and Royal Guardsman. Thomas sold thousands of these bad-sounding, unreliable amps before the

Vox AC30 Top Boost

The top ten all time classics

public caught on. Because it lost the biggest market for some of the hottest products of the 1960s and 1970s, the Jennings Musical Instrument company eventually collapsed. Tom Jennings never realized the full commercial potential of his innovations.

The easiest way to distinguish an original Vox from its transistorized clone is to plug in; the amp that sounds like all the Beatle songs is the British amp, while the one that sounds like a Lawrence Welk record gone bad is the Thomas Organ Vox. Fortunately, the name plate provides a fast means for visually identifying a US-made transistor Vox: a ring surrounds the Vox logo. Visually identifying authentic tube Vox amps is a little tougher, but originals have a JMI (Jennings Musical Instruments) logo on the control panel and/or the ID plate. If it doesn't say JMI on the amp, it's probably one of reissue AC30s made over the last decade by companies eager to cash in on the amp's popularity.

Interestingly, the transistor Vox amps made by Thomas Organ used Bulldog speakers. If you can find an old Super Beatle cabinet cheap enough, buy it for the four 12" speakers inside; they'll make any amp sound better.

If you're lucky enough to own an original AC30, here's a tip from the late steel guitarist/inventor Red Rhodes. The original rectifier was a GZ34 (or 5AR4 in the US), and when these English amps are plugged into our 60 Hz current (it's 50 Hz in the UK), it drives the tubes a little too hard and quickly wears them out. Try substituting a GZ32 (5U4 in the US) to lower the voltages a bit; it won't affect the tone, and makes the amp more reliable. It's also advisable to have the filter caps checked and replaced with fresh ones, if necessary. The Bulldog speakers are generally very reliable and, since they have Alnico magnets, stay powerful for years. However, they do occasionally blow. If that happens, do not change to a different speaker; try to have them reconed as original as possible (no chrome dust covers, please). A reliable reconer should be able to deal with them. Otherwise, Celestion now offer a great-sounding reissue, called the Alnico Blue. It's expensive, but probably comes closer to the original than any modern speaker available.

The 1965 Marshall JTM45 and Park 50

Power	45W RMS into 8 or 16 ohms
Tubes	One 6L6 or KT66 duet output (later models used EL34s), three 12AX7 preamps, one GZ34 rectifier (earliest models)
Speakers	Four Celestion 16-ohm G12-20s in a 4x12" 16-ohm cabinet. The combo has two 12" 25-watt Celestions
Special features	Two channels (bright and normal), bass, middle, treble, and presence.

Okay, okay, so I can't pick just one Marshall for an all-time favorite. Instead, I'd like to pick several from the 'glory years' – 1963 to 1972 – and count them as two. Jim Marshall's earliest amplifiers were rated at 45W and were direct knock-offs of the Fender 4x10" tweed Bassman, right down to the U.S.-made surplus transformers, 6L6 output tubes, three 12AX7 preamp tubes, and a tube rectifier. Understandably, the sound of these is very similar to the Bassman. Marshall's cabinets were much better suited for bass, however, because he used four 12" speakers in a closed-back design. Built on a chassis of hand-bent aluminum, early JTM45 amps had no serial numbers; the first 100 or so can be identified by small, red enamel-filled logos with the name 'Marshall' in block letters. That evolved into a gold plastic plate with block Plexiglas lettering, and then to the familiar white plastic script logo, although early versions were plated with either silver or gold.

As their popularity grew, Marshalls gradually began to take on more of their own identity and distinct sound. Responding to guitarists' demands, JTM45s gradually began to sound brighter as Marshall began employing Europe's most common power tube, the EL34. Most of these amps had one channel designed for bass, with a second, brighter channel for guitar. The earliest JTM45 combos had a great blues tone, again very reminiscent of tweed Fenders. As Fender began building more sophisticated amps, Marshall stuck with the raw designs pioneered by the 1950s Tweed series, a formula that proved to be right on target for the type of sound that musicians came to demand during the late 1960s and early 1970s. That sound, favored by such artists as Eric Clapton and Jimmy Page, propelled Marshall to a dominant position in the rock market, particularly in Europe.

This success, however, had a downside. In 1965, desperate to build a factory large enough to meet demand, Jim struck a deal with the Rose-Morris company, which agreed to give him the necessary cash in exchange for marketing rights and sole distribution of Marshall products for 15 years. This meant shutting the door on many music shops that had helped establish the company. To get around the problem, Jim decided to produce a line of amplifiers under another name: Park. Some time in late 1965, Jim and Johnny Jones, a drummer pal, formed a distribution company for these 'other Marshalls.' Park amps were built in Jim's factory with Marshall components, though the design and appearance differed slightly. For instance, Park amps had the same four inputs and two channel volumes as the Marshall, but the inputs were separated by the channel volumes. Jim used Park to experiment with designs that the Rose-Morris people didn't want to market. This resulted in some really interesting models specially built for keyboards or for really hot solo guitar. One of my favorite amps is a red '66 Park 50W top, which sounds like a perfect mix between a Marshall and a tweed Fender. It really screams. Park was quietly closed in the late 1970s, a couple years before Jim Marshall's 15-year Rose-Morris contract ended. Snap one of these up if you get a chance.

Marshall JTM45

Park 50

96 THE TUBE AMP BOOK

The top ten all time classics

The Marshall Super Lead 100

Power	100W RMS into 4, 8 or 15 ohms
Tubes	One EL34 quartet output, three 12AX7 preamp
Speakers	Celestion G12H-30s (16 ohms), usually in a 4x12" 16-ohm cabinet
Special features	Same as JTM45 and Park 50.

Marshall Super Lead 100

The next significant bit of Marshall history unfolded in 1966 with the release of the first 100W amps. The earliest models had 'Super 100' screened on a white plastic rear panel, but used leftover JTM45 front panels. These early amps used either two 50-watt power transformers or a huge, sideways-mounted power supply transformer that delivered over 600 volts to the EL34 output tubes (a very high power supply compared to today's Marshalls with around 450 volts on the tubes). This resulted in some really loud amps that were a little hard on speakers. Built with clear Plexiglas control panels with silk-screened black text on a gold background, these early 100-watters are known as 'Plexi' Marshalls, and are highly rated by players for their softer, yet powerful, tone. They aren't as bluesy-sounding as the early JTM45s or as bright as later Marshalls. Marshall switched to metal control panels sometimes in late 1969, ending the era of the 'Plexi'.

Marshall amps built between 1969 and 1976 were brighter and had slightly more gain. The company also gradually upgraded the Celestion speakers from the early 15W type (similar to the Vox Bulldog series) to the 30W G12H-30 series. All of the Celestions used during this gradual upgrade (which included 15W, 20W, 25W, and 30W models) were preceded by a 'G' for Guitar and '12' for 12" speaker, for example, G12-20). Jimi Hendrix liked Marshall bass cabinets because they used the Celestion 30H series – the 'H' standing for 'heavy.' They're very popular with heavy metal players for their raw, crunchy tone and are, in my opinion, the final improvement to the line.

This was also the era of the colored Marshalls. Dealers could special-order amplifiers in various colors of vinyl, including purple, orange, red, white, blue, and green. Deep Purple, the Guess Who, and many other great bands of the 1970s toured with these custom-colored Marshalls and created quite a stir. Custom colors were discontinued in 1976. That same year, Marshall introduced the master volume series and lost something in the process. Although they continued to make non-master volume models, pre-1976 amps seem to have more kick and are definitely more collectable.

If you own an early Marshall, or even the newer models, you can maximize performance by paying attention to a few simple rules. First, the quality of the tubes makes a world of difference. You should use good quality preamp tubes with low noise and no adverse matched output tubes. The company's US importer started putting the U.S.-made 6550 in Marshalls in the mid 1970s, while the UK-built EL34 is used everywhere else in the world. The EL34 is definitely preferable for guitarists who like that nice, smooth Marshall overdrive sound. The 6550 produces a harder, cleaner sound best employed by loud rhythm players or bassists. You can easily have your Marshall set up to use either tube by your local tech.

Second, the filter capacitors (the big blue cans mounted near the output tubes on your chassis) are famous for drying up, which causes a funny, out-of-tune low overtone when in the heavy overdrive solo mode. They must be replaced by a qualified repairman and need changing every five or six years, on the average. And no matter what anybody tells you, do not lower the voltage to your amp by using a Variac. Although this may produce some interesting distortion tones, it will adversely affect the life of your tubes, transformers, and other living things. To operate correctly, your amp wants the proper voltages at all times. Finally, be sure you've got your Marshall set for the proper impedance and that your cabinets are correctly wired; your amp will love you for it, and it will sound better and last longer. Check the impedance selector to ensure it doesn't get loose and make a faulty contact, the most common cause of blown output transformers and about a $200.00 repair.

The 1973 Hiwatt Custom 100

Power	100W RMS into 4, 8 or 16 ohms
Speakers	Four cast-frame, 8-ohm 12" Fanes
Special features	Four inputs; two volumes; bass, middle, treble, and master volume.

Hiwatt Custom 100

There were many other British amps besides Marshall, most of which were Marshall 'wannabes.' Hiwatt was one of the exceptions; it took the concept of a powerful British stack to a new level. Hiwatts were loud and rather clean, compared to Marshalls. If you've ever played one that's set up right, you'll remember the experience. The sound is tight and solid at the low to medium volume settings; when turned up full, it really moves some air.

It's impossible for me to think of Hiwatt without picturing Pete Townshend pounding out power chords with that swinging windmill style he made famous. Pete was one of the first players to recognize the value of Hiwatt amps, although many of the top English bands toured with walls of Hiwatts, which rose to fame as the up-market alternative to Marshall. This was because of an extreme commitment to quality by Hiwatt's founder and chief engineer, Dave Reeves.

Dave started the company in the late 1960s in a town just south of London. The line initially consisted of 50W and 100W tops coupled with the traditional 4x12" cabinets. The early Hiwatt cabinets had a straight front and no casters, sitting on simple wooden skids. Their chassis were heavy duty, and the volume and tone pots were the best available. The transformers were custom-made by Partridge and

The top ten all time classics

were switchable to operate at four different line voltages to adapt for a tour in any country. They could drive speaker loads of 4, 8, and 15 ohms, as well as a 100-volt line system.

When I first opened up an original Dan Reeves Custom 100, the sheer quality of assembly took my breath away. All components were mounted on thick fiberglass trim strips (no printed circuit boards for Dave), with the connecting wiring harnesses neatly hand-braided and laid out at perfect right angles.

You get the feeling that these amps were built to last forever and, indeed, I've seen very few of these early amps in for repair. Hiwatt amps of this vintage had a well-deserved reputation for reliability – and they really had to; just ask Pete's road crew. Hiwatt added several different items to its line over the years, such as 200W tops and 50W combos, called Bulldogs, but the meat of the line was always the 50W and 100W tops. Most of the early amps had model numbers starting with 'DR,' the designer and owner's initials.

Hiwatts have their own distinct tone, which is great for that hammering, bell-like rhythm track. I was lucky enough to pick up a vintage Custom 100 on a recent trip to England, and I highly recommend that serious players and collectors find one for themselves soon. The price on these amps is still relatively low, even in the United States; they can be found for under $500.00 in good working order. But increasing appreciation of vintage Hiwatts indicates that will undoubtedly change.

Tragically, Dave Reeves died in 1981 at the peak of his success, and control of the company passed to three of his employees. Hiwatt's popularity gradually declined, and the company was eventually sold to Eric Dixon, who launched a new phase of the Hiwatt story. Dixon's amps were redesigned with new features such as built-in reverb, switching channels, and high-gain overdrive distortion. These new amps kept up with current market trends, but still retained the classic Hiwatt look. The ideas behind the amps were basically sound, but certain design features caused problems.

For example, the EL34 power tubes were mounted on circuit boards, rather than employing the traditional chassis-mounted system. Though this provides faster and more economical assembly, tubes don't like the vibration that occurs on a printed circuit board, and circuit boards tend to warp with extreme heat.

Hiwatt has passed into other hands since then, and has even been owned by two different companies for a time – Rick Harris of Music Ground for the UK, and Fernandez for the rest of the world. All indications are that the English Hiwatts being built today have largely returned to the standards set by Dave Reeves 30 years ago, though the name can also appear on small, cheap, transistorized combos built in the far east, so watch out. Get your hands on the right one from the newer makers, however, and you've got an amp with a feel and sound not far off the great original models.

The 1978 Ampeg SVT

Power	320W RMS into 2 ohms
Speakers	Eight 10" Special Ampeg Bass Speakers in each cabinet
Special features	Two channels, including a bright channel with midrange boost with three selectable tones, bass boost, and bright switches; the SVT power section allows for adjustment of bias current and phase inverter balance.

Ampeg SVT

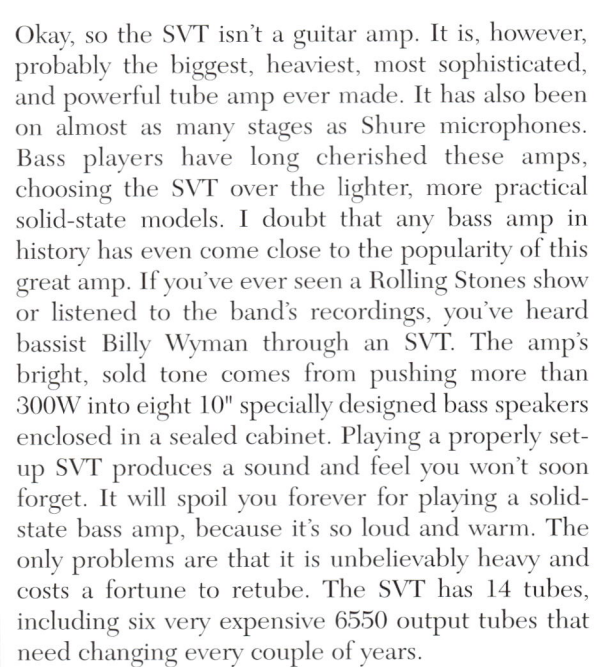

Okay, so the SVT isn't a guitar amp. It is, however, probably the biggest, heaviest, most sophisticated, and powerful tube amp ever made. It has also been on almost as many stages as Shure microphones. Bass players have long cherished these amps, choosing the SVT over the lighter, more practical solid-state models. I doubt that any bass amp in history has even come close to the popularity of this great amp. If you've ever seen a Rolling Stones show or listened to the band's recordings, you've heard bassist Billy Wyman through an SVT. The amp's bright, sold tone comes from pushing more than 300W into eight 10" specially designed bass speakers enclosed in a sealed cabinet. Playing a properly set-up SVT produces a sound and feel you won't soon forget. It will spoil you forever for playing a solid-state bass amp, because it's so loud and warm. The only problems are that it is unbelievably heavy and costs a fortune to retube. The SVT has 14 tubes, including six very expensive 6550 output tubes that need changing every couple of years.

The flexibility of the preamp is great and uniquely Ampeg. There are two channels feeding different tone sections, and the tone variations of the lead channel are wonderful, with a true middle-boost section that uses a transducer element and switches for three different levels of mid boost. In addition, there is a bass control with a switch than can select either a bass cut or an ultra-low boost. There are also a treble control and a bright switch for each channel.

The amp top provides a very sophisticated system for adjusting the bias, a most important adjustment for proper sound and reliability in any tube amp. It consists of two bias voltage adjustments and an overall balance adjustment, all of which are monitored by a digital volt meter plugged into two test points. All of these adjustments and test points are located behind a removable panel on the front of the amp. Even though there are detailed instructions for adjusting the SVT, I do not recommend doing this unless you are a qualified technician. For the qualified tech, however, the SVT is easy to adjust.

Another great design feature of the SVT is the 5W, 10-ohm resistor connected between the tube's cathode and ground, which the company calls a 'blow-out' protection circuit. If a single power tube shorts out, this resistor smokes and essentially disconnects the defective power tube from the circuit without affecting the rest of the tubes or blowing a fuse. The amp simply loses a sixth of its power but keeps right on playing. Truly a great

Modern classics

engineering job from the Ampeg team that developed the SVT.

Perhaps the most noticeable feature of the SVT is its unique cabinet design, a radical departure from all other popular bass cabinets of the time. The speakers have a 2" voice coil and relatively shallow cones, and when placed in a sealed cabinet, they produce a bass response above all the fundamentals on the bass. As a result, no particular string is favored. Most 15" and 18" ported cabinets of the day were bassier-sounding but lost output on the low E or even low G compared to the relatively even response of the SVT cabinet. This was because the low resonance (the limit of the cabinet's efficiency, or "how low can you go") of other ported cabinets didn't reach to the bottom fundamentals of the electric bass. The upper strings had the fundamental, while the lower strings didn't, and thus seemed to lose volume. The SVT cabinet, however, has a resonance that 'rolls off' at a much higher frequency and ignores the fundamentals of all four strings, which makes it seem more balanced. These lost low fundamentals aren't really practical anyway, because they caused vibration problems onstage, generally interfere with the other instruments and microphones, and can't be heard in the audience. The SVT is, by comparison, tight and balanced, and therefore fits better into the overall sound of the group. It has great definition, and you can hear all the notes the bassist is hitting.

Interestingly, this type of bass cabinet design has recently seen a resurgence with the Trace Elliot and the SWR designs, which also employ 10" speakers in sealed cabinets. The SWR even has a tweeter for an extended top end, adding to that solid bottom tone. These designs are enjoying much success in the marketplace today, but the SVT showed the way.

The SVT was initially developed for Ampeg in the early 1970s by an outside company called Uni Music.

The first SVTs used a weird output tube, the 6146, which was quickly changed to the more common 6550. Ampeg changed hands several times over the years. The first company to buy Ampeg was Selmer, which in turn was purchased by Magnavox. Magnavox moved the company from Linden, New Jersey, to a small corner of the Magnavox TV factory in Tennessee.

Magnavox continued to make the SVT virtually unchanged until it sold the company to MTI in the early 1980s. MTI contracted to have the amps made in Japan and offered a few of the most popular original Ampeg models, including the B-15 Portaflex and the SVT. The Japanese-built SVTs were exact copies of the original designs, though their components were slightly below the quality level of the US-made amps, which gave them a bit of a bad reputation for not being as good as the originals. The Japanese amps are still good values in the used marketplace, however, and parts are pretty much interchangeable with the US-made SVTs.

The latest chapter in the Ampeg story has taken a turn upward. The St. Louis Music company, headed by Gene Kornblum, acquired the Ampeg name and built a limited edition of 500 SVTs using the original designs and component suppliers from the old Linden era of Ampeg. Each amp has a plaque mounted on the rear to indicate its order of manufacture within the 500. These amps are no longer available and are fast becoming collectors' items. St. Louis Music has since offered an evolved SVT II that has some updated features, such as graphic EQ and other modernized items. They have also recently offered reissues of other popular bass models, some new designs, and have even offered Ampeg guitar amps again. I'm sure they will enjoy great success with their new SVT, but my heart will always favor the original. I guess I'm just a hopelessly vintage kinda guy.

Modern classics

While I have openly selected my favorite vintage amps here, there are dozens of tube amp builders today offering limited productions of great amps, many of which will rival the vintage classics for tone – and are often built just as well, or better. Many of them have been given brief entries in the preceeding 'Amp Companies' chapter – companies like Kelley, Victoria, THD, Kendricks, Top Hat, Dr Z, Trainwreck, Matchless, Soldano, Rivera, VHT, and plenty of others I am forgetting – but they deserve a few extra words as a group for the quality of work they are putting out (or in some cases, did put out until succumbing to the perils of the small business).

Generally, each of these companies is the product of one guy with a vision of how to build the world's greatest tube amp, no holds barred. They employ individual designs and construction techniques that, combined with variations of tube types and materials, give them each a unique personality.

I've used many of these amps myself, some we've worked on or re-Groove Tubed, usually for one of our Friends and Relations folks. They all sound and function a little different, but my taste runs to the understated, simpler designs that offer great tone, lotsa touch, and few knobs. I have some personal favorites that I have detailed elsewhere.

If you're looking for a unique and unusual tube amp like this, however, you will also need a rich aunt, because these kinds of limited production, hand-built amps will usually cost more than a Boogie, and far more than your off-the-shelf Fender or Marshall.

Modern classics

Like custom made guitars, they cost more, but the tone is usually worth it.

It's my opinion that the amp is at least half the sound of rock'n'roll guitar – and if a hand-built classic Paul Reed Smith guitar (my favorite) can fetch several thousand dollars, then so can the best of the hand-built tube amps available today!

The common thread that binds these 'modern classic' amplifiers is quality construction with attention to fine detail. The designs are simple; most of them are copies of real basic 1950s and 1960s production amplifiers. No fancy bells or whistles, no MIDI, just simple tube circuits, most originally available from the RCA tube applications handbook as used by Leo Fender and other early amp companies. The thing that makes these modern classics special is the correct blend of a simple tube circuit design with the right materials and construction know-how to achieve the ultimate tone recipe. In other words, it's craftsmanship – in the tradition of the Martin guitar or the Steinway piano. These new models are mostly hand-built with personal pride attached to every amp, and the attention to detail pays off as these amps almost always achieve their primary design goal: great guitar tone!

Another commonality to these amps is their relatively high cost, simply stated: they ain't cheap! Fact is, low-production, hand-built amps like these take time and money to produce, just like a hand-made Martin guitar. Frankly, I do not know of any budget (dare I say cheap?) tube amp for sale today that both sounds good, and is build to last – unless you're talking about a used Super Reverb. Just compare one of the many budget tube amps (including the US-made variety) to any of the modern classic tube amps, and you'll hear what I mean; not bad for the money, but not a great amp. While the 'Big Boys' spend a fortune on artist endorsements and ad campaigns, they spend much less on design, materials and workmanship.

Unfortunately, you just can't assembly line mass-produce quality tube amps these days like some kind of karaoke system. Mass producing tube amps in the 1960s was a snap. All the support systems were there to provide high-quality tube amp components: sockets, transformers, and most importantly, plenty of high-quality tubes! Nowadays, it's real hard work to reproduce the quality levels of 1960s tube amps. The guys who build these modem classics are definitely not getting rich selling their amps. But perhaps more important than getting rich, they are happy craftsmen doing what's important to them: building their ultimate guitar amp with no compromise. These guys are modern tone pioneers fighting the strong currents of mediocrity that are everywhere these days. Their accountants must hate them, but as a player, I love these guys!

I encourage you to seek out one of these modern classics before they become extinct like so many of the original classic amps detailed throughout this book. We should support them. They are keeping tube amps honest and true to the original idea: great tube amps make for great music. And who knows, any of these guys could be the next Jim Marshall, who, by the way, can still build a great amp!

Technical articles

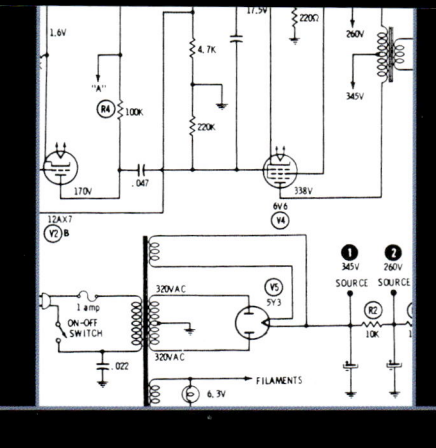

Maintenance and modification of the world's most popular tube amps

Technical

Tech articles

Behind the tube mystique
By Aspen Pittman (Reprinted from Guitar Player Magazine)

Guitarists both today and yesterday are often linked by one piece of equipment: a tube amp. In fact, the tube amp is currently enjoying its greatest popularity with musicians, even though there have been great strides in transistor amp technology over the past 25 years. Guitarists prefer tube amps. Why do designs built around tubes sound different from those following the solid-state approach? Simply, tubes work differently. This article explains their construction, function, and applications.

Tube Construction and Operation

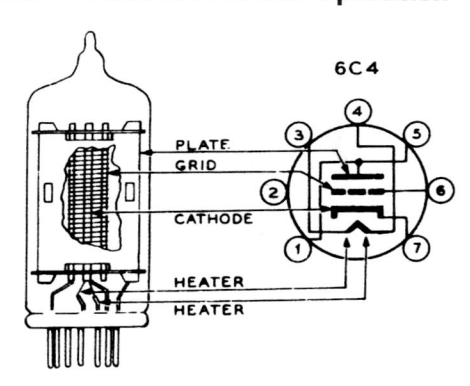

FIG. 1 Inside a basic tube are four active electronic elements as shown.

A tube is an electronic device consisting of a minimum of four active elements: a heater (filament), a cathode, a grid, and a plate. All of these are sealed in a glass enclosure with its air removed – a vacuum – to prevent the parts from burning. The location of a simple tube's parts is shown in **Fig. 1**. The filament is heated in order to warm the cathode. Once heated, the cathode begins to emit electrons, which flow from the cathode (which is negatively charged) toward the plate (which is positively charged). The grid's purpose is to control this flow. If the grid were absent, this movement of the electrons would be uncontrolled, much like water rushing from a faucet that is opened all the way (**Fig. 2**).

Theory of Operation

When a small signal is applied to the grid, it causes a larger change in the current that flows between the cathode and plate. In effect, it acts as a valve (in fact, the British refer to a tube as a "thermionic valve" – or just valve for short). A portion of the amp's electronic circuitry, the grid bias control, adjusts the proper voltage setting

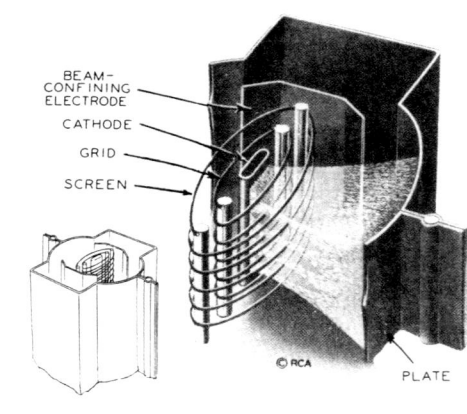

FIG. 2 Electrons flow at a controlled rate from the cathode to the plate.

of the grid. The amount of bias varies from tube to tube, depending on its sensitivity, and it acts to keep the tube "idling." When the grid bias is properly set, the tube is balanced to the circuit, and therefore produces a clean, powerful signal. Proper biasing also extends the tube's life.

For optimum performance, the bias setting should be checked whenever power tubes are changed – preferably by a qualified technician using an oscilloscope. A bias adjustment is a relatively simple operation, and can be performed for a minimal bench charge (typically $15.00 to $20.00). Some symptoms of improper bias setting include the amp running too hot, excessive hum after it's been on for a short while, or distortion that just doesn't sound right. The amp doesn't necessarily have to sound bad for its tubes to be incorrectly biased, and these symptoms may indicate other problems. However, if your amp is behaving in an extraordinary manner, a trip to the shop may head off damage to it, regardless of the cause.

Tube Functions in the Amp

Let's look at a common example of how a tube works in an amplifier. Imagine a small guitar amp with no volume or tone controls: just a guitar input, one tube, an output transformer and a speaker. The guitar's pickup produces a small voltage, the result of the string vibrating in the pickup's magnetic field. In general, this signal is applied to the grid, which in turn causes a large current flow from the cathode to the plate. Because of this, a correspondingly large voltage now appears at the plate. This plate is connected to an output transformer, which matches the impedance to that of the speaker. (Because there is a great disparity between impedances of the tube amplification circuit and speaker, the transformer must act as a buffer to interface the two components.) Thus, a small,

> **WARNING!**
> *Potentially lethal voltages are present in guitar amps, even when they are switched off and disconnected from the mains. Filter caps store voltage which can remain in the amp and be discharged through you. You should never open up your tube amp for any reason unless you are familiar with safe procedures for discharging these caps – and keeping them discharged – without doing harm to yourself or any of the components within your amp. If you do not understand these procedures and/or have no prior experience with them, never open up a tube amp – take it to a qualified professional. The authors of these technical chapters, the editor, and the publisher assume no responsibility for any damage whatsoever to persons or property that may result from undertaking the work outlined in these articles.*
>
> *There is a simple procedure to discharge the caps in amps with standby switches, which you should follow before opening up this type of amp (while this or other appropriate procedures should be used before working on any other amp):*
>
> *With the speaker attached, turn on the power switch on the amp, wait a minute and then turn on the standby (you should always wait a minute before turning on the standby to protect your power tubes by letting them warm up gradually). Wait a couple of minutes and then turn off the power switch on the amp, leaving the standby on. Wait a couple of minutes and then turn off the standby and disconnect the amp from the wall. This will reduce the DC voltage inside the amp to less than 10 volts – but you should check this with a multimeter before proceeding. If further current remains, you must bleed it off with a specially prepared bleeder resistor. Again, if you don't already know how to do this, don't try it yourself.*

low-power signal from a guitar's pickup can produce a high-powered signal to drive the speakers.

Naturally, amps don't all sound alike. This is due to variations in the types and quality of tubes used, as well as in the specific circuit design of the amp. In other words, some tubes amplify more than others under similar conditions. Also, the amount of gain a tube produces varies with the circuit design. This is why different amps can sound very different, even though they use the same tube types.

Technical articles

Additionally, certain amps use completely different types of tubes. A good example of this is the English-type Marshall using European EL34s in its power amp section, compared to the US type, which employs American-made 6550 power tubes. The US and English styles sound and play very differently, reflecting the character of their power tubes. The English EL34 tubes yield more distortion than their American counterparts, although they produce roughly the same amount of volume. With internal bias modification (which mostly involves changing some resistance values), any US Marshall amp can use European EL34s, and vice versa.

Multi-Stage Amps
Larger and more complex amps have many stages of tube amplification: preamp stages, signal-processing stages, and power amp stages.

THE PREAMP. The preamp stage is much like a mixer in a PA system, which must amplify an incoming mike or guitar signal to line-level strength before the signal can be processed with effects for tone shaping. Likewise, a tube amp must preamplify a guitar's signal so that it can be further processed. This is the first gain stage of the tube amp.

SIGNAL-PROCESSING STAGES. An example of a signal-processing stage is the reverb section, where the signal is diverted through a reverb spring system, returned by another gain stage, and finally blended with the original signal. Tone controls and second gain stages (often employed for an overdrive effect) are other examples of signal-processing stages.

POWER AMP STAGES. The power amp section takes the preamp's signal and amplifies it many times to a level that can drive the speakers. All tube amps with power ratings of 10 watts or more (other than a few rare, larger single-ended or dual single-ended amps) employ a push/pull power amp design. This means that the power tubes work as a team to amplify the signal and drive the speakers. Practically all transistor amps employ a push/pull configuration, as well. The output tubes all share in the sound, so for maximum efficiency it is desirable to use tubes that operate as similarly as possible. Also, for efficiency it is desirable to use tubes that operate as similarly as possible. Also, for efficiency, use power tubes of the same make – manufacturers' specifications for tubes bearing the same stock number may vary over a broad range. And, if one power tube is bad, it is advisable to change all of them. Having one fresh, powerful tube and three old ones, for example, can create an imbalance in the push/pull effect, resulting in inefficient operation. The power amp section is only as strong as its weakest link. So, if one tube out of four is faulty or varies from the others in its performance character, the overall sound of the amp will be limited.

The process of output tube matching dates back almost as far as tube amps. The military began matching certain properties of tubes to produce longer field life and higher performance. Later, top audio companies such as McIntosh developed a system to match power tubes for use in audio amps, and the companies would guarantee performance specifications only when their matched tube sets were used. Unfortunately, because of the mechanical nature of the device and the extreme operating temperatures that exist within the tube (around 700 degrees Fahrenheit), it is impossible to specifically "manufacture" matched power tubes. However, once the tube is made, it can be performance tested for various parameters and matched into sets containing other tubes with identical characteristics.

Limitations
Since a tube is a mechanical assembly of parts that forms an electronic device, it is subject to some mechanical problems and limitations. Tubes wear out in direct proportion to how hard they are worked (due to the circuit design) and how often and loud you play your amp. Vibration and jarring shorten the tube's life as well. Ideally, a tube could be built so that no vibration existed between its mechanical elements. However, in practice this doesn't happen. So when the tube is vibrated (usually by the speaker), the elements shake, resulting in an additional signal being amplified. This phenomenon is commonly referred to as "microphonics." The construction methods and materials used in the chassis and cabinet may actually serve to create pleasing microphonics that give the amp a distinctive, desirable sound. However, adverse tube microphonics can be a big problem when the elements of the tube rattle or ring, producing a signal all by itself. A tube with this problem is unsuitable for use in music amps, much like a faulty guitar pickup or a bad microphone is undesirable for most musical purposes.

When to Replace Tubes
So, when should you change your tubes? Chances are, your power tubes are worn out when your amp starts sounding weak, lacks punch, makes funny noises, has its power fading up or down, or loses highs or lows. If gain in one channel hums, lacks sensitivity to touch, or generally feels as if it's working against you, a preamp tube could be malfunctioning, and is in need of replacement. In both cases, though, the tubes may not be at fault. Unless you are skilled in specific troubleshooting, regard the high-voltage circuits found in amplifiers as extremely dangerous (read and re-read our "Warning" printed at the beginning of the Technical Articles section). Take the amp to a professional for diagnosis and repair.

Unfortunately, you can't simply pull your tubes out and take them to the drugstore or local electronics outlet to evaluate them on one of the tube-testing machines designed for TVs. This is because of the high voltage levels at which guitar amplifier tubes are driven. Amp tubes can be powered with 450 volts or more, whereas the testing machine provides only about 150 volts; this difference can completely foul up a diagnosis. Tube-for-tube replacement and a before-and-after comparison is often the most reliable test.

Good-sound, non-microphonic preamp tubes are the exception, not the rule. Quality preamp tubes, along with matched sets of power tubes, are a little harder to find, and you may pay more when you do locate fl them. However, you can expect improved sound and longer life, so there is a payoff.

Tubes, Transistors and Distortion
No tube primer would be complete without an explanation of how tubes distort in a way that is different from transistors. Tubes distort uniquely because as the signal emitting from the plate approaches its maximum potential, the tube gradually begins to react less and less to the original input signal. This results in a type of compression of the signal and produces a soft clipping. Clipping occurs when the input signal increases, but the maximum power has been reached. Thus, the signal becomes cut off, or "clipped." Transistors, on the other hand, react exactly the same to the input signals right up to their maximum power; then they stop quickly, creating a sharp clipping. These different types of clipping produce different series of harmonics (overtones). When the transistor amp clips, it produces more odd-order harmonics (and in its worst case can sound hollow and dry), whereas tube distortion produces even-order harmonics. Tube distortion generally sounds warmer. Various types of transistor and tube distortion are possible, depending on the amp's design.

In the case of a tube amp, preamp and power amp tubes have different distortion characteristics due to the difference in both their tubes and their circuit design. For example, relying on a master volume distortion circuit by itself will yield less sensitivity to variations in a player's touch than if the amp is attenuated – has its volume limited – after its power stage (that is, with a power attenuator). This is due to the contribution of the output transformer to the amp's sound, as well as to the difference in sonic qualities between different power tubes compared to preamp tubes. Leaving some of the distortion to the power amp section rather than relying mainly on the preamp section gives a broader range of sensitivity. In addition, the nature of the tube allows the player to vary his touch, producing different tone responses from the amp according to the manner in which he plays.

There are many variables in tube amp designs, and each has its characteristic sound

Technical articles

and quirks. Regardless of what type of amp you use, you will find that, like strings on guitars or oil in an automobile, tubes do wear out. Amps are not maintenance-free, and as they age, they undergo changes. The tubes are subjected to wear and tear, some of the electronic parts lose their initial properties and pots and jacks get old. Bad tubes can cause premature failure of other parts, such as the output transformer, speaker, and other vital components. If your amp sounds bad, weak, or otherwise not up to par, don't just hope the problem will go away. Get it fixed! Keep on top of the maintenance, replace the tubes when necessary, and get the most from your amp.

WHY CHANGE TUBES?

There are two good reasons to change tubes in your amp. First, because the tube has simply burned out, causing a malfunction with the amp. Second, and perhaps the more important reason, is to improve the performance of your amp. Your tube amp's sound will deteriorate gradually as its tubes wear down. If you've got a good guitar, you don't wait until the strings break before you change them, do you?

The effect the tubes have on your amp is much like your guitar has with its strings – the harder and more often you play, the faster the tubes will wear. So, just as your strings sound dead long before they may actually break, your tubes will lose power and tone long before they finally burn out. This wear down process is so gradual, you often won't notice it until you actually change to new (and hopefully better quality) tubes. It's much the same as when you change to new strings, the first play is like a brand new experience!

The most common symptoms of worn tubes would be: excessive noise, microphonic ring or squeal, loss of power, loss of highs, mushy lows, and erratic changes in output levels. Let's look at the three stages of your tube amp – the preamp, the power amp, and the rectifier stage (only in older amps and lower powered amps).

The rectifier stage

The only other tube you might find in your amp is a rectifier tube. This tube converts AC wall electricity to the DC electricity used inside your amp to power the preamp tubes and the power tubes. Normally, the rectifier will only be found on older amps or lower power amps such as Fender Princetons and Deluxe Reverbs. The guitar's signal never actually passes through this tube so the rectifier has no direct effect on the sound of your amp. However, since it acts as a power supply tube for the other tubes, it can "sag" when the demand for power is great (for instance, when you turn it up or pluck the string hard), and the amount of sag induced – or the lack of it – can play a big part in the feel and dynamics of an amp. (See the full section below which compares different types of tube rectifiers.)

Higher power amps all have an improved rectifier section that's made with solid-state diodes that will deliver more power without any sag. Marshalls, Boogies, and Twin Reverbs all have solid-state rectifiers. If your tube rectifier fails, the amp's pilot light will stay on but no sound whatsoever will be heard. Many types of plug-in tube rectifier replacements are available, and these usually improve the overall tightness of the amp's sound and will in some cases increase the power output.

In many cases we recommend this conversion to the solid-state model, but caution the owner of any real old amp to have it installed by a tech who can monitor the amp's performance, since it may have several capacitors or resistors that are ready to fail. The increased voltages from many solid-state rectifiers might cause these older components to fail sooner. Once they have been replaced, the amp will have improved performance and the rectifier need never be replaced again. If your amp is a truly collectible vintage piece, however, you will probably want to stick with a quality replacement tube rectifier of the correct type.

A WORD ABOUT CHANGING TUBES

Rule 1
Turn off the amp and unplug it from the AC outlet. Allow tubes to cool for five minutes.

Rule 2
Follow the tube location guide for correct placement of the various models (this is usually found on the inside wall of the cabinet). If no guide exists and the model numbers are not indicated next to their corresponding sockets, then assume the tubes are presently in correct position and replace them one at a time so as not to forget their placement.

Rule 3
When changing the power tubes, be sure to release the spring clamps holding the tube's base in its socket. Now grasp the top of the bottle and slowly work the tube out of its socket with a gentle, circular motion. Replace the new tube also by holding the top of the tube and pushing with a gentle circular movement. Notice the power tubes have a locator notch in the center of the plastic piece – this assures the proper location of the pins and will prevent you from inserting it wrong. Never force the tube into the socket until you've made sure this locator notch is in correct alignment.

Rule 4
When changing preamp tubes, use the same gentle circular motion described above to remove and replace the tubes. The proper location of the tubes is insured by a gap in the pin sequence – make sure the gap is correctly aligned with the blank space in the socket. This gap is usually indicated by a notch or nipple in the base of the socket where the shield attaches. If a pin on the tube is bent, it probably won't fit into the socket. If this occurs, cake apart a common ball point pen and use the front half of the pen (with the small hole) to insert over the bent pin and straighten it out by gently bending it back. Don't try bending the pin with your fingers as this will likely crack the glass base. ∎

TUBE REFERENCE GUIDE
"What's the best tube for my amp?"

"What's the best tube for my amp?" This is the Big Question, maybe even the reason you bought this book. I hate to disappoint you, but the quick answer is, "It just depends!" (You didn't really think this would be so easy, did you?) However, I would hope the complete answer you seek is contained somewhere in this chapter.

It was this daily occurring Big Question that prompted me to write the 1st edition of *The Tube Amp Book* nearly 17 years ago. I was time-selfish (and a bit lazy) and simply wanted to avoid answering the same open-ended questions over and over again. I decided it would be easier to objectively quantify the differences between tubes and print a small paperback guide to answer this and other commonly repeated questions. That 32 page 1st edition *TAB* had grown to over 800 pages by my 1995 4th edition, and the latest version is now in your hands (I don't now how many pages it will end up at, but the pages are already twice the size of the old *TAB*). So, the Big Question has obviously continued to dominate conversations of players and engineers worldwide. In the end, of course, the Big Answer might not be the same for everyone, as everyone is different. More than ever, this Big Question needs an answer... so we have tried, more than ever, to provide you with the resource to answer it for yourself!

Many tubes have come and gone in the last 17 years, and with the progressive editions of the *TAB* I have tried to keep up with the changing tube availability. However, our last edition is now about seven years out of date with regard to tubes. So it was really time to update and also expand in more detail just because this subject is of growing interest to tube amp lovers everywhere.

So, what is the best tube? Or, how about the best guitar, or speaker? Naturally, no two guitarists will agree on what sounds "the best." But nearly all players can hear the differences between two 6L6 types in the same Fender Super Reverb amp, while most of their audiences can not. I am convinced this keen hearing is a result of genetic breeding that has blessed (or cursed?) us guitar players with "Dog

Technical articles

Ears" that can hear the fine differences between speakers, guitars and in this case tubes. This also insures we players will always have something to talk (argue?) about, if only to keep our wives and/or girlfriends bored to tears (don't you just love to do that?).

As I discuss each tube available today (and some not so available), I will also provide both a subjective overview of each tube, and also a more detailed technical report. The "Dog Eared Opinions" were compiled by myself and a few other Dog Eared expert types here at GT. The detailed technical report section on each tube, or body of tubes, are the result of hard measured data using our own specially designed comparative test amps and tube testing systems. Some of you will want the quick read so you can can get back to playing your amp, while I know there is a growing second group who wants more information so they can delve deeper into the tube mystic.

That said, let's get back to that opening question – to which I usually reply:

"What kind of amp do you use?"
"What do you want 'more' or 'less' of?"
"What style(s) of music do you play?"

The answers to these questions help me dial in a meaningful Big Answer for the player. In fact, there simply isn't *a* "best" tube no more than there is *a* "best guitar." Perhaps I can elaborate a bit on this Q&A process so you can help answer that Big Question for yourself. Here's what I need to know first.

"What kind amp do you use?"

Almost every amp made was designed on a test bench somewhere with a guy playing a guitar and tinkering with the circuit design until he got it just right. Then the prototype was passed along for the production department to figure out how to build the thing. Fender and Marshall are designed in much the same way as are Boogie, Ampeg or any of the smaller boutique amps out there. In almost every case, the designer uses tubes currently available. Back in the '50s these tubes were off shelf RCA and GE tubes; today they would be off shelf Russian or Chinese origin. I say "off shelf" because no major tube amp manufacturer can afford to buy hand-selected premium tubes, and they usually want the lowest cost component to keep their prices affordable. So, because tube frequency response in a preamp tube can vary so much from one type to another, and also because there is a wide variance in power tube output ratings and responses, the designer will usually optimize his design for the tube he used during the development process.

Therefore, if you want that stock sound you bought the amp for, then usually that original type will work out best. But also if you are looking for something more, then you at least know your starting point, and the following tonal and performance descriptions will help you find a better replacement.

For example, Fender in recent years has been using mostly the 12AX7R for most of its preamp tube requirements, and the 6L6R for the power section (a key to suffice letter codes will follow). If we try these tubes in a vintage Fender which was designed years ago when US-made tubes were the standard, then the newer "R" types may sound a bit brittle and/or dark as compared to what a nice pair of NOS 6L6s – or a reissued 6L6GE – might sound like. In a new Fender amp, however, which was designed using the newer Russian type tubes, the Russian types sound just fine, as the newer amp was originally "voiced" for that tube. Of course an NOS tube, or another higher quality tube might also sound great in a new Fender – in fact they might sound even better to your ears, but if you are going for a stock sound, as the designer intended, then the newer Russian tube will sound fine and save you some money.

Remember, while we can try to make the selection process easier for you, you are the best judge of what's the best tube (to your ears), so there is no substitution for simple experimentation with both your preamp and power tubes, in your own amp, played they way you play. In the case of experimenting with preamp tubes it's real easy, as these do not need any technical assistance to change preamp tubes (just replace them carefully, and don't bend the pins!). With power tubes, however, it's not that easy. We strongly advise against just "plugging in" new tubes (unless you are replacing the exact same GT tube with a similar GT rating number) as this process usually requires a bias adjustment, and that usually requires taking the amp chassis out of it's wood cabinet, which should only be done by a trained technician who has the expertise and audio test equipment to correctly set the bias on the new tubes. The exceptions to this advice are, of course, cathode-biased Vox-types and some smaller-powered tweed Fenders and similar amps, where the bias is set, so you can simply plug in a new set of matched output tubes.

By the way, biasing an amp is a similar process to tuning up an engine. It's not that complicated if you have some basic training and tools. We also have Bias manuals and kits available for those who want to learn, and we even have a chapter in this book that can get you started.

"What do you want 'more' or 'less' of?"

Here I am trying to find out if the player's amp is too bright (and so recommend a softer, warmer preamp tube), or if it's too loud (and so recommend a lower rating on our GT Power Tube Gain:Distortion 1-10 rating system). If it's an EQ request, then the solution is often found in changing the critical 1st preamp tube in the gain stage of your preamp section (also, see Mark Baier's chapter on new and NOS preamp tube characteristics, which follows). This tube is usually labeled V1 on the schematic or tube use chart, or if you do not have either of those for reference, V1 will usually be the preamp tube farthest from the Power Transformer (it's placed there to keep the hum at the lowest possible levels). The Power transformer is the big one nearest the AC cord/input. Also, an easy way to find V1 is it is almost always the tube closest to the input jack. Most of the really dramatic differences between preamp tubes will be observed by sampling the tubes in the V1 stage.

Although all 12AX7 preamp tubes are electrically equivalent, and therefore easily interchangeable without any adjustments, they will all sound (and feel) different, and some are very different! In fact they can be as different as 8 Ohm 12" speakers, or 6-string guitars, which are also electronically the same but, as we know, can have a world of difference in tone and response.

So, if I hear a player wants his amp brighter or darker, or wants less gain for a softer touch, or more crunch, this narrows the search down to the preamp tubes. Also, if there is a "problem" – for example, a microphonic ring issue on a certain note, or the amp is high in hum or noise – a faulty preamp tube (usually in V1) will generally be the source of the problem, and a simple substitution can often cure it in a jiffy.

However, if I learn a player wants the amp to have more (or less) clean headroom before distortion sets in, then I'd try a different power tube, or try a different rating number of the same power tube he likes. For example, if the player wants to go for early distortion with more compression for, say, blues lead tone, I might recommend a GT Gain:Distortion power rating of #1-3 and/or a power tube type with less output. But if he's looking for more clean headroom before distortion and/or more power in general, I would suggest a GT Gain:Distortion power rating of #8-10 and/or a power tube type with more output.

These GT Gain:Distortion #1-10 ratings are a nice secondary benefit of our own proprietary dynamic GT matching process which is done under full voltage simulated conditions and with a dynamic signal burst as a part of the process. It should be noted there are many other tube vendors on the market using the older traditional static testing systems to measure one or two electrical performance parameters, which is helpful in some ways but not the total picture we want to see for a more musically meaningful matching. Either way, if you are replacing tubes of the same type from any reputable tube vendor who has tested and rated his matched tubes, chances are good you will not have to rebias your amp. But if you change power tube *type*, regardless of any rating similarities, you should have your amp rebiased.

The primary reason we match power tubes is

Technical articles

to improve the sound and performance of our amp. A properly matched set of power tubes is noticeably better than the same tubes "off shelf." At GT, we match tubes so your amp will have more tonal balance, to increase sustain by eliminating weird phase cancellations unmatched tubes have, and finally to run the amp's output section cooler and thereby maximize usable tube life.

There are also significant differences in how much power is produced from tubes of the same production batch and in the same circuit. With our GT Rating System for power, our matched power tube sets give me another alternative to dial in the exact tone and response the player is seeking. Lower-rated/powered tubes will distort sooner and offer a wider range of distortion. The higher-rated power tubes will have more headroom, bass response and generally more punch overall. Tubes in the middle, say #4-7 have the normal characteristics and these are usually the choice and recommendation for most player. Indeed, after almost 25 years of experience in this field, more than 70 per cent of my customers prefer the middle numbers.

These differences between power tube types, and the rating system differences are easily seen in my Power Tubes Comparison chart. Higher power isn't always better; many times the lower-powered 6L6 tube will have the exact tone the player is looking for. I remember a local studio guitar gunslinger spending hours in our listening room sampling different 6L6 tubes in his old Fender 4x10" Bassman. When he finally emerged he had picked the least expensive, lowest powered Chinese "Coke Bottle" shaped 6L6 power tube over the more modern high-powered types, and also over the classic NOS GE tubes we had at that time (which were *much* more expensive). So it is still a matter of trying and listening, and a matter of personal application and/or taste. What I try to do by asking these basic questions is to just narrow down the choices. In the end you will still need to draw your own conclusions.

"What style(s) of music do you play?"

For example, if you are a blues player 100 per cent of the time, and you've got a Marshall amp, I would steer you toward a weaker power tube with a lower rating number and perhaps a 12AXC7C preamp tube for its rounder, fatter tone. However, if you are a Heavy Metal player with the same amp, I might rather suggest a more aggressive preamp tube, say the ECC83S and/or a higher powered EL34 power tube such as the GTE34Ls, with a medium or high performance rating. If you play various styles, then I'd go for something more in the middle of the choice ranges. However, many professional players we deal with today carry several 12AX7 types along with them. They will pop in a tube suited for blues like a 12AX7C, and later change to another suited for hard rock, like an ECC83S. They will interchange them depending on which band they're playing with that night, room acoustics, and even for getting more sounds out of a recording session. Remember, the 12AX7 preamp tube is a "user serviceable part" – you can change them without any bias adjustment, so it's a safe and inexpensive way to "mod" your amp.

The fact is, there is no perfect tube complement in a Fender or Marshall amp that suits *everyone's* taste. That's why there are 100s of different amps or guitars to choose from (thank God!). Unfortunately, there are not even a dozen choices available for tubes from any one type. We have listed in the following section all the known types available on the market today. Additionally, we have found through the years that our rating system of #1-10 within a particular power tube type is a valuable asset in dialing in the exact performance a player is seeking from his amp.

Unfortunately, tube factories' quality control is no better today than it was when we started testing US made tubes 25 years ago. In fact it is even worse. So pretesting and matching (in the case of power tubes) is more important than ever.

More recently, we have also developed a preamp testing and rating system for selecting higher and lower gain tubes, and tubes with faster and slower rise times (attack, or touch) which we call our Special Applications Group. The SAG testing and rating systems for preamp tubes allow us to predetermine the "touch" of the preamp stage, ie fast aggressive pick attack or softer, warmer touch for a blues player. Additionally SAG Matched Phase Inverter tubes (MPI) are specially selected to have identical output and rise time features for the phase inverter section that allows for better output tube performance. These measurements must be done on a sophisticated vacuum tube curve tracer that measures tubes dynamically and displays the tube's individual response curve. You can learn more about our SAG tubes elsewhere in following chapters in this book.

Now let's separate the discussion between the two main choices you need to consider when changing tubes in your amp: the preamp tubes and the power tubes. Each tube amp section contributes to the total sound and response of the amp, but in very different ways. First let's look at the various preamp tubes.

The tubes in your amp's preamp section:

Preamp tubes take the small signal from your guitar pickup and amplify it to a "line level" (or to the signal level of a typical FX unit or PA mixer's output). Additionally, the preamp tubes are the active components of the various features of your amp, like the EQ stage and other features such as gain boost stages, reverb send and return and tremolo (in higher-quality amps, anyway; more affordable models often

Note On NOS Tubes

Many times these NOS tubes on today's market are "left overs" from previously rejected production lines, and/or "pulls" which have been removed from scrapped surplus gear. Tubes like this may have had many hours of use, and also may not have been tested for noise and/or microphonics... and therefore may be worse than current manufacture tubes that are readily available for much less money and, in some cases, pretested to assure highest quality. So buyers of NOS tubes for big money should request a verified quality test before buying, and/or be sure they are in original boxes and not "used". However, if you're lucky enough (or rich enough) to find some pristine Mullard ECC83s or some GE Clear top 6L6 tubes, then you will definitely like what you hear. These tubes, made back in the vacuum tube's golden days of the '50s and '60s, are much like anything made back then... very good! (You already know about the Fender Strats or Gibson Les Pauls made in this time period.) The only other thing I could caution an NOS buyer about would be the alarming trend to "duplicate" old boxes and even silkscreen tubes themselves to disguise, for example, a modern Ei 12AX7/ECC83 from Yugoslavia as an original Telefunken 12AX7. As the prices for originals rise, so do the unscrupulous vendors... so buyer beware! Now that you are forewarned, you can start hunting the fleamarkets or eBay for these NOS beauties – but expect to pay a premium (and watch out for me peaking over your shoulder at the fleamarket – you might have to act fast!).

use solid state driver circuits for these features). Usually, each special feature is controlled and influenced by a particular preamp tube. Most all preamp tubes are dual triodes, and most often the dual triode is in the 12AX7 family (aka: ECC83 or 7025). The current field of newly made tubes is quite wide; at least six 12AX7 types are currently on the market, coming from four factories in four countries, and more are expected after this book will go to print as we are currently working on a Mullard 12AX7 exact reissue to be made in the Chinese Shuguang tube factory. There are also many variations of NOS or past production 12AX7 tubes still available on the market. But NOS tubes will usually cost more and, in my experience, they are not always worth it.

Understanding our preamp tube rating system:

There are at least three critical parameters in evaluating the general differences between manufacturers of the following preamp tubes.

Guitar Amplifier Sites

http://www.diyguitaramp.com/tech.html

http://users.chariot.net.au/~gmarts/ampsmain.htm

http://www.drtube.com/guitamp.htm

http://en.wikipedia.org/wiki/Guitar_amplifier

http://www.edaboard.com/profile.php?mode=register

Technical articles

These are the average gain, output, and quality variance spread between the worst and best tubes from any random batch of 1,000 preamp tubes we process on a daily basis.

To be sure, there are the other "problem" issues such as microphonics, hum and noise which we test to screen out when selecting the tubes that will become Groove Tubes premium stock, but it would be unfair to just talk about our final GT selected "cream of the crop" tubes from any of these various factories because you may not have access to Groove Tubes and/or may choose to buy tubes from another tube vendor in their original factory boxes, or as we call them, "off shelf" tubes.

The Quality spread is a good indicator of what your "odds" are in getting a premium quality tube from a batch of "off shelf" tubes. However, the gain and output will be valid references even if you chose Groove Tubes or another premium selected tube. The gain and output characteristics are more typical of the design, while the QV is more typical of that factory's attention to manufacturing detail, processing, and general quality control. Unfortunately, you will notice that none of the following preamp tubes will have a very high QV rating. This is because with only four factories making preamp tubes today, there is little incentive to make them better or toss out problem tubes with high hum, noise or microphonics; most factories knowingly sell sub par tubes, it's just a fact of life in this business. Actually, these market conditions were the reason I was able to offer value to musician who just didn't want the hassle of buying 10 preamp tubes to find (maybe) five good ones. We've made a pretty good living over the last 24 years doing their QC for them, and helping define the musically meaningful differences between all types of tubes as they are used in today's guitar and bass amplifiers.

GAIN: Gain will be shown as a percentage of meeting the design specification for any of the various types. In the case of a 12AX7, the design specification for Gain should be 100 Mu. So if the reading is 93 per cent, it means these tubes on average hit 93 Mu. Note: In fact, few examples of 12AX7 tubes made today meet the original design spec for a 12AX7 tube. Do not be alarmed, as a low-gain tube doesn't mean it is unusable, as most amp designers take this into consideration when designing and and leave plenty of tolerance for a lower gain tube. In fact, if you place a high gain tube in some circuits it may have a tendency to have more microphonics and noise than a lower again tube would, and so may not work as well as they stock tube. You really never know until you try a preamp tube in your amp. (This is one reason we recommend having several types of preamp tubes on hand: it's cheap insurance you will get the tone you are looking for at any particular gig or session.)

OUTPUT: Output (current) will also be shown as a percentage of meeting the optimum spec. So, if an optimum spec for a 12AX7 is 1.2 milliamps of output current, an average reading of 1.2 milliamps will show 100 per cent. Once again, few tubes made today, on average, will meet the original output spec. It should be noted that very few tube vendors today take the time, or have the equipment, to test for output current. However, we feel output current is among the most meaningful parameter for predicting how a given tube circuit will perform. Complex modern circuit designs work better with a tube that has a strong current output to push the signal through the maze of the front-end components and extensive wire harnesses, while older, simpler designs are less sensitive to the variances in output current for a preamp tube.

QUALITY VARIANCE (QV): A lower percentage number indicates a comparatively tighter production tolerance (that is, fewer rejected tubes). In general, it means more tubes will pass any quality selection process, especially so for our GT performance testing systems. Tube types with a tighter Quality Variance spread, or lower QV%, will also usually have more uniform gain and output characteristics. Remember that these rating are for stock "off shelf" tubes as we receive them from the factories around the world. After our preselection processing, most of these types will have much higher Gain and Output ratings, and also the QV% would be a comparative 0 per cent.

LAST NOTE ON THE RATING PARAMETERS: Almost all tubes fall within "usable" gain and output specs, although they will not meet original design specs set by RCA or GE in the Golden Age of vacuum tubes. In those days, there were literally dozens of factories competing for massive amounts business, as tubes were used for everything from car radios to clocks. So I mention this to assure you that most "off self" tubes are "OK" and will not cause the amp to sound bad or have problems in most amp circuits (unless you have a very high-gain amp). However, tubes that fall short in output current may have a lackluster tone with little punch, even though they are technically "usable." Also, a strong output rating may be more important to the player than a high gain rating, while QV percentage can be used as an "odds" indicator on the Roulette Wheel of the "off shelf" world of tube vendors.

Tube Origins, what the model name suffixes mean:

I have listed the following tube types for comparison in this chapter by Groove Tubes catalog name, and where applicable the other common names this tube is known by from other tube vendors. Our GT model name also uses a suffix code letter to denote country of origin. Since today there are really only six tube factories in just five countries that produce any significant quantity of tubes, it is easy to cross reference these to other vendors, assuming they also indicate the country of origin.

These suffixes and countries of origin are:

"R" is from the Reflector factory in Russia. Most Sovtek labeled tubes come from here.

"R2" (12AX7R2) indicates the second evolution of a preamp tube from Reflector. In power tubes, "R2"(6L6R2, 6550R2, EL34R2) can also indicate Svetlana production, the second Russian factory making the 6L6, EL34 and 6550 fin St. Petersburg.

"S" is from the former Tesla factory, now called JJ, located in Slovakia.

"Y" is from the Electronska Industri (EI) factory in Serbia, formerly in Yugoslavia.

"C" is from the Shuguang factory located in China.

"GE" USA: Groove Tubes, San Fernando California, but only the 6L6GE and 6CA7GE and in very limited "boutique" productions.

12AX7-C
Gain: 93% Output: 92% QV: 17%

This is a warm and linear tube with a round, fat tone. It is well suited to rock, blues, and jazz. Therefore it is perhaps the most versatile of the 12AX7/ECC83/7025 family. There are many Chinese variants that have been produced over the last 10 years, and many tube vendors sell the older versions, so be sure to look for the new ones here. This is possibly the best of the current 12AX7 family with regard to "meeting original specs," and for my vintage amps it is my personal favorite. However, if you have a modern amp designed around the Russian tubes, then it may have too much gain and/or output, so will have more noise/hum and/or adverse microphonics. If you can buy a version selected to reduce these characteristics, however, then you might be very satisfied with this tube in a modern amp.

12AX7-R (aka: Sovtek 12AX7WA)
Gain: 86% Output: 77% QV: 42%

This tube comes from the Reflektor factory in Russia. It is perhaps the most commonly used tube by many amp makers because it is reliable and quiet. The main reason it is quiet is its generally lower gain and output. It will sound dark and harsh in a vintage amp, but will sound much better in a more modern amp that was designed around this tube's characteristics. Generally speaking, this is a "stock" tube that is

Technical articles

often changed out for a higher gain and/or output tube by knowledgeable guitarists. But if you want few problems and good reliability, this is the tube for you. If "tone" is your main criteria, look elsewhere.

12AX7-R2 (aka: Sovtek 12AX7LPS)
Gain: 83% Output: 83% QV: 42%

This tube has a relatively a long plate structure. It is brighter than most of the others (though not as bright as a 7025), but it is also more expensive in most catalogs. Its long plate structure can be more prone to microphonics in combo amplifiers. These tubes have gone through a rollercoaster quality ride in QC as they have entered the market over the last few years. We have had batches that were all bad, ones that seemed good but then died suddenly, and other batches that sailed through QC with a high yield and never had any problems. That said, it isn't so uncommon with a new design to have some stability problems.

12AX7-R3 (aka: Sovtek/Electro Harmonix 12AX7EH)
Gain: 87% Output: 83% QV: 17%

This tube is the newest tube made in the Reflector factory. It has a shorter plate structure, and has a flat, or linear tone. It is not quite as warm or linear as the 12AX7C, but is just about as versatile. It is a quiet tube, and works nicely in many amplifiers.

ECC83-S (aka: JJ ECC83)
Gain: 85% Output: 112% QV: 58%

This tube has terrific output and is a very strong tube by any standard, usually rating well over the original 12AX7 spec for Output. But they are pretty low on the Quality Ratio scale. That can be good or bad just depending on the amp design you put it into… but usually this is a great tube for most musical styles in most of the amps out there, and a personal favorite of mine for hard rock or bass preamps. It has a stronger midrange response than many, with a bit of roll off on the high end compared to GT7025, but has more highs that most others. While all tubes (and especially this ECC83) need to be purchased from a trusted tube vendor that tests tubes pretty extensively if you want a good result, these are the masters of current drive. In a modern amp such as a Bogner, Diezel, Rivera, or Mesa Recto series, these are the tubes that will push signal through those complex front ends.

7025-Y from the Ei factory
Gain: 90% Output: 46%
Tolerance Spread: 25% – 60%

Long, smooth plates, the most bright of the current tubes. These are the characteristic sound of the Fender Tolex years; they are very articulate. In V1 and V2 of a Fender Tolex-era amp, these give the original sound signature. These tend to be too bright in Marshall Plexi-era amps for some tastes when used with a Strat or Tele, but if you tend to load these amps down with pedals in the front end, they can help resolve the loss of treble from the pedals. Long plates can tend to be microphonic in combo amps. The Ei factory was the OEM for many "premium" tube folks in the past, such as Siemens and Telefunken at times, so if the plate structure looks familiar, this is because the West German companies gave their old tooling to Ei and also technical assistance to start them up when they were phasing out tube manufacturing in their own country.

5751-R New manufacture, aka Sovtek (also several NOS types are still around)
Gain: 70% Output: 1.2 milliamps
QV: N/A (not enough test results to publish)

The original 5751 is a low-gain 12AX7 type, and is used in V1 by some folks like SRV to change the ratio of preamp distortion to power tube distortion. However, this current manufacture from Russia is so far outside of NOS sample specs on curve traces, that this tube is a 5751 in name only. Perhaps down the road this will change. If you use these new 5751s and like them, that is great, but they are not like a vintage 5751. When there is a good new 5751 available, we will let you know.

12AT7-Y
12AT7-C
2AT7-A (USA NOS JAN spec)
6201-M (NOS, similar low noise/microphonic type used for tube microphones)
Gain: 60-70 Output: 10.0 milliamps

These are commonly used as phase inverter and reverb drivers in post-tweed Fender amps, and amps styles after the Fender circuits. They can be safely used in the first gain stage (V1) of an amplifier but they will drop the gain a bit. This substitution for a 12AX7 type in V1 may make some amps a bit quieter with regard to background noise, and can yield more clean headroom. This substitution is also one easy way to change the percentage of output distortion to preamp distortion in non-master volume amps. The "blackface" Fender amps evolved to use the 12AT7 as a phase inverter or "driver" tube for the power section, while the "tweed" Fender amps (along with the early Marshall amps which were evolved from the tweed amp design) used a 12AX7 for the phase inverter. This was a major design change for Fender, and a big reason why post-1960 Fender amps sound tighter and cleaner than a more aggressive Marshall amp. You can get some of that juicy tweed-era drive in your Blackface or modern Fender amp simply by substituting the 12AX7 for the 12AT7 in this position. Conversely, you can clean up your vintage Marshall (but lose some drive) by using the 12AT7 instead of its standard 12AX7 in the final driver stage of your preamp (the tube next to your power tubes). Substituting one in for a 12AX7 in the V1 stage will most likely leave you unimpressed, however, as most players like a hotter preamp drive stage.

12AY7 (Various brands of NOS tubes are still around)
Gain: 44 Output: 3.0 milliamps

This tube was used in first gain stage (V1) of very early Fender tweed-era amps, and is basically interchangeable with a 12AX7 type. It has about 1/2 the gain of a 12AX7, but much greater current output, so it can change the sound and feel of your amp if you decide to try a substitution. In modern Marshall type amps such as a DSL or TSL series, this tube in V2 will bring the touch, feel, and sonic qualities closer to Plexi era Marshall amps. This tube is also known as a 6072 or 6072M, which denotes a selected low noise version useful in studio equipment and some tube microphones.

12AU7-Y Current production
Gain: 16-18 Output 10 milliamps

These are commonly used as phase inverter in McIntosh Hi-Fi amps, some Ampeg amps, and will yield the most clean headroom in many amps while imparting a softer touch. In the latest Fender Pro Series amps such as the Pro Reverb, Concert, and Twin, this tube in V2 (or a 12AY7), will drop the gain of the high gain channel and impart a tone that is less intense or "buzzy" in its character. This substitution (for a 12AX7) will allow a much wider range in the sweep of the volume and gain controls of this channel.

108 THE TUBE AMP BOOK

Technical articles

EF86/6267-R (aka: Sovtek)
Gain: 2000 **Output** 3.0 milliamps
QV: 50% (or higher!)

A 9-pin pentode, used in the first gain stage of some newer amps such as the Matchless D/C30, Bad Cat Black Cat, Dr Z Route 66, and others, and there are some used in rare vintage Vox amps that inspired the use of this tube by the modern designers. The current versions of this tube from the Russian factory are very inconsistent, to be kind. Their output is way above spec, sometimes 50 per cent over what is to be expected from an old Mullard version that Vox used. Also, they have transconductance that is not close to the original's either. These factors make finding a low-noise version of this tube very difficult, and expensive if a company like ourselves can only find one in 10 that is "usable" in a Matchless amp. These need to be closely screened and tested using very sophisticated test gear that is generally out of reach for most internet tube vendors or smaller vendors. I really wish these modern amp designers would choose tubes we can still get in decent quality. But many players demand that "cranked" tone quality of these tubes in a Class A amp, so I understand the dilemma. Once you find a good one, however, I cannot argue with their unique and juicy tone.

Power Tubes

Many people believe that a 6L6 is a 6L6, or that an EL34 is an EL34. Not true! Just like preamp tubes, the design and consequent performance of any of these common power tubes listed below will differ very much. Although the differences you will notice are not so much an "EQ" thing like preamp tubes, but more of a feel or response thing. The power output from two 6L6 tubes can vary as much as 20 per cent, or moreso in the case of a random testing of EL34 or KT88 tubes. And each of these various examples from any of the particular power tube families listed below will have wide differences in their sonic signature, too. We try our best here to again give you the benefit of our "Dog Eared" opinions compiled from nearly 24 years of testing and grading power tubes from all the factories making tubes today.

There will be rating number next to the power tube types listed below that shows the tube's performance in "like" circuits, using the same plate voltages and bias voltages. The output is in milliamps. A tube with a higher reading is more powerful than a tube with a lower reading. It is as simple as that. This does not mean a higher number is "better" – there are many other aspects regarding tone. This is just a simple power measurement to show which tube is stronger. This does not indicate or mean that a higher current tube is better; as always, one piece of the puzzle is only that, and there are more pieces to consider... including what your application is and what amp you are retubing. This is simply a basic power output scale.

We also will show you the measured rating differences in frequency areas in our Power Tube Comparison chart that we have been compiling for more than 15 years. Notice the differences in the power output of the various tube families at the three common guitar related frequencies where we take output readings: Low, Mids, and Highs; these best indicate how they might affect the sound in your amp. Once again, more power isn't always better – in fact some players want their amps to act like lower-powered amps (and there are other tricks to do this in another part of this book). Basically, consider the broad choices in power tubes for your amps to be a blessing, because if one doesn't turn you on, perhaps changing to another one will! There are usually several variations of each family of power tubes, and these all sound and behave differently. It just takes a bit of experimenting, which is a little more complicated than preamp tube experimentation because each change will usually require a bias adjustment that usually must be performed by a qualified technician.

Lastly, please do not "mod" or sell your amp until you have tried at least two or three power tube options (unless you just totally hate it: then sell it and buy one you like to begin with). I can't tell you how many "modded" marshalls have been totally ruined by some hack with a soldering iron and a little bit of "knowledge"... when all the player really needed was a low noise, selected preamp tube and/or a properly biased set of new power tubes. We commonly charge guys $100s of dollars in labor "reversing" some hair-brained amp mod on an otherwise great sounding Marshall or Fender amp, and when they get it back with a fresh set of matched power tubes and some low noise, fat sounding 12AX7s... they're ecstatic! So your first stop, and safest route to modifying your amp's tone is always trying some different preamp or power tubes!

That said, let us proceed to describing the power tubes on the market today.

Tube Origins: Model Names and Suffixes

I have listed the following tube types for comparison in this chapter by Groove Tubes catalog name, and where applicable the other common names this tube is known by from other tube vendors. Our GT model name also uses a suffix code letter to denote country of origin. Since today there are really only six tube factories in just five countries that produce any significant quantity of tubes, it is easy to cross reference these to other vendors assuming they also indicate the country of origin.

These suffixes and countries of origin are:
"R" is from the Reflector factory in Russia. Most Sovtek-labeled tubes come from here.
"R2" (6L6R2, 6550R2, EL34R2) can also indicate Svetlana production, the second Russian factory making the 6L6, EL34 and 6550 in St. Petersburg.
"S" is from the former Tesla factory, now called JJ, located in Slovakia.
"Y" is from the Electronska Industri (EI) factory in Serbia, formerly in Yugoslavia.
"C" is from the Shuguang factory located in China.
"GE" USA: Groove Tubes, San Fernando, California, but only the 6L6GE and 6CA7GE and in very limited "boutique" productions.

The 6V6 Family
Applications: *Used in tweed and black face Fender Champ, Princeton, Deluxe and other sub-25W amps; vintage Gibson, Supro, and other amps.*
NOS reference tubes:
USA RCA "blackplate" 6V6A, Output 43.7 mA
USA Tung-Sol 6V6A, Output 45.4 mA

6V6-R (aka: Sovtek and/or Electro-Harmonix)
Average Output: 51.9 mA
This is a newer tooled Russian 6V6. This tube holds up well to plate voltages of 450+ volts, and will do very well in amps such as the Fender Deluxe Reverb. It is reliable but perhaps a bit harsh or aggressive in tone as compared to an

---

Matching Power Tubes

The idea to match tubes is very valid. This is because power tubes are usually working in "teams" either as a set of two or four in class A amp design like a Vox AC30, or as opposing pairs in the "push-pull" Class A/B designs most common in our Fender or Marshall style guitar amps. The idea is that if we can put two (or four) tubes together that sound exactly alike, then the sound will be balanced and even. While this is true in theory, the trick is to measure tubes in a way that shows what they will sound like when you are playing your guitar through them... and that's the rub. As we say, that's what makes a horse race!

In fact there are several "industry accepted" and/or common methods to measure and thereby "match" power tubes. The most common methods measure one or two static parameters like simple output current in milliamps, for example. These measurements are interesting when comparing differences between different manufactures because it shows off differences in design or component quality. We show some of those types of

continued on next page

Technical articles

continued from previous page

measurements here to compare the general differences in output power between tubes in the 6L6 or EL34 family of tubes. But power measurements can not predict how a tube will sound, no more than measuring the power capacity of speakers tells us how they sound. There are many other factors that must be considered when "matching" two items for a similar tone. So, considering that there are many considerations of what makes a tube's individual sonic signature – and power output is only one piece of the puzzle – how best to "match" tubes is an ongoing controversy. Many tube vendors offer static power "matched", or some other "static" (or non-interactive way) measurement to show they have matched tubes. However, (and here comes the controversy) there are at least two basic problems with conventional and simple output power matching as used by most tube vendors and amp techs today:

1) Tubes "wear out" at different rates, very differently from one to another. This is a result of the variables in the construction material, the amount of vacuum that is sealed in at the final production stage, and other factors of design and assembly. So, as tubes get some hours of use on them, these original small differences in output measurements change and are no longer "matched." If an EL34 is supposed to have an average output equivalent to 25 watts in a Marshall 50W amp, then matching two up for identical "outputs" of 24.3 watts on Day 1 will show that these same tubes, after just 10 hours of use, could be as much as 2W apart, or 10 per cent off! This is a moving target at best, and no stable way to predict how they will wear out.

2) And this is the more important point: power tubes of the same type and from the same production runs have a small range from most power output to least power, less than 10 per cent in most cases. One EL34 might measure as low as 23 watts, and the best might be around 25 or 26 watts. As some of you know, these slight differences of power are not audible with human ears, but can only be measured in milliamps with a fine scale metering system. The technical formula for hearing differences in volume is that it takes double the power to hear a 3dB of change in volume, and a 3 dB of change is almost unnoticeable. It's like moving the volume control on your hi-fi amp up one notch. For example, can you say a Marshall 100W amp is twice as loud as a Marshall 50W amp? Or is a 200W PA twice as loud as a 100W PA? Certainly not, and that's because of the God-given physics of how we are made. Human ears, even those pesky musicians' "Dog Ears" cannot hear these small differences in power between tubes. So if we can't hear small differences in power, matching tube in a static power system just doesn't work. I know, because when I got the idea to start Groove Tubes 25 years ago, that was the first "straight out of the book" method I tried... and I couldn't tell any improvement over a random set of tubes picked off the shelf. However, I knew every set of tubes I tried, even from the same master pack of 100 tubes, would change the way my amp sounded and played; some sets were more alive and balanced, others sounded dead, and these were brand new tubes right out of the box!

The secret to our dynamic testing systems evolved over the next few months, and I had a lot of head scratching going on from pretty smart engineers until we found our unique recipe that gave me the results I was looking for. I am not going to reveal exactly how we do it, but I will describe the general concept: we match tone, not power. We reasoned that tubes were not being used for power, or we would have long ago switched over to those better power devices called transistors that produce gobs of power with far less cost. No, we stuck with tubes because of their tone and response. We liked the way we could pick the string and create a progressive timbre of tone, from round and fat to crunchy to shreddy... just depending on how they responded to the dynamics of our playing style, or our touch. This dynamic signal coming from our guitar was our first clue. We needed to find a way to measure the gradual change in "tone" that occurred as the signal increased rapidly, like a blink of an eye fast. Changes in tone, or distortion, can be easily heard. While human ears can barely hear a 100 per cent difference in "power," our ears can easily hear a 5 per cent distortion change! In fact, we measure distortion in fractions because it is so noticeable to human ears. So rather than measure one or two small pieces of the puzzle, we developed a way to stand back and measure the Big Picture. We found a relationship of how fast the gain was increased to the degree of distortion that gain change made, kinda like picking a string, only done on a computer in a repeatable manner. Now we were on to something big, as we measured vast differences in this Gain:Distortion ratio system, and when we put a set of tubes together with common characteristics like this, Bingo!, these sets were amazingly musical, with even frequency response, long sustain, and even ran the transformers 15 to 20 per cent cooler! Like balancing tires, the machine just ran smoother and all six notes of a full chord would ring out the same and sustain equally too! When players started using and touring with Groove Tubes matched in this manner, they old us that the tubes were lasting longer, more than twice as long, and they stayed musical to the end! When we measured these sets again after 100s of hours, we were amazed to find they were still closely matched in our dynamic measuring systems.

We discovered the fact that these variations if the Gain:Distortion ratios between tubes of the same type were the result of the way the tubes were hand assembled – that is to say, one grid was a little closer to the cathode than another tube's grid – the result of the variance in any hand-assembled product like a guitar or crystal glass. The human factor crated a "snowflake" syndrome: no two were exactly alike nor could they ever be made to be. The physical tolerances were just to large for a microscopic electron to move the same way in every bottle as it passed from the cathode, through the grids, and onto the plate. An added dimension of this form of testing was that our test results grouped tubes into matched sets that had various distortion differences: early, normal, and late distortion characteristics. This meant in addition to offering better sounding tubes, we could also offer choices of different sounding tubes... and these choices would be repeatable! A low-number set will make the amp act like a lower powered amp as it will distort earlier in its power range and give a wide rainbow of distortion tones with more compression as the payer plays harder. A higher rating set would stay clean longer as you turned up the volume, and demonstrate more attack. The ones in the middle acted pretty much like you'd expect, or normal. We decided to use a simple 1-10 scale so that players could try different sets, and once they found the perfect fit for their application, they could always duplicate that sound when it was time to change tubes. Furthermore, if they would replace their worn tubs with tubes of the same type, rated in the same GT system, they didn't need to reset, or rebias the amp.

This sounds like a quick discovery, but it actually took almost a year and the help of many talented engineers and "Dog Eared" musicians like myself to dial in the perfect recipe... and we still use it today. This early tube adventure would be the basis of a long and successful company by providing consistent sets of matched tubes, so that players could actually hear a difference when they switched to GTs. When I think about this process now, I just have to thank God for providing all the right clues and friends along the way – it was my first miracle.

Technical articles

6V6-C
Average Output: 47.2 mA

NOS GE or RCA 6V6 tube. They are only available from Sovtek, various vendors and also from GT in matched sets of duets, quartets and sextets.

This 6V6 is recently redone off new tooling. This tube also holds up nicely to higher plate voltages and so has been a more reliable tube as compared to previous Chinese versions. I would describe the tone as more like the classic NOS GE 6V6. It has more of a softer, rounder clean tone (as compared to the GT 6V6-R) and a nicely compressed distortion when pushed hard, which is generally preferred by vintage players. They are available from various vendors and also from GT in matched sets of duets, quartets and sextets.

The 6L6/5881/KT-66 Family

Applications: *The classic "big Fender" tube, used in the tweed Bassman, Super, Pro and Bandmaster; blackface Super Reverb and Twin Reverb, and similar; also used by Mesa/Boogie, vintage Gibson amps, and countless modern makes; KT66 used in the original Marshall JTM45.*

NOS reference tubes:
USA Tung-Sol 5881, Output: 76.1 mA
USA General Electric 6L6, Output 72.0 mA
British MO Valve KT66, Output 82.5 mA

GT6L6-C (aka: Shuguang 6L6)
Output: 69.0

This is a relatively new offering, on the latest tooling from the Chinese Shuguang tube factory. This version has marked improvements over their earlier 6L6 designs as it is more rugged than in the past and has a new, larger bottle too. I have always liked the sound of the Chinese power tubes, but as they have been lighter weight they tended to have a shorter lifespan (maybe 20 per cent shorter) than say a Russian 5881WXT type. As with lighter strings, they may mean changing more often but the tone could make it a worthwhile trade-off. This new tube has a slightly softer, warmer tone compared to the more commonly used Russian variety, but is definitely more "musical" and is well suited for all styles of playing. They are available from various vendors and also from GT in matched sets of duets, quartets and sextets.

6L6-CB
Average Output: 71.6 mA

This is our old favorite Chinese made 6L6 with a coke bottle (CB) shape. It has light weight component construction and a softer vacuum, so it's quicker to distort, but with a warm soft tone. A great blues tube for smaller venues when you want to tone it down a bit. In a 40W Fender Super Reverb these will start to break up nicely at "3" on your volume with a humbucker style guitar. They are becoming scarce as they are not made on a full time basis but still might be available from various vendors and also from GT in matched sets of duets, quartets and sextets.

6L6-S (aka: JJ 6L6)
Average Output: 81.7

This is a real robust 6L6, from the JJ factory, formerly Tesla. This new design is kinda a "hot rod" 6L6 with a real high output; it's one of the most high powered options for a Fender amp owner. The high power manifests itself in a tighter bass response, and a crystal clear top end too. I would generally say this tube's power and rather stiff response makes it good for clean style jazz players and high powered rock players. The processing quality of JJ tubes is very good, so we usually get a long-life performance from these tubes, too. The only downside is that the stock pins in the JJ tube's bases are flat style, and so can be hard to push into a socket and perhaps might spread the socket because the solder builds up at the end of the pins. To correct this, we ship US made tapered pins which JJ uses when making the bases for our tubes. This is the only significant technical difference between a stock off shelf JJ 6L6 tube and the GT6L6-S, except of course the GT tubes are matched by our exclusive testing systems and available in duets, quartets and sextets.

6L6-R (formerly GT6L6-B; aka Sovtek 5881WXT)
Output: 71.0

This is a sturdy tube, one of the more physically robust of the 6L6 options. This is the stock tube in the more recently manufactured Fender mid and lower priced amps. It has a relatively low output, and so in vintage Fender amps can sound a bit lifeless at low volumes and also a bit harsh when pushed into distortion. This is the tube most often changed for one of the several other 6L6 types when a player is looking for something more out of his amp. They are available from Fender aftermarket sales as GT matched duets and quartets, as "off shelf" from various vendors and also from GT in matched sets of duets, quartets and sextets.

6L6-R2 (aka: Svetlana 6L6)
Output: 73.5 mA

This is one of the newer models on the market. The initial production and introduction of this tube just prior to 2002 produced a good, stable tube and the design was something like a Sylvania STR-387 copy. These have a tight sound, much like the tubes Fender used in the late 70's and 80's. However, more recently, the quality has been erratic, and we have noticed some customers complaining on the internet about the tube performance and quality issues. This is likely due to changes in the manufacturing and processing and we have noticed a lower vacuum and some material inconsistencies. And remember, there is a lot of "hear-say" in the web these days, so always consider the source. Our GT customers of this tube have been very happy, but this is because we pretest and burn in all our power tubes before we match and package them, weeding out any problems like this. That said, however, when you get a good 6L6 from this factory, they are truly a great sounding tube, with good power and very much like an NOS Sylvania or Phillips 6L6 type. Just be careful where you buy these, and we suggest a vendor that will test for low vacuum, gas leakage, and/or grid leakage. They are available from Svetlana, various vendors and also from GT in matched sets of duets, quartets and sextets.

5881-A (aka: NOS, JAN Philips origin 1987)
Output: TBA mA

This NOS small-bottle 6L6 variant was made for compact military use but has the same plate structure as the large bottle Sylvania/Philips production. There are still many 1,000's on the market so I expect they will be fairly available into the year 2004 or so. We always buy a large cache of any decent quality NOS US power tube when they come up; this is no exception. But as the prices for these are relatively high, we do not sell so many anymore. That's probably because there has been a big improvement of the commonly available power tubes from the

THE TUBE AMP BOOK 111

Technical articles

remaining tube factories, coupled with the fact that the later NOS power tubes are not nearly as good as the NOS from the '50s or '60s. Of course they work out great in Fender Blackface amps and any modern design as well.

This is a typical example of a whole group of "special application" tubes made for commercial and military purposes. Our example here is the small-bottle 6L6, which was ruggedized, mostly for military use. It has a smaller bottle that could be used in more compact field radios, an increased thickness of the mica insulator to make it less likely to break down under heavy use and transportation, and usually more getter flash and/or a higher vacuum to improve life expectancy. Otherwise these ruggedized tubes usually had the same plates, grids, cathode, heater as the standard 6L6 from the company making it. Therefore the power, tone and response would be very similar, if not exactly like, the "consumer" version. Of course the tube companies could charge a lot more from the government, who spec'd these additional modifications, so they were less available to the public and more scarce on today's NOS market.

Amp companies like Fender rarely used these versions because of the increased cost, unless reduced space was a consideration, as in some of those early "short" Showman heads of the early '60s that used the GE 5881/6L6WGBs for a while. Today, and yesterday, and for our guitar market, these "Mil Spec," or JAN (joint Army Navy spec) 6L6s have a life expectancy and durability advantage over most anything on the current market, but their tight, strong tone may not please a blues player as much as the much-less expensive, warmer 6L6 from China. The tone of this particular model shown here, a late '70s/early '80s production from Phillips (the evolution of the earlier Sylvania production), has a similar character to the more common Phillips/Sylvania consumer 6L6 from that era. That is, in a nutshell, a strong output powered tube with solid, even tone, and great dynamics or attack. Although not as soft and warm sounding as, say, the 6L6GE or 6L6C, it's a great tube with a classic USA style 6L6 tone and will last for many hours more than current production Russian tubes for example. Available from GT while supplies last in matched duets and quartets.

KT66-C aka: Shuguang KT66
Output: 91.6 mA

The GEC (UK origin) KT-66 was used in the Marshall JTM-45 amps, and the first 100W Marshall amps as well. Its tone can be heard being demonstrated by Eric Clapton on the John Mayall Blues Breakers recordings.

The original KT66 is the European equivalent to the US 6L6 design and is absolutely interchangeable for the 6L6 in all circuits. As the Chinese tube factory did design a very close copy, they do look and sound pretty similar to the originals, but performance wise they fall a bit short. Not that that makes them undesirable, as any KT66 design has and interesting tonal quality that is not so much GE-style 6L6 tight and clean, but more towards an EL34 with a crunchy breakup and softer overdrive as compared to a US made 6L6. This tends to give a Fender a slight British tone and feel, which can sometimes be a good thing. The obvious comparison is to the other KT66-HP which we designed and make in one of the Russian factories, and IMHO these Chinese sound very close to ours, but are lower in power and how much plate voltage they can handle. These are only good to about 450 volts B+ maximum, while the KT66-HP has a stronger bass response and can handle plate voltage of up to 550vdc, so can be used in the more aggressive boutique amps with higher plate voltages. In amps like these, and also in older vintage Fender Twin Reverbs with the higher plate voltages, the KT66-HP will sound and perform way better. They are available "off shelf" from various vendors or from GT in matched sets of duets, quartets and sextets.

KT66-HP Groove Tubes Exclusive
Output: 85.3 mA

This tube is our own design and exclusive tooling which is built under contract for us by the Reflector factory in Russia. We tried to design an exact copy of the GEC tube, and even made several of the components especially for this project. It has a different sound and performance than other Russian tubes (of Russian design) which are made in the same factory. For example, the Russian power tubes in general are a bit dark and harsh on the distortion tone, but this tube is exactly like an original MOV KT66, with a fat, solid, bell-like tone that is also very warm on the breakup point and beyond... very creamy! In blind listening tests, no one has been able to consistently tell the difference between these and the original; "Dog Ears" are the final test. Notice it also has a very high output, the result of a redesign after the first year's production that allowed the tube to handle even higher plate voltages than the original. (Incidentally, the first version, which we called just the GTKT66, still sounds great but these are not recommended for amps with plate voltages above 450vdc, and anyway we haven't made this variety since 1999). As compared to the KT66-C, these will handle plate voltages up to 550 volts.

After we introduced this amp several amp makers really fell in love with it, including Mike Soldano, and the Dr Z amp company even designed a model around it: his Route 66 amp!

BTW, for you trivia freaks, the "KT" originally meant "Kinkless Tetrode," and this tube was developed by the MO Valve Company of UK to get rid of the 6L6 "kink" in the response curve. This is why the mids of this tube are smoother and more linear than a 6L6 or 5881.

They are available only from GT in matched sets of duets, quartets and sextets.

GT6L6-GE Groove Tubes Exclusive
Output: 76.2 mA

This is our baby... our little bundle of joy...and our first home made "boutique" tube from our little tube factory here in San Fernando, California. Some said making this tube here again would be impossible – and they were almost right – but it just took a little longer than expected! After four years and that many hundreds of thousands of dollars, we finally pulled it off! We were blessed to find a large section of the original GE manufacturing and processing machinery, along with the original production drawings and processing schedules (the recipe!). We even discovered some original GE plate material, enough to make about 25,000 tubes... so we can only technically claim about 90 per cent US components.

The background of this tube is impressive, and it's a part of music history: it was the tube used exclusively by Fender in their Golden Years, from 1953 (the tweed amps) through 1978 (the blackface amps and beyond) before they switched to the ruggedized Sylvania STR tube. I also like the Sylvania 6L6 STR387 Fender used from 1978 through 1985; it also has good power and life, but also has a harder edge to its tone. By contrast, the GE "clear top" has a that cool, vintage Fender tone most would agree is the standard by which all 6L6s are measured. Also, this was the tube used in the Jimi Hendrix Fender Showman heads at the 1968 Hollywood Bowl concert (not always a Marshall guy in concert!). Hendrix's Fender studio recording amps had these tubes, too. Many folks don't realize that Hendrix's classic guitar amp tone often came from a Fender blackface amp in the studio world, and these amps always had the GE 6L6 clear top.

Knowing this history, when the chance came to purchase the GE 6L6 and 6CA7 production line along with original materials from our good friends at Richardson Electronics (who also gave us valuable technical support and advice), we jumped at. We then went to great lengths to duplicate the manufacture and processing recipes of the GE factory. IMHO, we got as close as humanly possible these days, considering we

112 THE TUBE AMP BOOK

Technical articles

are operating in a world that is technically 22 years beyond the time this line was turned off, back in 1980. In our own blind listening tests, the GT Dog Eared types couldn't tell any difference at all, and most of the outside reviews by our Dog Eared customers agree. And when compared to any other current 6L6 tube on the market today, *all* the Dog Ears pick this tube every time. The only close rival might be the KT66HP or GT6L6S type, and this always comes down to taste. Notice the higher output power; in this case you can really feel the difference in this tube, especially in the bass response. It's solid!

Of course all this investment and restartup of a US production line make it, bay far, the most expensive 6L6 in our catalog, almost double the cost of the Russian. However, if you've got that old (or new) blackface Twin Reverb you may not have a choice once you hear these tubes: it's like Hendrix all over again.

Myles Rose, owner/operator of Guitar Amplifier Blueprinting, said this about the GT 6L6GE tube: "I have been running my set of these GT6L6GEs from February 2002 at 105% output, 24 hours a day, seven days a week. They currently have over 6700 hours on them (as of Feb. 2003), and they have only dropped 3 to 4 milliamps during this period. I believe this is due to a very high vacuum, and USA parts, cathode coatings, etc. While these are more expensive than most non-USA 6L6s due to materials and labor costs being much higher, their cost is offset with a very long life, and great sonic qualities.

"I have found they typically have at least a 15-degree wider sound stage image in any amplifier that uses 6L6 tubes. The (curve) traces on this tube are duplicates of the traces of the original General Electric tubes, and they sound very close, too. This is not surprising, as these are made on the original tooling from the last plant that made these in the USA, and they use the same formula for plate materials and coatings. The mica spacers also come from the original GE source. I coined these tubes 'NVM,' for 'New Vintage Manufacture' (as opposed to NOS for New Old Stock)."

To be honest, shortly after Myles posted this report this on his internet site (www.guitaramplifierblueprinting.com), we hired him to head up our internet technical support department here at Groove Tubes. The GT 6L6GE is only available from Fender aftermarket sales and GT in matched sets of duets, quartets and sextets.

The EL34 family

Applications: *Used in Marshall, Hiwatt, Orange, Selmer, larger Vox and most other Euro tube amps; also a favorite of Rivera amps until the Quiana model, and of larger Matchless designs.*
NOS reference tube(s):
USA GE 6CA7/EL34, Output: 108.2 mA
British Mullard EL34, Output: 93.1 mA
Euro Siemens EL34, Output: 89.8 mA

EL34-C (aka: Shuguang EL34)
Output: 91.0 mA

This latest Chinese EL34 has similar response curves and sonic performance to the classic Siemens EL34s which were used in the late '60s to mid '70s Marshall amps. We really like this tube, and it's also the least expensive EL34 on the market today, which is nice if you like the sound and are a guitarist on a budget. Although in our past experience the Chinese tubes have had the "trade off" of a low price for a shorter useable lifespan, the more recent production tubes from the only remaining Chinese tube factory, Shuguang, seem to have improved and are lasting about 20 per cent longer than tubes made just a few years ago. Seems they got the message from my earlier books!

EL34-R (aka: Sovtek EL34-EH)
Output: 89.0 mA

The Russian Reflector factory makes this fairly rugged version. This model has been used for many years by various companies including Marshall (however they have more recently switched to the EL34R2 built by Svetlana). It has an even or flat tone, not too bright or aggressive. This tube was designed to be "Mullard Like", but it's a bit harsher on clipping (as with most Russian power tubes). The current Chinese EL34 offerings are closer to the tone and feel of the original Siemens tubes in older Marshall amps. They are available from Sovtek, "off shelf" from various vendors, and from GT in tested and matched sets of duets, quartets and sextets.

EL34-R2 (aka: Svetlana EL34)
Output: 86.3 mA

Svetlana (Russian) produces this 25-watt version, and it is the current OEM tube of Marshall amps. We like its overall tone as it's not as harsh on breakup as the other Russian EL34 and has sweeter highs, too. It's closer to the classic Seimens EL34 in style. When you get a good production run of these, they are very nice tubes indeed… but I should caution that the Svetlana tubes we've been processing over the last few years have been more inconsistent than those from other factories. The output has varied more than 15 per cent from one batch to the next, however the output measurement above is from a "good" batch, and lately they have been improving their consistency. To be safe, try to buy these from a vendor that can test for grid and/or gas leakage issues, and ones with reasonably high output, then the QC issue is not a factor. They are available from Svetlana, "off shelf" from various vendors and also from GT in our tested and matched sets of duets, quartets and sextets.

GTE34Ls Groove Tubes Exclusive
Output 97.1 mA

This product started life 12 years ago as an exclusive GT design which we partnered on in the old Tesla factory, but continues to be produced today in the newly upgraded JJ factory located in Slovakia. JJ also makes their own E34L version but our "s" version has a neat trick that makes it a better sounding and longer lived tube. GT developed an additional component, much like a "heat sink" assembly, and attached it to the seam of the plate assembly to dissipate the increased heat and power this "hot rod" EL34 produces. The result was to improve power up to a 30-watt tube (most EL34 tubes average 25 watts, so this makes it about 15-20 per cent hotter than a stock EL34). Please note, however, that the bias should be checked and usually adjusted higher for best results if replacing a normal EL34. The GTE34Ls has very high output (industry's highest power rating, by far) and it's tone has a strong midrange, to go with a crystalline top end and the big bass response from that extra 15-20 per cent Output power. A preferred Marshall replacement tube for many of our GT regular pros, such as Billy Gibbons, Joe Perry, and Joe Walsh, to name a few. This is the stock amp tube for Bad Cat, Matchless and recently Bogner too. There is a similar stock JJ E34L available from various vendors, but this model is only from GT in matched duets and quartets and sextets.

GT6CA7GE Groove Tubes Exclusive
Output: 97.9 mA

A strong beam pentode with an active beam forming element, and this is the main difference from a common EL34. Although it is a direct replacement for any EL34 amp requirement, this tube has a much higher vacuum than an EL34, and much stronger internal construction. To make it, Groove Tubes follows a similar production approach as with the GT6L6GE,

THE TUBE AMP BOOK 113

Technical articles

using the original GE machines, materials, and processing recipe. These will run at 800+ plate volts, and so can be substituted for many 6550/KT88 applications. Of course, you should beware that there are several "6CA7" tubes being made today which are little more than repackaged EL34 assemblies in a larger bottle. These "fakes" often do not even have the performance of the same companies' EL34 offerings and do not trace or perform in any way as the original USA GE 6CA7 of the past. Look for a first release of the GT6CA7GE as early as Summer 2003, as a Groove Tubes exclusive, and matched in our exclusive duets, quartets and sextets.

The 6550/KT88 Family

Applications: *some USA-distributed Marshalls; Ampegs; a variety of higher-powered vintage and contemporary amps.*
NOS reference tubes:
USA GE 6550a: Output TBA mA
British MO Valve Gold Lion KT88: Output TBA mA

6550-C (aka: Shuguang 6550A)
Output: 107.3 mA

Chinese made and slightly weaker than the original US-made version. However, these can actually sound better to some of the Dog Eared types as they distort easier (earlier) in Marshall style amps compared to the GE6550A. These have been made in two types over the recent years: a coke bottle shape (distorts faster), and the straight bottle with more power. Typically used in amps where a lot of power or clean headroom is desired. Very good value and also good tone; not expected to last as long as the Russian versions, but they have a more musical tone to many ears. They are available "off shelf" from various vendors and from GT in tested and matched sets of duets, quartets and sextets.

6550-R (aka: Sovtek 6550A)
Output: 92.5 mA

Russian made by Reflector factory. Oldest Russian version and a mainstay of many OEM amp manufacturers. Although this tube is not popular in the guitar amp world, mostly because they are clean and late to distort, it is a rugged and long-life option of you need a solid performer. I would characterize the tone as a bit dark with a harsher breakup as compared to the other Russian 6550 or the Chinese versions. They are available from Sovtek, "off shelf" from various vendors and also from GT in tested and matched sets of duets, quartets and sextets.

GT6550-R2 (aka: Svetlana 6550A)
Output: 92.5 mA

Made by Svetlana, this is the latest Russian version of the USA classic high-powered pentode and is the new favorite of many OEM amp manufacturers. Although this tube is not popular in the guitar amp world, mostly because they are clean and late to distort, it is a rugged and long life option of you need a solid performer. I would characterize the tone as warmer than the other Russian 6550, but not as warm as the Chinese versions. They are available from Svetlana, "off shelf" from various vendors and also from GT in tested and matched sets of duets, quartets and sextets.

KT88-C (aka: Shuguang KT88)
Output: 103.6 mA

Earlier version of this Chinese KT-88 made by Shuguang factory. Very nice and strong power and good guitar tone, too; warmer than 6550 types. This tube is my preferred KT88 for a softer tone, and finds a home in amps such as the ampeg SVT bass maps, the Marshall/Park 75 amps, the rarer Marshall Major amps, VHT and other higher-powered tube amps. They are available "off shelf" from various vendors and from GT in tested and matched sets of Duets, Quartets and Sextets.

KT88-C2
Output: 103.6 mA

Latest "improved" version of this Chinese KT-88 is also made by Shuguang factory. Very nice and strong power, and great guitar tone, too. Warmer than 6550 types. This version is more true to the original MO Valve Gold Lion (it's hard to tell them apart) and was particularly made for the hi-fi market (an affordable savior for McIntosh amps!). This tube also finds a home in amps such as the Ampeg SVT bass maps, the Marshall/Park 75 amps, the more rare Marshall Major amps, VHT and other higher powered tube amps. They are available "off shelf" from various vendors and from GT in tested and matched sets of duets, quartets and sextets.

GTKT88-SV Groove Tubes Exclusive
Output: 97.3 mA

This product started life 12 years ago as an exclusive GT design which we partnered on in the old Tesla factory, but continues to be produced today in the newly upgraded JJ factory located in Slovakia. JJ also makes their own KT88 version but our "SV" version has a neat trick that makes it a better sounding and longer lived tube. GT developed an additional component, looking much like a "heat sink" assembly (as with our GTE34L), and attached it to the seam of the plate assembly so as to disipate the increased heat and power this "hot rod" KT88 produces. The result was to improve power up to a 60-watt tube (most KT88 tubes average 50 watts, so this makes it about 20 per cent hotter than a stock KT88 without the heat sinks). However, please note the bias should be checked and usually adjusted higher for best results if replacing a normal KT88/6550. The GTKT88SV has very high output (industry's highest power rating, by far) and its tone is a strong midrange, to go with a crystalline top end and the big bass response from that extra 20 per cent power. They are only available from GT in matched sets of duets, quartets and sextets.

KT90 (aka El KT90 - El KT99)
Output: 114.8 mA

This is a most unusual looking KT88-equivalent tube, as it has a top-evacuated bottle (like a big preamp tube). It does have a redesigned and beefy plate structure that makes it a powerful option for this family. We have always liked the sound too, kinda a cross between the EL34 and the KT88, so a good powerful Heavy Metal tube for sure. The inconsistent availability of this tube, along with the changes in materials and processing over the years, has made it the most "unknown" of the large power tubes – you never know if you're going to get a good run or a "dud" run – and the politics of the region don't help a smooth production flow ether. However, we have a small but loyal fan base for this tube among our pro customers, but not nearly as many as for the KT88SV which is our best seller in this family.

The EL84 Family

Applications: *the classic Vox tube of the AC15 and AC30, and popular with modern amps that follow those designs, such as Matchless, Bad Cat, TopHat, and many others; also found in some vintage Gibson amps, smaller Marshalls, and some recent 15W Fenders. Currently the world's most popular tube for sub-30W amps.*
NOS reference:
Euro Phillips EL84: Output: 46.5 mA
USA GE 6BQ5: Output: 48.0 mA

EL84-R (aka Sovtek EL84-EH)
Output: 44.6 mA

Made by Reflector in Russia (as the Sovtek). Reliable, but not as articulate as the others. Darker sounding in most amps. The stock tube

Technical articles

EL84-Y (aka: EI EL84)
Output: 53.0 mA

in most Fender, Vox, and many other EL84-based amps. Not as bright as the EL84-S.

Brighter than the S or R version and inconsistent material and quality. usually this can be a dark-plate tube, but recently they have been using a shinny Stainless steel for the plates. While this is probably OK for a power tube (not a preamp tube, however), it looks so different as to turn off the vintage customer. We, however, like its strong output power and crisp tone; more highs than other EL84 tubes, too.

GEL84-S (aka: JJ EL84)
Output: 45.3 mA

We think this is probably the best of the current EL84s, as judged by most of our Dog Eared customers. It has good power balanced with a warmer, more even tone than either of the other EL84s made today. In fact, it is a better rock tube too because of its solid midrange response. More articulate mids and highs than the EL84R.

The 7027a Family
NOS reference tube:
USA Sylvania 7027a, Output TBA mA

7027 (aka: JJ 7027)
Output: TBA mA

Common only to early Ampeg amps like the V4 and VT22 series. This was a powerful pentode similar to the EL34 and 6L6, kind of a cross between the two in tone and response. It has a very hi-fi tone that gave Ampeg their unique sonic signature. This tube by the old Tesla factory, now called JJ, is the only option today unless you are blessed to uncover a small stash of the early GE or Sylvania 7027a, which I believe are a better choice if cost is no object.

The 7591a Family
NOS reference tube:
GE 7591a Output: TBA mA

GT7591a (aka: Sovtek 7591a)
Output: 58.4 mA

Today this tube is only made by Reflector, Russia. Used in many hi-fi amps in the past and a few smaller combo guitar amps from Ampeg. A very clear, even tone, and smooth, soft distortion. Available from Sovtek, "off shelf" vendors, and from GT in tested and matched duets, quartets, and sextets. ∎

REPORT ON EXISTING TUBE FACTORIES

In the case of Russia there are two factories, one in the west in St. Petersberg called Svetlana, and the other in the middle of the country, in Saratov, called Reflector. At Groove Tubes, we use the R and R2 suffix to differentiate between these factories and also to denote evolving designs from the same factory. Here's the Groove Tubes suffix guide and a bit of story behind these factories:

"R" is from the Reflector factory in Russia. Most Sovtek labeled tubes come from here.
"R2" (ie 12AX7R2) indicates the second evolution of a preamp tube from Reflector.
"R3" (ie 12AX7R2) indicates the third evolution of a preamp tube from Reflector.

I have been to the Reflector factory on many occasions, and even partnered on a few designs. It is a massive industrial complex contained in a four-square-block area and is almost 100 per cent vertically integrated – they even blow their own glass! Unfortunately, as tube demand for consumer and military applications has dwindled to a small fraction of former times, this 50-year-old factory today operates at less than 5 per cent capacity. We partnered with Reflector a few years back to produced our KT66HP, which is completely made on their premises. I am happy to report that this tube is still in production at the time of this writing as an exclusive GT tube type. We also seriously considered buying Reflector when the government wanted to liquidate them a few years back, but eventually decided against it. Fortunately, another US tube vendor, Sovtek, headed by Mike Matthews of Electro-Harmonix pedals fame, bought a large share of the factory and has kept them producing their full line of tubes. Many thanks go out to Mike for keeping this fine factory in business and guiding them into to the 21st century. They make a rugged and consistently good product used by Fender, Marshall and Ampeg, to name a few, and probably supply more than 50 per cent of the world's tube demand.

"R2" (ie 6L6R2, 6550R2, EL34R2) can also indicate the second Russian factory made 6L6, EL34 and 6550 from Svetlana in St Petersberg.

This is also a former high-volume tube production factory responsible for many of the Russian tube designs and established many years ago when tubes were the mainstream component in many electronics. The Russians were still making all-tube TVs and Radios when we in the West were buying Sony Walkmans! They are also a large, integrated tube factory with all component and processing done "on site," but because of lack of sales potential they were "dark" for many years. The management has changed hands several times over recent years. Production (and quality) has been sporadic, and today they also are running at less than 5 per cent capacity. They currently are supplying Marshall its power tubes, and have opened a Svetlana distribution company based in the USA. We truly hope the new management and marketing partners can stabilize this factory, as they produce some of the best sounding and highest quality power tubes – although as I mentioned, quality has been inconsistent in recent years.

"S" is for tubes from the former Tesla factory, now called JJ, located in Slovakia.

Tesla had many different tube factories all around the former Czechoslovakia, but the largest receiving tube factory was located near the old border of today's Czech Republik and Slovak Republic. I was fortunate to partner with four other equal partners in the rebuilding and restarting of the Tesla factory about 10 years ago. This was after the Berlin Wall had been torn down and Czechoslovakia had become a free country and so began to "privatize" the government factories. After a hard couple of years, our privatization plan for Tesla was interrupted by the dividing of the country into the now Czech Republic and Slovak Republik, and so our "company" disappeared overnight!

I lost touch with the factory and several of my former partners during this time, only later to find one of the partners had "rescued" the Tesla factory machines and tools out of a liquidation sale and transferred them to a new location inside of the newly formed Slovak Republic. Jan Jurco, who was one of my our five original Tesla

THE TUBE AMP BOOK

Technical articles

partners, is now the owner and operator of the Tesla legacy, now renamed after his initials. I am happy to report that two of the tubes we had developed while at Tesla are still made exclusively today for Groove Tubes by JJ, the GTE34Ls and GTKT88SV. Mr. Jurco has also developed many new types, and greatly improved all the former tubes to a new level of performance. His tubes are extremely high on Output ratings, usually meeting or exceeding original specifications in these areas a of gain and output. This makes them great for souping up vintage amps, but also less suitable to high-gain amp designs, where the increased gain and output can make them literally too hot and so they can have hum, noise and microphonic issues in some designs originally made with the Russian type tubes. That said, however, this is perhaps one of my favorite tube factories, and Jan Jurco is a great old friend and provider of some of Groove Tubes' most popular tubes.

"Y" is for tubes from the Electronska Industri (EI) factory in Serbia, formerly in Yugoslavia.

This factory located in Nis, Serbia, was originally started with the help of the German electronic companies of Seimens and Telefunken. The Germans were phasing out tube production in the '60s and transferred their machines, processing technology, and designs to this location. Therefore EI is one of the most (relatively) modern tube factories in the world. Although they are dependent on some raw components from outside, their clean rooms for assembly and overall machine quality are second to none. Unfortunately, the politics of this region are volatile and war has taken quite a toll on this jewel of a tube factory. This was one of the factories partially owned by the war criminal and former president Milosovik, so for many years we were unable to import any products from Serbia. Thankfully we have a relative peace there now, but they still do not have "Most Favored Nation" status, which means for a US company like GT, we pay more that 40 per cent import duty and freight. Also, EI has had inconsistent quality due to inconsistent material availability and financing. Still, despite war, lack of cash and material, they have courageously struggled against all odds (and near misses by Coalition bombing) to continue production.

One of our favorite sounding 12AX7 types, the GT7025, comes from this factory, and although the QV is very wide, when you get a good one it is most reminiscent of an original Telefunken; in fact many unscrupulous tube vendors have been relabeling the EI with the Telefunken logos to sell at stupid NOS prices to unsuspecting hi-fi and MI customers. The easy way to tell the difference is that original Telefunken tubes have a small diamond logo on the glass in between the pins at the bottom. As Europe becomes more stable in this region (we pray), I look for EI to regain their former prominence in the world tube market. The designs are very good, and they are completely capable – they just need financing, some new marketing, and our prayers. I almost thought we'd lost them during the bombing and during the sanctions; I am happy to report they are still alive and the future looks bright.

"C" is for tubes from the Shuguang factory located in China.

This is the last receiving tube factory left in China, and they are a very old and esteemed factory by Chinese standards. Although recent investment and restructuring has made Shuguang a primarily CRT factory (they are a major supplier of TV screens and computer monitors in China), the old receiving tube line is still intact and making great sounding tubes, in my humble opinion. The tubes we have seen over the last 24 years have been mostly power tubes like 6L6, EL34 and KT88, but just recently they have tooled up to produce a 12AX7, and after many evolutions and retooling, they make what I think is the warmest sounding 12AX7 in the world – perfect for vintage amp lovers, as it is much like an old GE 12AX7.

Like all current preamp tube production lines, consistent quality is a problem, and the Shuguang 12AX7 is no exception. You may hear one that sounds terrible with high noise or hum, and another from the same batch is perfect! Over the years, the Chinese tubes got a bad reputation for flaming out on power tubes, or humming on preamp tubes, and all I can say is don't give up on them! We at GT of course weed out those imperfect tubes so our customers don't suffer the usual 5-out-of-10 syndrome… but all in all we have found their QC about the same as most the other tube factories.

Compared to their preamp tubes, however, their power tubes are a bit under-built with regard to materials, and the vacuums are a bit soft, so they do not last as long as, say, a Russian 6L6. On the other hand they sound more like the original US types like GE I guess because they copied those designs. When I first visited Shuguang, I was shocked to see a room full of what I thought were RCA grid winders, but on close inspection I noticed they were just really nice "Chinese" copies! Hey, if they can make really close Rolex copies, how much harder can it be to make an RCA grid winder? I also think Chinese tubes in general have a great tone because their cathode nickel is among the best ever made, very pure, and this is a factor in tube performance. Like good woods are important to a guitar, the cathode nickel is a "secret recipe" and key ingredient to what a tube will sound like. Again, I think they just did a good copy of the US formulas… and why not!

"GE" Groove Tubes production, San Fernando, California, USA: limited production of the 6L6 and 6CA7 tubes to the original USA General Electric specs.

I mention this last because it is the most recent tube factory, that has only recently gone "on line", and because we supply a small fraction of the overall market. In fact, before May of 2001, these tubes didn't exist except in NOS stockpiles and at escalating prices on eBay. This GE 6L6 was the factory tube of Fender, Silvertone, Ampeg and many other US companies from the early '50s through the late '70s, when Fender and others switched to Sylvania and imported tubes. We were very blessed to have purchased key elements of the original GE 6L6 and 6CA7 production line from Richardson Electronics, including many hundreds of pounds of the original plate materials. I will not bore you with the four year saga and the sad tale of selling my vintage amp collection in order to finance this incredible project, but needless to say there was a lot of sacrifice and many failures before we shipped our first tube in the early months of 2002.

However, certainly we can not compare our little "boutique" tube line to any of the above mentioned "mint condition fully integrated original" tube factories in Russia or eastern Europe, where almost all the processing and component production is done in house. We had to be creative, and so have enlisted the help of many companies to develop a practical production plan for the USA today. We had the help of several retired experts like John Mark of RCA, John Gummer of GE, George Graham of Richardson Electronics (still producing tubes in the US for commercial and military) and, lastly, Charles Widner, who recently restarted the Western Electric factory for the hi-fi market and found a large cache of original GE plate materials, which I use in the 6L6 production.

We were also blessed to find a few of the original companies who supplied GE components, such as the mica insulators, and a local company who had the original four slide machines that stamp out intricate GE plate designs. Without these blessings falling into my lap, there is *NO WAY* I could have succeeded. In truth, to bring these processes in house would be economically impossible just to make a few tubes for the guitar market.

Thankfully, we had the original GE processing formulas, and a lot of expert help with the critical processing needed to make tubes. Also, you have to want to make tubes real bad: like restoring that old car (or guitar), it simply can not be done for profit… it has to be a love thing. After so many years of dreaming about GE clear top 6L6s and RCA black plates this was a dream come true. I just wanted to make a few select tubes for a few select customers. It's kinda like restoring an old car: it always costs more than it's worth, but the joy of completing the task is more payment than I have ever imagined.

Technical articles

Although I had partnered in both production and design many times over my 24+ year tube adventure in several of the above mention factories, I never fully appreciated how very difficult it is to produce a vacuum tube; it is certainly the hardest undertaking I have ever done, or could imagine.

OK, that's the while story on the few remaining antique tube factories that supply 100 per cent of the market today. If there is any other factory out there, I do not know about it and would be glad to learn about another I have missed. Frankly, there are only two primary reasons we even still have tubes today:

1) There were a few of the factories "left over", from a hundred or so that were producing tubes in the '50s. They were still viable businesses in the '80s simply because these countries were so "backward", both economically and technologically, and hadn't progressed at the rate of speed the West had. That's what eventually won the Cold War: Levis, rock'n'roll, and Walkmans! And...

2) The critical "Dog Eared" musician types who refused to believe the often repeated marketing lines that said: "Finally, a transistor amp that sounds just as good as your old tube amp!" I wished I had a nickel for every time I have read that line, but I never played any transistor or modeling amp that could stay up with my old Fender Deluxe Reverb or Marshall 100 Super Lead... and I don't expect to play one that can in my lifetime. It was this pressing market demand that woke up a few of the remaining slumbering tube factories to provide a product that just wouldn't die.

While early tube production from these various "New East" factories were much less reliable in quality and tone than the classic "Old West" tubes (RCA, GE, Tung-sol, Amprex, Telefunken, Mullard, MO Valve), they have consistently improved. They have invested in new tooling, copied the old classic designs closer and closer, and now I can safely predict that the music world's tube supply for the foreseeable future is secure. In fact, there are more tubes on the market today for our guitar and bass amps than when I started almost 25 years ago! Even my mom doesn't question any more why I am in the tube business, or that *anyone* could still be in the tube business these days. Now that's real progress! ■

THE WIDE WORLD OF PREAMP TUBES
Alternate Types and Interchangeability
By Myles Rose

The preamp tubes are the most critical, least expensive, and most overlooked tubes in your amp. Unlike power/output tubes, which are routinely matched when they are sold (in different ways, some much better than others), preamp tubes are tested, at best, to: (a) make sure they work, (b) ensure they are not microphonic. These little tubes, however, have a crucial effect on the tonal signature of your sound, and are interchangeable without adjustment or the need of an amp tech.

In testing, I have found that some suppliers don't seem to test their preamp tubes at all, as I have sometimes even found that one side of the triode is dead. Since most warranty preamp tubes for up to six months and longer, they possibly figure that it is cost-effective to just send them out as they get them in, and if there is a problem, it is cheaper to just give the customer another tube. This is of little comfort to somebody that either has to make another trip to their music store or, worse, box up the bad tube and ship it back to the supplier, and then wait for its replacement. This is one reason to consider a proven supplier when you buy preamp tubes.

Today's amplifiers, whether modern high-gain types or boutique amplifiers, have one thing in common: the preamp tube in the first gain stage (usually V1 and/or V2) sets the tone and initial gain structure of the amplifier.

Amp Design Relative To Preamp Tubes
Today's modern amps get just about all of their characteristics in the preamp section. How the gain stages are set up, how the EQ is set up, gain structure, and tone stacks – all are the main aspect of the sound character of the amplifier.

Amps made by Mesa/Boogie, Fender, Marshall, Bogner, Peavey, and others all use the same Sovtek, Svetlana, JJ/Tesla, Electro-Harmonix, and other power tubes from the same factories. In spite of the same output sections, and in many cases the same range of B+ voltages on the plates of the output tubes, these amps sound different. This is all because of different designs, primarily in the front end, or initial gain section of the amplifier.

Inconsistencies
Today's newly made preamp tubes are very inconsistent compared to the tubes of the 1940s – '60s. The medical and military sectors in the west have little if any need for tubes today. They are primarily used today in audio applications. The needs of the high-end audiophile are more easily met than those of the musician, as their tubes are not subjected to the same stresses as those in a guitar amplifier, they use less of them, and they last much longer. There are high-end audio suppliers that will match tubes and hand select them, at much higher costs (check out a Western Electric 300B matched pair for example). They pop their tubes in, and 10 years later all is still just fine.

Tubes for the guitar and bass player for use in the preamp section are a different story. The tubes today are very inconsistent. You contact your local tube supplier, plunk down your money, and the roulette wheel is set into motion.

To show the inconsistency, we went through a batch of over 100 tubes that were from the Electro-Harmonix 12AX7EH, ECC83, 7025; Sovtek 12AX7WA, LP, LPS; Chinese 12AX7C (old tooling and new tooling); and a few others. Basically, the standard 12AX7 spec that applies to 12AX7/ECC83/7025 tubes has a reference of 1.2 mA at 250 volts with a -2 volt bias.

Some people like to use those little references that say if you want less gain than a 12AX7, use a 12AT7, as it has only 70 per cent of the gain of a 12AX7 etc. These little tips are cute, but with the wide range of inconsistency out there, they are not all that useful, as it is still a matter of chance. The 12AT7 has a different current capacity than a 12AX7, so if you are just looking for less gain, then you may even find you get it with just a different 12AX7 – even from the same brand, same date code, and same batch – just by swapping around tubes that are already in your amplifier. With today's inconsistent offerings, the old tables of gain cannot be used with much accuracy.

In the tubes we went through (keeping in mind our 1.2 mA/1600 transconductance industry standard spec) we found our samples ranged from 0.7 mA to 1.6 mA. When you take into account that the amplification factor of a 12AX7 is 100, you can see that there is a dramatic difference in these tubes. Looking at a 1.6 mA tube, we see a 33 per cent factor of increase over the standard. This is a large number. A 1.0 tube versus a 1.2 tube will turn the gain you loved in your 5150 into something less than what you used to know and love it for. You sit dumbfounded... "How can this be? I just put in new tubes, the same as what I had before?"

Gain Versus Output in Preamp Tubes
Think of your days back in science class, where you built or saw one of those big ball devices, that created a half a million volts, made a great light show, and you could touch it, and have the class' hair stand on end! Gain in a guitar amp is much the same. You can have a 20W amp with high gain, and shred all day long – it's just as "gainy" as a 100W monster. It is the wattage, or output, that differs.

Today's 12AX7s as an example, all have about the same gain, which in this case should be about 100. Most today fall below this, in the range of maybe the mid 80s, with some samples going up to maybe 110. The big difference in tubes, though, is current output. A typical 12AX7 is expected to put out 1.2 milliamps at a given test voltage. Today's tubes in general, put out as little as 70 per cent of that in 80 per cent of the cases. A tube at 0.8 milliamps has a full 30 per cent less output than what is expected. This is like a 50 watt amp putting out 30 or so watts. Some tubes are better than others from various

Technical articles

manufacturers. Some examples of this are tubes like the widely used, and perhaps the most popular, Sovtek 12AX7WA. This is a sturdy tube, generally free of microphonics, with acceptable gain. Part of the reason they are quiet is they tend to be lower in gain than many other 12AX7s, but are also much lower on output current. It is much the same as with power tubes; where in the same amp, one duet of output tubes will put out 50 watts, while another set gives only 45 watts. Thus, the Sovtek 12AX7WA is quiet, due to lack of output and gain in many amp circuits. These tubes are sturdy and inexpensive, and help a lot of amps make it through the warranty period. On the other side of the coin are the JJ ECC83 tube. This is a part of the 12AX7 family, but different construction, plate materials, cathode coatings, and other factors give this tube a bit more gain than most others. This can be 100-120 Mu, rather than the 85 or so of a 12AX7WA Sovtek. The big difference in the ECC83 in general though, is its current output – at times over 1.5 milliamps, or in some cases, *three times more* than a Sovtek 12AX7WA.

Like I have said in the past, preamp tubes are a crapshoot. You buy your tubes and take your chances. Some folks, like Groove Tubes, screen and test for microphonics, noise, *and* low output. The ones that do not pass all tests are rejected; in some cases, this amounts to over 50 per cent of the factory run. There is still a range of differing specs found even in these more rigidly tested tubes, but the spread is much tighter. In any of these cases, the end user still does not know what the tube is actually doing. The SAG area at GT takes tubes, and runs them through another process, where all the specs are recorded and traces are performed. These traces show rise time, and other factors, and knowing all of these specs can help you to better tailor your preamp's tonality, feel and performance. All of this analysis has resulted in a number of special-applications preamp tubes: the SAG-MPI's (matched phase inverters), and other "kits" such as the MHG (Marshall High Gain) kits, which can be used in any 12AX7 based amp, not just Marshalls. I guess I should have called this a "High Power" kit, which would have been more accurate actually. There are Fender High Gain Kits, and Fender Soft Touch Kits, and the SAG generally tailors preamp tubes for specific uses and tastes.

You want even *more gain* from your Triple Rectifier or Bogner? Look at those first gain stage preamp tubes, and get some tube vendor to measure them for you. If you have a 1.1 in there, and put in a 1.3, you will hear the difference in gain *immediately*. This is not a subtle change that only the "experts" can hear. Leave the settings on the guitar and amp the same, swap the tube, and listen again.

When we see a transconductance of 1200 versus the 1600, the way the tube reacts is different too, in this case, its rise time is about 25 per cent slower. This might be just the ticket for a blues player, looking for some nice initial compression on the pick attack, but it may not be the sound for a metal or speed player. Transconductance in the testing, ranged from 1060 to 1790. Remember, 1600 is the industry standard.

There is one other aspect of preamp tubes. Unlike power tubes, where one tube is one tube, a preamp tube is two tubes in one bottle. There is an A side and a B side. The are independent units sharing only the heater. In one channel of a Fender blackface amp, as an example, the inputs use one side of the tube, then comes the tone stack, then the other side of that tube is used for the second gain stage. In the Normal channel V1 is used, and in Vibrato channel V2 is your first gain stage. Since these are two different circuits with the tone controls between them, it is not as much of a concern whether or not the A and B side of the tube are close to the same performance, or "matched". But, anytime we use that tube in the phase inverter position (or driver) position of the amp, which is the driver for the power tubes, then having the two sides matched is important. This matching subject is covered in more detail a few paragraphs below, so I won't elaborate on this here.

New vs Vintage Amp Needs
NOS tubes are sought after by folks that have original amps like Fender tweeds and the like. If you want the original sound, feel, and character of these amps, then NOS is probably the best way to go. Getting NOS tubes for your amp to be correct is much easier in some ways than getting decent new tubes from a dealer who doesn't test and match thoroughly (though the NOS will certainly be more expensive). There are some good folks that deal with NOS tubes, but there are plenty of others just selling old tubes as they get them – and these expensive, supposedly "new" old stock tubes might be just as dead as poorly selected new tubes when you get them.

When it comes to new tubes for you modern or classic amp, or new boutique amp, here is the problem: run-of-the-mill preamp tube suppliers guarantee the tubes to work, and not be microphonic. That is about all they can do. Going through tubes that retail for less than $20 in most cases, one at a time to measure them, is beyond reason economically. Other folks do offer these checking and testing services, however, at an additional cost in most cases. To my mind, the money is well spent. When you want a nice, high-gain tube for your Rivera or Demeter, putting in a tube that is 30 per cent down from spec is not the ticket! At that point, what can you do with that tube? Take it back? Why? It works. The store or vendor never stated it would do anything more than "work." Perhaps they will exchange it, and now you start the process over? And over. And over.

Recently I was working with a fellow named Tom Dunn. He plays a 5150 II. We performed a blueprinting session of his amp, and found after going through maybe two dozen of his own tube stock that he had picked out his tubes by ear, and placed them in the most advantageous positions in his amp. You can do this by ear, if you have the ears of this guy, and also the time (he did this over many months)… and the tube stock.

Your first gain stage in your amp is its soul, sound, and character. We talked here about gain, and a little about rise time, which is a subject in itself. We did not get much into "sound," such as the articulation and definition that comes from NOS tubes like the Mullards and Telefunkens. If you have an older amp with a more moderate gain structure, and want it to sound closer to magic, then this is the way to go. In a modern amp, the design is not aiming to get a lot of the articulation from the output section. Today's designs look for two or three or more stages of gain, channel switching – which we did not have on the older amps of yesterday – and flexibility. The only flexibility we had when I was the age of most of you, was a high-gain input and low-gain input… or turning the reverb on or off.

All I can suggest, is try to find a tube vendor that can supply you with the tubes you need with some degree of classification. This way, if you have a 1.3/1670 tube in there now, and you want to tone it down a bit, then maybe go for a 1.1 – it will make a difference. If you want tonal changes in color, rather than gain and compression, then you want to go with a little stash of tubes, depending on your use for the day or evening. Most of my clients keep the following:

NOS JAN 12AX7A: Most often general use.
12AX7C: Chinese 12AX7 – to take off some amp edge or brightness
12AX7EH: Electro-Harmonix – for general use
ECC83: for the Marshall sort of sound
7025: for the Fender '60s and '70s sound
5751: for blues and less aggressive attack (and perhaps less gain as compared to an in-spec 12AX7).

For more detailed assessment of a range of current production and NOS preamp tubes, see the piece by Mark Baier of Victoria Amplifiers that follows the end of this chapter.

Gain/Output and Matching
The first basic point to remember is that unlike a power tube such as a 6L6, EL84, EL34, 6550, and others, a preamp tube is a "dual triode" in most cases. This includes tubes such as a 12AX7, ECC83, 7025, 12AT7 and others. In regard to matching of two sides of a preamp, many folks feel that in a balanced circuit which, in a way, "sums" the two sides of the tube, that balancing is not necessary. The high end audio industry recognizes the need for a balanced phase inverter or drivers, unlike some in the musical instrument industry.

Technical articles

In this crucial position, however, this balance of two sides of a triode is necessary for optimal operation. I will use an example to try to illustrate. Taking a twin engine airplane, let's say we have a typical light twin with two 300 HP engines designed to cruise at 250 knots. We have each engine at 1800 rpm. One engine is at 24" of manifold pressure, while the other is at 22" of MP. The airplane structure itself is the "balance," much like the balanced circuit that some amp builders feel will negate the need for a balanced phase inverter. The plane may fly just fine, but we need to add a bit of rudder trim, and our fuel consumption will be higher, and overall performance and balance of the airplane (or amp) will suffer.

Now in the "art" of balancing a dual triode, there are many folks that will perform "matching" for a few dollars of additional cost. In almost all cases, this "match" is a gain match, or a current match, or a transconductance match. Due to very high inconsistencies in today's preamp tubes, matching for any of these factors is an improvement over an unknown tube. This sort of matching at least makes sure that we do not have a 200 HP engine on one side of our airplane and a 400 HP engine on the other side. These numbers seem like wide examples? Not in the least. A typical new preamp tube these days at a given bias and voltage, in the 12AX7 family, will run from a milliamp to three times that. They typical average is only about 70 per cent of what is expected as standard spec, by the way.

Now, in our airplane example, our horsepower may be way off with two different HP engines; in the stock amp, and with the easiest form of balance as above, at least our HP is the same, which is of help. But, we still have the factor that one of the engines is running at a different RPM or Manifold Pressure. This is the "time" component that is missed in most matching. You can balance the circuit all you want, from an amp designer's standpoint, to compensate for voltages or current, but you cannot balance the time component, all one can do is average, in a manner.

A true balanced dual triode is just like two output tubes. We want their characteristics balanced in many aspects. This is why we do not plug in one EL34 and one 6L6, even though this would work, and make some sound come out of the amp. (In fact, in this example, these two totally different tubes would probably be closer in characteristics than the typical untested new dual triode of today.) What is required here is to select for output, and *time*. This can only be done on very sophisticated equipment such as a vacuum tube curve tracer. The two curves – measured at all voltages of operation, and with a signal applied to the tube – are compared and sought to be as close to identical as can be achieved. This is very costly as it is very time consuming. At times, only 1 in 50 tubes will make the grade. Remember, it is this little tube that drives your final output stage.

The folks in the high end audio industry know the difference in balanced or matched inverters, and many amp makers do also. There are still a lot of "amp" folks out there that want to fight the points here, mostly because they have not taken the time component into their thinking. If they have tried "matched" phase inverters from simple sources with simple current match methods, this may be one reason that there was not as much difference as they had expected.

Now, the question of gain versus output in preamp tubes, especially as it applies to the PI/driver stage, is one that still eludes a lot of folks. Using a nice tube in a current drive stage will yield little improvement in tone, as this stage is looking for drive and current more than gain. So, and RFT or Mullard used here, for example, would work well in one sense, in that it is a nice tube and probably more equally matched on the two sides of the triode – just because it was made better in the first place. This would be the only advantage.

Now, if you could find a cheap tube like a Sovtek 12AX7WA (if the phase inverter was an AX7 and not an AT7, as an example), but this "cheap tube" was matched on the A and B sides, and had good output of at least 1.0mA per side, then you'd be in better shape than with a mismatched expensive tube like an RFT, Mullard, etc. Keep your great tubes for the first amp stages. This is where gain is important more than output.

Considering again the example of some suppliers matching dual triodes for a single factor: some match for gain, some for transconductance. This is *much* better than not matching at all. But, for the best results, tubes should be looked at for a number of factors... gain, output, transconductance, etc. I also trace tubes for rise time, as this is critical when a speed metal player wants less or more pick attack noise, slower or faster response, or a jazz player wants more compression on the front end.

As you can already see, no doubt, it's a lot more complicated that it seems in a lot of cases. In the "old days," amps were made with components, such as resistors and capacitors, that were +/– 20 per cent. In reality, the norm or these parts were pretty close to spec. When it comes to preamp tubes, there are specs too, but nobody lists them or warrants them. Unlike most resistors, tubes do vary – a lot – and +/– 50 per cent is *very* common. This is one of the reasons that one amp sounds great and another just okay, even two of the same make and model.

The first gain stage in an amp is critical, and a matched output stage is critical. They both have a different set of requirements that need to be addressed. Sometimes its gain, sometimes output, sometimes TC, sometimes rise time, sometimes curves, linearity, etc.

In the course of my work blueprinting amplifiers, I have had to explain over and over about a mis-matched output section and its impact to many. In any class A/B amplifier, because of the NFB (negative feedback loop, often tapped into by something we know as the Presence control), any disparity between the upper part of the sine wave (produced by half of the output tubes), and the lower part of the sine wave (produced by the other half of the output tubes), is cancelled out by the NFB circuit by design. This is the reason some notes "sing" when your amp is pushed in the output section (rather than pushing the input in a master volume amp), and other notes do not have the same magic. The reason some of the great blues players have that tone, is that their amps are taken care of by people that know how to adjust or deal with some of the issues that cause this lack of luster.

Since no tubes are even close to identical, this cancellation is always going on. The object is to limit this as much as possible. The most common way people match an output section, is to use good quality matched tubes. The industrial spec for a match here can be as high as +/– 20 per cent. A good match by a lot of tube vendors is +/– 10 per cent. I believe that even the untrained ear can hear the difference when a output section matched within +/– 5 per cent is used. In the amps I set up for the folks that retain me, my spec is less than +/– 2.5 per cent.

The most overlooked and misunderstood part of the output section is the 12AX7/ECC83 (Marshall or Vox style, or tweed Fender style) or 12AT7 (blackface Fender style) Phase Inverter tube. This is the tube that drives the output tubes. A lot of folks that specialize in making amps sound great don't understand this, but fix this accidentally. They tend to use very good tubes, such as JAN spec 5751s – where the match is closer – as well as closer matched tubes in the output section. They also use tubes that sound good in the first gain stage positions, rather than the common Sovtek WA tubes which most manufacturers use (because they are sturdy, not as expensive, and ship well without developing microphonics).

When I scope an amp in the lower frequency region, the vast majority of the time, the upper and lower parts of the sinewave are not even close to equal. This is more disparent than just a slightly mismatched set of output tubes. At this point, I install a matched phase inverter/driver. Matched phase inverters and output tubes are one of the reasons some amps "sing" and others are pedestrian compared to their brothers and sisters.

If you seem to have a lot of dead spots, try a new phase inverter tube. This is usually the preamp tube that is the closest to your output tubes. It is a trial and error process, but you may get lucky. If you want to be sure, however, buy one that has been matched for you, for all the factors dscussed above.

By the way: this is *extremely critical* with dropped tunings, seven-string guitars when

Technical articles

played in the lower ranges, and even more critical for bass players who use tube amplification.

Myles Rose is proprietor of Amplifier Blueprinting in Los Angeles, California, and head of the technical support department at Groove Tubes. ∎

THOUGHTS ON 12AX7 TYPE TUBES
By Mark Baier

I'm always a bit surprised by how little the average guitar plinker knows about the glass bottles that powers his amp. Sure, much has been written by self-proclaimed gurus as to the physics of vacuum tubes, or how to turn a BF Twin into a '51 Super, but how much discussion is ever given to the *sonic* differences between a GE and Sovtek 12AX7? I'm gonna invent new and exciting sonic adjectives for the occasion! You'll want to rip your hair out trying to decipher what "chewy and refined with a touch of lacy brilliance" actually means! When I'm done writing this, I'm gonna compose my monthly article for *Wine Spectator*...

At this point it's probably a good idea to note that I did not conduct an extensive, dedicated listening test using a specific model amp and guitar. These opinions are based on years of playing many guitars and amps of various manufacturers and vintages. The focus of this assessment will be with a '50s Fender tweed-type amp in mind. Aside from being my favorite kind of amp to play tone-wise, they make a good platform for evaluating tubes due to their simple circuit topologies.

Preamp Tubes

GE 12AX7: Most commonly available as JAN type, mid '80s vintage. Has a soft top end – not edgy or piercing. Nice, solid mids and lows. Lots of depth and texture to the tone. Older orange box versions seem to be a bit more brilliant. Can be a noisy tube; this would be a problem for high gain amps. Can sound grainy in some amps. Moderate amount of gain. Good NOS American tone... for years these were standard issue in your Fender amp. If you're using the recent JAN stuff, pack some spares: these were an end run situation, and QC was a bit slack. Standard Phase inverter tube in a new Victoria.

ECG 12AX7: Like the GE mentioned above, this tube is most commonly found as JAN examples. With the Wild West style of capitalism that is the rule in the tube biz these days, these JAN stocks are the most reliable source of actual NOS US and British tubes. These tubes are visually identical to the '60s and '70s Sylvanias. Like a vintage Sylvania, they display a lot of gain and midrange/top-end brilliance. They sound great in a cathode follower stage. Real creamy when driven. A preamp tube that can cut through the mix. Not as textured as a '60s RCA, but possessing a more authoritative body. Can be a problem tube; prone to microphonic behavior and fizzy static that comes and goes at will. A nice vocal tube that will color a soundstage with its usage. We use them as a standard tube in our 3x10" and 4x10" amps to give some bite to the darker Mojo 10" Alnico speakers we use.

RCA 12AX7/7025: Through the years, RCA produced a few different types of 12AX7s. Typical late '50s types would have a large, ribbed plate that is dark gray/black in color, branded 12AX7 and 12AX7A. This vintage is rich and appealing sounding with a very musical midrange bloom. Round, warm bass with a compelling, lacy top end (don't say I didn't warn you). Very similar in construction and tone to contemporaneous Tung-Sols.

Existing examples are increasingly rare and unpredictable in performance. Can be a noisy tube. You know that "noise" test point on your TV-7/DU? It was put there because of this tube. Expect to screen this tube before using; average samples will frustrate you with uneven performance. Early-mid '60s–'70s examples of the RCA 12AX7 are my favorite in this family of tubes. They have a shorter ribbed plate structure and the actual plate is lighter in color. Very broad, even response. No frequency group is accentuated... everything sounds even and in perfect harmony. Most true NOS examples exhibit goodly amounts of gain, though they are not as forward as Sylvanias. Particularly fine with Fender guitars. Bouncy and expressive. Perhaps the best sounding hi-fi tube around, right up there with Telefunkens; expect to pay up for real, tested, NOS examples. I dearly love these in our 80212s. They tickle the ceramic magnet 12"s we use beautifully. The 7025 versions are screened, tested versions that were specified by many manufacturers because of their low noise, audio intention. Physical characteristics of this tube are virtually identical to the classic Mullard type. I've seen Mullard, Brimar, Amperex and GECs that appear to be identical to the RCA 7025. Most European examples have a seam at the top of the envelope. Performance is interchangeable as well. I suspect that the lauded Mullard "M" series of tubes are hyper-tested examples of this classic RCA structural platform.

Telefunken ECC83/12AX7: Known through two versions, ribbed and flat plate. Has achieved mythical status, certainly perceived as the benchmark 12AX7 by hi fi nuts, and with good reason. Even scope pulls can test way above minimum and still sound terrific... It's not unreasonable for this tube to last for more than 100,000 hours! Lots of *Oomph* when used in a guitar amp... rich and three-dimensionally complex. Its note decay is very musically textured and fine. Breaks up with an encompassing, balanced presentation. Transparent sounding, giving a broad, even response. Richly toned without sounding muggy. Lots of air and space, a delicate authority is achieved.

Very commonly found in old hi-fi sets and industrial test gear. Not as common in American guitar amps like Fender, but seen quite often in off brands like Guild and Silvertone. Often rebranded with the amp manufacturers name. These work great in higher gain amps like Marshall, where noise would preclude the use of a more "colorful" tube.

Ei ECC83/12AX7: A new Eastern European type currently being produced in Serbia. This tube shows lots of promise. Extremely similar to the Telefuken in construction. Can be discerned by the seam on the top of the envelope; the Tele doesn't have this. Sonically, the Ei is a bit more colorful than the Tele. Can be edgy and harsh sounding, with a very forward midrange. Terrific tube in higher gain rock amps like a Matchless. Must be carefully selected for noise and microphonics. Aggressive sounding with a definite bite. Not a subtle, refined tube. Expect to reject 50 per cent for noisy operation. Not as musical as a GE or RCA, but still very good. Many are being rebranded by fast-buck hustlers as Telefukens. Watch it!

Sovtek 12AX7: Available in a number of versions. Early types are 12AX7WA and WB. Recent examples labeled WXT and WXT+. Higher distortion and gain than RCA and Mullard types. This characteristic endears them to users looking for a distorted, aggressive attack. Not as quiet as the Chinese 12AX7, but possessing a lower noise floor than common US made GE and ECG types. Very reliable, long-lasting tube. Lacks real voice and character when utilized in old style Fender tweed and BF era amps. Somewhat lifeless and flat sounding in amps where tube disposition can be discerned. Great tube for amps that derive their tones through circuitry manipulation, ie multi-stage cascading gain amps, built-in effects amps, "lead" amps with buttons and knobs that say things like "pull thick," "drive," "solo" and "thrash." Not microphonic or particularly noisy. Really a "plug it in and forget about it" tube. Good for guys and gals who don't want to be bothered by fiddling around endlessly with their tubes...

Shuguang 12AX7: Chinese made, current production in limbo. Last batch reputedly made in mid 1997. Very quiet, well-made 12AX7. Reputedly made using Mullard or MOV equipment. Rich toned, if a bit lacking in low end and mids. Nice sense of space and air around notes. Fairly high gain, good for modern amps. Its low noise floor makes it good for critical high-gain applications. The downside to this tube is its longevity – or lack thereof. These things wear

Technical articles

out quickly, getting smudgy and flat; as if someone threw a blanket over the tone. Easily recognizable by the shiny metal stiffeners between plate structures. Always house branded; common guitar brands include Mesa, GT, Fender, and many others. Designer versions can get pricey. Recent versions of this tube include a shiny plate type dubbed a 7025. Some folks really swear by this variant. If cathode is the same as other Chinese versions, it's gonna have the same longevity problems. More musical than the Sovtek, but not as compelling and textured as an NOS GE or RCA. Very common as recent Fender and Mesa OEM type.

GE 5751: Originally a 12AX7 with beefed up internal elements for operation in action-man military and industrial applications. Five star versions are highly prized for their careful parameter control. Most examples in stock these days are the JAN stuff from the mid '80's. Very good side to side matching. Very tubey sounding tube. If you're trying soften up a harsh modern amp, this may be the "go to" tube. Don't over use it; too many 5751s in the tone zone will deaden things, as there is less presence and air to their top end. Nice and clear sounding – kind of a soft brilliance. I like them best as input amplifiers replacing he 12AY7 called for in '50s Fenders.

ECG 5751: I don't have a lot of experience with this one… I would expect a greater reject rate than the GE's based on my experience with all the ECG stuff.

RCA 5751: A very nice tube, but very hard to find. Be careful these days when purchasing NOS examples. I suspect that RCA didn't make a whole lot of these; consequently, you risk buying used stuff unless you know what you're doing. When you do find them, you'll be treated to a rich, full sounding tube that is detailed and three-dimensional. Very musical. The GE sounds thinner when compared directly… the RCA has more body and depth.

This list hardly represents a complete assessment of every preamp tube available, but is intended to address to sonic characters of the most commonly found types. The basic structural platforms are the GE, ECG/Sylvania, RCA (two types), Telefuken, Reflector(Sovtek), and Shuguang. The 12AX7 type tubes available today will no doubt be an example of one of these. I hope my impressions of the differences will be of value to you and enlighten you to the tonal manipulation possible with this benchmark tube type.

Mark Baier is founder and proprietor of Victoria Amplifier Co, and is a noted philosophizer and essayist on all things of tonal concern to guitarists.

VACUUM TUBE RECTIFIERS

By Myles Rose

People have many opinions on rectifiers in vacuum tube guitar amplifiers. Some feel that solid state rectifiers are more reliable. This is true in one sense, but gives vacuum tube rectifiers an "implied" unreliability. I find vacuum tube rectifiers rarely fail. If they do fail, it is usually a case of physical damage, or the rectifier being bad in the first place, and failing in newly made amplifiers where the rectifier failed due to bad manufacturing. In the case of the Chinese rectifiers, when new, I have seen a 50 per cent failure rate. A bad rectifier will usually show itself in one of two ways: (1) power light comes on, but *no* sound at all comes from the amp; (2) when you flip the amp off standby and into play when all seemed fine, your fuse will blow.

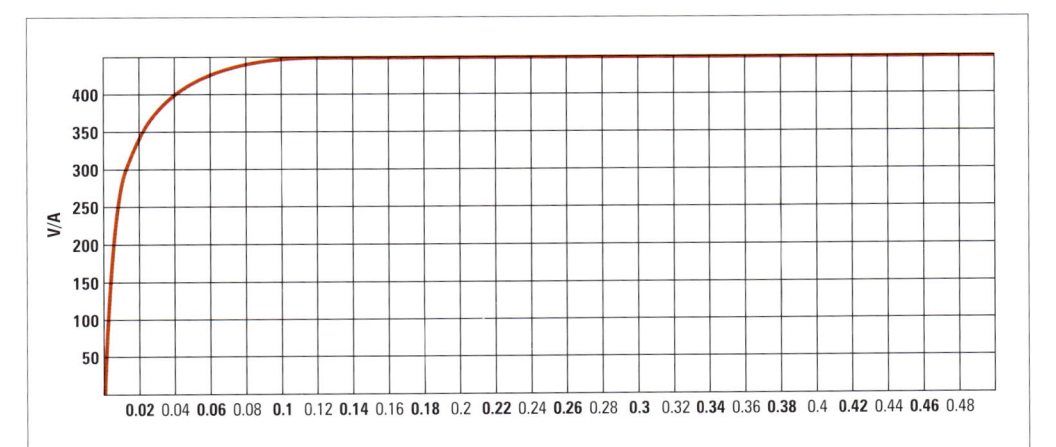

Solid state diodes: With the solid state rectifier, we had full power in 0.1 seconds. Our voltage was 449.08 as previously stated.

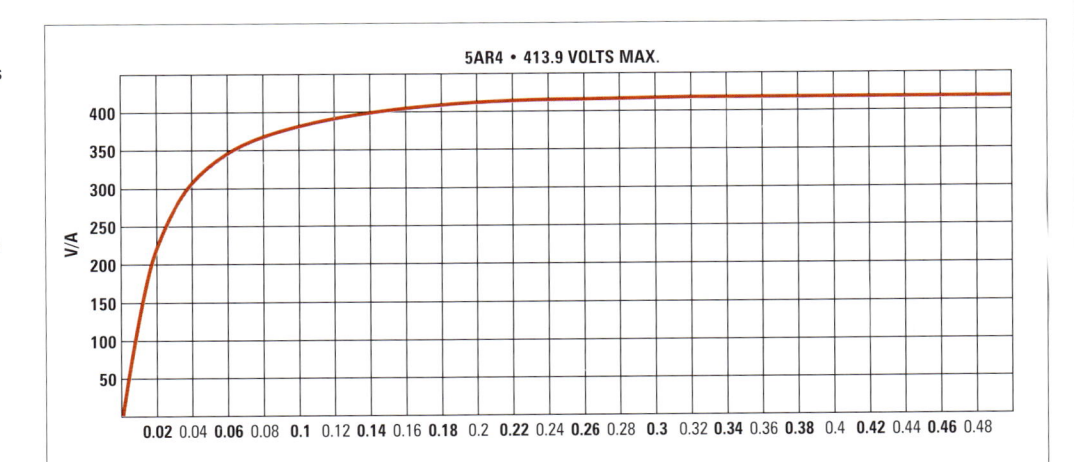

5AR4 • 413.9 VOLTS MAX.

5AR4: A 5AR4 vacuum tube rectifier came next. This is one of the strongest vacuum tube rectifier units used today. It took about two times as long as the diodes, 0.2 seconds, for our maximum voltage to be developed, in this case 413.9 volts. This is a fairly large drop from the earlier solid state rectifier – but is pretty standard for the drop between solid state and tube rectifiers.

There are rectifiers with 4 pins and 5 pins, of the same type. The four pin rectifier may be indirectly heated, and not work well in some amp designs, so if a 5 pin rectifier comes out of your amp, put a 5 pin rectifier back into your amp. The rectifier converts the AC line current coming into your amplifier to the DC voltages that are needed. A solid state rectifier is easier to fit into an amp design, can provide more power, does not run as hot, and is less expensive than its tube counterparts.

The Marshall JTM-45 used a tube rectifier, but when Marshall came out with the 50W version of the amp, a solid state rectifier was a change in the design. Most Fender amps over 40W use a solid state rectifier.

If you have an older amp with a tube rectifier and want to replace it with a solid state replacement, be sure to first check your amp and make sure all the other components of the amp are in good condition. The solid state rectifier is capable of higher voltages, and weak

THE TUBE AMP BOOK 121

Technical articles

5U4-G • 357.06 VOLTS MAX.

[Chart: V/A vs. time (0.02–0.48 seconds), curve rising to ~340]

5U4-G: Going to a 5U4-G, commonly used in some Mesa Rectifier Series amps, we found that it took 0.44 seconds to develop full voltage. This is a long time compared to the solid state device, and twice as long as the 5AR4. This is almost a half a second, and most people can hear something that is 1/2 a second long in duration or delay. Our high voltage was 357.06 volts, not far short of 100 volts less than the solid state device.

5R4-GYB • 530.93 V

[Chart: V/A vs. time (0.02–0.48 seconds), curve rising to ~320]

5R4-GYB: The 5R4-GYB, often used in place of the 5U4-G, was the next tested. It showed about the same rise time as the 5U4-G above, but had a maximum voltage of 330.93 volts. This makes a nice rectifier in some Fender Deluxe type amps that have been made with everything from the little 5Y3 to a GZ34 in the past. It's a great blues rectifier in these amps.

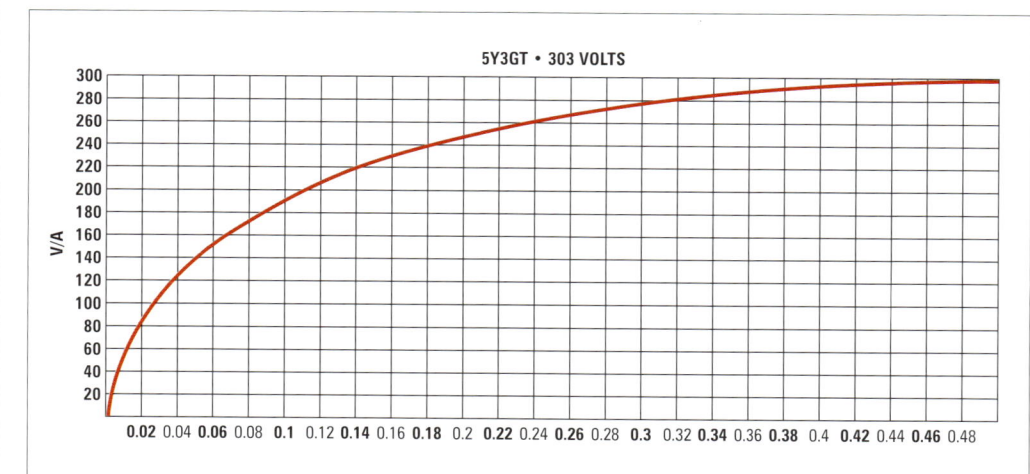

5Y3GT • 303 VOLTS

[Chart: V/A vs. time (0.02–0.48 seconds), curve rising to ~300]

5Y3GT: Lastly, we tried the 5Y3GT, a common and popular rectifier in smaller vintage amps. Rise time was almost a full half a second at .44 seconds. Our maximum voltage was 303 volts.

resistors or caps will wear out and blow faster after its installation.

As you will see in the charts below, different rectifiers have different characteristics. In many amps, you can get a different sound and feel by replacing the rectifier with different types. Matchless is one example that lets the user choose various tube rectifiers as part of their features. A solid state rectifier will give very fast rise time and response, as the voltages are produced very quickly. A vacuum tube rectifier will yield more to the player's touch dynamics, sound warmer and less harsh by some folks feelings, and give the compression and sustain in a much different way than its solid state brother.

With a tube rectifier, when the player initially hits a loud note or chord there is voltage sag, in some cases a *lot* of sag. As the note or chord starts to decay, the voltage then builds, and what you have in essence is a built-in compressor/sustain device. If you look at the charts, you will see how fast the voltage is developed with a solid state rectifier versus a tube rectifier.

Using the same circuit, we replaced the rectifier section. We used the same voltage input in all cases, but in the case of the 5Y3GT rectifier in the last test, our transformer output of 333v had to be reduced to 330v, as the 5Y3GT would have been pushed just a touch beyond its design limits. This change had negligible results on the final outcome.

The tests were measured over a period of 500 milliseconds, or half a second. Using the commonly configuration of 1N4007 diodes, it's easy to see – in comparison with the vacuum tube rectifiers – the difference in rise time to get the voltage we are after. Also, this is a somewhat less apparent factor, but in this case with the solid state rectifier, we had 449.08 volts available for our B+ voltage, the highest voltage in the group. This is also something to keep in mind when you replace a tube rectifier with a solid state replacement, as your output tube bias will probably need to be checked. In the case of a class A amplifier such as a Vox AC15, AC30, a Matchless amp, most Carr amps, or others, the bias is automatically taken care of, but in an older class A amp you may also want to be concerned about the output transformer and capacitors if they are old or original. The amp will be running at higher voltages, and in some cases *much* higher voltages, as you will see later.

Now you have a little more information on rectifiers, and how the changing of types in a rectifier equipped amplifier can change the sound of the amp, the power of the amp, and the characteristics of touch, feel, and sustain. In most cases, the NOS rectifiers are better than those of today. The Chinese units still have a long way to go to reach the specs of those built in the USA and western Europe in the past. The Russian units are the best of the new ones, and

Technical articles

the JJ factory is currently working on a re-release of the GZ34 which shows promise. Mesa Boogie products currently use mostly the Chinese variety.

As a safety note: in rectifier swapping, you can generally go "down" but not always "up". This means you can put a 5U4 or 5Y3 in an amp that had a 5AR4, but it is not recommended to put a 5AR4 or 5U4 in an amp that had a stock 5Y3. The increase in voltages may be too much for the other components in the amplifier – particularly if it is a vintage amp.

Rectifier substitutions & cross reference:
GZ34 = 5AR4
GZ32 = 5U4 = 5V4GA
GZ31 = 5U4GB
GZ30 = 5Y3GT

Typical Rectifier voltage drops
5Y3   –60 volts @ 125 mA
5U4GB   –50 volts @ 275 mA
5U4   –44 volts @ 225 mA
5V4   –25 volts @ 175 mA
5AR4   –17 volts @ 225 mA

OUR PHILOSOPHY ON AMP MODIFICATIONS

One of the most appealing aspects of vacuum tube amplifiers, which you lose with solid-state designs, is the organic nature of the technology. You can get *into* tube circuitry – change capacitors, resistors and other components – to personalize the sound. After all, this is no more than amp designers do, and it's the difference from one amp to another.

Solid-state technology has all these same components, but because they operate at very low voltages, they can be combined in integrated circuit chips. This is a very *rigid* technology: it can do a lot, but what you get is what you get. It's difficult to modify the chip – or, for that matter, a solid-state amp.

Vacuum tube technology operates at high voltages. Therefore it is **VERY DANGEROUS**. A very real risk is damaging or destroying your amp. If that's not bad enough, you can seriously damage or destroy **YOURSELF**. The moral is: always make sure that any work done on your amp is done by a qualified technician. (*Again, read our* **Warning** *at the front of this Technical section; if you aren't experienced in this stuff already, don't even open up your tube amp – take it to a professional.*)

There are lots of guys like this around, and more all the time. In fact, some of them have become highly skilled and creative. They are able to offer a significant range of sonic and functional variations from stock designs. We know most of them, because sooner or later any guitar tech with aspirations of being a designer has to come to terms with tubes in his life. The more aware tech will automatically install good tubes on any significant modification, perhaps in the same spirit in which a dentist will insist on cleaning your teeth. Basic health begins with good hygiene.

The real truth is that changing your tubes is the most basic modification you can do or have done to your amp. Another real truth is that very often you can get what you're looking for in your amp without further effort – or expense. There's a whole range of differences you can explore, simply by plugging and unplugging. A good tech or dealer can give you a fresh look at your own music and capabilities in this way.

It makes sense to try this first, since many mods are not easily reversible. Then if you want something more, you ought to consider this: amp modification is an art form, just like playing guitar or bass. There are guys who are skilled, talented and creative. There are guys who are still learning and make mistakes, and guys who are never gonna learn – just like guitar and bass players. They all do it a bit differently, and often have strong convictions about their art. Does that sound familiar?

Once you've decided to modify your amp, a whole new world of complexity opens up to you. Often it's probably neither necessary nor worth it. Remember that almost *any* mod may reduce the value of your amp, and many deliberately sacrifice reliability and/or tube life to achieve its sonic effect.

For some it may be magic. At Groove Tubes, we're all inevitably friends and relations with top designers and mod shops. We exchange tips and stories back and forth in a way which quietly helps musicians generally. We've begun a policy of testing some of their mods ourselves, as part of our tube development process. We have discovered some which are specifically meant to *increase* reliability and/or tube life. We publish many of these on the following pages.

We also have published what we think are the best of the master volume mods for Marshall (there are lots of really bad ones) because we think enough players can benefit from this knowledge, although I'm not a real fan of the master volume-type distortion. However, this is a personal decision to modify your amp and not for us to offer opinions. Just make sure your tech is qualified, and that the sound you're after is one you want to live with, not just an attractive effect.

Of course, all these mods are different from model to model and manufacturer to manufacturer. A Marshall is not a Fender, and a Boogie is no longer a Bassman. As a result, there are literally hundreds of mods we've heard about. In particular, we advise avoiding any unnecessary modifications to collectible, vintage amps (or any that might soon *become* collectible), which will almost certainly devalue them instantly. Before undertaking any mod, consider what you want from the result and research what other good tube amp might already exist out there that will do this for you. The simplest answer might be to buy a different amp that already has the sonic characteristics you desire – and to sell yours to someone else who will like it just the way it is.

A WORD ABOUT BIASING YOUR AMP

The bias control on an amp is much like the idle control on an engine. That is to say, there is an optimum point of bias for an amplifier that allows for good sound and maximum tube life. This optimum bias point will be different with a new set of tubes no matter which tubes you buy, so the bias should always be checked when changing power tubes. (Preamp tubes are self-biasing.)

We have described below some symptoms of an amp that is biased incorrectly. If you suspect your amp is either "under-biased" or is "over-biased" after installing your new tubes, turn the amp off and take it to a qualified serviceman for a bias adjustment. Do not attempt to bias the amp yourself, as you may make things worse – the bias control is not a user-serviceable adjustment.

The charge for a bias adjustment should cost no more than a minimum bench charge ($10 to $20) so it's cheap insurance for your new tubes and you'll get the most sound out of your amp.

Under-Biased Amp
Amp is "idling too high" so tubes are running very hot, causing them to burn out fast and possibly to short out. The tubes' plates (large gray metal housing) will glow red from heat and the amp will lack punch and might hum.

Over-Biased Amp
Amp is "idling too low" so tubes are running cool. Amp will sound dirty at any level and will sound low on power. This type of distortion is called "crossover" distortion. Crossover distortion is a non-musical type of distortion and isn't as pleasing to hear as "harmonic distortion."

Correctly-Biased Amp
Amp will sound clean and tight at low to medium levels. When pushed to maximum, amp will produce an even harmonic distortion – musical distortion, if you will. When installing good matched, tested, correctly biased tubes, expect an overall tonal improvement along with better balance and sustain.

Biasing an amp with a Variable Bias Control
The proper method of biasing a tube amplifier requires a signal generator, an oscilloscope, a volt meter, and preferably a dummy load resistor. First, remove power tubes and measure the bias voltage at the grid (usually pin 5). Adjust the voltage to the largest negative voltage and

Technical articles

install power tubes – the tubes are now at the over-biased position and are running cool with lots of cross-over distortion. Now apply a 2000-cycle signal to the input of the amp and connect the proper impedance load to the output. Turn volume of amp up to 70 per cent output, or just prior to clipping, and get a picture of the signal on the scope. Notice the notch indicating cross-over distortion in the signal. Adjust the bias control gradually until this notch just disappears. The amp is now properly biased.

Biasing a Preset Fixed-bias Amp

A preset fixed-bias amp may be adjusted – it's just more trouble than an amp that has a variable control in the bias circuit (note that the term "fixed bias" refers to all grid-biased designs, though some are variable and some are preset at the factory). Some of the most common preset fixed-bias amps are Boogies, Hiwatts, some Ampegs and some Fenders. All Marshalls and most Fender do have adjustable bias controls. So why doesn't every amp-maker use an adjustable bias? Good question: the answers we get from the companies aren't very good ones. It's probably because they wanted to save money in production. If the company can buy tubes with uniform performance, then they can set a bias point that will be pretty close for the average power tube. Unfortunately, when the player has to change tubes, he has to find a set of tubes that are identical to the last set of the tubes or the amp won't be properly biased.

In order to change the bias of a fixed-bias amp, you should consult a qualified tech who can change the resistor values in the bias circuit to achieve proper bias voltage. This can be done by using a substitution box and changing just one resistor, or a capacitor, in some Ampeg amps. (If you are using a matched set of Groove Tubes and the amp is running too hot, you might simply change to a lower performance number and solve the problem. Likewise, if the amp is running too cool, try a higher number and the amp should perform correctly. This is simpler than modifying the bias circuit, but you are limited to the range of Groove Tubes performance ratings you may choose from. If you do need to change the bias point to accommodate your chosen set of Groove Tubes, at lease you can get that same rating time after time and not have to worry about biasing your amp again.)

About the bias probe

There is a new device available called the Bias Probe. It is an inline adapter which is plugged in between the amplifier and an output tube and directly monitors the total current flowing through the tube. This reads out directly on a digital voltmeter (DVM) and may be adjusted easily to the required settings which are provided in the Bias Probe manual. A correctly-biased amp will have an exact current range the tube will draw, so therefore the measurements made with the Bias Probe are repeatable and quick to get.

This method offers the best of both worlds: it's fast, accurate, and requires a minimum of test equipment. Perfect for the band on the run or the music shop that doesn't have a full-time tech on staff and wants to be sure a customer's amp is correctly biased when selling a new set of tubes.

The Bias Probe is a product of JHD Audio and is available through them, or you may order one directly from us. Any DVM that reads a 200-MV range will do nicely.

BOOGIE BIAS POT INSTALLATION
By George Sayer, Amp Labs, NYC

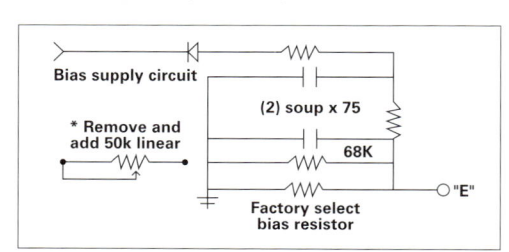

[Bias supply circuit diagram: * Remove and add 50k linear, (2) soup x 75, 68K, Factory select bias resistor, "E"]

The majority of Mesa/Boogie's amps can be easily rebiased for all variations of matched output tubes available by simply installing a 50k linear trim pot in place of the fixed resistors used in factory production (* above). I have found that their are often much preferable tubes to use in Boogies of the past to those instaled as stock, and changing to a variable bias circuit will allow you to experiment with some of these. In some cases, especially the Simul-Classes, the individual 220k bias feed resistors at each tube socket may need to be varied slightly, say 180k or so, to accommodate the difference between 6L6/5881s and EL34s – but usually just the overall bias control will suffice.

Tube life and amplifier output can be greatly maximized with the installation of the variable bias resistor (a trim pot). Let your Boogie live life to the fullest – get it biased!

More info or questions? Call George at Amp-Labs, NYC, (516) 618-2568.

GT SURVIVAL TIPS TUBE AMPS

Tube amps are simple, and so they are easy to keep running smoothly. If you neglect to follow a few simple rules, however, you can buy yourself some expensive trouble. Here are some things you can try that will put your amp in top condition and keep it there.

TIP #1 – Speaker Impedance

The proper matching of the impedance between your tube amp and speaker is extremely important. Improper matching will cause severe tube wear and is a common cause of early tube failure. Some amplifiers are more sensitive to this than others. Among the most sensitive are Marshall amps. Pay attention that the Marshall's impedance selector is on 16 ohms when you're running a common 16-ohm Marshall cabinet, and reduce it to 8 ohms when adding a second identical cabinet. Always check your cabinets by measuring with a volt meter on the ohm scale (these meters read low, i.e.: an 8-ohm cabinet might read 6 ohms while a 4-ohm cabinet could read 3 ohms). Another way to determine the impedance of your cabinet is to read the individual speaker impedance and note how they are wired.

If there are two 8-ohm speakers wired in parallel (output + to both + 's and output - to both -'s) then the cabinet will be a 4-ohm load. If the two speakers are wired in series (output + to speaker #1 +, #1 - to #2 +, #2 - to output -) then the cabinet will have a 16-ohm load. In other words, parallel wiring halves the impedance of the speakers while series wiring will double it.

Find out the specified output impedance of your amp by asking a service station to check it on a volt meter, or perhaps your local dealer. The common amps are: Marshall, variable 4, 8, and 16 ohms; Fender Deluxes and Princetons, 8 ohms; Fender Twins and Dual Showmans, 4 ohms; Fender Super/Reverbs and 4x10" amps, 2 ohms.

Beware the dangers of using a power attenuator with your Marshall, as most power attenuators do not match impedances closely enough for these amps – though some newer types by THD and Marshall themselves are significant improvements on older designs. Using a power attenuator might let your Marshall distort at lower levels, but at the expense of much more rapid output tube wear – premature failure of the output tubes is common in Marshalls used with power attenuators. Fender amps are not as sensitive to power attenuators as Marshalls, because of differences in design in the output section. However, since the tubes are putting out full power into the attenuator, they will wear out quicker than if they were just coasting at a moderate output level. If you like the sound you get with the attenuator, be prepared to spend a little more on power tubes.

TIP #2 – Power Tube Replacement

The regular replacement of power tubes is normal in amps with regular use. Just when to change them can vary with the type of use the amp gets and how often it's used. Most players should change their tubes once a year if they play moderately loudly and fairly often. As the output tubes wear out, both the bass and treble responses of the amp will begin to suffer. This power loss from worn-out tubes isn't always noticeable because it occurs gradually over time, and because power level differences aren't easily noticed. It takes twice the power for the

Technical articles

ear to hear just 3 db more, and that's just barely audible! Worn tubes will usually have poor, mushy bass response. Regular power tube replacement will guarantee consistent and reliable performance. It's cheaper in the long run.

Tip #3 – Driver/Phase Inverter Tube Replacement
The driver tube operates in conjunction with the power tubes to form the power-amplifier section of the amp. The best power tubes will sound bad with a weak driver tube, as this is the tube that controls the output tubes. If it cannot control the output tubes, the amp can't sound its best. This will show up particularly at higher-power playing, or when playing the amp distorted. *I advise people to replace the driver tube whenever replacing the output tubes!*. In most amps, the driver is the smaller tube (12AT7, 12AX7, 7025, 12AU7 or similar), which is adjacent to the output tubes.

TIP #4 – Retensioning Tube Sockets
NOTE: Because the tube sockets are connected to the very highest voltages in the amplifier, we suggest that the following work be done only by those having the proper knowledge of electrical safety as regards working on tube amps. When tubes are changed again and again over time, the sockets' female parts begin to stretch and not make good tight contact with the tube pins. This can lead to arcing and intermittent connections between the tube and the amp. This condition can be aggravated by the vibration from your speakers and so may occur on certain notes on your guitar or keyboard. You can correct this by replacing the socket (last resort) or by retensioning the socket with a large safety pin, jeweler's screwdriver, or small ice pick. Use a tool with an insulated handle if at all possible. First: disconnect the amp from the AC outlet and allow the amp to drain off any voltage by leaving your speakers hooked up to the amp with the standby "ON." This takes just a few minutes and could save an awful experience later. Now remove the tubes and notice the contacts located inside each pin hole of the socket. These contacts spread the pin hole, and you want to gently push the contact to close the hole just slightly – do not push the contacts in so far that the tube will not re-insert. After you've re-tensioned all the contacts, replace the tubes and notice how much tighter they are held.

You may also find corrosion on the contacts. Try spraying a little contact cleaner on a tube and inserting it into the socket a few times. This will improve the connection to the tube and prevent future corrosion.

TIP #5 - Capacitors and Resistors
The most common problem we see in tube amps (other than tubes) is worn-out capacitors and bad resistors. What follows are some common symptoms of bad resistors and capacitors, why they can go bad, and how to locate and fix the problem.

NOTE: Tube amplifiers contain high voltages which may be lethal, even if the amp has been off for some time. We do not recommend that you open your amp, or try to perform any repair operations unless you are properly trained in electronic servicing. Again, there are large voltages present in your amplifier that can kill, even with your amp unplugged from the wall. Having said all that, you may now read on.

A common result of cheap tubes failing is that they will take out a screen grid resistor with them (usually located across the inside of the tube socket, or near by). These take the heat when the tube shorts and can fall out of specification easily. This will cause improper function of any power tube you place in the faulty socket – if the resistor is open, the tube may as well not be in the socket! In any case, the amp will not be reliable until the screen grid resistor(s) have been replaced. Fender amps usually have a 1-watt 470-ohm screen grid resistor, while Marshalls generally use 5-watt 1000-ohm resistor for this purpose. The screen grid resistors can be checked using an ohm-meter to measure their resistance. The measurement should be within 10 per cent of its marked value.

Another common source of poor sound quality would be worn-out filter capacitors in the output or supply stage of the amp. This is especially common in amps over ten years old. These are fairly large components and are often mistaken for "metal tubes" at first glance. These caps filter out the 60-cycle hum from the power source and through the years they dry out and filter less and less. As the 60-cycle hum is now present in your audio output, it will create an odd harmonic that will seem to follow your notes up and down the scale. It's almost like having somebody singing off-key all the time. In addition, since the amp is now producing sub-harmonic notes, the power is sapped and the overall response of the amp will become weak and sound mushy.

Inspection of filter caps can usually determine if they are bad. These large metal cylinders are easy to spot. Fender amps have them on the underside of the chassis, between the transformers, covered by a 4" x 6" metal pan. It is therefore not usually necessary to remove the amp chassis from the wood cabinet. As in our tube socket retensioning tip, first disconnect the amp from the AC outlet and allow the amp to drain off any voltage by leaving your speakers hooked up to the amp with the standby "ON." Remove the pan and "drain" the remaining voltage from the capacitor by touching a screwdriver from the hot side of the caps to ground, or attach a bleeder resistor to do this even more safely. Now inspect the top side (or positive) of the part, looking for a broken or swollen seal. This can look like a little bubble about to pop, or it could have already burst and have powder coming out. Capacitors have this relief seal to expose when they go faulty. Be sure to replace them with the same value (or greater value) and make sure they are placed with the proper polarity.

Marshalls have their filter capacitors placed upright on the chassis held at the base with a clamp. The chassis must be removed from the wood cabinet to inspect the filter caps. Observe the same procedure for inspection of the capacitors. It should be mentioned that if you replace your filter caps, unless you are using pre-formed caps, many techs advise you should connect your amp to a Variac and power the amp up slowly to allow the caps to charge and form properly. ■

WHAT NOT TO DO!
Reprinted from Tom Mitchell's excellent book How to Service Your Own Tube Amp. The entire book and an accompanying video is available from the GT Catalog.

It is always surprising to me how much misinformation (or total lack of information) there is regarding the safe and proper methods of operating and maintaining audio equipment. To set the record straight on some of the most prevalent misconceptions, the following is a list of what *not* to do.

Do Not Void Your Warranty
Tampering with the innards of an amplifier will void the warranty. As long as the warranty is in effect, you can get your amp fixed for free – so why waste a valuable benefit like that? This is not to say you can't perform your own maintenance such as cleaning pots and contacts, changing tubes, retensioning the tube sockets, and so on. Remember, doing your own repair work and modifications will void the amp's warranty. After the warranty expires, you're on your own. Then you can feel free to start tinkering.

WD-40 Is Not A Cleaner
Neither is Gunk, rubbing alcohol, or water (the use of any of these products will cause serious damage). If you want to clean a potentiometer, switch, or contact, use "tuner cleaner" or a spray that leaves a silicone lubricant behind. Radio Shack 64-2315 tuner cleaner, or Cramolin R-2 (and R-5) are excellent choices (Cramolin being the preferred type). When cleaning tube sockets, however, I recommend using a contact cleaner that leaves no residue behind, such as Radio Shack 64-2322, Rawn contact cleaner, or Caeon 27 (Caeon being the preferred type).

The lubricant in tuner cleaner, combined with the high voltages present in tubes, attracts dust

Technical articles

and airborne contaminants, many of which are organic in composition. This means that their chemical compositions are based on the element carbon, which is used in the construction of resistors. In a short period of time, the foreign material will build up and bridge together adjacent pins on the tube socket, causing a phenomenon called "arc." Watching this happen is very much like watching a small fireworks display, the result being a destroyed tube, burned up tube socket, and possibly other problems as well. The point here is, use a cleaner that leaves nothing behind. The converse of this rule is also true. Do not use non-lubricating contact cleaner in potentiometers. It will dissolve the grease in the pot bushing that allows the shaft to rotate. The result is frozen pots. Use tuner cleaner in pots, use contact cleaner in tube sockets... PERIOD.

Do Not Set The AC Voltage Selector Incorrectly

Not all amplifiers have selectable line voltages. For those that do, these are the facts. The power transformer is designed to provide a specific output voltage from the secondary, given a specific primary (or power line) voltage. The circuitry in the amplifier is designed to operate at this pre-determined voltage. Incorrect setting of the selector can cause voltages far below or far above those required for normal and safe operation.

I have heard the surprising (and dangerously incorrect) notion that setting the selector at 220 volts with a line voltage of 120 volts gives you a good distortion sound. Doing anything like this (or, even more dangerous, selecting 120 volts with 220 volts on the power line) can wreak havoc with your output tubes and transformer due to incorrect bias, and incorrect B+ (power supply). If you plan to do this, prepare for massive failure that even an experienced technician will have difficulty servicing. If you feel like doing this, take a deep breath and wait until the feeling passes. In other words, the rule is this: if you plan to monkey with your voltage selector... DON'T. Use your multimeter to find out what the line voltage from the wall socket is, set your selector to the nearest voltage below what the meter reads, and leave it there.

Do Not Use A Variac

A variac is a variable AC source (vari-AC). It is a special type of transformer with a large knob which permits the operator to select an output voltage anywhere between zero and 130 volts AC, given a 120 volt AC input. The consensus on using a variac with an amplifier is that you can get great distortion at 90 volts (or 130 volts). The facts here are the same as setting the voltage selector incorrectly. If you can afford to buy a new amplifier every week, go ahead and use one of these, but be advised that you can get a better sound out of your equipment by installing new tubes often, checking the bias, and performing regular maintenance check-ups. A variac is used by technicians to test amplifiers by slowly bringing the AC power up and monitoring the amp's symptoms on test equipment. The improper use of variacs is one of the things that keep technicians in business, due to the massive failures that occur from their use by non-technicians. The best way to use a variac is to not use a variac.

Do Not Set The Impedance Selector Incorrectly

Matching the impedance of your speakers to the output impedance of your amplifier is a critical adjustment. Some people have the misguided notion that putting an 8 ohm load on an amp with the impedance selector set to 16 ohms somehow increases the output power (by some sort of "magic," I suppose), or that running a 16 ohm load with the selector at 4 ohms gives you a "good" sound at a lower volume. Regardless of the myths, one thing that doing these things will do is to cause the power tubes to work too hard (due to the impedance mismatch), resulting in premature tube failure and possibly output transformer damage as well. If the amplifier was designed to operate at a specific output impedance, by all means make sure that the speakers used are the correct impedance. The amplifier was designed to operate at its peak efficiency when the output is properly loaded.

Do Not Mix Speaker Cabinet Impedances

When connecting more than one speaker cabinet to an amplifier, make sure that they are all the same impedance. When connecting two similar cabinets together to the same amplifier, the impedance is divided by two (this is only true when the impedances are the same). This formula does not apply to speaker cabinets of dissimilar impedance, as the method of calculation is different. For example, connecting an 8 ohm cabinet and a 16 ohm cabinet together to a common amplifier will result in a combined impedance of 5 1/3 ohms. Since there is no such impedance selection available on any amplifier, an impedance mismatch would occur, resulting in a potential for damage to the amplifier's output circuit. Choose your impedances wisely.

Do Not Operate An Amp Without A Speaker Attached

One of the fastest and surest ways of causing serious damage to an amplifier's output circuit is to operate it without a load (speaker) attached to the secondary of the transformer. When the impedance selector is properly set and the correct load is attached to the speaker jack, the impedance required by the plates of the output tubes is correctly balanced by the load (speaker) that is connected to the secondary of the output transformer. In the absence of a load on the transformer secondary, an imbalance occurs.

Bear in mind that the speaker is actually a part of the amplifier circuit, and removing it is as potentially damaging as removing any other component from the circuit inside the amplifier. As the speaker operates, it reacts with the circuitry within the amplifier. Without the reaction from the speaker the output circuit becomes extremely unstable. Output tubes, sockets, associated components, and the output transformer will most likely become smoldering casualties when the amplifier is operated without a load.

Never Use Guitar Cable As Speaker Cable

This is a problem that is not really an obvious one. Everyone does this at one time or another without realizing the potential for damage. Another way of describing guitar cable is by its more appropriate name, "shielded cable." Typical shielded cable has a single insulated wire in its center completely surrounded by a flexible metal braid. This braid is the cable's "ground" and shields the internal connector (called the "hot" lead) from external interference (usually a 50-60 cycle hum). While this is ideal for reducing noise for guitar inputs, this construction gives rise to problems when used in an output circuit. Owing to the large amount of conductive surface area provided by the shield, it follows that this type of cable has a large amount of internal capacitance.

Given this fact, the longer the cable, the larger the capacitance. A typical 20 foot long guitar cable can add sufficient unwanted capacitance to the output circuit to interfere with the performance of the amplifier. Since this capacitance reacts with the impedance of the speaker, an impedance mismatch occurs and the potential for circuit damage is present. An additional problem exists. In some less expensive shielded cable, the internal "hot" lead is very narrow gauge wire. Speakers operate at high current levels, and since this small wire is designed to operate with mike levels signals (.025 volts is typical) it is not capable of handling speaker level (in excess of 100 volts peak-to-peak in typical of a 100W amp into a 16 ohm speaker). In some extreme cases, the insulation on the "hot" lead in the cable literally gets hot, the insulation melts, and the "hot" lead shorts to the shield, effectively shorting out the amplifier. Amplifiers were not designed to run into a dead short, and damage can (and most likely will) occur. Use quality speaker cable that is at least 16 gauge. Twelve gauge is preferred (the smaller the gauge number, the thicker the cable), and keep the cable length as short as possible, just long enough to reach from the amp to the speaker cabinet.

Never Use Speaker Cable As Guitar Cable

Fortunately, if you do this, no damage will occur. Speaker cables are not shielded, and so are not constructed to block out external interference. This is just a performance tip. If you plug your guitar into your amp and you hear a loud hum, but the guitar still works, you probably mistakenly used a speaker cord in place of a guitar cord. This is a mistake that you will probably only make once (more often if your guitar and speaker cables look alike).

Do Not Exceed Fuse Ratings
When a fuse blows, your amp is trying to tell you something. A blown fuse means you have a bad amp, not a bad fuse. A fuse is a protection circuit designed to interrupt the flow of current when the fuse's current rating has been exceeded. If you power your amp up and the fuse blows, replace the fuse with a new fuse with the same current rating. If it blows again, the amp needs immediate service. If the fuse doesn't blow, you should perform some routine maintenance checks to the amp (bias incorrectly set, loose tube sockets, etc). If you don't have the right type of fuse, do not use an automotive 30 amp fuse, or wrap foil around the bad fuse. This invites trouble. In fact, if you decide that you are going to do it in spite of this advice to the contrary, invite some friends over with marshmallows and wienies because your amp is going to go up in smoke.

No Soldering Guns
Use a soldering iron, not a soldering gun. A soldering gun has a blunt tip, is clumsy to use, and concentrates too much heat on the work being soldered to be practical for electronic work. Using a soldering gun can cause permanent damage to circuit boards and components. Always use a soldering iron.

Disconnect AC When Not Measuring Voltage
It seems so obvious, and yet it's easy to forget to do this (and remember: even with the amp switched to "off," 120 volts is still entering the chassis at the power switch point unless you have unplugged it from the wall). Of course if you do forget, you run the risk of having a "shocking" reminder never to do it again. In fact, you may not get another chance to make this mistake twice, since you run the risk of being electrocuted.

Isolate Yourself From Earth/Ground
When working on electronic equipment, perform your work on a wooden workbench, wear rubber-soled shoes (Converse high-tops will do nicely), and perform your work in a room with a wooden or carpeted floor. Doing all of these things reduces the risk of fatal electrocution, should you forget to disconnect the AC from the equipment being serviced. Still use extreme caution for these procedures only reduce risk, they do not eliminate it.

Probe With One Hand
Keep your other hand in your pocket or behind your back. Also make sure that the hand operating the test probe never touches the chassis. The only thing that should be making an actual physical contact with the circuitry is the tip of the test probe. If you were using two hands, with one hand holding the amplifier chassis, which is ground, and the other hand holding the test probe and your hand slipped and touched the power supply, your body would complete the connection, and the power source would discharge directly through your heart. The rest would be history, and so would you. You should use the recommended clip leads to avoid having to use two hands to make voltage measurements. Clip one lead to the chassis of the amplifier and leave it there. Now you only need to use one hand and one probe to take your voltage readings.

Check Power Supply Filter Capacitors For Stored Charge
Even with the AC disconnected, the potential for electrocution is still present. It is the function of capacitors to store charge, and in the case of the power supply filter capacitors, they do an amazingly good job. Although some amplifiers have resistors attached directly across the capacitors to give the stored charge a way to "bleed off" (these are called "bleeder resistors"), it is not unusual to see an amplifier that does not have these resistors. If there are no bleeder resistors in the power supply, as in some Marshall 50W amps, it is common to find a stored charge of many hundreds of volts still in the capacitor even though the amplifier has been off for hours. To remedy this, technicians use a power resistor with clip leads on both ends to bleed off the charge.

Use caution, however. Even after the charge has been bled off, the charge may reappear after the bleeder is removed, even though you did not turn the amplifier back on. This is due to "soakage." The dielectric in the capacitor "remembers" the charge that the capacitor used to have, and it causes the charge to return. Monitor the filter caps periodically, and leave the bleeder attached until you power up again.

Observe Component Polarities
Electrolytic capacitors and rectifiers are both polarized components. That is to say that they will only function properly in a circuit when the current is flowing through it in one direction. This also means that if the component leads are reversed in the circuit, catastrophic failure is certain to occur. Electrolytic capacitors are labeled with a positive terminal and a negative terminal. If it is installed backwards in an amplifier circuit, which has voltages in excess of 400 volts, count on an explosion. Rectifiers are also polarized components. They have a band at one end and this band tells you which direction to install it in the circuit. A rectifier permits current to flow through in one direction, but blocks current flow in the opposite direction. If installed in reverse it will definitely result in failure not only to the rectifier, but most probably to other circuits as well. If you must replace one of these types of components, make sure the new one goes in the same way that the old one came out.

When Removing Wires, Leave Clues For Yourself
Whenever I need to replace a transformer, I clip the wires off where they attach in the circuit so that about half an inch of insulation is left. Transformer leads are color-coded, and this saves me a considerable amount of time when I reinstall the replacement. It is simple to find where the new wire attaches, since a small piece of color-coded insulation remains where the old wire used to be. Wiring to tube sockets is usually color coded as well, and this trick works when a tube socket needs to be replaced.

Before Disassembly, Draw A Picture
Although the phrase has been over-used, a picture is worth a thousand words. If you are going to remove a tube socket for example, draw a picture of what it looks like before you begin. Show where the grid resistors attach, how the wiring looks, color codes, and any other relevant details. When you begin to reconstruct the circuit, you know in advance how it is supposed to look when the job is complete.

Use Your Sharpie
One tool that I use very often is my "Sharpie" indelible ink marker. The ink from this pen resists heat extremely well, and I use it for writing directly on tubes. With tubes installed in their sockets, I draw a "register mark" on them that starts about half an inch from the base and extends off the tube, on to the socket, and ends up on the chassis. When I must remove a tube for some reason, the register mark saves me time when I plug the tube back in. I line up the mark on the tube with the mark on the socket and I can push it right into place without having to fumble around trying to align it properly. I also write the date that I installed the tube, so that there is no mistaking how old it is when it comes time to replace it.

A Sharpie can be used to make a reference mark on a reverb tank, so that the connecting lines get reconnected properly, and it can also be used to mark the polarity of a rectifier or capacitor on a circuit board to facilitate replacement, assuring me that the new part goes in observing the correct polarity. Of course, a Sharpie can be used as a prod to check for loose components, since it is made of plastic and is non-conductive. By using a little imagination, a Sharpie can definitely become a time saving tool.

Do Not Handle Hot Tubes With Your Bare Hands
This is another tip that seems so obvious, and yet everyone is guilty of doing this at some time.

Technical articles

Grabbing a hot tube with your bare hands is a good way of building up calluses very quickly, if blisters don't bother you much. This hurts about as much as grabbing a hot soldering iron with your bare hands, which is also a little foolhardy. If you must remove a tube while it is still hot, do so with a towel, or an oven mitt. The preamp tubes don't get nearly as hot, but caution is still recommended. Either protect your hands before grabbing a hot tube, or wait about ten minutes for the tube to cool off. Handling a tube while it is still hot can also cause damage to the tube if it is handled roughly. Hot tubes are fragile, so handle them with care.

Do Not Measure VOLTS With A Multimeter Set To OHMS
A multimeter is a sensitive instrument and can easily be damaged if used incorrectly. Measuring high voltage with your multimeter inadvertently set to a low ohms setting can (and usually will) cause permanent damage. Always anticipate what you will be measuring and set your multimeter accordingly before you take a reading.

MAINTENANCE CHECKLIST
By Tom Mitchell

Disconnect The AC Line
Before disassembly, the AC line must be disconnected. Doing so will permit you to avoid electrocution.

Check For Stored Voltage In The Filter Capacitors
Power supply filter capacitors can store charge for days. If these capacitors have no way to "bleed off" the stored charge, it is quite possible that days later many hundreds of volts are still present. This stored charge is potentially lethal. Make sure you discharge the capacitors through a bleeder resistor.

Check For Loose Connections
Try moving each component with a non-conductive probe (like the Sharpie that you've heard so much about). Physically move each interconnecting wire and check for bad connections on either end. Finally, make sure that all screws, bolts, nuts, and jacks are tight, since ground connections to the chassis are typically made through them.

Check For Faulty Or Damaged Components
Visually inspect each component for signs of cracking or over heating. Electrolytic capacitors loose their electrolyte through the positive end. Check for signs of leakage.

Check For Arc On Tube Sockets
When a power tube goes bad, it often burns the socket that it is plugged into. The black carbon residue left behind is called "arc." This arc must be removed because it causes a conductive path between two adjacent pins on the tube. In extreme cases, replacement of the entire socket is recommended.

Tighten The Tube Socket Pin Connections
The connections in the tube socket become slack with time and do not make a good connection, especially when tubes are replaced frequently. Loose tube sockets are frequently the reason for socket arc. Remove the tubes. Using a jeweler's screwdriver, apply a small amount of pressure on the pin contacts until they move in towards the center slightly, effectively reducing the diameter of the contact. Repeat with all the other contacts. Spray some non-lubricating contact cleaner into the socket and replace the tubes.

Clean Potentiometers, Contacts, And Switches
Spray tuner cleaner into each potentiometer and rotate each shaft no less than five times (more if very scratchy). Spray cleaner into each input and output jack and insert and remove a plug into each jack no less than five times. Finally, spray cleaner into all switches (except power and standby) and operate no less than five times.

Check Bias
At this point it will be necessary to reattach the AC and power up the amplifier while disassembled. Make sure that the speaker is attached before powering up. Check the bias (or balance) and readjust, if required.

Reassemble
Make sure that you disconnect the AC power and speaker line prior to reassembly.

THE AMPLIFIER SIGNAL CIRCUITS
Reprinted from Electric Guitar Amplifier Handbook by Jack Darr

The first thing to do in servicing an electric guitar amplifier is to look it over carefully. Find out what it isn't doing, and, just as important, find out what it is doing. As in all electronics work, the diagnosis is the hardest part. First, look for what is working; this will give you an idea as to what is not.

Troubles in amplifiers will fall into three classes: it will be dead, weak, or sound funny. This last class covers hum, oscillation, motorboating, and similar things; in other words, the amplifier is making the noises itself when it should not. The first thing to decide is in which class this trouble falls. Make the easiest possible test: turn the instrument on and listen to it. A normal reaction in a functioning amplifier is a slight rushing sound in the speakers called "blow" (as if you were blowing very softly into a microphone). Some slight hum is also normal, especially if the input connections are open and the volume controls are turned up. All tubes should show a little light in the top, but none of them should get red hot or show any flashing between the elements. If you see the latter, turn off the amplifier immediately; there is a short somewhere in the power supply. Also, if you hear a loud hum, smell smoke, or see smoke coming from under the chassis, turn it off.

Take the easiest problem first – the completely dead amplifier. Nothing happens when you turn it on. This means that some part has completely broken down, and it is easy to find. Simply check out all circuits in the amplifier, beginning with the power supply. It doesn't take long to find a bad component with the proper tests. These tests are listed in later paragraphs. First, look at the statistical order of failures in this kind of electronic equipment. The author's experience in actual repair operations indicates certain troubles are more likely than others. The experienced technician checks them in the order of frequency of occurrence, and he finds the trouble faster. Likelihood of failures come in this order:

1. Tubes
2. Power supply
3. Components – resistors, capacitors, and controls
4. Cables – plugs and wiring between the guitar and the amplifier
5. Transformers – output transformers, speakers and power transformers. Remember this list, and use it; it will make the repair job a lot faster. If you find a completely dead amplifier, the first thing to look for is a bad tube. The second most likely source of failure is something in the B+ power supply, and so on in the order given.

In checking electric guitars, break the complete system down into three parts: the amplifier itself (including the speakers), the connecting cables, and the guitar (including the pickup and controls on it). Here is how to check it out. First, pull out all cables to the instruments and mikes. Turn the amplifier on, and listen for any signs of trouble. See if you can hear the normal blow or hum that means it is alive. If you do not, check at any one of the inputs. Turn its volume control all the way up, and touch the hot terminal of the jack. If the amplifier is working, you ought to hear a very loud buzz or honk noise in the speaker. There are two easy ways to make this test: One, plug in one of the cables, and touch the tip of the phone plug; this is always connected to the hot terminal on the jack. Two, make up a special test plug with the hot wire brought out to where you can touch it with a fingertip. Try this on an amplifier you know is working, and you will recognize the sound the next time you hear it.

Technical articles

If you hear a loud buzz, chances are the amplifier is all right, so go to the connecting cables. Plug them into the amplifier one at a time, and touch the center conductor of each cable. Again, the loud buzz says this section is functioning. Before going on, however, flex each of the cables near both connectors while touching the center conductor. Any static or break in the buzzing sound indicates there is a problem in the connector. Look over each one carefully for poorly soldered connections, broken wires, and strands of wire shorting across the connector. If these occur, you will need to repair or replace the cable.

When the connecting cables have been eliminated as the source of trouble, only the pickup remains. About all that can be done here is to substitute a new unit, and discard the original or have it rewound. Since these usually are sealed units, it is probably quickest to replace the pickup if replacements are readily available. If you don't hear a loud buzz when you touch a hot input terminal, the amplifier is dead. Check to be sure that all volume controls are turned on in the channel you are testing. (In all servicing, you must watch out for the obvious; it is easy to overlook. For instance, if you are not careful, you may take the amplifier out of the case looking for a dead stage, and then find out that the master gain control had been turned off. Don't laugh – it has happened!)

If you can't get a sound through the amplifier from one input jack, try another one: try them all, in fact, before you pull the amplifier chassis out of the case. It may be that one channel is dead and the others are all right. If no sound is heard at all, pull the amplifier. More detailed testing will have to be performed.

Set the amplifier upside down on the bench, and make sure that the speaker is still connected. In high-powered amplifiers you can overload the output tubes and burn up a very expensive output transformer in about one minute if the amplifier is turned on without the right load (the speakers) connected. In some amplifiers you may have to rig up extension wires, but this is easy. Transistor amplifiers are even more critical than the high-powered tube-types, on this point. There are two basic circuits used: the output transformer type and the transformerless type. Each has its own individual requirements. With the class A single-transistor type, do not turn the amplifier on with the speaker open. These will stand a short across the output, but operating the amplifier with the output open may blow the output transistor(s).

The output-transformerless types are exactly the opposite. They can withstand an open circuit in the output, but a short across the speaker terminals will blow both output transistors in a fraction of a second. When you hook up extension cables to the speakers or to a dummy load resistor, be very sure that there are no dangerously exposed bare wires, such as wires twisted together for extensions. Use only well-insulated test leads with insulated alligator clips, etc.

The reasons for this circuit peculiarity will be discussed in detail in the section on power-output testing.

Checkout Procedures

Here is a step-by-step method of testing that will show you where the trouble is in the least possible time. This is based on actual field experience in repairing these amplifiers, so follow it as closely as you can.

1. Check the B+ voltage – most of the troubles will be found in the power supply.
2. Check the amplifier, stage by stage, for voltage on plates and screen grids. Use a DC voltmeter for testing, set on a scale that is at least twice the maximum voltage you expect to find. For example, if there is about 250 to 300 volts of B+, then use a 500-volt scale to save damaging the meter.

When servicing, always start at the output – the speaker and output tubes – and work your way back toward the inputs. Why? Because this is the fastest way! No matter what you find in the early stages of the amplifier, you can't tell if the basic trouble is fixed unless the output tubes and speakers are working. So, start at the output. Check back through the circuit, fixing all troubles as you find them, and when you get to the input, the amplifier will be working.

When you make voltage measurements, watch the meter reading and listen also. When you touch the voltmeter prod to the plate of an amplifier tube, you will hear a small pop in the speaker if everything is working past that point. This pop won't be very loud when the plate of the power-output tube is touched, but it will be on the control grid. So, if you get the right voltage on the plate and screen, touch the grid with the prod. This should give you a louder pop, for you have gone through the circuitry of the tube, which amplifies the tiny disturbance you make when you touch the grid with the prod.

Signal Tracing

As you go toward the input, you will hear louder and louder pops. This is one of the oldest methods of troubleshooting known – it was worked out back in the early days of radio where they called it "the circuit disturbance test." It is still just as good as it was then, since it works every time. As you go along, watch for the stage where there is no pop. That is where the trouble is!

Fig. 4-1 shows a partial schematic of a typical amplifier. The idea is to check the signal path by listening for pops as the voltages are being checked. This path starts at the input and goes all the way through in a plain series circuit. Each time the signal passes through an amplifier stage, it gets louder (more amplification). Anything that breaks the chain will stop the signal right there. You can see the test method: Start at the output and work back toward the input. The numbers in Fig. 4-1 show the correct sequence. For this test it is assumed that the power supply has been checked and the right B+ voltage found at the filter output (X). If the output of the power supply is correct, the power supply itself must be all right, and the trouble must be somewhere in one of the amplifier circuits.

As an example, look at a typical case of trouble. Suppose you get good loud pops all the way up to and including test point 6. At point 7 you get a pretty weak pop, and hardly any at all on 8 (the control grid). This means that the trouble is somewhere in the preamp stage, and everything from there on is normal. The first step is replacing the tube – this is done simply because it is easiest and the problem could be a bad tube. If this doesn't help, leave the new tube in, at least until the trouble is found.

Next, measure the DC voltages around the tube plate and screen. Assume that the plate

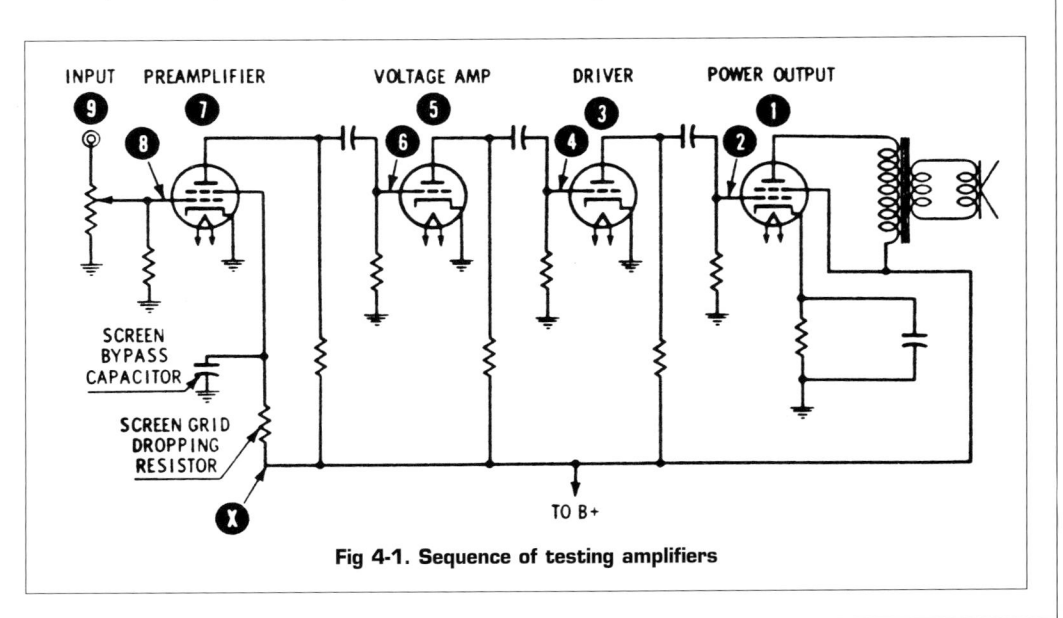

Fig 4-1. Sequence of testing amplifiers

THE TUBE AMP BOOK 129

Technical articles

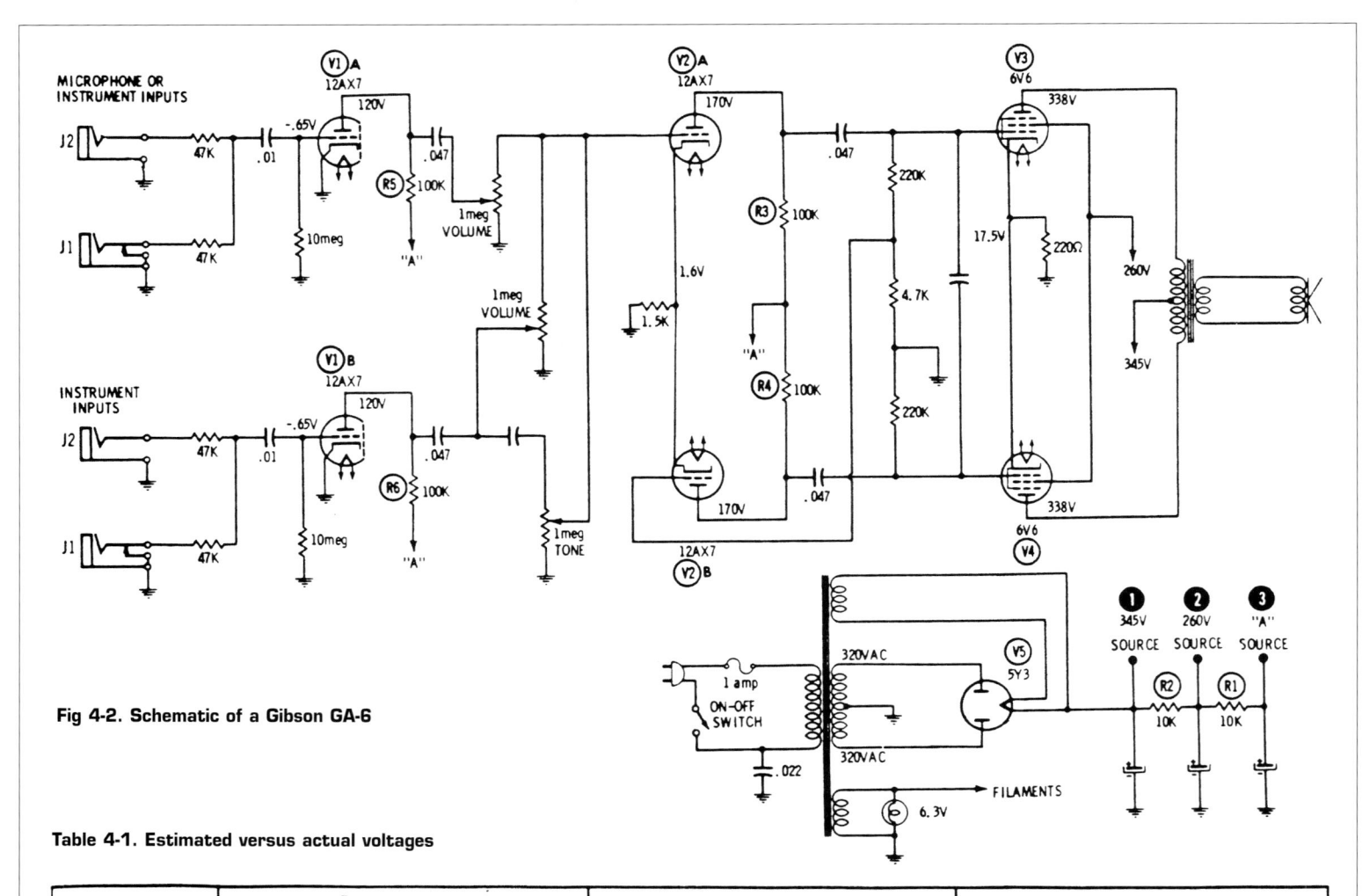

Fig 4-2. Schematic of a Gibson GA-6

Table 4-1. Estimated versus actual voltages

Tube	Plate Voltage Estimate	Plate Voltage Actual	Screen Voltage Estimate	Screen Voltage Actual	Cathode Voltage Estimate	Cathode Voltage Actual
V1	120V	120V				
V2	160V	170V				
V3, V4	345V	345V	270V	260V		
V5					350V	345V

voltage is nearly all right, but there is no screen grid voltage at all. This means that there is one of two troubles: an open screen-grid dropping resistor, or a shorted screen bypass capacitor. Either one will give the same symptom. Now you start to eliminate. (All of this work is a straight process of elimination. Just keep testing until you find the bad part, once the defective stage has been isolated.)

First, measure the supply voltage at point X (the supply end of the resistor) to be sure that it is there. When you look at the schematic, it would seem that the earliest checks of the power supply output would also check out this point. Remember, however, there are wires connecting the various common points in the chassis, so these wires have to be eliminated as possible points of failure, at least indirectly. Assuming the proper voltage is present at X, the fault must be in the screen grid dropping resistor or the screen bypass capacitor. Turn the set off, and take a resistance measurement with an ohmmeter from the screen-grid tube pin to ground. If the capacitor is shorted, there will be a zero reading here – a dead short. If the capacitor is good, you will get a reading of the resistance of the screen dropping resistor plus the resistance to ground through the power supply. A normal reading here is something like one to two megohms. In this kind of circuit the screen dropping resistor is usually 820,000 ohms to 1.2 megohms.

This resistor could be open, so you take your next measurement directly across the resistor itself; the reading should be the rated value. All resistors are color coded to tell what size they are supposed to be. The ohmmeter reading must agree within 10 per cent of this. If this resistor reads completely open, there is the trouble! Replace it with another of the same size and wattage, and turn the set on. The screen grid voltage reading will be normal. The input will now pop as loud as it should, and the input jack will give a very loud buzz or honk when touched with a fingertip – if the resistor was the only source of trouble.

Determining voltages without service data

In the previous section, you made a voltage analysis of the amplifier using information gained from the schematic diagram of the amplifier. However, at times schematics are hard to find. Now see what can be done if you must test an amplifier circuit without this information. Fortunately, most of the amplifiers are conventional and since they use the same basic circuits, you can use a model amplifier for comparison. It has been done for many years. Servicing is easier if you have the service data, of course, but you can still test an amplifier and find the trouble if you know what each stage is supposed to do and how it does it. That is the

130 THE TUBE AMP BOOK

Technical articles

reason for so much detail in the first section. How can this information be used in checking an unknown amplifier?

Fig. 4-2 shows the schematic of a commercial amplifier. This one isn't actually unknown, but it will serve as an example. What should the normal voltages be? Incidentally, there is a very valuable feature in your favor when checking voltages in vacuum-tube amplifiers: tolerance. A tube voltage can be inside a certain range, and still be all right. For example, a tube plate voltage rated at 100 volts can measure from 90 to 110 volts and still operate without affecting the performance of that stage. This is a 10 per cent tolerance; many voltages have 20 per cent or even slightly more. The only voltage that is really critical is the grid bias.

When you start on the unknown amplifier, the first thing, as always, is the supply voltage. Check the B+ voltage, at the filter input (point 1 on the schematic). How much should it be? A very accurate idea can be arrived at by measuring the ac voltage on the plates of the rectifier, and converting. With a normal load it can be assumed that the rectified voltage will be 10 to 20 per cent above the rms voltage on the plates. In this one you will find about 320 volts rms on the plate, so an added 10 per cent will give about 350 volts at the rectifier cathode for a guess.

In the circuit shown, a 10,000-ohm resistor (R2) is used as a filter choke, giving a fairly large voltage drop. The circuit indicates that the plates of the output tubes are connected directly to the rectifier output (filter input) through the primary winding of the output transformer; their plate currents will not flow through the filter resistor. This connection provides more voltage on the output tube plates; it also results in more hum. However, this hum is canceled out in the push-pull output transformer, so this circuit is a practical arrangement to get a bit more plate voltage and consequently more output. Here the power-tube plate voltage will be very close to the voltage found on the rectifier cathode, or about 345 volts, since the only drop is in the output transformer.

What should the voltage be at the filter output (point 2)? There screen current for two 6V6s is being drawn through this resistor and also the plate currents of the first two tubes, 12AX7 twin-triodes. From the tube manual, screen current is about 4.0 mA for the pair of 6V6s, and 1.0 mA each should be a fair average value for the first four triodes. This gives about 8 mA current, which, by Ohm's law, is an 80-volt drop across 10,000 ohms or 270 volts at point 2.

When you examine the B+ circuits further, you will find another 10,000-ohm filter resistor (R1). This one carries only the plate currents of the preamplifier triodes (previously estimated at 4 mA for the four), so the drop across it is 40 volts. This gives an estimated 240 volts at point 3. Plate currents in voltage amplifier stages average from about 0.5 mA to 1.5 mA. The 1500-ohm cathode resistor connected to tube V2

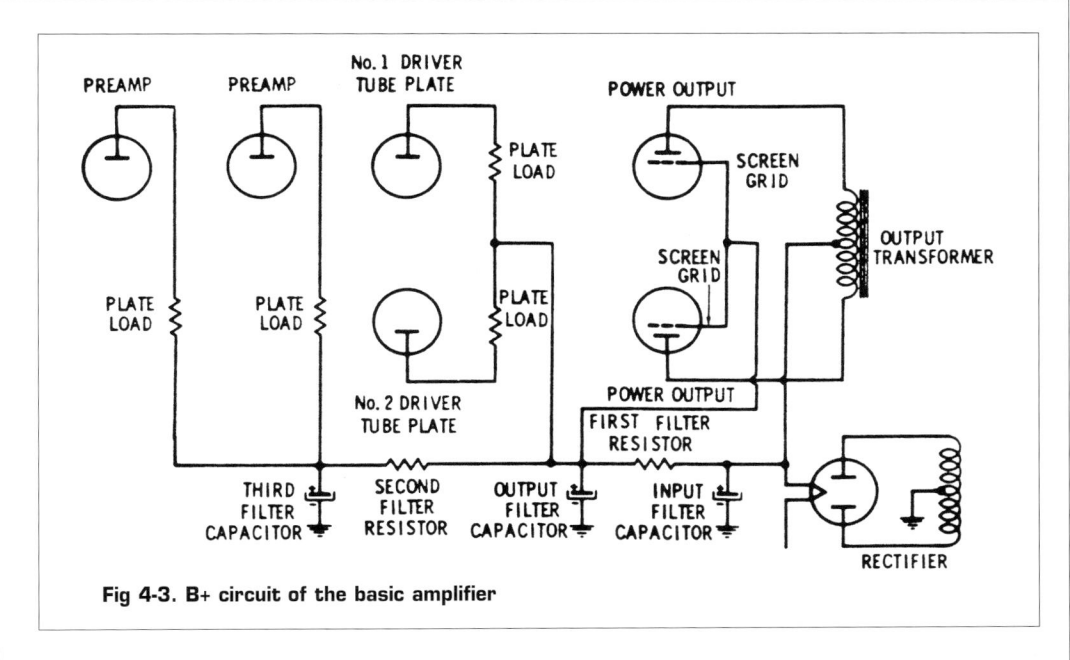

Fig 4-3. B+ circuit of the basic amplifier

makes the negative bias higher than on V1, and reduces the tube current. If the plate current in V2 is assumed to be 0.8 mA and 1.2 mA is assumed for V1, the drop across load resistors R3 and R4 will be 80 volts, and across R5 and R6 it will be 120 volts. Since the estimate for point 3 came out 240 volts, the plate voltage on V2 will be 160 volts, and on VI it will be 120 volts. Thus, approximate readings for all points in the B+ circuit have been obtained.

When estimates are compared with the manufacturer's published data (Table 4-1), they turn out to be reasonably close – within 10 per cent, in fact. Things will not always work out this well, but you see that it is possible to estimate all B+ values using some educated guesses and a tube handbook.

How do you know the size of the plate load resistors? They are color-coded so you can tell at a glance. Just find the plate connections on each socket, and look at the color coding on the resistor connected there. Any electronics handbook will tell you what the colors mean. You can also get an idea of what the normal plate voltage should be from the typical operating conditions table given for each tube in the tube manual.

Fig. 4-3 shows the complete B+ supply circuit for typical amplifier stages, beginning at the first place where DC voltage appears, the rectifier cathode or filter input. Learn this circuit; it applies to all amplifiers. If it is a bigger amplifier, there will be more; smaller ones will have less. It is always the same basic circuit. You can lift it out of the amplifier, mentally, and follow it through to see if there is any trouble in the plate voltages.

Localizing the trouble

Performance tests provide an easy way to find out just where trouble is. In other words, see just how much of the whole amplifier is working, and then concentrate on checking the part that isn't. It is easy to do. Turn the amplifier on, and make voltage and pop tests through the circuit, beginning at the output. The first time you go through a stage and it doesn't pop, there it is.

Take a typical trouble and see how to pin it down. For instance, assume the output stage and the B+ supply in Fig. 4-2 are all right, but either the amplifier does not work, or it has a very bad tone. On pop tests you find that the plate of the upper driver tube (V2A) has a pretty weak pop, and the grid of the same tube has hardly any at all. Obviously, something is wrong, but what?

Check the plate voltage; instead of the normal 170 volts or so, there is about 50 volts. This pinpoints the trouble as being somewhere in the driver stage. The first thing to check is the tube, so replace it – simply because this is the easiest thing to do, and experience has shown that tubes cause a lot of troubles. However, the results are the same, so the tube must have been all right.

To proceed, look at the B+ supply circuit in Fig. 4-2. Note that the plate voltage of this tube is fed through a 100,000-ohm plate-load resistor (R3). Turn the amplifier off, and measure the resistance of this resistor. If it has opened up or increased in value, the symptoms would be exactly what have been described. However, it checks right on the nose at 100,000-ohms, so go on to consider other possibilities. The supply voltage at the bottom or line end of the resistor is all right, because it measures the same as the screen-grid voltage on the power-output tubes checked earlier.

Summarizing the situation, the load resistor is all right, the tube is all right, but still there is not enough plate voltage. The only condition that can cause these symptoms is too much plate current being drawn through the load resistor, since it will also cause too large a voltage drop. The plate voltage is dropping across the resistor instead of across the tube. A tube draws too much plate current when the grid bias is wrong,

Technical articles

so measure the voltage on the control grid. It ought to be zero; there is no bias voltage fed to the grid from any external source, and the 1.0-megohm grid resistor goes directly to ground.

To measure grid voltages you must use a high-impedance meter – a vtvm or high-resistance vom, since this, like all grid circuits, is a very high impedance. A low-resistance meter will cause the voltage present to be incorrect, since the meter itself acts as a shunt.

Assume that there is about 5 volts positive on the grid. This is definitely wrong. No grid in this type of amplifier ever reads positive if it is in good shape. It will be either zero or slightly negative. A 5-volt positive bias on a grid will cause the tube to draw a very heavy plate current; thus, the plate voltage will drop very badly because of the excess drop across the plate-load resistor.

Where could this voltage come from? Only a one-megohm resistor and a coupling capacitor are connected to this grid. The resistor goes straight to ground, so this is not a very likely source of voltage; however, the coupling capacitor is connected to the plate of the preceding tube, and this tube as about 120 volts positive on its plate. This is a likely suspect.

In all cases a capacitor must be a completely open-circuit to DC. The capacitor is used to transfer the AC signal voltages from the plate (output) of one stage to the grid (input) of the one following; it must always block any DC from getting through. (Although the correct name for these is coupling capacitors, you will find them called blocking capacitors in some cases.)

From the symptoms that have been assumed, it looks as if the capacitor must be leaking DC onto the grid. To make sure, disconnect the grid end of the capacitor, and hook the DC volts probe of a vtvm to the open end. Now turn the amplifier on. If the capacitor is leaking you'll read a positive voltage on the open end. This should be zero, of course, since a good capacitor is a completely open circuit for dc. A normal capacitor with good insulation will give just a very slight kick of the meter needle as it charges up. Then this reading will slowly leak off through the input resistance of the meter. If you have any residual reading, any voltage showing at all after the first charge has leaked off, the capacitor is bad and must be replaced.

Fig. 4-4 shows how this test is made. With the capacitor hooked to the grid resistor as in the original circuit (Fig. 4-2), you will probably read 5 to 6 volts dc. With the capacitor disconnected, you may read as high as 35 to 40 volts positive dc on it if it is leaky. The input resistance of the vtvm (11 megohms average) is much higher than the 1-megohm grid resistor. If you use a vtvm for this test, set it on a low DC volts scale. If you use a vom, set it on a voltage scale that will carry the maximum voltage to be read. In this case it is the 120 volts on the preceding tube plate. You can't blow up a vtvm with a voltage overload, but you can damage a vom, so be careful. After the first charging kick, set the meter to a lower voltage scale. For the final test use the lowest scale available; even one volt positive through a coupling capacitor means it must be replaced.

You cannot make a leakage test with a common ohmmeter. The actual leakage through these capacitors is very small. If you could measure it, the resistance would go up to almost 100 megohms (far above the capacity of a service ohmmeter), but the capacitor will still leak enough to cause a lot of trouble. The voltage test is sure and fast, so use it.

Capacitor leakage is a very common trouble; that is why it is used as an example. It will cause loss of volume, a very bad distortion, and even damaged tubes if the leakage is bad enough. All of these problems result from the change in the grid-bias voltage. The amplifier tubes are driven into a very nonlinear part of their operating range, and the tone suffers very severely as a result. In fact, after a little practice you will almost be able to identify the problem by listening to the amplifier. Leaking coupling capacitors give the tone a characteristic muffled sort of sound that is easy to spot. Now examine the process that you went through and the methods you used to find the trouble. Can you see the orderly steps in the example just given? The amplifier was examined one stage at a time, until a stage that was not doing its job was located. You stopped right there, found that trouble, and fixed it, before going any farther.

You used a process of elimination to find the defective component. In electronics work there are always several things that can cause any given trouble. Did you notice that things were eliminated one at a time, until the faulty item

Fig 4-4. Testing a coupling cap for leakage

was reached? First the tube (it is the easiest), then the plate supply voltage, next the plate load resistor, and finally the real villain, the leaky coupling capacitor were checked out. There are only a certain number of parts in any circuit that can cause any given trouble. Patiently eliminate them one at a time, and eventually you will find the right one. You may find it the first time; on the other hand, you may have to go all the way, as you did in the example. Just keep on until you find it. Later in this book there are more elaborate tests using complicated equipment. However, you will find that in this, as in all other electronics work, the majority of the troubles can be located and fixed with only very simple test equipment plus a good bit of plain old common sense. This is because a very large percentage of troubles are simple ones – a dead tube, a burned resistor, a leaky capacitor, and so on. Even the more complicated troubles will have very simple causes.

Always remember the process of elimination, and use it. If you know how each circuit works, you can quickly find the one that is not working, and start from there. ∎

MODS FOR AMPEG

Conversion Of 12DW7 To 12AX7

Courtesy of the MTI company when they owned Ampeg.

Most of the common Ampeg bass and guitar amps used 12DW7 preamp tubes, which makes things a little sticky since the tube has been discontinued. Have no fear, however, because we have the official Ampeg changeover operation, and it's pretty simple to change over and use the very common 12AX7 type tube.

For Models VT22, VT40, V4, V4B, and V2 simply plug in the 12AX7 in place of the single 12DW7 in the preamp section.

For the SVT bass amp, change Resistor R-25 to a 4.7K, and Resistors R-7 and R-6 to 220K. Then simply plug in the 12AX7 tube (also fine are ECC83 and 7025). ∎

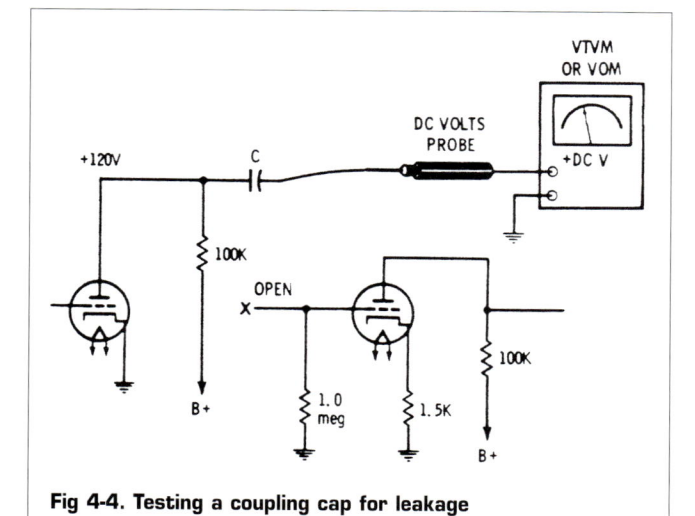

IMPROVING THE CATHODE BIAS CIRCUIT

Courtesy of Ken Fischer, Trainwreck Circuits

Most of the common Ampeg amps use a cathode bias system, unlike the Fender-style amps where the grid supply is adjusted to set the proper bias. The majority of these cathode bias amps use cathode resistors and bypass capacitors that are of too low wattage and voltage. A good rue of thumb is to double the wattage rating of the resistors and double the voltage rating of the bypass capacitors. Keep the

Technical articles

bypass caps away from the heat of the cathode resistor: if this cap shorts out, your tubes will be history.

Conversion Of Ampeg Amps From 7027a To EL34

Many Ampeg amps built in the '60s and '70s used a 7027A power tube. This tube has been discontinued for some time now, and so players with these amps are going to be in big trouble. These amps are the V4, V4B, VT22, and VT40 among others. Therefore we have engineered a few possible substitutions that can be easily made by any qualified tube amp tech. You can choose to modify your amp to use either EL34s or 6550s, depending on the sound and performance you're looking for .

7027A to EL34: Take the amp to a qualified tech and show him the schematic in this book along with the following instructions. Change the following resistors R41, 42, 47 & 48 (from 470 ohms 1W) to 1000 ohms 5W. Next, change resistor R49's value to a correct bias point while looking at a scope picture indicating the crossover distortion notch.

7027A to 6550: Use the same procedure as the above EL34 conversion and additionally change resistor R50 to a value of 82,000 ohms (82K). You will still have to change the value of R49 to the proper level to adjust the bias point correctly.

Converting An SVT From 6164B/8289A To 6550

The early SVT can be converted to use 6550 output tubes, resulting in a substantial reduction in re-tubing cost, with relatively little effect on performance.

The modified amplifier (prototype) produced more than 280 watts RMS into 4 ohms at below 3 per cent THD, line voltage maintained at 177 volts, 60 Hz. This is a reduction in output of only -0.30 dBR. A slightly softer clipping occurred at full output, and the amplifier biased slightly cooler. The modification consists of four stages:

1) Rewire the tube sockets for the different basing
2) Derive a new screen supply
3) Rescale the bias range
4) Change driver tube type

CAUTION: Be very careful! This amp has high voltages – much higher and of much higher current capability than most amps – it only takes one mistake!

Parts And Equipment Required:

To perform this modification, you will need:
– two 12AT7 tubes
– six 6550 (matched set preferred)
– six 1000-ohm 10W resistors

Several each of resistors in the range of 62K to 100K, 1/4W 5 per cent. Soldering iron, rosin core solder, DVM or VOM, scope and signal generator, 4-ohm 250W (or greater) dummy load, some degree of skill

Preparation:

Remove the power amplifier chassis from the cabinet, and using compressed air or a vacuum cleaner and soft brush, thoroughly clean the dust from the cabinet and power amplifier.

Remove the output tubes and drive tubes, and place the chassis bottom up, with the output tube sockets away from you.

Rewiring:

1) Move the gray or white wire from pin 1 (or 6, some amps) to pin 8, (6-sockets).

2) Move the orange wire from pin 3 to pin 4 (6-sockets).

3) Unsolder the six plate-cap wires from the circuit board, and using #22 or heavier-stranded hookup wire, wire each pad to its corresponding socket at pin 3 (6-sockets). Make sure that the wire that you use is rated for at least 600 volts DC.

4) Leave pin 5 alone (6 tubes).

5) Remove six 22-ohm 10-watt resistors from the PCB, and replace them with six 1000-ohm 10W resistors. These resistors should be R-31, 33, 45, 28, 39 and 42 on your schematic.

6) Looking at the foil side of the PCB, locate the common end of the six resistors just replaced. They are interconnected by an "I"-shaped trace, which has a trace branching to the screen supply ("E" on your schematic). The trace connecting to the screen supply must be cut. Be sure that you cut only the trace that feeds power to these six resistors!

7) Solder a piece of hookup wire on the foil side of the PCB, between the junction of the six resistors that were isolated in step 6 and the B+ source pad, which is located on the edge of the PCB to the left of the plate lease. Several red wires, including the red/white out transformer center tap lead, are attached here.

8) It is now necessary to change the values of R21 and R22, to allow the amp to be properly biased. The original value of these resistors is 22K. They are located on the end of the PCB which is away from the power tubes, and away from the bias pots. A good starting value for the new resistors is 82K (1/4W 5 per cent is fine). The final values in the prototype amp were 68K and 82K. The different values result mainly from the poor tolerance of the bias pots.

9) Replace the two 12BH7 driver tubes with 12AT7s.

10) With the output tubes out of the amp, adjust the bias pots for maximum bias (greatest negative voltage) at pin 5 of the 6550s.

11) Install a set of tubes (preferably an old set of "pulls") and power up the amp. (It will be necessary to jumper around the power-supply connectors to turn the amp on.) Monitor the bias test points, and verify that the correct bias voltage at the test point can be achieved – it should fall near center rotation of the bias pot. If the test point reads too high, increase the value of R21 or R22 as appropriate, to bring the bias within range. If it reads too low, reduce the value of R21 or R22.

12) Now it's time to see if the amp still works. Connect a suitable test load to the amplifier, and apply and input signal, while monitoring the output on a scope. Adjust the symmetry (balance) control for symmetrical clipping, and verify that the amplifier will output near full power into load. If it produces a proper output waveform at better than 200W, chances are you made no mistakes.

13) Remove the test power tubes, set the bias back to maximum bias, and replace the output tubes with the new set. (In an amplifier with this many output tubes, it is very important that the tubes used be matched.)

14) With a load connected, and no input, power up the amp and adjust the bias to produce 0.70 volts at test points.

15) Apply signal, and adjust the symmetry control for symmetrical clipping.

16) Put the amp back together – it's done!

MODS FOR FENDER

POST-TWEED AMPS NOISE REDUCTION

Courtesy of Ken Fischer, Trainwreck Circuits

On the post-tweed vintage Fender amps that use 1/2-watt preamp resistors: if the amp makes crackling, popping, or hissing noises, remove the first-stage preamp tube(s). If the noise stops, try a fresh replacement (two tubes in a two-channel amp). If the noise returns, it is not a case of defective preamp tubes. Instead, change all of the 100K-ohm preamp resistors and the chances are 95 per cent that the noises will go away. This is a very common problem with Fender amps.

THE TUBE AMP BOOK 133

Technical articles

Fender preamp mod for warmer, cleaner sound

The preamp section of many Fender amps uses a 250 pf disc cap, a .047 and a .1 ceramic cap. Try replacing the 250 pf disc cap with a silver mica cap of the same value and changing the .047 and the .1 with .02 or .022 polypropylene caps. This will improve the warmth and dynamic response of most Fenders. However, don't throw away the old parts in case you prefer the original tone to this mod. After all, it's still just a matter of taste. ∎

FENDER AMP CONVERSION TO EL34
Courtesy of Ken Fischer, Trainwreck Circuits

Any Fender amp that uses 6L6 power tubes can be converted to use EL34 tubes. While the useful life of the EL34 is somewhat shorter than the stock 6L6, the change in tone may make some players feel it is worth spending a little extra for this conversion. After all, beauty is in the ear of the beholder. (Be aware that, before performing this mod, you should ascertain that your power transformer's heater supply should be able to handle the extra current draw, as EL34s require 600 mA per tube than 6L6s.)

Step 1: Between pins 4 and 6 on each output socket is a 470-ohm resistor. Replace this with a 1000-ohm, 5-watt resistor.

Step 2: Between pins 1 and 5 on each outputs tube socket is a 1500-ohm resistor. Disconnect it from pin 1 and solder it so that one end remains on pin 5, and the other end stands straight up. Connect the wires that ran to pin 1 to the ends of the 1500-ohm resistors that are standing free.

Step 3: Connect pin 1 to pin 8 with a solid wire.

Step 4: Increase the value of the resistor in series with the bias diode to reduce to bias voltage to obtain proper bias. This will vary from amp to amp and, of course, on the tube grading number.

Note: If you wish to use 6L6 tubes after this mod, simply adjust the bias supply. ∎

MAXIMIZING SILVERFACE FENDER AMPS
How to Maintain, Mod & "Blackface" Your '70s Fender
By Brinsley Schwarz
Reprinted from The Guitar Magazine (UK)

Compared to most hand-wired, point-to-point tube amps on the market your bog-standard '70s silverface Fenders are going for a song, and they can be as robust and toneful as the boutique boys – with just a little technical know-how. Here are some simple ways that you – or your favorite qualified amp tech – can make the most of these built-like-a-brick-shithouse bargains.

During the later-mid 1960s, Fender (CBS) made alterations to their range of amplifiers which, in the opinion of most, left them as pale imitations of their predecessors. In these articles we're going to look at how to rework these amps to earlier, better standards and also at what modifications can be made to make them a little more versatile and suitable to modern use.

The amps we're concerned with are known as "silverface" because of their silver control panels. Introduced in 1968, they are valued far less than the earlier blackfaces (though CBS bought Fender in 1965, changes during the first transitional years were gradual). The two ranges were basically similar, mostly the same circuitry, layout and transformers. But the differences, although minor, are crucial and, combined with a downgrading in component quality, resulted in a loss of tone and playability which cost Fender's reputation dear. Fortunately, most of the changes are reversible and silverfaces can be made to live.

Inevitably, in any article about amplifiers there's bound to be some tech talk, but I'll try to keep it simple. The intention is not to persuade you all to start digging around inside your amps – although that's how I got started – but rather to help you to understand a little of how things work, how and why changes should be made and what's possible, and to help you get the best from your amp. Some of this information can apply to all tube amps, but I'll be concentrating on these old Fenders, mostly with two channels, reverb and vibrato (but not including 70W and 130W versions, which are beyond help). I'll also be including some general *Tips*, which will be indicated in *italics*.

See the *Warning!* at the start of this book's technical section, which tells you about the dangers of messing with amplifiers. They can be extremely dangerous even when switched off and disconnected from the wall power. There is a simple procedure to avoid the dangers, so please read the warning until you understand it, and ideally refer back to a more detailed resource on this topic. We won't be opening up the amp this issue, but to undertake the steps in parts two and three you'll need to digest and understand these safety instructions thoroughly. If you don't understand safe procedure already, don't open up *any* guitar amp... take it, and this book, to a pro who can do the work for you.

Valve Job

To start with let's go over the exterior layout of the amp and how you can get the most from what's there. I'll always be looking at the amp from behind and the right way up. Over on the far right are the two preamp valves (tubes). The furthest to the right is V1, the preamp tube for

> **TIP:** *If you don't use the normal channel you are wasting a tube – it's on and using itself whether you are using it or not – so take it out and keep it as a spare. Also, the Normal and Vibrato channels share a small bit of circuitry and removing one of the tubes makes the other one a bit more lively.*

the Normal channel. This should be a 12AX7 (or its European equivalent ECC83).

V2 is the Vibrato channel preamp tube and is also a 12AX7. Whichever channel you use, choose its tube with care. Not all 12AX7s are the same, different makes have different characteristics – just as different makes of guitar strings have – and the 12AX7 comes in many guises. It's worthwhile investigating this, and if possible buying from a reputable supplier of quality, tested tubes, as the right tube can make all the difference, just as the right pickup can on your guitar (see the *Tube Reference Guide* near the front of this book's technical section).

Next is V3, this is the reverb send tube and should be a good quality 12AT7 (or its European equivalent ECC81) if you want good reverb.

V4 is the reverb return and mix tube. Here the vibrato channel gets a little top boost and is mixed with the reverb signal. Use a good quality 12AX7; a lot of noise problems can arise in this part of the amp and a good tube here helps.

V5 is the vibrato tube, the one that creates the vibrato. A working 12AX7 is sufficient as this is not a factor in "tone," merely function.

Last of the preamp tubes is V6 which is the "phase inverter" or "driver tube." These amps are push/pull, that is to say that one power amp tube (or pair of tubes) amplifies the top half of the sound wave (positive) while the other one (or pair) amplifies the bottom half (negative). The driver tube and its circuitry takes the preamp signal and splits it into two signals, one positive and one negative phase, for the two halves of the power amp stage. Always ensure your driver tube is a good one, and change it whenever you change the power tubes. If your amp turns on and lights up but there is no sound from anywhere, check the driver tube first. That's it for the preamp tubes. This is the first place where your tone is shaped and using the right, not necessarily the most expensive tube can help. You can change your amp quite a lot here.

To dabble a little and find out what different

> **TIP:** *The reverb driver is in fact a little class A amplifier on its own, giving about a watt into 8Ω. Try plugging a small speaker into the reverb send socket, remove the driver tube (V6), or turn down the master volume control if you have one, and you've got a one watt practice amp.*

Technical articles

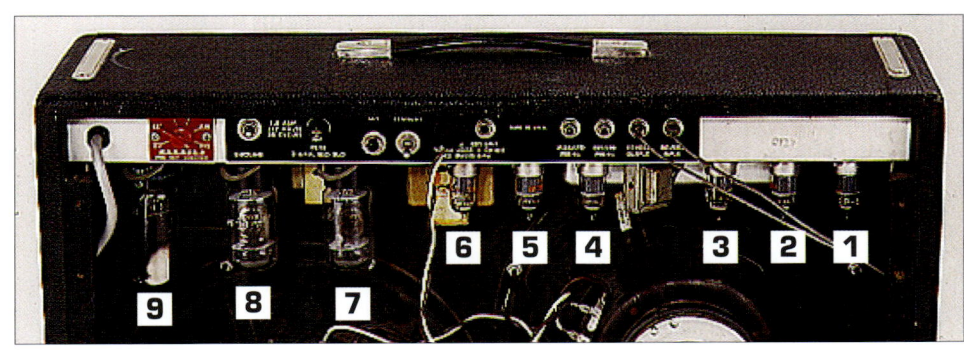

Diagram 1: Tube Line-up
1. Normal ch. preamp tube
2. Vibrato ch. preamp tube
3. Reverb send tube
4. Reverb return/mix tube
5. Vibrato tube
6. Phase inverter/driver tube
7. Power/output tube
8. Power/output tube
9. Rectifier

TIP: *Preamp tubes are usually covered with metal cans and you can remove these for better tone. I know, you're thinking 'more mumbo-jumbo', but here's why it's true: inside the tube there is an element called the cathode which sits at the centre of another element called the plate. The cathode gives off electrons and the plate uses a positive charge to attract them. This is part of how the tube works. Okay, now put a metal can over the tube. Connected to ground via the chassis, the can is negative with respect to the plate. The electrons, which are your tone, are attracted to the positive plate but repelled by the negative can and are thus confused, just like your tone. Try this: take the cans off your tubes, turn your amp on and play, listening critically. Now put the cans back (take care as the tubes are hot) and play again. Case proved... though if this causes your amp to becomes noisy, you'll have to put the cans back one at a time until the noise goes.*

tubes do, try a 12AY7 or a 12AU7 in V1 for a warm, ultra clean tone, try a Chinese 12AX7 in V2 for grainy gain and an ECC81 (high spec 12AT7) in V6 for an efficient and balanced driver stage. Experiment – preamp tubes aren't expensive and you can always use them somewhere if you don't want to stick with the changes.

Power Ranger
Next along the back of your amp come the power tubes. These amplify the preamp signal and provide the power (watts) to drive the speakers. A lot has been said already about matched power tubes and I don't want to dwell on the debate here, but there are some relevant points. Power tubes can come matched and graded. Grading enables you to choose whether your power stage has clean headroom or early drive and keep it that way by renewing with the same grade replacements. Matched sets can last longer, since they are balanced with respect to bias and current and so are all hopefully working optimally together.

Last of the tubes is the rectifier tube, on the extreme left. All amps have a rectifier – it can be tube or solid state. The rectifier changes AC voltage into DC, which most of the amp uses. Tube and solid state rectifiers sound different. Put very simply, here's how this works: your amp is turned up loud but you're not playing, everything is on idle, then you play a note, the power tubes' demand for current shoots up; the solid state rectifier says "no problem" and the tube turns full on, the front end of your note plays loud and then decays naturally as it dies away.

But the tube rectifier doesn't work like that; the demand for current isn't met immediately, the tube can't turn full on instantaneously, and so the the front end of your note sounds a little squashed. Then as the note decays and the demand for current relaxes, the rectifier can produce and the note swells, almost getting louder as it decays. This is a simple description of compression (also referred to as "sag"), and amps with tube rectifiers have this squashy, compressed and singing quality when turned up. It's simple to convert tube rectification to solid state, there are even plug-in units available. But going the other way round, although possible, is much more complicated.

Any Old Iron
The rest of the amp consists of the power/mains transformer, situated on the left; the choke, which is a small filtering transformer; the output transformer, which matches the power tubes to the speakers; and the reverb drive transformer, which is the little one between V3 and V4. You don't need to know much about these, just that they should be clean and free from rust, so don't store your amp in the garden shed, where moisture and extreme temperature changes can get at them. For the health of your output transformer, recommended speaker impedances should be adhered to.

Lastly, let's locate the the filter cap box,

TIP: *If your amp works but seems to overdrive or distort early, you could have a broken choke.*

which is the large flat box behind the preamp tubes. Filter capacitors can be extremely dangerous and have been the subject of much discussion and controversy. Their renewal or upgrade will be one of the subjects covered below when I'll go through how to "blackface" your silverface amp.

One final word... silverface Fenders are potentially really good amps. They are hand wired and easy to work on. In the USA upgrading and modifying these amps has been commonplace for years and has caused their value to rise slightly, but they can still be had cheaply. Advertised prices in the UK have soared recently and I believe they are too high. It's easy to spend a lot of money on the right tubes and speakers, upgrades and mods to get the amp working properly. Current prices may be almost fair for an example in mint condition with rust-free transformers and good speakers, but many of these amps have been badly mistreated and are in need of lots of TLC. The price you pay should reflect this and not the faulty salesman's logic "it's Fender and it's old so it must be worth lots of money."

Blackfacing
Now we are ready to move on. Far from merely stopping at maximising your '70s era silverface Fender amp's stock potential, let's dive into "re-converting" your all-tube tone tank into one of the most hallowed guitar amps of all time: a pre-CBS "blackface" Fender combo.

This will involve a simple rewire of one area of the amp and the straight replacement of some components elsewhere. These are simple mods and old Fenders are easy to work on, but if you have even the smallest doubt about your abilities take your amp to a qualified tech. Also, there's a lot of information coming up that will enable you to have "just a better amp" or "a completely different one." You might want to read it *all* before you decide how far you want to go and before you start buying up parts. When I started learning this stuff by messing around with my Twin Reverb, I did little bits at a time as I read and found out more, often undoing and chucking away what I'd done a few days before. I tried out a lot of stuff before I got to the amp I have now and what I'm doing in these articles is giving you what I found to work and be worthwhile, in the simplest terms. Of course some of you may want to find out more and there are plenty of sources from which to learn. Some of you won't, but I hope you'll all learn something useful.

Biased Opinion
To begin with, let's examine how to rewire the

THE TUBE AMP BOOK

Technical articles

DIAGRAM 1(a) SILVERFACE WIRING

This board not always in this position and sometimes wire to the changed pot has resistor on it – remove resistor change wire as diagram 1b

Resistors and wiring connections which will be changed in layout below are colored green (not always these values but change them to fig 1b values anyway)

DIAGRAM 1(b) BLACKFACE WIRING

Change to this pot wiring

VIEWED FROM REAR OF AMP

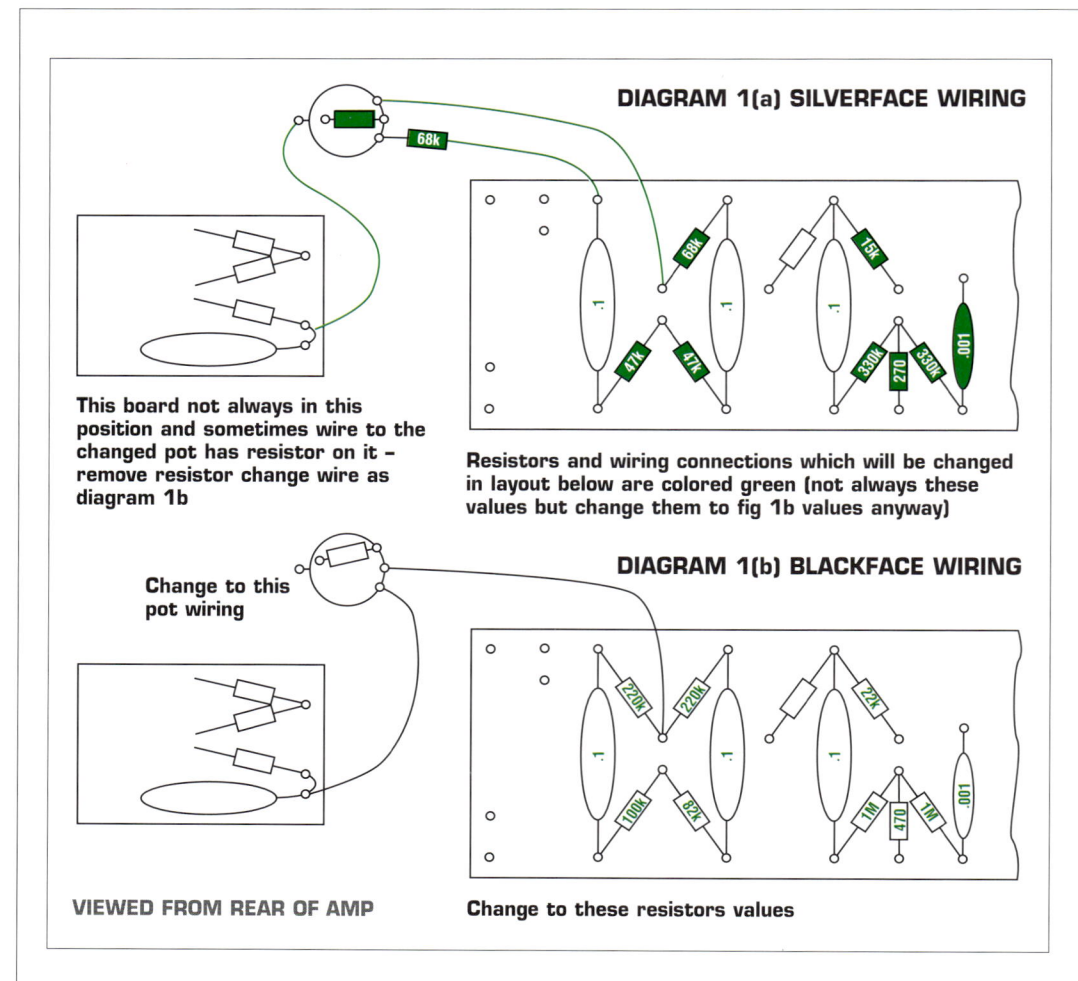

Change to these resistors values

driver and bias circuit. The bias circuit in silverface amps does not allow the amp to be rebiased, only for the bias to be balanced between the two halves of the power stage. This is not ideal and does not allow you to get the best – or even what you may choose to get – from your amp and the power tubes you select. What is bias? Well, tubes were not designed just to use in audio amps, they were used for dozens of other functions – oscillators, non-audio amplifiers, switches, etc. They are general purpose electronic devices that have to be 'pushed' into a particular operating area to become audio frequency amplifiers. Bias is that push, a negative voltage applied to the tube to make it work the way we want it to. There are different ways to bias an amp, and different, often strongly held, opinions on which way is best and/or correct. I always bias by the "measured current" method.

Whatever method you choose to use you need to rework your amp to the earlier spec. The driver circuit (powered by V6) is also not very ideal, it's weak and flat sounding; the "blackface" design is stable with more drive and gain. **Diagram 1** (a & b) shows the differences, and you can see you have to change just seven or eight resistors and one wire. The new – that is, *older* – circuit should liven up your amp and give you more gain, and the bias circuit will allow the bias to be adjusted correctly or to your taste. While you are doing this is a good time to upgrade the four caps in the circuit as well.

Upgrading Coupling Caps

The other real problem with these amps is that CBS/Fender seemed to have embarked on a component downgrading policy. Throughout any tube amplifier there are capacitors which are either for tone shaping or for what is known as "coupling." Your tone goes through these caps and is effected by their quality. Coupling caps are used between tube stages, they separate DC voltage (power that the tubes use in order to work) which they don't allow to pass, and AC voltage (your guitar signal) which they let through. To start with, Fender changed from the really good blue tubular caps found in most blackfaces, to brown "chocolate drop" caps, which aren't very good at all. Later, after the damage to their reputation from this and other changes became apparent, they changed again to somewhat better (blue drop) caps. Both these latter types can be improved upon and it's okay to just replace them with the same value, but better, caps.

I use two makes as upgrades: Sprague, also known as "Orange drops," and Xicon. Sprague caps have been used in many top range American amps for years and are, I guess, the favorites. Spragues sound warm, tight and controlled; Xicons are grainier with a little more top end (many techs and players are also fond of Mallory types, though these are getting more

TIP: *However, you can also use this opportunity to change the amp's sound for the better as well, by altering some of the values slightly... see the mods nearer the end of this chapter.*

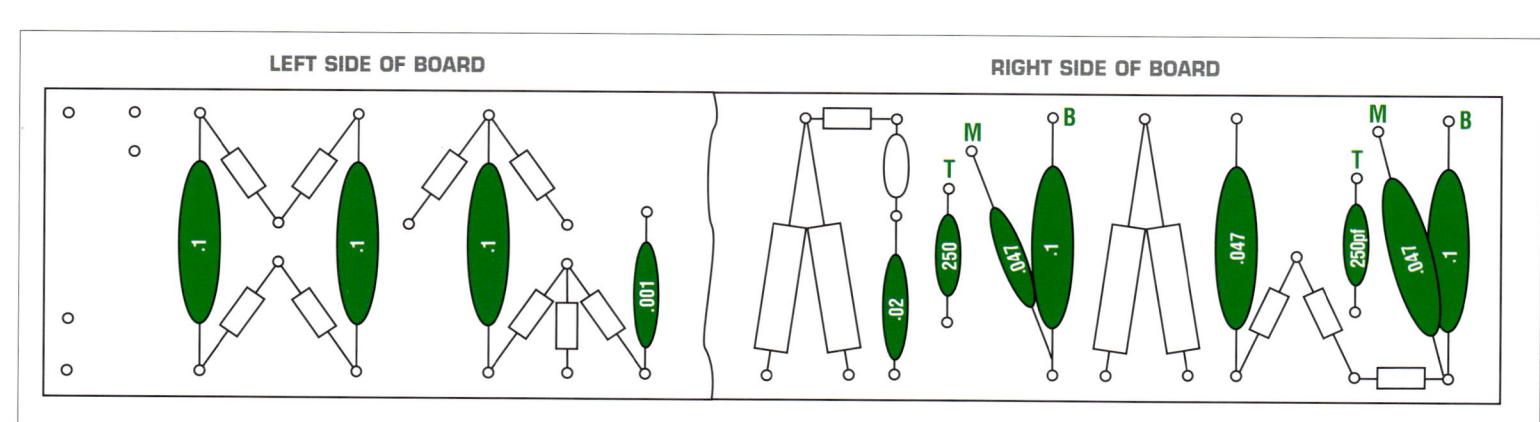

LEFT SIDE OF BOARD   **RIGHT SIDE OF BOARD**

DIAGRAM 2 CAP UPGRADE

VIEWED FROM REAR OF AMP

All caps colored green to be upgraded and can also be modified. BMT = Bass, Middle, Treble. Treble caps (and bright switch caps, not shown) should be changed to silver mica – Sprague, Xicon or other quality signal cap

136 THE TUBE AMP BOOK

Technical articles

difficult to find, and some more expensive NOS caps, too). Whatever you use, the caps must be rated for at least 600 volts. Changing the coupling caps and the bass and middle tone caps to Sprague or Xicon, and the treble control and bright switch caps to silver mica, will give better, clearer, warmer tone and a sweeter top end. This means a complete recap, between 15 and 18 caps. (See **Diagram 2** for the rest of these).

Filter Caps

Last in the component quality upgrade are the filter caps. These are not in the soundpath at all, and are used to "clean" the DC voltage that the amp uses. So how do they affect the sound? Surely if they're working that's good enough? Opinions differ here and there are no straightforward answers. Filter caps used in guitar amps are "electrolytic," and they go bad with time. Their average lifespan to be working within correct spec is probably around 10 years, so if your amp has original caps (light brown cylinders with "Mallory" printed on them) they probably aren't working as they were intended to when new. And here is the rub: if they are working and your amp works and sounds OK, then why change them at all? Well, all I can say is I have replaced the filter caps in lots of old Fenders with Sprague electrolytics and the amp always sounded better for it – tighter bottom end, more punch and quite a lot quieter. To spot a broken cap, look for stuff oozing from either end, little "blisters" in the rubber end seal, or broken/loose wires. Even if none of these are present, if they are still the originals in a 30-year-old amp, you can pretty much assume they have dried up.

There is another reason for considering a filter cap renewal and this applies to all old amps. Cap failure can cause damage to your mains transformer. These cost far more than filter caps and in some cases it may prove extremely difficult to locate the right part. I'm not trying to scare everyone with an old amp into new filter caps – many are out there working away just fine – it's just another consideration.

Messing with these things is dangerous, so as an added precaution to the *WARNING!* printed in this book, here is a second procedure to guard against an extremely unpleasant shock. Get an insulated alligator clip lead, cut it in half and solder a 1k to 10k, 2 watt resistor between the two halves and tape or heatshrink wrap the join so that the whole thing is insulated. Having followed the procedure outlined for bleeding off voltages in an amp with a standby switch (like these Fenders), carefully remove the filter cap box, attach one clip to the positive (+ve) end of the large cap at the left end of the board (usually 70 microfarads or "µFs" and put the other clip on the chassis. Repeat this with each of the 5/6 caps. This will make them safe. To double check, use a voltmeter set on DC volts between the +ve of each cap and the chassis to read the volts left in the cap; this should read zero. If you're in any doubt whatsoever get this done by a tech. We can't emphasise caution strongly enough here!

If you want to do this yourself, then obtain the correct value and correct voltage replacements and observe the correct polarity when installing the new caps. Electrolytic caps have positive and negative sides, the positive side is usually marked on the caps as '+ve' or simply '+', and they won't work at all (sometimes they'll self distruct) if fitted the wrong way around.

If you want to go further than merely replacing what's there, here are some things to consider. Filtering is done in stages which are separated first by the choke and then by resistors. The first stage is usually made up of two large caps in series (generally 70µF). You can change the sound of the whole amp by altering these two caps, as shown in **Diagram 3**. Changing up to 100µFs will tighten up the bottom end and add a little punch. Upping them to 220µFs will give lots of extra punch, especially good if you're going to include an overdrive channel.

Next comes three 20µF caps separated by two big resistors. The last cap feeds the two channel preamps and you can make the preamps a little more open sounding by changing this cap down to 10µFs. The resistor keeps the stages apart, but also drops the voltage a little and by increasing the value of the last resistor you can lower the voltage in the preamp stages. This produces a "browner" sound, more vintage and more ready to get gainy. Tube rectifiers, however, don't like to see big filter caps and so you should stick as close to the original values as possible in these amps.

I don't think that it's worth doing a cap job at all unless you're going to use really good caps; once again I think Spragues are best. Mallorys are great too but not as easy to get. European and Japan makes function correctly but don't sound as good.

Resistor Upgrade

Lastly, some of the upgrades and mods will require changing some resistors, and here's some more amp voodoo stuff. There are different types of resistor, they can be made from different materials, and we need to use the right ones. Old Fenders have "carbon composite" resistors; they're dark brown and quite large by modern standards, and these are not readily available – although they can be bought from specialist suppliers in the US. Modern resistors are "carbon film" or "metal film," "enamel" or "ceramic." Some would insist that carbon comp sound best but carbon film are also good. Ceramic resistors are best in just one place: this is to replace the 1 watt, 470 ohm resistors between pins 4 and 6 of the power tubes. New 4 watt 470 ohm ceramics are better and safer here. Also, size is important. Resistors need to be at least 1/2 watt and also of a similar size to those being replaced. Obviously, as technology demands smaller components, large resistors have all but disappeared and are now relatively expensive, but you can still find them and it has been proved to me that they sound better… stands to reason, doesn't it? Big components = big sound, little components = ???, correct?

Mod Squad

Now that we've got our silverface back to blackface specs, it should be sounding a lot better already. But why stop there? Let's take a look at some of the simple modifications that can be done to these easy-to-work-on '70s Fender amps, and also discuss some more comprehensive ones that would be too complex to explain fully in the space available here. As I've said before, I'm not necessarily advocating that you all start modifying your amps, but I hope to maybe awaken you to some possibilities. Also, understanding a little of how and why amps work should help you to get more out of yours and to make more informed purchases in the future.

Also, note that while these silverface amps aren't yet up to the highly collectable status of Fender's blackface, brownface and tweed models, they're becoming increasingly desirable nonetheless. Even if you (and, possibly, other players) consider the mods discussed here to be "improvements," any alteration of an amp from original could detract from it's resale value. On the other hand, because these amps are simple to work on, these modifications are mostly reversible. Ultimately, you have to weigh it up for yourself: if these mods appeal to you and

DIAGRAM 3 FILTER CAPS UPGRADE
READ WARNINGS – OBSERVE POLARITY!

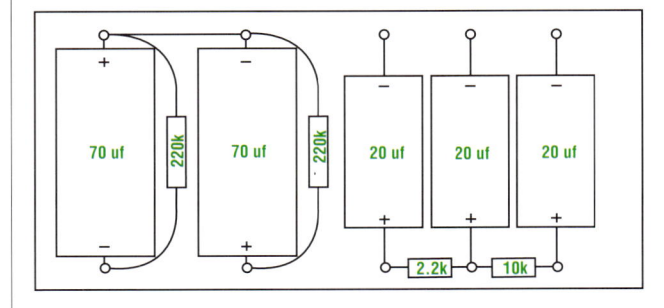

Replace 70 ufs with 80 ufs for same spec or 100 ufs for firmer low end. Blackface spec has 1k resistor not 2.2k and 4.7k not 10k. All resistors 2 watt.

VIEWED FROM REAR OF AMP

Technical articles

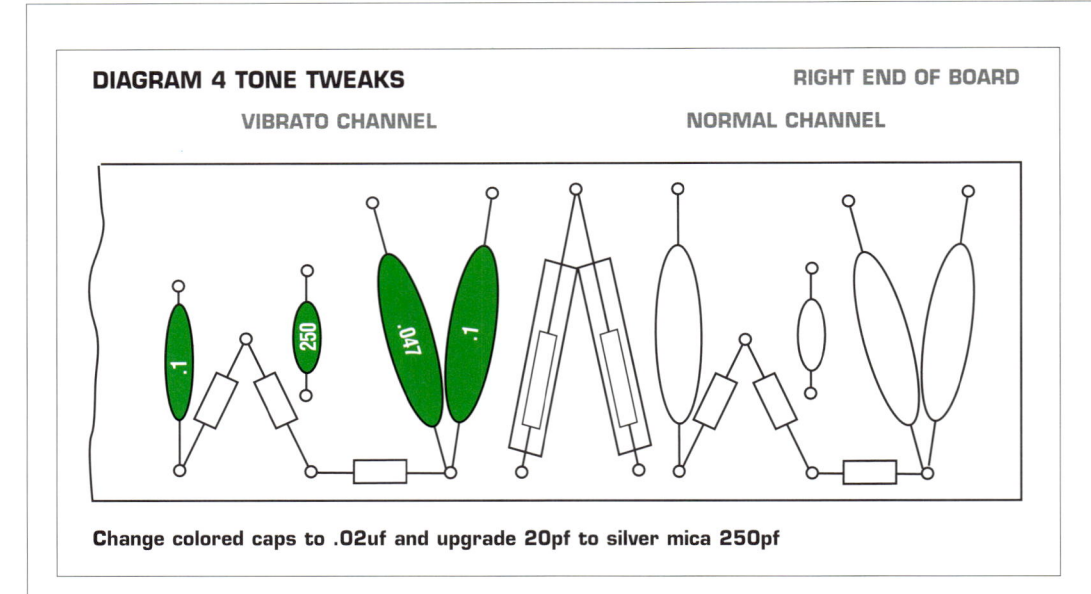

DIAGRAM 4 TONE TWEAKS — RIGHT END OF BOARD
VIBRATO CHANNEL — NORMAL CHANNEL

Change colored caps to .02uf and upgrade 20pf to silver mica 250pf

you plan to keep the amp for a while – possibly a lifetime – then the rewards become self evident.

Tone Tweaking
Since the Normal channel of your amp is probably lying dormant, this is the obvious candidate for modifying, leaving you with one original channel (the Vibrato one) and one altered one. The two channels come together later on in the amp so you can use a passive A/B box to switch from one channel to the other.

As I hinted earlier, there are some very simple changes that you can make to component values that will produce a more open tone from your Fender. The original tone and coupling caps are usually big value ones and these can give a sound that is mid-light or a bit bottom/top heavy, almost as though you're playing through PA speakers. **Diagram 4** shows what needs to be changed, just three caps. The bass tone cap is usually a .1µF, the mid control a .47µF and the coupling cap varies. Changing these to .02µF caps (often seen as .022µF), the values used in older Fenders and Marshalls, will give a more natural guitar sound that feels easier to play. Personally, I prefer this even as an original circuit in Fenders, so in a clean channel you could use .02 caps throughout, and I would suggest that you try it in your Vibrato channel, leaving the Normal channel for something a little more adventurous.

Hot Mod
There is enough space on the circuit board for any one of the many preamps that use just one tube. Flicking through diagrams in the back pages of this very book will reveal old designs used by Fender, Gibson and Marshall, to name but a few. Some of these would involve a different layout than the one in your silverface but the one I've chosen to show you is very simple to install. It produces a Marshall type voicing and much more gain – and with a humbucker it can give mucho drive and crunch. **Diagram 5** shows how. The normal channel circuitry is all at the extreme right hand end of the board; as you can see you only need to change five caps, four resistors and one pot. You also need to build on a little extra one-cap/one-resistor circuit on a tag terminal between pin 8 of V1 and ground. As with all of this work, make sure you clean out all of the old wiring and desolder the eyelets so that you start afresh with a clean board. As I have already said, there are many designs to choose from that will fit into the normal channel space. There's no need to be restricted by the control layout either, and simpler preamps – like those found in tweed Fenders or Marshall's 18 – 20W series, will work and sound good and, if you're into it, you can mess around until you find something that you really like.

Another mod that has proven extremely popular with my customers is to make this into a tweed Deluxe channel, so I will give this in more detail further below.

Effects On Both Channels
Okay, now you've got two different and very usable channels, so you'll want reverb and vibrato to work on both. No problem, you just have to move a few parts and some wire. Over on the left side of the board there are two 220k resistors (red/red/yellow stripes), which mix the two channels together as they go into the driver circuit at V6. The left hand one is usually marked with an X on Fender drawings and has a wire going all the way over to the output coupling cap of the normal channel, also marked with an X (the wire twists around several other wires on it's way across the board). Remove this 220k resistor and the wire but leave the other resistor with two wires going to it in place. Move both channels' treble caps, fit the two new 220k resistors and rewire, all as shown in **diagram 6**. Now you have reverb and vibrato on both channels.

Negative Feedback Mods
One more area that you can change to big effect is the feedback resistor. This is the only resistor in the driver circuit that we left untouched in the 'blackface' mod. Feedback is used to control the power stage and to stop it from running away or oscillating. As standard in Fenders from the '60s, the effect of the feedback circuit is quite heavy. Raising the value of the resistor will give you the effect of increased power-amp gain or lowered headroom. The feedback resistor in these amps is usually 820Ω. I usually raise this to 5.6k to make the power stage a little more lively, but you can go as far as you like, however if your amp develops noise or stops making any sound at all (this due to inaudible runaway

DIAGRAM 5 MARSHALL/HIGH GAIN — RIGHT END OF BOARD
VIBRATO CHANNEL — NORMAL CHANNEL

Optional change cap to .68uf 245 volt

Normal channel to Marshall voicing and more gain. Change colored components and wire as shown. Also change 10k middle pot to 25k

VIEWED FROM REAR OF AMP

* These two restrictors can be any value between 150k and 220k to vary amount of gain lift

Build 1.5k resistor and .68uf cap (25 volt) from Pin 8 to single tag grounded by valve base securing nut and bolt

138 THE TUBE AMP BOOK

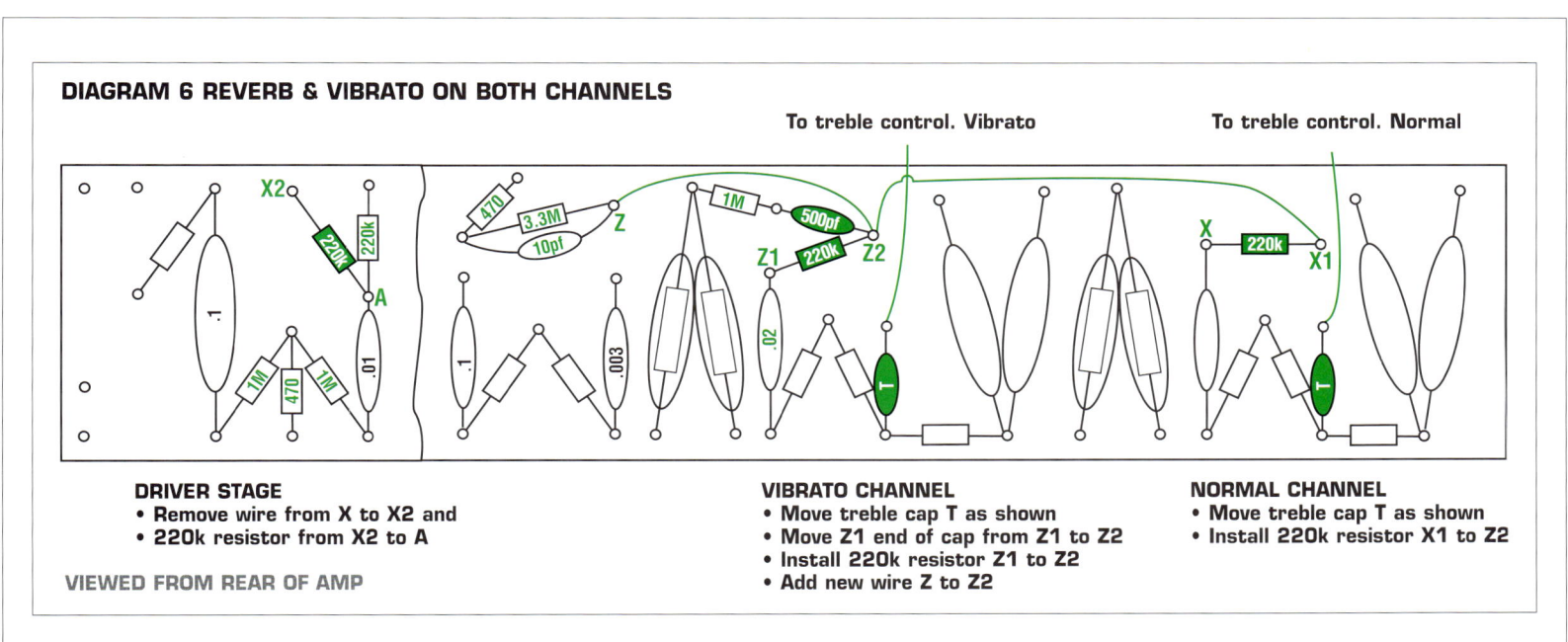

DIAGRAM 6 REVERB & VIBRATO ON BOTH CHANNELS

DRIVER STAGE
- Remove wire from X to X2 and
- 220k resistor from X2 to A

VIEWED FROM REAR OF AMP

VIBRATO CHANNEL
- Move treble cap T as shown
- Move Z1 end of cap from Z1 to Z2
- Install 220k resistor Z1 to Z2
- Add new wire Z to Z2

NORMAL CHANNEL
- Move treble cap T as shown
- Install 220k resistor X1 to Z2

oscillation/feedback) you'll need to back down the value until it's stable again.

More Extreme Measures

The mods that I've gone through have all been quite simple to do, and yet they can make a big difference to your amp. Since these Fenders are so easy to work on, you can go much further, with modifications that would be too complicated to go into here. But I can talk about them a little – and if you have sufficient experience yourself or know a good local amp tech, you can follow these up on your own initiative.

If you dump the vibrato you get another tube to play with and you can use this in conjunction with the vibrato channel preamp tube, V2, to get a tweed Bassman or JTM 45 channel. These are three-stage preamps but with two tubes you've got 4 stages so you can add another stage to this to get big overdrive, too. If you dump reverb as well that gives you more tubes to play with and you've got enough space to include a tube parallel effects loop, which is essential if you run a rack digital effects unit, as your tube dry sound will be kept out of the digital soundpath and so remains intact. There'd also be room to put in a five-stage Soldano type overdrive channel. Of course your Fender will never sound exactly like a Soldano or a '58 Bassman or a JTM 45, but it can take on those characteristics, with good crunch or heavy duty overdrive.

Speaker Upgrades

The one remaining thing to be discussed is speakers. These were not fabulous when new in silverfaces, and by now they are usually very tired, broken or have been replaced, often without much consideration for tone. There are different ways of looking at replacements, depending on the type of amp you want. Remember, you can make your amp quite a lot louder by using efficient speakers and also alter the tone with ceramic or alnico magnets. Electrovoice speakers are at least 3 db more efficient than most guitar speakers – that's about the same difference between 50W and 100W. Two EV10s in a Vibrolux, for example, will produce more volume and enable you to turn up without the speakers dying. The new Celestion Century speakers are also extremely efficient – lightweight, too. If you have a small amp which isn't quite loud enough to use live but you really love the sound and you can't find a bigger, louder amp that sounds the same, then consider using more efficient speakers. The volume boost may well be enough to enable you to use the amp on stage.

Jensens are being built again and are readily available. These would hopefully give your amp that vintage sound with good dynamics and tone. Also, speakers like these really jump and move air when you play a note, which makes them great for recording. There are some good vintagey replacements available – Mojo, Kendrick, Weber and Naylor spring to mind – many of which have had rave reviews in guitar mags – but I normally turn to Celestions. They make a sort of "better quality" range which includes the Classic Lead 80, Greenback, Vintage 30 and Modern Lead 70. These are reasonably priced and very useful as replacements in silverface Fenders. Classic Lead 80s are my favourites in Twin and Pro Reverbs; they are clear and can take the clean power. Vintage 30s are also good, although maybe a little loose in the low end and, to counteract this, a good trick is to use one Vintage 30 and one GT12-100. The latter has good firm bottom end and combines well with the grainy top of the Vintage 30. Celestion also make a Vintage 10 which sounds good and is powerful (though sadly they are apparently discontinuing their great sounding G10L-35), but I guess the Jensen 10" speakers would be my favourites for Super Reverbs. The main thing is to realise that speakers are a very important part of your amp and, if you need to change them, a little thought and perhaps a little extra expenditure could make a big difference.

Over the years most manufacturers, for whatever reasons, have not always fitted the best components available (speakers included), but sometimes those parts that have been readily available have luckily turned out to sound really good (Leo Fender's early use of alnico-magnet Jensen speakers, for example). The silverface era was *not* one of these lucky times, but with some effort and a little cash you can make yourself a fine amplifier.

Super Deluxe: Tweed-Style Preamp Mod

This is a great, easy way to add a raw, gainy new voice to an already great silverface Fender which you have already improved and "blackfaced," and is an especially nice mod for one of the most popular of the silverface range, the Deluxe Reverb. Although the preamp section of the Deluxe is virtually identical to all of the other two channel models in the range, the amp is different in two very important ways. The first is that it uses two 6V6 power tubes; the second is that it runs on lower voltages (that is to say, the voltages used inside the amp after they have been supplied by the mains transformer are lower). As a result of these differences, the Deluxe is quite a hot little amp and is a longtime favorite with Telecaster players. Small and not too loud at about 22W, it's an ideal amp for cranking up full, great for small blues and country gigs, and the ideal recording amp too.

This mod is for the Normal channel, which most people ignore completely, and will also work really well in a blackface Deluxe. (Both of these models are now becoming more valuable, the blackface extremely so, and any alterations will probably reduce the value of your amp from the "it's old, original and Fender" viewpoint, but if the work is done carefully, the old parts and wiring can be kept and easily restored. And if you're not using that Normal channel anyway…)

The design I'll tell you about here is a modified version of the preamp in a tweed

Technical articles

DIAGRAM 7 SILVERFACE DELUXE 'NORMAL' CHANNEL

DIAGRAM 8 TWEED CHANNEL IN SILVERFACE DELUXE

→ Always means to ground

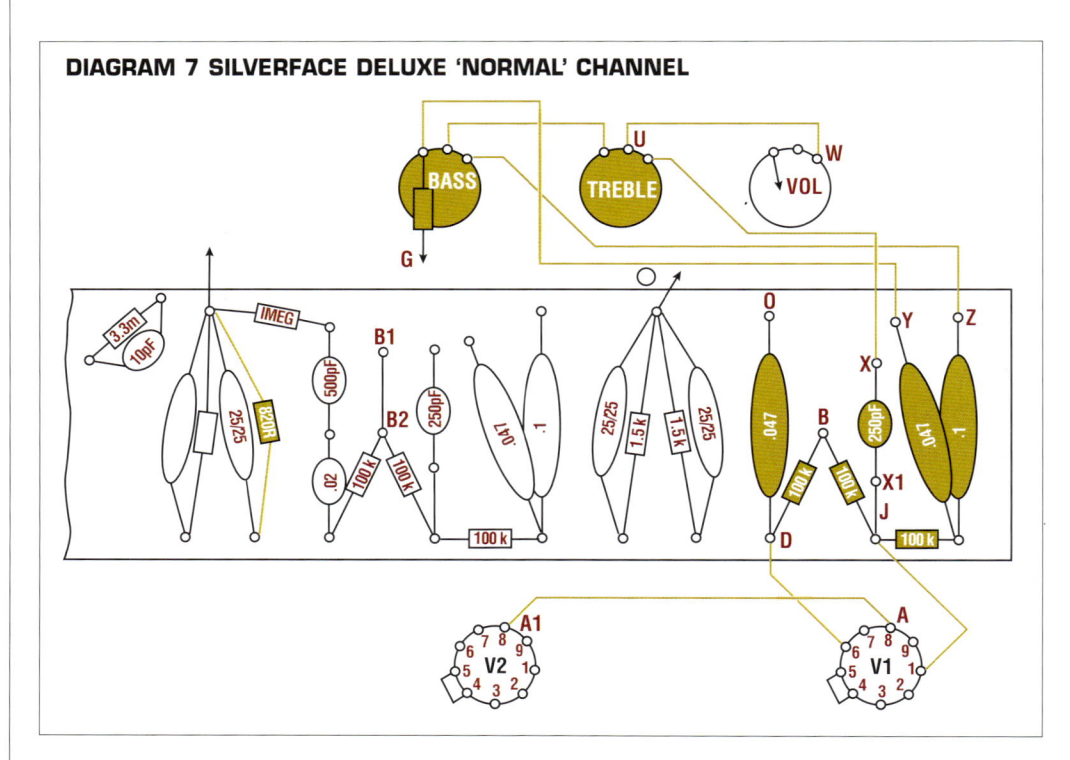

Fender Deluxe from the '50s. The main difference between that and the later Deluxes – as far as we are concerned – is that the earlier series had just a single tone control as opposed to the treble and bass controls later introduced. This is a big difference; the circuitry involved in the two-control design (and three-control design on the larger amps, in which this mod will also fit) robs the amp of a lot of midrange volume. The tweed Deluxe circuit has very little mid loss and gives a fat, hot sound which combines well with the strong driver stage found in all blackface Fenders to give anything from a warm, clean blues tone to a big, growly overdrive. Of course vintage Fender amp enthusiasts will already realise that there are many more differences between the tweed and later versions of the Deluxe, but by mimicking this preamp we are going part way toward the tweed sound, and providing a great new voice option in a simple-to-build package.

Procedure

The circuit is easy to build and requires just a few parts. The reverb and vibrato can be added to work on this channel as well (see the mod above) and the normal channel's three volume, treble and bass controls will become the new gain, tone and master vol controls.

Diagram 7 shows the original amp layout, with all the parts to be changed indicated (there may be minor positional differences, but all the layouts have basically the same circuitry). **Diagram 8** shows the new layout with all the new parts in place, and **diagram 9** shows inside the filter cap box, which is the large metal box on the underside of the amp, from where you need to run one new wire. Be careful with these caps! They can be very dangerous – so do read and follow the **WARNING!** printed elsewhere.

The Normal channel uses just one preamp tube, and the gain and tonal characteristics of the new circuit can be altered dramatically by the type of tube used. The original narrow-panel (5E3 circuit) tweed Deluxe used a 12AY7 and you can experiment with this tube or even a 12AU7 for warm and clean sounds, but I build this as a hot, growly, blues overdrive channel and for that, a 12AX7 is best.

The addition of a master volume control makes the channel very versatile. Of course, it will all depend on your guitar and pickups, but the sound should be lively and hot even with decent single coil pickups, and with good humbuckers you should be in Robben Ford territory. You can switch between the two channels with a simple passive A/B box, and don't forget you can place effects pedals before and/or after an A/B box for varied applications.

As I've already said, this circuit works well in a Deluxe, but I've also had good results from it in the other two-channel Fenders, even in a Twin Reverb, but because these amps differ from each other, mainly due to voltages, rectifier type and output transformer size, the mod will sound different from model to model.

How To: Preparation

Read these instructions several times before you start, locate all of the parts, eyelets and wiring in your amp to be sure you can follow the instructions, and read the WARNING! again.

Diagram 7 shows the board layout at the far right side of the amp (looking in with the back of the amp chassis facing you). This is the area you are going to be working on. Not all the wiring is shown, just what needs to be removed or de-soldered; leave the rest in place. (If your amp has just one point at X or X1, then you need a slightly different layout; you can probably find it in the schematics included in this book, or on the web, if you can't work it out for yourself. The one used here, however, is the most common.)

1. Locate the points on the board marked O and B. You'll find a wire coming from each point: the two wires are twisted together.

   The wire from point O goes all the way over to a 220k resistor on the left end of the board. De-solder this from point O, untwist it from the other wire and leave it hanging over the front of the amp. You'll need it later.

   The wire at point B goes leftwards, usually

140 THE TUBE AMP BOOK

Technical articles

to point B1 (but sometimes to point B2). Remove this wire altogether, leaving everything else soldered to points B1 and B2 in place. **Note:** not all amps have both B1 and B2; older models may have just point B2. With these, remove the wire from B to B2.

2. Remove the wire from V1 pin 8 to V2 pin 8, points marked A and A1 on diagram 1.

3. Remove all the colour-indicated parts from the board and de-solder and lift out the colour-indicated wiring from the board. Clean out all of the eyelets with a solder sucker. Remove the wire from V1 pin 1 altogether.

4. Remove the bass and treble controls. Keep as much of this assembly together as you can; you should only need to de-solder the three wires at the board – points X, Y and Z – and the wire from tag U to tag W at the volume control. Cut the resistor on the bass control midway between the resistor body and the solder point on the amp chassis (G). You should now be able to remove the two controls with their wiring intact, which will be helpful if you should ever want to restore the amp to original.

5. Leave connected the wire from the volume control's middle tag to V1 pin 7 and the wires to/from V1 pins 2, 3, 4, 5 and 9. You should now have cleared and cleaned the area to be worked on and you can start to rebuild.

Rebuilding
Read this section through several times before beginning to acquaint yourself with the parts and new layout. **Diagrams 8** and **9** apply now!

Parts List
R1 – 68k 2W (27k to 68k, see text)
R2 & R3 – both 100k 1/2W
R4 & R5 – both 1.5k 1/2W

C1 – 8μF or 10μF at 450volts, Sprague electrolytic
C2 & C3 – both .02μFs at 600v, Xicon
C4 – .68μFs at 25v (.68 to 4.7μF, see text)
C5 – optional change: leave original 25/25 in place or reduce to anything down to .68 μFs (see text)
C6 – 500pf at 600v, silver mica
C7 – .0047μF at 600v, Xicon

P1 & P2 both 1meg audio (log) high-quality pots. Use US-made Alpha or CTS pots, and use good quality wire, rated for at least 600 volts.

Steps
1. You'll notice on the parts list that some of the components have multiple values. That's because you have some choices and you can alter the way the channel sounds with different values.

R1 provides what is known as B+ voltage to the tube. The higher the value, the lower the voltage: low voltages give a "browner" sound (ie more distorted and easily overdriven). So if you're going to build my blues overdrive channel then use a larger value resistor. Once again, different Fender models supply different voltages to this preamp tube, and you're looking for something like 200 to 230 volts at point B (**diagram 8**).

C4 and C5 are the tube's cathode bypass caps. These are not necessary at all for the tube to work but they add gain and so they're almost always used (switching one or more of these caps in and out of circuit is often called a "fat" switch). They also control the frequency response of the tube: the larger the value, the lower the frequency. The stock value in most Fenders is 25μFs (microfarads) and this is as large as is needed for more than the full range of the guitar. You can keep both of these to the smallest value of .68μFs; this will provide less bass response, but if you want to keep one large for more bass then it should be C5, and that's why I have left the stock 25/25 in place. C4 can be anything from .68 to 4.7 μFs. Using a .68 cap here keeps the tone from getting too overblown and heavy (when overdriving with humbuckers on a Les Paul, for example) and that's why I've suggested that value. But it's up to you, and if you're not using the channel for gain, or you play a Telecaster, then you could try something larger.

2. Locate the filter cap cover underneath the amp behind the preamp tubes and remove it. Inside you'll see five filter caps. Find point F on **diagram 9** and run a new wire from this point through the grommet G and to point F on **diagram 8**. You may have to remove everything from F on the filter cap board, clean out the eyelet and resolder all the parts and the new wire all at the same time.

3. Now solder in R1, and the new wire, at point F. Run a wire connecting the points E – this provides C1's ground – and solder C1 at point E. C1 will have +ve (positive) and –ve (negative) ends. –ve is at point E (to ground).

4. Moving on, solder R1, R2, R3 and C1 at point B and continue on through the circuit. You'll find this method of moving from eyelet to eyelet is the easiest way to work, as the parts are fixed at one end first and everything is soldered into each eyelet at the same time – and it also ensures you don't leave anything

DIAGRAM 9 TWEED IN SILVERFACE DELUXE

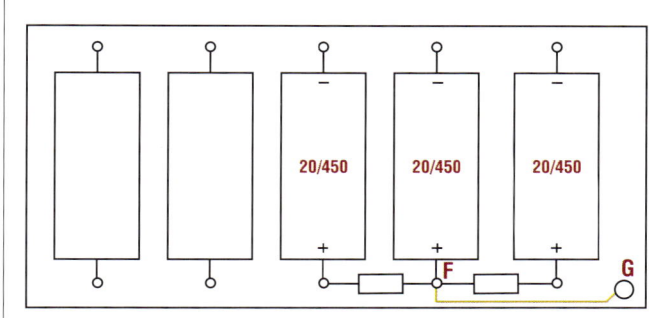

20/450 20/450 20/450
F G

Filter cap box: run new wire from point F through grommet G to point F on diagram 8

out! Make sure the electrolytic caps are installed the right way around, +ve toward the tube, –ve to ground.

5. The final task is to connect the wire that you left hanging over the front of the amp. This should still be twisted around other wires as it comes from its 220k resistor over on the left end of the board. Bring it level with the new master volume control and, still running down the centre of the board, bend it towards the master vol, cut it to length, and solder it to the middle tag of the pot as on **diagram 8**.

6. **Diagram 8** shows all the parts and wiring to be installed. Less solder is better than more; solder joints should be shiny – give all parts a wiggle once the solder is set and be sure that wires soldered to the tube pins are really soldered, not just hooked over them. They can be deceptive!

After you've checked your work (I still do this at least twice), you're ready to turn on. You can check the voltage at point B if you like (use a digital voltmeter set at 1000 volts DC and be careful); anything around + 200 to 230 volts is good for a hot, fat blues tone. This done, you're ready to put the amp back together and play.

Parts should be available from popular electronics suppliers, by mail or over the internet. I can also offers parts kits, containing absolutely all components needed for this modification, from my company Grumble Amplifiers at a cost of £27.50 (approx $45 plus extra for shipping).

Brinsley Schwarz – a seminal figure in London's pre-punk "pub rock" scene of the 70s – is the former lead guitarist with the Brinsley Schwarz band, Nick Lowe, and Graham Parker's band The Rumour, and is now a guitar tech at London's Chandler Guitars and proprietor of his own amp mod and custom building operation Grumble Amplifiers in the UK, contact: (011-44)-208-647-9771.
Schematic diagrams by Steve Bailey

Technical articles

WHERE IT ALL BEGAN

Essential Ingredients of the Tweed Tone

By Mark Baier of Victoria Amplifier Company

Much has been written about Leo Fender's vision, and his impact on the music business and popular culture continues to loom large in our lives. Over the years, the Stratocaster and Telecaster have become icons overshadowing an equally important aspect of the man and his work – the guitar amplifier. First and foremost a radio repairman, Leo Fender created amplifiers that were born of his vision to provide reliable, sonically superior products for professional use. Years of repairing the radios and instrument amplifiers of the day led Leo to develop products that were more reliable, more electronically advanced, and just plain cooler than anything else available.

To fully appreciate the guitar amplifiers we call "tweed," it's important to understand how the amplifiers evolved throughout the late '40s to the end of the tweed era in 1960. The earliest amps Leo built were very simple in design and construction, using common radio, phono and PA designs of the day to produce low-powered steel guitar practice amps like the Princeton and Champion models and the larger professional-oriented Pro and Dual Professional models. These early models had relatively low gain circuitry and used octal style tubes such as the 6SJ7, 6SL7, and 6N7. These octal tubes have a sonic signature all their own, typified by a recessed top-end response and a clear, precise, clean low gain tone. When driven hard, the early tweed amps have a loose feel that is somewhat squashy, with a strong midrange component. Maximum wattage ratings were modest, as the 6L6s of the day were not rated for the kind of power that they would be later in the '50s. The individual circuit elements were also dictated by the tubes employed at the time. The now obsolete 6SJ7 pentodes and 6SL7 medium mu dual triodes require different methods of operation than later designs, further contributing to the early tweed sound.

The demands being placed on Leo and his design staff were considerable, and musicians in the field provided Fender with valuable input on how the amplifiers were performing and how they might be improved. With the convention of the time calling for the entire band to plug into the same amp, increasing power must have been a paramount goal. Thanks to the brilliant minds at RCA and Western Electric, new tubes and new circuit designs were readily available, affording Fender the opportunity to meet customer demand. Tube manuals and reference materials of the day made RCA and Western Electric engineering efforts available to anyone who knew what to do with it, and Leo knew his way around. No doubt, a well-worn *Radiotron Designers Handbook* was never far from his reach.

The availability of preamp tubes in the 9-pin miniature envelope further influenced Fenders amplifiers. These tubes hit the market in the late '40s and came replete with new RCA circuit designs to be utilized. During the '50s, Fender would mix and match the different types of circuits such as input amplifiers, tone shaping circuits, driver and phase inverters, etc, in a constant evolutionary process. By the time Fender was using the 12AY7 and 12AX7 as preamp tubes in all of their amplifiers, obsolete designs like grid leak bias and paraphase inverters gave way to more modern circuits, which made for more stable operation and greater power with less distortion – now we are getting somewhere!

The Phase Inverter

By the late '50s, Fender was using the "long tailed pair" phase inverter on the Bassman and Twin models, finally allowing the true potential of the 6L6WGB/5881 to be realized. This inverter was loosely adapted from the classic Williamson/Mullard high quality amp topology, and Fender's derivation of it has become the standard ever since. Remember that while Fender was employing common, readily available tube amp circuits, they were always tinkering with them, trying to optimize performance. It speaks volumes that 50 years later, these tube circuits are still being employed by virtually everybody in the business. The relevance of Leo Fender and his take on musical instrument amplifiers cannot be understated.

The sonic consequence of the different circuit elements is subtle, but apparent, for those with discerning ears, and it is most evident in the phase inverter. Early paraphase and the later self-balancing paraphase types are easily driven into clipping. This results in loss of top end and intense harmonics being generated in the midrange band. This also accounts for the onset of the distorted tone of the amp at a modest volume. The 6L6s in these late '40s early '50s models (5B, 5C, 5D models) were being asked to amplify an already distorted signal, so don't expect a lot of high-power 6L6 growl here. The driven note of these amps will be more saturated and compressed, giving them a spongy feel at higher volumes. The cathodyne inverter, also known as a split-load inverter, first appears in 1955 in the "E" series of amplifiers. (*Put simply, the letter in the Fender circuit designations used at this time indicated the evolution of the design as it changed through the years, while the number following it indicated a specific model.*) This elegant design uses one 12AX7 to deliver a truly balanced signal to the power tubes. We are now able to get more volume and headroom out of those 6L6s, which, throughout the '50s, are evolving themselves, increasing available power and wattage.

The improvement is not perfect, however, as the cathodyne inverter requires a driver stage in front of it to deliver the current needed. The result is that the inverter can still be driven into clipping, imparting its signature on the distorted character of the amp. Sonically, the effect is different than the paraphase, which has a squashed midrange vibe. The cathodyne produces a sweet and even compression that rolls off the top end somewhat. The harmonics leap out of the amp, yielding a rich and musical distorted note. The tweed amps employing this inverter (all the E series amps) seem to find themselves described as "sweet," "creamy" and "rich." The amount of phase inverter distortion can be controlled in amps using this circuit, because the signal coming off the tone controls is feeding the driver. When the tone controls are full on, the driver is seeing a hot signal, causing it to go into clipping, producing the distortion we all love. One can achieve a driven tone at a more modest volume than power tubes will allow by manipulating the tone controls to push the inverter over the edge before the power tubes start breaking up. If you want to clean things up, backing the treble control off to about "7" will do just that. Now, with the phase inverter holding together, we can hear more of the power tubes' contribution to the distorted tone of the amp. But to truly hear the sound of that 6L6 snarl without the phase inverter coloring things, they need to be driven by that most sultry of phase inverters, the "long tailed pair."

The long tail inverter as used by Fender in the late 1950s has become the industry standard, finding use at one time or another in virtually every brand of guitar and hi-fi amp made since 1960. It is far from a simple design, using a complex cathode coupled topology that practically eliminates distortion from the phase splitter. Another big plus for the long tail is the ability to generate a higher peak-to-peak voltage swing. This allows for more efficient power amplifier design (AB2), in turn increasing wattage. Fender tweaked the book values described in the classic *Mullard Tube Circuits for Audio Amplifiers*, but the theory of operation is the same. Leo merely played with the gain and balance until it met his requirements of low distortion, stable operation and low parts count. Leo Fender will always be remembered as a man who knew the value of economy and thrift, be it in design or manufacturing.

The Output Transformer

The models employing this new inverter were the late '50s "F" series amps. The 5F6 Bassman and 5F8 Twin represent the pinnacle in guitar amplifier design of their day, or of any day. Nearly every significant guitar amplifier designed post-1960 is derivative in some way or another of these seminal amplifiers. For the first time, a performer was able to get 40W (or 80W) of undistorted power, and after years of playing

Technical articles

soft, distorted amplifiers, this was loud, sucka! One important innovation found on the 5F6 and 5F8 models that permitted the extra wattage were the large-core interleaved output transformers (part #'s 45249 & 45268). Two features stand out on these transformers: the larger, heavier lamination (also called the stack, or core), and the hi-fidelity seven-layer interleaved primary and secondary windings. The larger stack translates to an honest 40 watts of power before the low end hits the ceiling – pretty important when designing a bass amp, which is what the "Bassman" was all about, right? The interleaved windings improve high frequency response. The windings are leaved together, causing them to be in closer physical proximity with each other, creating better inductive coupling. The general rule is easy to remember: the low-end response of any audio transformer is dictated by the size of the stack. You need that steel there to swallow and reproduce those low frequencies; basic physics. The upper frequencies depend on the windings and the engineering behind them. There are many ways of winding high-end transformers, (just ask anybody investing in 300Bs), and a respectful discussion of it is beyond the scope of this article, but suffice it to say that Leo Fender was fitting his amplifiers with the finest transformers ever used in a guitar amp.

While the larger core transformers of the late '50s Bassman and Twin models dramatically improved bass response, Fender used the physical principals of transformer operation to intentionally restrict bass response in other models. The E and F series of the Pro, Super, and Bandmaster were fitted with a smaller 28W piece of iron. This seems a bit odd when you consider that the tubes are capable of producing 40W. In fact, the original Triad engineering data specifically mentions restricted low frequency response as a design criterion. It all makes sense when you consider the wattage rating of the speakers used in these amps. As Jensen literature of the day rates the P10R as a 9-watt speaker, Leo certainly foresaw the imminent demise of a lot of them at full volume. Speaker failure can wreak havoc in a tube amp, and the use of the smaller transformer acts to restrict the amount of punishing low-end getting through to the speaker, thus making the amplifiers more reliable and enduring.

These transformers (part #'s 45216 & 45217) were not interleaved like their larger cousins. They are what is known as a layer-wound transformer. The primary is wound around the bobbin and the secondary is wound around that; one layer on top of the other. This is what gives these amps their smooth top-end, which is slightly recessed when compared to the Bassman and Twin models. These two types of Fender transformers, the larger 40W and smaller 28W types, remain unchanged to this day, and have been used by Fender ever since.

The Cabinet

Another important aspect to all things tweed is the cabinet and its effect on amplifier performance. In the late '40s and early '50s, Fender made the beautiful "V" front cabinets, featuring two discreet baffle boards mounted at an angle using a large chrome strip to connect them in the middle. The cabinets themselves were made of soft, resonant western pine, and the baffles, like all tweed-era amps, were thin and lightweight, allowing a lot of vibration to be transferred to the cabinet. This has the effect of the cabinet becoming a passive radiator of sound, ever so slightly contributing to the dynamic character and feel of the amplifier. In the V-front, the cabinet is oversized, giving these amps a deep and rich low-end. This period also saw the manufacture of the "TV" front cabinets. In these, the baffle is mounted to the front face and is secured evenly, with none of it "floating." This arrangement couples the baffle and speaker to the cabinet nicely, allowing the thin front face to vibrate, creating a lively, open sound. Though none of these amps features reverb, the resonant quality imparted by the cabinet design and materials hints at it, and this is a subtle yet significant element of tweed tone.

The wide panel amps of the early '50s (series D) and narrow panel models of the mid to late '50s (series E & F) feature baffle boards that are attached to the cabinet at the top and bottom but not at the sides. This is commonly referred to as a floating baffle. The effect of the cabinet resonance is not as pronounced as the TV and V front types, but is still of consequence. Much of the percussive and vocal qualities found in tweed amps are attributable to these lively pine cabinets and their light, thin baffle boards.

Design Intentions

Any assessment of tweed amplifiers would be flawed without some thought about the intentions of Leo Fender and his designers. Before Leo ever made a Telecaster, he was building steel guitar amplifiers and PA equipment. In an amplifier of this kind, undistorted clean tone is the goal, and all engineering efforts served to increase output and lower distortion. Throughout the evolution of the Fender amplifier, this has always been the guiding principle. The selection of the 12AY7 as the input amplifier tube (V1) of choice in the tweed amps is obviously a step to delay the onset of clipping in the preamplifier.

It is also worth noting that Fender amplifiers were made to have Fender guitars plugged into them. As their guitars sported pickups that are rather low output, Fender employed higher gain preamplifiers to complement these guitars. Imagine the look on the face of the first player to plug a P-90 equipped Gibson into a '52 Pro. BB King spoken here! These were also amps designed for serious professional use. Leo Fender's years of repairing other companies' amplifiers served to educate him about what was going to hold up and what wasn't. Consequently, the construction quality was without equal. The use of the vulcanized fibre eyelet board served two purposes. The first was to shock mount the critical components from the abuses of the road. True point-to-point constructed amplifiers, where the electronic components are soldered directly to their destinations, from point "A" to point "B" as it were (between a volume pot lug and a tube pin, for example), don't hold up well to being dropped out of the van 300 nights a year. Abuse will catch up with that old Gibson amp, cowboy! The other is ease of service and manufacture. The aforementioned Gibson amp looks like a rat's nest inside – there are resistors and capacitors going every which way in apparent disorder. The Fender, with the neatly laid out eyelet board and perfectly dressed yellow wire connecting everything to the tube sockets and pots, looks like a work of art by comparison. The icing on the cake are the masterfully drawn schematic and layout drawings provided with every Fender amp. The guy who drew these schematics, many of which are reproduced in this book, was the Michelangelo of electronic draftsmen. It's a shame he didn't draw everybody's schematics.

By early 1960, Leo Fender once more took the road less traveled and reinvented the guitar amplifier for yet another generation. The Tolex covered amplifiers of the '60s ushered in a new concept in cabinet construction, one which would allow Fender to fully realize his vision of supremely roadworthy amplifiers. The superior tonality of a Fender amp was by now a given, and the innovations of reverb and tremolo were built upon the sonic bedrock that is all things tweed. The electronic design elements that had been firmly established by 1958 were the root system. The thicker and tighter baffle boards of the Tolex amps reduced cabinet vibration, in turn focusing more energy to Promised Land – clean, undistorted guitar tone. The completely revamped chassis attachment, a vast improvement on the tweed U-frame chassis, would set the new standard, one that endures to this day. By late 1963, form and function found their final and perfect union with the introduction of the AB763 series, and the old Fender tweed amps were overlooked for years.

Thanks to the vintage guitar madness of the 1990s, the product of a hard-core nucleus of enthusiasts (you were one of them, Aspen!), a new appreciation was created for the old dusty tweed workhorses. Amplifiers that were considered only as "used amps" 10 years ago now command top dollar, and continue to be appreciated as the magnificent tools they always were. That old '59 Bassman... she may be bruised and a bit smoky, but with a little respect (and a new set of tubes) will still dare us to make love to our guitars like never before. ■

"Mark would like to thank Les Plopa (KB9 RBY) for technical assistance in writing this article"

THE TUBE AMP BOOK 143

Technical articles

Modified Hiwatt bias circuit
(To existing bias supply; Caps are 100 MFD @ 100 VDL; Diodes are 1N4007)

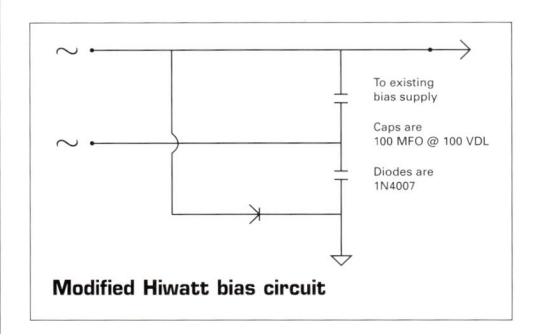

MODS FOR HIWATTS

HIWATT bias doubling circuit

Hiwatt, Orange and Sound City amps often do not have enough bias supply to run reliably. Red Rhodes has a simple solution as shown below. Each side of the sine wave is rectified by a single diode (1N4007) and used to charge the two caps (100MFD @ 100VDC) which are connected in series. The overall negative bias voltage is thereby doubled.

MODS FOR MARSHALLS

100-WATT TO 50-WATT

Many players who have 100W Marshalls are playing in clubs that are smaller and won't allow for high volume guitar levels, levels that make Marshalls sing better. One easy solution is to remove two of the power tubes and thereby reduce the power by about half. This is a quick and easy method, but you must follow two important rules for this to be a safe mode to run your amp in.

Rule One: Your 100W Marshall has four power tubes that are split into two halves in a circuit design called Class A/B or often called a push/pull circuit. Simply stated, the left half pushes the top side of the sine wave (the note) and the right half pulls the bottom side of the sine wave. You may pull out two tubes from your output stage *but* you must take only one from each half, one from the two left tubes and one from the two right tubes. I usually just take out the outside two and leave the inside two. Remember if two of your quartet of output tubes are powering the amp and the other two are on the shelf, you're wearing them out at different rates – so alternate them occasionally so that their life expectancy stays about the same.

Rule Two: Your output tubes have an impedance just like your speakers do. When you set the speaker impedance on the back of your Marshall, you're actually setting the interfacing impedance between your output tubes and your speaker, via the output transformer. If you remove two tubes from your output stage, you have changed the impedance of the output stage. A stock 100W Marshall with two cabs should have the impedance selector set at 8 ohms with all four power tubes installed. If you remove two tubes, change the impedance to half of this, or 4 ohms. If you are running only one cab at 16 ohms, change the selector to 8 ohms. In other words, you need to lower by half the impedance at the selector. This will then produce the proper impedance between the tubes and your speakers.

Marshalls have almost always used the EL34 type power tube in their 50W and 100W amps. However, the amps they exported to the USA from the mid-'70s until mid-1986 have been sold with the USA-made 6550 power tube. Although Jim Marshall and his company preferred the sound and response of the EL34, the US distributor chose the 6550 for its increased reliability and lifespan. When they changed distribution again, they standardized the whole line to the EL34 worldwide. There still are plenty of the 6550 amps out there and guys are still wondering why their amps sound so much different from the "old Marshalls." Well, those old Marshalls had EL34s in them which sound completely different (See the Marshall chapter in the Amplifier Companies section for more detail on this).

We offer the following two modifications to convert your 6550 Marshall into an EL34 Marshall with a smoother, warmer distortion. We also offer a mod to convert UK EL34 Marshall into a 6550 model, popular with some heavy metal players who like it real loud and crunchy.

Marshall with 6550s to Marshall with EL34s

To modify a USA Marshall (6550s) to the EL34 tube type, the bias needs to be decreased by about 10 volts. This may be accomplished in two ways. The first way follows the Marshall factory schematic and is the way it is done on amps made in the factory for the two different markets, 6550 output tubes for the USA and Japan, and EL34s in Europe and the rest of the world.

The first method replaces three resistors in 50W models, and four in the 100W amps. Notice the two schematics of the 50W and 100W amps: the resistors that set the range of the bias pot are two 220Ks and one 56K for the European amp while they are changed to two 150K and a 47K for the USA. amps. There is also a change on the 100-watter from a 27K to a 15K near the 1N4007 diode.

To convert your USA Marshall to use EL34 tubes, start by locating these resistors in the bias circuit that set the voltage range of the bias pot, and change them to their alternate values noted on the schematic. The resistors are located toward the right end of the PC board, when the amp is out of the wood, upside down, and the controls are toward you. Three are located directly above the bias pot, the other is near the rectifier diodes which are at the right. When the modification is complete, the bias reading at pin 5 should vary with the bias pot between -36 volts and -45 volts for the 100W Marshall and show a range between -34 volts and -40 volts for the 50W model. It goes without saying that the proper point of bias would still need to be adjusted with actual set of EL34s in the amp. This should be done using a signal generator and a scope, adjusting the proper bias by observing the cross-over distortion in the sine wave at the most over-biased point and rotating the bias pot the other direction until the sine wave just straightens out. If you do not have the luxury of a scope and signal generator, we would also recommend setting the bias with a Bias Probe available from Hunt, Dabney and Associates. This device monitors the total current flowing through the tube while a manual provides listings of most amps and the proper flow that would be measured in a correctly biased amplifier.

The second mod for using EL34 tubes in an amp made for 6550s is a simpler and faster method we've used here at Groove Tubes, and this may be easier for you.

With the amp out of its cabinet, upside down, with the controls facing you, locate the 47K resistor which is directly above and in line with the bias pot. Solder another 47K resistor in parallel with this. With the bias pot set to "max" and the output tubes out of the amp, check to see that at least -38 volts is present on the grid of the socket (pin 5). If so, install the tubes and bias the amp. If there is more than -38 volts, that's okay, but much less may result in not being able to bias the amp properly. The value of the added resistor may be adjusted to set that voltage, with a lower value lowering the available bias.

Another important item to change when changing 6550s to EL34s or vice-versa is the feedback wire that connects to the speaker impedance selector from the output transformer. It is usually the purple wire and should be connected to the 4-ohm leg of the switch if you are using 6550s and the 8-ohm leg if using EL34s.

Marshall with EL34s to Marshall with 6550s

To modify a European Marshall for 6550s, the bias needs to be increased by about 10 volts. There are two acceptable ways to accomplish this. The first is the "official Marshall" method. The second is our own method, which is faster and easier. This may be done by locating the resistors in the bias circuit that set the voltage range, and changing them to their alternate values, which are screened on the PC boards under the parts. The two resistors in question are located toward the right end of the PC board, when the amp is open, upside down, and the controls are toward you. One is located directly above the bias pot, the other is near the rectifier diodes which are at the right. Follow the enclosed schematic to confirm the correct values for either a 50W or the 100W Marshall.

144 THE TUBE AMP BOOK

When correctly modified, the bias pot will now supply a range somewhere between -44 volts and -55 volts for the 100W Marshall and a range between -38 volts and -48 volts for the 50W model. Naturally, the bias must be set for the specific set of 6550s to be used, preferably with a scope and signal generator to set the exact amount of bias. Another important item to change when changing 6550s to EL34s or vice-versa is the feedback wire that connects to the speaker impedance selector from the output transformer. It is usually the purple wire and should be connected to the 4-ohm leg of the switch if you are using 6550s and the 8-ohm leg is using EL34s.

Marshall master volume
Courtesy of Ken Fischer, Trainwreck Circuits

This modification is to the phase inverter section of your Marshall. Locate the two 220K resistors that split apart at the bias feed on the circuit board. One of these is the positive while the other is the negative feed to the output tubes' grids. They are tied together at one end and split to form "V" on the circuit board. Follow the diagram and remove these resistors, replacing them with a dual pot. The value of this dual pot should be between 100K and 250K.

This master volume modification will return the amp to normal "pre-mod" performance specs when the dual pot is turned all the way up and still produces a good master volume. The actual gain will remain unchanged, however, and the distortion level from a Marshall preamp into the master section will vary with the vintage of the amp (later ones have more gain than the very early ones). The amount of distortion produced by the master mod can be increased by increasing the gain from the preamp through additional modification – *however*, this type of mod will change the tone of your amp, possibly decrease the tube life, and increase the noise level of your amp. Therefore, try the dual pot first and see if that's enough before radically changing your Marshall.

Transient suppression MOD (tube saver MOD)

There are two minor modifications that can be made to Marshalls, and most other amps, that will greatly increase reliability. They both reduce the transient spikes that eat up power tubes – usually present when the amp is really cranked way up. These spikes don't do much to improve your sound and they make the amp unstable, while damaging the power tubes.

Mod 1: Find a metal oxide varistor with a rating of 250 volts. You can also use two 130-volt varistors wired in series. These 130-volt varistors can be found at Radio Shack if you're not near a professional electronics store. Place the varistor(s) between the primary leads of the power transformer.

Mod 2: Find 6 1N4007 diodes. Install three on each side of the power output tubes that have the plate winding attached to them. Three to side A and the other three to side B. The plate winding will be attached to pin 3 on all Marshall as well as most other amps. Attach the three diodes in series from pin 3 to ground, making sure they are in the right polarity. The right polarity for the 1N4007 diodes is the band pointing toward the plate of the tube or pin 3.

Trainwreck amps and Mesa/Boogie amps are the only amps we know of that come standard with both of these circuits.

Power transformer early 50W bias/standby MOD
Courtesy of Ken Fischer, Trainwreck Circuits

On early Marshall 50W amps, the bias supply is connected to the cold side of the standby switch. When the amp is switched to play, the output tubes run full plate voltage with no bias for the few seconds it takes for the bias voltage to build. This can result in blown HT fuses and shortened tube life. The solution is to move the bias feed to the hot side of the standby switch. For visual guidance, please refer to the Marshall diagrams contained elsewhere in this book.

Hum reduction tip
The physical location of the wires running through your amp chassis can increase or decrease the amount of hum you hear. This little tip will make a big improvement in lowering the hum in all Marshalls. Open the amp up and, looking at the circuit board, find the first two preamp tubes farthest away from the power tubes. Notice the green wires that come from pins 2 and 5 on the first tube and from pin 2 on the second tube. These wires carry the low level signal through the first gain and tone stage and are therefore sensitive to picking up hum. Simply pull these wires up and away from the chassis and the other wires feeding the socket (these carry high voltages). This will greatly reduce the hum picked up on these wires. Leave the other wires flat against the chassis.

VOX AC30 CHECK-UP
By David Petersen
Reprinted from The Guitar Magazine (UK)

There are a number of relatively simple jobs a player can undertake to ensure his or her vintage AC30 is running in tip-top condition, and perhaps save larger maintenance and repair bills further down the road. Since the AC30 was conceived mainly as a more powerful version of Vox's then-flagship amp the AC15 – the basic sound of which most players were already happy with – it naturally shares many key circuit features with its predecessor. It kept the EL84 output tubes (but this time using two pairs for the extra power) running in Class A mode and, importantly, unlike most of the all-pervasive Fender amps of the time, the Vox design had no "negative feedback" circuit.

Feedback circuits were originally introduced to suppress harmonic distortion of all kinds – although, conversely, this type of distortion is something which can be vital in the right proportions to good, rich guitar tone. Negative feedback can also reduce background noise level – but it tends to rein in desirable output stage distortion as well. The AC30's "open loop" operation thus accounts for much of the amp's unique tonal character, allowing the unusually high second harmonic content of the EL84 tubes to expand the tone in a natural way. In other words: they sound good.

Condition Check
The thing about AC30s is this: the open loop design we just mentioned relies heavily on both the preamp and power amp sections running sweetly. This amp uses the gain factors of both the single-tube input stage of each channel and the simple power output stage to achieve sufficient sensitivity for its purpose, and deterioration in either of these circuits will lead to inadequate gain. Feedback designs like Fender and Marshall can afford to lose a few dB of gain before any audible problems: AC30s, on the other hand, have no real margin of performance.

The usual cause of trouble is ageing EL84 tubes – either because they're just plain ancient or, in the case of more recently fitted ones, because the life expectancy of newer tubes compares poorly to those made in the '50s and '60s by European makers who had to conform to stiff standards such as those of the British Valve Association (BVA to its friends – as marked on those precious Mullards you occasionally come across). It may be necessary to replace modern output tubes every 18 months; with the old BVA-marked products it was more like every five years, assuming about 1000 hours running time a year – with proportionally longer or shorter replacement periods according to how much you play your amp.

Other symptoms of tube ageing include uneven background noise (be careful, though: this can have at least two other causes) and high hum level. One or more output tubes going "low emission" can cause this last problem by creating an imbalance in the push-pull output stage that limits its normal hum cancelling ability. Another sign of this is a tendency to "run red" in one or more output tubes, which emit an unusual glow easily seen through the vent slots. Confusingly, the overheating tube is not always the bad one – its glow can mean that its neighbour has given up and left it to do all the work. Either way, don't leave the amp running like this, as "red" is an alarm signal which says something expensive is about to happen.

Technical articles

Tube Job

Revalving an AC30 is not unduly difficult, and the cathode-bias design – which evens out differences in the operating point of a reasonable range of EL84s – makes it easier (it also means that the amp has no provision for bias adjustment, unlike a grid-bias design such as Marshall or Fender). Best choice is a matched quartet of EL84s chosen for medium or high gain and matched under Class A operating conditions. Anyone buying tubes for AC30s should ask if this has been done.

The rectifier is the tube with the unenviable task of sorting the incoming voltage, changing it from alternating current (AC) into direct current (DC). The Mullard GZ34 rectifier fitted to JMI amps is a rugged old tube and you'll often find them in good condition 35 or 40 years down the road, surprising considering the load they're under. The GZ34 is immediately identifiable: it's the biggest tube in the amp. It's hard to find a good replacement these days, but if you manage to find a real NOS (new old stock) Mullard, be ready to pay any affordable price for it against the day when the existing one fails. This unwelcome event is usually heralded by the failure of the control panel mains fuse. The preamp tubes do not require frequent replacement, particularly if they are BVA, but substituting new good quality ECC83s (12AX7s) in the first stage and driver positions can give a tired-sounding AC30 a new lease of life. Viewing the chassis from the rear, these are the fourth and fifth tubes from the left in a Normal model, or fourth, fifth and sixth in a Top Boost.

Heavy Weather

The environment is the AC30's next biggest enemy. The amp runs hot – no way around this – but check that nothing blocks any ventilator when the amp is on and that the rear of the unit is open to the air. At the other end of the thermal scale, don't leave it stored anywhere you yourself wouldn't be comfortable, with particular emphasis on the trunk of your car and outside storage areas, sheds, unheated garages, and so forth. Long spells of damp can cause corrosion of the contacts between tubes and sockets, and it's not uncommon for one or more sockets to require replacement – an awkward job for the inexperienced, and quite expensive if done by a professional.

A little preventative tip here might save trouble later: **with the amp unplugged from the wall** (once again, read – and understand – our **WARNING!** at the start of the Technical section before opening up your AC30 or any guitar amp) take off the rear cover by removing the six screws and slide the amp carrier panel out rearwards – not all the way, but enough to allow access to the tubes. Be careful not to tug at the speaker wires: if they're tight, unscrew the terminal block screws on the chassis and disconnect them, noting carefully which wire goes to which terminal. Standard connection is negative (black or blue) to center, positive (red or yellow) to lower terminal, a connection worth checking anyway as lots of amps have it wrong.

Now, one by one, remove each tube, dip its pins in a small container of light machine oil (3-In-1 is fine) sufficiently to cause a droplet to appear on each pin, then re-insert it in its holder. This is a very effective protection against corrosion and can sometimes even be used to cure an amp that would otherwise need replacement tubes, but don't be too liberal with the oil, or clouds of smoke may billow forth from the amp next time you use it – somewhat alarming, though it goes away after a bit. This trick is also good against noisy, suspect tube contacts. Now reconnect the speakers and put everything back as it was. Procedures are a little different for the current issue AC30, but a newer amp is unlikely to have corrosion problems yet anyway.

Dipping each tube's pins in light machine oil then reinserting them can help cure minor corrosion problems in tube sockets, a common problem in older amps.

Speaker Mind

Speaker problems are relatively uncommon. A legendary – and valuable – component of an original, vintage AC30, the "Blue" (and to a slightly lesser extent, the largely similar "Silver") is easily capable of handling 40W of guitar input in each speaker, perhaps 20W of bass guitar, and the other types of speakers used at different times were never far behind (not that you will want to temp fate by trying this either with your originals or the reissues: both are expensive units!). As the AC30 has a very gradual onset of distortion with not too many fast speaker-killing squared waveforms in its output, the speakers in these combos have an easy life. But poor storage conditions can cause corrosion of the voice coil or its connections, and a sudden return to hard service after years of inactivity, as sometimes happens when the amp is sold by its previous long-time-no-play owner to an enthusiastic new one, can cause problems.

As the 2x12" speakers are wired in series, if one unit "open circuits" (fails) it will mute the other, resulting in a sudden and total loss of output. **Do not** try to continue playing if this happens as it could damage the output tubes, or worse – and more expensively – the output transformer.

Find out which speaker is open-circuit by bridging the terminals on one and then the other with a piece of wire or a pair of sharp-nosed pliers (no dangerous voltages here). If you hear the usual background noise return to the cab when you bridge a speaker, then the one you've bridged is the dead one. Fortunately Celestion still issue re-coning kits for "Blues" and G12Ms, and there are a few good speaker repairers still working in the USA and the UK, so there should be no problem getting your AC30 speakers restored to the proper standard.

If you want to replace your vintage speakers for new – either because they can't be fixed or 'cos you'd rather gig with newer reproductions, storing the original units safely to preserve the resale value of the amp – Celestion now produces a highly regarded reissue, called the Alnico Blue. They sell for big money, but are considered one of the finest-sounding lower-wattage speakers available today. Weber Vintage Sound Technology in the USA also builds two good Vox replacement speakers with genuine alnico magnets, which they call the Blue Dog and Silver Dog, and which cost somewhat less, and a lot of players are impressed with these.

If the red pilot lamp fails or goes intermittent ("on the blink"), this might require no more than carefully tightening the screw-fit lamp readily seen from the back of the chassis **with the amp unplugged from the mains** and the rear cover removed. If the amp is obviously working but the lamp won't light, throw in a replacement 6-volt LES bulb.

The Big Stuff

Other jobs are best left to the experienced technician as they involve knowledge of electronics and/or work on the chassis in the powered-up state. Examples of jobs you may need to take to your local tech include: replacement of the main filter electrolytic can in

Technical articles

the power chassis, recommended for any JMI model as these components lose performance whether used or not after about 20 years; identifying and replacing resistors whose values have drifted by more than 20 per cent, often resulting in low, noisy or even dead preamp channels; restoration of tremolo; replacement of the input jack array.

There are many other possibilities – these are just the common ones – but if it takes a little time and expense to put your AC30 into top shape, the result nearly always pays off handsomely in musical terms… and it's inspired some of the biggest names in our business.

David Petersen is one of the UK's most respected Vox amplifier specialists, is co-author (with Dick Denney) of the book The Vox Story, *and is chief amplifier technician at Chandler Guitars in London.*

THE LAST WORD ON CLASS A
By Randall Aiken

Which is better, class A or class AB?
From a guitar amplification standpoint, neither class of operation is "better," they are just different. You shouldn't get too hung up on the "class A" designation, because most of the push-pull amplifiers that are supposed to be class A aren't really class A at all, they are just cathode-biased, non-negative feedback class A/B amplifiers. Operating class is not the reason for the tonal differences between these amplifiers.

What is the difference, then?
The cathode biasing and lack of negative feedback are two of the main differences between the Vox clones and the Marshall/Fender style amps. The typical Marshalls and Fenders used a fixed-bias output stage with negative feedback from the output back to the phase inverter input, while the Vox clones use a cathode-biased output stage and no global negative feedback. In addition, the output tubes and preamp stage/phase inverter configurations contribute greatly to the tonal signature of these amplifiers.

Cathode biasing vs fixed biasing
In a cathode-biased amplifier, the bias voltage is developed across a cathode resistor that is bypassed with a big electrolytic capacitor. In a class AB amplifier, as the current through the tube increases, the average voltage across the cathode resistor changes, which modulates the plate current, creating a bit of "sag" and a dynamic change in the harmonic structure of the note that changes while playing. This occurs because the plate current in a class AB amplifier is not continuous for the entire AC cycle. The tube goes into cutoff for a portion of the cycle, which means that the average DC level of the signal on the cathode will shift, changing the operating point of the tube, with the resulting dynamic tonal changes. The average value of a sine wave is zero, but the average value of a clipped sine wave, such as occurs when the plate current is cut off for some percentage of time, is not zero. The current in a true class A amplifier is constant, so it doesn't exhibit this bias shift, unless driven to clipping, where all bets are off. This is why a cathode bypass cap is not necessary in a true class A output stage – the plate currents are equal and out of phase, unless there is an imbalance in the output transformer, the output tubes, or the drive signals (it is a good idea to use one anyway, for these reasons). The fixed-bias amplifier maintains the bias at a more constant level, so it doesn't have the constantly changing operating point that varies with the output level.

The effect of global negative feedback
The use of global negative feedback does several things: it flattens and extends the frequency response, it reduces distortion generated in the stages encompassed by the feedback loop, and it reduces the effective output impedance of the amplifier, which increases the damping factor. All of these things affect the tone in some manner.

The flattened, extended frequency response obviously changes the tonal character by removing "humps" in the output stage response and producing more high and low-end frequencies. The distortion reduction makes the amp sound cleaner and more "hi-fi," up to the point of clipping. Perhaps the main difference for the "feel" is the increased damping factor produced by the negative feedback loop. The decreased effective output impedance causes the amp to react less to the speakers. A speaker impedance curve is far from flat; it rises very high at the resonant frequency, then falls to the nominal impedance around 1kHz, and again rises as the frequency increases. This changing "reactive" load causes the amp output level to change with frequency and changes in speaker impedance (a dynamic thing that changes as the speakers are driven harder). Global negative feedback generally reduces this greatly. This can be good or bad, depending upon what you are looking for.

Negative feedback makes the amp sound "tighter," particularly in the low end, where the speaker's resonant hump has the most effect on amplifier output. This is better suited for pristine clean playing or a tight distorted tone, while a non-negative feedback amp has a "looser" feel, better suited to a bluesy, dynamic style of playing. The other disadvantage of a negative feedback amplifier is that the transition from clean to distorted is much more abrupt, because the negative feedback tends to keep the amp distortion to a minimum until the output stage clips, at which point there is no "excess gain" available to keep the feedback loop operating properly. At this point, the feedback loop is broken, and the amp transitions to the full non-feedback forward gain, which means that the clipping occurs very abruptly. The non-negative feedback amp transitions much more smoothly into distortion, making it better for players who like to use their volume control to change from a clean to a distorted tone.

There is an output stage topology that is kind of in between, called "ultralinear" operation. This uses local negative feedback to the screen grids of the output stage by means of a tapped output transformer primary. This increases the damping factor and makes the amp a bit tighter without the use of a global negative feedback loop (you can use global negative feedback with ultralinear output stages, but you may not like the tone as much). The Dr Z Route 66 amplifier uses an ultralinear output stage. There is also a triode output stage, which has even higher damping factor than ultralinear, but some players feel that it sounds too "compressed" and midrangey, while others like it. Part of the reason for the midrange emphasis is the increased input capacitance of triode mode over pentode mode because of the Miller effect, which in effect, multiplies the grid to plate capacitance by the gain of the tube. This increased capacitance rolls off the high frequencies.

The impact of output tubes and preamp stages
The other main difference is the use of EL84 tubes vs EL34 or 6L6 variants. The EL84 tube generates more distortion than an EL34 or 6L6 (even when running clean, particularly without negative feedback), and has a nice overdrive tone. These distortion products tend to color the tone more, even at low levels where the amp is perceived to be "clean." In addition, the preamp stages and phase inverters are completely different, which has a huge effect on the amp tone. The details of the preamp differences are beyond the scope of this article, but they can be seen by comparing the schematics.

Does true class A operation require any particular current or bias point?
True class A operation does not have to be above any particular current rating or dissipation. It depends on the tube type, the power supply voltage, the reflected impedance, and the required operating point. However, in general, when a class A power amplifier is designed, the bias point is chosen to correspond with the spot on the plate curves at the intersection of the load line, the plate voltage, and the maximum dissipation curve that gives maximum symmetrical swing in both directions before clipping. This means that the tube is biased right at maximum plate dissipation, which is okay, because the dissipation is

Technical articles

maximum at idle in a class A amplifier, and does not increase with applied signal, as it does in a class AB or class B amplifier.

This is not to say that that is the only current and voltage that will work. If you lower the plate voltage by 100V, you will find another "optimum" spot where these lines intersect. If you change the reflected load impedance, you will find yet another optimum spot. There is, however, an upper limit on the voltage that can be applied where you can no longer bias for symmetrical swing about the idle point without exceeding the plate dissipation ratings. This is the limiting voltage for that tube in true class A operation *at the max recommended tube ratings*. If you choose to run the tube over ratings, as seems to be the case in SF Fender Champs, you can bias the tube to a point that is running class A, but is above the maximum dissipation curve. Although this seems to work with some tubes, it is not a recommended practice.

This holds true for both single-ended and push-pull designs. In push-pull class A, the bias point and plate supply voltage is the same as for single-ended, but there is a phase inverter and a center-tapped transformer, which are used to increase power and reduce distortion (even-order harmonics are canceled, and power supply hum is canceled in a balanced push-pull amp). Power is twice that of single ended (for a two-tube push-pull vs. a single tube single-ended, etc).

To get a better feel for this, take a set of plate curves for a given tube, and draw a load line representing the reflected impedance (it has a slope corresponding to the negative reciprocal of the reflected load impedance, and passes through the intersection of the bias current and plate voltage lines), and draw a curve representing the plate dissipation (it will be a parabolic shape, with each point equal to the current that corresponds to the plate dissipation divided by the plate voltage). The load line should just touch the plate dissipation curve at the selected plate voltage (for max power out – if you want less than max power, it can be below the dissipation curve). The current corresponding to this point will be the required bias current, and the dissipation will be maximum at that point. All tube signal swings will occur on the load line (assuming a purely resistive load – reactive loads generate elliptical load lines), so you can find the plate voltage swing for a given grid voltage swing, and you will see that you will have to either change the plate voltage or the reflected load impedance, or both, in order to get the optimum class A bias point. Don't forget that the actual plate voltage swings both above and below the supply voltage, and the center of the swing is the actual plate supply voltage. This is kind of confusing at first, because it isn't intuitive that you could get a 400V peak with only a 250V supply (ie, a swing from 100V to 400V, centered around 250V). The "extra" voltage comes about because of the nature of how the output transformer works.

Does biasing at max dissipation guarantee class A operation?

Just because you are biased at max dissipation does not mean you are class A! You must be in the region where the voltage swing is symmetrical and biased in the center of the range, where plate current flows for all unclipped output. Biasing to a high voltage and low plate current whose product equals the maximum plate dissipation might not allow this, because, although you are at max plate dissipation, the bias point is such that plate current will flow for an appreciably less time on the negative signal swing (cutoff) than it will on the positive signal swing (saturation), and *no* load line can be found that will allow symmetrical swing, or it will be in such a non-linear portion of the curves as to be unusable. This is because the plate voltage is too high, and the max allowable current without exceeding dissipation limits is too low. The same thing can occur on the other end of the scale, where you can reduce the plate voltage to a point that the max dissipation current will exceed the maximum allowable plate or cathode current ratings of the tube. There is an optimum area of the curves that will become apparent when you start drawing load lines and picking bias points. It is a bit of an iterative process, so the tube manufacturers make it easy for you by listing typical class A operating conditions in the data sheets.

In theory, you can take a class AB push-pull amplifier and convert it to class A push-pull operation, *however*, you would, in nearly all cases, have to reduce the plate voltage to be able to bias the tubes into the class A region, because the whole reason for going to class AB is to get higher power, so the plate voltage is run higher and the idle current lower than what is allowed in class A. Once again, you have to look at the plate curves for the particular tube to determine where the allowable class A region is. If you simply bias a class AB amp to max dissipation at idle, you will find that as you apply a signal, the tubes will dissipate more power, and they will start to glow a lovely cherry red color, and something will croak. In addition, the power supply and/or output transformer may not be able to handle the extra current required for true class A operation, so, unless you know the ratings of the trannies, it is best not to attempt this, even if you lower the supply voltage.

Are those class A amplifiers I see advertised really class A?

There is much debate raging in the marketplace about "class A" amplifiers, and whether or not they are truly class A, or just class AB amplifiers unscrupulously marketed to the unsuspecting public as class A. The truth is that most, if not all, are in reality cathode-biased, non-negative feedback class AB amplifiers, contrary to what the manufacturer's literature may say.

What is the difference between class A and class AB, and why is it a problem for so many people?

The fundamental problem is in how class AB is defined, and how people interpret it. The people who say a class AB amp is "class A at lower volumes" are technically wrong, but for the right reasons. If you were to define class A as being only conduction for a full 360 degree phase angle, you would be correct. However, there is more to the definition of amplifier classes than that.

The defining factor in determining whether or not an amplifier is class A, class AB, or class B *has* to be made at the full output before clipping, otherwise, the class definitions have no meaning whatsoever. It is indeed, a very black and white thing, and depends on the bias point on the characteristic curves, and the load line, among other things.

If, at the full undistorted output, the plate current flows in each tube for a full 360 degrees of the input conduction cycle, the amplifier is class A. However, if the amplifier is biased such that the plate current cuts off for an appreciable time during each cycle at this full undistorted output power, it is then a class AB amplifier. If it is biased such that each side is in cutoff for half the input cycle, it is a class B amplifier. Note that cutoff does not mean that the output of the amplifier is clipped, or distorting. Cutoff refers to plate current cutting off on one side of a push-pull pair for a portion of the cycle, while the other side continues to function. The output waveform is still a clean, unclipped sine wave, because the transformer sums the two "halves" of the input signal into one composite signal. Effectively, one tube amplifies the "upper half" and the other tube amplifies the "lower half." This is done to provide higher efficiency and greater output power. In a class AB amplifier each tube amplifies a bit more than half the signal, in order to reduce the distortion that occurs at the zero crossings of the waveform, which is called "crossover distortion."

Here is where the problem comes in: because a class AB amplifier is biased so that the plate current flows for the entire cycle at lower output levels (which is done to reduce crossover distortion), many people claim it is a "class A amplifier at lower volumes." This is simply not true. It is operating in conditions *similar* to class A, but is not a class A amplifier by any means. It is still a class AB amplifier, no matter what you choose to call it.

Now, what are the differences, you might ask? Well, for one, the Class AB amplifier is biased in a more non-linear portion of the characteristic curves, which means it has more distortion than a true class A amplifier. Also, the efficiency will be greater than is theoretically possible with a class A amplifier at these levels. There is a very real difference in tone and operating conditions between a true class A 10W amplifier running at say, 1W, and a 10W class AB amplifier running at 1W. Same output level, same overall power level, *but* a different class of operation, different amount of distortion, different efficiency, *and* a different tone, even though neither one of them is in cutoff for any portion of the output cycle at that low level. This is due to the bias point differences and load line differences. The differences become even more apparent when the amplifiers are run at their full undistorted output power. The true class A amplifier will have no crossover distortion, while the class AB amplifier will. The average plate current for the true class A amplifier will not change, or will change very little, from idle to full output power, while the average plate current in a class AB amplifier will increase dramatically. This will lead to "sag" in the power supply that doesn't exist in the true class A amplifier, which again results in a tonal change.

As you can see, there is indeed such a thing as a "true class AB" amplifier, just as there is a "true class A" amplifier, and the class definitions are not at all ambiguous, except to those who don't understand them, or choose to ignore them for marketing advantage.

One more thing: What if you push the class A or class AB amplifier into clipping? Does it then become a class AB/ B, C, or D amplifier? No, of course not. It is simply the same class amplifier it was to begin with, but driven into clipping. A class A amplifier driven to clipping is still a class A amplifier by definition. This is why amplifier classes are defined the way they are. Otherwise, the class designations would have no meaning. Any amplifier can be driven beyond its limits into a full-clipped square wave output (unless it is limited), but that doesn't make it a class D switching amplifier, now does it?

Which one should I buy?

The bottom line is this: don't worry about whether an amp is class A or not; find out if it is cathode-biased or fixed-biased, and whether is uses global negative feedback or not, whether it uses a pentode, triode, or ultralinear output stage, and what type of output tubes are used. These parameters will give an idea of the "feel" of the amp, but in the end, you still must play the amp and use your ears to tell you which one is best suited for your playing style. Don't make a decision based on technical specs alone – or marketing speak – because you might miss out on a great-sounding amplifier!

Randall Aiken is proprietor of Aiken Amplification, www.aikenamps.com.

REPLACING OUTPUT TRANSFORMERS

By Doug Conley

The output transformer of a guitar amplifier is the key to its successful performance. Occasionally they go bad and must be replaced, and the technician or DIY'er must understand several things. The output transformer does three things. First, it transforms the modulated AC, superimposed on the plate voltage, into a separate signal, removing the several hundred volts DC from the speaker circuit. It also has to turn the differential voltage – the two-phase signal with each phase 180 degrees out of phase – to a single-ended current with one end connected to ground for feedback to the preamp. Most importantly, it acts like a transmission, converting a high voltage, low current signal to one of low voltage and high current so it can power the speaker, which is low impedance.

If you need a new output transformer, usually you are best to obtain an exact replacement from the factory. If this isn't possible, perhaps if the maker is out of business, usually you can find a good substitute if you know three things.

The transformer must have the proper impedance ratio. This is like the gear ratio of a transmission and is expressed as an impedance ratio, say, 1500 to 1, or 6000 to 4. This impedance ratio is *not* the same as the voltage, or turns ratio: it is the *square* of it. (Or the turns ratio is the square root of the impedance ratio.) If the amplifier requires a 4000 ohm plate-to-plate impedance and has 4,8 and 16 ohm outputs, you need a 4K to 4-8-16 transformer. This means its secondary has taps to provide the correct "gear reduction" so that if a 4,8 or 16 ohm load is used, the tubes have the same load provided.

Some manufacturers don't publish the needed figure, so then you either guess, using a similar transformer from a model with the same output tubes, power output, and plate voltage; or you get a similar amp with a good transformer and measure. Put a signal generator, at 1 KHz, across the secondary feeding a known speaker load, set it at a nice even voltage, and then measure the voltage across the plates of the output tubes connections (with the amp unplugged and the tubes out!). Divide it by the input voltage, square it, and multiply by the speaker impedance. For example, you have an amp with 4 ohm and 8 ohm outputs. Put a signal generator across the 8-ohm output and set it to, say, .25 volts. Connect your meter across the anode pins of the output tubes, and you measure 2.5 volts, so you have a 10:1 voltage ratio, or a 1000 to 1 impedance ratio. You need an 8000 to 4 and 8 ohm transformer.

Also, the amount of power the transformer can handle must be known. The transformer must be capable of handling at least as much power as the old one. Power handling is pretty much a function of weight, so, get a transformer from an amp of the same or a little more power, and one that weighs as much or a little more.

Also, the configuration of the transformer is important, in that you must have a push-pull transformer for push-pull circuits, and should have one designed for a single tube for the one-tube, single-ended models. These have specially designed cores to handle the unbalanced DC currents. Also, if you need "ultra-linear" screen taps – which are taps on the primary to drive the screen grids – be sure to get them.

You won't see them on guitar amps, but some hi-fi and PA amps have feedback tertiary windings, cathode windings, or – McIntosh was notorious for this – bifilar windings for cathode and plate. If you have such a beast, there's no substitute: either you get that exact transformer, or you're sunk.

Exact replacement transformers are color-coded the same way and you can just wire them in. Otherwise, you must make a diagram of where everything goes. Ohm the windings out: the primary has the high resistance and the secondary the low. By following the relative resistance of the taps, it should be obvious what's what.

On feedback amps, you must first hook up the secondary, determining which are the various impedances of output. The two with the most resistance are common and high ohms: the lowest ohm tap – assuming you have 4/8/16 or 2/4/8 – will be this one's center tap. Because the impedance is the square of the voltage, the center tap is 1/4 of the outside legs, and the half impedance gets 7/8 the voltage. The middle tap will be between the center tap and one (the high) tap.

Then hook up the primary – the center tap goes to B+, and the outer legs to the plates. Ultralinear taps *must* go to the tube whose plate is nearest them. Then hook up a dummy load and power up, with no signal and all the amp gain controls down. Slowly start turning them up, monitoring the output on a scope or signal tracer. If the amp howls, you have the primary reversed! Change the plate leads, and the screen taps with them.

If you get the factory exact replacement, this will be done for you: the color code will reflect proper phasing. If you get a different factory unit, get a schematic from the amp it goes in, which should reflect proper phasing – if the amp has the same number of stages as the one in the schematic! I use the new color code to set the secondaries and then phase the primary by trial. You can establish that the feedback is wrong, if the amp oscillates, by disconnecting the feedback loop.

Also, expect that if you change the transformer, the distortion sound of the amp will change, too. With vintage amps, the old transformers were far inferior technically to what is currently produced, and so will not distort as

Technical articles

much. You could try getting one from the companies selling "exact reissues" of tweed Fenders: they are deliberately compromised. ■

SPEAKERS AND CABS
By Dave Hunter
Reprinted from the book Recording Guitar and Bass (Backbeat Books 2002), by Huw Price, edited by Dave Hunter

It can be far too easy to get wrapped up in all the potential ingredients that contribute to your electric guitar tone – pickups, strings, wood types, preamps, tubes, power amps, and so forth – without pausing to consider where the sound comes from. Simple: your speakers. Whatever components play a part in the signal chain that shapes every note you play, the humble speaker is still what takes the sound to the listener, or indeed the microphone. Admittedly, an amount of DI recording has always occurred, and better sounding direct-to-board units are available all the time – digital amp emulators, tube preamps with speaker simulators, and so forth – but the majority of recordings are still made by miking up amps. (In fact, in a miked amp, the speaker's effect as a low pass filter and the resultant attenuation of high frequencies plays a big part in what we have come to consider a "great guitar sound," which is why straight DI'd guitar often sound strange and "unnatural" to us, even if it is a more realistic rendering of the electric guitar's natural sound.)

Amp techs who know their onions have long declared that your speakers (and speaker cabinet design) are responsible for 50 per cent of your tone, but the speaker is often the last thing a player considers in an effort to overhaul an unsatisfactory sound. Replacing a stock speaker with a different type, even of the same size, can instantly alter your amp's sound more than any other single component change – and be the quickest, simplest, and potentially cheapest way to convert a mediocre combo into a tone machine that sounds great and is easier to play.

It's not always just a matter of installing a "better" speaker, however, so knowing a little about the general tonal characters of a number of speaker types can prove valuable to any guitarist. That said, cabinet type and design can itself play a major part in shaping your sound. Therefore, let's take a brief look at both of these major ingredients: first driver types, then cabinet designs.

Vintage, Or Low-Powered Speakers
You can split guitar drivers very broadly into two categories: "vintage," or lower-powered types; and "modern," or higher-powered types. Within these, they can generally also be divided into "British" and "American" sounding units, though most larger manufacturers today cross the boarder between the two.

In the '40s, '50s and early '60s, guitar amps rarely carried speakers rated higher than 15W to 30W power handling, and indeed guitar amps in the early days rarely put out more than the upper figure. These were fine singly in the recording studio, or in multi-driver cabs at dance hall volumes, but push them hard and they started to "break up" in sonic terms, adding a degree of speaker distortion to the amp's own distortion when played near the peak of its operating capacity (or well below it in some cases... gotta' love those tweed Deluxes!). As guitarists found themselves in bigger and bigger venues, requiring higher clean volume levels, amp builders sought out more robust speaker designs – but these cost more than the cheaply built, lower-rated drivers, so weren't universally employed even so. For their own part, a lot of players who weren't seeking absolute "clean clean" enjoyed the edge, bite and apparent compression that a little speaker distortion added to their sound, and lower-powered drivers – with all their gorgeous, inherent "flaws" – quickly became a big part of the foundations of the rock'n'roll sound.

In the USA, Jensen were far and away the most respected – and for a time, most used – manufacturer of lower-powered speakers, and their 12" P12R (15W), P12Q (20W) and P12N (30W), 10" P10R (15W), P10Q (20W) played a huge part in the signature sounds of classic amps from Fender, Gibson, Ampeg, Silvertone and others (note that original Jensen spec sheets often give even *lower* power ratings for these! The reissued units, however, are universally higher). Each of these models has its distinctive characteristics, but together they are broadly characterized by bell-like highs, somewhat boxy but rather open and transparent mids, and juicy, saturated lows (to the point of flapping, farting out and all-round low-frequency freak out in the lesser-rated models). All of which combines to produce great sweet, tactile clean sounds when driven a little, and gorgeous, rich overdrive when driven a lot.

All were built using alnico magnets, paper cones and paper voice coil formers, too, which universally prove essential ingredients in the vintage driver formula. Depending on costs and availability, most of the same manufacturers also fitted units from Utah, Oxford, CTS and others, which more often than not shared some of the Jensens' characteristics (provided they were alnico-magnet designs), but were rarely as revered by players.

Many of the more popular Jensen models are available again from the revitalized maker's Vintage Reissue range. These capture at least some of the originals' tonal characteristics, though materials are not 100 per cent exact matches between the new and old units (though the alnico magnets remain on the "P" models). Eminence also builds a number of vintage-styled drivers (and are often, in fact, the manufacturer behind many own-label speakers from Kendrick, Fender, THD and others), and the smaller builder WeberVST also offers some highly respected vintage-repro units.

British vs. American
On the other side of the Atlantic, Elac, Goodmans and Celestion were building 10" and 12" speakers that weren't a world away in design from the Jensens of the USA. Using pulp-paper cones and alnico ring magnets to achieve power handling conservatively rated at from 12W to about 20W in the early units, these appeared most famously as the Goodmans Audiom 60 and (Celestion-built) Vox "Blue Bulldog" G12 in the Vox AC15s and AC30s of the late-'50s and early '60s. By far the more famous and highly sought-after of the pair, the Celestion unit has sweet, rich, musical mids and appealing (if not overwrought) highs, with not tremendous low-end reproduction but an extremely flattering tonality overall, and great dynamic range. In short, they're one of the most beloved guitar speakers of all time. It was (and in Celestion's very impressive reissue form, is) a highly efficient speaker, too, offering 100dB (measured @ 1W/1M), versus figures ranging from less than 90dB to around 96dB for similar Jensen units. This means that a pair in a 2x12" cab topped with a 30W Vox AC30 chassis makes for

---

SPEAKER DISTORTION

Distinct from amplifier distortion, speaker distortion occurs when a driver is pushed to near its operating limits, where the voice coil and paper cone begin to fail to translate the electrical signal cleanly, and thereby produce a somewhat (or sometimes severely) distorted performance. Put simply, the paper cone begins to flap and vibrate beyond its capacity, and introduces a degree of fuzz into the brew.

To keep the concept clear in your head, imagine two different amps: amp A is a 60W Mesa/Boogie MkII with a 100W EV driver, running with the cascading gain preamp cranked, but the master volume down to about 2/10; the sound you hear is definitely distorted, but that is coming entirely from the amp (preamp, in fact) and the speaker is operating well within its limits. Amp B is a 15W Vox AC15 running flat-out into a 15W Celestion Alnico Blue speaker; the sound you hear is part distortion from the floored amp, and part distortion from a speaker hit with very nearly more power than it can handle, especially on the dynamic peaks... and the result is a wilder, more uncontrollable sound – but a great one, by many standards.

150 THE TUBE AMP BOOK

a pretty gutsy combo indeed, despite the apparently low numbers on paper.

Celestion evolved the British driver *du jour* into the ceramic-magnet G12M "Greenback" (rated 20W to 25W) in the mid '60s, when they were generally found in multiples of four inside the classic Marshall 4x12" cabs that helped broadcast the rock message to the masses in ever larger arenas, at ever greater volumes. The Greenback is warm, gritty and edgy, with a none-too-firm bottom end but plenty of oomph when tackling the output in numbers, especially in a closed-back cab (and remember, four of them together can take 100W, and better share the low-frequency load without flapping out). This speaker, as much as any amp, typifies the "British sound" sought after by so many blues-rock guitarists. In the late '60s and '70s the slightly higher-rated G12H took on a heavier magnet to give a tighter low-end response, but otherwise carried on with the Brit sound tradition. Fane, too, deserve an honorable mention for some good sounding lower-powered speakers they have produced over the years, as well as the sturdy drivers that helped give many Hiwatts, for example, their legendary big, bold sound.

Modern, Or High-Powered Speakers

As many higher-powered guitar amps evolved to cope with the larger concert halls, builders sought speakers that could take the full punch and transmit it relatively uncolored, as undistorted as possible. All contemporary manufacturers offer a few models of this type, but early classics came from the American makers JBL and EV. The former helped to make early 100W Fender Twin Reverbs into some of the loudest combos in the States, while EVs were Randall Smith's preferred choice of speaker to take the brunt of the cascading-gain blast in his compact (1x12") but powerful early Mesa/Boogie combos. JBLs present a rounded midrange with an edge of bark and nasal honk, and ringing, occasionally piercing highs. Various popular EV models down the years have tended to be muscular, balanced and aggressive, while still very musical and fairly "hi-fi" in a guitar context – keeping their punch together well when under stress – and remain the top choice of a number of rock soloists.

Today, Celestion's Classic Lead 80 and G12H-100, a number of Eminence models and others offer the similar power-handling capabilities and firm-yet-musical response in bigger amps. While an ability to handle massive power levels sounds like a desirable characteristic for *any* speaker, you have probably already perceived the trade-off; firm, robust drivers barely flinch when hit with the full-whack from lower-powered amps like, say, a Fender Deluxe Reverb, a Vox AC15, or a TopHat Club Royale – all of them great recording amps, as it happens – and the tone resultant from this partnership can therefore be somewhat tight, dry and constipated. Getting back to that old speaker distortion… hey, in many circumstances we like it, and when it comes to achieving a characterful semi-clean, crunch or distortion sound at lower volume levels it can be a real boon. For mega-watt rockers, however, who need a firm sound on the big stage, for either bold clean playing, high-gain distortion, or gut-rumbling low-string riffs, the advent of high-powered drives was – and remains – a godsend.

The alnico camp includes such all-time classics as the Jensen P10R, P12R, P12Q and P12N; JBL D120 and D130; and the Celestion Alnico Blue (aka Vox "Blue Bulldog"). Other than the P12N and the JBLs these are all sub-30W drivers. It's not impossible to built a high-powered alnico unit, however, and in addition to the robust, big-magneted JBL drivers, Britain's Fane still offers a 12" alnico-magnet speaker rated at 100W. About as expensive as Celestion's pricey (but low-wattage) Alnico Blue, the Fane can be a great sounding unit in the right amp, too, but needs to be hit with pretty high volumes to start giving up the good stuff. In low-powered combos, therefore, it can sound a little dark and tight – though that might be what some players want to achieve.

As for the ceramics camp, its own classic models are probably loved by as many players as are the alnicos mentioned above. Among them are the Celestion Greenback, "H," Classic Lead 80, and Vintage 30 models; Jensen "C" series drivers; and all of the much-loved EV models.

Efficiency And Sensitivity

Even if they have put some thought into the general "sound" of their speakers, the majority of guitarists fail to consider the effect that speaker sensitivity – that is, the speaker's efficiency in translating wattage into sound on the air – will have on the volume of their amp. As covered earlier in this chapter, the speaker plays an enormous part in determining the decibels (dB) an amplifier is capable of producing, and is as important as the amp's wattage rating in generating that thing we call volume.

Sensitivity is not related in any way to power handling ability. There are 15W speakers rated at 100dB (measured @ 1W/1m), and 100W speakers rated at 97dB. Remember what we have discussed elsewhere: every doubling of an amp's power creates only a 3dB increase in output. Hit these respective 15W and 100W speakers in turn with the signal from, say, a good 15W tube amp and the lower-powered speaker will give you a significantly higher audible volume (and probably sound better in the process if you're a rock'n'roll fan, with more speaker distortion and coloration added to the brew). Hit the pair of them in turns with a 60W Mesa/Boogie MkII cranked to full, and the sensitive 15W speaker will blow; it's important to remember that high efficiency does not equate with high power-handling capabilities! Stick in one of the new Celestion Centuries, however, rated at an astonishing 102dB and capable of handling the Boogie's full 60W, and you've got a loud little combo.

What all this means is that you can make your amp of choice far more muscular by fitting a speaker(s) with a higher sensitivity rating, for example, converting a 20W combo fitted with a Celestion G12M-25 "Greenback" (97dB) that just won't cut it at gigs into a club player's dream by changing for a 100dB Celestion G12H-30. As much as this would seem to be universally a "good thing," a more sensitive speaker might make make your much-loved recording combo too loud for optimum studio use. Alternatively, you just might prefer the sound of its original, less efficient driver (or find equal reason to swap down from a loud, sensitive speaker to a juicy, softer, insensitive one). Referring back to our beloved tweed Fender Deluxe, Tremolux and the like: these originally came with Jensen P12Rs and P12Qs, which weren't highly efficient volume producers, but sounded delightful.

Speaker Size

Size matters, but bigger isn't always better. Far and away the majority of guitarists pump their air via 12" speakers, but plenty of great amps carry 10" drivers as well: the tweed Fender Bassman, Vibrolux and Super Amp; the

---

ALNICO MAGNETS

Players have raved for years about the great mojo of alnico-magnet speakers versus ceramic-magnet units, and while the science behind this is difficult to quantify in such limited space, there is certainly something to it. As used in speakers as well as in more vintage-style (or simply higher-end, these days) pickup magnet assemblies, alnico (an alloy of aluminum, nickel and cobalt) is generally regarded as the "musical magnet." Thanks to the relative scarcity of cobalt – and high prices from the early '60s onward – it's also an expensive alternative, especially when you need to gather together enough of it to manufacturer a big magnet for a high-powered driver.

As a rough rule of thumb, alnico speakers tend to be musical, sweet, and harmonically rich without harshness; ceramic speakers are characterized by muscular, aggressive, punchy performance (though both of these are generalizations, and either type – if well designed – can also possess elements of the other's characteristics). In the early days of ceramics, it was chosen because it was relatively affordable. This also translated into ceramics being the magnet of choice for heavy, firm, high-powered modern units.

Technical articles

blackface Fender Tremolux and Super Reverb; the Vox AC10 and the Matchless Lightning '15 210. A few great combos carry 15"s, namely Fender's tweed Pro, brownface Vibrosonic and blackface Vibrosonic Reverb. All of the above – including countless combos and stacks carrying 12"s – are stunning guitar amps.

It's a common misconception that 10"s are inherently "bright and trebly," while 15"s are "bassy and woofy." It ain't necessarily so. A decent 15" speaker designed for guitar amp use will be capable of producing as much audible treble as any 10" driver – or at least more than you probably want to hear anyway, at full whack on the amp's treble control. On the other hand, a pair of 10"s offer more speaker-cone surface area than a single 12", and are capable of reproducing a more solid, forceful fundamental, provided their frequency response dips to the lowest-produced notes on the 6-string guitar (as most do).

The impact of cone size centers much more around attack and response times – that is, the delay between plucking a string on the guitar and the speaker getting audible airwaves into motion. At one end of the scale, 10"s have a faster response due to the shorter distance the cone has to travel, which results in crisp, articulate notes and a speedy attack, with a lot of definition at the interior of the note. This can also translate to seemingly more lively highs, though not necessarily *more* high frequency content as such. At the other end of the scale, 15"s – with their much larger, deeper cones – are slower to get moving and pumping the air, the result being a slightly less articulate attack. As you would expect, 12"s fall between the two, offering a good compromise. (We're talking tiny fractions of a second here, but it all contributes to the resultant tone.) Singly, 10"s do have trouble giving enough oomph to your lows, while 15"s can get to "flappy" for some playing styles (the very reason Mr Fender changed his Bassman from a 1x15" to a 4x10" in the mid '50s. Yeah, it's a classic guitar amp – but the change to a 4x10" succeeded in making it a better bass amp in its day, too). Hence the broad popularity of the 12".

Cabinets

In addition to the sound of the drivers themselves, the cabinet you bolt them into plays an important part in shaping your sound. Cabinet design and resultant tonality could make a book in itself, but there are a few major factors and formats worth understanding to begin to pin down your requirements. These include open vs closed-back designs, single or multiple drivers, cabinet wiring, baffle construction, and wood types.

Open vs Closed Backs

Simply bolting a piece of wood across the back of a box or leaving it open can be one of the single greatest tone-influencing aspects of a cabinet's design. Open-backed cabs accentuate the higher frequencies, with a wider, more "surround-sound" dispersion. They generally offer a broad, well-rounded, transparent, and relatively "realistic" sounding frequency response. Their low end response, however, tends to be somewhat attenuated, due to the partial phase cancellation resultant from soundwaves from both the front (driver pumping forward) and back (driver pumping backward) of the cab reaching the listener – or microphone, when recording – at the same time. This isn't usually enough to cause the alarming phase cancellation problems that can play havoc with some multi-miked recording sessions, but it does influence the overall sound signature of open-backed cabs... sometimes for the better, sometimes for the worse, depending on what is required for the type of music you play. For the same reasons – put simply – the deeper the dimensions of the open backed cab, the better its low-end reproduction.

The same phenomenon means that if you stand anywhere but straight in front of an open-backed combo, it's likely to sound louder than a similar but closed-back amp, with a more "omnidirectional" sound projection. Again, depending on your requirements, this can make it either easier or harder to achieving your desired sound.

Closed-back cabs offer a tighter, fuller low-end response, with more directional sound projection which comes straight out from the front of the speakers. Sometimes this comes with slightly spongy, compressed-sounding mids (though often this element is accentuated by the drivers popularly included in such cabs), and slightly attenuated highs. A closed-back cab will also be relatively quieter than a similarly loaded and powered open-back cab if you stand to the side or, obviously, behind it. The closed-back cab is particularly suited to that chest-thumping *chunk* and low-end thud required of a lot of heavier rock styles.

In a nutshell, you can think of the "open-back" sound as the gritty, edgy, full-throated wail of a Fender tweed Deluxe; the bright, bold twang of a Twin Reverb; or the juicy, sweet mids of a Vox AC30. Think of the "closed-back" sound as the bowel-rumbling blast from a Marshall Plexi, or the foundation-rattling roar of a Mesa/Boogie Triple Rectifier stack.

Single vs. Multiple Drivers

The number of speakers in your cab obviously plays a big part in determining how much air you're going to move, but other aspects of the way that running through a 1x12", 2x10" or 4x12" will affect an amp's tonality are less obvious. To understand this concept thoroughly, some understanding of phase cancellation is necessary, though there isn't space here to go into this subject in detail. In short, the more speakers you bolt into a cab, the more potential you create for phase cancellation between drivers. Even speakers of the same make and model react slightly differently to the signal presented them, and of course they are in slightly different positions relative to the ear of the listener, so the variations in sound waves they produce inevitably results in some phase cancellation.

This slight "blurring" of the sound from phase cancellation in multi-speaker cabs can indeed be a good thing in some circumstances; it contributes to many classic sounds and might be exactly what you want to achieve. It can also sometimes make life harder in the studio, in particular... or easier, if miking up each of two speakers in, say, a Bluesbreaker combo and blending them right creates just the sound you're looking for. Yet again, it's all about the right sound for the right performance – but it's worth knowing in advance which formats might be more difficult to work with than others.

Some makers take advantage of the complementary interaction of two or more speakers in a cab by mixing entirely different types of drivers. Matchless use one juicy, soft Celestion Greenback and one tighter, brighter Celestion G12H in their D/C-30 cab, with excellent results. The designers at Trace Elliot selected a Celestion Vintage 30 12" and a 10" Vintage 10 for the Gibson Super Goldtone combo, and it helped to create a great-sounding new amp. This is something you can try yourself. If you are tempted to replace both speakers in a twin-style cab, try replacing just one first. That might be enough to change the sound for the better, and the mix of speaker types might make for a richer, more complex sound besides. (Of course, always mix speakers of the same impedance. That is something you do not want to mix'n'match.) On the other hand, the speakers might "fight" each other sonically and the sound might be worse than before (and improve again when two of the same new speaker are installed). Sadly, like with guitar pickups, trial-and-error is often the only way to tell which replacements speakers will best suit your amp and playing style.

Related to this, with any multiple-driver cab, yet another factor has its say in your sound, namely...

Cabinet Wiring

We're talking series vs parallel wiring here, or indeed a combination of the two. This can be a very complex issue (hey, like so much we have covered up until now!), but it's not hard to comprehend at least enough of the subject to know roughly how it affects your tone. All speakers interact with the amp's output transformer and, via that, its output tubes, to create different degrees of damping and resonance according to impedance and how hard you're driving them (very roughly speaking). For this reason, the way you wire together two or more speakers will affect your tone.

Technical articles

Speakers wired in parallel tend to damp and restrain each other somewhat more, offering a slightly tighter response and smoother breakup – all of which is highly desirable in some circumstances. Speakers wired in series tend to run looser, with less damping, resulting in a more raw, open sound – which is, yep, highly desirable in other circumstances. If you have a 2x12" or 2x10" cab with two 8 ohm speakers and an amp with 4 ohm, 8 ohm and 16 ohm outputs you can test it both ways and decide which you prefer, and even install a switch on your cab to flick between the two. Two 8 ohm speakers in parallel will create a 4 ohm load, while in series they'll create a 16 ohm load, so set your amp accordingly.

Be aware, however, that if one speaker in a parallel set-up blows, the other will keep functioning and the amp (namely the output transformer) will remain safe for a time, though you should still shut it down pronto and replace the blown speaker. When one speaker of a series-wired pair blows, your output will die instantly because each speaker relies on the other to complete the signal flow, so dive for that power switch before your OT starts to fry! This is most likely why Fender wired all of their earlier multi-speaker cabs in parallel, rather than because of anything having to do with tonal differences between the two arrangements.

Baffle Construction

The way in which the speaker baffle (the board at the front of the cab to which the speaker is bolted) is affixed to the cabinet, as well as its thickness and wood type, will also affect the overall sound. Thinner, more loosely fixed baffles will vibrate more, and pass more vibrations into the rest of the cabinet too, creating cab resonance and standing waves that add to the sound of the speaker itself. Vintage cabs, as a rule, tend to be built this way more than modern cabs. The classics of the breed include Fender's wide and narrow-panel tweed amps from the '50s, like the 5E3 Deluxe, the 5F6A Bassman, and their brethren. These are built with "floating baffles," attached to the cab's front panels by a pair of screws at each of two sides of the baffle only. When cranked, the baffle adds its own resonance and the whole cab begins to sing – and the amp reacts that much more like an instrument in itself. For the right style of music, this can sound great. In other circumstances, the tone can be perceived as woolly or blurry. (For an even better look at the difference between the various Fender tweed-style cabs, check out Mark Baier's chapter Where It All Began earlier in this section.)

Fixing the baffle tightly to all sides of a sturdy, thick-wooded cab tends to restrain cab vibration and baffle resonance, and translates a greater proportion of the pure speaker tone into the resultant sound. In general, this can mean a somewhat more controlled response and tighter lows. Fixing a baffle more tightly to a lightweight, more loosely built cab, however, can really get the whole cabinet jumping!

Wood Types

All of this interacts, too, with the type of wood from which the cab and baffle are built. Broadly speaking, plywood and chipboard offer less cabinet resonance than do solid woods (well, less and lesser respectively, if you will). This can be desireable for some designs, where the amp builder has factored in precisely what he wants from his amp and drivers, and doesn't want too much unpredictable cab sound to get in the way. On the other hand, the resonance that does occur in a chipboard or MDF cab – and their is always some – can sometimes sound dead and "unmusical."

Solid wood cabs, usually built from pine or similar soft woods, as used in the early Fender, Gibson and Vox cabs and others, offer a more uniform, musical resonance and, well, woodier tone. To some ears this can be slightly indistinct and unfocused, a little "woolly;" to others, it is sonic beauty incarnate. Either way, it's a contributory factor to some of the greatest sounds of rock'n'roll history.

Good, Better, BEST?

So which of all of the above is best for you? Sorry, but only you can decide. Little of what we have covered here – vintage vs modern or American vs British speakers, fixed vs floating baffle, series vs parallel wiring – is a "better/worse" dichotomy. It's all a matter of choosing the right tool for the job, and the one which best suits your sonic tastes and playing style.

The importance of all this lies, as ever, in understanding the vast range of tonal choices available to you. Never ignore the major role that speaker selection and cab design play in the tonal recipe, and you'll be better able to achieve the results you want from your amp. Experiment, mix'n'match, and see what you can come up with. Whereas previously you were convinced you needed to lay down another wad of cash for a vintage Marshall half-stack to get the sound your band requires on a couple tracks, you might discover that injecting your 12W Fender Princeton Reverb through the studio's in-house, closed-back 4x12" pulls off an impressive JTM45 impersonation – while distorting a lot quicker and sweeter, and being easier to record to boot. Cool, huh?

SMALL AMPS vs BIG AMPS IN THE STUDIO

By Huw Price
Reprinted from Recording Guitar and Bass (Backbeat Books 2002) by Huw Price, edited by Dave Hunter

I want to explode a few myths that have built up surrounding amplifiers. The most common of these is that you need a big amp to achieve a big sound in the studio; often the opposite is true. Legend has it that Jimmy Page – commonly associated with Marshall stacks and Les Pauls – was actually more inclined towards Telecasters and Supro amps in the studio. Billy Gibbons is also known to regard his 15W tweed Fender Deluxe as his all-time favourite amp.

There is a major difference between a guitarist's requirement for live performance and studio work. This was particularly true in the late '60s and early '70s, when the PA systems were woefully inadequate. Musicians had to provide the volume if they wanted to be heard, and to be able to hear themselves. This is why Vox AC15s, AC30s, Fender Bandmasters and even Marshall 50-watters no longer cut the mustard. I believe that a culture of machismo built up over high-powered amps, and it lingers to this day.

A high powered Marshall or Hiwatt can sound amazing when it's turned up to "patent applied for" in a big room. The problem is that nowadays, most studios are small and the volume can be too much for the room. It's also very difficult to achieve good results with big, cranked amps when they have to share the room with the drums, bass and vocals. This is when you hear the dreaded "you'll have to turn down" from the engineer. Personally, I hate saying that to guitarists because we all know that the amp will not sound as good when you wind it back below the "sweet spot."

Anyone with experience of traditional tube guitar amps – ones with one or two volume knobs – knows that the amp has to be turned up to a certain level before it comes to life. Anything below this sounds flat and dull; what's more, it is harder to play. The real magic in tube guitar amps comes not only from overdriving the preamp tubes but also from distorting the power tubes. The distortion characteristics of power tubes are different to preamp tubes. The little fellas tend to have a fizzy, more compressed sound whereas the big ones have a more rounded and musical tone. Overdriven power tubes are also more responsive to the players touch. When you play light or turn your guitar volume pot down, they sound clean. Turn up or play hard and they distort. In other words, you can play the amp as well as the guitar. Your amp is not there simply to make your guitar louder or to act as a high volume buzz box; it can be an instrument in itself.

Okay, let's think how to avoid the dreaded "turn it down." It's no use refusing point blank – the engineer will not be asking you because he doesn't like your sound, it's probably because he has problems with spillage or rattling snares. It is more than likely that everyone else in the group would prefer you to turn down, too. So how do you get that great cranked-up tone without getting thrown out of the studio (or your group)? One simple answer springs to mind: use a smaller amp.

This is not a joke; many top guitarists have

Technical articles

known for years that 4W, 10W or 15W amps can sound huge on record. Quite often they keep quiet about it in interviews, thinking their image might be blown if they admit that their monster tone was made with a Fender Champ. A lot of the great guitar sounds of the '50s and '60s were made with little Fender amps, Gibson amps and also budget stuff like Supros and Danelectros.

Many people feel that those small-amp sounds are exciting because the amp is operating way beyond its design parameters. If you do this, random and unpredictable things start to occur – many of which can result in some fantastic guitar sounds on record. When you turn up a Marshall JCM800 or Mesa/Boogie Dual Rectifier it produces a different sound – preamp distortion. These amps are designed to do this and the circuits are operating quite comfortably. This isn't necessarily worse, but it isn't perhaps as ragged or exciting, either.

Don't just take my word for it. Here's how Stones guitarist Keith Richards put it to *Guitar Player* magazine way back in November, 1977:

"I never use an amp in the studio that I use on stage. I mean, stage amps are far too big. Probably the biggest mistake inexperienced players make is thinking that to have a lot of volume in a studio you need a huge amp. It's probably the opposite. The smaller the amp, the bigger it's going to sound, because it's already going to sound like it's pushed to the limit. Whereas you can never push a Marshall stack to its limit in the studio because it will always sound clean."

The point I'm making (with a little help from Keef) is this: why bother having a really cool 100W Marshall, Hiwatt, Soldano, Mesa/Boogie or Fender amp if you can never turn the volume past three and a half? I've seen guitarists spend hundreds on distortion pedals, compressors, graphic equalizers, power soaks, etc, etc, all in an attempt to emulate the sound their amp produces naturally when it's cranked up. If you still can't get the "big amp = big sound" thing out of your head, look at it like this: in the studio, whatever amp you play through will be captured by a microphone diaphragm that may be less than 1" in diameter, transmitted down a hair-thin wire, recorded as imperceptible magnetic particles on a tape or as digital code… and eventually reproduced through speakers that might be anything from the 1/2" diameter of headphone drivers to the 4" to 8" diameter of the average home hi-fi units. In this environment, physical size no longer matters; sonic size is everything.

Tube Rectification

Another wonderful thing about cranking up amps that have tube rectifiers – as so many smaller amps do – is that the amp will compress naturally. This is often referred to as "sag." At high volumes the demand on the power supply can be too much and volume doesn't rise in proportion to the physical effort put into the playing. Consequently, the guitar is easier to record because it has no transient peaks. Power tubes themselves will also induce some compression-like characteristics when pushed hard, but the effect is more noticeable in tube-rectified amps. The Fender Bassman reissue comes with a tube rectifier for recording and a solid-state rectifier for gigging. On the other hand, if you want a tight, firm, immediate response in all circumstances, an amp with solid state rectification might be more suited to your requirements.

Tube rectifiers also allow for a gentler start when you power up, which is healthier for the tubes if there is no standby switch. Smaller amps also allow engineers to use their high quality condenser microphones very close to the speaker without worrying too much about damaging capsules. The microphones will sound happier with far less tendency to distort. Ribbon mics can also be used with reduced risk of damage, and these can be *great* mikes for recording electric guitar – though a big, powerful amp will blow them as quickly as you can whack out an open E minor. I also advocate the use of low-powered amps for clean sounds. I achieved a monster sound with a Neumann U67 placed right against the speaker of a Fender Champ on a recent project; the amplifier was turned to 1.

None of the above is intended to imply that a big amp can't sound great, too. But you will stand the best chance of doing a large amp justice if you record it in a large, properly designed live room in a professional studio – and fewer of us have access to those these days for any kind of affordable money. Even so, a big amp is still going to be more of a challenge to mike up and record. For me, small amps have the magic in the studio. They consistently prove a lot easier to work with, and a lot better sounding when put to the task of recording.

In concluding this section on small amplifiers, let me ask you to remember that a 100W amplifier is not twice as loud as a 50W amplifier – it is only 3dB louder. Or as Nigel Tufnel might say: "That's three louder." The relationship between power and volume is not linear. Each doubling of power produces only a 3dB increase in volume. Speaker efficiency is often neglected in this equation, too (as discussed in the preceding chapter on *Speakers And Cabs*, and the following chapter on *Power Modeling*). Typical values for guitar speakers are around 90dB, which represents an efficiency of less than 10 per cent. To assess speaker efficiency, a speaker is fed with 1W of amplifier power and an SPL measurement is taken at a point 1 metre away. A speaker efficiency of 10 per cent means that 90 per cent of the amplifier's power is wasted – usually turning to heat. If you like the tone of your little amp but feel the need for more volume, try buying a more efficient guitar speaker. A unit with an efficiency rating of 96dB produces a 6 dB increase in volume versus an 90dB-rated speaker. The perceived volume increase is the same as bumping up the power of your amplifier from 25W to 100W. Check the ratings of a few speakers, and you might be surprised.

POWER MODELING
By Myles Rose

The vast majority of the time when I first walk into a venue where I will be listening to music for the night, I can generally tell if the performance will be a memorable one. I can usually tell from the equipment setup, and not to brag, but I have about a 90 per cent track record. The 10 per cent of the time I am mistaken, it is easy to explain. I did not know the performer or group, and had no idea what to expect; I was invited as a guest, and already knew the music was not to my particular taste; or a few other reasons.

The big tip-off is amplifier compliment – amplifier power that is. There are folks that have blinding-fast technique. Speed metal players, fast articulate players, folks with speed as their underlying strong point. Frankly, this is not my to personal taste. I am generally impressed for ten minutes, but then my attention is lost. I generally ask myself, are these folks practicing, or just looking for the right note? One note played with feeling – and that has tone – is worth 100 64th note triplets from my point of taste.

The biggest problem, however, is amplifier power. When I see a 100W amp on the stage of a 150-seat venue, I know that I am in for trouble… most of the time. If it is a jazz player looking for a clean sound, then I am safe. If it is a speed player, well, then it is what I expect. If it is a rock or blues band, then I know I am in for a very one-dimensional performance most of the time. I know, with good prospects, a few other things: the player does not understand amps or tone, perhaps his main rig is broken and this 100-plus-watt amp was borrowed, or they are into a hi-fi sort of sound with little or no dynamics.

A 100W amp, or even a 50W amp, will not distort in its output section at rational volume levels. Folks who are known for great tone and who are revered as great players generally stay at around 50W or less, even in the largest venues. They are looking for a particular sound, tone and feel. They let the stadium sound systems do the rest. If you cannot turn your amp, and most amps, to at least 6 or so on the volume, you will never tap the soul of most tube amps.

Folks also do not understand "loudness." Many think a 100W amp is *twice* as loud as a 50W amp. This is not all the case. Double your wattage, and all you gain is 3dB. Sure, folks talk about "headroom," and think this is a huge requirement. Folks that actually need headroom are clean players… rhythm players, jazz players for some styles, and pedal steel players to name a few. An amp with a lot of headroom is a hi-fi amp. It will be clean, and not have the dimension or touch dynamics of a lower-powered amp. If you want more loudness out of

Reference

Amp Volume Shell Game

This is a test I do a lot of times for folks to show there are differences in power tubes – generally to demo high quality tubes such as the Groove Tubes 6L6GE versus the Sovtek 5881WXT (which is a very popular tube which comes stock in a lot of amps).

Take two Fender 50W Bassman amps, or any two amps which are the same. In one amp, install a duet of GT-6L6GE tubes, leaving the stock tubes in the other. Connect them to like speaker cabinets. With a signal generator and a 1000 Hz or 400 Hz tone, and a dB Spl meter, set the amps where the non-GE-tubed amp will output a level of 80dB one meter on axis. On the GE tubed amp, set the level to 70db, -10db down from the first amp. The tone controls should be set the same.

Have a player (hopefully with a small audience, as there will be more income potential), play for a few minutes. Switch back and forth using a common A/B switch. Do not A+B the amps, only use amp A or B at any given time.

Bet the folks in the audience which amp is louder; 90 per cent or more of the time, the GE-equipped amp will be chosen. Proceed to show them – via the signal generator and sound pressure level meter – that the GE amp is -10db down in actual loudness from the other amp. Let the folks in the audience play with the amps and meter as you collect your winnings. It is also nice to show at the end of this demo that the GE-equipped amp also has a 15 to 20-degree wider soundstage image.

The reason for this? Picture playing a nice classical piece of music at 80dB using a small, hand-held transistor radio, and then the same piece at the same level on a very high-end stereo system with large speakers. The level is the same, but the harmonic content is so much larger with the big stereo that the human brain is fooled into thinking it is louder. It has to do with the Fletcher-Munchen curve, and some other physics. Bottom line is: with better and more articulate tubes, there is just a lot more complexity in the recipe, spices in the soup – so to speak – rather than just my grandmother's boiled chicken in water.

a 50W amp, double your speaker area, or go with a more efficient speaker. Going from an 83dB speaker to a 95dB speaker is almost the same gain in volume as going from 50W to 200W in amp power.

Then there are those folks that have the great idea of pulling two tubes out of their 100W amp to turn it into a 50W amp. You can do this, and relatively safely if you do it right (the procedure is outlined in the *Mods For Marshalls* chapter of this section); but to put it bluntly and get a lot of argument, this is not the best idea. A great amp is made up of many components. Power transformers, output transformers, capacitors, and other parts make up the design. If one takes a Marshall 100W Super Lead, and pulls two of the tubes, and properly sets the impedance selector, turning the amp into a 50W amp, what actually happens? Well, we have a 100W power supply that is now even less taxed than before. The "50W" Marshall will now have less dynamics, less feel, less touch sensitivity. It will be a nice, clean, hi-fi, 50W amp. It's 100W power supply will never reach saturation. It's output transformer will never be pushed. It will actually be cleaner than it was as a 100W amp. The only distortion you will get is when the output tubes are at their limit, and this will be an unbalanced sound, although some might think this is just to their own tastes.

Modeling amps? Some folks think that modeling amps have some strong points such as a lot of sounds for the dollar or in a given space, or for recording. Some think in a live venue, modeling amps can have limitations. To my way of thinking, this live aspect is sort of a "yes and no." I see many folks with 50W amps in small clubs, where the soul of the amp is never tapped. Put an amp like a Line 6 Vetta, Fender Cyber Deluxe or Cyber Twin, or Vox modeling amp out there, and you may be surprised. I may still prefer the *proper* wattage tube amp by far, but I will take the modeling amp every time over the wrong tube amp. Why? Modeling amps allow a degree of touch dynamics and tonal ranges to be captured at most any level. You have all sorts of controls for this ability. A Fender tweed Bassman in a small venue will never be able to be cranked to its level of tone potential for some music styles. A Line 6 Vetta may pull off the "tweed sound" of the virtual Bassman at low volumes in a much more convincing and pleasing manner, at least to my tastes. Better yet, though, use a tweed Deluxe!

To this end, all I can suggest is, listen to amps, and play them. See how they react to your touch. If this is not a part of your music and style – such as many folks that start the song at 110db and end it at 110db – then most any amp will work pretty well. As you develop an ear for different tone aspects, and fingers and touch that can give you at least two more playing dimensions, then you will move to the next step of being a better player, and also have a more heightened ability as a listener.

Reference section

Tube/amp replacement guide

A listing of preamp and power tube complements for most popular models from common amp manufacturers.

ACOUSTIC

Model	Tubes
160	2-7025, 2-12AT7, 1-6L6 Quartet
160	3-7025, 1-12AT7, 1-6L6 Quartet
164/165	2-7025, 2-12AT7, 1-6L6 Quartet
164/165	3-7025, 1-12AT7, 1-6L6 Quartet
G-100T	3-7025, 1-12AT7, 1-6L6 Quartet
G-60T	2-7025, 1-12AT7, 1-6L6 Duet
Tube 60	2-7025, 1-12AT7, 1-6L6 Duet

ADA

MB-1	2-12AX7
MP-1	2-12AX7
T-100S	2-12AX7, 1-6CA7 Quartet

ALEMBIC

Preamp	2-7025
F|X	1-12AX7

AMPEG

Jet-12 – early 2-6K11, 1-7591 Duet
Note: some older Jet-12s used 6BK11s instead of 6K11s. This tube is no longer available, and there is no adequate substitute.

Jet-12	2-7025, 1-7591a Duet
Jet-12R	2-7025, 1-6U10, 1-7591a Duet
GS-12R Rocket II	2-7025, 1-6U10, 1-7591a Duet
G-12 Gemini 12	3-7025, 1-6CG7, 1-7199, 1-7591a Duet
GV-15 Gemini – early	4-7025, 1-6CG7, 1-7199, 1-7591a Duet
GV-15 Gemini	4-7025, 1-6CG7, 1-7199, 1-6L6 Duet
GV-22 Gemini 22	4-7025, 1-6CG7, 1-7199, 1-7027 Duet
SB-12	2-7025, 1-6L6 Duet + 1-5AR4
B-25	2-7025, 1-7100, 1-7027 Duet + 1-5AR4
B-22X	4-7025, 1-6CG7, 1-7199, 1-7027 Duet + 1-5AR4
B-42X	4-7025, 1-6CG7, 1-7199, 1-7027 Duet + 1-5AR4
GV-12	1-7025, 1-6U10, 1-12DW7, 1-7591 Duet
AC-12	1-7025, 1-6U10, 1-12DW7, 1-7591 Duet
AX-44C	1-7025
AX-70	1-7025

THE TUBE AMP BOOK 155

Reference

B-12XT	4-7025, 1-7027 Duet + 1-5AR4
VT-40 – late	1-7025, 2-12DW7, 1-6AN8, 1-6CG7, 1-7027 Duet
VT-40	3-7025, 1-6K11, 1-12DW7, 1-6AN8*, 1-6CG7, 1-7027 Duet
VT-22 – later	1-7025, 3-12DW7, 1-6CG7, 1-6AN8*, 1-7026 Quartet
VT-22	3-7025, 1-6K11, 1-12DW7, 1-6AN8*, 1-6CG7, 1-7027 Quartet
VT-60	4-7025, 1-12AU7, 1-6L6 Duet
VT-120	4-7025, 1-12AU7, 1-6L6 Quartet
V-2	3-7025, 1-12DW7, 1-12AU7, I-6K11, 1-6CG7, 1-7027 Duet
V-3	4-7025, 1-12AT7, 1-6550 Duet
V-4	3-7025, 1-12DW7, 1-12AU7, I-6K11, 1-6CG7, 1-7027 Quartet
V-4B	2-7025, 1-12DW7, 1-12AU7, 1-6KI1, 1-7027 Quartet
V-5	3-7025, 1-6550 Quartet
V-7	4-7025, 1-12AT7, 1-6550 Quartet
V-9	1-7025, 4-12DW7, 1-6CG7, 2-12BH7, 1-5550 Sextet
B-15 early	2-6SL7, 1-7199, 1-6L6 Duet + 1-5AR4
B-15N	3-6SL7, 1-6L6 Duet + 1-5AR4
B-15S	2-7025, 1-12DW7, 1-12AU7, 1-7027 Duet
B-18	2-6SL7, 1-7199, 1-7027 Duet
SVT – early	1-7025, 4-12DW7, 1-6C4, 2-12BH7, 1-6550 Sextet

Note: Early SVTs had 6146 power tubes and have converted to 6550s. The conversion is shown in the Tube Cross-Reference Guide. We recommend this conversion as quality 6146s are hard to find and 6550s are superior. In 1982, SVT began using 5-7025, 1-6C4, 2-12BH7, 1-6550 Sextet.

SVT-II	4-7025, 2-12AU7, 1-6550 Sextet
SVT-IIP	Preamp 3-7025

AUDIO RESEARCH

SP-3 Series	8-7025
SP-6 A and B	6-7025
SP-6E	4-7025, 2-ECC88
SP-8	4-7025, 2-ECC89, 1-12BH7
SP-10	12-ECC88, 1-ECC81, 2-6L60S Duet
D75A and D76	4-7025, 8-6CG7, 1-6550A, 1-6550 Quartet

D79, D150, D-70/115/250 – Call the Groove Tubes Sales Department with your serial number.

BEDROCK

1200	5-12AX7, 1-EL34 Duet
1400	3-12AX7, 1-EL34 Duet
1600	6-12AX7, 1-EL34 Quartet

BENSON

300	3-7025, 1-EL34 Duet
400	3-7025, 1-EL34 Quartet

CARR

Hammerhead	2 – 12AX7, 1 – EL-34 Duet
Rambler	3 – 12AX7, 1 – 12AT7, 1 – 6L6GC Duet
Slant 6V	4 – 12AX7, 2 – 12AT7, 1 – 6V6GT Quartet, 1 – 5AR4
Slant 6V Double Power	4-12AX7, 2 – 12AT7, 1 – 6L6GC Quartet
Imperial	3 – 12AX7, 1 – 12AT7, 1- 6L6GC Quartet
El Moto	4 – 12AX7, 1 – EL-34 Quartet

Note: When an effects loop option is ordered for either the Slant 6V or El Moto please add 1 – 12AU7 to the tube compliment.

CARVIN

VTR 2800 50-watt	1-12AT7, 1-6CA7 Duet*
VTR 2800 100-watt	1-12AT7, 1-6CA7 Quartet*
VTX 100	3-7025, 1-6L6 Quartet
X-B12 30-watt	3-7025, 1-6L6 Duet
X-T12 60-watt	3-7025, 1-6L6 Duet
X-V112 100-watt	3-7025, 1-6L6 Quartet
X-VI12 – later	3-12AX7, 1-EL34 Quartet
X-V212 100-watt	3-7025, 1-6L6 Quartet
X-V212 – later	3-12AX7, 1-EL34 Quartet
X-V212E 100-watt	3-7025, 1-6L6 Quartet
X-60 60-watt	3-12AX7, 1-EL34 Duet
X-60B 60-watt	3-7025, 1-6L6 Duet
X-100 100-watt	3-7025, 1-6L6 Quartet
X-100B 100-watt	3-7025, 1-6L6 Quartet
X-100B – later	3-12AX7, 1-EL34 Quartet

Carvin amps cannot use EL34 tubes since pin 1 and 8 are not wired together. See the 6L6-to-EL34 mod in the Servicing and Modification Section under Fender amps in Aspen Pittman's The Tube Amp Book.

CHAMPION

R&R 10-watt	1-7025, 1-6V6 Duet
R&R 20-Watt	1-7025, 1-EL34 Duet

CONN

Strobe Tuner 2-6AQ5, 1-6X4, 1-5879, 1-12AT7, 1-12AU7

CONRAD JOHNSON

Premier One	4-6CG7, 1-5751, 2-6550 Sextets
Premier Three	2-7025, 5-5751, 2-5965
MV-75A	2-6CG7, 1-5751, 2-6550 Quartet

CORNFORD

Harlequin	two 12AX7s and one EL84
Hellcat	five 12AX7s and an EL84 quartet
Hurricane	four 12AX7s and two EL84s
M.K.50 H	four 12AX7s and two 5881/6L6s
R.K.100	four 12AX7s and an EL34 quartet

DEMETER

TGA-3	6-12AX7, 1-6550 Quartet
TGP-3	6-12AX7
VTBP/201	6-12AX7
VTDB-2	1-12AX7

DR Z

Carmen Ghia V1 GT-ECC83S, V2 5751, One duet GT-EL84S #5-6 (JJ EL-84), 5Y3 Rectifier

MAZ-18 Jr.	V1 & V2 GT-ECC83S, V3 GT-GT12AT7C, V4 GT-12AX7R, V5 GT-12AX7R2, One duet GT-EL84S #5-6, 5AR4 (GZ34) Rectifier
MAZ-18 Jr.	"NR" Non-reverb V1 & V2 GT-ECC83S (JJ ECC83S), V3 GT-12AX7R2, One duet GT-EL84S #5-6, 5AR4 (GZ34) Rectifier
MAZ-38 Sr.	V1 & V2 GT-ECC83S, V3 GT-GT12AT7C, V4 GT-12AX7R, V5 GT-12AX7R2, One quartet GT-EL84S #5-6, 5AR4 (GZ34) Rectifier
MAZ-38 Sr.	"NR" Non-reverb V1 & V2 GT-ECC83S (JJ ECC83S), V3 GT-12AX7R2, One quartet GT-EL84S #5-6, 5AR4 (GZ34) Rectifier
Mazerat	V1 GT-ECC83S, V2 5751, One quartet GT-EL84S #5-6, 5AR4 (GZ34) Rectifier
Prescription	V1 GT-ECC83S, V2 & V3 GT-12AX7C, One quartet (4) GT-EL84S #5-6 5AR4 (GZ34) Rectifier
Prescription Extra Strength	V1 GT-ECC83S, V2 EF86 / 6267, V3 GT-12AX7R2 or 12AX7M, One quartet (4) GT-EL84S #5-6 5AR4 (GZ34) Rectifier
Delta 88	V1 EF86 / 6267, V2 12AX7LPS (GT-12AX7R2) or GT-12AX7M, One Duet of #5 or #6 GT-KT88SV
Z-28	V1 EF86 / 6267, V2 GT-12AX7 or 5751, One duet (2) GT-6V6, 5AR4 Rectifier
Route 66	V1 EF86 / 6267, V2 GT-12AX7 or 5751, One duet (2) GT-KT66HP, 5AR4 Rectifier
KT-45	V1 EF86 / 6267, V2 GT-12AX7 or 5751, One duet (2) GT-E34LS, 5AR4 Rectifier
SRZ-65	3 12AX7, one duet GT-E34LS, GZ34 (5AR4) rectifier.
6545	V1 and V2 12AX7 (GT-ECC83S), V3 EF86 / 6267, V4 12AX7, One duet (2) GT-E34LS, some with 5AR4 some solid state.

DIAZ

CD-100	3-7025, 1-12AT7, 1-6L6 Quartet

DYNACO

PAS Series	4-7025, 1-12X4
Stereo 70	2-7199, 1-5AR4, 1-EL34 Quartet
MK III	1-6AN8, 1-5AR4, 1-6550 Quartet

FENDER

Bandmaster VOS	1-12AY7, 2-7025,1-6L6 Duet + 5U4
Bandmaster OS	6-7025,1-6L6 Duet
Bandmaster NS	3-7025, 1-12AT7,1-6L6 Duet
Bandmaster/Reverb	4-7025, 2-12AT7,1-6L6 Duet + 5U4
Bantam Bass	2-7025, 1-12AT7,1-6L6 Duet + 5U4
Bassman 4-10 OS	1 1-12AY7, 2-7025,1-6L6 Duet + 5U4
Bassman 4-10 OS	2 1-12AY7, 206925,1-6L6 Duet + 2-5U4
Bassman 4-10 OS	3 2-12AY7, 1-7025,1-6L6 Duet + 2-5U4
Bassman 4-10 VOS	1-6SC7, 1-6SL7,1-6L6 Duet + 5U4
Bassman Top 50wt	3-7025, 1-12AT7,1-6L6 Duet + 5U4
Bassman Top 70wt	3-7025, 1-12AT7,1-6L6 Duet
Bassman Ten	70wt 2-7025, 1-12AT7, 1-6L6 Duet
Bassman 100	2-7025, 1-12AT7,1-6L6 Quartet

Reference

Bassman 135	2-7025, 1-12AT7,1-6L6 Quartet	Tremolux OS	4-7025,1-6L6 Duet + 5AR4**	Clavioline-keybrd.	2-6SN7, 1-6J5, 1-6J7
Blues Jr.	3 12AX7, 1 EL84 Duet	Tremolux Top	3-7025, 1-12AT7,1-6L6 Duet -t- 5AR4	Clavioline-amp.	1-6J5, 1-0A2,1-6V6 Duet + 5V3
Champ OS	1-6SJ7, 1-6V6 + 5Y3	Twin VOS	3-6SC7, 1-6J5,1-6L6 Duet + 5U4	Duo-Medalist	3-6EU7, 1-7025, 2-12AU7,1-7591 Duet
Champ OS	1-7025, 1-6V6 + 5Y3	Twin VOX	3-12AY7, 2-7025,1-6L6 Duet + 2-5U4	Falcon	3-6EU7, 1-7025, 2-12AU7,1-7591 Duet
Champ 12 (1990)	1-7025, 1-6L6			Falcon	4-7025,1-EL84 (6BQ5) Duet
Concert	6-7025,1-6L6 Duet + 5U4	Twin OS	1-12AY, 2-7025,1-6L6 Quartet +5AR4	GA-Custom	3-6SJ7, 2-6SQ7, 2-6J5,1-6L6 Duet + 5T4
Concert	6-7025,1-6L6 Duet	Twin Cream/Brown	6-7025,1-6L6 Quartet		
Concert '83	5-7025, 2-12AT7,1-6L6 Duet	Twin Reverb	4-7025, 2-12AT7,1-6L6 Quartet	GA-1RT-1	1-7025, 1-6BM8 + 5Y3
Deluxe #5C3	1-6SC7, 1-12AY7,1-6V6 Duet + 5Y3	Twin Reverb	'83 5-7025, 2-12AT7,1-6L6 Quartet	GA-1RVT	1-7100, 1-6EU7, 1-6BM8 + 5Y3
Deluxe #5D3	1-12AY7, 1-7025,1-6V6 Duet + 5Y3	Vibrolux VOS	2-7025,1-6V6 Duet + 5Y3	GA2-RVT	4-6EU7, 1-12AU7,1-6V6 Duet + 5Y3
Deluxe #5E3	2-7025,1-6V6 Duet + 5Y3	Vibrolux OS	3-7025, 1-12AT7,1-6L6 Duet + 5AR4	GA-4RE	2-6EU7, 1-12AU7
Deluxe Brown	3-7025,1-6V6 Duet + 5Y3	Vibrolux Brown	4-7025,1-6L6 Duet + 5AR4	GA-5	1-7025, 1-6V6 + 5Y3
Deluxe Reverb	4-7025, 2-12AT7,1-6V6 Duet + 5Y3	Vibrolux Reissue	5 12AX7, 1 12AY7, 1 6L6 Duet	GA-5T	2-6EU7, 1-6AQ7 + 6X4
Deluxe Reverb '83	5-7025, 1-12AT7,1-6V6 Duet	Vibro Champ	2-7025,1-6V6, 5Y3	GA-6	1-12AY7, 1-6SL7,1-6V6 Duet + 5Y3
Dual Professional	5 12AX7 1 6V6GT 1 6L6 Quartet	Vibrosonic		GA-6 newer	2-7025,1-6V6 Duet + 5Y3
Dual Showman	3-7025, 1-12AT7,1-6L6 Quartet	Vibroverb Brown	4-7025, 2-12AT7,1-6L6 Duet + 5U4	GA-8	1-6EU7, 1-6C4, 1-6BQ5 + 6CA4
Dual Showman Head (no reverb, 1990)	4-7025, 1-12AT7,1-6L6 Quartet	63 Vibroverb Re	5 12AX7, 1 12AY7, 1 6L6 Duet	GA-8T	1-7025, 2-6BM8 + 5Y3
		Vibro-King	5-12AX7, 1-6V6 EL-84 early models, 1-6L6 Duet	GA-9	1-6SJ7, 1-6V6 + 5Y3
Dual Showman Rev	4-7025, 2-12AT7,1-6L6 Quartet			GA-14	2-7025,1-6V6 Duet + 5Y3
Harvard	1-7025, 1-6V6 + 5Y3	Fender 30	4-7025, 2-12AT7,1-6L6 Duet + 5U4	GA-15	1-7025, 1-6SL7,1-6L6 Duet -t- 5Y3
Harvard	1-7025, 1-6AT6,1-6V6 Duet + 5Y3	Fender 75	3-7025, 2-12AT7,1-6L6 Duet	GA-15RVT	2-6EU7, 1-12AU7,1-EL84 (6BQ5) Duet
Hot Rod Deluxe	3 12AX7, 1 6L6 Duet	Fender 100 P.A.	4-7025, 2-12AT7,1-6L6 Duet	GA-16T	2-7025,1-6V6 Duet 4- 5U4
Hot Rod Deville	3 12AX7, 1 6L6 Duet	Fender 160 P.S.	3-7025,1-12AT7, 1-12AU7, 1-6CA8, 1-6L6 Sextet	GA-17RVT	2-6EU7, 1-7025, 2-6AQ5 + 6CA4
Musicmaster Bass	1-7025,1-6V6 Duet			GA-18T	2-6EU7, 1-6CA4,1-EL84 (6BQ5) Duet
Princeton VOS	1-6SL7, 1-6V6 + 5Y3	Fender 300 P.S.	2-7025, 1-12AT7, 1-6V6, 1-6550 Quartet	GA-19RVT	3-6EU7, 1-7199,1-6V6 Duet + 5Y3
Princeton OS	1-7025, 1-6V6 + 5Y3			GA-20	1-6SL7, 2-6SJ7,1-6V6 Duet + 5Y3
Princeton	2-7025, 1-6V6 + 5U4	Fender 400 P.S.	6-7025, 1-12AT7, 1-6L6, 1-6550 Sextet	GA-20T	1-12AY7, 1-7025, 1-5879, 1-6SQ7, 6V6 Duet + 5Y3
Princeton Rev	3-7025, 1-12AT7,1-6V6 Duet + 5U4				
Princeton Rev '83	3-7025, 1-12AT7,1-6V6 Duet	'59 Bassman (reissue)	3-7025, 1-6L6 Duet	GA-20RVT	3-6EU7, 2-12AU7,1-EL84 Duet + 5Y3
Pro VOS	3-6SC7,1-6L6 Duet + 5U4			GA-25	1-6SJ7, 2-6J5,1-6V6 Duet + 5Y3
Pro OS	2-12AY7, 1-7025,1-6L6 Duet + 5U4	Tone Master (new model)	3-7025,1-6L6 Quartet	GA-25RVT	4-6EU7, 1-12AU7,1-6V6 Duet + 5Y3
Pro Brown	6-7025,1-6L6 Duet			GA-30	1-6SC7, 2-6SJ7,1-6V6 Duet + 5Y3
Pro	3-7025, 1-012AT7,1-6L6 Duet + 5AR4	Vibro King (new model)	5-7025, 1-EL84,1-6L6 Duet	GA-30 Invader	2-7025,1-6V6 Duet + 5Y3
				GA-30RV Invader	3-6EU7, 1-12AU7,1-6V6 Duet + 5Y3
Pro Reverb	4-7025, 2-12AT7,1-6L6 Duet 4- 5U4	65-London	18 watts, 2-12 or Head, Duet of EL84, 2-12AX7, 1-EF86, 1-EZ81	GA-30RVT Invader	4-6EU7, 2-12AU7,1-7591 Duet + OA2
Pro Series Concert	7-12AX7, 1-12AT7, 1-6L6 Duet			GA-35RVT Lancer	1-7025, 2-6EU7, 2-12AU7, 1-7591 Duet + OA2
Pro Series Pro Rev	7-12AX7, 1-12AT7, 1-6L6 Duet	65-Marquee Club	36 watts, 2x12 or Head, Quartet of EL84, 2-12AX7, 1-EF86, 1-GZ34		
Pro Series Twin	7-12AX7, 1-12AT7, 1-6L6 Quartet			GA-40 (early)	1-6SN7, 3-6SJ7, 1-6V6, 1-6V6 Duet + 5AR4
Quad Reverb	4-7025, 2-12AT7,1-6L6 Quartet	(OS=Old Style, VOS=Very Old Style)			
Rack Amp RPWI	2-7025, 1-12AT7,1-6L6 Quartet	**some were made with 6BQ5/EL84 power tubes		GA-40 Les Paul	2-5879, 1-7025, 1-6SQ7, 1-6V6 Duet + 5Y3
Rack Guitar Preamp	2-6C10, 4-7025, 2-12AT7				
Rack Bass Preamp	2-6C10, 4-7025	**GARNET AMPS**		GA-40T	3-6EU7, 1-12AU7,1-7591 Duet 4- 5AR4
Rack 200 wt amp	2-12DW7, 2-12BH7, 2-7025, 1-6550 Sextet	BTO	1-12AX7, 1-6AU6, 1-6SN7, 1-6AN8, 6CA7 Quartet	GA-45RVT Satum	4-6EU7, 1-12AU7, 1-6CG7, 1-6L6 Duet + OA2
				GA-46 Accordion Pre	3-7025, 2-5879
Reverb Unit OS	1-7025, 12AT7, 1-6K6	Lil' Rock	1-12AX7, 1-6SN7, 6V6 Duet,		
Reverb Unit	3-7025, 1-6V6	Pro	2-12AX7, 1-6AU6, 1-6SN7, 1-6AN8, 6CA7 Duet	GA-46 Accordion Amp	1-6SN7, 1-6V6,1-6550 Duet + 5AR4
Showman OS	6-7025,1-6L6 Quartet				
Showman	3-7025, 1-12AT7,1-6L6 Duet + 5U4	Pro Vocal	5-12AX7, 1-6SN7, 6V6 Duet	GA-50	2-6SJ7, 2-6J5,1-6L6 Duet + 5V4
Super 2-10	V front 2-12AY7,1-6V6 Duet + 5U4	Pro 200	2-12AX7, 2-12AU7, 6CA7 Duet,	GA-50T	3-6SJ7, 1-6SN7, 1-6SL7, 1-6L6 Duet + 5V4
Super 2-10	1-12AY7, 2-7025,1-6L6 Duet + 5U4	Pro 400	2-12AX7, 2-12AU7, 6CA7 Quartet		
Super VOS	3-6SC7,1-6L6 Duet + 5U4	Pro 600	2-12AX7, 2-12AU7, 6CA7 Sextet	GA-55	2-12AY7, 1-6SC7,1-6L6 Duet + 5V4
Super OS	1-12AY7, 10-7025,1-6L6 Duet + 5U4	Session Man	4-12AX7, 1-12AU7, 6CA7 Duet,	GA-55RVT Ranger	4-6EU7, 1-12AU7, 1-6CG7, 1-6L6 Duet + OA2
Super Brown	5-7025,1-6L6 Duet				
Super Reverb	4-7025, 2-12AT7,1-6L6 Duet + 5U4	**GIBSON**		GA-60	2-6EU7,1-7591 Duet + 5AR4
Super Reverb	4-7025, 2-12AT7,1-6L6 Duet	Atlas Medalist	2-6EU7, 1-6C4,1-6L6 Duet	GA-70 Ctry-Western	1-7025, 1-12AY7, 1-12AU7, 1-6L6 Duet + 5V4
Super Champ '83	1-7025, 1-12AT7, 1-6C10,1-6V6 Duet	Atlas IV	2-6EU7, 1-6C4,1-6L6 Duet		
Super Twin	2-7025, 1-12AT7, 1-6CX8, 1-6C10, 1-6L6 Sextet	BA-15RV	3-6EU7, 1-12AU7,1-6V6 Duet + 5Y3	GA-75 Recording	2-6EU7, 1-6CG7,1-6L6 Duet + 6C4
		Bass 25	2-6EU7, 1-7591 Duet	GA-75W	1-7025, 2-6SC7, 1-6SJ7, 1-6L6 Duet + 5V4
Super Six	4-7025, 2-12AT7,1-6L6 Quartet	Bass 50	2-6EU7,1-EL34 Duet		
Super 60 '90	2-7025, 1-12AT7,1-6L6 Duet	BR-3	2-7B4, 1-6J5,1-6V6 Duet + 5Z4	Super Medalist	2-6EU7, 2-12AU7, 1-6AU7, 1-7025, 1-7591 Duet
Super 112 '90	2-7025, 1-12AT7,1-6L6 Duet	BR-6	1-6SL7, 1-6SN7,1-6V6 Duet + 5Y3		
Super 210 '90	2-7025, 1-12AT7,1-6L6 Duet	BR-6F	1-6SJ7, 1-6SN7,1-6V6 Duet + 5Y3	Thor Bass Amp	2-6EU7, 1-6CA7 Duet
Tonemaster	3 12AX7, 1 6L6 Quartet	BR-9	1-6SN7,1-6V6 Duet + 5Y3	Titan I, II, & III	3-6EU7, 2-12AU7, 1-OA2, 1-6FQ7,1-6L6 Quartet
Tremolux VOS	1-12AY7, 207025,1-6L6 Duet + 5AR4				

THE TUBE AMP BOOK 157

Reference

Titan Medalist	Same as above
Hawk	3-7025,1-EL84 (6BQ5) Duet
Les Paul TV or JR	1-6SJ7, 1-6V6 + 5Y3
Skylark T	1-6X4, 2-7025, 1-EL84 (6BQ5)
Skylark	1-7025, 1-6X4, 1-EL84 (6BQ5)
SG Systems 100	1-7025,1-8417 Duet
SG Systems 200	1-7015,1-8417 Sextet
EH-150	3-6SQ7, 1-6N7, 1-5U4,1-6L6 Duet
Echoplex	2-6EU7, 1-6C4
XFL-3	2-12AX7, 1-12AT7
XFL-60/60	4-12AX7, 2-12AT7,1-6L6 Duet or 1-EL34 Duet

GT ELECTRONICS

Model Tubes – Matched power sets indicated in bold.

STP-B	1-7025
STP-G	2-7025, 2-ECC-83, 1-6V6 Duet

NOTE: The following Groove Tubes Electronics products accept a wide variety of tubes without modifying the amp. The owner can customize his own amp and change the sound characteristics as desired. To select the combinations best for you, refer to the Groove Tube Catalogue of Tube Types.

TRIO	5 preamp tubes – choose 7025 or 12AX7
STP-G	* 3 preamp tubes – choose 2-7025, 1-12AX7
	* 1 phase inverter – choose 12AX7, 12AT7 or ECC-83
	* 2 output tubes – choose 1-6V6 Duet or 1-EL34 Duet
STA-1 Preamp	* 4 preamp tubes – choose 12AX7 or 12AX7
	* 2 phase inverters – choose 12AX7, 12AT7 or ECC-83
	* 2 output Duets – choose 2-EL34 Duets, 2-6L6 Duets, 2-6550 Duets, or 2-KT88 Duets – or any combination of these Duets.
D-75 Dual amp	* 2 preamp tubes – choose 12AX7, 7025 or ECC83
	* 2 phase inverters – choose 12AX7 or 12AT7
	* 2 output Duets – same as the STA-1 Preamp. Choose any combination of the four basic types – 6L6, EL34, 6550 or KT88.
D-120 Dual amp	* 2 preamp tubes – choose 12AX7, 7025 or ECC83
	* 2 phase inverters – choose 12AX7 or 12AT7
	* 2 output Duets – choose KT88, 6550 or KT88
D-75 Studio	4-7025, 2-Duets of 6550, KT88 or KT90
Soul-o 75	* 4 preamp tubes – choose 7025, 12AX7 or ECC83
	* 1 phase inverter – choose 12AX7-or 12AT7
	* 1 Duet – choose 6L6, EL34, 6550, KT88 or KT90
Soul-o 150	Same as Solo 75 except has 1 Quartet of the above choices.

HAMMOND

B-3 Preamp	1-12BH7, 2-6AU6, 2-6C4, 1-6X4, 1-7025, 1-12AU7

HIWATT

30W	4-7025,1-EL84 Quartet
30WR	4-7025,1-EL84 Quartet
50	4-7025,1-EL34 Duet
100	4-7025,1-EL34 Quartet
200	4-7025,1-EL34 Sextet
400	5-7025,1-KT-88 (6550) Sextet
C520	4-7025,1-EL84 Duet
D50L	4-12AX7,1-EL34 Duet
D50LR	4-12AX7,1-EL34 Duet
D50LRC	4-12AX7,1-EL34 Duet
D100L	4-12AX7,1-EL34 Quartet
D100LR	4-12AX7,1-EL34 Quartet
PRE-1	3-12AX7
PW50	4-12AX7,1-EL34 Quartet
S50	4-12AX7,1-EL34 Duet
S50LC	4-12AX7,1-EL34 Duet
S100L	4-12AX7,1-EL34 Quartet

KASHA

KA-150	3-12AX7,1-6550 Quartet
Rockmod-1	4-12AX7
Rockmod-2	5-12AX7
Rockmod-3	5-12AX7

JIM KELLEY

Single Channel	2-7025,1-6V6 Quartet
Single Channel w/R	3-7025, 1-12AT7,1-6V6 Quartet
Ft. Act. Chn. Swt.	3-7025, 2-12AT7,1-6V6 Quartet

KMD

GV-60	Transistor pro, w/ 1-6L6 Duet
GV-100	Transistor pro, w/ 1-6L6 Quartet
GV-100S	Transistor pre, w/ 1-6L6 Quartet

LANEY

A50	4-ECC83,1-EL34 Duet
A100	4-ECC83,1-EL34 Quartet
PT-50	4-ECC83,1-EL34 Duet
PT-100	4-ECC83,1-EL34 Quartet
PT-30	3-7025,1-6V6 Duet
PT-50 MV	3-7025,1-EL34 Duet
PT-50 AOR	4-7025,1-EL34 Duet
PT-100 MV	3-7025,1-EL34 Quartet
PT-100 AOR	4-7025,1-EL34 Quartet
ST-30	4-7025,2-6V6

LEGEND

G-50 and G100	3-7025 w/ Transistor amp

LESLIE

Model 147	1-12AU7, 1-OC3,1-6550 Duet
Model 122	1-12AU7, 1-OC3,1-6550 Duet

MARANTZ

Model 7	6-ECC83
Model 8	2-6CG7, 2-6BH6,1-EL34 Quartet
Model 9	1-6CG7, 2-ECC88, 2-EL34 Quartets

MARKLEYAMPS

40 SR	1-7025, w/ Transistor amp
80 SR	1-7025, w/ Transistor amp
40 DR	1-7025, w/ Transistor amp
80 DR	1-7025, w/ Transistor amp
150 DR	1-7025, w/ Transistor amp
T-60 wt.	3-7025,1-6L6 Duet
T-120 wt.	3-7025,1-6L6 Quartet
CD-40	3-7025,1-6L6 Duet
CD-60	3-7025,1-6L6 Duet
CD-120	3-7025,1-6L6 Quartet
CD-212	4-7025,1-6L6 Quartet
RM100 MT	2-7025,1-6L6 Quartet
Preamp DR	1-7025
Preamp DB	1-7025

MARSHALL

Artiste 50	4-7025,1-EL34 Duet
Artiste 100	4-7025,1-EL34 Quartet
10 Watt Trem	1-7025,1-ECL86 Duet
18 Watt Trem	3-7025,1-EL84 Duet, 1-EZ81
18/20 Watt	2-7025,1-EL84 Duet, 1-EZ81 (early only)
DSL50	4xECC83, 2xEL34
DSL100	4xECC83, 4xEL34
DSL201	4xECC83, 2xEL84
DSL401	4xECC83, 4xEL84
EL84 20/20	2xECC83, 1xECC82, 4xEL84
EL34 50/50	2xECC83, 2xECC81, 4xEL34
EL34 100/100	2xECC83, 2xECC81, 8xEL34
JCM600	4xECC83, 2xEL34
JCM601	4xECC83, 2xEL34
JCM602	4xECC83, 2xEL34
JMP-1	2xECC83
JTM45	3-7025,1-6L6 Duet, 1-GZ34 (5AR4)
JTM45OS	3xECC83, 1xGZ34, 2xKT66
JTM310	3xECC83, 2x5881
JTM312	3xECC83, 2x5881
JTM600	4xECC83, 2xEL34
JTM610	4xECC83, 2xEL34
JTM612	4xECC83, 2xEL34
JTM615	4xECC83, 2xEL34
JTM622	4xECC83, 2xEL34
MF350	2xECC83
TSL60	4xECC83, 2xEL34
TSL100	4xECC83, 4xEL34
TSL122	4xECC83, 4xEL34
TSL601	4xECC83, 2xEL34
TSL602	4xECC83, 2xEL34
1959SLP/1959SLPX	3xECC83, 4xEL34
1974X	3xECC83, 1xEZ81, 2xEL84
1987X/1987XL	3xECC83, 2x5881
2061X	2xECC83, 2xEL83
2203X	3xECC83, 4xEL34
2203ZW (Zakk Wylde Sig)	3xECC83, 4x6550
2204	3xECC83, 2xEL34
2555SL (Slash Signature)	3xECC83, 4xEL34
2555	3xECC83, 4xEL34
6100	7xECC83, 4x5881
1962	3-7025,1-6L6 Duet, 1-GZ34 (5AR4)
50 Watt	3-7025,1-6550 Duet (EL34 in UK)*
50 Watt w/Trem.	4-7025,1-6550 Duet (EL34 in UK)*
100 Watt	3-7025,1-6550 Quartet (EL34 in UK)*

158 THE TUBE AMP BOOK

Reference

100 Watt w/Trem.	4-7025,1-6550 Quartet (EL34 in UK)*	Blue Angel	5-7025,1-6V6 Duet, 1-EL84 Quartet
200 Watt (Major)	3-7025,1-KT88 Quartet (6550-USA)	Dual Rectifier	5-7025,1-EL34 Duet,
Model 2000	6-7025,1-6550 Sextet (sometimes EL-34)	Road King	1-6L6 Quartet, 2-5U4
Model 2001	3-7025, 2-12AT7,	Tigris	4-7025, 2-6V6 Duets, 2-EL84
	1-6550 Octet (Worldwide)	Quartets	
Model 2100	3-12AX7,1-EL34 Quartet	Baron	4-7025, 2-6L6 Sextets
Higain			
Model 2103	3-7025,1-6550 Quartet (EL34 in UK)*		
Model 2203	3-7025,1-6550 Quartet (EL34 in UK)*	**METALHEAD**	
Model 2205	5-7025,1-6550 Duet (EL34 in UK)*	Elan MK II	5-12AX7
Model 2210	5-7025,1-6550 Quartet	Elan MK III	5-12AX7
	(EL34 in UK)*		
Model 2500 Higain	3-12AX7,1-EL34 Duet	**MITCHELL**	
Model 2501 Higain	3-12AX7,1-EL34 Duet	Pro 100	4-7025,1-6L6 Quartet
Model 2502 Higain	3-12AX7, 1-EL34 Duet	Pro 100 w/EQ & Rev	5-7025,1-6L6 Quartet
Model 3203	1-7025,1-EL34 Duet	Pro 100 w/EQ & Rev	3-7025, 1-12DW7,1-616 Quartet
Model 4001 Studio 15	2-7025,1-6V6 Duet*	Deluxe	4-7025,1-6L6 Duet
Model 4010 1-12"	3-7025,1-6550 Duet (EL34 in UK)*		
Model 4100 Dual Rev	3-12AX7,1-EL34 Quartet		
Model 4101 Dual Rev	3-12AX7,1-EL34 Quartet	**MOJAVE**	
Model 4102 Dual Rev	3-12AX7, 1- EL34 Quartet	Coyote	3 12AX7, EL84 Duet, GZ34/5AR4
Model 4104 2-12"	3-7025,1-6550 Duet (EL34 in UK)*	Peacemaker 100 watt	4 12AX7, EL34 Quartet
Model 4210	5-7025,1-6550 Duet (EL34 in UK)*	Plexi 45	3 12AX7, 6L6 / KT666 Duet,
Model 4203	1-7025,1-EL34 Duet		GZ34/5AR4
Model 4500 Dual Rev	3-12AX7, 1-EL34 Duet	Scorpion 50 watt	4 12AX7, EL34 Duet
Model 4501 Dual Rev	3-12AX7, 1-EL34 Duet	Sidewinder	3 12AX7, EL84 Quartet, GZ34/5AR4
Model 4502 Dual Rev	3-12AX7, 1-EL34 Duet		
Model 9001	3-7025		
Model 9005	4-7025,1-EL34 Quartet	**MUSICMAN**	
* In 1986, Marshall officially dropped the 6550 tube,		RD 50	1-7025,1-6L6 Duet
replacing it in all applications with the original EL-34 tube.		RD 65	1-6L6 Duet* (Transistor Preamp)
		RD 100	1-6L6 Duet* (Transistor Preamp)
McINTOSH		RD 112	1-6L6 Duet (Transistor Preamp)
C-22	6-7025	RD 120	1-6L6 Duet (Transistor Preamp)
MC-30	-7025, 1-12AU7, 1-12BH7,1-6L6 Duet + 5U4	RP 65	1-6L6 Duet* (Transistor Preamp)
MC-40	1-7025, 1-12AU7, 1-12BH7,1-6L6 Duet 4- 5AR4	RP 100	1-6L6 Duet* (Transistor Preamp)
MC-60	1-7025, 1-12AU7, 1-12BH7,1-6550 Duet	RP 115	1-6L6 Duet (Transistor Preamp)
MC-75	1-7025, 1-12AU7, 1-12BH7,1-6550 Duet	Sixty-Five	1-7025,1-6CA7 Duet
MC-240	3-7025, 2-12AU7, 2-12BH7, 2-6L6D 4- 2-5U4	One-Thirty	1-7025,1-6CA7 Quartet
MC-260	3-7025, 2-12AU7, 2-12BH7, 2-6550D + 2-5AR4	75	1-6L6 Duet (Transistor Preamp)
MC-275	3-7025, 2-12AU7, 2-12BH7, 2-6550 Duets	150	1-6L6 Quartet (Transistor Preamp)
MC-3500	2-7025, 2-6DJ8, 1-6CG7, 1-6BL7,1-6L6 Octet	*early model used 6CA7, check amp	
MESA/BOOGIE		**MUSITECH**	
Subway Blues	4-7025, 1-EL84 Duet	Stereo 240	5-7025, 1-12AT7, 1-5AR4
Subway Rocket	4-7025, 1-EL84 Duet		1-6L6 Duet + 1-EL34 Duet
F-30	4-7025, 1-EL84 Duet	Stereo 215	5-7025, 1-12AT7,
F-50	4-7025, 1-6L6 Duet		1-6V6 Duet + 1-EL84 Duet
F-100	4-7025, 1-Quartet		
Nomad 45	5-7025, 1-EL84 Quartet		
Nomad 55	5-7025, 1-6L6 Duet	**NOMAD**	
Nomad 100	5-7025, 1-6L6 Quartet	50 chn. swt.	4-7025, 1-12AT7,1-EL34 Duet
Dual Caliber 100	6-7025, 1-6L6 Quartet	100 chn. swt.	4-7025, 1-12AT7,1-EL34 Quartet
20/20 Stereo	3-7025, 2-EL84 Duets	50 reverb	6-7025, 1-12AT7,1-EL34 Duet
Stereo 2:50	3-7025, 2-6L6 Duets	100 reverb	6-7025, 1-12AT7,1-EL34 Quartet
Rectifier Stereo 2:100	3-7025, 2-6L6 Duets		
Formula Preamp	5-7025		
V-Twin Preamp	2-7025	**ORANGE**	
V-Twin Rack	2-7025	80	3-7025,1-EL34 Duet
Rectifier Recording	6-7025	120	3-7025,1-EL34 Quartet
Preamp		AD5	1-12AX7, 1-EL84
Heartbreaker	7-7025, 1-6L6 Quartet, 1-5AR4	AD15/12	2-12AX7, 1-EL84 Duet + 5AR4
Maverick	6-7025, 1-EL84 Quartet, 1-5AR4		

PEAVEY			
Classic 60	2-12AX7,1-6L6 Duet		
Classic 60/60	3-12AX7,1-6L6 Duet		
Classic 120	3-12AX7,1-6L6 Quartet		
Classic 120/120	4-12AX7, 2-6L6 Quartet		
Classic & VTX	1-6L6 Duet (Transistor Preamp)		
SDuce	1-6L6 Quartet (Transistor Preamp)		
Heritage	1-6L6 Quartet (Transistor Preamp)		
Mace	1-6L6 Sextet (Transistor Preamp)		
MX	1-6L6 Quartet (Transistor Preamp)		
Roadmaster (early)	1-6550 Quartet (Transistor Preamp)		
Roadmaster (newer)	4-7025, 2-12AX7,1-6L6 Sextet		
Rockmaster	3-7025,1-6L6 Quartet		
Rockmaster (newer)	3-12AX7,1-6L6 Quartet		
Encore 65	3-7025, 1-12AT7,1-6L6 Duet		
TG RAXX	4-12AX7		
TB RAXX	4-12AX7		
Triumph 60	4-12AX7, 1-12AT7,1-6L6 Duet		
Triumph 120	4-12AX7, 1-12AT7,1-6L6 Quartet		
Butcher	3-7025,1-6L6 Quartet		
Vintage (early)	2-6C10,1-6L6 Quartet		
Vintage (newer)	1-6L6 Quartet (Transistor Preamp)		
VTA 400	3-7025, 1-6AN8,1-6550 Quartet		
VTA 800	3-7025, 1-6AN8, 2-6550 Octet		
VTB 300	2-7025, 1-6AN8,1-6550 Quartet		
VTG 300	2-7025, 1-6AN8, 1-6550 Quartet		
POLYTONE			
Fusion	4-7025, 1-12AU7, 1-12AT7,1-6L6 Quartet		
RANDALL			
RGTES	4-12AX7, 1-12AT7,1-6L6 Quartet		
RGT100	4-12AX7, 1-12AT7,1-6L6 Quartet		
RCT100HT	4-12AX7, 1-12AT7,1-6L6 Quartet		
RISSON			
Clone	3-7025,1-6550 Quartet		
CTA	2-7025, 1-12AT7,1-8417 Quartet		
ETA-150	3-7025,1-6550 Quartet		
ETA-100	3-7025,1-6550 Duet		
LTA 120	3-7025, 1-12AT7,1-6550 Duet		
LTA 120 Reverb	3-7025, 2-12AT7,1-6550 Duet		
LTA 200 Reverb	3-7025, 2-12AT7,1-6550 Quartet		
RIVERA			
Chubster	5-12AX7, 1-EL34 Duet		
Quiana	5-12AX7, 1-6L6 Duet or 1-6L6
Quartet			
TBR-I Rudolph Schenker	8-12AX7, 2-KT77 Duet		
TBR-1 Ted Nugent	6-12AX7, 1-12AT7, 2-KT77 Duet		
TBR-I	7-12AX7, 2-EL34 Duet		
TBR-1M	7-12AX7, 2-EL34 Duet		
TBR-1SL	7-12AX7, 2-EL34 Duet		
TBR-2	7-12AX7, 2-6550 Quartets		
TBR-2M	7-12AX7, 2-6550 Quartets		
TBR-2SL	7-12AX7, 2-6550 Quartets		
TBR-2 John Sykes	8-12AX7, 2-6550 Quartets		
TBR-2 Jerry Garcia	5-12AX7, 2-12AU7,
	2-6550 Quartets		
TBR-2B	5-12AX7, 2-6550 Quartets		
TBR-3 Hammer 120	4-12AX7, 2-EL34 Duets		

THE TUBE AMP BOOK 159

TBR-4 Preamp	6-12AX7,1-EL84 Duet
TBR-5 Hammer 320	4-12AX7, 2-6550 Quartets
TBR-6 Preamp	4-12AX7
TBR-7 Power Amp	4-12AX7, 2-5881 Duets
M60 Combo	5-12AX7, 2-EL34 Duet
M100 Combo	5-12AX7,1-EL34 Quartet
M100 Combo (J. Garcia)	5-12AX7,1-6550 Quartet
S120 Combo	6 12AX7, 2-EL34 Duets

ROLAND

Bolt 30	1-12AT7,1-7391 Duet
Bolt 60	1-12AT7,1-6L6 Duet
Bolt 100	1-12AT7,1-6L6 Quartet

SILVERTONE

50 Watt w/Reverb	7025, 2-6CG7,1-6L6 Duet
100 Watt w/Reverb	4-7025, 2-6CG7,1-6L6 Quartet

SOLDANO

Astroverb	5 12AX7 and one duet EL84
Avenger	(available with either 6L6 or EL34 transformer, not interchangeable) 4 12AX7 and one quartet 6L6 or EL34
Hot Rod 50	4 12AX7 and one duet 6L6
Hot Rod 50+	5 12AX7 and one duet 6L6
Lucky 13 50 watt	6 12AX7 and one duet 6L6
Lucky 13 100 watt	6 12AX7 and one quartet 6L6
Reverb-O-Sonic	5 12AX7 and one duet 6L6
SLO 100	5 12AX7 (4 without loop) and one quartet 6L6
Space Box reverb unit	2 12AX7
Supercharger GTO pedal	2 12AX7

Discontinued products:

Hot Rod 100+	5 12AX7 and one quartet 6L6
Hot Mod	one 6C10 or 6K11/6Q11
SL60	4 12AX7 and one duet 6L6
SL105	3 12AX7 and two duets 6L6
SM100	2 12AX7 and one quartet 6L6
SP77	4 12AX7
Surf Box	5 12AX7 and one EL84
X88	6 12AX7
X99	5 12AX7

SOUND CITY

Bass 150	2-7025, 1-12AT7,1-6550 Quartet
Concord	3-7025, 2-12AT7,1-EL34 Quartet
GT-50	1-7025, 1-12AT7,1-EL34 Quartet
PA-400	3-7025, 1-12AT7,1-EL34 Quartet
PA-200	3-7025, 1-12AT7,1-6550 Quartet
MK-IV 120	4-7025, 1-12AU7,1-EL34 Sextet
LB-50 Plus	3-7025, 1-12AT7,1-EL34 Duet
LB-200 Plus	4-7025, 1-12AT7,1-6550 Quartet
50 watt	3-7025,1-EL34 Duet
100 watt	3-7025, 1-12AT7,1-EL34 Sextet
200 watt	3-7025, 1-12AT7,1-6550 Quartet

SPECTRA (DEAN MARKLEY)

30 T	3-7025,1-6V6 Duet
60 T	3-7025,1-6V6 Duet
120 T	3-7025,1-6L6 Quartet
2-12 T	4-7025,1-6L6 Quartet

SUNDOWN

Formula 50	4-7025,1-6L6 Duet
Rebel 50	3-7025,1-6L6 Duet
Artist Combo	5-7025,1-6L6 Quartet
Rebel 100	3-7025,1-6L6 Quartet
Artist 30 Combo	5-7025,1-6L6 Duet
SD1012C	5-7025,1-6550 Duet
SD1000H	5-7025,1-6550 Duet

SUNN

Coliseum PA	3-7025, 1-6AN8,1-6550 Quartet + 5AR4
Model A	4-7025,1-6550 Quartet
Model A212	4-7025,1-6550 Duet
Model T	4-7025,1-6550 Quartet
Sceptre	1-7025, 1-12AU7, 1-6AN8*, 1-6550 Duet**
Sentura	1-7025, 1-12AU7, 1-7199,1-6CA7 Duet
Sonic	1-7025, 1-12AU7, 1-6AN8*, 1-6550 Duet**
Solos I	2-7025, 1-12AU7, Transistor powered
Spectrum I	1-7025, 1-7199,1-6CA7 Duet
Solarus (190L)	1-7025, 1-12AU7, 1-6AN8*, 1-6550 Duet**
100S	1-7025, 1-12AU7, 1-6AN8*, 1-6550 Duet**
200S (190B)	1-7025, 1-6AN8*, 1-6550 Duet**
1000S	1-7025, 1-12AU7, 1-6AN8*, 1-6550 Quartet
1200S (350L)	1-7025, 1-12AU7, 1-6AN8*, 1-6550 Quartet
2000S (350B)	1-7025, 1-6AN8*, 1-6550 Quartet

* later models may use 7199 ** early models may use 6CA7

SWR

SS-180	1-7025
PB-200	1-7025
SM-400	1-7025
SM-400 (newer)	1-12AX7
Baby Blue	1-12AX7
Redhead	1-12AX7
Studio 220	1-12AX7

THD

4-10	1-12AY7, 3-7025,1-6L6 Duet
2-10	1-12AY7, 3-7025,1-6L6 Duet
V-Front	1-12AY7, 3-7025,1-6L6 Duet
Tweed Head	1-12AY7, 3-7025,1-6L6 Duet
4-10 Reverb	1-12AY7, 3-7025, 1-12AT7,1-6L6 Duet
2-10 Reverb	1-12AY7, 3-7025, 1-12AT7,1-6L6 Duet
V-Front Rev	1-12AY7, 3-7025, 1-12AT7,1-6L6 Duet
Tweed Head Rev	1-12AY7, 3-7025, 1-12AT7,1-6L6 Duet
50W Rack Head	1-12AY7, 4-7025, 1-12AT7,1-6L6 Duet
100W Rack Head	1-12AY7, 4-7025, 1-12AT7, 1-6L6 Quartet

NOTE: 7025 cab be substituted for 12AY7 for increase in gain.

EL34s, 6CA7s, KT88s and 6550s can be substituted for 6L6s with a simple bias-voltage adjustment.

TRAINWRECK

Liverpool-30 wt.	3-7025,1-EL34 Quartet
Express-22 wt.	3-7025,1-6V6 Duet, some with 1-EL34 Duet
Liverpool Rocket	3-7025, 1-5AR4, 4-EL84
Express-36wt.	3-7025,1-EL34 Duet

THUNDERFUNK

50B	5-7025, 1-12AT7,1-6550A Duet
100B	5-7025, 1-12AT7,1-6550A Duet
50ELS	6-7025, 2-12AT7,1-EL34 Duet
100ELS	6-7025, 2-12AT7,1-EL34 Quartet
50LS	6-7025, 2-12AT7,1-6550A Duet
100LS	6-7025, 1-12AT7,1-6550A Quartet

TRACE ELLIOT

Hexavalve	3-ECC83, 1-6550 Sextet
Quatravalve	3-ECC83, 1-6550 Quartet
TWinvalve	3-ECC83, 1-6550 Duet
VA350	3-ECC83, 1-6550 Sextet
VR350	3-ECC83, 1-6550 Sextet
GP12XV	3-ECC83

TRAYNOR

Custom Valve 40	3-12AX7, 1-6L6 Duet
YBA-2	2-7025,1-EL84 Duet
YBA-1 & 4	2-7025,1-6CA7 Duet
YBA-3	2-7025,1-6CA7 Quartet
YGL-3 (3A or MK-3)	4-7025, 1-EL84,1-6CA7 Quartet
YGL-3 older	5-7025,1-6CA7 Quartet
YGM-3 & 4	4-7025,1-EL84 Duet
YRM-1 (SC) & YVM-1	4-7025,1-6CA7 Duet
YRM-1 older	4-7025, 1-EL84,1-6CA7 Duet
YSR	6-7025,1-6CA7 Duet

TUBE WORKS

901	1-12AX7
902	1-12AX7
903	1-12AX7
904	1-12AX7
910	1-12AX7
913	1-12AX7
922	2-12AX7
924	2-12AX7

TUSC

CF-50	1-12AT7,1-6L6 Duet
CF-100	1-12AT7,1-6L6 Quartet
DF-50	1-12AT7,1-6L6 Duet
DF-100	1-12AT7,1-6L6 Quartet

VHT

G2000S	2-12AX7, 2-12AT7, 2-EL34 Duet
G2150C	2-12AX7, 2-12AT7, 2-EL34 Quartet
G2150S REV4	3-12AX7, 2-6550A Duet
G2150S REV5	2-12AX7, 2-12AT7. 2-6550A Duet
G2250S	2-12AX7, 2-12AT7. 2-6550A Quartet
G2300SB	2-12AX7, 2-12AT7,1-EL34 Quartet, 1-6550A Quartet
50C	5-12AX7,1-EL34 Duet
100C	5-12AX7.1-EL34 Quartet
100S	5-12AX7,1-6550A Duet

Preamp   5-12AX7
NOTE: Use only the 6550 "A" version of Groove Tubes for VHT output circuits

VICTORIA
518-T	5Y3GT OR 5V4GT, 6V6GT, 12AX7
20112-T	5Y3GT, 2-6V6GT, 1 12AX7, 1 5751 OR 12AY7
DOUBLE DELUXE	5Y3GT OR 5AR4/GZ34, 4-6V6GT, 1-12AX7, 1-5751 OR 12AY7
35210-T	5U4GB, 2-6L6GCGE, 2-12AX7, 1-5751 OR 12AY7
35115-T	5U4GB, 2-6L6GCGE, 2-12AX7, 1-5757 OR 12AY7
35310-T	5A4R/GZ34, 2-6L6GCGE, 2-12AX7, 1-5751 OR 12AY7
45410-T	5AR4/GZ34, 2-6L6GCGE, 2-12AX7, 1-5751 OR 12AY7
80410-T	5AR4/GZ34, 4-6L6GCGE, 2-12AX7, 1-5751 OR 12AY7
50212-T	5AR4/GZ34 OR 5U4, 2-6L6GCGE, 3-12AX7, 1-5751 OR 12AY7
80212-T	5AR4/GZ34, 4-6L6GCGE, 3-12AX7, 1-5751 OR 12AY7
VICTORI-ETTE	5AR4/GZ34, 2-EL84, 3-12AX7, 1-12AT7
VICTORILUX	5AR4/GZ34, 4-EL84 OR 2-6L6GCGE/6550A, 3-12AX7, 1-12AT7
SOVEREIGN	5AR4/GZ34, 2-EL34 OR 6L6, 2-12AX7, 1-12AT7, 1-6BM8, 1-EF86

VOX
AC-4	1-7025, 1-6267, 1-6V4, 1-EL84
AC-10	1-7025, 1-6267, 1-6U8A, 1-7189 Duet
AC-15	2-7025, 1-12AT7, 1-EL84 Duet + 5AR4
AC-15 '82	2-7025, 1-12AT7, 1-EL84
AC15TB/TBX/TB2	5xECC83, 1x5Y3, 2xEL84
AC-30	4-7025, 1-12AU7, 1-EL84 Quartet 4- 5AR4
AC-30 w/Reverb	6-7025, 1-12AU7, 1-EL84 Quartet + 5AR4
AC-30 w/Reverb '82	6-7025, 1-12AT7, 1-EL84 Quartet
AC-30 Top	7-7025, 1-12AU7, 1-EL84 Quartet
AC-30	6-7025. 1-EL84 Quartet Standard '90
AC-30	6-7025, 1-EL84 Quartet Limited '90
AC30 Custom Classic	3xECC83, 1xGZ34, 4xEL84
AC30TB/TBX (1992-2004)	5xECC83, 1xECC82, 1xGZ34, 4xEL84
AC-50	3-7025. 1-12AU7.1-EL34 Duet -+- 5AR4
AC-100	1-7025. 2-12AU7, 1-EL34 Quartet
Cambrudge Reverb	3-7025. 1-12AU7, 1-EL84 Duet + EZ81
V125	6-7025, 1-12AT7, 1-EL34 Quartet
v-125	3-7025. 1-12AT7. 2-7189, 1-EL34 Quartet
Concert 501	5-7025. 1-EL84 Quartet
Concert 502	5-7025. 1-EL84 Quartet
Concert 100	4-7025. 1-EL43 Quartet
AC-30 TB '86	5-ECC82, 1-EL84 Quartet
AC-30 TB/RV '86	7-7025. 1-EL84 Quartet

NOTE: Additional tube sets available for the following tube Hi Fi manufacturers: Audible Illusions, Beard, Berning, Bogen, Counterpoint, Fisher, Harmon-Kardon, Music Reference, H.H. Scott.

Tube/spec sheets

RCA 6550 Push-Pull AF Power Amplifier — Class A1 spec sheet, Radio Corporation of America, Electron Tube Division, Harrison, N.J. DATA 2, 5-62.

Tube/spec sheets

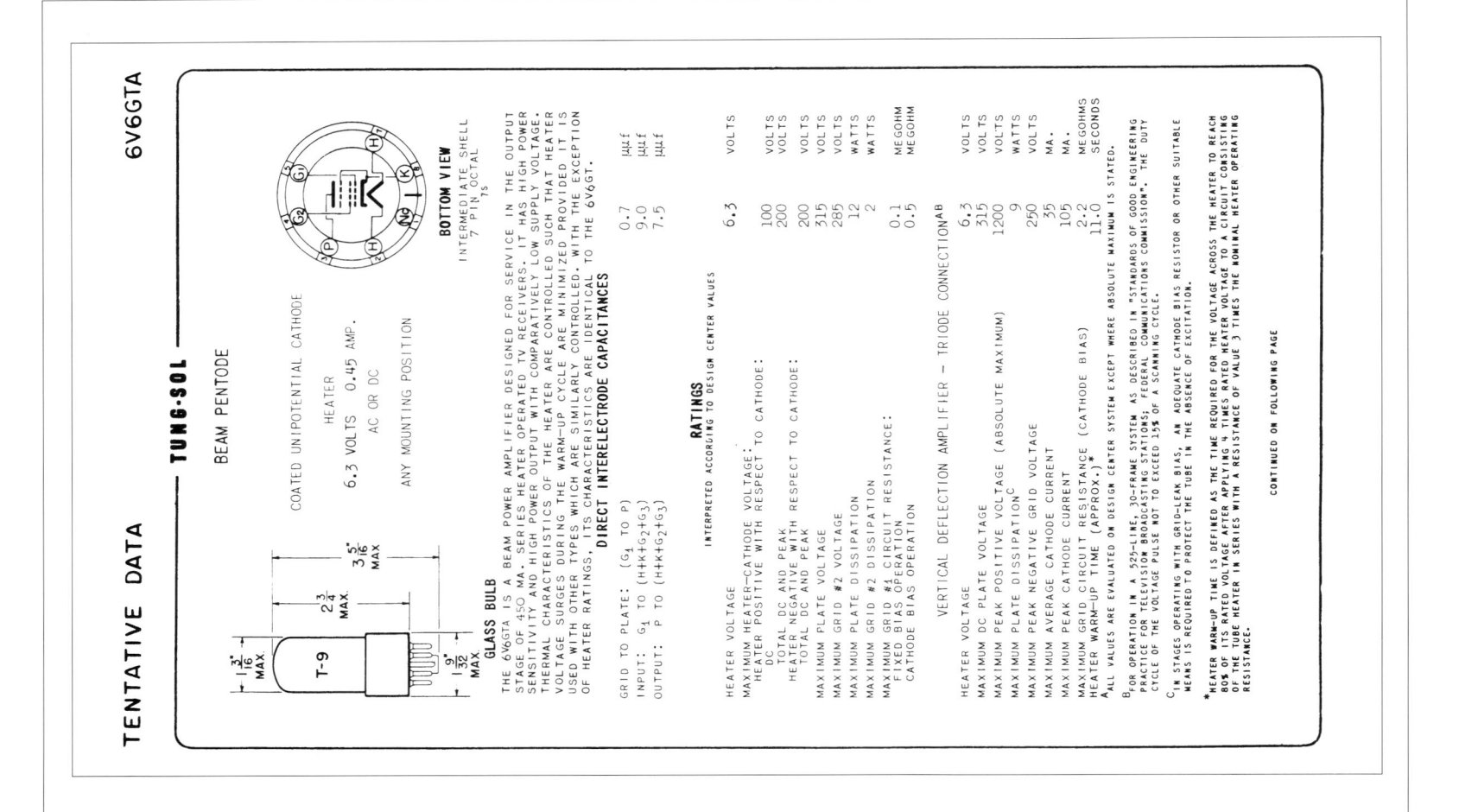

6V6GTA — TUNG-SOL

TENTATIVE DATA

BEAM PENTODE

COATED UNIPOTENTIAL CATHODE

HEATER
6.3 VOLTS   0.45 AMP.
AC OR DC

ANY MOUNTING POSITION

GLASS BULB
T-9
1 3/16 MAX
2 3/4 MAX
3 5/16 MAX
9/32 MAX

BOTTOM VIEW
INTERMEDIATE SHELL
7 PIN OCTAL
7S

The 6V6GTA is a beam power amplifier designed for service in the output stage of 450 ma. series heater operated TV receivers. It has high power sensitivity and high power output with comparatively low supply voltage. Thermal characteristics of the heater are controlled such that heater voltage surges during the warm-up cycle are minimized and provided it is used with other types which are similarly controlled. With the exception of heater ratings, its characteristics are identical to the 6V6GT.

DIRECT INTERELECTRODE CAPACITANCES

GRID TO PLATE: (G₁ to P)	0.7	µµf
INPUT: G₁ to (H+K+G₂+G₃)	9.0	µµf
OUTPUT: P to (H+K+G₂+G₃)	7.5	µµf

RATINGS
INTERPRETED ACCORDING TO DESIGN CENTER VALUES

HEATER VOLTAGE	6.3	VOLTS
MAXIMUM HEATER-CATHODE VOLTAGE:		
HEATER POSITIVE WITH RESPECT TO CATHODE:		
DC	100	VOLTS
TOTAL DC AND PEAK	200	VOLTS
HEATER NEGATIVE WITH RESPECT TO CATHODE:		
TOTAL DC AND PEAK	200	VOLTS
MAXIMUM PLATE VOLTAGE	315	VOLTS
MAXIMUM GRID #2 VOLTAGE	285	VOLTS
MAXIMUM PLATE DISSIPATION	12	WATTS
MAXIMUM GRID #2 DISSIPATION	2	WATTS
MAXIMUM GRID #1 CIRCUIT RESISTANCE:		
FIXED BIAS OPERATION	0.1	MEGOHM
CATHODE BIAS OPERATION	0.5	MEGOHM

VERTICAL DEFLECTION AMPLIFIER – TRIODE CONNECTION[A][B]

HEATER VOLTAGE	6.3	VOLTS
MAXIMUM DC PLATE VOLTAGE	315	VOLTS
MAXIMUM PEAK POSITIVE PLATE VOLTAGE (ABSOLUTE MAXIMUM)	1200	VOLTS
MAXIMUM PLATE DISSIPATION[C]	9	WATTS
MAXIMUM PEAK NEGATIVE GRID VOLTAGE	250	VOLTS
MAXIMUM AVERAGE CATHODE CURRENT	35	MA.
MAXIMUM PEAK CATHODE CURRENT	105	MA.
MAXIMUM GRID CIRCUIT RESISTANCE (CATHODE BIAS)	2.2	MEGOHMS
HEATER WARM-UP TIME (APPROX.)*	11.0	SECONDS

[A] ALL VALUES ARE EVALUATED ON DESIGN CENTER SYSTEM EXCEPT WHERE ABSOLUTE MAXIMUM IS STATED.

[B] FOR OPERATION IN A 525-LINE, 30-FRAME SYSTEM AS DESCRIBED IN "STANDARDS OF GOOD ENGINEERING PRACTICE FOR TELEVISION BROADCASTING STATIONS; FEDERAL COMMUNICATIONS COMMISSION." THE DUTY CYCLE OF THE VOLTAGE PULSE NOT TO EXCEED 15% OF A SCANNING CYCLE.

[C] IN STAGES OPERATING WITH GRID-LEAK BIAS, AN ADEQUATE CATHODE BIAS RESISTOR OR OTHER SUITABLE MEANS IS REQUIRED TO PROTECT THE TUBE IN THE ABSENCE OF EXCITATION.

*HEATER WARM-UP TIME IS DEFINED AS THE TIME REQUIRED FOR THE VOLTAGE ACROSS THE HEATER TO REACH 80% OF ITS RATED VOLTAGE AFTER APPLYING 4 TIMES RATED HEATER VOLTAGE TO A CIRCUIT CONSISTING OF THE TUBE HEATER IN SERIES WITH A RESISTANCE OF VALUE 3 TIMES THE NOMINAL HEATER OPERATING RESISTANCE.

CONTINUED ON FOLLOWING PAGE

6V6GTA — TUNG-SOL

TENTATIVE DATA

CONTINUED FROM PRECEDING PAGE

TYPICAL OPERATING CONDITIONS AND CHARACTERISTICS

CLASS A₁ AMPLIFIER – SINGLE TUBE

HEATER VOLTAGE	6.3	6.3	VOLTS
HEATER CURRENT	0.45	0.45	AMP.
PLATE VOLTAGE	180	250	VOLTS
GRID #2 VOLTAGE	180	250	VOLTS
GRID #1 VOLTAGE	−8.5	−12.5	VOLTS
PEAK AF GRID #1 VOLTAGE	8.5	12.5	VOLTS
ZERO-SIGNAL PLATE CURRENT	29	45	MA.
MAXIMUM-SIGNAL PLATE CURRENT	30	47	MA.
ZERO-SIGNAL GRID #2 CURRENT	3	4.5	MA.
MAXIMUM-SIGNAL GRID #2 CURRENT	4	7	MA.
PLATE RESISTANCE (APPROX.)	50 000	50 000	OHMS
TRANSCONDUCTANCE	3 700	4 100	µMHOS
LOAD RESISTANCE	5 500	5 000	OHMS
MAXIMUM-SIGNAL POWER OUTPUT	2	4.5	WATTS
TOTAL HARMONIC DISTORTION (APPROX.)	8	8	PERCENT

CLASS A₁ AMPLIFIER – PUSH-PULL
UNLESS OTHERWISE SPECIFIED, VALUES ARE FOR TWO TUBES.

HEATER VOLTAGE	6.3	VOLTS
HEATER CURRENT	0.45	AMP.
PLATE VOLTAGE	250	VOLTS
GRID #2 VOLTAGE	250	VOLTS
GRID #1 VOLTAGE	−15	VOLTS
PEAK AF GRID #1 TO GRID #1 VOLTAGE	30	VOLTS
ZERO-SIGNAL PLATE CURRENT	70	MA.
MAXIMUM-SIGNAL PLATE CURRENT	79	MA.
ZERO-SIGNAL GRID #2 CURRENT	5	MA.
MAXIMUM-SIGNAL GRID #2 CURRENT	13	MA.
PLATE-TO-PLATE LOAD RESISTANCE	10 000	OHMS
MAXIMUM-SIGNAL POWER OUTPUT	10	WATTS
TOTAL HARMONIC DISTORTION	5	PERCENT

CLASS A₁ AMPLIFIER – TRIODE CONNECTION

HEATER VOLTAGE	6.3	VOLTS
HEATER CURRENT	0.45	AMP.
PLATE VOLTAGE	250	VOLTS
GRID VOLTAGE	−12.5	VOLTS
PLATE CURRENT	49.5	MA.
TRANSCONDUCTANCE	5 000	µMHOS
AMPLIFICATION FACTOR	9.8	
PLATE RESISTANCE (APPROX.)	1 960	OHMS
GRID VOLTAGE FOR Ib = 0.5 MA. (APPROX.)	−36	VOLTS

162 THE TUBE AMP BOOK

6EU7 — TUNG-SOL

TENTATIVE DATA

TUNG-SOL ELECTRIC INC., ELECTRON TUBE DIVISION, BLOOMFIELD, NEW JERSEY, U.S.A., JUNE 1, 1960 PLATE #5869

TWIN TRIODE
MINIATURE TYPE

UNIPOTENTIAL CATHODE

HEATER
6.3±10% VOLTS   0.3 AMP.
AC OR DC
ANY MOUNTING POSITION

GLASS BULB

BOTTOM VIEW
SMALL-BUTTON NOVAL
9 PIN BASE
9LS

THE 6EU7 IS A HIGH-MU TWIN TRIODE IN THE 9 PIN MINIATURE CONSTRUCTION. IT IS ESPECIALLY DESIGNED FOR USE IN HIGH-GAIN RESISTANCE-COUPLED LOW-LEVEL AUDIO-AMPLIFIER APPLICATIONS, SUCH AS PREAMPLIFIERS FOR MONOPHONIC AND STEREOPHONIC PHONOGRAPHS, AND MICROPHONE AMPLIFIERS. THE BASING ARRANGEMENT ENABLES THE CIRCUIT DESIGNER TO OBTAIN GOOD ISOLATION BETWEEN CHANNELS WHEN THE TUBE IS USED IN A STEREO SYSTEM.

DIRECT INTERELECTRODE CAPACITANCES
WITHOUT EXTERNAL SHIELD

	UNIT #1	UNIT #2	
GRID TO PLATE	1.5	1.5	µµf
GRID TO CATHODE AND HEATER	1.6	1.6	µµf
PLATE TO CATHODE AND HEATER	0.2	0.2	µµf

RATINGS
INTERPRETED ACCORDING TO DESIGN MAXIMUM SYSTEM

AMPLIFIER - CLASS A₁
VALUES ARE FOR EACH UNIT

HEATER VOLTAGE	6.3±10%	VOLTS
MAXIMUM PLATE VOLTAGE	330	VOLTS
MAXIMUM GRID VOLTAGE:		
NEGATIVE BIAS VALUE	55	VOLTS
POSITIVE BIAS VALUE	0	VOLTS
MAXIMUM PLATE DISSIPATION	1.2	WATTS
MAXIMUM PEAK HEATER-CATHODE VOLTAGE:		
HEATER NEGATIVE WITH RESPECT TO CATHODE	200[A]	VOLTS
HEATER POSITIVE WITH RESPECT TO CATHODE	200	VOLTS

[A] THE DC COMPONENT MUST NOT EXCEED 100 VOLTS.

CONTINUED ON FOLLOWING PAGE

6EU7 — TUNG-SOL

TENTATIVE DATA

CONTINUED FROM PRECEDING PAGE

TYPICAL OPERATING CONDITIONS AND CHARACTERISTICS

CLASS A₁ AMPLIFIER

HEATER VOLTAGE	6.3±10%		VOLTS
HEATER CURRENT		0.3	AMP.
PLATE VOLTAGE	100	250	VOLTS
GRID VOLTAGE	−1	−2	VOLTS
AMPLIFICATION FACTOR	100	100	
PLATE RESISTANCE (APPROX.)	80,000	62,500	OHMS
TRANSCONDUCTANCE	1,250	1,600	µMHOS
PLATE CURRENT	0.5	1.2	MA.

EQUIVALENT NOISE AND HUM VOLTAGE
REFERENCED TO GRID, EACH UNIT

AVERAGE VALUE• RMS   1.8   µVOLTS

• MEASURED IN "TRUE RMS" UNITS UNDER THE FOLLOWING CONDITIONS; HEATER VOLTAGE OF 6.3 VOLTS AC; CENTER TAP OF HEATER TRANSFORMER GROUNDED; PLATE SUPPLY VOLTAGE, 250 VOLTS DC; PLATE LOAD RESISTOR, 100,000 OHMS; CATHODE RESISTOR, 2700 OHMS BYPASSED BY 100-µf CAPACITOR GRID RESISTOR, 0 OHMS; AMPLIFIER COVERING FREQUENCY RANGE BETWEEN 25 AND 10,000 CPS.

B. DESIGN-MAXIMUM RATINGS ARE LIMITING VALUES OF OPERATING AND ENVIRONMENTAL CONDITIONS APPLICABLE TO A BOGEY ELECTRON DEVICE OF A SPECIFIED TYPE AS DEFINED BY ITS PUBLISHED DATA, AND SHOULD NOT BE EXCEEDED UNDER THE WORST PROBABLE CONDITIONS. THE DEVICE MANUFACTURER CHOOSES THESE VALUES TO PROVIDE ACCEPTABLE SERVICEABILITY OF THE DEVICE, TAKING RESPONSIBILITY FOR THE EFFECTS OF CHANGES IN OPERATING CONDITIONS DUE TO VARIATION IN DEVICE CHARACTERISTICS. THE EQUIPMENT MANUFACTURER SHOULD DESIGN SO THAT INITIALLY AND THROUGHOUT LIFE NO DESIGN MAXIMUM VALUE FOR THE INTENDED SERVICE IS EXCEEDED WITH A BOGEY DEVICE UNDER THE WORST PROBABLE CONDITIONS WITH RESPECT TO SUPPLY-VOLTAGE VARIATION, EQUIPMENT COMPONENT VARIATION, EQUIPMENT CONTROL ADJUSTMENT, LOAD VARIATION, SIGNAL VARIATION, AND ENVIRONMENTAL CONDITIONS.

OPERATING CONDITIONS AS RESISTANCE COUPLED AMPLIFIER
EACH UNIT

PLATE SUPPLY VOLTAGE	90			180			300			VOLTS
PLATE LOAD RESISTOR	0.1	0.22	0.47	0.1	0.22	0.47	0.1	0.22	0.47	MEGOHM
GRID RESISTOR (OF FOLLOWING STAGE)	0.22	0.47	1.0	0.22	0.47	1.0	0.22	0.47	1.0	MEGOHMS
CATHODE RESISTOR	4700	7400	13000	2000	8500	6700	1500	2800	5200	OHMS
PEAK OUTPUT VOLTAGE	6	9	11	25	34	39	57	69	77	VOLTS
VOLTAGE GAIN	35[C]	45[D]	52[E]	47	59	66	52	65	73	

C AT 2 VOLTS (RMS) OUTPUT
D AT 3 VOLTS (RMS) OUTPUT
E AT 4 VOLTS (RMS) OUTPUT

NOTE: COUPLING CAPACITORS SHOULD BE SELECTED TO GIVE DESIRED FREQUENCY RESPONSE. CATHODE RESISTORS SHOULD BE ADEQUATELY BYPASSED.

Tube/spec sheets

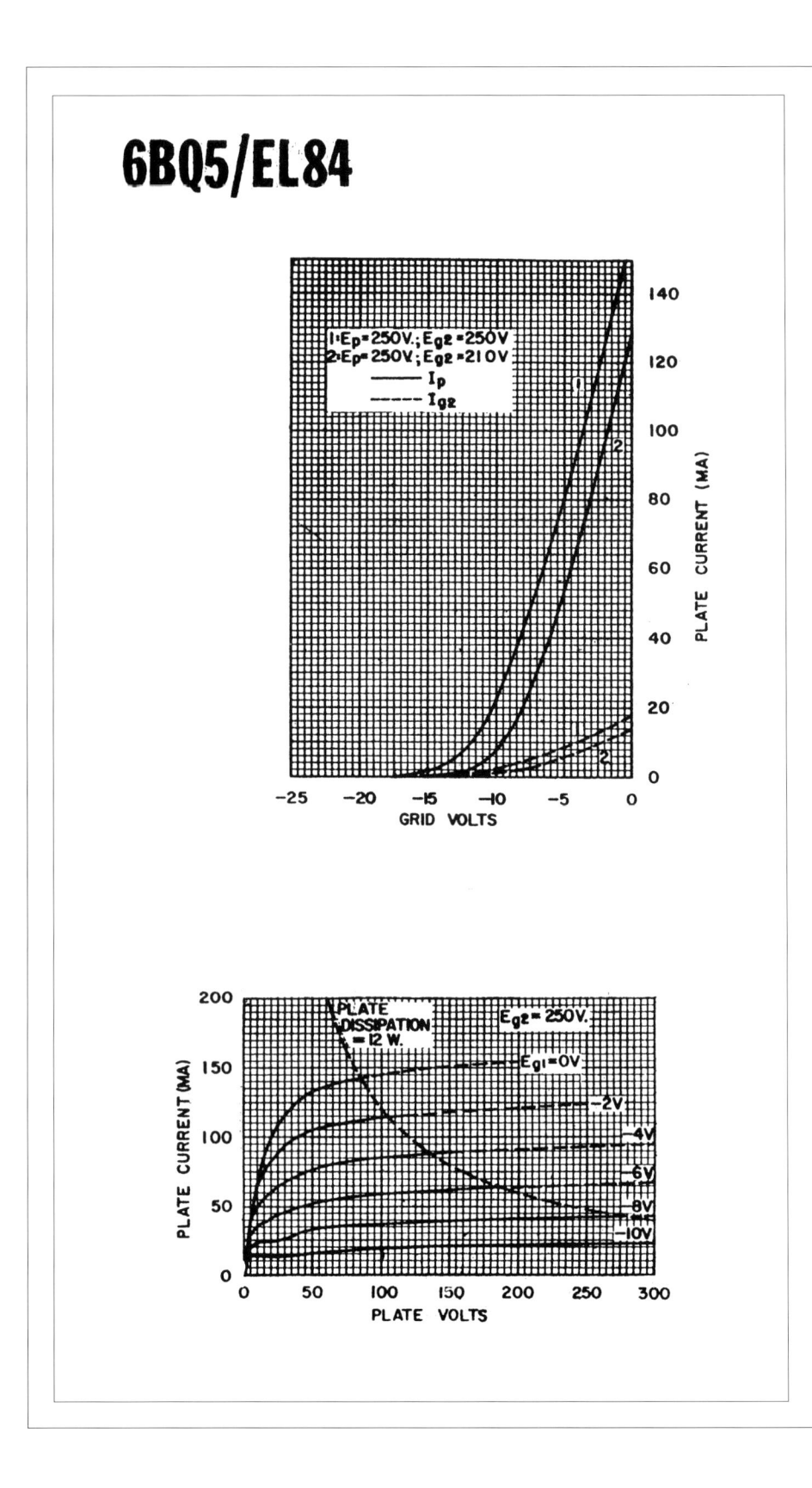

6BQ5/EL84

Class A - Triode Operation
(Screen grid connected to plate)

Plate Voltage	250 volts
Common Cathode Resistance	270 Ω
Plate Load Resistance	3.5 KΩ
Plate Current (zero signal)	34 mA
Plate Current (max. signal)	36 mA
Input Signal Voltage (rms)	6.7 volts (rms)
Power Output	1.95 watts
Percent Distortion	9.0 %

Class AB - Triode Operation
(Two tubes, push-pull. Screen grid connected to plate)

Plate Voltage	250	300 volts
Common Cathode Resistance	270	270 Ω
Plate to Plate Load Resistance	10	10 KΩ
Plate Current (zero signal)	2x20	2x24 mA
Plate Current (max. signal)	2x21.7	2x26.0 mA
Input Signal Voltage (rms)	8.3	10 volts (rms)
Power Output	3.4	5.2 watts
Percent Distortion	2.5	2.5 %

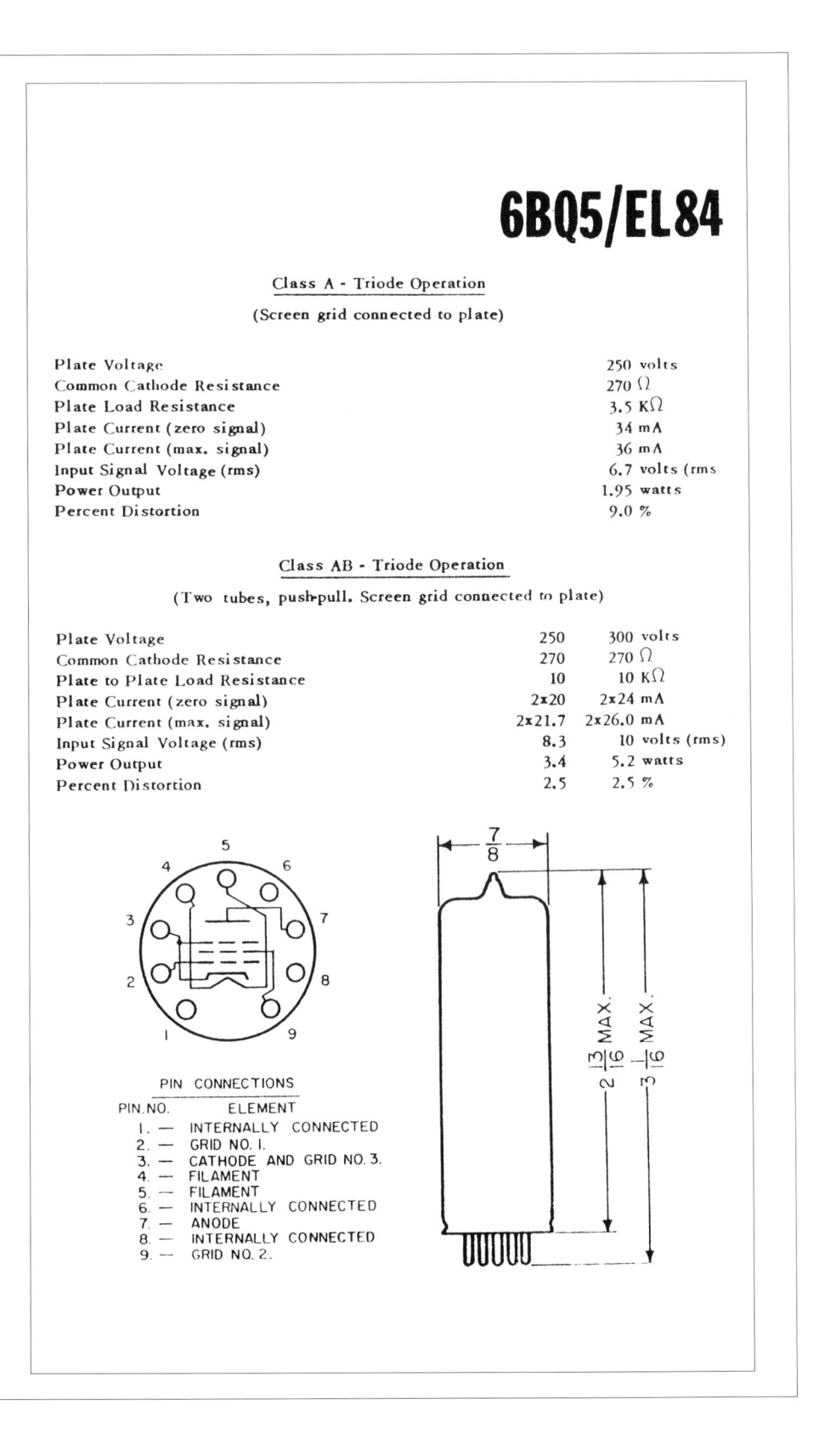

PIN CONNECTIONS

PIN NO.	ELEMENT
1.	INTERNALLY CONNECTED
2.	GRID NO. 1.
3.	CATHODE AND GRID NO. 3.
4.	FILAMENT
5.	FILAMENT
6.	INTERNALLY CONNECTED
7.	ANODE
8.	INTERNALLY CONNECTED
9.	GRID NO. 2.

164 THE TUBE AMP BOOK

Tube/spec sheets

6BQ5 — TUNG-SOL — TENTATIVE DATA

TYPICAL OPERATING CONDITIONS AND CHARACTERISTICS — cont'd.

CLASS A, ONE TUBE — cont'd.

PLATE VOLTAGE	250			VOLTS
GRID #2 VOLTAGE	210			VOLTS
GRID #1 BIAS	-6.4			VOLTS
CATHODE RESISTOR	160			OHMS
PLATE LOAD RESISTANCE	7000			OHMS
INPUT A.F. VOLTAGE (RMS)	0	0.3	3.4	3.8	VOLTS
PLATE CURRENT	36	---	36.6	36.5	MA.
GRID #2 CURRENT	3.9	---	7.3	8.0	MA.
TRANSCONDUCTANCE	10400			µMHOS
PLATE RESISTANCE	40000			OHMS
AMPLIFICATION FACTOR OF GRID #2 WITH RESPECT TO GRID #1	19			
MAX. SIGNAL POWER OUTPUT[B]	0	0.05	---	4.7[C]	WATTS
TOTAL HARMONIC DISTORTION[B]	---	---	10	---	PERCENT
SECOND HARMONIC[B]	---	---	1.8	---	PERCENT
THIRD HARMONIC[B]	---	---	9.3	---	PERCENT

CLASS B, TWO TUBES

PLATE VOLTAGE	300		VOLTS
GRID #2 VOLTAGE	300		VOLTS
GRID #1 BIAS	-14.7		VOLTS
LOAD RESISTANCE, PLATE TO PLATE	8000		OHMS
INPUT A.F. VOLTAGE (RMS)	8	10	VOLTS
PLATE CURRENT	2×10	2×37.5	2×46	MA.
GRID #2 CURRENT	2×1.1	2×7.5	2×11	MA.
MAX. SIGNAL POWER OUTPUT	0	11	17	WATTS
TOTAL HARMONIC DISTORTION	---	3	4	PERCENT

CLASS AB, TWO TUBES

PLATE VOLTAGE	250		VOLTS
GRID #2 VOLTAGE	250		VOLTS
COMMON CATHODE RESISTOR	130		OHMS
LOAD RESISTANCE, PLATE TO PLATE	8000		OHMS
INPUT A.F. VOLTAGE (RMS)	0	8	10	VOLTS
PLATE CURRENT	2×31	2×37.5	2×36	MA.
GRID #2 CURRENT	2×3.5	2×7.5	2×4	MA.
MAX. SIGNAL POWER OUTPUT	0	11	17	WATTS
TOTAL HARMONIC DISTORTION	---	3	4	PERCENTS

CONTINUED ON FOLLOWING PAGE

6BQ5 — TUNG-SOL — TENTATIVE DATA

CONTINUED FROM PRECEDING PAGE

TYPICAL OPERATING CONDITIONS AND CHARACTERISTICS

CLASS A, ONE TUBE

HEATER VOLTAGE	6.3				VOLTS
HEATER CURRENT	0.76				AMP.
PLATE VOLTAGE	250				VOLTS
GRID #2 VOLTAGE	250				VOLTS
GRID #1 BIAS	-7.3				VOLTS
CATHODE RESISTOR	135				OHMS
PLATE LOAD RESISTANCE	5200				OHMS
INPUT A.F. VOLTAGE (RMS)	0	0.3	3.4	4.3	4.7	VOLTS
PLATE CURRENT	48	---	---	49.5	49.2	MA.
GRID #2 CURRENT	5.5	---	---	10.8	11.6	MA.
TRANSCONDUCTANCE	11300				µMHOS
PLATE RESISTANCE	38000				OHMS
AMPLIFICATION FACTOR OF GRID #2 WITH RESPECT TO GRID #1	19				
MAX. SIGNAL POWER OUTPUT[B]	0	0.05	4.5	5.7	6.0[C]	WATTS
TOTAL HARMONIC DISTORTION[B]	---	---	6.8	10	---	PERCENT
SECOND HARMONIC[B]	---	---	3.0	2.0	---	PERCENT
THIRD HARMONIC[B]	---	---	5.8	9.5	---	PERCENT

PLATE VOLTAGE	250				VOLTS
GRID #2 VOLTAGE	250				VOLTS
GRID #1 BIAS	-7.3				VOLTS
CATHODE RESISTOR	135				OHMS
PLATE LOAD RESISTANCE	4500				OHMS
INPUT A.F. VOLTAGE (RMS)	0	0.3	3.5	4.4	4.8	VOLTS
PLATE CURRENT	48	---	---	50.6	50.5	MA.
GRID #2 CURRENT	5.5	---	---	10	11	MA.
TRANSCONDUCTANCE	11300				µMHOS
PLATE RESISTANCE	38000				OHMS
AMPLIFICATION FACTOR OF GRID #2 WITH RESPECT TO GRID #1	19				
MAX. SIGNAL POWER OUTPUT[B]	0	0.05	4.5	5.7	6.0[C]	WATTS
TOTAL HARMONIC DISTORTION[B]	---	---	7.5	10	---	PERCENT
SECOND HARMONIC[B]	---	---	5.7	5.0	---	PERCENT
THIRD HARMONIC[B]	---	---	4.5	8	---	PERCENT

PLATE VOLTAGE	250				VOLTS
GRID #2 VOLTAGE	250				VOLTS
GRID #1 BIAS	-8.4				VOLTS
CATHODE RESISTOR	210				OHMS
PLATE LOAD RESISTANCE	7000				OHMS
INPUT A.F. VOLTAGE (RMS)	0	0.3	3.5	4.2	5.5	VOLTS
PLATE CURRENT	36	---	---	36.8	36	MA.
GRID #2 CURRENT	4.1	---	---	8.5	14.6	MA.
TRANSCONDUCTANCE	10000				µMHOS
PLATE RESISTANCE	40000				OHMS
AMPLIFICATION FACTOR OF GRID #2 WITH RESPECT TO GRID #1	19				
MAX. SIGNAL POWER OUTPUT[B]	0	0.05	---	---	5.6[C]	WATTS
TOTAL HARMONIC DISTORTION[B]	---	---	4	10	---	PERCENT
SECOND HARMONIC[B]	---	---	1.7	---	---	PERCENT
THIRD HARMONIC[B]	---	---	---	8.7	---	PERCENT

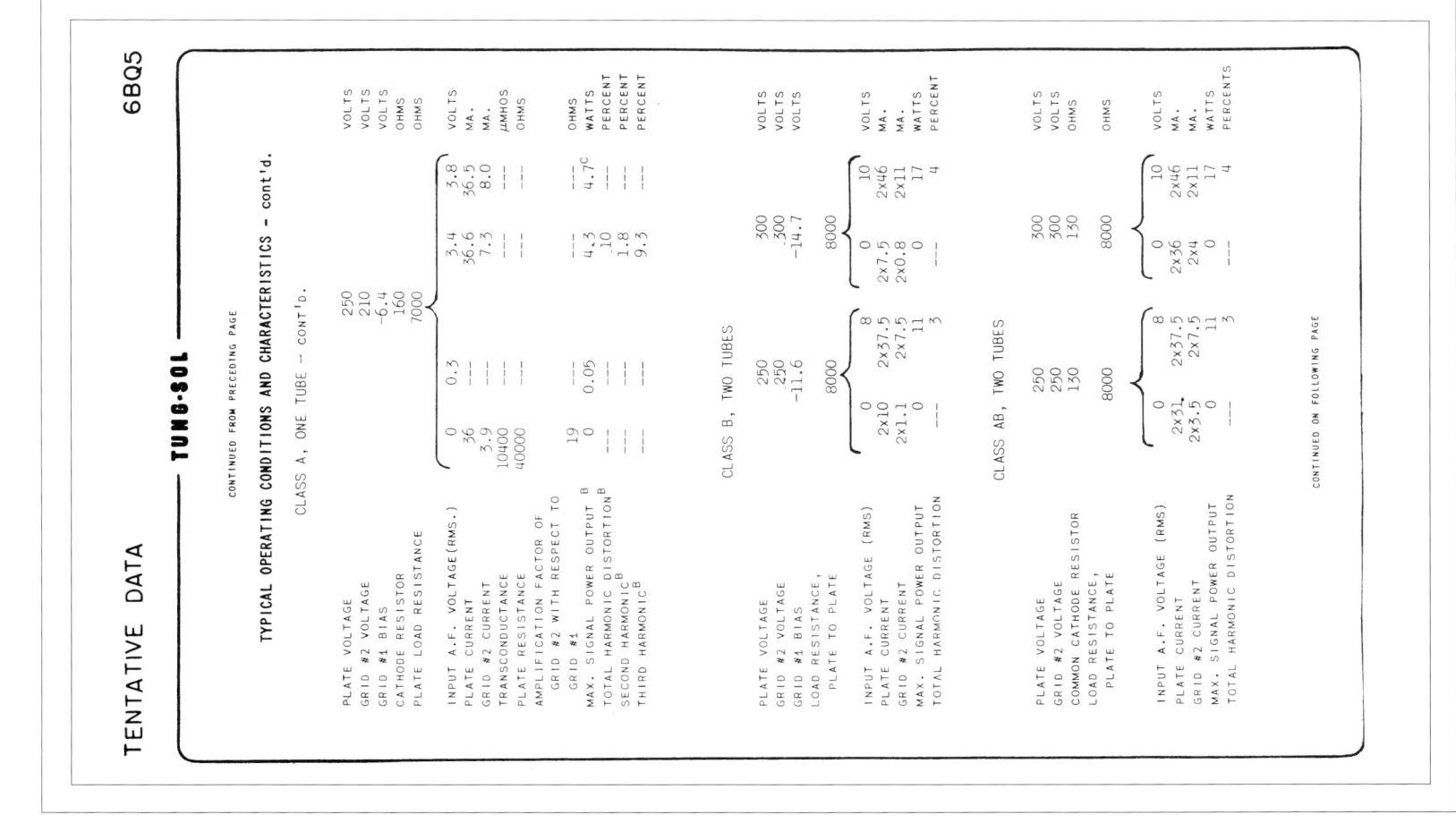

THE TUBE AMP BOOK 165

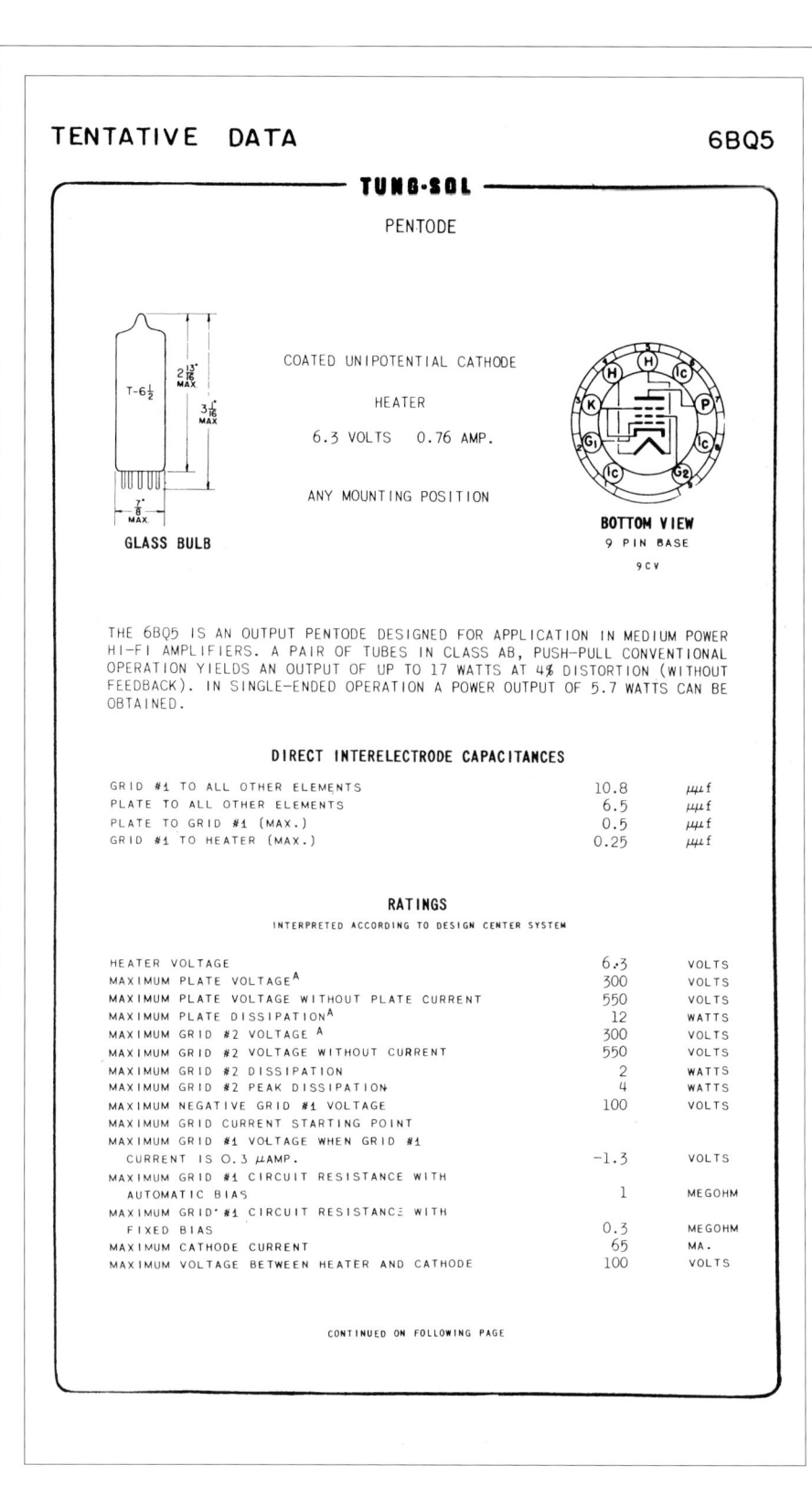

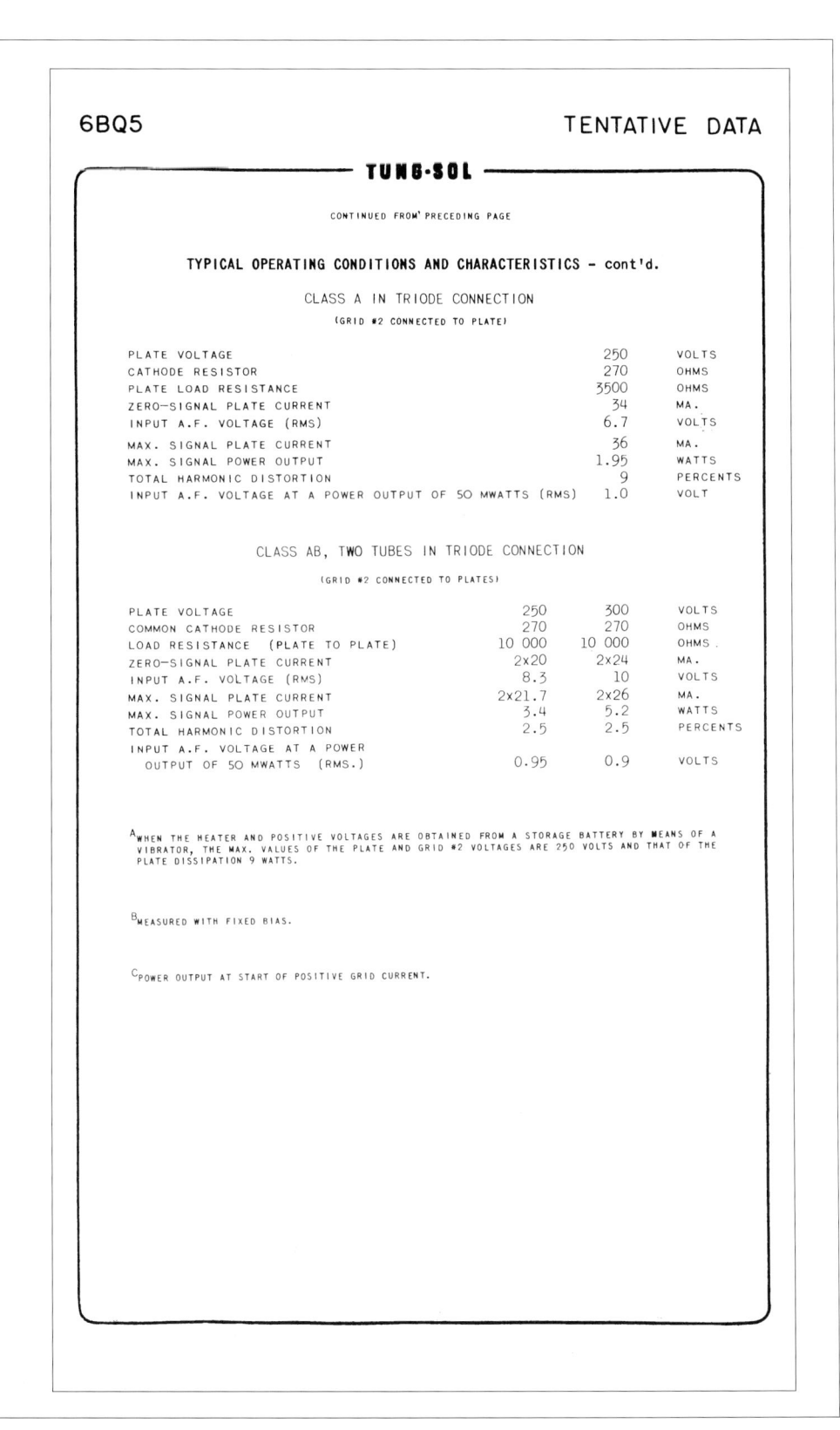

Tube/spec sheets

TENTATIVE DATA — TUNG-SOL — 6BQ5

PENTODE

COATED UNIPOTENTIAL CATHODE
HEATER
6.3 VOLTS   0.76 AMP.

ANY MOUNTING POSITION

GLASS BULB
T-6½

BOTTOM VIEW
9 PIN BASE
9CV

THE 6BQ5 IS AN OUTPUT PENTODE DESIGNED FOR APPLICATION IN MEDIUM POWER HI-FI AMPLIFIERS. A PAIR OF TUBES IN CLASS AB, PUSH-PULL CONVENTIONAL OPERATION YIELDS AN OUTPUT OF UP TO 17 WATTS AT 4% DISTORTION (WITHOUT FEEDBACK). IN SINGLE-ENDED OPERATION A POWER OUTPUT OF 5.7 WATTS CAN BE OBTAINED.

DIRECT INTERELECTRODE CAPACITANCES

GRID #1 TO ALL OTHER ELEMENTS	10.8	μμf
PLATE TO ALL OTHER ELEMENTS	6.5	μμf
PLATE TO GRID #1 (MAX.)	0.5	μμf
GRID #1 TO HEATER (MAX.)	0.25	μμf

RATINGS
INTERPRETED ACCORDING TO DESIGN CENTER SYSTEM

HEATER VOLTAGE	6.3	VOLTS
MAXIMUM PLATE VOLTAGE	300	VOLTS
MAXIMUM PLATE VOLTAGE WITHOUT PLATE CURRENT[A]	550	VOLTS
MAXIMUM PLATE DISSIPATION[A]	12	WATTS
MAXIMUM GRID #2 VOLTAGE[A]	300	VOLTS
MAXIMUM GRID #2 VOLTAGE WITHOUT CURRENT	550	VOLTS
MAXIMUM GRID #2 DISSIPATION	2	WATTS
MAXIMUM GRID #2 PEAK DISSIPATION	4	WATTS
MAXIMUM NEGATIVE GRID #1 VOLTAGE	100	VOLTS
MAXIMUM GRID CURRENT STARTING POINT CURRENT IS 0.3 μAMP.	-1.3	VOLTS
MAXIMUM GRID #1 CIRCUIT RESISTANCE WITH AUTOMATIC BIAS	1	MEGOHM
MAXIMUM GRID #1 CIRCUIT RESISTANCE WITH FIXED BIAS	0.3	MEGOHM
MAXIMUM CATHODE CURRENT	65	MA.
MAXIMUM VOLTAGE BETWEEN HEATER AND CATHODE	100	VOLTS

CONTINUED ON FOLLOWING PAGE

TENTATIVE DATA — TUNG-SOL — 6BQ5 — TENTATIVE DATA

CONTINUED FROM PRECEDING PAGE

TYPICAL OPERATING CONDITIONS AND CHARACTERISTICS – cont'd.

CLASS A IN TRIODE CONNECTION
(GRID #2 CONNECTED TO PLATE)

PLATE VOLTAGE	250	VOLTS
CATHODE RESISTOR	270	OHMS
PLATE LOAD RESISTANCE	3500	OHMS
ZERO-SIGNAL PLATE CURRENT	34	MA.
INPUT A.F. VOLTAGE (RMS)	6.7	VOLTS
MAX. SIGNAL PLATE CURRENT	36	MA.
MAX. SIGNAL POWER OUTPUT	1.95	WATTS
TOTAL HARMONIC DISTORTION	9	PERCENTS
INPUT A.F. VOLTAGE AT A POWER OUTPUT OF 50 MWATTS (RMS)	1.0	VOLT

CLASS AB, TWO TUBES IN TRIODE CONNECTION
(GRID #2 CONNECTED TO PLATES)

PLATE VOLTAGE	250	300	VOLTS
COMMON CATHODE RESISTOR	270	270	OHMS
LOAD RESISTANCE (PLATE TO PLATE)	10 000	10 000	OHMS
ZERO-SIGNAL PLATE CURRENT	2x20	2x24	MA.
INPUT A.F. VOLTAGE (RMS)	8.3	10	VOLTS
MAX. SIGNAL PLATE CURRENT	2x21.7	2x26	MA.
MAX. SIGNAL POWER OUTPUT	3.4	5.2	WATTS
TOTAL HARMONIC DISTORTION	2.5	2.5	PERCENTS
INPUT A.F. VOLTAGE AT A POWER OUTPUT OF 50 MWATTS (RMS.)	0.95	0.9	VOLTS

[A] WHEN THE HEATER AND POSITIVE VOLTAGES ARE OBTAINED FROM A STORAGE BATTERY BY MEANS OF A VIBRATOR, THE MAX. VALUES OF THE PLATE AND GRID #2 VOLTAGES ARE 250 VOLTS AND THAT OF THE PLATE DISSIPATION 9 WATTS.

[B] MEASURED WITH FIXED BIAS.

[C] POWER OUTPUT AT START OF POSITIVE GRID CURRENT.

Tube/spec sheets

6CA7/EL34

TENTATIVE DATA

Class AB₁ Audio Amplifier
Distributed Load Connection

Maximum Ratings (Design Center Values)

Plate and Grid No. 2 Supply Voltage	500 V
Plate Dissipation	25 W
Grid No. 2 Dissipation	8 W
Cathode Current	150 mA
Grid Current Starting Point - Grid No. 1 Voltage when Grid No. 1 Current is 0.3 μA	−1.3 V
Grid No. 1 Circuit Resistance	500 KΩ
External Resistance Between Heater and Cathode	20 KΩ
Voltage Between Heater and Cathode	100 V

Typical Operation (Fixed Bias - Two Tubes Push Pull)

Plate Supply Voltage	500 V
Grid No. 2 Supply Voltage	(See Note 1)
Grid No. 1 Bias	−44.5 V
Plate to Plate Load Resistance	(approx.) 7000 Ω
Plate and Grid No. 2 Current (Zero Signal)	2×57 mA
Plate and Grid No. 2 Current (Max Signal)	2×112 mA
Input Signal Voltage (rms)	32 V
Power Output	60 W
Harmonic Distortion	2.5 %

Note 1:
Screen voltage is obtained from taps located at 43% of the plate winding turns. An unbypassed resistor of 1 KΩ in series with each screen grid is necessary to prevent screen overload.

PIN CONNECTIONS

NO.1- GRID NO.3
NO.2- HEATER
NO.3- PLATE
NO.4- GRID NO.2
NO.5- GRID NO.1
NO.6- N.C.
NO.7- HEATER
NO.8- CATHODE

Revised 9/60

12AY7

TUNG-SOL

DOUBLE TRIODE

MINIATURE TYPE

COATED UNIPOTENTIAL CATHODE

HEATER

SERIES	PARALLEL
12.6 VOLTS	6.3 VOLTS
150 MA.	300 MA.

AC OR DC

FOR 12.6 VOLT OPERATION APPLY HEATER VOLTAGE BETWEEN PINS #4 AND #5. FOR 6.3 VOLT OPERATION APPLY HEATER VOLTAGE BETWEEN PIN #9 AND PINS #4 AND #5 CONNECTED TOGETHER.

WHEN OPERATING FROM AN AC HEATER SUPPLY, DO NOT USE THE 12.6 VOLT CONNECTION IF LOW-HUM CAPABILITIES ARE TO BE REALIZED.

ANY MOUNTING POSITION

BOTTOM VIEW
SMALL BUTTON
9 PIN BASE
9A

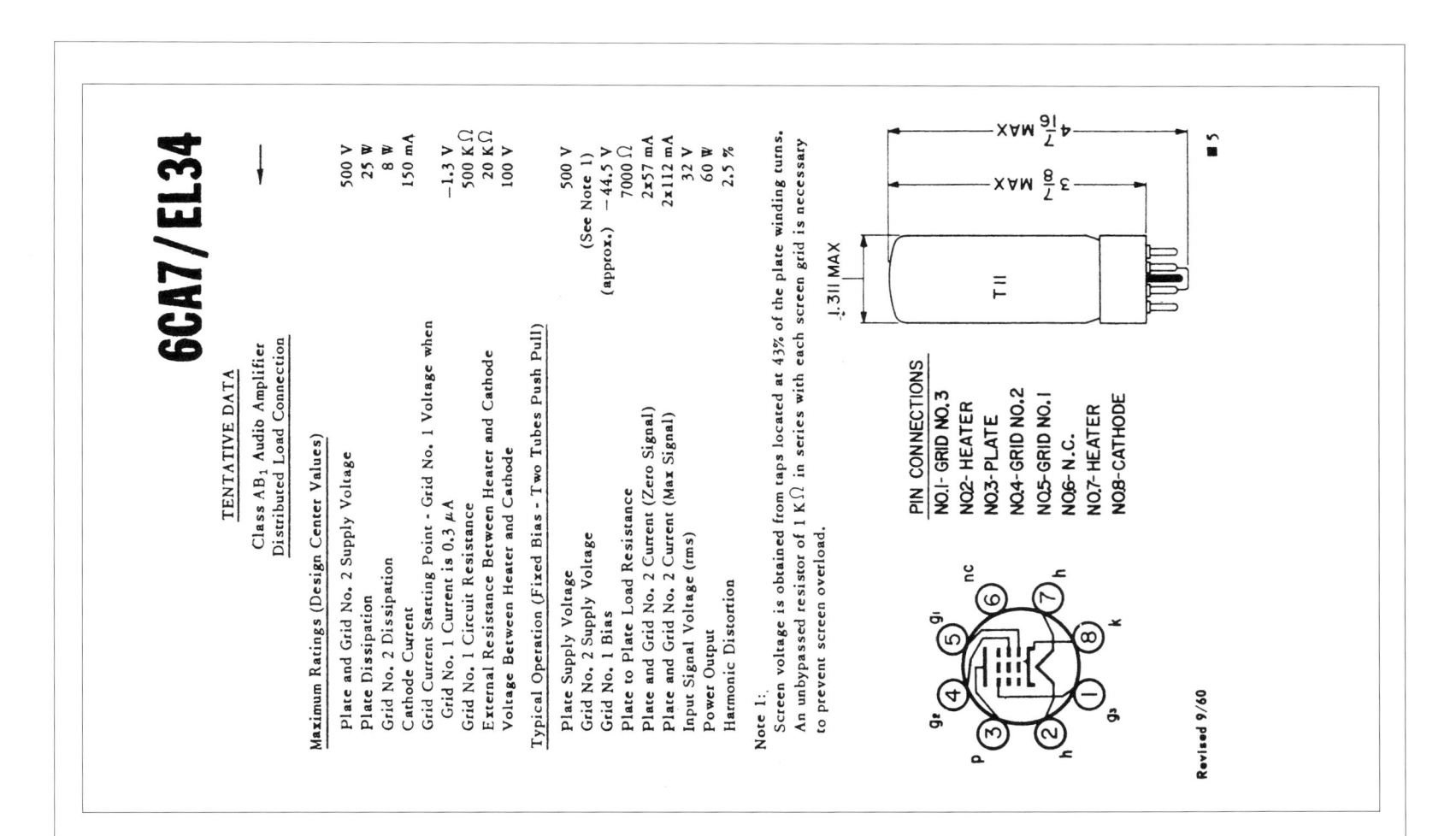

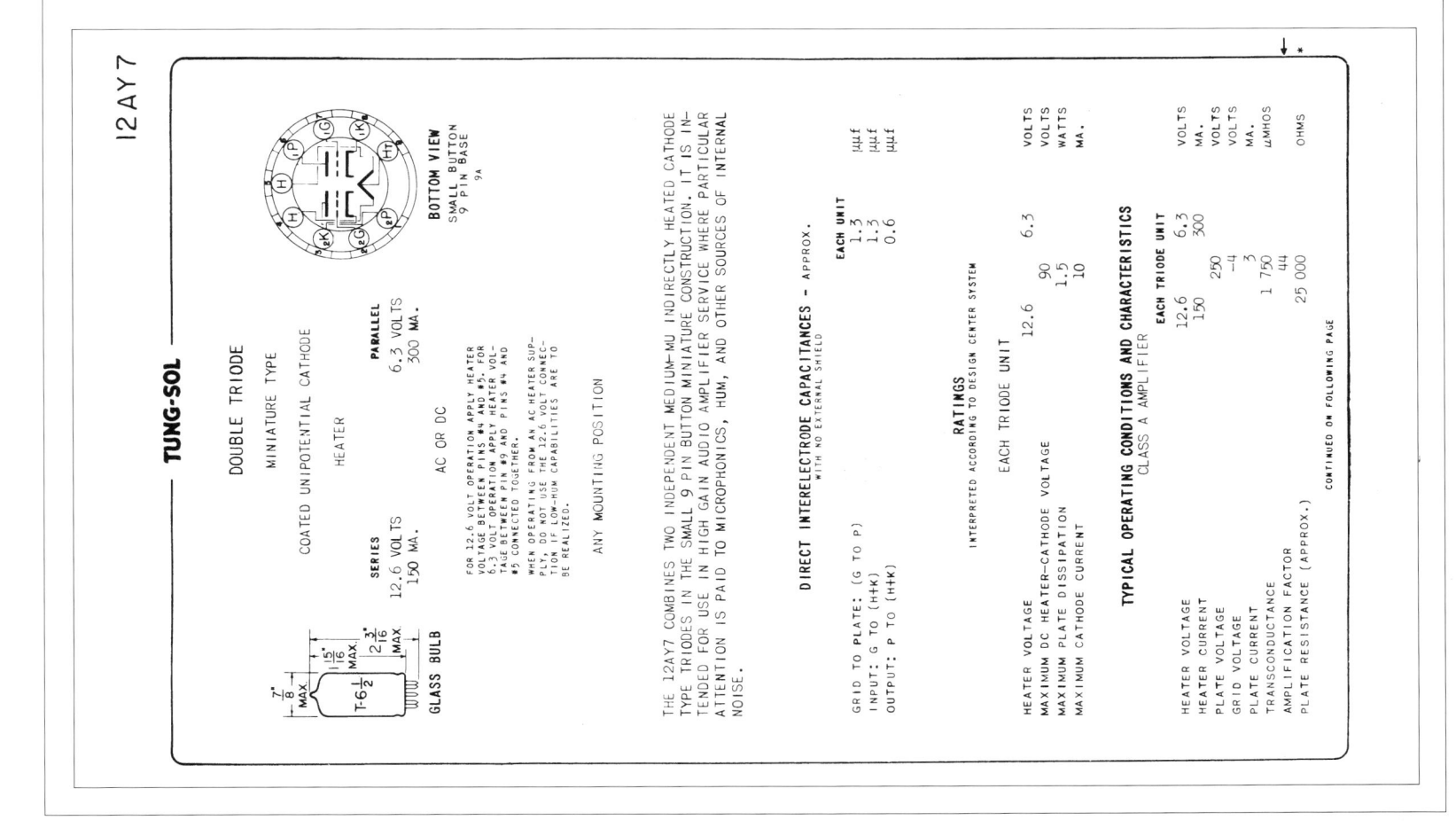

GLASS BULB T-6½

THE 12AY7 COMBINES TWO INDEPENDENT MEDIUM-MU INDIRECTLY HEATED CATHODE TYPE TRIODES IN THE SMALL 9 PIN BUTTON MINIATURE CONSTRUCTION. IT IS INTENDED FOR USE IN HIGH GAIN AUDIO AMPLIFIER SERVICE WHERE PARTICULAR ATTENTION IS PAID TO MICROPHONICS, HUM, AND OTHER SOURCES OF INTERNAL NOISE.

DIRECT INTERELECTRODE CAPACITANCES - APPROX.
WITH NO EXTERNAL SHIELD

	EACH UNIT	
GRID TO PLATE: (G TO P)	1.3	μμf
INPUT: G TO (H+K)	1.3	μμf
OUTPUT: P TO (H+K)	0.6	μμf

RATINGS
INTERPRETED ACCORDING TO DESIGN CENTER SYSTEM

	EACH TRIODE UNIT	
HEATER VOLTAGE	12.6 6.3	VOLTS
MAXIMUM DC HEATER-CATHODE VOLTAGE	90	VOLTS
MAXIMUM PLATE DISSIPATION	1.5	WATTS
MAXIMUM CATHODE CURRENT	10	MA.

TYPICAL OPERATING CONDITIONS AND CHARACTERISTICS
CLASS A AMPLIFIER

	EACH TRIODE UNIT	
HEATER VOLTAGE	12.6 6.3	VOLTS
HEATER CURRENT	150 300	MA.
PLATE VOLTAGE	250	VOLTS
GRID VOLTAGE	−4	VOLTS
PLATE CURRENT	3	MA.
TRANSCONDUCTANCE	1 750	μMHOS
AMPLIFICATION FACTOR	44	
PLATE RESISTANCE (APPROX.)	25 000	OHMS

CONTINUED ON FOLLOWING PAGE

THE TUBE AMP BOOK 167

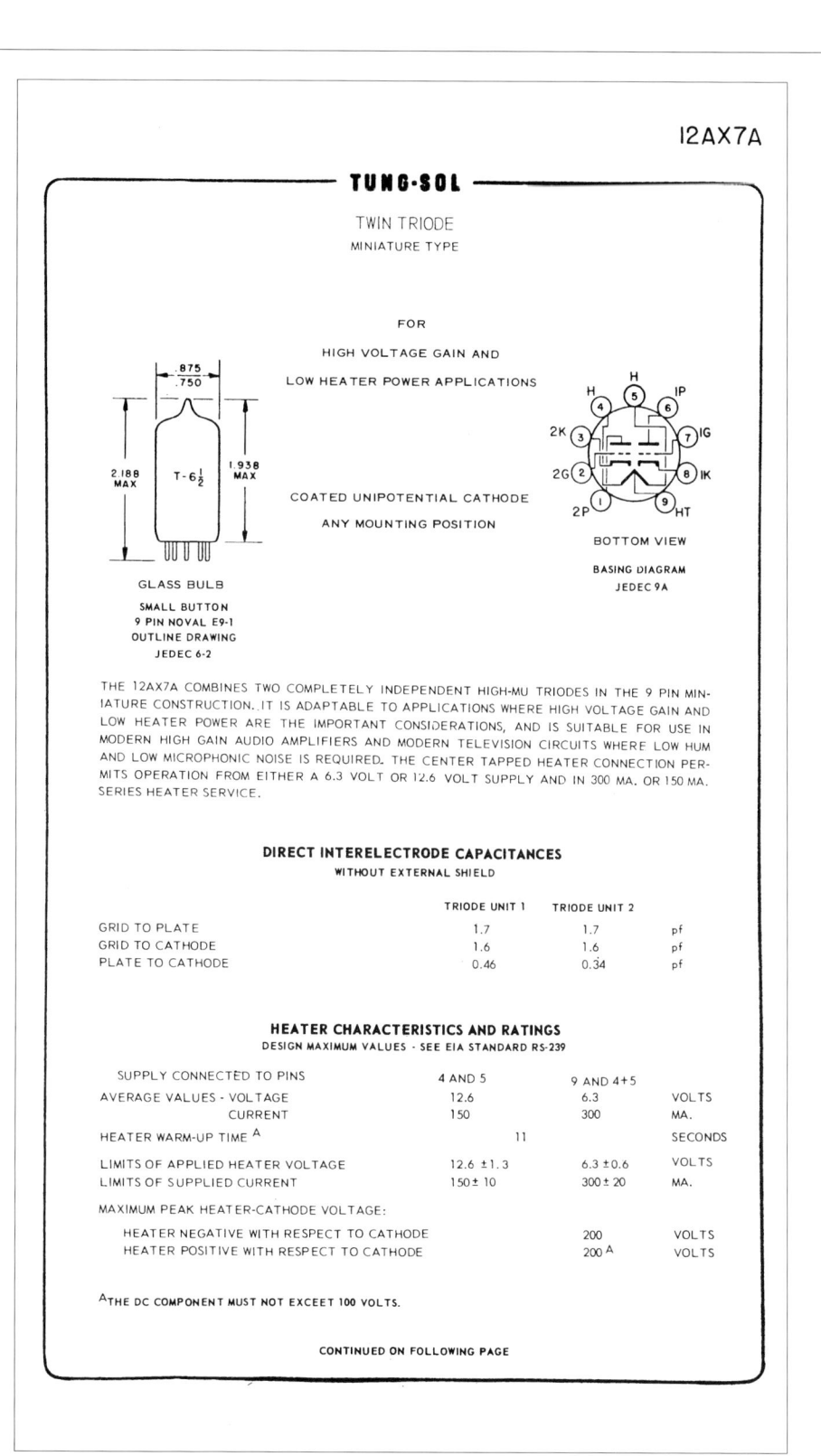

Tube/spec sheets

12AY7 — TUNG-SOL

CONTINUED FROM PRECEDING PAGE

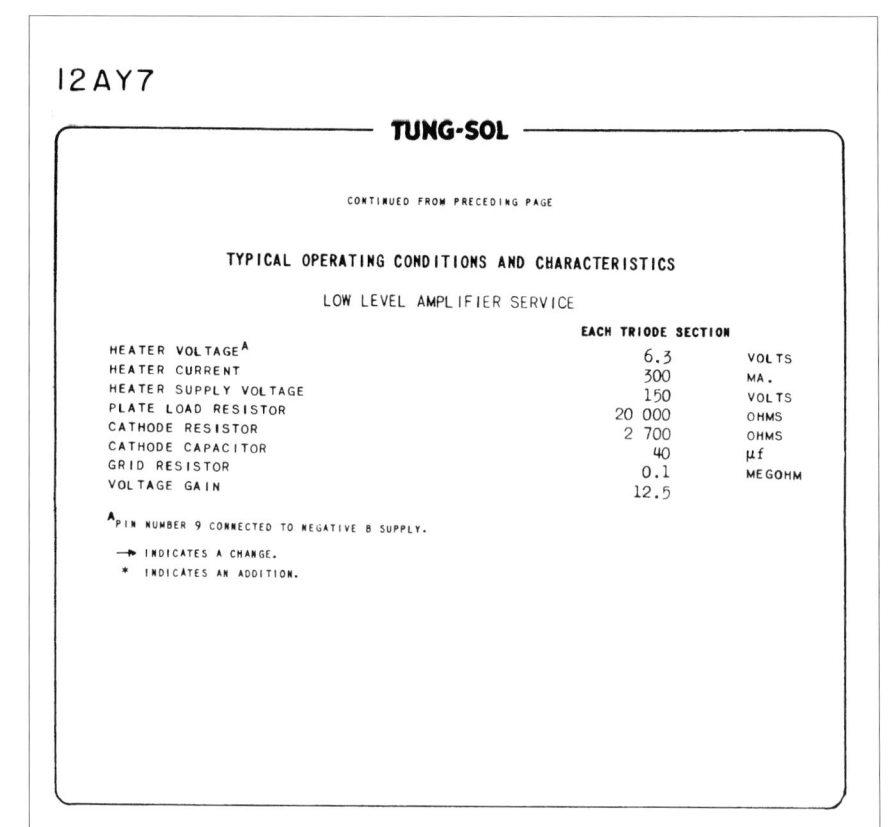

TYPICAL OPERATING CONDITIONS AND CHARACTERISTICS

LOW LEVEL AMPLIFIER SERVICE

EACH TRIODE SECTION

HEATER VOLTAGE [A]	6.3	VOLTS
HEATER CURRENT	300	MA.
HEATER SUPPLY VOLTAGE	150	VOLTS
PLATE LOAD RESISTOR	20 000	OHMS
CATHODE RESISTOR	2 700	OHMS
CATHODE CAPACITOR	40	µf
GRID RESISTOR	0.1	MEGOHM
VOLTAGE GAIN	12.5	

[A] PIN NUMBER 9 CONNECTED TO NEGATIVE B SUPPLY.

→ INDICATES A CHANGE.
* INDICATES AN ADDITION.

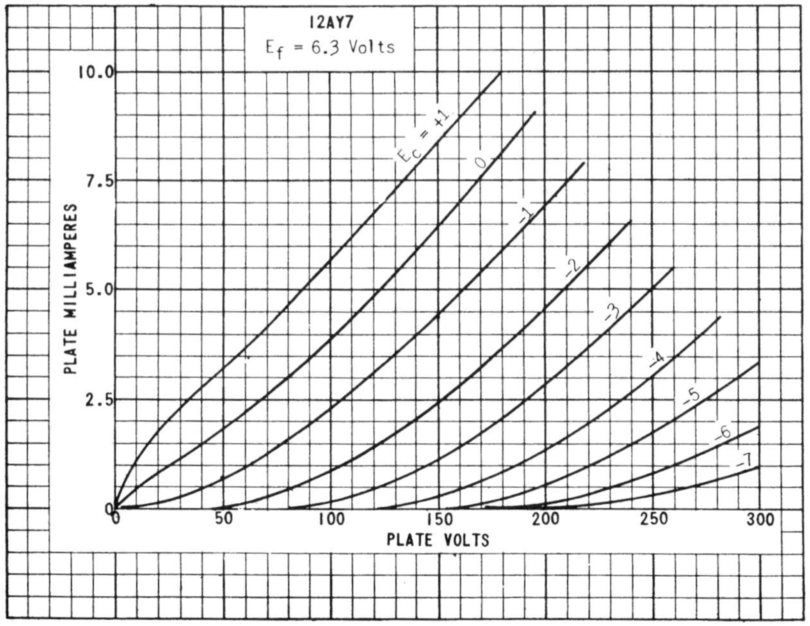

Plate characteristics curves: $E_f = 6.3$ Volts; Plate Volts (0–300) vs Plate Milliamperes (0–10.0).

12AX7A — TUNG-SOL

TWIN TRIODE
MINIATURE TYPE

FOR

HIGH VOLTAGE GAIN AND
LOW HEATER POWER APPLICATIONS

COATED UNIPOTENTIAL CATHODE
ANY MOUNTING POSITION

GLASS BULB
SMALL BUTTON
9 PIN NOVAL E9-1
OUTLINE DRAWING
JEDEC 6-2

Dimensions: .875/.750, 1.938 MAX, 2.188 MAX, T-6½

Basing diagram (bottom view) JEDEC 9A: pins — 2P, 2G, 2K, H, H, 1P, 1G, 1K, HT

THE 12AX7A COMBINES TWO COMPLETELY INDEPENDENT HIGH-MU TRIODES IN THE 9 PIN MINIATURE CONSTRUCTION. IT IS ADAPTABLE TO APPLICATIONS WHERE HIGH VOLTAGE GAIN AND LOW HEATER POWER ARE THE IMPORTANT CONSIDERATIONS, AND IS SUITABLE FOR USE IN MODERN HIGH GAIN AUDIO AMPLIFIERS AND MODERN TELEVISION CIRCUITS WHERE LOW HUM AND LOW MICROPHONIC NOISE IS REQUIRED. THE CENTER TAPPED HEATER CONNECTION PERMITS OPERATION FROM EITHER A 6.3 VOLT OR 12.6 VOLT SUPPLY AND IN 300 MA. OR 150 MA. SERIES HEATER SERVICE.

DIRECT INTERELECTRODE CAPACITANCES
WITHOUT EXTERNAL SHIELD

	TRIODE UNIT 1	TRIODE UNIT 2	
GRID TO PLATE	1.7	1.7	pf
GRID TO CATHODE	1.6	1.6	pf
PLATE TO CATHODE	0.46	0.34	pf

HEATER CHARACTERISTICS AND RATINGS
DESIGN MAXIMUM VALUES · SEE EIA STANDARD RS-239

	4 AND 5	9 AND 4+5	
SUPPLY CONNECTED TO PINS			
AVERAGE VALUES - VOLTAGE	12.6	6.3	VOLTS
CURRENT	150	300	MA.
HEATER WARM-UP TIME [A]	11		SECONDS
LIMITS OF APPLIED HEATER VOLTAGE	12.6 ±1.3	6.3 ±0.6	VOLTS
LIMITS OF SUPPLIED CURRENT	150± 10	300± 20	MA.
MAXIMUM PEAK HEATER-CATHODE VOLTAGE:			
HEATER NEGATIVE WITH RESPECT TO CATHODE	200	200	VOLTS
HEATER POSITIVE WITH RESPECT TO CATHODE	200 [A]	200	VOLTS

[A] THE DC COMPONENT MUST NOT EXCEED 100 VOLTS.

CONTINUED ON FOLLOWING PAGE

Tube/spec sheets

12AU7A

TUNG-SOL

TWIN TRIODE
MINIATURE TYPE

UNIPOTENTIAL CATHODE
FOR
AUDIO FREQUENCY AMPLIFIER
OR COMBINED OSCILLATOR AND
MIXER APPLICATIONS IN
T.V. RECEIVERS

ANY MOUNTING POSITION

GLASS BULB
SMALL BUTTON NOVAL
9 PIN BASE E9-1
OUTLINE DRAWING
JEDEC 6-2

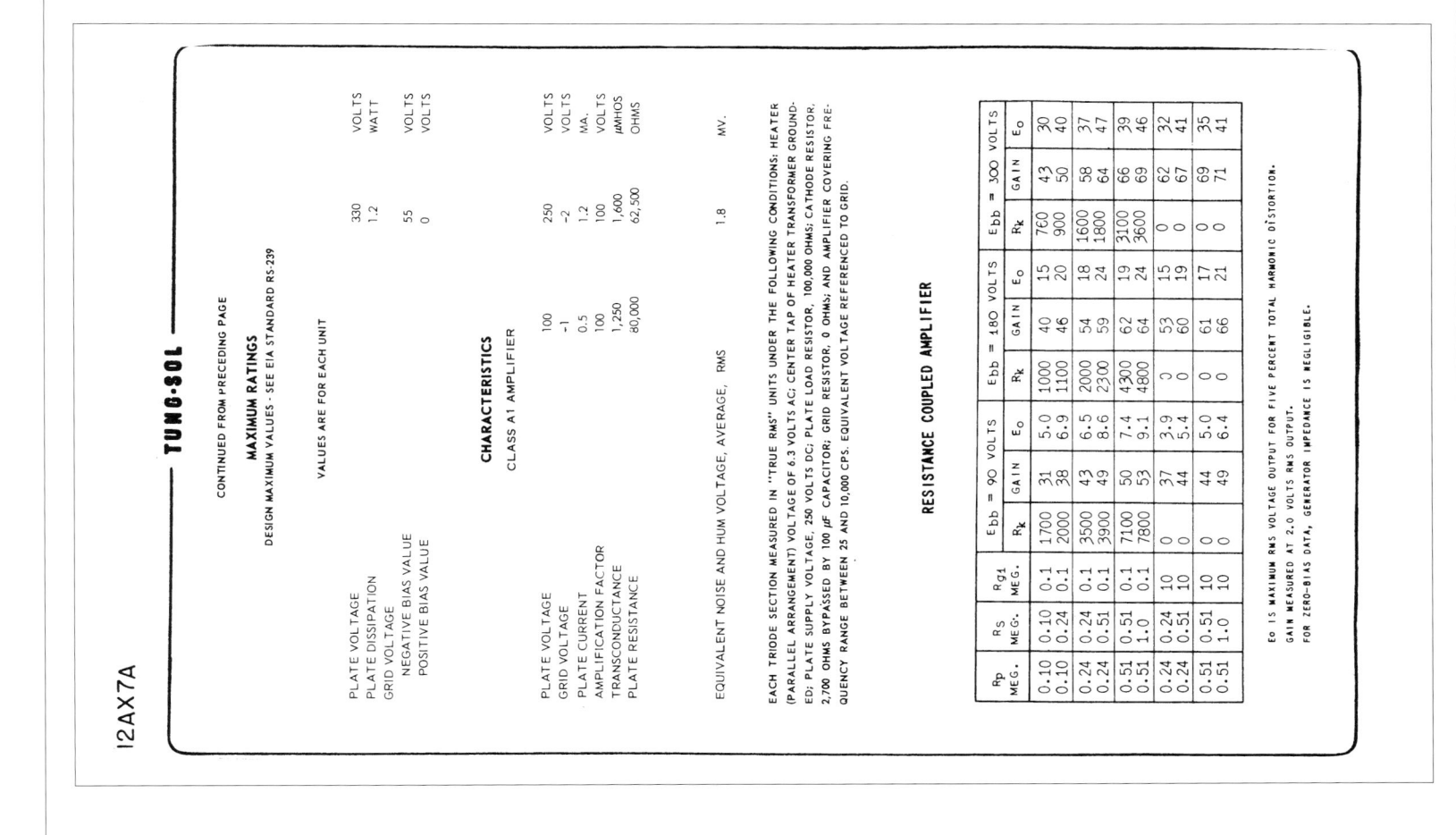

BASING DIAGRAM
JEDEC 9A
BOTTOM VIEW

THE 12AU7A COMBINES TWO INDEPENDENT MEDIUM-MU INDIRECTLY HEATED CATHODE TYPE TRIODES IN THE 9 PIN MINIATURE CONSTRUCTION. IT IS ADAPTABLE TO APPLICATION EITHER AS AN AUDIO FREQUENCY AMPLIFIER OR AS A COMBINED OSCILLATOR AND MIXER, EXCEPT FOR HEATER RATINGS IT IS IDENTICAL TO THE 7AU7 AND THE 9AU7.

→ DIRECT INTERELECTRODE CAPACITANCES

	TRIODE UNIT T1	TRIODE UNIT T2	
GRID TO PLATE	1.5	1.5	pf
GRID TO CATHODE	1.6	1.6	pf
PLATE TO CATHODE	0.50	0.35	pf

HEATER CHARACTERISTICS AND RATINGS
DESIGN MAXIMUM VALUES - SEE EIA STANDARD RS-239

AVERAGE CHARACTERISTICS		
HEATER IN SERIES	12.6 VOLTS	
HEATER IN PARALLEL	6.3 VOLTS	
HEATER SUPPLY LIMITS:		
VOLTAGE OPERATION		
HEATER IN SERIES	150	MA.
HEATER IN PARALLEL	300	MA.
	12.6±1.3	VOLTS
	6.3±0.6	VOLTS
MAXIMUM HEATER-CATHODE VOLTAGE:		
HEATER NEGATIVE WITH RESPECT TO CATHODE		
TOTAL DC AND PEAK	200	VOLTS
HEATER POSITIVE WITH RESPECT TO CATHODE		
DC	100	VOLTS
TOTAL DC AND PEAK	200	VOLTS

CONTINUED ON FOLLOWING PAGE

12AX7A

TUNG-SOL

CONTINUED FROM PRECEDING PAGE

MAXIMUM RATINGS
DESIGN MAXIMUM VALUES - SEE EIA STANDARD RS-239

VALUES ARE FOR EACH UNIT

PLATE VOLTAGE	330	VOLTS
PLATE DISSIPATION	1.2	WATT
GRID VOLTAGE		
NEGATIVE BIAS VALUE	55	VOLTS
POSITIVE BIAS VALUE	0	VOLTS

CHARACTERISTICS
CLASS A1 AMPLIFIER

PLATE VOLTAGE	100	VOLTS
GRID VOLTAGE	-1	VOLTS
PLATE CURRENT	0.5	MA.
PLATE VOLTAGE	100	VOLTS
AMPLIFICATION FACTOR	1,250	
TRANSCONDUCTANCE	1,600	µMHOS
PLATE RESISTANCE	62,500	OHMS
	80,000	

EQUIVALENT NOISE AND HUM VOLTAGE, AVERAGE, RMS   1.8   MV.

EACH TRIODE SECTION MEASURED IN "TRUE RMS" UNITS UNDER THE FOLLOWING CONDITIONS: HEATER (PARALLEL ARRANGEMENT) VOLTAGE OF 6.3 VOLTS AC; CENTER TAP OF HEATER TRANSFORMER GROUNDED; PLATE SUPPLY VOLTAGE, 250 VOLTS DC; PLATE LOAD RESISTOR, 100,000 OHMS; CATHODE RESISTOR, 2,700 OHMS BYPASSED BY 100 µF CAPACITOR; GRID RESISTOR, 0 OHMS; AND AMPLIFIER COVERING FREQUENCY RANGE BETWEEN 25 AND 10,000 CPS. EQUIVALENT VOLTAGE REFERENCED TO GRID.

RESISTANCE COUPLED AMPLIFIER

Rp MEG.	Rs MEG.	Rg1 MEG.	Ebb = 90 VOLTS Rk	GAIN	Eo	Ebb = 180 VOLTS Rk	GAIN	Eo	Ebb = 300 VOLTS Rk	GAIN	Eo
0.10	0.10	0.1	1700	31	5.0	1000	40	15	760	43	30
0.10	0.24	0.1	2000	38	6.9	1100	46	20	900	50	40
0.24	0.24	0.1	3500	43	6.5	2000	54	18	1600	58	37
0.24	0.51	0.1	3900	49	8.6	2300	59	24	1800	64	47
0.51	0.51	0.1	7100	50	7.4	4300	62	19	3100	66	39
0.51	1.0	0.1	7800	53	9.1	4800	64	24	3600	69	46
0.24	0.24	10	0	37	3.9	0	53	15	0	62	32
0.24	0.51	10	0	44	5.4	0	60	19	0	67	41
0.51	0.51	10	0	44	5.0	0	61	17	0	69	35
0.51	1.0	10	0	49	6.4	0	66	21	0	71	41

Eo IS MAXIMUM RMS VOLTAGE OUTPUT FOR FIVE PERCENT TOTAL HARMONIC DISTORTION. GAIN MEASURED AT 2.0 VOLTS RMS OUTPUT. FOR ZERO-BIAS DATA, GENERATOR IMPEDANCE IS NEGLIGIBLE.

THE TUBE AMP BOOK 169

Tube/spec sheets

12AU7

TUNG-SOL — TWIN TRIODE

MINIATURE TYPE

COATED UNIPOTENTIAL CATHODE

HEATER	SERIES	PARALLEL
	12.6 VOLTS	6.3 VOLTS
	0.15 AMP.	0.3 AMP.
	AC OR DC	

FOR 12.6 VOLT OPERATION APPLY HEATER VOLTAGE BETWEEN PINS 4 AND 5, FOR 6.3 VOLT OPERATION APPLY HEATER VOLTAGE BETWEEN PIN #9 AND PINS #4 AND #5 CONNECTED TOGETHER.

- .875" MAX
- 1.938" MAX
- 2.188" MAX
- T-6½
- GLASS BULB
- SMALL BUTTON 9 PIN BASE E9-1
- OUTLINE DRAWING JEDEC 6-2

Pin diagram (BOTTOM VIEW BASING DIAGRAM JEDEC 9A): 1 2P, 2 2G, 3 2K, 4 H, 5 H, 6 1P, 7 1G, 8 1K, 9 HT

THE 12AU7 COMBINES TWO INDEPENDENT MEDIUM-MU INDIRECTLY HEATED CATHODE TYPE TRIODES IN THE SMALL 9 PIN BUTTON CONSTRUCTION. IT IS ADAPTABLE TO APPLICATION EITHER AS AN AUDIO FREQUENCY AMPLIFIER OR AS COMBINED OSCILLATOR AND MIXER.

DIRECT INTERELECTRODE CAPACITANCES

	WITH SHIELD[A]	WITHOUT SHIELD	
TRIODE UNIT 1			
GRID TO PLATE: (G TO P)	1.5	1.5	pf
INPUT: G TO (H+K)	1.8	1.6	pf
OUTPUT: P TO (H+K)	2.0	0.40	pf
TRIODE UNIT 2			
GRID TO PLATE: (G TO P)	1.5	1.5	pf
INPUT: G TO (H+K)	1.8	1.6	pf
OUTPUT: P TO (H+K)	2.0	0.32	pf

[A] EXTERNAL SHIELD #315 CONNECTED TO CATHODE OF UNIT UNDER TEST.

RATINGS

(INTERPRETED ACCORDING TO DESIGN CENTER SYSTEM)

EACH TRIODE UNIT

	CLASS A1[B] AMPLIFIER	VERTICAL[B] DEFLECTION AMPLIFIER	
MAXIMUM HEATER-CATHODE VOLTAGE:			
HEATER NEGATIVE WITH RESPECT TO CATHODE:			
TOTAL DC AND PEAK	200	200	VOLTS
HEATER POSITIVE WITH RESPECT TO CATHODE:			
DC	100	100	VOLTS
TOTAL DC AND PEAK	200	200	VOLTS
MAXIMUM PLATE VOLTAGE	300	300	VOLTS
MAXIMUM PEAK POSITIVE PLATE VOLTAGE	—	1200	VOLTS
MAXIMUM PLATE DISSIPATION:[C] (ABSOLUTE MAXIMUM)			
EACH PLATE	2.75	2.75	WATTS
BOTH PLATES	5.5	5.5	WATTS
MAXIMUM PEAK NEGATIVE GRID VOLTAGE	—	250	VOLTS
MAXIMUM PEAK CATHODE CURRENT	20	20	MA.
MAXIMUM GRID CIRCUIT RESISTANCE		60	MA.
FIXED BIAS OPERATION	0.25	—	MEGOHM
CATHODE BIAS OPERATION	1.0	2.2	MEGOHMS

[B] FOR OPERATION IN A 525-LINE, 30-FRAME SYSTEM AS DESCRIBED IN "STANDARDS OF GOOD ENGINEERING PRACTICE FOR TELEVISION BROADCASTING STATIONS," FEDERAL COMMUNICATIONS COMMISSION. THE DUTY CYCLE OF THE VOLTAGE PULSE MUST NOT EXCEED 15 PERCENT OF A SCANNING CYCLE.

[C] IN STAGES OPERATING WITH GRID-LEAK BIAS, AN ADEQUATE CATHODE BIAS RESISTOR OR OTHER SUITABLE MEANS IS REQUIRED TO PROTECT THE TUBE IN THE ABSENCE OF EXCITATION.

CONTINUED ON FOLLOWING PAGE

12AU7A

TUNG-SOL

CONTINUED FROM PRECEDING PAGE

→ MAXIMUM RATINGS

DESIGN MAXIMUM VALUES - SEE EIA STANDARD RS-239

VALUES ARE FOR EACH UNIT

	CLASS A1 AMPLIFIER	VERTICAL DEFLECTION OSCILLATOR	
PLATE VOLTAGE	330	330	VOLTS
PLATE DISSIPATION:			
EACH PLATE	2.75	2.75	WATTS
BOTH PLATES	5.5	5.5	WATTS
CATHODE CURRENT	22		MA.
PEAK NEGATIVE PULSE GRID VOLTAGE		440	VOLTS
AVERAGE CATHODE CURRENT		22	MA.
PEAK CATHODE CURRENT		66	MA.
MAXIMUM CIRCUIT VALUES:			
GRID CIRCUIT RESISTANCE:			
FOR FIXED BIAS, GRID-RESISTOR BIAS, OR CATHODE-BIAS OPERATION	2.2	2.2	MEGOHMS

	HORIZONTAL DEFLECTION OSCILLATOR	VERTICAL DEFLECTION AMPLIFIER	
DC PLATE VOLTAGE	330	330	VOLTS
PLATE DISSIPATION			
EACH PLATE	2.75	2.75	WATTS
BOTH PLATES	5.5	5.5	WATTS
PEAK POSITIVE-PULSE PLATE VOLTAGE		1200	VOLTS
PEAK NEGATIVE-PULSE GRID VOLTAGE	660	275	VOLTS
AVERAGE CATHODE CURRENT	22	22	MA.
PEAK CATHODE CURRENT	330	66	MA.
MAXIMUM CIRCUIT VALUES:			
GRID CIRCUIT RESISTANCE:			
FOR FIXED BIAS, GRID-RESISTOR BIAS, OR CATHODE-BIAS OPERATION	2.2	2.2	MEGOHMS

TYPICAL OPERATING CHARACTERISTICS

CLASS A1 AMPLIFIER - EACH UNIT

PLATE VOLTAGE	100	250	VOLTS
GRID VOLTAGE	0	-8.5	VOLTS
AMPLIFICATION FACTOR	→19.5	17	
PLATE RESISTANCE	6250	7700	OHMS
TRANSCONDUCTANCE	3100	2200	µMHOS
PLATE CURRENT	11.8	10.5	MA.
GRID VOLTAGE (APPROX.) FOR PLATE CURRENT OF 10 µAMP.		-24	VOLTS

→ INDICATES A CHANGE.

170 THE TUBE AMP BOOK

Tube/spec sheets

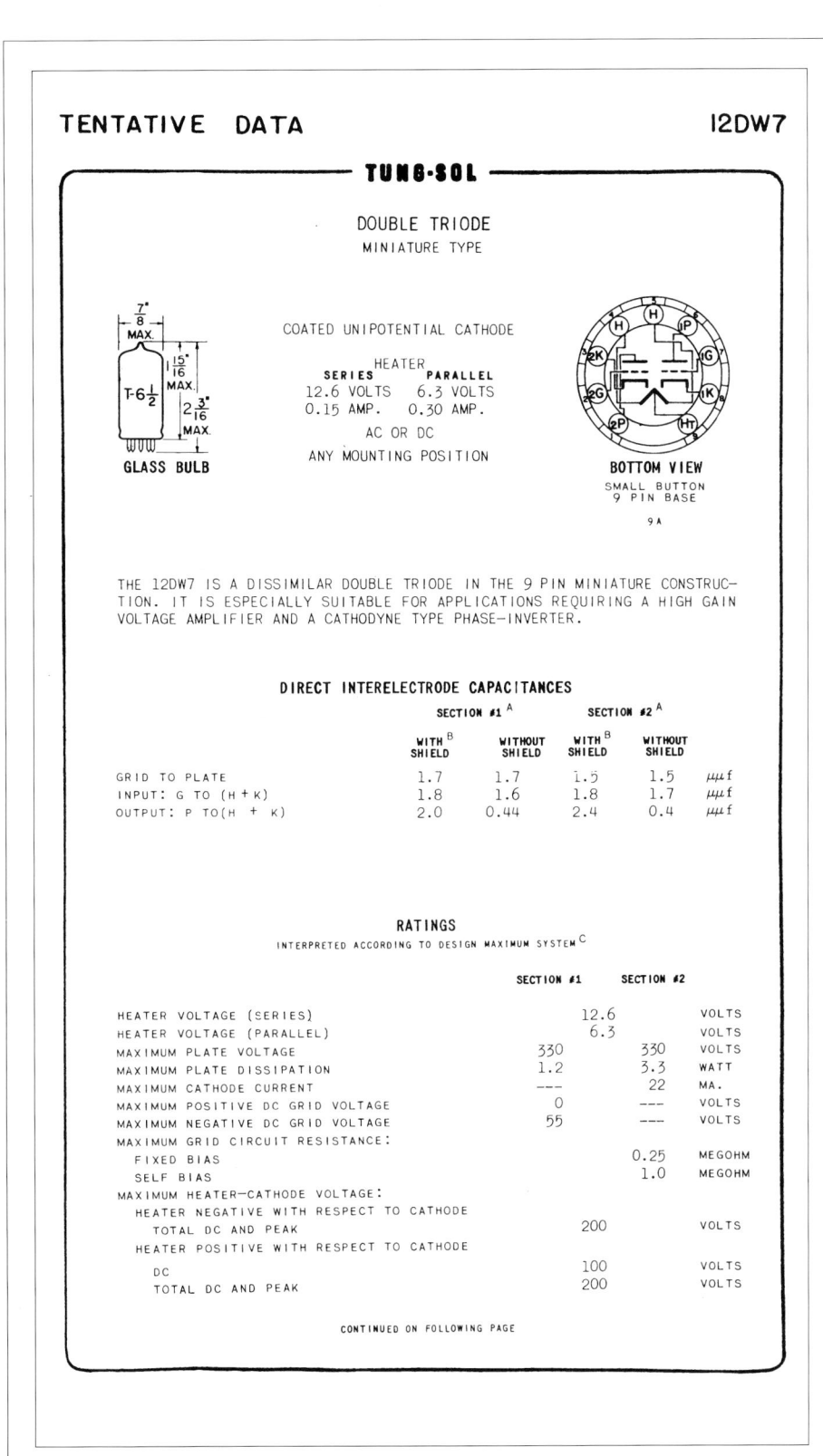

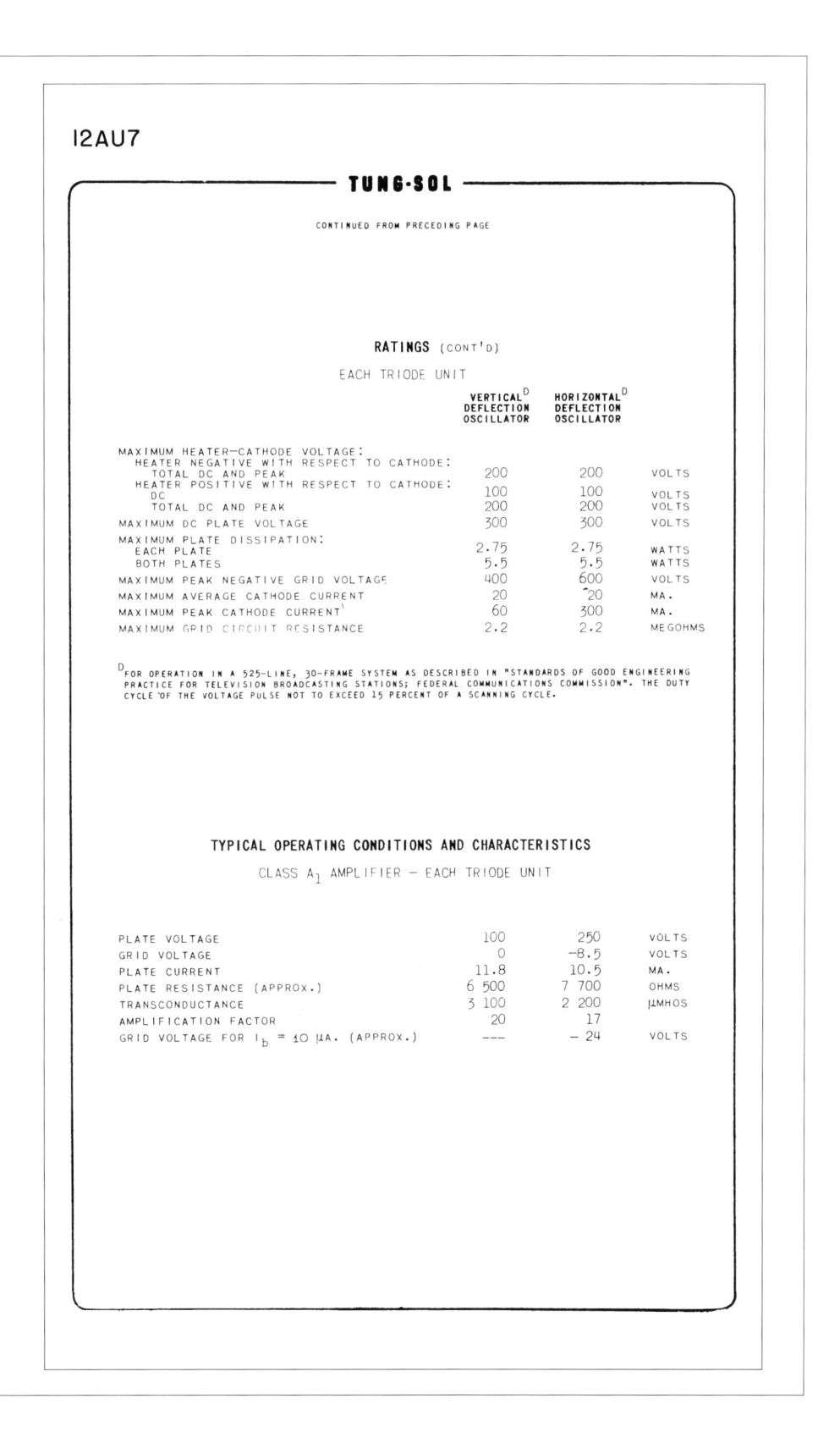

12DW7

TENTATIVE DATA

TUNG-SOL

DOUBLE TRIODE
MINIATURE TYPE

GLASS BULB

COATED UNIPOTENTIAL CATHODE

	SERIES	PARALLEL
HEATER	12.6 VOLTS	6.3 VOLTS
	0.15 AMP.	0.30 AMP.

AC OR DC

ANY MOUNTING POSITION

BOTTOM VIEW
SMALL BUTTON
9 PIN BASE
9A

THE 12DW7 IS A DISSIMILAR DOUBLE TRIODE IN THE 9 PIN MINIATURE CONSTRUCTION. IT IS ESPECIALLY SUITABLE FOR APPLICATIONS REQUIRING A HIGH GAIN VOLTAGE AMPLIFIER AND A CATHODYNE TYPE PHASE-INVERTER.

DIRECT INTERELECTRODE CAPACITANCES

	SECTION #1 [A]		SECTION #2 [A]		
	WITH [B] SHIELD	WITHOUT SHIELD	WITH [B] SHIELD	WITHOUT SHIELD	
GRID TO PLATE	1.7	1.7	1.5	1.5	μμf
INPUT: G TO (H + K)	1.8	1.6	1.8	1.7	μμf
OUTPUT: P TO (H + K)	2.0	0.44	2.4	0.4	μμf

RATINGS
INTERPRETED ACCORDING TO DESIGN MAXIMUM SYSTEM [C]

	SECTION #1	SECTION #2	
HEATER VOLTAGE (SERIES)		12.6	VOLTS
HEATER VOLTAGE (PARALLEL)		6.3	VOLTS
MAXIMUM PLATE VOLTAGE	330	330	VOLTS
MAXIMUM PLATE DISSIPATION	1.2	3.3	WATT
MAXIMUM CATHODE CURRENT		22	MA.
MAXIMUM POSITIVE DC GRID VOLTAGE	---	---	VOLTS
MAXIMUM NEGATIVE DC GRID VOLTAGE	0	---	VOLTS
MAXIMUM GRID CIRCUIT RESISTANCE:			
FIXED BIAS	55	---	MEGOHM
SELF BIAS	0.25	1.0	MEGOHM
MAXIMUM HEATER-CATHODE VOLTAGE:			
HEATER NEGATIVE WITH RESPECT TO CATHODE			
TOTAL DC AND PEAK		200	VOLTS
HEATER POSITIVE WITH RESPECT TO CATHODE			
DC		100	VOLTS
TOTAL DC AND PEAK		200	VOLTS

CONTINUED ON FOLLOWING PAGE

12AU7

CONTINUED FROM PRECEDING PAGE

TUNG-SOL

RATINGS (CONT'D)
EACH TRIODE UNIT

	VERTICAL [D] DEFLECTION OSCILLATOR	HORIZONTAL [D] DEFLECTION OSCILLATOR	
MAXIMUM HEATER-CATHODE VOLTAGE:			
HEATER NEGATIVE WITH RESPECT TO CATHODE:			
TOTAL DC AND PEAK	200	200	VOLTS
HEATER POSITIVE WITH RESPECT TO CATHODE:			
DC	100	100	VOLTS
TOTAL DC AND PEAK	200	200	VOLTS
MAXIMUM DC PLATE VOLTAGE	300	300	VOLTS
MAXIMUM PLATE DISSIPATION:			
EACH PLATE	2.75	2.75	WATTS
BOTH PLATES	5.5	5.5	WATTS
MAXIMUM PEAK NEGATIVE GRID VOLTAGE	400	600	VOLTS
MAXIMUM AVERAGE CATHODE CURRENT	20	20	MA.
MAXIMUM PEAK CATHODE CURRENT [1]	60	300	MA.
MAXIMUM GRID CIRCUIT RESISTANCE	2.2	2.2	MEGOHMS

[D] FOR OPERATION IN A 525-LINE, 30-FRAME SYSTEM AS DESCRIBED IN "STANDARDS OF GOOD ENGINEERING PRACTICE FOR TELEVISION BROADCASTING STATIONS; FEDERAL COMMUNICATIONS COMMISSION". THE DUTY CYCLE OF THE VOLTAGE PULSE NOT TO EXCEED 15 PERCENT OF A SCANNING CYCLE.

TYPICAL OPERATING CONDITIONS AND CHARACTERISTICS
CLASS A_1 AMPLIFIER – EACH TRIODE UNIT

PLATE VOLTAGE	100	250	VOLTS
GRID VOLTAGE	0	-8.5	VOLTS
PLATE CURRENT	11.8	10.5	MA.
PLATE RESISTANCE (APPROX.)	6 500	7 700	OHMS
TRANSCONDUCTANCE	3 100	2 200	μMHOS
AMPLIFICATION FACTOR	20	17	
GRID VOLTAGE FOR I_b = 10 μA. (APPROX.)	---	-24	VOLTS

THE TUBE AMP BOOK 171

Tube/spec sheets

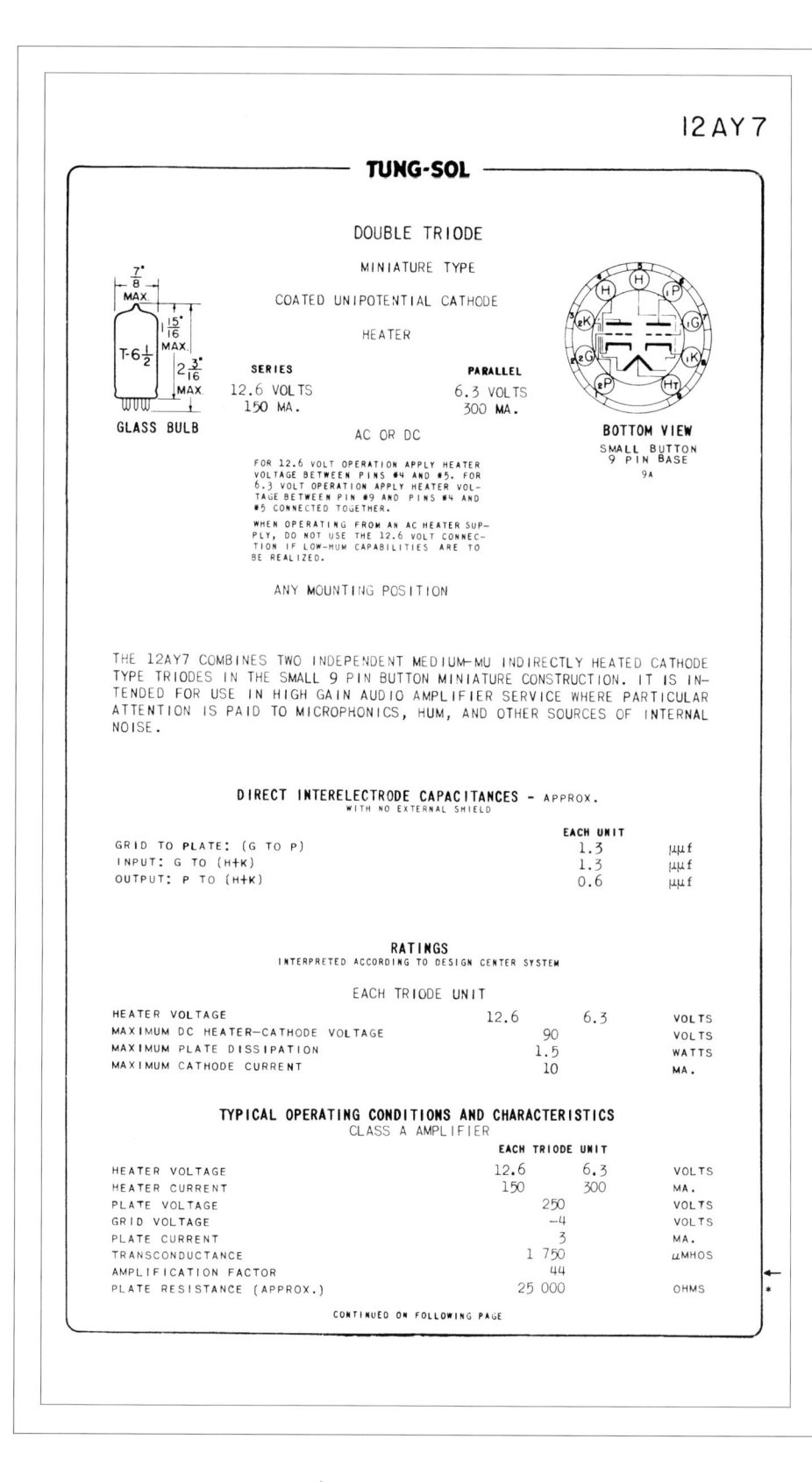

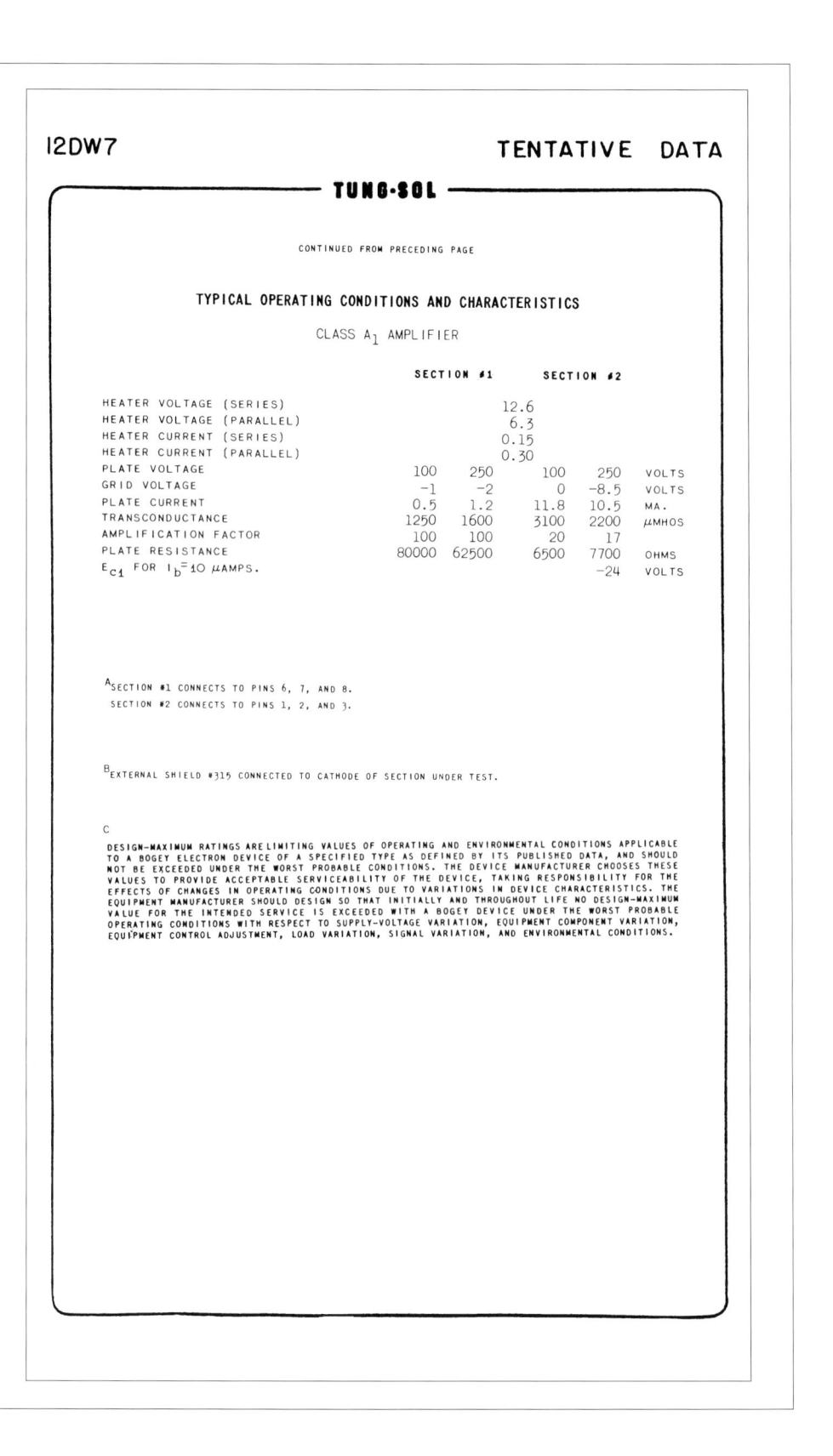

12DW7 — TUNG·SOL — TENTATIVE DATA

CONTINUED FROM PRECEDING PAGE

TYPICAL OPERATING CONDITIONS AND CHARACTERISTICS
CLASS A₁ AMPLIFIER

	SECTION #1	SECTION #2
HEATER VOLTAGE (SERIES)	12.6		VOLTS
HEATER VOLTAGE (PARALLEL)	6.3		VOLTS
HEATER CURRENT (SERIES)	0.15		
HEATER CURRENT (PARALLEL)	0.30		
PLATE VOLTAGE	100	250	100	250	VOLTS
GRID VOLTAGE	−1	−2	0	−8.5	VOLTS
PLATE CURRENT	0.5	1.2	11.8	10.5	MA.
TRANSCONDUCTANCE	1250	1600	3100	2200	μMHOS
AMPLIFICATION FACTOR	100	100	20	17	
PLATE RESISTANCE	80000	62500	6500	7700	OHMS
E_{c1} FOR $I_b = 10$ μAMPS.				−24	VOLTS

A SECTION #1 CONNECTS TO PINS 6, 7, AND 8.
SECTION #2 CONNECTS TO PINS 1, 2, AND 3.

B EXTERNAL SHIELD #315 CONNECTED TO CATHODE OF SECTION UNDER TEST.

C DESIGN-MAXIMUM RATINGS ARE LIMITING VALUES OF OPERATING AND ENVIRONMENTAL CONDITIONS APPLICABLE TO A BOGEY ELECTRON DEVICE OF A SPECIFIED TYPE AS DEFINED BY ITS PUBLISHED DATA, AND SHOULD NOT BE EXCEEDED UNDER THE WORST PROBABLE CONDITIONS. THE DEVICE MANUFACTURER CHOOSES THESE VALUES TO PROVIDE ACCEPTABLE SERVICEABILITY OF THE DEVICE, TAKING RESPONSIBILITY FOR THE EFFECTS OF CHANGES IN OPERATING CONDITIONS DUE TO VARIATIONS IN DEVICE CHARACTERISTICS. THE EQUIPMENT MANUFACTURER SHOULD DESIGN SO THAT INITIALLY AND THROUGHOUT LIFE NO DESIGN-MAXIMUM VALUE FOR THE INTENDED SERVICE IS EXCEEDED WITH A BOGEY DEVICE UNDER THE WORST PROBABLE OPERATING CONDITIONS WITH RESPECT TO SUPPLY-VOLTAGE VARIATION, EQUIPMENT COMPONENT VARIATION, EQUIPMENT CONTROL ADJUSTMENT, LOAD VARIATION, SIGNAL VARIATION, AND ENVIRONMENTAL CONDITIONS.

12AY7 — TUNG-SOL

DOUBLE TRIODE
MINIATURE TYPE
COATED UNIPOTENTIAL CATHODE

HEATER

SERIES	PARALLEL
12.6 VOLTS	6.3 VOLTS
150 MA.	300 MA.

AC OR DC

FOR 12.6 VOLT OPERATION APPLY HEATER VOLTAGE BETWEEN PINS #4 AND #5. FOR 6.3 VOLT OPERATION APPLY HEATER VOLTAGE BETWEEN PIN #9 AND PINS #4 AND #5 CONNECTED TOGETHER.

WHEN OPERATING FROM AN AC HEATER SUPPLY, DO NOT USE THE 12.6 VOLT CONNECTION IF LOW-HUM CAPABILITIES ARE TO BE REALIZED.

ANY MOUNTING POSITION

GLASS BULB: T-6½, 7/8" MAX, 15/16" MAX, 2 3/16" MAX

BOTTOM VIEW
SMALL BUTTON 9 PIN BASE 9A

THE 12AY7 COMBINES TWO INDEPENDENT MEDIUM-MU INDIRECTLY HEATED CATHODE TYPE TRIODES IN THE SMALL 9 PIN BUTTON MINIATURE CONSTRUCTION. IT IS INTENDED FOR USE IN HIGH GAIN AUDIO AMPLIFIER SERVICE WHERE PARTICULAR ATTENTION IS PAID TO MICROPHONICS, HUM, AND OTHER SOURCES OF INTERNAL NOISE.

DIRECT INTERELECTRODE CAPACITANCES — APPROX.
WITH NO EXTERNAL SHIELD

	EACH UNIT	
GRID TO PLATE: (G TO P)	1.3	μμf
INPUT: G TO (H+K)	1.3	μμf
OUTPUT: P TO (H+K)	0.6	μμf

RATINGS
INTERPRETED ACCORDING TO DESIGN CENTER SYSTEM

	EACH TRIODE UNIT		
HEATER VOLTAGE	12.6	6.3	VOLTS
MAXIMUM DC HEATER-CATHODE VOLTAGE	90		VOLTS
MAXIMUM PLATE DISSIPATION	1.5		WATTS
MAXIMUM CATHODE CURRENT	10		MA.

TYPICAL OPERATING CONDITIONS AND CHARACTERISTICS
CLASS A AMPLIFIER

	EACH TRIODE UNIT		
HEATER VOLTAGE	12.6	6.3	VOLTS
HEATER CURRENT	150	300	MA.
PLATE VOLTAGE	250		VOLTS
GRID VOLTAGE	−4		VOLTS
PLATE CURRENT	3		MA.
TRANSCONDUCTANCE	1 750		μMHOS
AMPLIFICATION FACTOR	44		
PLATE RESISTANCE (APPROX.)	25 000		OHMS

CONTINUED ON FOLLOWING PAGE

Tube/spec sheets

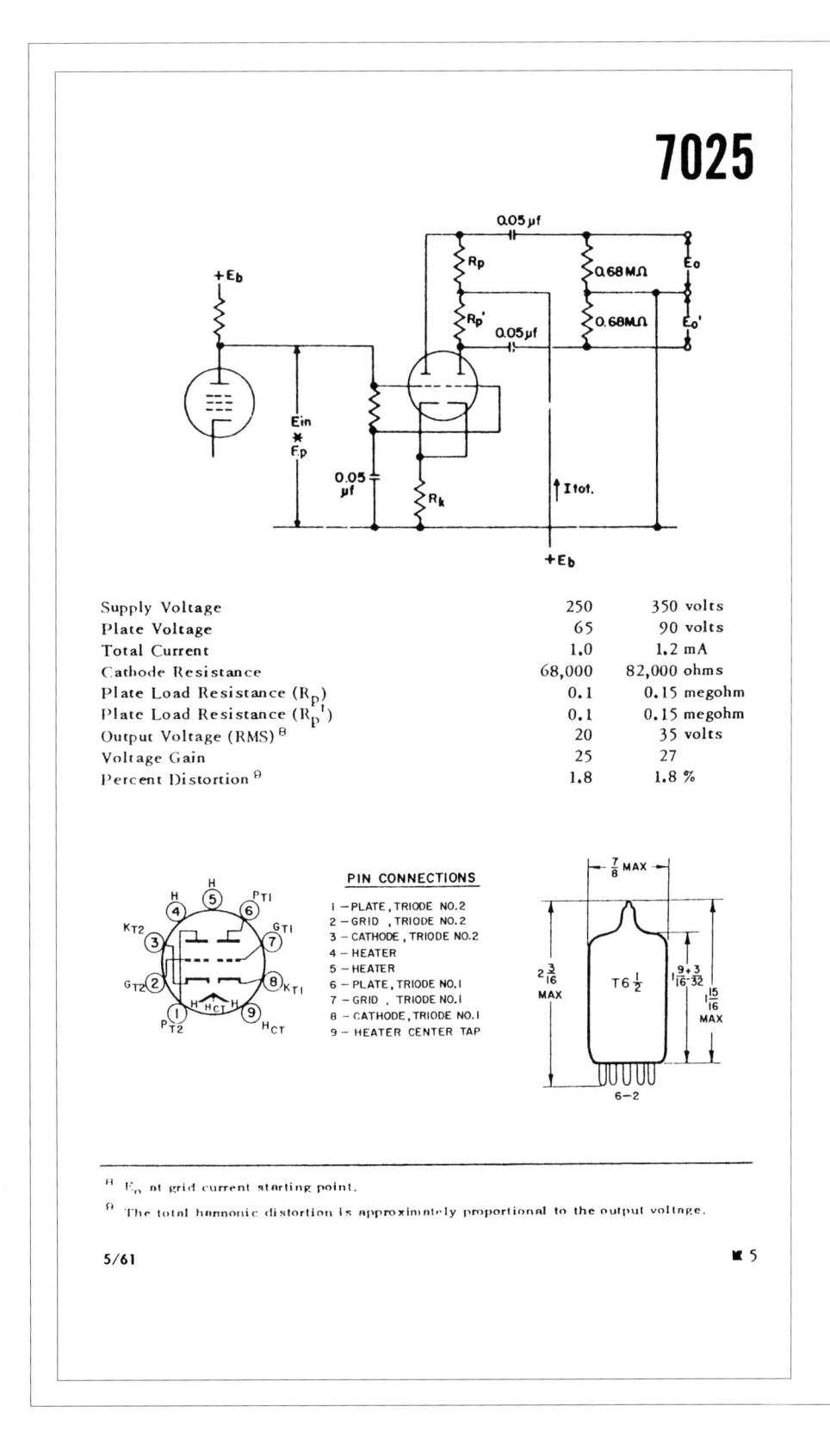

12AT7 — TUNG-SOL

CONTINUED FROM PRECEDING PAGE

TYPICAL OPERATING CONDITIONS AND CHARACTERISTICS

CLASS A_1 AMPLIFIER – EACH TRIODE UNIT

HEATER VOLTAGE	12.6	6.3	12.6	6.3	VOLTS
HEATER CURRENT	150	300	150	300	MA.
PLATE VOLTAGE	100		250		VOLTS
CATHODE BIAS RESISTOR	270		200		OHMS
PLATE CURRENT	3.7		10		MA.
PLATE RESISTANCE	15 000		10 900		OHMS
TRANSCONDUCTANCE	4 000		5 500		µMHOS
AMPLIFICATION FACTOR	60		60		
GRID VOLTAGE (APPROX.) FOR I_b = 10 µA.	−5		−12		VOLTS

TYPICAL CIRCUIT FOR CONVERTER OPERATION AT 100 MEGACYCLES

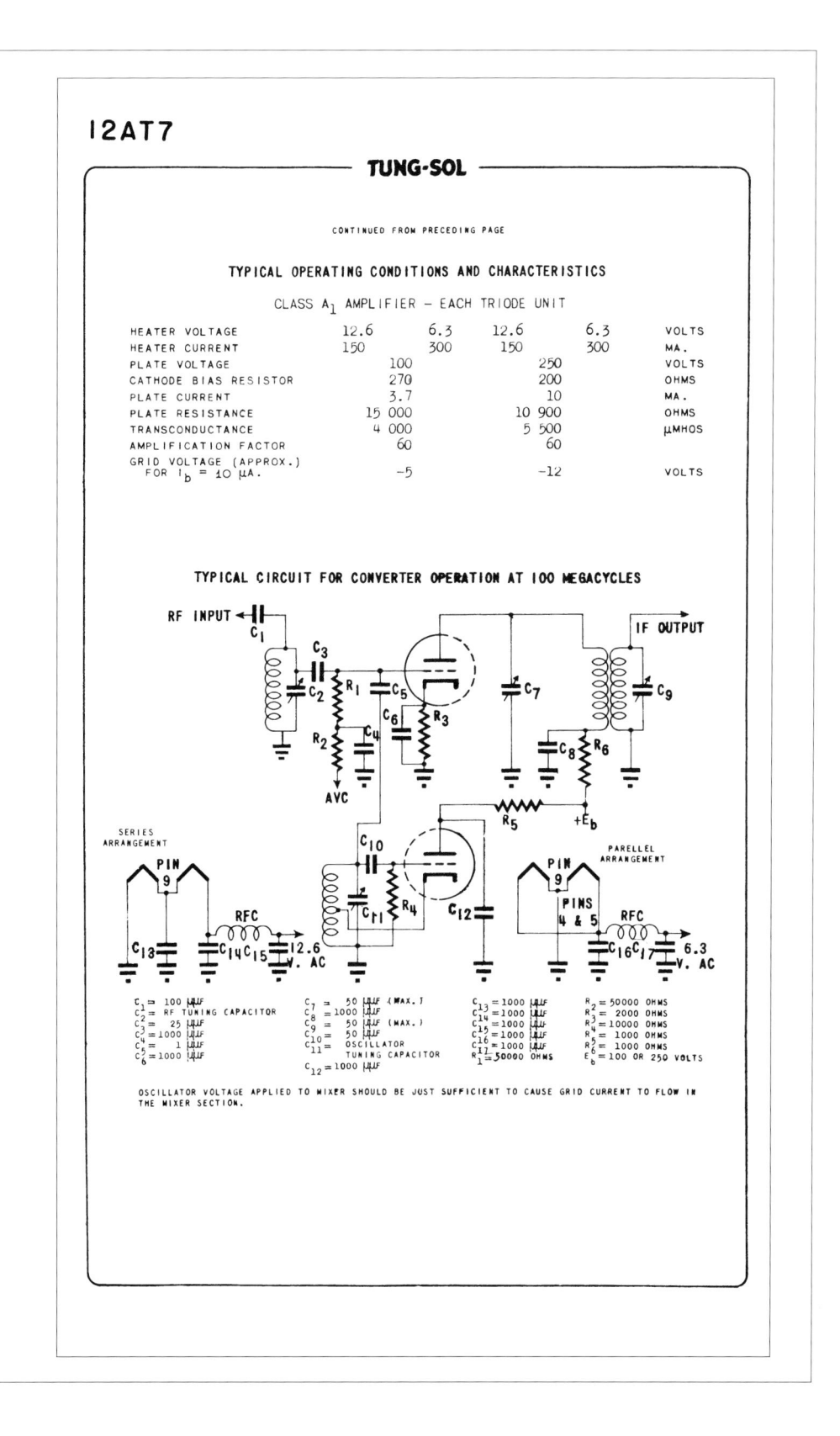

C_1 = 100 µµF
C_2 = 25 µµF
C_3 = 1000 µµF
C_4 = 1 µµF
C_5 = 1000 µµF
C_6 = 1000 µµF
C_7 = 50 µµF (MAX.)
C_8 = 1000 µµF
C_9 = 50 µµF (MAX.)
C_{10} = OSCILLATOR TUNING CAPACITOR
C_{11} = 1000 µµF
C_{12} = 1000 µµF
C_{13} = 1000 µµF
C_{14} = 1000 µµF
C_{15} = 1000 µµF
C_{16} C_{17} = 1000 µµF

R_1 = 50000 OHMS
R_2 = 2000 OHMS
R_3 = 1000 OHMS
R_4 = 1000 OHMS
R_5 = 50000 OHMS
R_6 = 1000 OHMS
E_b = 100 OR 250 VOLTS

OSCILLATOR VOLTAGE APPLIED TO MIXER SHOULD BE JUST SUFFICIENT TO CAUSE GRID CURRENT TO FLOW IN THE MIXER SECTION.

7025

Supply Voltage	250	350 volts
Plate Voltage	65	90 volts
Total Current	1.0	1.2 mA
Cathode Resistance	68,000	82,000 ohms
Plate Load Resistance (R_p)	0.1	0.15 megohm
Plate Load Resistance (R_p')	0.1	0.15 megohm
Output Voltage (RMS) [8]	20	35 volts
Voltage Gain	25	27
Percent Distortion [9]	1.8	1.8 %

PIN CONNECTIONS
1 – PLATE, TRIODE NO.2
2 – GRID , TRIODE NO.2
3 – CATHODE , TRIODE NO.2
4 – HEATER
5 – HEATER
6 – PLATE, TRIODE NO.1
7 – GRID , TRIODE NO.1
8 – CATHODE, TRIODE NO.1
9 – HEATER CENTER TAP

[8] E_o at grid current starting point.
[9] The total harmonic distortion is approximately proportional to the output voltage.

5/61

Tube/spec sheets

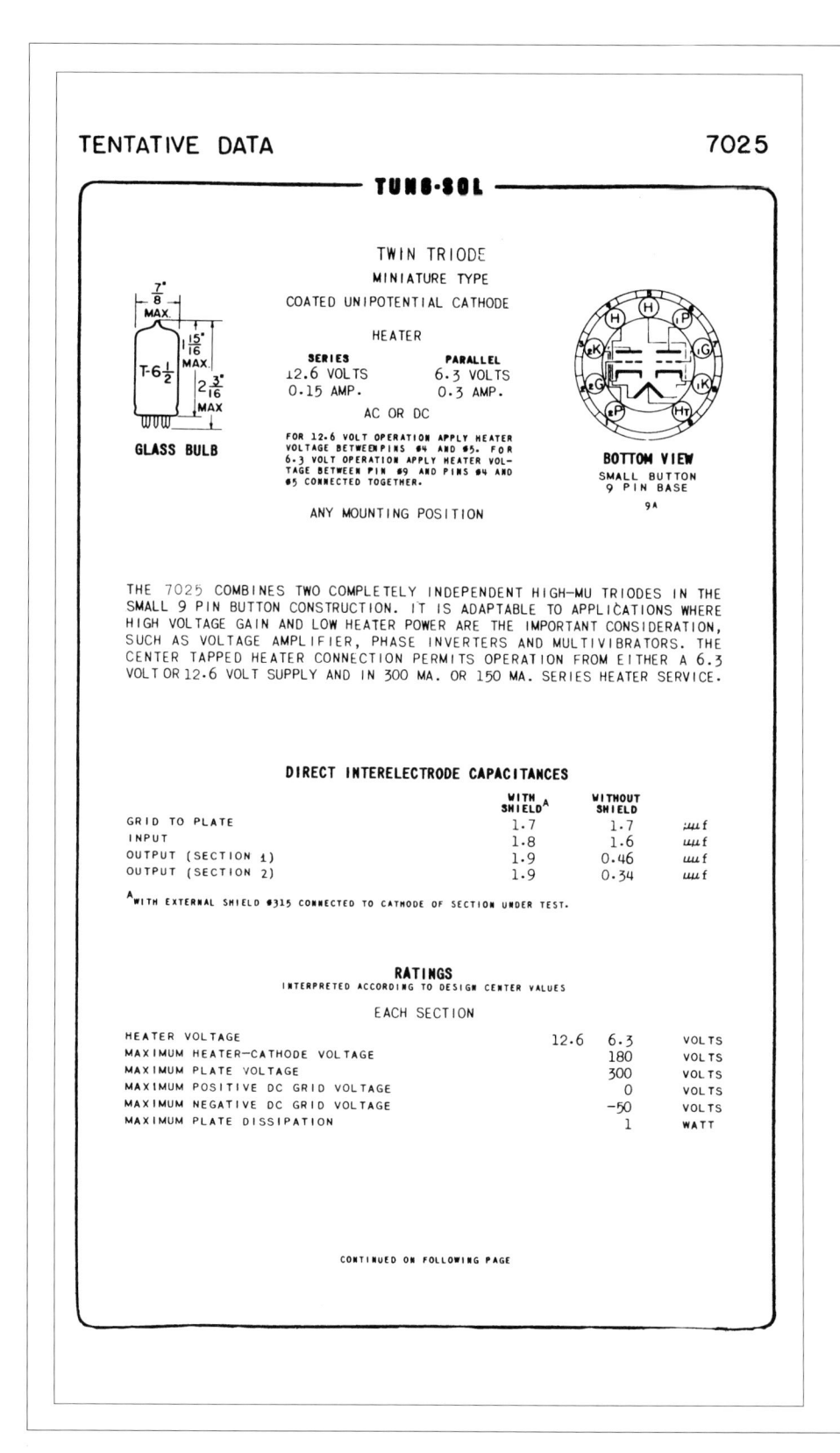

7025

[Average plate characteristics chart: Plate Current (milliamperes) 0–6 vs Grid Volts −4 to 0, with curves for Ep=100V and Ep=250V. Dated 5/61.]

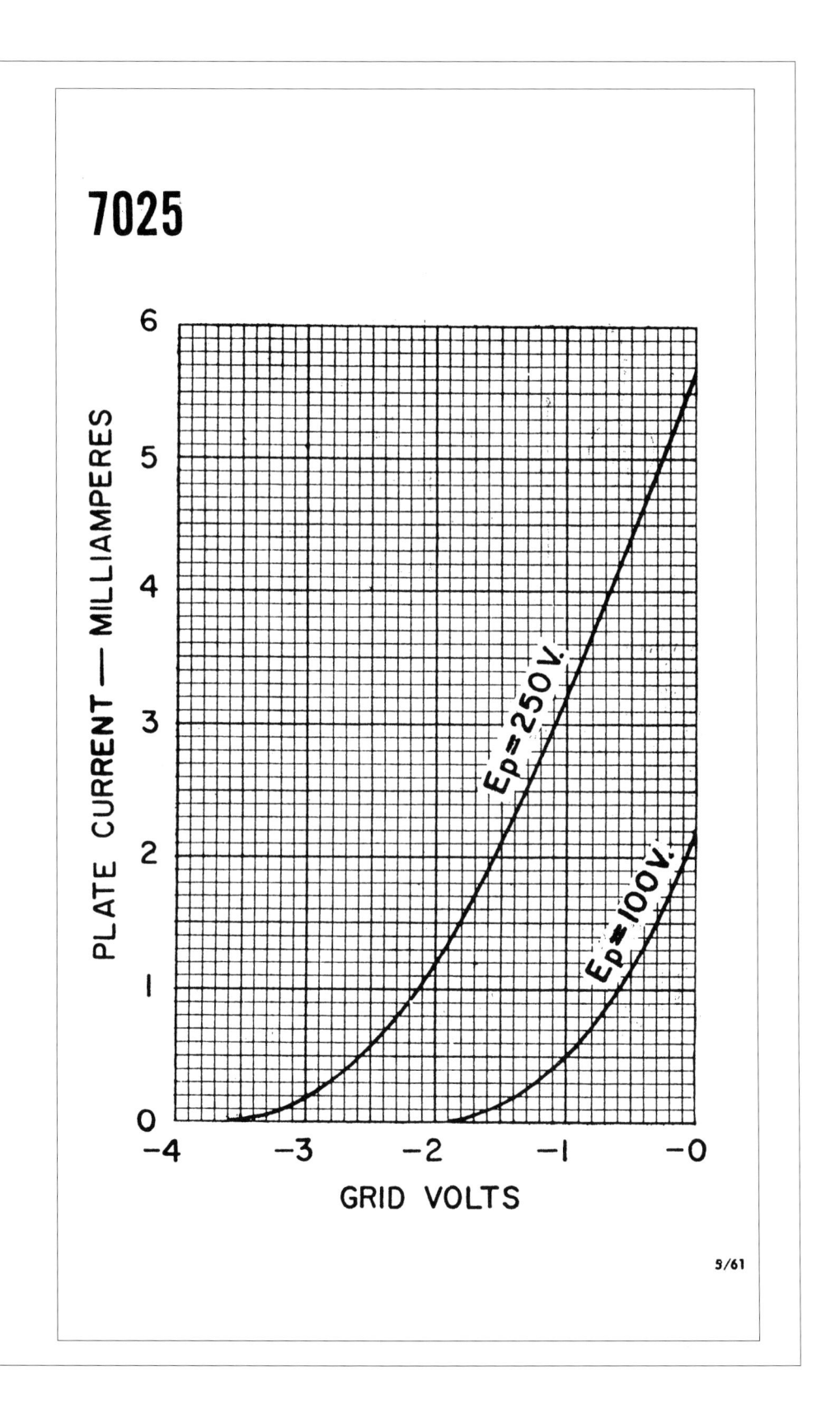

TENTATIVE DATA — TUNG-SOL — 7025

TWIN TRIODE
MINIATURE TYPE
COATED UNIPOTENTIAL CATHODE

HEATER

SERIES	PARALLEL
12.6 VOLTS	6.3 VOLTS
0.15 AMP.	0.3 AMP.

AC OR DC

FOR 12.6 VOLT OPERATION APPLY HEATER VOLTAGE BETWEEN PINS 4 AND 5. FOR 6.3 VOLT OPERATION APPLY HEATER VOLTAGE BETWEEN PIN 9 AND PINS 4 AND 5 CONNECTED TOGETHER.

ANY MOUNTING POSITION

GLASS BULB T-6½
- 7/8" MAX
- 15/16" MAX
- 2 3/16" MAX

BOTTOM VIEW
SMALL BUTTON
9 PIN BASE
9A

THE 7025 COMBINES TWO COMPLETELY INDEPENDENT HIGH-MU TRIODES IN THE SMALL 9 PIN BUTTON CONSTRUCTION. IT IS ADAPTABLE TO APPLICATIONS WHERE HIGH VOLTAGE GAIN AND LOW HEATER POWER ARE THE IMPORTANT CONSIDERATIONS, SUCH AS VOLTAGE AMPLIFIER, PHASE INVERTERS AND MULTIVIBRATORS. THE CENTER TAPPED HEATER CONNECTION PERMITS OPERATION FROM EITHER A 6.3 VOLT OR 12.6 VOLT SUPPLY AND IN 300 MA. OR 150 MA. SERIES HEATER SERVICE.

DIRECT INTERELECTRODE CAPACITANCES

	WITH SHIELD[A]	WITHOUT SHIELD	
GRID TO PLATE	1.7	1.7	µµf
INPUT	1.8	1.6	µµf
OUTPUT (SECTION 1)	1.9	0.46	µµf
OUTPUT (SECTION 2)	1.9	0.34	µµf

[A] WITH EXTERNAL SHIELD #315 CONNECTED TO CATHODE OF SECTION UNDER TEST.

RATINGS
INTERPRETED ACCORDING TO DESIGN CENTER VALUES

	EACH SECTION	
HEATER VOLTAGE	12.6 6.3	VOLTS
MAXIMUM HEATER-CATHODE VOLTAGE	180	VOLTS
MAXIMUM PLATE VOLTAGE	300	VOLTS
MAXIMUM POSITIVE DC GRID VOLTAGE	0	VOLTS
MAXIMUM NEGATIVE DC GRID VOLTAGE	−50	VOLTS
MAXIMUM PLATE DISSIPATION	1	WATT

CONTINUED ON FOLLOWING PAGE

Tube/spec sheets

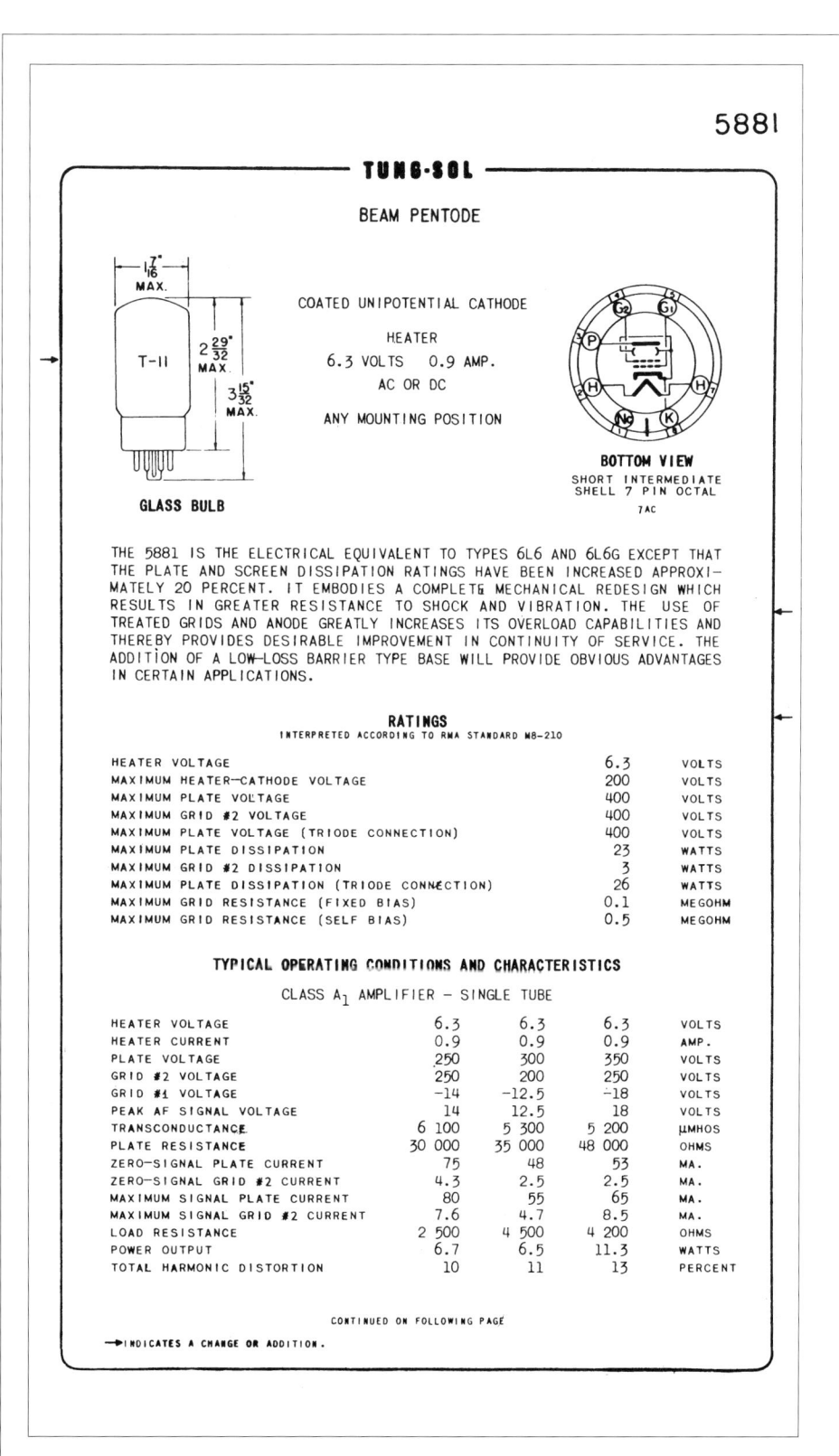

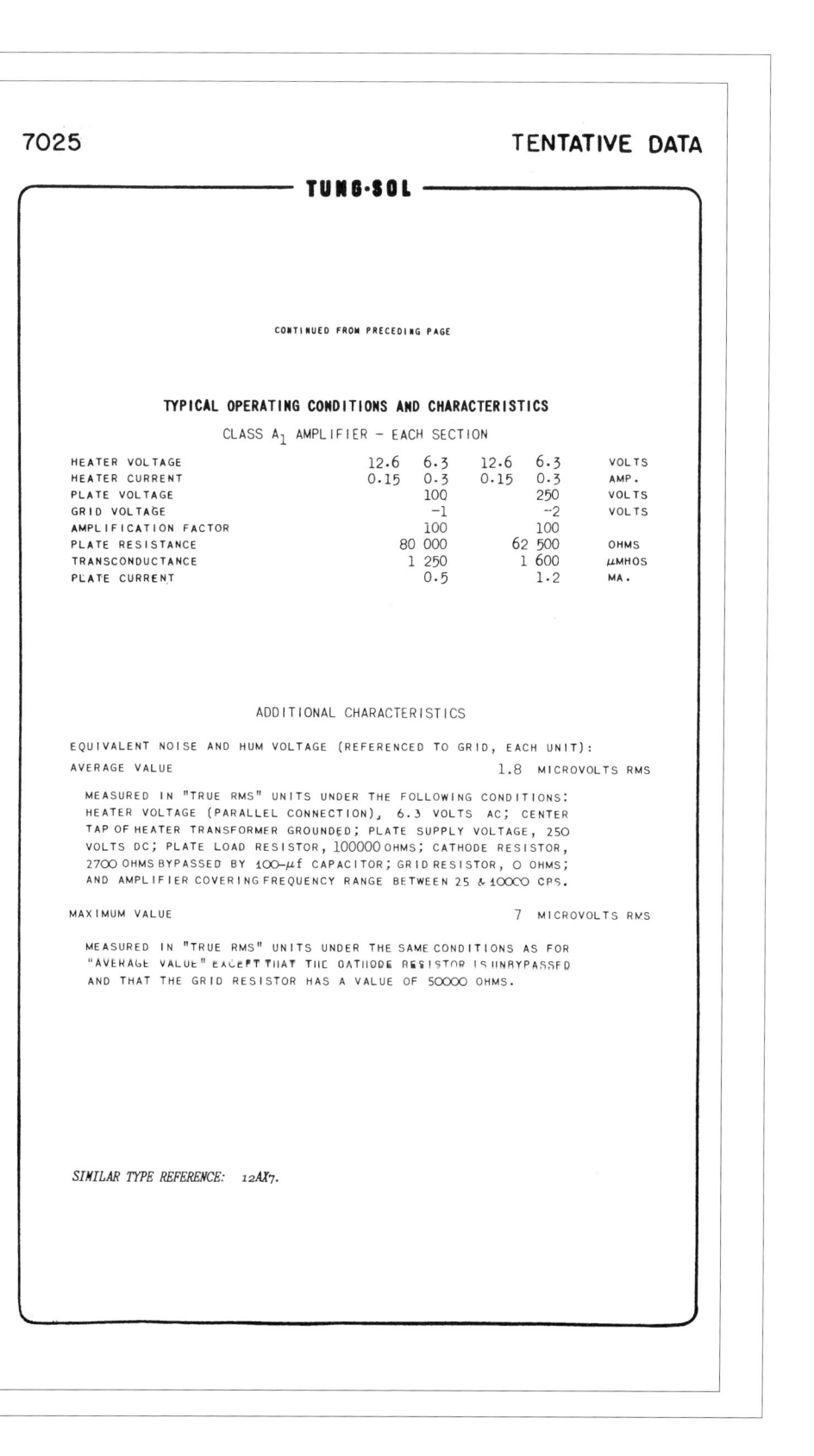

7025 — TUNG-SOL — TENTATIVE DATA

CONTINUED FROM PRECEDING PAGE

TYPICAL OPERATING CONDITIONS AND CHARACTERISTICS
CLASS A₁ AMPLIFIER - EACH SECTION

HEATER VOLTAGE	12.6	6.3	12.6	6.3	VOLTS
HEATER CURRENT	0.15	0.3	0.15	0.3	AMP.
PLATE VOLTAGE		100		250	VOLTS
GRID VOLTAGE		-1		-2	VOLTS
AMPLIFICATION FACTOR		100		100	
PLATE RESISTANCE	80 000		62 500		OHMS
TRANSCONDUCTANCE	1 250		1 600		μMHOS
PLATE CURRENT	0.5		1.2		MA.

ADDITIONAL CHARACTERISTICS

EQUIVALENT NOISE AND HUM VOLTAGE (REFERENCED TO GRID, EACH UNIT):

AVERAGE VALUE .... 1.8 MICROVOLTS RMS

MEASURED IN "TRUE RMS" UNITS UNDER THE FOLLOWING CONDITIONS: HEATER VOLTAGE (PARALLEL CONNECTION), 6.3 VOLTS AC; CENTER TAP OF HEATER TRANSFORMER GROUNDED; PLATE SUPPLY VOLTAGE, 250 VOLTS DC; PLATE LOAD RESISTOR, 100000 OHMS; CATHODE RESISTOR, 2700 OHMS BYPASSED BY 100-μf CAPACITOR; GRID RESISTOR, 0 OHMS; AND AMPLIFIER COVERING FREQUENCY RANGE BETWEEN 25 & 10000 CPS.

MAXIMUM VALUE .... 7 MICROVOLTS RMS

MEASURED IN "TRUE RMS" UNITS UNDER THE SAME CONDITIONS AS FOR "AVERAGE VALUE" EXCEPT THAT THE CATHODE RESISTOR IS UNBYPASSED AND THAT THE GRID RESISTOR HAS A VALUE OF 50000 OHMS.

SIMILAR TYPE REFERENCE: 12AX7.

5881 — TUNG-SOL

BEAM PENTODE

COATED UNIPOTENTIAL CATHODE

HEATER
6.3 VOLTS 0.9 AMP.
AC OR DC
ANY MOUNTING POSITION

GLASS BULB
T-11
$\frac{7}{16}$" MAX.
$2\frac{29}{32}$ MAX.
$3\frac{15}{32}$ MAX.

BOTTOM VIEW
SHORT INTERMEDIATE SHELL 7 PIN OCTAL
7AC

THE 5881 IS THE ELECTRICAL EQUIVALENT TO TYPES 6L6 AND 6L6G EXCEPT THAT THE PLATE AND SCREEN DISSIPATION RATINGS HAVE BEEN INCREASED APPROXIMATELY 20 PERCENT. IT EMBODIES A COMPLETE MECHANICAL REDESIGN WHICH RESULTS IN GREATER RESISTANCE TO SHOCK AND VIBRATION. THE USE OF TREATED GRIDS AND ANODE GREATLY INCREASES ITS OVERLOAD CAPABILITIES AND THEREBY PROVIDES DESIRABLE IMPROVEMENT IN CONTINUITY OF SERVICE. THE ADDITION OF A LOW-LOSS BARRIER TYPE BASE WILL PROVIDE OBVIOUS ADVANTAGES IN CERTAIN APPLICATIONS.

RATINGS
INTERPRETED ACCORDING TO RMA STANDARD M8-210

HEATER VOLTAGE	6.3	6.3	VOLTS
MAXIMUM HEATER-CATHODE VOLTAGE	200		VOLTS
MAXIMUM PLATE VOLTAGE	400	350	VOLTS
MAXIMUM GRID #2 VOLTAGE	400	250	VOLTS
MAXIMUM GRID #2 VOLTAGE (TRIODE CONNECTION)	400		VOLTS
MAXIMUM PLATE DISSIPATION	23		WATTS
MAXIMUM GRID #2 DISSIPATION	3		WATTS
MAXIMUM PLATE DISSIPATION (TRIODE CONNECTION)	26		WATTS
MAXIMUM GRID RESISTANCE (FIXED BIAS)	0.1		MEGOHM
MAXIMUM GRID RESISTANCE (SELF BIAS)	0.5		MEGOHM

TYPICAL OPERATING CONDITIONS AND CHARACTERISTICS
CLASS A₁ AMPLIFIER - SINGLE TUBE

HEATER VOLTAGE	6.3	6.3	6.3	VOLTS
HEATER CURRENT	0.9	0.9	0.9	AMP.
PLATE VOLTAGE	250	300	350	VOLTS
GRID #2 VOLTAGE	250	200	250	VOLTS
GRID #1 VOLTAGE	-14	-12.5	-18	VOLTS
PEAK AF SIGNAL VOLTAGE	14	12.5	18	VOLTS
TRANSCONDUCTANCE	6 100	5 300	5 200	μMHOS
PLATE RESISTANCE	30 000	35 000	48 000	OHMS
ZERO-SIGNAL PLATE CURRENT	75	48	53	MA.
ZERO-SIGNAL GRID #2 CURRENT	4.3	2.5	2.5	MA.
MAXIMUM SIGNAL PLATE CURRENT	80	55	65	MA.
MAXIMUM SIGNAL GRID #2 CURRENT	7.6	4.7	8.5	MA.
LOAD RESISTANCE	2 500	4 500	4 200	OHMS
POWER OUTPUT	6.7	6.5	11.3	WATTS
TOTAL HARMONIC DISTORTION	10	11	13	PERCENT

CONTINUED ON FOLLOWING PAGE.

→ INDICATES A CHANGE OR ADDITION.

Tube/spec sheets

TUNG-SOL 6L6GB

TENTATIVE DATA

CONTINUED FROM PRECEDING PAGE

CLASS A₁ PUSH-PULL AMPLIFIER – PENTODE CONNECTION
VALUES ARE FOR TWO TUBES

HEATER VOLTAGE	6.3	6.3	VOLTS
HEATER CURRENT	0.9	0.9	AMP.
PLATE VOLTAGE	250	270	VOLTS
GRID #2 VOLTAGE	250	270	VOLTS
GRID #1 VOLTAGE	−16	−17.5	VOLTS
PEAK AF GRID TO GRID VOLTAGE	32	35	VOLTS
TRANSCONDUCTANCE (EACH TUBE)	5 500	5 700	μMHOS
PLATE RESISTANCE (EACH TUBE)	24 500	23 500	OHMS
ZERO-SIGNAL PLATE CURRENT	120	134	MA.
MAXIMUM SIGNAL PLATE CURRENT	140	155	MA.
ZERO SIGNAL GRID #2 CURRENT	10	11	MA.
MAXIMUM SIGNAL GRID #2 CURRENT	16	17	MA.
LOAD RESISTANCE	5 000	5 000	OHMS
POWER OUTPUT	14.5	17.5	WATTS
TOTAL HARMONIC DISTORTION	2	2	PERCENT

CLASS AB₁ PUSH-PULL AMPLIFIER – PENTODE CONNECTION
VALUES ARE FOR TWO TUBES

HEATER VOLTAGE	6.3	6.3	VOLTS
HEATER CURRENT	0.9	0.9	AMP.
PLATE VOLTAGE	360	360	VOLTS
GRID #2 VOLTAGE	270	270	VOLTS
GRID #1 VOLTAGE	−22.5	−22.5	VOLTS
PEAK AF GRID TO GRID VOLTAGE	45	45	VOLTS
ZERO-SIGNAL PLATE CURRENT	88	88	MA.
MAXIMUM SIGNAL PLATE CURRENT	132	140	MA.
ZERO SIGNAL GRID #2 CURRENT	5	5	MA.
MAXIMUM SIGNAL GRID #2 CURRENT	15	11	MA.
LOAD RESISTANCE	6 600	3 800	OHMS
POWER OUTPUT	26.5	18	WATTS
TOTAL HARMONIC DISTORTION	2	2	PERCENT

CLASS AB₂ PUSH-PULL AMPLIFIER – PENTODE CONNECTION
VALUES ARE FOR TWO TUBES

HEATER VOLTAGE	6.3	6.3	VOLTS
HEATER CURRENT	0.9	0.9	AMP.
PLATE VOLTAGE	360	360	VOLTS
GRID #2 VOLTAGE	225	270	VOLTS
GRID #1 VOLTAGE	−18	−22.5	VOLTS
PEAK AF GRID TO GRID VOLTAGE	52	72	VOLTS
ZERO SIGNAL PLATE CURRENT	78	88	MA.
MAXIMUM SIGNAL PLATE CURRENT	142	205	MA.
ZERO SIGNAL GRID #2 CURRENT	3.5	5	MA.
MAXIMUM SIGNAL GRID #2 CURRENT	11	16	MA.
LOAD RESISTANCE	6 000	3 800	OHMS
POWER OUTPUT	31	47	WATTS
TOTAL HARMONIC DISTORTION	2	2	PERCENT

TUNG-SOL 5881

CONTINUED FROM PRECEDING PAGE

CLASS A₁ AMPLIFIER – SINGLE TUBE – TRIODE CONNECTION
GRID #2 CONNECTED TO PLATE

HEATER VOLTAGE	6.3	6.3	VOLTS
HEATER CURRENT	0.9	0.9	AMP.
PLATE VOLTAGE	250	300	VOLTS
GRID #1 VOLTAGE	−18	−20	VOLTS
PEAK AF SIGNAL VOLTAGE	18	20	VOLTS
ZERO-SIGNAL PLATE CURRENT	52	78	MA.
MAXIMUM SIGNAL PLATE CURRENT	58	85	MA.
AMPLIFICATION FACTOR	8	---	
TRANSCONDUCTANCE	5 250		μMHOS
LOAD RESISTANCE	4 000	4 000	OHMS
TOTAL HARMONIC DISTORTION	6	5.5	PERCENT
POWER OUTPUT	1.4	1.8	WATTS

CLASS A₁ PUSH-PULL AMPLIFIER
VALUES ARE FOR TWO TUBES

HEATER VOLTAGE	6.3	6.3	VOLTS
HEATER CURRENT	0.9	0.9	AMP.
PLATE VOLTAGE	250	270	VOLTS
GRID #2 VOLTAGE	250	270	VOLTS
GRID #1 VOLTAGE	−16	−17.5	VOLTS
PEAK AF GRID TO GRID VOLTAGE	32	35	VOLTS
TRANSCONDUCTANCE (EACH TUBE)	5 500	5 700	μMHOS
PLATE RESISTANCE (EACH TUBE)	24 500	23 500	OHMS
ZERO-SIGNAL PLATE CURRENT	120	134	MA.
MAXIMUM SIGNAL PLATE CURRENT	140	155	MA.
ZERO-SIGNAL GRID #2 CURRENT	10	11	MA.
MAXIMUM SIGNAL GRID #2 CURRENT	16	17	MA.
LOAD RESISTANCE	5 000	5 000	OHMS
POWER OUTPUT	14.5	17.5	WATTS
TOTAL HARMONIC DISTORTION	2	2	PERCENT

CLASS AB₁ PUSH-PULL AMPLIFIER
VALUES ARE FOR TWO TUBES

HEATER VOLTAGE	6.3	6.3	VOLTS
HEATER CURRENT	0.9	0.9	AMP.
PLATE VOLTAGE	360	360	VOLTS
GRID #2 VOLTAGE	270	270	VOLTS
GRID #1 VOLTAGE	−22.5	−22.5	VOLTS
PEAK AF GRID TO GRID VOLTAGE	45	45	VOLTS
ZERO-SIGNAL PLATE CURRENT	88	88	MA.
MAXIMUM SIGNAL PLATE CURRENT	132	140	MA.
ZERO SIGNAL GRID #2 CURRENT	5	5	MA.
MAXIMUM SIGNAL GRID #2 CURRENT	15	11	MA.
LOAD RESISTANCE	6 600	3 800	OHMS
POWER OUTPUT	26.5	18	WATTS
TOTAL HARMONIC DISTORTION	2	2	PERCENT

CONTINUED ON FOLLOWING PAGE

→ INDICATES A CHANGE OR ADDITION.

Tube/spec sheets

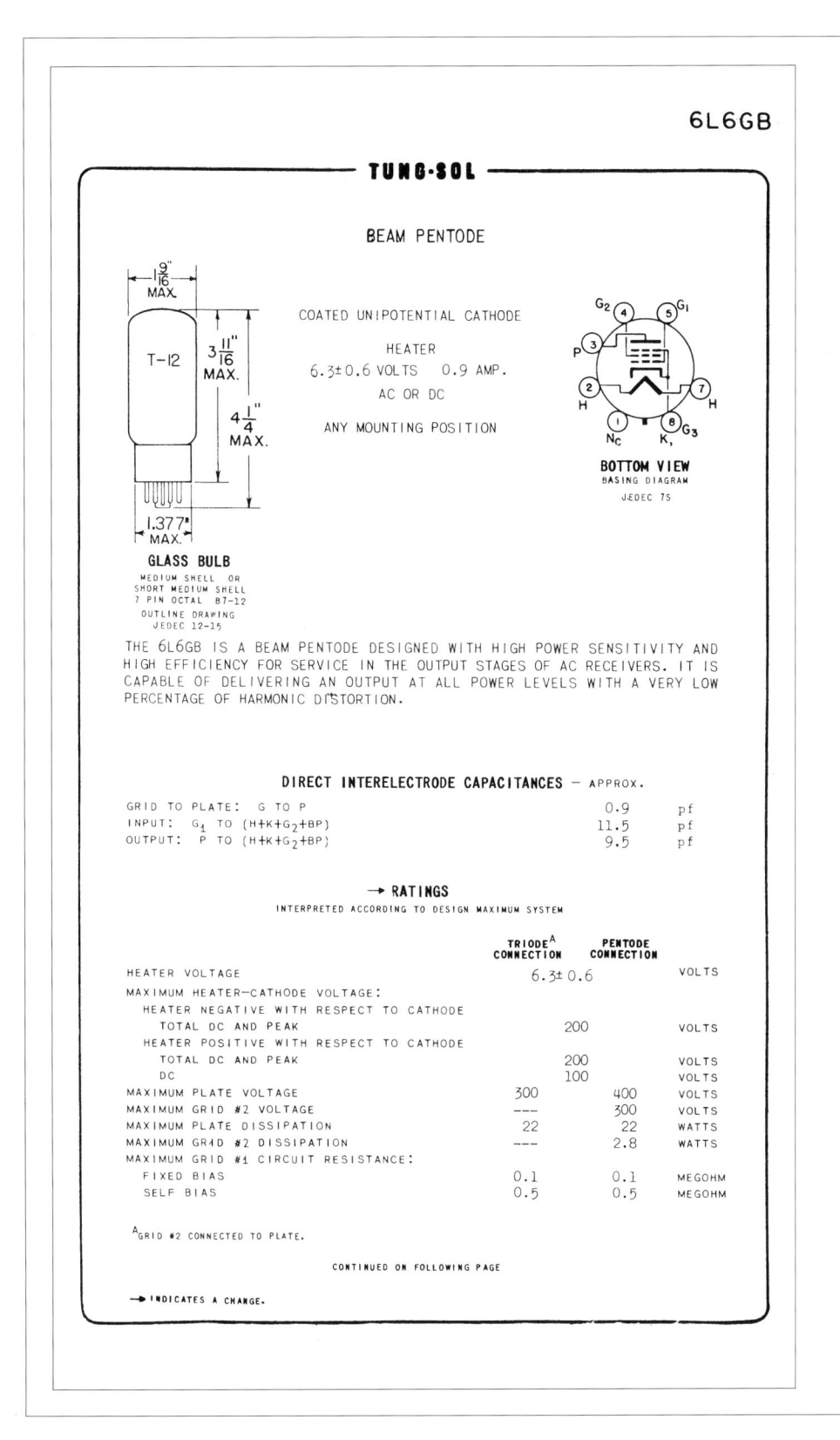

6L6GB

TUNG-SOL

BEAM PENTODE

GLASS BULB
MEDIUM SHELL OR
SHORT MEDIUM SHELL
7 PIN OCTAL 8T-12
OUTLINE DRAWING
JEDEC 12-15

COATED UNIPOTENTIAL CATHODE

HEATER
6.3±0.6 VOLTS   0.9 AMP.
AC OR DC

ANY MOUNTING POSITION

BOTTOM VIEW
BASING DIAGRAM
JEDEC 7S

THE 6L6GB IS A BEAM PENTODE DESIGNED WITH HIGH POWER SENSITIVITY AND HIGH EFFICIENCY FOR SERVICE IN THE OUTPUT STAGES OF AC RECEIVERS. IT IS CAPABLE OF DELIVERING AN OUTPUT AT ALL POWER LEVELS WITH A VERY LOW PERCENTAGE OF HARMONIC DISTORTION.

DIRECT INTERELECTRODE CAPACITANCES — APPROX.

GRID TO PLATE: G TO P	0.9	pf
INPUT: G_1 TO (H+K+G_2+BP)	11.5	pf
OUTPUT: P TO (H+K+G_2+BP)	9.5	pf

→ **RATINGS**
INTERPRETED ACCORDING TO DESIGN MAXIMUM SYSTEM

	TRIODE^A CONNECTION	PENTODE CONNECTION	
HEATER VOLTAGE	6.3±0.6		VOLTS
MAXIMUM HEATER-CATHODE VOLTAGE:			
HEATER NEGATIVE WITH RESPECT TO CATHODE			
TOTAL DC AND PEAK	200	200	VOLTS
HEATER POSITIVE WITH RESPECT TO CATHODE			
TOTAL DC AND PEAK	200	200	VOLTS
DC	100	100	VOLTS
MAXIMUM PLATE VOLTAGE	300	400	VOLTS
MAXIMUM GRID #2 VOLTAGE	---	300	VOLTS
MAXIMUM PLATE DISSIPATION	22	22	WATTS
MAXIMUM GRID #2 DISSIPATION	---	2.8	WATTS
MAXIMUM GRID #1 CIRCUIT RESISTANCE:			
FIXED BIAS	0.1	0.1	MEGOHM
SELF BIAS	0.5	0.5	MEGOHM

^{A} GRID #2 CONNECTED TO PLATE.

→ **INDICATES A CHANGE.**

CONTINUED ON FOLLOWING PAGE

THE TUBE AMP BOOK 177

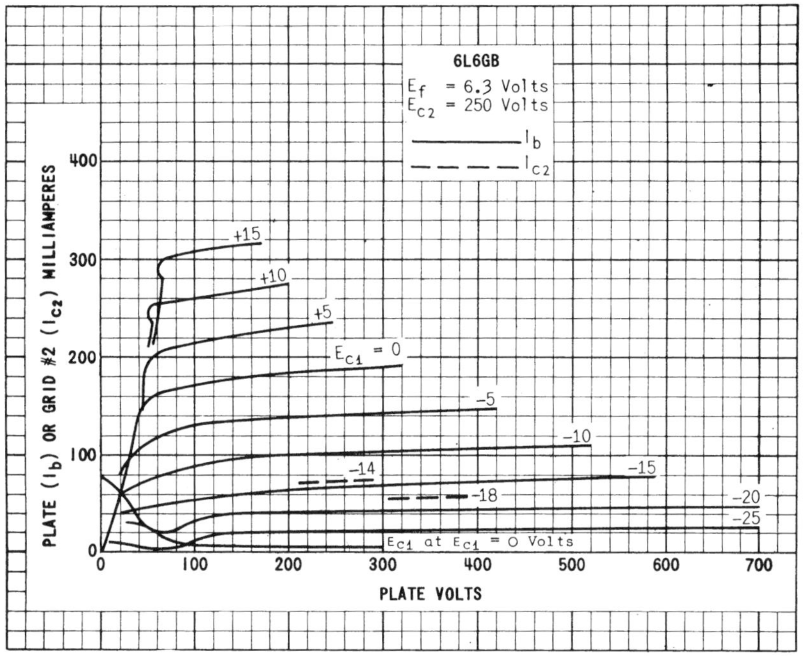

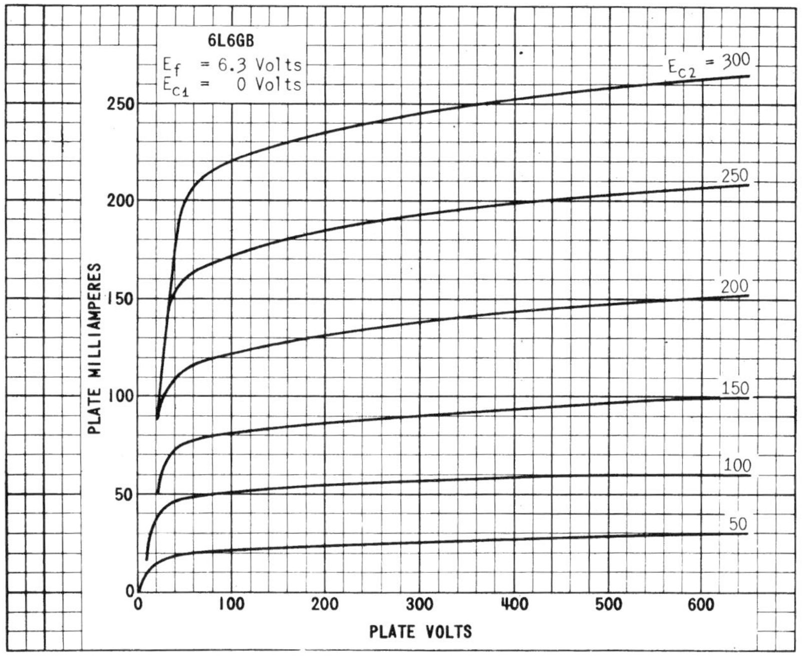

Tube/spec sheets

TUNG-SOL 6L6G, 6L6

CONTINUED FROM PRECEDING PAGE

CLASS A₁ PUSH-PULL AMPLIFIER – PENTODE CONNECTION

VALUES ARE FOR TWO TUBES

HEATER VOLTAGE	6.3	6.3	VOLTS
HEATER CURRENT	0.9	0.9	AMP.
PLATE VOLTAGE	250	270	VOLTS
GRID #2 VOLTAGE	250	270	VOLTS
GRID #1 VOLTAGE	−16	−17.5	VOLTS
PEAK AF GRID TO GRID VOLTAGE	32	35	VOLTS
TRANSCONDUCTANCE (EACH TUBE)	5 500	5 700	µMHOS
PLATE RESISTANCE (EACH TUBE)	24 500	23 500	OHMS
ZERO SIGNAL PLATE CURRENT	120	134	MA.
MAXIMUM SIGNAL PLATE CURRENT	140	155	MA.
ZERO SIGNAL GRID #2 CURRENT	10	11	MA.
MAXIMUM SIGNAL GRID #2 CURRENT	16	17	MA.
LOAD RESISTANCE	5 000	5 000	OHMS
POWER OUTPUT	14.5	17.5	WATTS
TOTAL HARMONIC DISTORTION	2	2	PERCENT

CLASS AB₁ PUSH-PULL AMPLIFIER – PENTODE CONNECTION

VALUES ARE FOR TWO TUBES

HEATER VOLTAGE	6.3	6.3	VOLTS
HEATER CURRENT	0.9	0.9	AMP.
PLATE VOLTAGE	360	360	VOLTS
GRID #2 VOLTAGE	270	270	VOLTS
GRID #1 VOLTAGE	−22.5	−22.5	VOLTS
PEAK AF GRID TO GRID VOLTAGE	45	45	VOLTS
ZERO SIGNAL PLATE CURRENT	88	88	MA.
MAXIMUM SIGNAL PLATE CURRENT	132	140	MA.
ZERO SIGNAL GRID #2 CURRENT	5	5	MA.
MAXIMUM SIGNAL GRID #2 CURRENT	15	11	MA.
LOAD RESISTANCE	6 600	3 800	OHMS
POWER OUTPUT	26.5	18	WATTS
TOTAL HARMONIC DISTORTION	2	2	PERCENT

CLASS AB₂ PUSH-PULL AMPLIFIER – PENTODE CONNECTION

VALUES ARE FOR TWO TUBES

HEATER VOLTAGE	6.3	6.3	VOLTS
HEATER CURRENT	0.9	0.9	AMP.
PLATE VOLTAGE	360	360	VOLTS
GRID #2 VOLTAGE	225	270	VOLTS
GRID #1 VOLTAGE	−18	−22.5	VOLTS
PEAK AF GRID TO GRID VOLTAGE	52	72	VOLTS
ZERO SIGNAL PLATE CURRENT	78	88	MA.
MAXIMUM SIGNAL PLATE CURRENT	142	205	MA.
ZERO SIGNAL GRID #2 CURRENT	3.5	5	MA.
MAXIMUM SIGNAL GRID #2 CURRENT	11	16	MA.
LOAD RESISTANCE	6 000	3 800	OHMS
POWER OUTPUT	31	47	WATTS
TOTAL HARMONIC DISTORTION	2	2	PERCENT

TUNG-SOL 6L6GB

CONTINUED FROM PRECEDING PAGE

TYPICAL OPERATING CONDITIONS AND CHARACTERISTICS

CLASS A₁ AMPLIFIER – PENTODE CONNECTION

PLATE VOLTAGE	250	300	350	VOLTS
GRID #2 VOLTAGE	250	200	250	VOLTS
GRID #1 VOLTAGE	−14	−12.5	−18	VOLTS
PEAK AF SIGNAL VOLTAGE	14	12.5	18	VOLTS
TRANSCONDUCTANCE	6 000	5 300	5 200	µMHOS
PLATE RESISTANCE	22 500	35 000	33 000	OHMS
ZERO SIGNAL PLATE CURRENT	72	48	54	MA.
MAXIMUM SIGNAL PLATE CURRENT	79	55	66	MA.
ZERO SIGNAL GRID #2 CURRENT	5	2.5	2.5	MA.
MAXIMUM SIGNAL GRID #2 CURRENT	7.3	4.7	7	MA.
LOAD RESISTANCE	2 500	4 500	4 200	OHMS
POWER OUTPUT	6.5	6.5	10.8	WATTS
TOTAL HARMONIC DISTORTION	10	11	15	PERCENT

CLASS A₁ AMPLIFIER – TRIODE CONNECTION[A]

PLATE VOLTAGE	250	VOLTS
GRID #1 VOLTAGE	−20	VOLTS
PEAK AF SIGNAL VOLTAGE	20	VOLTS
TRANSCONDUCTANCE	4 700	
PLATE RESISTANCE	1 700	OHMS
AMPLIFICATION FACTOR	8	
ZERO SIGNAL PLATE CURRENT	40	MA.
MAXIMUM SIGNAL PLATE CURRENT	44	MA.
LOAD RESISTANCE	5 000	OHMS
POWER OUTPUT	1.4	WATTS
TOTAL HARMONIC DISTORTION	5	PERCENT

[A] GRID #2 CONNECTED TO PLATE.

CONTINUED ON FOLLOWING PAGE

Tube cross reference guide

Tubes are often known by several correct model names. These names stem from different industries ordering tubes with additional quality standards or slight changes in the tubes' format, ie: "Glass Container" will be denoted as GC following the model number.

Another source of confusion is the European nomenclature. There is a corresponding name in Europe for nearly all US-made tubes and vice-versa. We've listed only the common tube models for instrument amplifiers found in our music industry. Occasionally, an amp manufacturer will list the same tube by two names, the first by a common name and the second with an industrial number. The industrial numbers will always be four digits and usually denote a superior quality (although these days, it rarely applies since the tube industry's quality has fallen off sharply).

The most common example of this is the dual channel Fender amps with reverb and tremolo. These amps will call for the 7025 and the 12AX7 in the same amp in different locations. In this case, the manufacturer is asking for a higher quality tube in the most critical gain stages of the two volume controls and the reverb recovery control section while the tremolo section can use any 12AX7, since this stage of the amp does not contribute to the audio quality of the signal.

What follows is our cross reference guide only for tubes known by several names.

USA	INDUSTRIAL	EUROPEAN
7025 = ECC83/12AX7	5871 = CT-6V6	ECC81 = GT-12AT7
12AT7 = ECC81	5881 = GT-6L6	ECC82 = GT-12AU7
12AU7 = ECC82	6087 = GT-5Y3	ECC83 = GT-7025/12AX7
12AX7 = 7025/ECC83	6135 = GT-6C4	ECC88 = CT-6DJ8
12AY7 = 6072	6136 = GT-6AU6	EL34 = GT-EL34/6CA7
12DW7 = 7247	6189 = GT-12AU7	EL37 = GT-6L6
5AR4 = GZ34	6201 = GT-12AT7	EL84 = GT-EL84
5U4 = GZ32	6626 = OA2	EZ81 = GT-6CA4
5Y3 = GZ30	6663 = 6AL5	GZ31 = GT-5U4
6CA4 = EZ81	6679 = GT-12AT7	GZ32 = GT-5U4
6CG7 = 6FQ7	6680 = GT-12AU7	GZ34 = GT-5AR4
6K11 = 6Q11/6C10	6681 = GT-12AX7	GZ37 = GT-5AR4
6U10 = 6AC10	6853 = GT-5Y3	KT66 = GT-6L6
6CA7 = EL34/KT77	7025 = GT-12AX7	KT77 = GT-EL34/6CA7
6L6 = 5881/KT66	7184 = GT-6V6	KT88 = GT-6550
6V6 = 7408	7247 = GT-12DW7	U50 = GT-5Y3
6550a = KT88/6550	7408 = GT-6V6	U51 = GT-5U4
ECC83 = 7025/12AX7	7543 = GT-6AU6	U52 = GT-5U4
EL34 = 6CA7/KT77	7729 = GT-12AX7	U54 = GT-5AR4
EL84 = 6BQ5	7730 = GT-12AU7	U77 = GT-5AR4

European Tube Nomenclature
(courtesy of Per Arvidsson, Tyreso, Sweden)

The first letter states the heater voltage or current:
A = 4 Volts
E = 6.3 Volts
D = 1.4 Volts-battery
G = 5 Volts
H = 150 m.A. series
K = 2 Volts - battery
P = 300 m.A. series
U = 100 m.A. series
V = 50 m.A. series
Q = Tetrode for transmitting power amp

The second letter states what type of tube it is:
A = Diode
B = Double diode
C = Triode
D = Tetrode
F = Pentode for small signals
H = Heptode
L = Pentode for power amps
M = Tuning indicator
Y = Rectifier
Z = Rectifier

The third letter states if the tube has a double or more functions. This letter (when present) has the same meaning as the second. Therefore, ECC83 means 6.3 V, Triode, Triode or EBF80 means 6.3 V, Double Diode, Pentode. The following numbers are just numbers! Sometimes the letters and numbers are reversed to indicate a special version, for example the E83CC means a "life long" version of the ECC83. There are, however, some tubes that do not follow these rules (ex: the British KT66 has a 6.3 V heater voltage, and not a 2 V battery voltage as the K should indicate. Rather than conforming to the standard code, the "KT" in this case stands for "kinkless tetrode").

Tube amp companies

Companies, like bands, come and go – even the most famous and influential. This is especially true of vacuum tube amplifier companies, and we're especially sensitive to it. As recently as the early '70s, essentially all musical instrument amps were tube type. By the mid '80s, about 50 per cent of guitar and bass amps were solid state (transistor) technology. Many companies didn't survive the transition. Even when the brand survived, the ownership frequently changed.

Groove Tubes maintains an ongoing effort to ease the musician's burden in this regard. We collect all the schematics we can find for vacuum tube designs, no matter how obscure, and publish the more popular in our Reference Guide. This edition has a much expanded offering, but if you need something you don't see here, call us; we may have it. Equally important, if you have a schematic which isn't in here, we'd sure like to have a copy of it. We don't pay any money for mem, but we keep improving the collection for everyone's benefit.

We have listed here the most up-to-date info on how to contact these companies and/or brands. If you have trouble locating them, a thorough web search can usually turn up up-to-date info these days.

Tube amp companies

ACOUSTIC AMPLIFICATION
Trademark recently bought by Samick Corp.,
18521 Railroad Street
City of Industry, CA 91748
626-964-4700
www.acoustic.mu
For info on out of production amps, some service available, contact:
ACOUSTIC – Doug Forbes
263 Chemeketa Street
Salem, Oregon 97301
(503) 315-8657
TOLL FREE 800-996-3684

ADA
ADA Signal Processors, Inc.
7303D Edgewater Drive
Oakland, CA 94621
(415) 632-1323
FAX: (415) 632-9358

AGUILAR
599 Broadway 7th Floor
New York, NY 10012
(415) 831-8200
info@aguilaramp.com
www.aguilaramp.com
Makers of mainly rack-mounted, high-powered tube amps for bass guitar.

AIKEN
Greenwood, SC, USA
(864) 993-8383
info@aikenamps.com
www.aikenamps.com
Highly respected custom-order builder of high-end, "updated vintage" style tube amps.

ALEMBIC
3005 Wiljan Court, Bldg. 4
Santa Rosa, CA 95407
(707)523-2611
FAX: (707) 523-2935
alembic@alembic.com
www.alembic.com
Makers of custom-order active guitars and basses and a limited range of amps.

ALESSANDRO
Alessandro High-End Products
Huntingdon Valley, PA
(215) 355-6424
hounddogcorp@msn.com
www.alessandro-products.com
See entry in main section.

ALLEN
1325 Richwood Road
Walton, KY 41094
(859) 485-6423
tonesavor@fuse.net
www.allenamps.com
Respected smaller maker of hand-built "blackface" style Fender tube amps and tube amp kits.

AMPEG
St. Louis Music Supply
1400 Ferguson Ave.
St. Louis, MO 63133
(314)727-4512
FAX: (314) 727-8929
www.ampeg.com
See entry in main section.

ASHDOWN ENGINEERING
Distributed in the USA by HHB Communications
1410 Centinela Ave.
Los Angeles, CA 90025-2501
(310) 319-1111
sales@hhbusa.com
www.ashdownmusic.co.uk
See entry in main section.

BAD CAT
2621 Green River Road Suite #105
Pmb #406
Corona, CA 92882
(909) 808-8651
badcatamps@earthlink.net
www.badcatamps.com
See entry in main section.

BEDROCK
1602C Concord Street
Framingham, MA 01701
(508) 877-4055

BENSON
Out of production, no info.

BLOCKHEAD
(845) 528-2229
blockheadamp@aol.com
www.blockheadamps.com
New York-based manufacturer of well respected, high-grade vintage Marshall reproduction style tube amps.

BOGNER
11411 Vanowen Street
North Hollywood, CA 91605
(818) 765-8929
www.bogneramplification.com
See entry in main section.

BRUNO
Ultra Sound Music Inc.
251 West 30th. Street - 6th. Floor
New York, NY 10001
ultramusic@nyc.rr.com
www.brunoamps.com
See entry in main section.

BUDDA
37 Joseph Court
San Rafael, CA 94903
(415) 492-1935
buddatone@budda.com
www.budda.com
See entry in main section.

CALLAHAM VINTAGE AMPLIFIERS
114 Tudor Drive
Winchester, VA 22603
(540) 955-0294
callaham@callahamguitars.com
www.callahamguitars.com
Makers of hand-built vintage reproduction tube amps.

CARR
433 W. Salisbury Street
Pittsboro, NC 27312
(919) 545-0747
info@carramps.com
www.carramps.com
Respected smaller builder of distinctive, American and British inspired hand-built tube amps.

CARVIN
12340 World Trade Drive
San Diego, CA 92128
(800) 854-2235
Long-time Californian supplier of quality, affordable direct-order tube and solid state guitar and bass amplifiers and PA systems.

CLARK
428 Center Street
West Columbia, SC 29169
(803) 467-6784
sales@clarkamplification.com
www.clarkamplification.com
See entry in main section.

CONN
C.G. Conn, Ltd.
1000 Industrial Parkway
Elkhart, Indiana 46516
(219) 295-0079
FAX: (219) 295-8613

CORNELL (DC Developments)
60 Eastcote Grove
Southend-on-Sea, Essex, SS2 4QB, UK
011-44 (0) 1702-610964
www.dc-developments.com
See entry in main section.

CORNFORD AMPLIFICATION
48 Joseph Wilson Industrial Estate
Millstrood Road
Whitstable, Kent CT5 3PS, UK
011-44 (0) 1227-280000
info@cornfordamps.com
www.cornfordamps.com
See entry in main section.

CRATE
A Division of St. Louis Music
1400 Ferguson Avenue
St Louis, MO 63133
(314) 569-0141
www.crateamps.com
Builder of popular, affordable tube and solid state amps.

DEMETER
15730 Stagg Street
Van Nuys, CA 91406
(818) 994-7658
info@demeteramps.com
www.demeteramps.com
Makers of modern-styled, multi-channel tube amps.

DIEZEL
Distributed by Salwender International
1140 North Lemon Street "M"
Orange, CA 92867
(714) 538-1285
www.salwender.com
German builder of powerful and expensive modern rock tube amps.

DR Z
17011 Broadway Avenue
Maple Heights, OH 44137
(216) 475-1444
drz@drzamps.com
www.drzamps.com
See entry in main section.

DUMBLE
California
See entry in main section.

DUNCAN, SEYMOUR
Seymour Duncan Corp.
601 Pine St.
Santa Barbara, CA 93117
(805) 964-9610
FAX: (805) 964-9749
Well-know manufacturer of replacement pickups, and a small range of guitar amplifiers.

EGNATOR
196 Oakland Suite D-209
Pontiac, MI 48342
(248) 253-7300
www.egnator.com

Tube amp companies

Makers of high-powered modern rock tube amps.

ENGL
Distributed in USA by JI Concept US Inc.
6440 Antigua Place
West Hills, CA 91307
(818) 610-2892
www.engl-amps.com
See entry in main section.

FARGEN
(916) 971-4992
www.fargenamps.com
Respected maker of small-run, hand-built tube amps.

FENDER
Fender Musical Instruments Corp
8860 E Chaparral Road Suite 100
Scottsdale, AZ 85250
(480) 596-9690
www.fender.com
See entry in main section.

GARNET
info@garnetamps.com
www.garnetamps.com
See entry in main section.

GIBSON
1818 Elm Hill Pike
Nashville, TN 37210
(615) 871-4500
FAX: (615) 871-4070
relations@gibson.com
www.gibson.com
See entry in main section.

GROOVE TUBES, GT ELECTRONICS,
GROOVE TUBES AUDIO
1543 Truman Street
San Fernando, CA 91340
(818) 361-4500
FAX: (818) 365-9884
info@groovetubes.com
www.groovetubes.com
See entry in main section.

HIWATT (Music Ground, UK)
51 Hallgate
Doncaster DN1 3PB, UK
011-44 (0) 1302-366803
musicgnd@aol.com
www.hiwatt.co.uk
See entry in main section.

HIWATT (Fernandes)
8163 Lankersheim Blvd
North Holllywood, CA 91605
(800) 318-8599
info@hiwatt.com
www.hiwatt.com

HAMMOND
Hammond Suzuki USA
733 Annoreno Drive
Addison, IL 60101
(630) 543-0277
hammondsuzuki@worldnet.att.net
See entry in main section.

HOFFMAN
190 Lakeland Drive
Pisgah, NC 28768
hoffmanamps@hoffmanamps.com
www.hoffmanamps.com
Respected small hand-builder of tube guitar amps, mainly now involved in amp parts/building supplies business.

HOLLAND
500 Wilson Pike Circle
Suite 204
Brentwood, TN 37027
(615) 377-4913
mike@hollandamps.com
www.hollandamps.com
Maker of quality, hand-built production and custom order tube guitar amps.

HUGHES & KETTNER
1848 S. Elmhurst Road
Mt Prospect, IL 60056
(800) 452-6771
info@hughes-and-kettner.com
www.hughes-and-kettner.com
See entry in main section.

K & M (Two Rock)
7880 Old Redwood Highway
Cotati, California 94931
(707) 664-0267
FAX (707) 664-0262
www.two-rock.com
See entry in main section.

KASHA
Kasha Amplifiers, Inc.
19441 Business Center Dr. #137
Northridge, CA 91324
(818) 772-4912
FAX: (818) 893-4331

JIM KELLEY AMPLIFIERS
Out of production. Contact an ex-Kelley
employee Wes Patterson
@ Pro Music Exchange
937 N. Main St.
Orange, CA 92667
(714) 744-9762
See entry in main section.

KENDRICK
110 West Pflugerville Loop
Pflugerville, TX 78660
(512) 990-5486
FAX (512) 990-0548
Maker of hand-built Fender reproduction style amps and other American-style tube guitar amps.

KITTY HAWK
Kitty Hawk Elektroakusdc GmbH
PO Box 65
5241 Gebhardshain, W. Germany
02662-6863
FAX: 02662-4463
German tube amp builder.

KMD AMPLIFIERS
Kaman Music Corporation
PO Box 507
Bloomfield, CT 06002
(203) 243-7941
FAX: (203) 243-7102

KOCH
Dist by Eden Electronics
Arnhemseweg 152
Amersfoort, 3817 CL
The Netherlands
011-31-33-4634533
Dutch maker of modern-style rock amps.

KOMET
1865 Dallas Drive
Baton Rouge, LA 70806-1454,
(225) 926-1976
kometamp@bellsouth.net
See entry in main section, under Trainwreck.

LANEY
PO Box 2632
Mount Pleasant, SC 29465-2632
(888) 860-1668
sales@laneyusa.com
www.laneyusa.com
See entry in main section.

LEGEND
Out of production, no info.

LESLIE
Hammond Suzuki USA
733 Annoreno Drive
Addison, IL 60101
(630) 543-0277
hammondsuzuki@worldnet.att.net
See entry in main section.

M&G
www.mandgamps.com
Respected maker of hand-built, smaller production-run and custom-order vintage Marshall-style reproduction amps.

DEAN MARKLEY
PO Box 507
Bloomfield, CT 06002-0507
(860) 247-7941
info@deanmarkley.com
www.deanmarkley.com
Guitar components company also known for offering a small range of guitar amps.

MARSHALL
Dist by Korg USA
316 South Service Road
New York, NY 11747-3201
(800) 872-5674
UK Address:
Denbigh Road
Bletchley
Milton Keynes, MK1 1DQ, UK
011-44 (0) 1908-375411
www.marshallamps.com
See entry in main section.

MATCHLESS
2105 Pontius Avenue
Los Angeles, CA 90025
(310) 444-1933
sales@matchlessamplifiers.com
www.matchlessamplifiers.com
See entry in main section.

MESA/BOOGIE
1317 Ross Street
Petaluma, CA 94954
(707) 778-6565
FAX: (707) 765-1503
info@mesaboogie.com
www.mesaboogie.com
See entry in main section.

METALHEAD
Metalhead Electronics
5707 Cahuenga Blvd
North Hollywood, CA 91601
(818) 980-1975
FAX: (818) 985-1624

MUSITECH
4040 Blackfoot Tr. Bay 110
Calgary, Alberta
T2G 4E6 Canada

MUSIC MAN
Ernie Ball, Inc.
151 Suburban Rd. Box 4117
San Luis Obispo, CA 93401
(805) 544-7726
FAX: (805) 544-7275
See entry in main section.

ORANGE
USA: PO Box 421849
Atlanta, GA 30342

Tube amp companies

(404) 303-8196
info@orangeusa.com

UK: 28 Denmark Street
London WC2H 8NJ, UK
011-44 (0) 207-240-8392
info@orangeamps.com
www.orangeamps.com
See entry in main section.

PEAVEY
Peavey Electronics
711 "A" Street
Meridian, MS 39301
(601) 483-5365
www.peavey.com
See entry in main section.

POLYTONE
Polytone Musical Inst. Inc.
6865 Vineland Ave.
North Hollywood, CA 91605
(818) 760-2300
Maker of widely used jazz combos.

RANDALL
A division of Washburn
444 East Courtland Street
Mundelein, IL 60060
(847) 949-0444
randall@randallamplifiers.com
www.randallamplifiers.com
Maker of a variety of tube and solid state amps, including large rock stacks.

REVEREND
27300 Gloede Unit D
Warren, MI 48088
(586) 775-1025
www.reverenddirect.com
Maker of modern-retro-design guitars and amps.

RISSON
Out of production, no info.

RIVERA
13310 Ralston Avenue
Sylmar, CA 91342
(818) 833-7066
rivera@rivera.com
www.rivera.com
See entry in main section.

ROLAND
5100 S. Eastern Avenue
PO Box 910921
Los Angeles, CA 90091-0921
(323) 890-3700
www.rolandus.com
Large music industry maker and distrubutor, though mostly know for their solid state guitar amps.

SAVAGE
12500 Chowen Avenue S
Suite 112
Burnsville, MN 55337
(952) 894-1536
savrok@prodigy.net
www.savageaudio.com
Respected maker of hand-built tube amps to mainly repro vintage British designs.

SOLDANO
1537 NW Ballard Way
Seattle, WA 98107
(206) 781-4636
www.soldano.com
See entry in main section.

SWR
9130 Flenoaks Blvd
Sun Valley, CA 91532-2611
(818) 253-4797
support@swrsound.com
www.swrsound.com
Maker of large, mostly solid state bass amps and acoustic guitar amps.

THD
4816 15th Avenue NW
Seattle, WA 98107-4717
(206) 781-5508
info@thdelectronics.com
www.thdelectronics.com
See entry in main section.

TOP HAT
920 E Orangethorpe Avenue
Suite B
Anaheim, CA 92801-1127
(714) 447-6700
TopHatAmps@aol.com
www.tophatamps.com
See entry in main section.

TRACE ELLIOT
A division of Gibson
1818 Elm Hill Pike
Nashville, TN 37210
(615) 871-4500
FAX: (615) 871-4070
relations@gibson.com
www.gibson.com
See entry in main section.

TRAINWRECK
Box 261
Colonia New Jersey 07067
(732) 381-5126
See entry in main section.

TRAYNOR
see Yorkville See entry in main section.

VERO
22436 S River Road
Joliet, IL 60431
(815) 467-7093
inro@veroamps.com
www.veroamps.com
Builder of respected modern-retro tube guitar amps with a wide range of custom-order options.

VHT
9130 Glenoaks Blvd
Sun Valley, CA 91352-2611
(818) 253-4848
info@vhtamp.com
www.vhtamp.com
See entry in main section.

VICTORIA
1504 Newman Court
Naperville, IL 60564
(630) 369-3527
www.victoriaamp.com
See entry in main section.

VOX
Dist by Korg USA
316 S Service Road
Melville, NY 11747
(516) 333-9100
www.voxamps.co.uk
See entry in main section.

WATKINS/WEM
"Southview" 3
Biggin Hill
London SE19 3HT, UK
011-44 (0) 208-679-5575
wem.watkins@btinternet.com
www.wemwatkins.co.uk
See entry in main section.

WIZARD
www.sneddon.net/ricwiz
See entry in main section.

YORKVILLE
4625 Witmer Industrial Estate
Niagra Falls, NY 14305
(716) 297-2920
info@yorkville.com
www.yorkville.com
Long-time Canadian builder of both quality and budget, solid state and tube guitar amps, including the Traynor line.

Glossary

GLOSSARY OF COMMON AMPLIFIER TERMS

By Randall Aiken

There are many terms commonly seen in literature describing guitar amplifier circuitry which can be somewhat confusing to a person with no prior background in electronics. This section will attempt to shed some light on these mysteries, and provide descriptions of common electronic terms, components, and circuitry in somewhat easy-to-understand form, although some terms represent concepts that are difficult to explain in simple language, so you may have to do some additional side reading to fully comprehend them. The material is presented in an alphabetical, glossary-style form, so there is some overlap in the definitions.

A – the symbol for amps, or amperes, which is a unit of current flow. Common prefixes are "m", for mA (10-3 amps), and "u", for uA (10-6 amps).

AC – Alternating Current. This is electric current that periodically changes the direction in which it flows. The most common form of an alternating current supply is the sinusoidal current that comes out of a wall outlet. It has no positive or negative terminals, because AC has no polarity, other than an instantaneous polarity that changes at a rate equal to the frequency of the current. Common household AC current is supplied at a frequency of 60Hz in the United States and some other countries, and 50Hz in other places in the world, most notably, England. "Hz" stands for "Hertz," which is the name of the unit for frequency, and means

Glossary

"cycles per second," indicating how many cycles, or changes from positive to negative, the AC waveform goes through each second. In some older literature, you may see the term "CPS," which stands for "cycles per second," used in place of "Hz." Alternating current does not have to be sinusoidal in shape; the square wave of a distorted guitar amplifier output is also AC, because it changes polarity periodically.

Active – a component that needs a power source to function, as opposed to a passive component. Examples of active components are tubes, transistors, opamps, etc. Also commonly used to refer to guitar pickups that have built-in preamps, which require batteries to operate.

Admittance – the reciprocal of impedance. $Y = 1/Z = G + jB$, where G = conductance, and B = susceptance.. The unit of admittance is the "mho," same as conductance.

Ali – the name given to the Marshall amplifiers that came after the Plexi's and had aluminum front panels.

Alnico – an alloy of aluminum, nickel, and cobalt which was commonly used in vintage speakers. It was replaced by cheaper and stronger ceramic (strontium ferrite) materials, but is making a comeback in "modern vintage" style speakers, such as the WeberVST Blue Dog and the Celestion Alnico Blue, among others.

Amplifier – the other half of rock'n'roll (thanks to Ritchie Fliegler for that one).

Anode – the "current collecting" element of an electron tube, also called the "plate." The anode usually has a large positive voltage connected to it in order to attract the negatively-charged electrons from the cathode element of the tube. If you look at a tube, this is the large greyish metal piece that encloses most of the other elements.

Attenuator – (a) a network that is used to reduce the amplitude of a signal. Typically, this is accomplished with two resistors, one in series with the signal and another from the output of the first resistor to ground. This attenuates the signal by an amount dependent upon the ratio of the resistor values.
(b) a device used to reduce the volume of an amplifier. It goes between the amplifier and the speakers, allowing a non-master volume amplifier to be cranked up to full power without being overly loud, in order to get the desired overdrive tone from the amplifier.

B – the symbol for susceptance, also the symbol for magnetic flux density.

B+ – the high voltage supply in a tube amplifier. The name is a holdover from the old days of battery-powered radios, which had an "A" supply for the filaments, a "B" supply for the high voltage, a "C" supply for the bias, and a "D" supply for the screen grids, if a separate supply was used. The conventions held when radios switched over to rectified AC supplies.

Back bias – a method of obtaining a negative bias voltage by means of a resistor or zener diode in the center tap of a full-wave rectifier circuit. The current in the center tap flows in the same direction for both half-cycles, so the voltage drop is the same for both. This full-wave-rectified negative voltage can be filtered and used as a negative bias supply. The downside is that all the plate current of the output stage flows through the back bias circuit, so it can be impractical for higher-powered amplifiers. Also, the resistive drop method should only be used for true class A amplifiers, because there can be a large difference between the idle and full-power current draw of a class AB or class B amplifier. The zener method is much more suitable, and in fact, creates a regulated bias voltage that is relatively independent of the current draw of the amplifier, provided it is above the minimum current necessary to keep the zener in the normal reverse breakdown region.

Bias – the amount of negative voltage applied to the grid of a tube with respect to the cathode, or the amount of idle current flowing in the tube when no AC signal is present on the grid pin.

Biasing – the term commonly used for the practice of setting the idle current in an output tube. Preamp tubes are biased as well, but they are biased only during the initial design of the amplifier and use what is known as "cathode biasing", and don't require rebiasing as part of general amplifier maintenance.

Blackface – the term given to older Fenders which had a black metal control panel. This era of Fender amps transitioned into the "silverface" amps, which had a silver metal control panel. The transition occurred at the time CBS bought the company, and some "improvements" were made to the circuitry of most of the amplifiers. These "improvements" are generally regarded as detrimental to the tone of the amplifier, which led to a practice known as "blackfacing" a Fender amp, which means converting the circuit back to the pre-CBS schematic.

Bridge rectifier – a set of four rectifiers arranged in a "square" or "diamond" shape (depending on how you look at it). The four diodes allow full-wave rectification without the need for a center-tap on the transformer.

Bypass cap – a capacitor that is connected from the power supply to ground. It "bypasses" the AC signals to ground, while passing the DC supply through. This is used to make the DC supply rail "clean," or free from AC noise. Usually bypass caps are relatively small, on the order of 0.1uF or so. Larger caps connected in the same manner are usually called "filter caps." This term is also used to refer to a capacitor connected across the cathode resistor on a tube. It bypasses the AC signal to ground without affecting the DC bias of the tube. This increases the gain of the amplifier stage. This capacitor can also be used to tailor the frequency response of the stage.

C – the symbol for capacitance

Cap – short for capacitor.

Capacitor – a device consisting of two parallel plates separated by an insulator, called the "dielectric." The capacitance is proportional to the area of the plates, and inversely proportional to the distance between them. Capacitors are used to block DC while passing AC. They are frequency-dependent devices, which means that their capacitive reactance, or "effective resistance" to AC increases as the frequency gets lower. This makes capacitors useful for tone controls, where different frequency bands must be passed, or for bypassing AC signals to ground while passing DC through for filtering purposes.

Capacitance – the "size" of a capacitor. The unit of capacitance is the Farad, but a one Farad capacitor would be quite large, indeed! The most common capacitors are sized in microfarads (μF, or mfd in very old texts – 10-6 farads), nanofarads (nF – 10-9 farads), and picofarads (pF – 10-12 farads).

Cathode – the "current generating" element of an electron tube. The heater heats the cathode to a very high temperature, causing it to emit electrons, which are then collected by the anode, or plate, which has a high positive voltage, which attracts the negatively charged electrons from the cathode.

Cathode biasing – a method of biasing a tube where the bias is generated by the voltage drop across a resistor in the cathode. The grid is referred to ground through a resistor, and the current flow through the cathode resistor produces a positive cathode voltage with respect to the grid, which is effectively the same as making the grid negative with respect to the cathode.

Chassis – the metal box that encloses the amplifier parts. It is usually made of steel, but occasionally aluminum is used. The transformers and choke are usually mounted on top, while the passive components are usually mounted inside the chassis.

THE TUBE AMP BOOK 183

Glossary

Choke – another term used for an inductor, most commonly an inductor used as a power supply filter.

Class A – an amplifier operating with the grid bias adjusted so plate current flows for the entire 360 degrees of the input waveform, by biasing the tube halfway between cutoff and saturation, in the most linear portion of the operating curves. The distortion is lowest in class A operation, but the efficiency is also very low. With the exception of single-ended amplifiers, the amplifiers most manufacturers call "class A" are actually cathode-biased class AB amplifiers.

Class A1 – class A operation where grid current does not flow for any portion of the input cycle.

Class A2 – class A operation where grid current flows for some portion of the input cycle.

Class AB – an amplifier operating with the grid bias adjusted so plate current flows for greater than 180 degrees, but less than 360 degrees of the input waveform, by biasing the tube above cutoff, but below the point required for class A operation. The distortion is higher at low signal levels than true class A, but the efficiency is higher, although not as high as class B, allowing more output power than class A for a given plate dissipation.

Class AB1 – class AB operation where grid current does not flow for any portion of the input cycle.

Class AB2 – class AB operation where grid current flows for some portion of the input cycle.

Class B – an amplifier operating with the grid bias adjusted so plate current flows for right at 180 degrees, by biasing the tube right at cutoff. The distortion is higher than class A or class AB, and there is usually a large amount of crossover distortion, but the efficiency is higher than class AB, allowing more output power for a given plate dissipation.

Class B1 – class B operation where grid current does not flow for any portion of the input cycle.

Class B2 – class B operation where grid current flows for some portion of the input cycle.

Combo – a guitar amplifier that has a built-in speaker.

Common cathode – the "standard" tube circuit where the cathode is connected to the "common" point on the circuit, usually ground, and usually through a resistor, which is often bypassed with a capacitor, placing it at "AC" ground potential.

Common grid – a tube stage which has the grid connected to the "common" point on the circuit, usually ground. This doesn't have to be a physical DC connection, it can be an AC ground, ie grounded through a capacitor.

Common plate – a tube stage which has the plate connected to the "common" point on the circuit, usually ground This doesn't have to be a physical DC connection, it can be an AC ground, ie grounded through a capacitor. This is the most often seen method of making a common plate stage, where the plate is connected directly to the power supply (the AC ground connection is through the power supply capacitors, which are essentially a short to ground for AC signals). This stage is commonly called a "cathode follower."

Control grid – a wire mesh element located between the cathode and plate of an electron tube which controls the flow of electrons between the two elements. The control grid draws no current, and as such, presents a high impedance to the driving circuit. Voltage variations on the control grid, with respect to the cathode, cause variations in plate current, which is the basis of amplification within the tube.

Coulomb – a unit of electron charge.

Coupling capacitors – capacitors which are used between stages in a guitar amplifier. They block the DC plate voltage of the previous stage, while passing the AC guitar signal on through.

Concertina phase splitter – the name given to the single-tube phase inverter in which the in-phase signal is taken off the cathode and the out-of-phase signal is taken off the plate, with equal-value plate and cathode resistors. This phase splitter configuration has excellent balance, but only unity gain. Also called a "split-load" phase inverter.

Crossover distortion – Crossover distortion is the term given to a type of distortion that occurs in push-pull class AB or class B amplifiers. It happens during the time that one side of the output stage shuts off, and the other turns on. Depending upon the bias point, there is a small amount of time where both tubes are in very non-linear portions of their operating curves, or even cut off entirely, and this "kink" in the transfer curves results in a distortion, or notch, at the zero crossing point of the reconstructed waveform.

Current – The term given to electron flow. The unit of current is the "amp," or "ampere," and indicates a current flow of one coulomb per second.

Cutoff frequency – The "corner point" of a filter, usually the point where the response is down –3dB compared to the midband signal level.

dB – decibels, the standard unit of measure of volume, or "loudness."

DC – Direct Current. This is electric current that flows in one direction only. The most common form of a direct current supply is a battery. The battery will have positive and negative terminals. If a circuit is connected between the two terminals, a current will flow in one direction only. The actual electron flow is from negative to positive, but "conventional" current flow is indicated as a current flow from positive to negative. This has been a source of confusion since the early days of electricity, and you will see both conventional and electron flow used in literature.

Decoupling – the process of isolating one stage of an amplifier from another. This is usually done by adding a resistor in series with the power supply to a gain stage and a large value electrolytic capacitor from the supply to ground after the resistor. Decoupling prevents oscillations and other noises that may occur due to unwanted feedback through the power supply connections. It also provides further filtering of the power supply to reduce ripple, producing a cleaner DC supply for the low-level preamp stages.

Decoupling capacitor – the large electrolytic capacitor used to filter the power supply after the decoupling resistor.

Decoupling resistor – the series resistor used to isolate one stage of an amplifier from another.

Dielectric – the insulating material used in a capacitor. Typical dielectric types used in amplifiers are: polystyrene, polypropylene, polycarbonate, polyester, and ceramic.

Diode – a two-element device which passes a signal in one direction only. They are used most commonly to convert AC to DC, because they pass the positive part of the wave, and block the negative part of the AC signal, or, if they are reversed, they pass only the negative part and not the positive part. This allows them to be used to generate a positive or negative DC supply. There are both solid-state and tube diodes. Since a diode will pass current in only one direction, they can also be used to "clip" the top or bottom part of a signal. Diodes are also commonly called "rectifiers" because they rectify the AC voltage, however, the term "rectifier" is usually reserved for diodes used in the power supply section of an amplifier, while "diode" is generally used in small signal, or low power applications, such as clippers.

Direct box – (also DI box, for "direct injection") a device that allows a guitar or amplifier to be connected directly into a mixing board without the use of a microphone. There are two basic

Glossary

types of direct boxes, those that go between the guitar and the amp, feeding a clean guitar signal to the board, and those that go between the amplifier output and the speakers, feeding the amp signal to the board. The latter usually contain some type of frequency compensation, or "speaker emulation" to give a sound similar to a miked speaker.

E – the symbol for electromotive force, or voltage

Effects loop – a circuit that allows insertion of external effects devices in the signal path of an amplifier. Noise performance is usually improved by using the effects loop rather than putting the effects in series with the guitar input.

Electron Tube – (also tube for short, or valve in the UK) the device used to make guitar amplifiers sound good! Actually, this is the name given to the amplifying devices in some guitar amplifiers. They consist of a glass tube containing several elements which are brought out to pins on the base of the tube. All of the air inside the tube is evacuated at time of manufacture, which keeps the filament from rapidly burning up.

Feedback – a circuit that allows a portion of the signal from a later stage in an amplifier to be "fed back" to an earlier stage, or within the same stage. Feedback can be voltage or current, negative or positive. Negative voltage feedback decreases gain, and is used to reduce distortion, flatten frequency response, increase input impedance, decrease output impedance. Negative current feedback increases output impedance, and is used in some solid-state amplifiers to obtain a more "tubelike" response. Positive feedback will increase gain, but can make a circuit oscillate if too much is applied. Sometimes a small amount of positive feedback is used to offset the reduction in gain caused by application of negative feedback.

Filament – the heating element in an electron tube, also called the "heater." The filament heats the cathode to a very high temperature, which "boils off" electrons, which are then collected by the plate. The filament can be seen as the glowing element through the holes in the plate of most tubes.

Filter – a circuit which is used to either block or reduce a range of frequencies. There are lowpass filters, which pass frequencies below a certain point, called the "cutoff frequency," highpass filters, which pass frequencies above the cutoff frequency, bandpass filters which pass frequencies above a lower cutoff frequency and below an upper cutoff frequency, bandstop filters, which pass frequencies below a lower cutoff frequency and above an upper cutoff frequency, and allpass filters, which pass all frequencies at the same amplitude, but which have certain phase or delay characteristics.

Filter caps – Filter capacitors. The term used for the large capacitors used to filter out the residual AC ripple in the power supply. The rectifier converts AC to pulsating DC, since it just allows current to flow in one direction. The output of the rectifier is a series of "humps," which must be "smoothed out" to become flat, ripple-free direct current. The filter caps store up the voltage on the positive rise of the pulsating rectified AC waveform, and hold it there while the rectified waveform goes down to zero. This charge, hold, charge, hold, etc. behavior is what smooths out the ripple. In general, the larger the capacitor, the less residual ripple there will be.

Fixed biasing – a method of biasing a tube or output stage by using a negative DC voltage on the grid with respect to the cathode. This name is sometimes confusing, because an amplifier may have a bias adjustment pot to adjust the negative grid voltage, but it is still called "fixed" biasing to differentiate it from "cathode biasing."

Flatness – the peak-to-peak deviation from the nominal voltage in the passband of an amplifier. Flatness is typically measured in dB. For example, if an amplifier has a passband "ripple" of + 0.5dB, it is said to have a "flatness" of + 0.5dB.

Frequency response – a measure of how "wide" a set of frequencies an amplifier will pass. Typically, this is specified as the frequency span between the lower and upper points where the amplitude of the signal has fallen off −3dB, or 0.707 times the midband voltage level. Closely related is the term "flatness," which specifies the deviation from center in the passband.

Full-wave rectifier – a rectifier that conducts on both positive and negative halves of the incoming sinusoidal signal. It produces a "pulsating" DC composed of single-polarity "humps" at twice the incoming AC frequency. The full-wave rectifier requires less filtering than a half-wave rectifier to produce the same degree of ripple in the output DC waveform.

Fuse – a component designed to protect electronic circuits, usually made of a thin piece of metal mounted in a glass or ceramic tube with metal end caps, that is designed to safely burn in two if the current passing through it exceeds the rated maximum.

G – the symbol for conductance.

Global negative feedback – negative feedback that is applied over several amplifier stages, as opposed to local negative feedback, which is applied on one stage only. An example of global negative feedback is the feedback loop in a Marshall or Fender amplifier, where there is a feedback path from the speaker output back to the phase inverter, through an attenuator composed of the "feedback resistor" and a resistor to ground on one side of the phase inverter.

Grid – the "control element" in a vacuum tube. The grid is normally biased negative with respect to the cathode. As the grid is made less negative with respect to the cathode, more current will flow from the cathode to the plate. As the grid is made more negative with respect to the cathode, less current will flow from the cathode to the plate. It usually only takes a relatively small grid voltage swing to control the plate current over it's entire range. Since the grid element controls of the current flow in the tube, it allows the tube to be used as an amplifier to take a relatively small input signal on the grid and generate a relatively large signal swing at the plate. The amount of signal voltage at the plate is equal to the current flowing through the tube multiplied by the resistance connected to the plate.

Grid leak biasing – The small amount of grid current in the tube generates a negative bias voltage across this resistor, which biases the tube to the proper operating point with respect to the cathode, which is grounded. This method of biasing is not very stable, and fell out of favor early on in the development of tube amplifiers. Most preamp stages now use cathode biasing as opposed to grid leak biasing.

Grid leak resistor – a very large resistor from the grid of a tube to ground, which is used to generate the bias voltage for the tube. See "grid leak biasing" for an explanation of how this works. This term is sometimes incorrectly used when referring to the grid-to-ground resistor in a cathode biased configuration, which is used to provide a DC ground reference for the grid circuit.

Grid resistor – the term usually given to a series resistor connected to the grid of a tube, also called a "grid stopper," but sometimes used to refer to the resistor connected from the grid of a tube to ground, which is also sometimes called a "grid leak" resistor.

Grid stopper – a resistor connected in series with the grid of a tube, usually right at the pin of the tube. It is used to prevent parasitic oscillations and reduce the chance of radio station interference by forming a lowpass filter in conjunction with the input capacitance of the tube.

Ground – The common "reference" point for the circuit. This is usually also connected to the chassis, but there can be independent circuit grounds and chassis grounds.

Glossary

H – the symbol for magnetizing force, also the symbol for the unit of inductance, the Henry.

Half-wave rectifier – a rectifier that conducts on only the positive or only the negative half of the incoming sinusoidal signal. It produces a "pulsating" DC composed of single-polarity "humps" at the incoming AC frequency, with a flat "dead time" during the time the input signal goes to the opposite polarity. The half-wave rectifier requires more filtering than a full-wave rectifier to produce the same degree of ripple in the output DC waveform.

Heater – the heating element in an electron tube, also called the "filament."

HT – stands for "high-tension," meaning high voltage. Occasionally the B+ fuse on an amplifier will be labeled "HT Fuse."

Hz – stands for "Hertz," which is the name given to the frequency of an alternating current. The units are in cycles per second. In some older literature, you may see this represented as "CPS," which, of course, stands for "cycles per second." A prefix of "k" or "M" is used to indicated kilohertz, or kHz, and megahertz, or MHz, which indicate thousands and millions of cycles per second, respectively.

I – the symbol for current

Impedance – a complex quantity containing both a resistance and a reactance. The symbol for impedance is "Z," and the unit of impedance is the ohm. $Z = R + jX$, where R is the resistance, and X is the reactance of the circuit, and j is the complex, or imaginary, operator, indicating multiplication by the square root of –1. Inductive reactances have positive imaginary components, and capacitive reactances have negative imaginary components. For example, an inductor of 1mH with a resistance of 8 ohms would have an impedance of (8 + j6.3) ohms at 1000 Hz. Since an impedance is a complex number, it has both a magnitude and a phase. Typically, when discussing amplifiers or speakers, impedances are referred to as the magnitude of the complex number, instead of the rectangular form as given in the definition. The magnitude of the (8+j6.3) example is 10.2 ohms, as calculated by the square root of the sum of the squares of the real and imaginary parts (the "length" of the resulting vector). The concept of imaginary numbers can be a bit confusing to those who haven't encountered it before. If you are interested in finding out more about this, check out a textbook on introductory circuit analysis, as they usually have a good treatment of the subject.

Inductance – the "size" of an inductor, not the actual physical size, but the "electrical" size. The unit of inductance is the Henry, or "H." Most power supply inductors, or chokes, are measured in henries, typically 2-20H. The inductance of a transformer primary may also be several henries. Smaller inductors are measured in millihenries (mH – 10^{-3} henries) or microhenries (µH – 10^{-6} henries).

Inductor – a circuit element consisting of a coil of wire would on a core material made of ferrous or non-ferrous material. An inductor resists changes in the flow of electric current through it, because it generates a magnetic field that acts to oppose the flow of current through it, which means that the current cannot change instantaneously in the inductor. This property makes inductors very useful for filtering out residual ripple in a power supply, or for use in signal shaping filters. They are frequency-dependent devices, which means that their inductive reactance, or "effective resistance" to AC decreases as the frequency gets lower, and increases as the frequency gets higher. This property makes them useful in tone controls and other filters.

IT – interstage transformer.

Jack – the input or speaker output connector on a guitar amplifier.

Jewel – the term commonly used to refer to the screw-on pilot light lens on Fender guitar amplifiers. These usually were red or green, but purple ones are purported to have real "mojo."

k – the prefix indicating "kilo" or thousands, as in a 10k resistor, which means ten thousand ohms.

K – the symbol for the cathode of an electron tube

L – the symbol for inductance.

LDR – light dependent resistor. Often used in referring to an optocoupler in which the active element is a photoresistor, whose resistance changes as current is passed through the lighting element, which is usually an LED or neon bulb.

LED – light emitting diode. These are semiconductor devices that emit light of various colors when an electric current is passed through them. They are typically used as indicators, but occasionally are used as clipping diodes because of their larger forward voltage drop when compared to a standard silicon diode.

Local negative feedback – feedback that is applied over one stage only, as opposed to global negative feedback, which is applied over several stages of amplification. An example of local negative feedback is a cathode follower, where the feedback signal is not so apparently derived by the current flowing through the cathode resistor, or a common-cathode stage with an unbypassed cathode resistor.

Long tail pair – a phase inverter topology that has a single resistor connected as a pseudo-current source from the junction of two tube cathodes, with the outputs taken off the individual tube plates, one in phase with the input signal, and the other out of phase with the input signal. The circuit gets its name from the "tail" resistor connected to the cathodes.

M – the prefix for mega, or millions, as in a 1M resistor, which means one million ohms.

Mains – the AC line voltage input (often more commonly used in British terminology). Occasionally the fuse on the AC input will be labeled "Mains Fuse."

Master volume – a second volume control, located at the end of the preamp section of a guitar amplifier, which allows the guitarist to turn the preamplifier up to the point of distortion, while keeping the overall volume low.

Microphonics – the tendency for a component to induce audible noise into the amplifier circuit when mechanically disturbed. Tubes are the most common microphonic component, and they will usually make an audible "thump" or "ring" when tapped. Occasionally, the problem is severe enough in combo amplifiers to cause uncontrollable feedback from the speaker to the tube, resulting in a "squealing" or "howling" noise when the volume is turned up loud. Although it is not commonly known, capacitors can also be quite microphonic. Different types have different levels of microphony, with ceramic types usually being the worst.

Miller capacitance – the effective multiplication of the plate-to-grid capacitance in a triode tube (or transistor) by the gain of the amplifying stage. Miller capacitance can decrease the frequency response of an amplifier stage by acting as a lowpass filter in conjunction with the source resistance of the preceding stage.

Modeling amp – a computer that is passed off as a guitar amplifier. Also see "solid-state."

Negative feedback – feedback in which a portion of the signal from a later amplifier stage is fed back to an earlier stage (or to the same stage) in such a manner as to subtract from the input signal.

Ohm – the unit of resistance or impedance.

Ohm's law – the fundamental relationship between voltage, current, and resistance. It is usually stated as: E = I*R, or V=I*R, where E or

Glossary

V = voltage (in volts. E stands for "electromotive force" which is the same thing as voltage), and I = current (in amps), and R = resistance (in ohms). The equation can be manipulated to find any one of the three if the other two are known. For instance, if you know the voltage across a resistor, and the current through it, you can calculate the resistance by rearranging the equation to solve for R as follows: R = E/I. Likewise, if you know the resistance and the voltage drop across it, you can calculate the current through the resistor as I = E/R.
A related equation is used to calculate power in a circuit: P = E*I, where P = power (in watts), E = voltage (in volts), and I = current (in amps). For example, if you measure 20V RMS and 2.5A into a load, the power delivered to the load is: P = 20*2.5 = 50W. This equation can also be rearranged to solve for the other two quantities as follows: P = E*I, E = P/I, and I = P/E. You can also combine the power equation with the first Ohm's law equation to derive a set of new equations. Since E = I*R, you can substitute I*R for E in the power equation to obtain: P = (I*R)*I, or P = I2R. You can also find P if you know only E and R by substituting I=E/R into the power equation to obtain: P = E*(E/R), or P = E2/R. These two equations can also be rearranged to solve for any one of the three variables if the other two are known. For example, if you have an amplifier putting out 50W into an 8 ohm load, the voltage across the load will be: E = sqrt(P*R) = sqrt(50*8) = 20V RMS.

Optocoupler – another name for optoisolator.

Optoisolator – a device which contains an optical emitter, such as an LED, neon bulb, or incandescent bulb, and an optical receiving element, such as a resistor that changes resistance with variations in light intensity, or a transistor, diode, or other device that conducts differently when in the presence of light. These devices are used to isolate the control voltage from the controlled circuit. Typical optoisolators are the Vactec and photoFET devices used in channel-switching amplifiers, as well as the "trem-roach" neon bulb/photoresistor package used in the tremolo circuit in some Fender amplifiers.

Oscillator – a circuit that produces a sustained AC waveform with no external input signal. Oscillators can be designed to produce sine waves, square waves, or other wave shapes. They are typically used as variable speed generators in tremolo circuits in guitar amplifiers.

OT – short for "output transformer."

Output transformer – a transformer used to match the low impedance of a speaker voice coil to the high impedance of a tube output stage.

Output transformers consist of at least two windings, a primary and a secondary. Some output transformers have multiple impedance taps on the secondary side, to allow matching to different speaker cabinets, typically 4, 8, and 16 ohms.

p – the prefix for "pica," or 1*10-12, as in a 100pF capacitor, which means 100x10-12 Farads. Originally the term "μμF" or "micro-micro Farads" was used.

Parasitic oscillation – an unwanted oscillation in a tube amplifier, often at supersonic, inaudible frequencies. Parasitic oscillations can cause all sorts of problems, including overheating output tubes and bad tone.

Passive – a component that doesn't need a power source to function. Examples of passive components are: resistors, capacitors, inductors, transformers, etc. Also used to refer to guitar pickups that don't have built-in preamps, and don't require batteries to operate.

PCB – short for "printed circuit board," or PC board. A piece of phenolic or glass-epoxy board with copper clad on one or both sides. The portions of copper that aren't needed are etched off, leaving "printed" circuits which connect the components.

Pentode – A five-element electron tube, containing a control grid, screen grid, suppressor grid, cathode, and plate as active elements, in addition to the filament.

Phase – the instantaneous "polarity" of an AC signal, or more correctly, the point in the rotation of the vector, measured in degrees, from 0 to 360 degrees total.

Phase inverter – a circuit that generates two output signals, each 180 degrees out of phase with the other. This is a bit of a misnomer, since it does more than just invert the phase of a signal, it actually generates two out-of-phase signals.

Phase splitter – another name for a phase inverter.

PhotoFET – an optoisolator in which an LED controls the turn on/off of a bilateral MOSFET device. These devices are commonly used as channel-switching devices.

Plate – the "current collecting" element in a vacuum tube. Also called the "anode." This is also the term used for each of the two terminals of a capacitor, which are on either side of the dielectric.

Plate dissipation – the amount of power dissipated in the plate element of a vacuum tube. At idle, or quiescent conditions, it is equal to the DC plate current multiplied by the DC voltage difference between the plate and cathode elements. When the tube is amplifying a signal, the average plate dissipation depends on several things, including the quiescent bias point, the amount of signal voltage between the plate and cathode, and the class of operation. Average plate dissipation can either increase, decrease, or remain the same at full power, depending on these things. In a class AB or class B amplifier, the power dissipation increases, because the signal swing above and below the quiescent point is not the same (the tube is in cutoff for a portion of the cycle) and in a true class A amplifier the plate dissipation remains the same, as the average signal change at the plate is zero, since it swings equally above and below the quiescent bias point.

Plexi – the name given to early Marshall amplifiers that had a gold Plexiglas control panel on the front and rear of the chassis. This was later changed in mid 1969 to gold aluminum front and rear panels, commonly referred to as an "ali-panel" Marshall.

Positive feedback – feedback in which a portion of the signal from a later amplifier stage is fed back to an earlier stage (or to the same stage) in such a manner as to add to the input signal.

Pot – short for "potentiometer."

Potentiometer – a variable resistor. It usually has three terminals: the two end terminals, across which the entire resistance appears, and a third terminal, the "wiper", which moves to a different spot on the resistor as the shaft is turned. In this manner, the resistance between the wiper and one end terminal gets smaller while, at the same time, the resistance between the wiper and the other end gets larger. This allows the potentiometer to be used as a variable voltage divider, for use in attenuators, such as volume controls or tone controls.

Power – the rate of doing work, equal to the voltage multiplied by the current in a circuit. In an amplifier, this work results in either heat or mechanical energy, such as moving the loudspeaker coil to produce sound.

Power amp – the high-level amplifying stage in a guitar amplifier. This is where the smaller preamp signal is converted into a high power signal necessary to drive the speakers to the desired output level.

Power transformer – a transformer used to convert the incoming line (or mains) voltage to a higher or lower value for use in the guitar amplifier. Typically, the power transformer will have at least one primary, but sometimes two or more, to allow use at 120V/240V/etc. mains

Glossary

voltages. There will also usually be a 6.3V filament winding, sometimes center-tapped to allow balancing the filament string symmetrically around ground for hum reduction. There is sometimes a 5V winding for use with a tube rectifier. This winding is eliminated when using a solid-state rectifier. There is also a third winding for generating the high voltage, or B+, as it is commonly called. This winding may be center-tapped, unless a bridge rectifier is used.

PP – push-pull.

PPP – parallel push-pull.

Preamp – the low-level amplifying stages in a guitar amplifier. This is where the tiny signal from the guitar pickup is amplified and shaped for the desired tonality before being sent to the power amplifier, which generates the high power signal needed to drive the speakers.

Presence – a control on a guitar amplifier that boosts the upper frequencies above the normal treble control range for added high-end. This control is usually a shelving type of equalizer, and is normally implemented as a lowpass filter inside the global negative feedback loop. By decreasing the amount of high frequencies that are fed back, the high frequencies at the output of the amplifier are boosted.

PSE – parallel single ended.

PT – power transformer.

PTP – point-to-point. A method of wiring an amplifier without using a PC board or – accurately speaking – without any form of circuit card or terminal strips, etc, where points in the circuit are directly connected by the components themselves. For example, an input jack connected to a preamp tube's input by the resistor's own leads, the tube's output connected to the volume pot's tag by a coupling cap's leads, etc.
Frequently, however, the term is used to refer to construction techniques in which the components are mounted on terminal strips or tag boards, and the wiring is put in by hand to make the circuit connections. Either technique is widely regarded as "sounding better" than PCB construction because of supposedly higher bandwidth, but this is a myth, as PCBs are regularly used into the MHz region. PTP wiring is generally better than PCB for guitar amps because of ease of maintenance and durability. Many manufacturers use cheap, single-sided PC boards without plated-through holes, which tend to pull up pads when a component is desoldered. Some even go so far as to not use a soldermask or silkscreen. This type of construction should be avoided, and is a good indication of a cheaply made amplifier. In short, amps employing hand-wiring/hand-builing techniques (PTP or otherwise) might be better sounding and more durable amps primarily because more care has gone into their construction and component selection.

Push-pull – In a push-pull amplifier, the power supply is connected to the center-tap of the transformer and a tube is connected to both the upper and lower end of the center-tapped primary. This allows the tubes to conduct on alternate cycles of the input waveform. A push-pull stage can be biased class A, where current flows in both tubes for the entire input cycle (but in opposite directions), or class AB, where current flows alternately in both halves, but less than a full cycle in each, or class B, where current flows only half the time in each tube. Most designs are biased class AB for best efficiency and power output with minimal crossover distortion (but not necessarily best "tone," although this is subjective). A push-pull stage requires at least two tubes to operate, but can have more connected in parallel with each side, resulting in an amp with four, six, or even eight output tubes for higher-power amps. This is called "parallel push-pull" operation, or PPP.

Q – the symbol for the "quality factor" or figure of merit for a reactive component, such as a capacitor or coil. Low reactive element Qs can affect the response of filters near the cutoff frequency. Also the symbol for "quality factor," or selectivity of a filter network, used to denote the relative "sharpness" of a filter. For instance, a high Q bandpass filter would be one that has a very narrow width and steep slopes on the sides. It is a measure of the ratio between the center frequency and the bandwidth of a bandpass filter.

R – the symbol for resistance.

RDH4 – "Radiotron Designer's Handbook, 4th edition" – the legendary "bible" of tube amplification, also known as "the big red book."

Reactance – the "imaginary" component of impedance, or the resistance to AC signals at a certain frequency. Capacitive reactance is equal to $1/(2*pi*f*C)$, and inductive reactance is equal to $2*pi*f*L$. The unit of reactance is the ohm.

Reactive load – a load that contains inductance or capacitance, either with or without resistance as well. An example of a reactive load is a loudspeaker which has an impedance that varies with frequency, unlike a purely resistive load, whose impedance is flat for all frequencies in the range of a guitar amplifier.

Rectifier – this is the same thing as a diode, but the term is usually reserved for diodes used in the power supply section of an amplifier.

Reflected impedance – the impedance seen "looking into" the primary of a transformer when the secondary is loaded with a specific impedance. The impedance on the secondary side is transformed by the square of the turns ratio of the transformer. For example, if a 2:1 turns ratio transformer has a 10 ohm load on the secondary, the impedance measured across the primary terminals will be 40 ohms, because the secondary impedance of 10 ohms is multiplied by 2^2, or 4.

Relay – an electromechanical switch, operated by passing current through a coil of wire wound around a steel core, which acts as an electromagnet, pulling the switch contact down to make or break a circuit. These are available in several types, including SPST (single-pole, single-throw), SPDT (single-pole, double throw), DPST (double-pole, single throw), and DPDT (double-pole, double-throw), and not as commonly, in multi-circuit configurations such as 3PDT or 4PDT (three and four poles, double-throw).

Resistance – the "size" of a resistor. The unit of resistance is the ohm. Resistors vary in size from fractions of an ohm to several million ohms. The prefix "k" is used for kilohms, or thousands of ohms, and the prefix "M" is used for megohms, or millions of ohms.

Resistive load – a load that contains no inductance or capacitance, just pure resistance An example of a resistive load is a dummy test load consisting of a single resistance equal to the output impedance of the amplifier under test. The resistive load has an impedance that is flat for all frequencies in the range of a guitar amplifier.

Resistor – a circuit element that presents a resistance to the flow of electric current. A current flowing through a resistance will create a voltage drop across that resistance in accordance with Ohm's law.

Resonance – a control on a guitar amplifier that boosts the lower frequencies at or below the normal bass control range for added low-end, also called "depth" or other names. This control is usually a shelving type of equalizer, and is normally implemented as a highpass filter inside the global negative feedback loop. By decreasing the amount of low frequencies that are fed back, the low frequencies at the output of the amplifier are boosted. Resonance is also the term given to an electronic circuit that contains both capacitive and inductive elements – there is a "resonant" point where the capacitive reactance equals the inductive reactance. Depending upon whether the elements are in series or parallel, this will result in a maximum voltage and maximum impedance across the elements (parallel resonance) or maximum current and minimum impedance through the elements (series

Glossary

resonance). If the circuit has resistance, either across the parallel resonant circuit or in series with the series resonant circuit, the maximim peak will be limited, and the bandwidth of the resonance will be broader. The relative "sharpness" of the resonant circuit is called the "Q," or "quality" factor. See the definition of "Q" for more details.

Reverb – a short, recirculating delay effect used on some guitar amplifiers. It is similar to echo, but instead of discrete, long delay repeats, it is a series of very short delays that add up to create a sense of spaciousness in the tone. A spring unit with a sending transducer at one end and a receiving transducer at the other end is usually used as the delay unit, although some amplifiers use an analog or digital delay line.

RMS – stands for "root mean square." It is a term used with AC voltages or currents to indicate the equivalent DC voltage or current. For a sine wave, the RMS value is equal to the peak-to-peak value divided by 2*sqrt(2), or 2.282, or the peak value divided by sqrt(2), or 1.414. You can also multiply the peak value by 0.707, which is the same as dividing by 1.414. The RMS value of the signal depends on the shape of the waveform. For instance, the RMS value of a square wave is not the peak value multiplied by 0.707, rather, it is equal to the peak value of the square wave.

Sag – a "drooping" of the power supply voltage in a guitar amplifier as a note or chord is played. This "drooping" causes a slight drop in volume, for an effect similar to a compressor. It adds "touch sensitivity" to the amplifier, and is one of the reasons tube guitar amplifiers sound subjectively better than solid-state guitar amplifiers.

Scaling – the process of shifting an electronic parameter up or down. For instance, a tone circuit that has a midrange boost/cut centered around 1kHz might be scaled to 800Hz to better suit the application. This would be an example of frequency scaling. Impedances may also be scaled up or down.

Schmitt phase inverter – a phase inverter configuration using two cathode-coupled tubes, with the first tube acting as a common cathode stage providing an out-of-phase signal at its plate, while the second tube operates as a common-grid stage, providing an in-phase signal at its plate. This type of inverter has moderately good balance, providing the plate resistor of the out-of-phase side is made slightly smaller than the in-phase plate resistor to compensate for differences in the amplification between the two stages. This phase inverter provides high gain.

Screen grid – a second grid element interposed between the control grid and the plate, to act as an electrostatic shield between them. This shielding action greatly reduces the input capacitance of the tube, which increases it's frequency response, and makes the plate current virtually independent of plate voltage. There is no screen grid in a triode, only in a tetrode or pentode.

Secondary emission – electrons in a vacuum tube may be moving at a sufficient speed to dislodge additional electrons when they strike the plate of the tube. These electrons emitted from the plate can reduce the current flow in the tube. A third grid element, called the "suppressor grid," is used to reduce the effects of secondary emission.

SE – single-ended.

Silverface – the name given to Fender amplifiers that have a silver control panel. The panel was changed from black to silver at the time CBS bought Fender. In addition, certain "improvements" were made to the circuitry at the same time. The general consensus is that these amplifiers don't sound as good as the blackface amplifiers, which has led to a practice known as "blackfacing" the amp, which means converting the circuitry back to match the blackface schematic.

Silkscreen – the name given to the "component identification" ink layer screened onto a printed circuit board. Also the name given to the lettering screened on the front and back of a guitar amp control panel. Name derives from the fine silk mesh screen template through which the ink is printed onto the component.

Single-ended – The term "single-ended", or SE for short, is given to an amplifier output stage configuration whose output transformer primary is not center-tapped. It has only two connections, one of which goes to the power supply, the other to the plate of the power tube. Tubes can also be paralleled for more power as in a push-pull stage, resulting in what is called "parallel single-ended" operation, or PSE. A single ended stage for guitar amplification is always biased class A. Old Fender Champs are a good example of a single-ended guitar amplifier. Higher power amplifiers are usually push-pull instead of single-ended, which allows higher efficiency and better frequency response with a smaller output transformer. Output transformers for single-ended amplifiers require an air gap to avoid saturation of the core due to the offset DC current in the transformer. This air gap greatly reduces the primary inductance. so the core must be made larger and the number of turns must be increased to obtain good low frequency response. A push-pull output transformer has no offset DC current flowing in the primary, because the DC bias current flows in opposite directions on each side of the primary, so it doesn't need an air gap, and can be made smaller. Single-ended output stages do not have the inherent even-harmonic cancellation and power supply rejection that push-pull output stages have, so the output tone is quite different, and the DC plate supply must be better filtered in order to keep the hum to a low level.

Solid-state – a component that has been specifically designed to make a guitar amplifier sound bad. Compared to tubes, these devices can have a very long lifespan, which guarantees that your amplifier will retain its thin, lifeless, and buzzy sound for a long time to come.

Solder mask – a coating on a PC board, usually a dark green or dark blue, but occasionally a yellowish color, which is designed to insulate and protect the copper traces and keep them from shorting together during the wave soldering process. The soldermask is "masked out" at solder pads, to allow for soldering component leads.

Speaker – a transducer designed to reproduce audio frequencies. There are many different models of guitar speakers, each with its own particular power handling capability and tone.

Speaker emulator – a device composed of filters that are designed to emulate the response of a loudspeaker, commonly used for direct recording applications.

Split-load phase inverter – the name given to the single-tube phase inverter in which the in-phase signal is taken off the cathode and the out-of-phase signal is taken off the plate, with equal-value plate and cathode resistors. This phase splitter configuration has excellent balance, but only unity gain. Also called a "Concertina" phase splitter.

Star ground – a preferred amplifier circuit grounding system, where all the local grounds for eachstage are connected together, and a wire is run from that point to a single ground point on the chassis, back at the power supply ground. Sometimes multiple star points are used for lower hum and noise levels in the amplifier.

Suppressor grid – a grid in a pentode vacuum tube that is used to minimize secondary emission from the plate, by virtue of its negative charge, which repels electrons emitted and returns them back to the plate. It eliminates the "kink" in the characteristic curves of a tetrode.

Susceptance – the reciprocal of reactance, measured in mhos.

Switch – a device that opens and closes an electric circuit.

Taper – the rate at which the resistance of a

Glossary

potentiometer changes as the shaft is rotated. There are several common tapers used in guitar amplifiers. There is linear taper, which means that the resistance changes linearly as the pot shaft is rotated, ie, the resistance at midpoint is half the total resistance from end to end. Another common taper is log taper, short for logarithmic taper, which means that the pot changes in a logarithmic fashion as the shaft is rotated, ie, the resistance at 1/10 rotation is half the total resistance from end to end. You may hear people occasionally mistakenly call this "analog taper," but there is no such thing. There is also a reverse log taper. The taper is chosen for the application. A volume control, for instance, will be a log taper, because the ear hears sound in a logarithmic fashion, and the volume must change accordingly to be perceived as linearly changing as the pot is turned. Depending upon the type of tone circuit, the pot used may be log or linear. If all the "action" occurs at one end of the pot, chances are the wrong type of pot is being used in the circuit.

Tetrode – A four-element electron tube, containing a control grid, screen grid, cathode, and plate as active elements, in addition to the filament.

Tolex – the original DuPont trade name given to the vinyl covering used on most guitar amplifiers, such as Marshall or Fender style vinyl (hence the capital "T" if it is genuine Tolex from DuPont). Purple and red Tolex have the best tone.

Tone – the characteristic sound of an amplifier.

Tone control – a potentiometer used for controlling the tone of an amplifier. This may be a single control or there may be multiple tone controls, commonly called a "tone stack."

Tone stack – The term used to describe the tone controls in a guitar amplifier. There are four main types of tone stacks used in most common guitar amplifiers. They are the Marshall style, the Fender style, the Vox style, and the lesser used Baxandall, or James style. These tone stacks vary in their construction, consisting of either a bass and treble control, or bass, mid, and treble controls. Some amplifiers have a tone stack consisting only of one control, usually a treble cut control, but sometimes it will be a single control that cuts treble at one end of the rotation, and cuts bass at the other end. These types of control are usually labeled "tone," or "cut."

Transconductance – the ratio of the tubes plate current to its grid voltage. The unit of transconductance is the "mho," which is measured in amps/volt, and is not surprisingly "ohm" spelled backwards, because one ohm is equal to one volt divided by one amp, so the unit of resistance, the ohm, is a volt/amp. Transconductance is one "figure of merit" for a tube. Higher transconductances mean higher gains and greater amplification from the tube.

Transformer – a device for changing levels of AC signals, or for changing impedances of circuits. It consists of a minimum of two coils, the primary and the secondary, wound on the same core. The core material can be ferrous (magnetic, such as iron), or non-ferrous (non-magnetic, such as an air core). Transformers used in guitar amplifiers are invariably wound on iron cores. An ideal transformer has no losses, it merely steps a voltage up or down in proportion to the turns ratio between the primary and the secondary. This is useful in converting the voltage from a wall outlet, typically 120V or 240V, into a higher voltage for the tube plate supply, typically 400V or more, and a lower voltage for the tube filament, typically 6.3V or 12.6V. The transformer will also "reflect back" to the primary the impedance which is connected to the secondary, in proportion to the square of the turns ratio. That is, if you have a 20:1 transformer with a 16 ohm impedance connected to the secondary, it will "look like" a 6.4K ohm impedance on the primary side. This is useful in matching the plate of a tube, which is very high impedance, typically on the order of several thousand ohms, to a speaker, which is very low impedance, typically on the order of 4, 8, or 16 ohms.

Transient response – the response of a circuit to a step waveform. An amplifier cannot perfectly reproduce an input step waveform because of the limited bandwidth and non-constant phase response of the amplifier. The transient response may indicate some "overshoot" or "undershoot" of the signal transition, or possibly some "ringing" or damped sinusoidal oscillations at the transition.

Tremolo – a circuit that periodically varies the amplifier output level at a rate and depth set by controls on the amplifier. The terms vibrato and tremolo are sometimes used interchangeably.

Triode – a three-element electron tube, containing a grid, cathode, and plate as active elements, in addition to the filament.

Tube – short for "electron tube".

Tweed – the name given to the covering on old Fender amplifiers which preceded the introduction of the Tolex vinyl covering.

u – (actually μ, the Greek letter "mu") the prefix for "micro", meaning one millionth, as in a 1uF capacitor, which means one millionth of a Farad. A lower-case "u" is usually used nowadays.

Ultralinear – the term given to the amplifier configuration developed by Hafler and Keroes, which uses taps on the output transformer to provide a negative feedback signal to the screen grids of the output tubes. This gives an operating point somewhere between that of a pentode and a triode. This form of operation was given a bad name due to a particularly sterile-sounding Fender amplifier that had an ultralinear output stage and far too much global negative feedback. A few of the misinformed amp "guru" types immediately denounced all ultralinear operation as sounding bad. The stigma has endured to this day, although this is slowly changing, with the help of amp makers like Dr Z, who are willing to experiment with different output topologies to produce a better sounding amplifier. Ultralinear operation, when used without global negative feedback, can sound quite good, as the local negative feedback provided by the screen taps increases the damping factor, lowering output impedance, and "tightening up" the bass, without the use of global negative feedback.

V – the symbol for voltage. Common prefixes are "m," for mV (10-3 volts), and "u," for uV (10-6 volts), and "k", for kV (103 volts).

Vactec – the common name given to the Vactrol optoisolator device used for channel switching. The name is printed on the Vactrol because the company that invented them was named Vactec, later EG&G Vactec.

Vactrol – an optoisolator device used for channel switching in many modern amplifiers (see Vactec above), such as Soldano and Mesa. It is a single package combining an LED and a photoresistor, which changes resistance from very high (essentially an open circuit) to very low (essentially a short circuit) as the current through the LED is turned on and off. It is used as a substitute for relays, to avoid the "clicks" and "pops" that can occur when they are used for channel switching.

Vacuum tube – Another name for "electron tube."

Valve – the British term for "tube," the full term for which is "thermionic valve."

Variac – the trade name for a brand of variable AC transformer. There are other brands, but this term is generically used to describe all of them. A variac allows adjustment of the incoming AC mains voltage. The better ones have meters for voltage, current, or both, and fuses for protection.

Vibrato – a circuit that periodically varies the pitch of a note. True pitch-shifting vibrato is not usually found on a guitar amplifier. The terms vibrato and tremolo are sometimes used interchangeably.

Voltage – the term for electric force. Voltage is the energy per unit charge created when positive and negative charges are separated.

Volume control – a potentiometer used for controlling the volume of an amplifier. Best setting is usually on "10" or higher.

W – the symbol for watts. Typical prefixes are "m," for thousandths, as in "mW," or "milliwatts," "k," for thousands, as in "kW," or "kilowatts," and "M," for millions, as in "MW," or "megawatts."

Watt – a unit of power (see W above). Contrary to popular belief, more is not always better.

X – the symbol for "reactance."

Y – the symbol for "admittance."

Z – the symbol for "impedance"

Copyright © 1999, 2000, 2001, 2002 Randall Aiken. Reprinted with approval from Aiken Amplification.

About the author

Richard Aspen Pittman is a native Californian who began performing music in local bands at age 11. He went to work for the Guitar Center in Hollywood at age 18 and his interests in music broadened to include the equipment and production side of the music business. He started out as a rookie salesman, which meant he would set up the guitars out of the box and sweep the floor. However, a short six months later, Aspen had been promoted from floor sweeper to floor manager of the Guitar Center, in charge of all sales and the buying of used equipment (now that's called "vintage gear"!). His GC customers over the next seven years included the very top names in the business and gave him the opportunity to help produce countless recording sessions and road tours, always in the quest for the ultimate sound. In 1972, Aspen left the retail business to join Acoustic Control Corporation as the head of West Coast sales, Artist Relations, and worked with their engineering team to develop new products. Over the next seven years, Aspen worked with dozens of top recording artists, helped develop many new product lines, and eventually became Acoustic's Director of Marketing worldwide.

In 1979, Aspen left Acoustic to form the consulting firm of Aspen and Associates, whose accounts included many of the top musical equipment manufacturers and performing groups in Southern California.

Later that year, Aspen enlisted the help of an engineering team to help him develop an idea he had: to improve the performance of tube guitar amps through a special tube testing and matching process. This process led to the founding of Groove Tubes, whose list of believers now includes virtually every top guitar player in the business. Aspen next enlisted a team of top engineers, players and scientists to develop another one of his ideas, the Speaker Emulator, which allows the output of a tube guitar amp to directly interface with a recording console or a PA mixer. Aspen formed GT Electronics to produce a series of products based around this patented idea as well as other tube gear designed for stage applications. Some of those products are mentioned in this book. Aspen went on to form Groove Tubes Audio to develop tube-related products for the studio, such as tube condenser microphones and direct boxes. All these separate divisions finally became consolidated under the Groove Tubes Llc company in 1998. This was the same year Fender approached Aspen to add Groove Tubes to all new-production Fender tubes, and as of January, all tube-based Fender amps come shipped with Groove Tubes.

Aspen also invented FATHEADs, a simple concept of adding mass to the headstock of any guitar and thereby increasing sustain and improving the harmonic balance of overtones. FATHEADS are a thin plate of bell brass that is used like a washer between the back of the headstock and the tuning gears. No modification to the guitar is required (tuning gears hold the FATHEAD to the back of the headstock) and the mass of the headstock is roughly tripled with the addition of a FATHEAD. FATHEAD was awarded a US patent in 1989. This idea later developed into the FATFINGER, which also added mass but clamped on any headstock in seconds to accomplish much the same results. Aspen was awarded yet another patent for FATFINGER in 2002.

Aspen co-invented yet another breakthrough in the SFX stereo technology. An old friend, Drew Daniels, who was a notable audio engineer, brought an idea to Aspen for development. The idea was to develop a system that could produce a stereo sound from a single speaker cabinet. The process involved an electronic encoder to manipulate a stereo signal for a specially built speaker cabinet that used two very different speakers to produce a 360 degree stereo sound field from the single cabinet. A few months later, Aspen had not only developed the system, but licensed it to the Fender company, who trade marked the name SFX (Stereo Field Expansion) technology and used this system in their award winning Acoustisonic amp. This Fender amp became a best seller and won "Product of the Year" at the NAMM show in 1999, and went on to win several other awards around the world. Fender is planning several other SFX amps and Groove Tubes also manufactures several custom SFX systems for guitar and keyboards. SFX was awarded a US Patent in 2001, bringing Aspen's US patent total to four!

Aspen's latest tube adventure started in 1998 when he was able to purchase much of the original General Electric tube production line for the legendary GE 6L6 and 6CA7 vacuum tubes. Aspen spent the next four years and a considerable amount of money setting up Groove Tubes Manufacturing, including buying a new facility in San Fernando, California. Aspen released the GT6L6GE at the 2002 NAMM show, and it was a faithful reproduction of the GE 6L6 down to the original NOS special plate materials. The tube has gone on to draw wide approval from players and engineers alike. The next plan is to recreate the 6CA7/EL34 in 2003, and the process is already well under way (this was the tube Aspen sold to Eddie Van Halen for the vintage Marshall he used on his first few albums).

Through all his business successes, Aspen has remained a player with a passion for collecting all types of classic tube amps and tube-based studio mikes and signal processors. His personal amp and mike collection has hundreds of examples of vintage tube amps from Ampeg to Vox, and mikes from RCA ribbons to Neumann tube models. Aspen has also authored several articles about tubes and tube amps for music publications, including *Guitar Player* magazine, so it seemed only logical that Aspen would be the one to write "the book on tube amps" as a reference guide for countless musicians who really need this information to keep their amps in top playing condition.

Aspen resides near his San Fernando, California, Groove Tubes factory with his wife and business partner Sigrid, two nasty blue & gold macaws, and two good dogs. He regularly attends Christian music worship groups at several nearby churches and spends his leftover time collecting and restoring watches, guns, guitars and of course tube amps!

THE TUBE AMP BOOK

Acknowledgements

STROKES FOR MY FOLKS

I started Groove Tubes with my eight-year-old daughter, two good dogs, and a bunch of good family and friends. We've grown to employ quite a few of those original family and friends, but I'd like to think most of these beautiful folks like to be a part of Groove Tubes just for the fun we have around here. God knows they're not in it for the money. Our common goal is an interest in improving the sound and performance of tube amps everywhere via the Groove Tubes system. I couldn't have gone this far with out these friends and family so I would like to dedicate this page to my GT folks. If you see any of these people on the street, run up to them and shake their hand – favors and/or cash tips are encouraged. There's a long list of good people who really helped me research and write the various editions of *The Tube Amp Book*, culminating in this *Deluxe Revised Edition*. Many couldn't be there for the Groove Tubes Family picture, so I'd like to take up a little space to express my deepest thanks to them for helping with my dream. In no special order:

Sigrid "Sigi" Pittman: God's blessing who became my wife and partner in Groove Tubes for the last 10 years. She really covers my back so I could spend time writing this book, again and again.

Dick Rosmini: My audio mentor that got me into classic recording gear. Dick was on the first PortaStudio development team. He helped me develop the Speaker Emulator and many GT Audio products. We lost Dick a few years back, a real loss for the GT family.

Ken Fischer: He gets the award for our favorite service guru and for most unique sense of humor. He has since made quite a name for himself as Trainwreck amp designer and builder.

Bob Nash: Ran Modern Music of Dudley, England… a legendary amp technician and gentleman who supplied many of my UK schematics.

Mike and Barbara Cooper: My English GT family. Mike's a walking encyclopedia on the British amp companies, and Babs ran our first GT UK.

Red Rhodes: Best tube amp repair expert I ever knew and responsible for lots of the little tips in the Service sections. Headed our guitar amp engineering department through our early years and was one of the world's best steel players. We also lost Red a few years back, but his Trio preamp and Dual 75 amp are still being produced.

John Sprung: Owner of Parts Is Parts (formerly American Guitar Center) in Wilmington, VT, fellow fanatic of those fabulous old tube amps.

Rick Harrison: Owner of Music Ground, Doncaster, England and very helpful with photos, catalogs and finding me many vintage beauties.

Sam Hutton: Restored Fender amps in Orange, California, and worked at Fender for years. We lost Sam too, another sad day for the GT family.

Ritchie Fleigler: Long-time friend and fellow amp nut, wrote his own book called *Amps! The Other Half of Rock & Roll*. Currently a big guy in Fender but has been a true friend and supporter of GT from the earliest days when he was at ESP and Marshall.

I would also like to thank the following supportive folks:
My mom, Kay
My Dad, Jerry
My Daughter, Autumn
Rick Benson, GT's own Hollywood insider
Myles Rose (GT guy who does Tech Support for our customers and on our web site – another tube guru nut working well below his market value to be around glowing filaments)
Mitch Colby (Marshall and Vox USA, fellow amp nut good friend of GT)
Howard Reitzes (helped me from the start, played keys in Iron Butterfly)
Marty Mehterian (keeps me rollin', does everything mechanical for GT)
Paul Patronete (long time GT employee, great player too)
Russ Jones (early founding partner in GT, long time friend and supporter)
Craig Swancy (Craig's Music, first GT family member from Texas, shucks!)
Norman Harris (Norms Rare Guitars, CA, long-time GT supporter, donated many guitars and amps for the photos sections of the book)
Jeffery "Skunk" Baxter (our GT Tubie Brother)
Mitch Margolis (GT braintrust – S45 and Vipre)
Drew Daniels (GT braintrust guru type – SFX)
Jimmy Wiggle (GT braintrust guru type – Single and S30)
Polo Corona (GT braintrust and has run our GT amp line for 15 years!)
Frank He (GT braintrust guru type – makes most of the GT mics)
Jaco Pastorius (old pal who helped me form and name the company)
Bob Weir (one of our first GT amp fans, turned Jerry onto our amps)
Jerry Donahue (one of our 1st Soul-o artists, incredible player and guy)
John Mark (former head of RCA tubes, helped me start up the tube factory)
George Graham (Richardson Electronics, helped me get the GE machines to make tubes)
Jan Jurco (owner of JJ tube factory, formerly my partner in Tesla… makes great tubes!)
Doug Chandler (1st GT dealer in UK was Chandler Guitars, now Doug partners our Groove Tubes Europe operation, along with Tina Sharpe)
Charlie Chandler and Paula Chandler and all at Chandler Guitars, Kew, London (loyal GT dealer, and let us run rampent through the shop snapping cool tube amp photos)
Nigel Osborne (publisher of this book – whose leap of faith produced this miracle)
Dave Hunter (Backbeat's talented editor who really helped me craft this 5th edition, 'nd he cun spel riel gud)

The Groove Tube
Schematic Archive

THE GROOVE TUBES SCHEMATIC FILE

This edition of The Tube Amp Book - like everything else about it - contains more schematic diagrams than ever before. The selection of those printed in the book itself has altered somewhat, however: the majority of the most significant designs are still here, though many more of these, and most of the more obscure diagrams, are contained on the accompanying CD.

The universal use of CDs for data storage and retrieval in recent years has made this by far the best way to bring you the most schematics possible, without further obliterating the world's forests to do so. While it is undoubtedly useful to have as many as possible reproduced in print form for easy reference, those on the CD will reveal larger sized, high-resolution diagrams that should be easier to read and use. They are handy for printing off to make your own modifications, alterations or notes right on the page, or to pin up over the workbench when you service the amp - all of which should save plenty of amp techs from having to mark, mar or otherwise deface this edition of the book. To that end, all of the diagrams printed in the book are also contained on the CD.

For a quick guide to which makes are here in print and which are on the CD only, scan the index below. A more detailed interactive index, with full model listings (alphabetized as accurately as possible) is on the CD itself. Be aware that if - for example - these book pages contain a schematic diagram for, say, a particular Fender amp, but not its accompanying layout diagram, the latter might well be contained on the CD (the two have been printed as near to each other as possible for the major models, but space issues make it impossible to do so for all such pairs). There may be a handful of diagrams published in Edition 4.1 which are not contained here - no doubt we have lost a file or two over the years - but most readers will agree it is a greatly improved resource. Likewise, there are a handful of "unidentified" diagrams for which we have no make/model reference, so check these as well if what you are looking for is not here.

Ampeg

Model	Page
B-12NB & B-15NB	4
J12B JET	4
B-12NF & B-15NF	5
R-12-R REVERB ROCKET	5
B-12N & B-15N	6
B-18N	6
B-12X	7
B-12XT	7
B-22X	8
B-25	8
B-25B	9
B-42X	9
ET-1	10
ET-2B	10
G-12	11
J-12D	11
SBT PREAMP	12
SB-12	12
SBT-12 (REV I)	13
SUPER ECHO TWIN	13
SVT (REV A)	14
SVT PREAMP	14
SVT PREAMP (REV B)	15
SVT PREAMP (REV C)	15
SVT POWER AMP (REV D)	16
SVT V9	16
SVT MODEL 6146	17
SVT MODEL 6146B	17

Bogen

Model	Page
CHB-20A	18
E30	18

Danelectro

Model	Page
REVERB BOX 9100	19

Epiphone

Model	Page
EA-10 DELUXE	20
EA-25 CENTURY	20
EA-26RVT	21
EA-30 TRIUMPH	21
EA-300RVT	22
EA-35T	22
EA-50T	23
EA-72	23

Fender

Model	Page
BANDMASTER 5E7	24
BANDMASTER 6G7	24
BANDMASTER AC568	25
BASSMAN 5E6	25
BASSMAN 5E6-A	26
BASSMAN 5F6	26
BASSMAN 5F6	27
BASSMAN 5F6-A	27
BASSMAN 5F6-A	28
BASSMAN 6G6-A	28
BASSMAN 6G6-A	29
BASSMAN 6G6-B	29
59 BASSMAN REV A	30
59 BASSMAN REV E	30
BLUES DE VILLE	31
BLUES JUNIOR	31
BASSMAN AC568	32
CHAMP-AMP 5E1	32
CHAMP-AMP 5E1	33
VIBRO-CHAMP AA764	33
VIBRO-CHAMP AA764	34
CONCERT 6G12	34
DELUXE 5C3	35
DELUXE 5D3	35
DELUXE 5D4	36
DELUXE 5E3	36
DELUXE 5E3	37
DELUXE 6G3	37
DELUXE REVERB-AMP AB763	38
DELUXE REVERB-AMP AB763	38
DUAL SHOWMAN REVERB AA270	39
DUAL SHOWMAN REVERB AA7E9	39
HARVARD 5F10	40
HARVARD 6G10	40
PRINCETON 5C2	41
PRINCETON 5E2	41
PRINCETON 5F2	42
PRINCETON 5F2	42
PRINCETON 5F2-A	43
PRINCETON 5F2-A	43
PRINCETON AA964	44
PRINCETON REVERB AA1164	44
PRINCETON REVERB AA1164	45
PRO-AMP 5C5	45
PRO-AMP 5D5	46
PRO-AMP 5E5	46
PRO-AMP 5E5	47
PRO-AMP 6G5	47
PRO-AMP 6G5-A	48
PRO-AMP 6G5-A	48
PRO-AMP AA763	49
PRO REVERB AA165	49
PRO REVERB AA165	50
PRO REVERB AA1069	50
REVERB 6G15	51
REVERB 6G15	51
SHOWMAN 6G14-A	52
SHOWMAN 6G14-A	52
SHOWMAN AB763	53
SHOWMAN AB763	53
SUPER-AMP 5C4	54
SUPER-AMP 5D4	54
SUPER-AMP 5E4-A	55
SUPER-AMP 5F4	55
SUPER-AMP 5F4	56
SUPER REVERB AA1069	56
SUPER REVERB AA1069	57
TREMOLUX 5E9-A	57
TREMOLUX 5E9-A	58
TREMOLUX 5G9	58
TREMOLUX 5G9	59
TREMOLUX 6G9	59
TREMOLUX 6G9-A	60
TREMOLUX AA763	60
TREMOLUX AA763	61
TREMOLUX AB763	61
TWIN-AMP 5C8	62
TWIN-AMP 5C8	62
TWIN-AMP 5D8	63
TWIN-AMP 5D8	63
TWIN-AMP 5E8-A	64
TWIN-AMP 5F8	64
TWIN-AMP 5F8-A	65
TWIN-AMP 5F8-A	65
TWIN-AMP 6G8	66
TWIN-AMP 6G8	66
TWIN-AMP 6G8-A	67
TWIN REVERB AA270	67
TWIN REVERB AA270	68
TWIN REVERB AA769	68
TWIN REVERB AB568	69
TWIN REVERB AB763	69
VIBRASONIC 5G13	70
VIBRASONIC 5G13	70
VIBROVERB AB763	71
VIBROVERB AB763	71
VIBROLUX 5E11	72
VIBROLUX 5F11	72
VIBROLUX 6G11	73
VIBROLUX AB763	73
VIBROLUX AA270	74
CBS 100W	75
75	75

Garnet

Model	Page
BANSHEE	76
BTO OVERDRIVE CIRCUIT	76
BTO POWER AMP	77
G15T GNOME	77
G45-90-100 PRE	78
G90TR POWER SUPPLY	78
G100 PREAMP	79
G200R ENFORCER	79
G250TR & G250D PREAMPS	80
G250TR & G250D POWER SUPP	80
G250TR & G250D REVERB & PRES	81
G250TR, G250D, G200S G G200	81
HERZOG RANDY BACHMAN	82
H-ZOG	82
LIL ROCK B90L	83
MACH 5 PREAMP & REVERB	83
PA90	84
PA90 MIXER	84
PRO BASS 190 STINGER FUZZ & PRE	85
REVERB UNIT	85

Gibson

Model	Page
ATLAS IV	86
ATLAS MEDALIST	86
BA-15RV	87
BR-1	87
BR-3	88
BR-4	88
BR-6F	89
BR-9	89
EH-100	90
EH-100	90
EH-125	91
EH-150 LATER	91
EH-160	92
EH-185 (6J7 PRE)	92
EH-185 (6SQ7 PRE)	93
EH-195	93
FALCON	94

READING SCHEMATICS

REPRINTED WITH PERMISSION FROM TOM MITCHELI'S EXCELLENT BOOK
"HOW TO SERVICE YOUR OWN TUBE AMP."

Reading schematic diagrams is really no big deal. The hardest part is remembering what each of the symbols means. Once you are comfortable with recognizing the symbols, a schematic diagram becomes nothing more than an electronic roadmap, or a sort of technical shorthand. Of course, as with all new things, you become more proficient by practicing. If you do not know how to read a schematic diagram, it will not impeded your ability to troubleshoot. However, if you do know how, you will find troubleshooting to be a little easier.

This chapter includes a reference table of the most common parts used in amplifier circuits. Next to each schematic symbol is a sketch of the component itself. This reference table should prove helpful in your attempts to interpret schematic diagrams. It is beyond the scope of this book to teach schematic reading. To do so would require a course in basic electronics, including hands-on lab experiments. I advise that you go to a junior college book store and buy a textbook on "Introduction to AC and DC circuits" or a similar title along those lines. In fact, I have seen good introductory electronics books in stores like Waldenbooks and B. Dalton. This type of book will build on what you have learned here. If you are interested in taking electronics seriously, it will be $15 well spent.

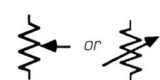

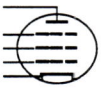

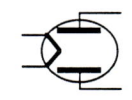

RESISTOR — **POTENTIOMETER** — **CAPACITOR** — **ELETROLYTIC CAPACITOR** — **TRIODE** — **DUAL TRIODE** — **PENTODE** — **TUBE RECTIFIER** — **RECTIFIER** — **NPN TRANSISTOR**

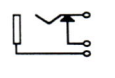

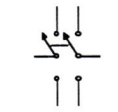

PNP TRANSISTOR — **SPEAKER** — **INCANDESCENT LAMP** — **NEON LAMP** — **FUSE** — **PHONE JACK** — **SWITCHING PHONE JACK** — **SPST TOGGLE SWITCH** — **DPST TOGGLE SWITCH** — **DPDT TOGGLE SWITCH** — **GROUND CONNECTIONS** (circuit ground, A.C. ground)

CIRCUIT WIRING (connection, no connection) — **SHEILDED CABLE** — **IRON CORE INDUCTOR** — **TRANSFORMER** — **POWER TRANSFORMER** (AC 110V 120V 200V 225V 245V, power supply, filament) — **OUTPUT TRANSFORMER** (plate, center tap, plate, 16 ohms, 8 ohms, 4 ohms, common)

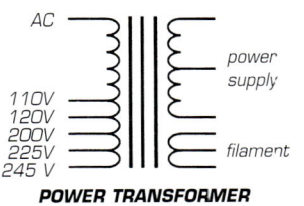

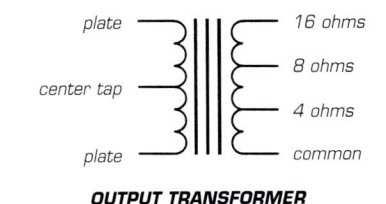

READING RESISTOR COLOR CODE

RESISTORS HAVE COLOR BANDS ON THEM THAT TELL HOW MUCH RESISTANCE THEY HAVE. THE COLOR BAND THAT CORRESPONDS TO THE FIRST DIGIT IS THE ONE THAT IS CLOSEST TO ONE END OF THE RESISTOR. THIS NUMBER COMBINED WITH THE SECOND NUMBER ARE MULTIPLIED BY THE NUMBER IMPLIED BY THE THIRD BAND.

BELOW IS A LIST OF SOME OF THE MOST COMMON RESISTOR VALUES FOUND IN AMPLIFIER CIRCUITS, AND THEIR COLOR CODES.

RESISTANCE	COLOR CODE
100 OHMS/10%	BROWN-BLACK-BROWN-SILVER
470 OHMS/10%	YELLOW-VIOLET-BROWN-SILVER
820 OHMS/10%	GRAY-RED-BROWN-SILVER
1 K OHMS/10%	BROWN-BLACK-RED-SILVER
1.5 K OHMS/10%	BROWN-GREEN-RED-SILVER
2.7 K OHMS/10%	RED-VIOLET-RED-SILVER
4.7 K OHMS/10%	YELLOW-VIOLET-RED-SILVER
5.6 K OHMS/10%	GREEN-BLUE-RED-SILVER
6.8 K OHMS/10%	BLUE-GRAY-RED-SILVER
10 K OHMS/10%	BROWN-BLACK-ORANGE-SILVER
15 K OHMS/10%	BROWN-GREEN-ORANGE-SILVER
47K OHMS/10%	YELLOW-VIOLET-ORANGE-SILVER
56 K OHMS/10%	GREEN-BLUE-ORANGE-SILVER
82 K OHMS/10%	GRAY-RED-ORANGE-SILVER
100K OHMS/10%	BROWN-BLACK-YELLOW-SILVER
220 K OHMS/10%	RED-RED-YELLOW-SILVER
470 K OHMS/10%	YELLOW-VIOLET-YELLOW-SILVER
1 MEG OHMS/10%	BROWN-BLACK-GREEN-SILVER

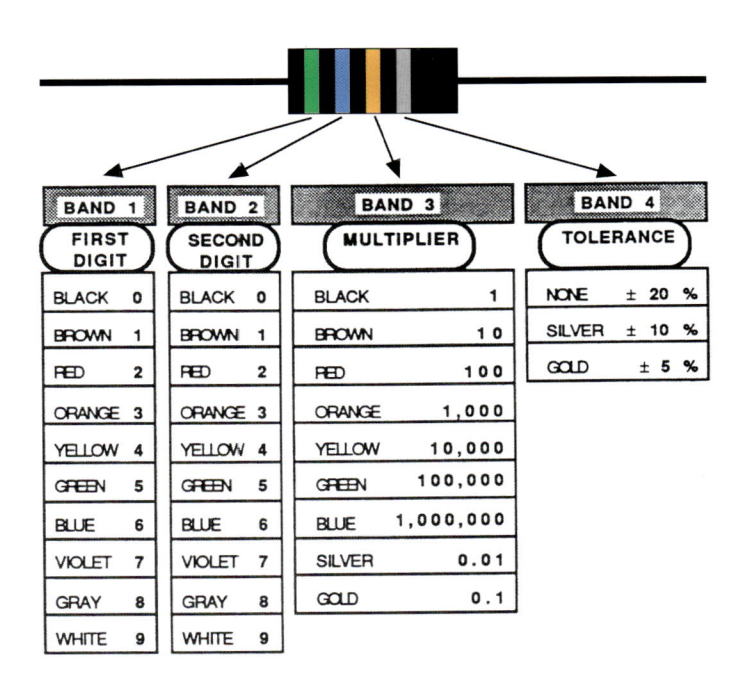

BAND 1 FIRST DIGIT	BAND 2 SECOND DIGIT	BAND 3 MULTIPLIER	BAND 4 TOLERANCE
BLACK 0	BLACK 0	BLACK 1	NONE ± 20 %
BROWN 1	BROWN 1	BROWN 10	SILVER ± 10 %
RED 2	RED 2	RED 100	GOLD ± 5 %
ORANGE 3	ORANGE 3	ORANGE 1,000	
YELLOW 4	YELLOW 4	YELLOW 10,000	
GREEN 5	GREEN 5	GREEN 100,000	
BLUE 6	BLUE 6	BLUE 1,000,000	
VIOLET 7	VIOLET 7	SILVER 0.01	
GRAY 8	GRAY 8	GOLD 0.1	
WHITE 9	WHITE 9		

PHYSICAL SIZE	POWER	LENGTH
	1/8 WATT	1/4 INCH
	1/2 WATT	3/8 INCH
	1 WATT	9/16 INCH
	2 WATT	11/16 INCH

The Tube Amp Book
DELUXE REVISED EDITION

by **Aspen Pittman**

Edited by **Dave Hunter**

Technical consultant **Myles Rose**

A BACKBEAT BOOK
First edition 2003
Published by Backbeat Books
600 Harrison Street,
San Francisco, CA 94107, US
www.backbeatbooks.com

Copyright 1986, 1988, 1991, 1993, 1995, 2003 Aspen Pittman, Except chapters: The Wide World of Preamp Tubes, Rectifiers, Power Modelling chapters copyright 2003 Myles Rose; Thoughts on 12AX7 Type Tubes, Where It All Began copyright 2003 Mark Baier; What Not To Do, Maintenance Checklist, copyright 2003 Tom Mitchell; The Amplifier Signal Circuits copyright 2003 Jack Darr; The Last Word on Class A, Amp Terms Glossary copyright 2003 Randall Aiken; Ampeg Cathode Bias and other Ampeg, Marshall, and Fender mods copyright 2003 Ken Fischer ; Maximizing Silverface Fender Amps copyright 2003 Brinsley Schwarz; Vox AC30 Check-Up copyright 2003 David Petersen; Replacing Output Transformers copyright 2003 Doug Conley; Speakers & Speaker Cabs and selected Amp Companies contributions copyright 2003 Dave Hunter; Small Amps Vs. Big Amps in the Studio copyright 2003 Huw Price

Archivist: **Aspen Pitman**
Editor: **Dave Hunter**
Technical consultant: **Myles Rose**
Digital retouching: **Phil Richardson**

Origination by Hong Kong Scanner Arts
Print by Colorprint Offset

03 04 05 06 07 5 4 3 2 1

GA2-RVT	94
GA-5T	95
GA-6 LANCER	95
LES PAUL GA-5 EARLY	96
LES PAUL GA-5 LATER	96
GA-6	97
GA-8 6V6	98
GA-15RVT	98
GA-17RVT SCOUT	99
GA-19RVT	99
GA-20	100
GA-20T	100
GA-20T	101
GA-20T	101
GA-25RVT	102
GA-30 (6SJ7 PREAMP)	102
GA-30RV	103
LES PAUL GA-40	103
GA-55RVT	104
GA-60	104
GA-75W	105
GA-77RET	105
GA-83S STEREO-VIB (PREAMP)	106
GA83S STEREO-VIB	107
GA-300RVT	107
HAWK	108
REVERB-12	108
SKYLARK/TREMOLO	109
THOR BASS AMP	109

Gretsch
CHET ATKINS	110

GT Electronics
TUBE DIRECT UNIT	111
STUDIO 220 PREAMP	111

Hammond
ORGAN REVERB AO-35 (EARLY A-100)	112
ORGAN AO-39	112
PWR-AMP AO-39 WIRING	113
ORGAN REVERB AO-44 EXP (A-100)	113

Harmony
H-204 (1995)	114
H-440	114

Hiwatt
DR PREAMP STAGE	115
DR 103-112 OUTPUT	115
DR201-203 OUTPUT	116
POWER SUPPLY 50W	116
POWER SUPPLY 100W	117
POWER SUPPLY 150W	117
STA 100-200 PREAMP	118
STA 400 SLAVE	118

Kay
K-505	119
K-520	119

McIntosh
175	120
275	120
MC-30 POWER AMP	121
MC-60 TYPE A 125	121
MC-225	122
MC-2100 POWER SUPPLY	122

Maestro
6V6 TREMOLO AMP	123
ECHOPLEX (1966)	123
ECHOPLEX EP-2	124
GA-78RV	124
M1-RVT	125
M-201	125
M-216RVT	126
MA-40RVT	126

Magnatone
180	127
480	127

Marshall
50W	128
50W MASTER VOL (MODEL 2204)	128
50W MASTER VOL (MODEL 2204 REV)	129
BASS AMP (MODEL 1992)	129
100W PA (MODEL 1968)	130
200W PA (MODEL 1966)	130
BASS AMP (MODEL 1992)	131
BLUESBREAKER REISSUE (MODEL 1962)	131
JCM 800 BASS (MODEL 1992)	132
JCM 800 LEAD (50 & 100W PWR)	132
JCM 800 LEAD (PREAMP)	133
JCM 800 REV (MODEL 4140/4145)	133
MAJOR 200W (MODEL 1963)	134
REVERB (2205/2810/4210 REV)	134
REVERB (2210 HEAD/4211 COMBO)	135
STUDIO 15 (MODEL 4001)	135
SUPER PA (MODEL 1963)	136

Mesa/Boogie
BOOGIE MARK I	137
BOOGIE MARK IIA	137
BOOGIE MARK IIB	138
BOOGIE MARK IIC & PREAMP	138
BOOGIE MARK III PREAMP	139
BOOGIE MARK III PWR/OUTPUT	139
BOOGIE BASS 400	140

Orange
120W GRAPHIC (EARLY MODEL)	141
120W GRAPHIC MKII	141
MODEL 125	142

Peavey
CLASSIC 30	143
CLASSIC 120	143
VAN HALEN 120 PREAMP (DIAG 1)	144
VAN HALEN 120 PREAMP (DIAG 2)	144

RCA Theremin
1929 THEREMIN	145

Randall
RGT-100	146

Revox
STEREO TAPE RECORDER	147

Rickenbacker
B16/16D	148
B SERIES GUITAR	148

Selmer
ZODIAC TWIN 30	149
ZODIAC TWIN 50	149

Silvertone
1396	150
1474	150
1482	151
1484	151
1485	152
2 X 6V6 AMP	152
2 X 6V6 AMP REVISED	153
OLD 2 X 6L6 TREMOLO AMP	153

Sound
BIG 12E & 25RT	154
STUDIO-20	154

Sound City
200W LE	155
BASS 150W	155
CONCORD	156
L-B 120 MARK IV	156
L-B PLUS 50	157
120W SLAVE	157

Sovtek
MIG 50 MASTER VOLUME	158
MIG 60 LEAD	158

Sunn
1200	160
2000S	160
MODEL A	161
MODEL T SUPER	161
SENTURA 1	162
SONIC II/1-40/200S/SORADO	162

Supro
AMP 1947	163
LATE '50S 1 X 6V6 AMP	163
REVERB	164
S6625	164
S6651	165
S6698 TREM/REV	165
STEEL GUITAR AMP 1946	166

Traynor
YBA-1 BASS MASTER	167
YBA-2 BASS MATE	167
YBA-3 POWER AMP	168
YBA-3 CUSTOM SPECIAL PREAMP	168
YGA-1 1A	169
YGM-2 GUITAR MATE	169

Univox
U-1011 LEAD AMP	170
U-1226 LEAD AMP	170

Valco
6650 TR	171
SUPER REVERB (MODEL 510-33)	171

Vox
AC10	172
AC15 MK II (1959)	172
AC15 MK III (1960)	173
AC30 REVERB (1978)	173
AC30 SILVER JUBILEE (VOX LTD)	174
AC30 (S.S. RECT)	174
AC30 TOP BOOST REISSUE (1994)	175
AC30 TWIN REVERB (W/MOD)	175
AC50/2 (GZ34 RECT)	176
AC50/4 MK 11 (S.S. RECT)	176
AC50/4 MK III	177
BERKELY II SUPER REVERB V-8	177
CAMBRIDGE REVERB V-3	178
V-15 (VOX LTD)	178

Wards Airline
GDR 8514A/8515/A	179
GDR 8517A	179
GDR-8518A	180
GDR 9012A	180
GIM 9151A	181
GVC-9058A	181

Watkins
1959 COPICAT MK II	182
DOMINATOR 35 MK IV	182
JOKER 30	183
W14T & W20T	183

Watkins/WEM
CLUBMAN MK 8	184
DOMINATOR 50	184
WESTMINSTER MK9	185

Western Electric
140A AMP	186
142A AMP	186
142A AMP (CONVERTED FOR 25W)	187
143A AMP	187

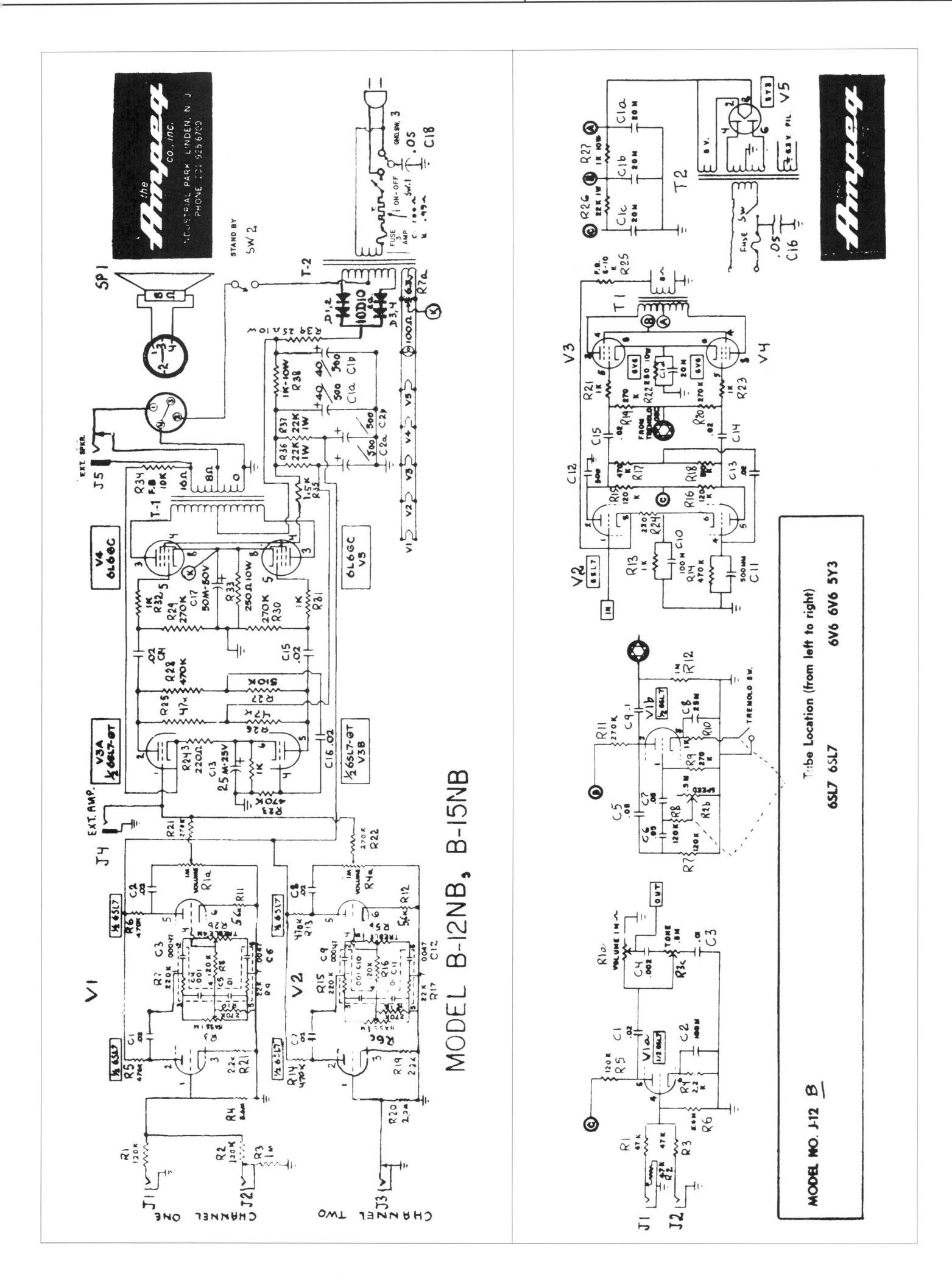

Ampeg

B-12NB & B-15NB
J12B JET

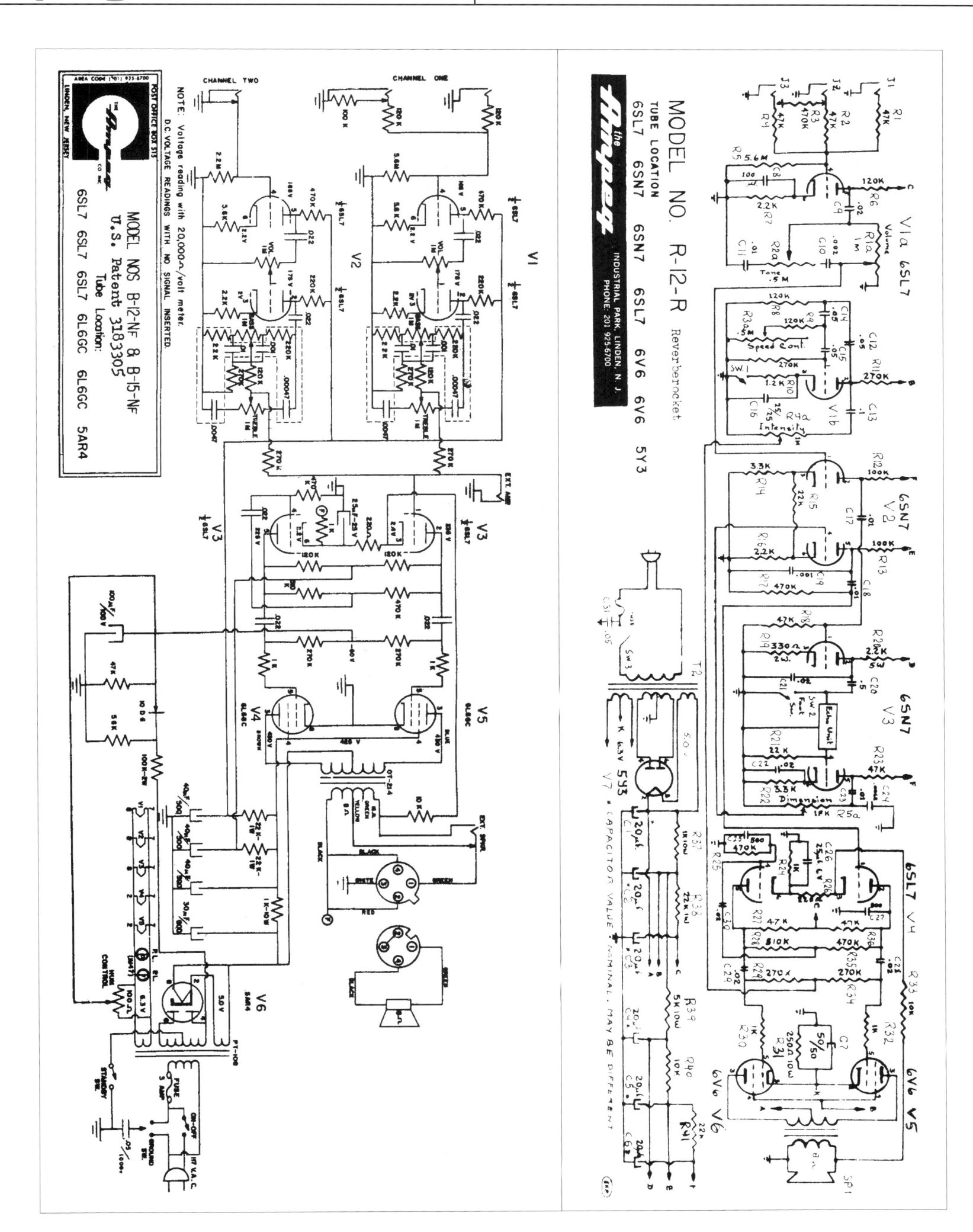

Ampeg

B-12NF & B-15NF
R-12-R REVERB ROCKET

B-12N & B-15N
B-18N

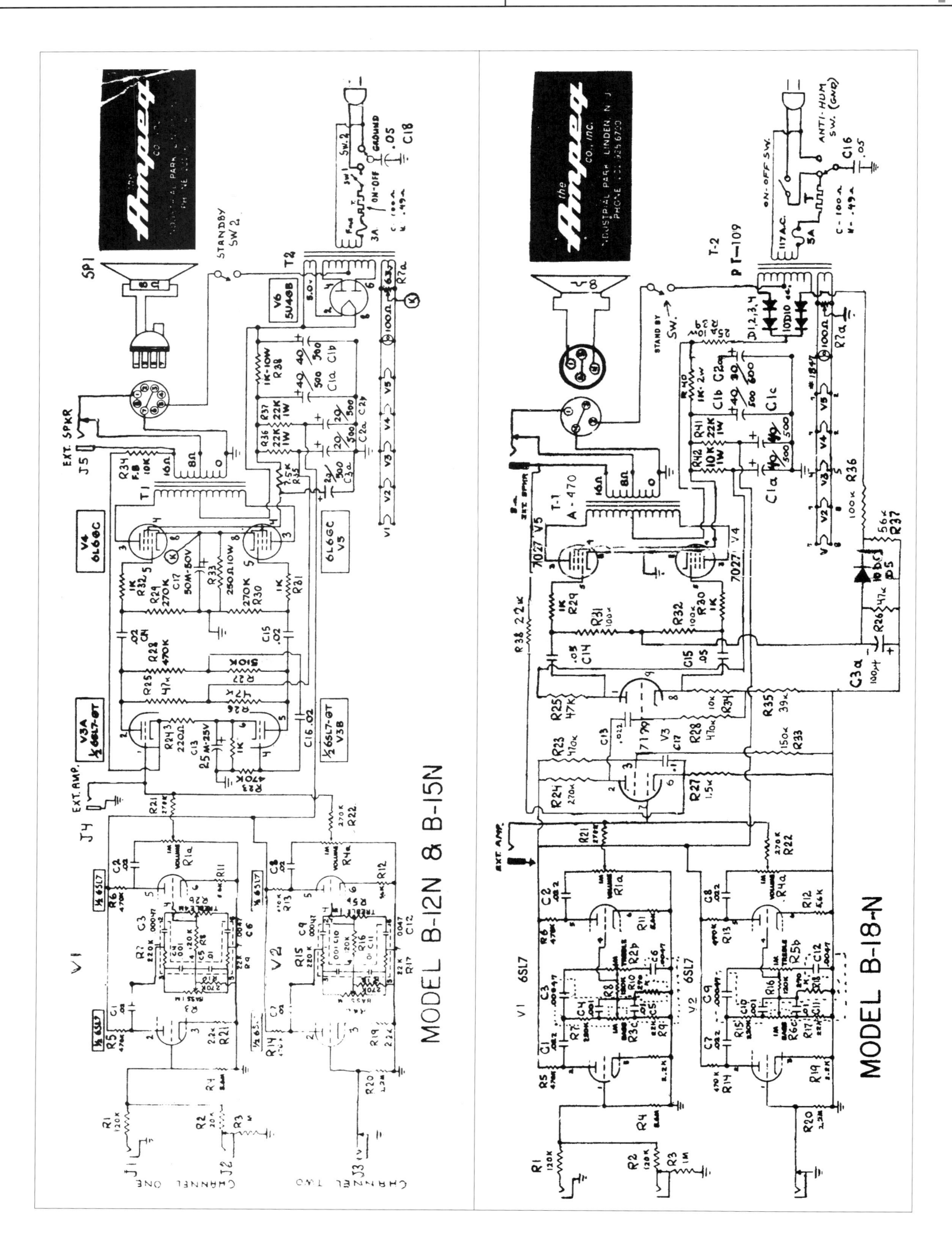

Ampeg

MODEL B-12N & B-15N

MODEL B-18-N

6 THE TUBE AMP BOOK

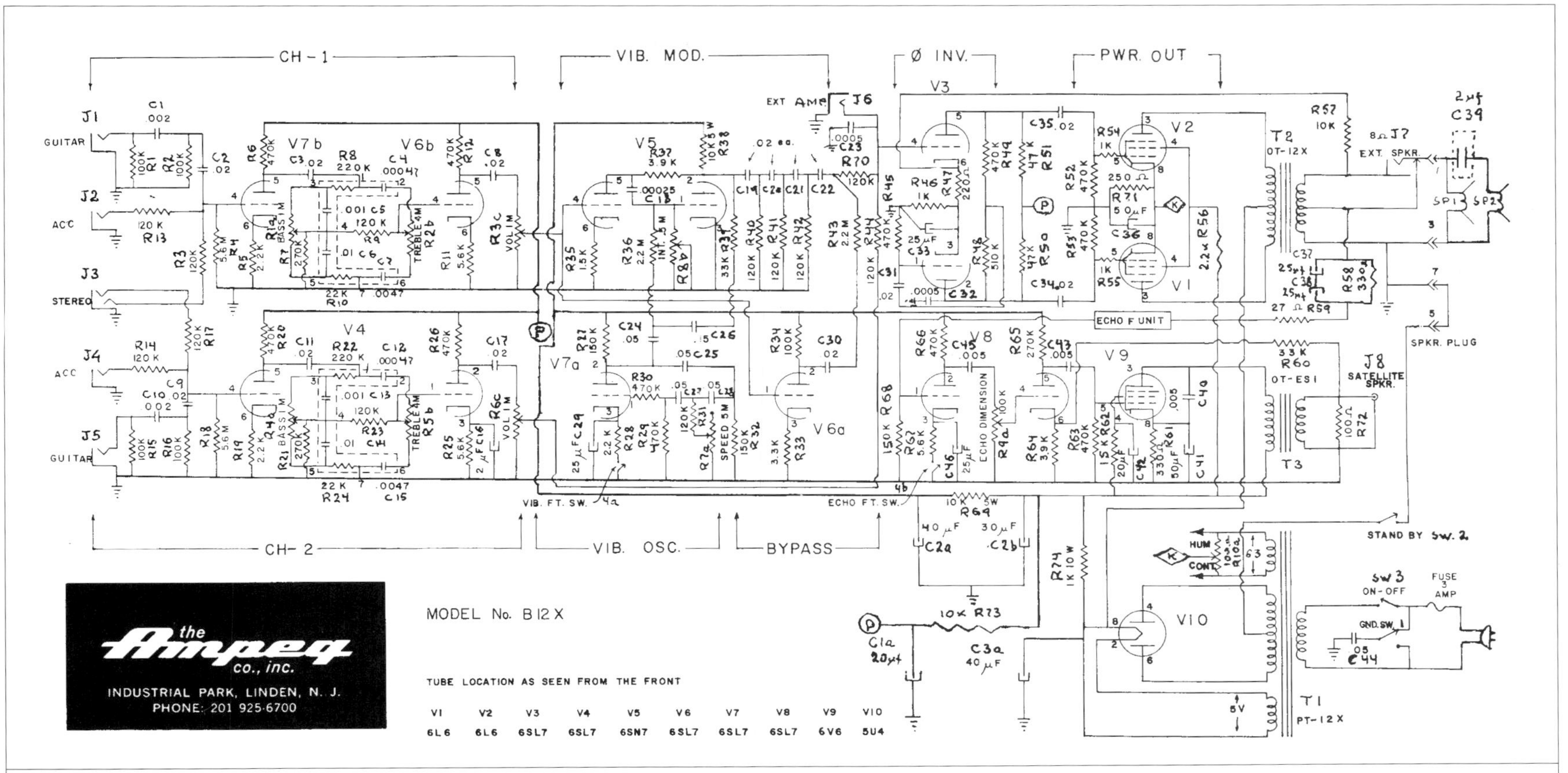
CH - 1
VIB. MOD.
Ø INV.
PWR. OUT
CH- 2
VIB. OSC.
BYPASS
J1 GUITAR
J2 ACC
J3 STEREO
J4 ACC
J5 GUITAR
EXT AMP J6
EXT. SPKR
SPKR. PLUG
J8 SATELLITE SPKR.
ECHO F UNIT
VIB. FT. SW.
ECHO FT. SW.
STAND BY SW. 2
SW 3 ON - OFF
FUSE 3 AMP
GND. SW. 1
HUM CONT.
OT-12X
OT-ES1
PT-12X
the Ampeg co., inc.
INDUSTRIAL PARK, LINDEN, N. J.
PHONE: 201 925-6700
MODEL No. B12X
TUBE LOCATION AS SEEN FROM THE FRONT
V1 V2 V3 V4 V5 V6 V7 V8 V9 V10
6L6 6L6 6SL7 6SL7 6SN7 6SL7 6SL7 6SL7 6V6 5U4

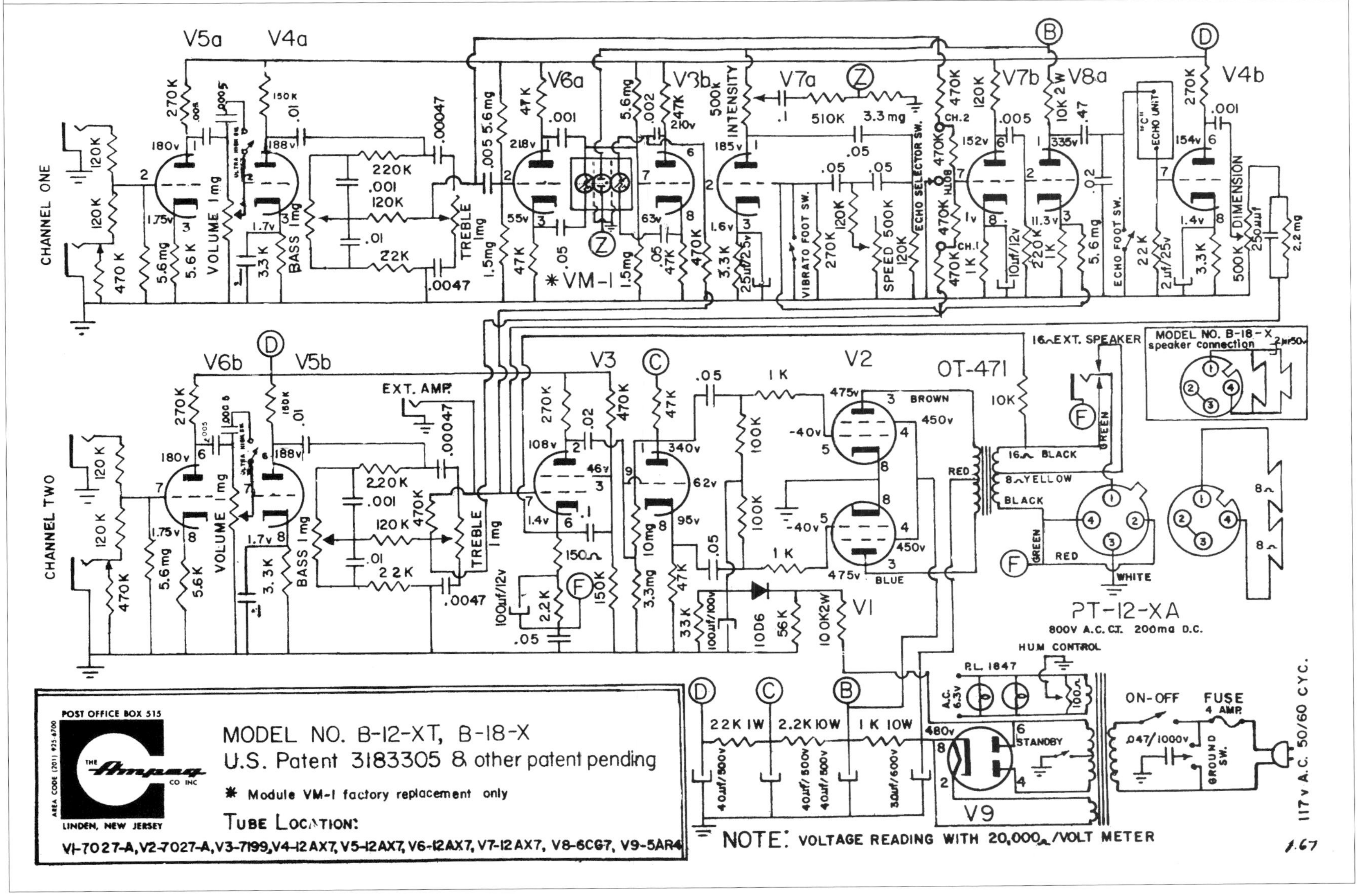
CHANNEL ONE
CHANNEL TWO
V5a V4a V6a V3b V7a V7b V8a V4b
V6b V5b V3 V2 V1 V9
VOLUME 1mg
BASS 1mg
TREBLE 1mg
*VM-1
INTENSITY
SPEED 500K
VIBRATO FOOT SW.
ECHO SELECTOR SW.
ECHO FOOT SW.
DIMENSION
"C" ECHO UNIT
EXT. AMP
OT-471
16Ω EXT. SPEAKER
MODEL NO. B-18-X speaker connection
PT-12-XA
800V A.C.C.T. 200ma D.C.
HUM CONTROL
ON-OFF
FUSE 4 AMP
STANDBY
GROUND SW.
117 v A.C. 50/60 CYC.
POST OFFICE BOX 515
AREA CODE (201) 925-6700
THE Ampeg CO INC
LINDEN, NEW JERSEY
MODEL NO. B-12-XT, B-18-X
U.S. Patent 3183305 & other patent pending
* Module VM-1 factory replacement only
TUBE LOCATION:
V1-7027-A, V2-7027-A, V3-7199, V4-12AX7, V5-12AX7, V6-12AX7, V7-12AX7, V8-6CG7, V9-5AR4
NOTE: VOLTAGE READING WITH 20,000Ω/VOLT METER
1.67

B-22X
B-25

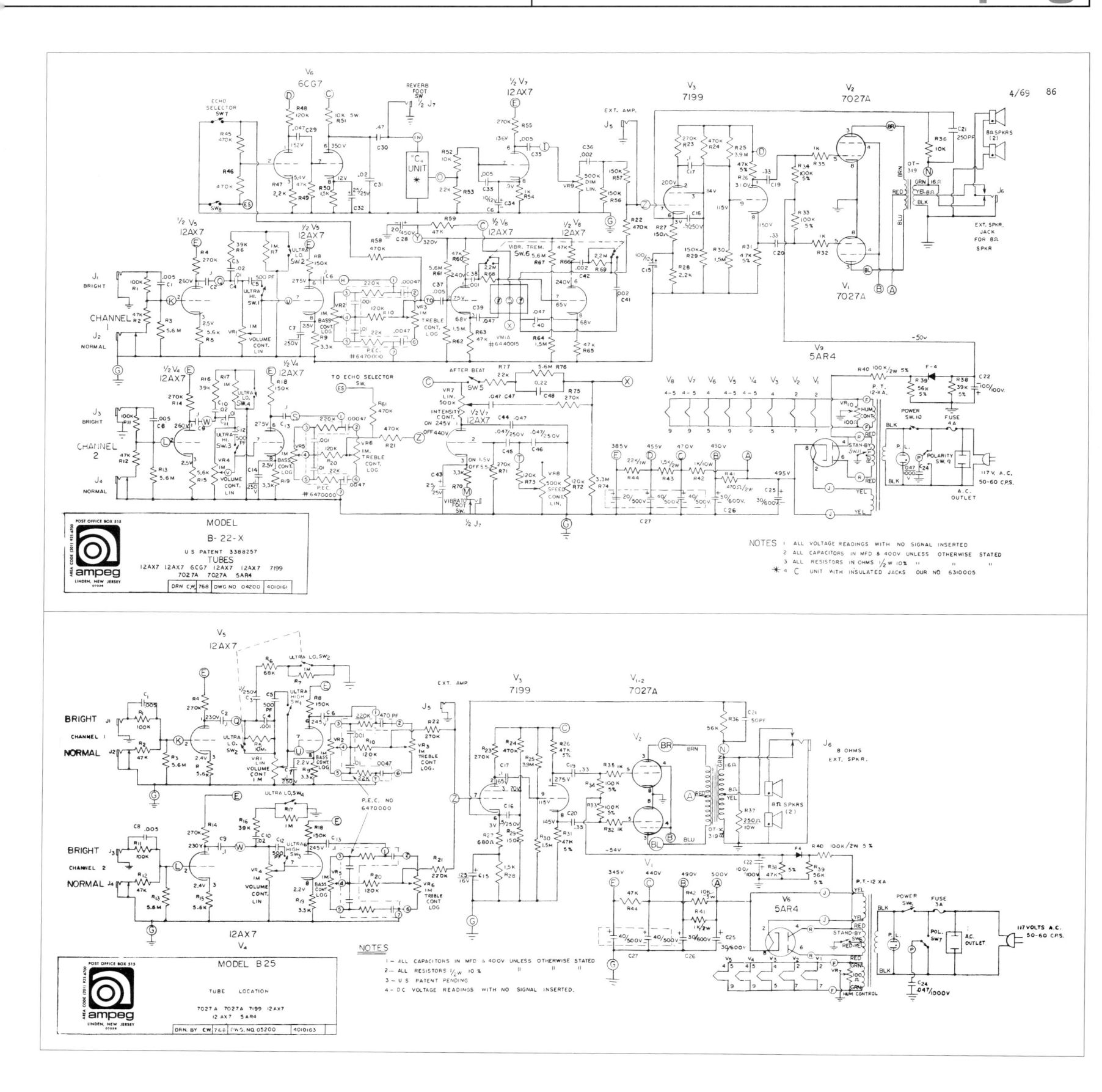

Ampeg

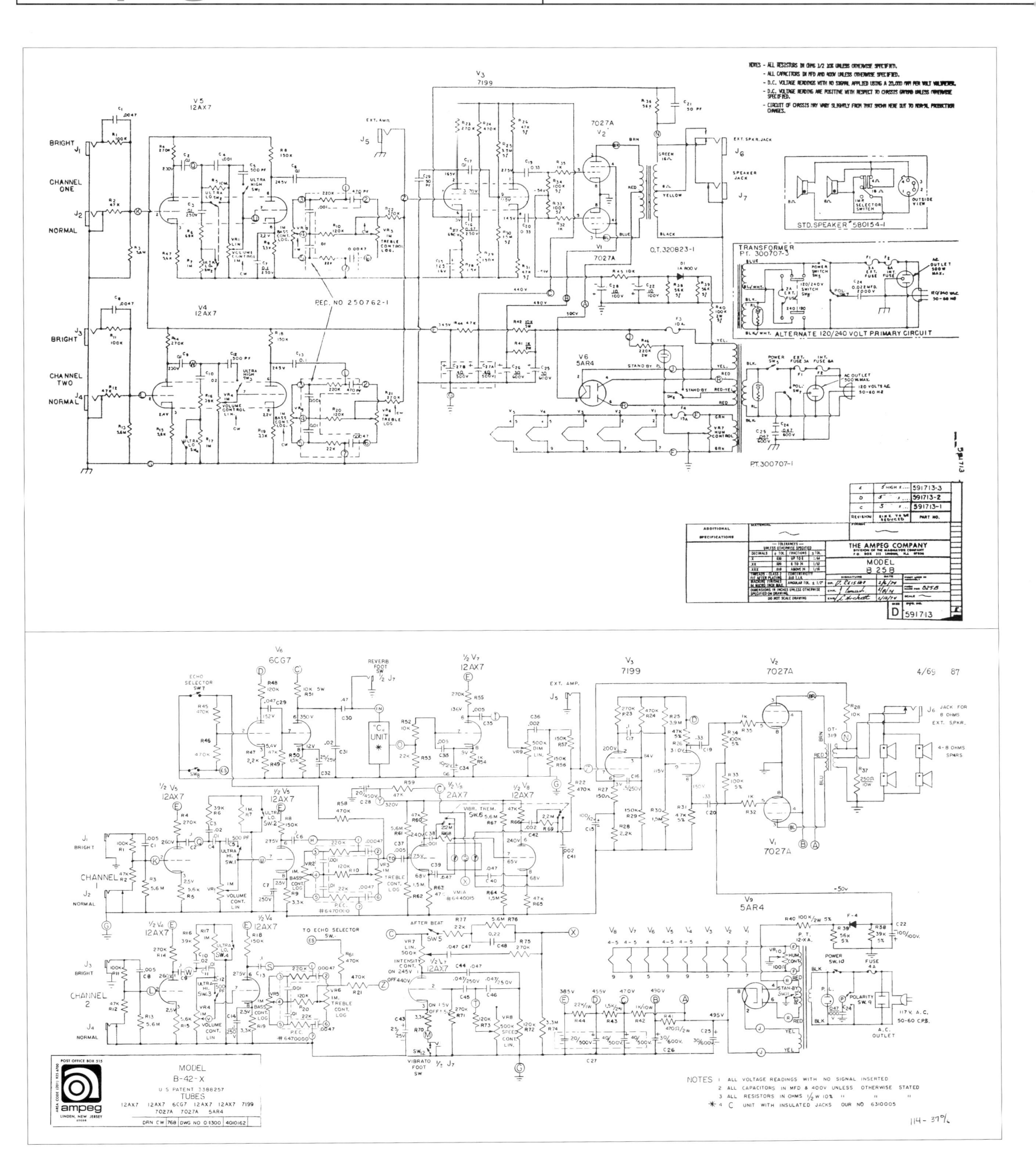

Ampeg

B-25B
B-42X

ET-1
ET-2B

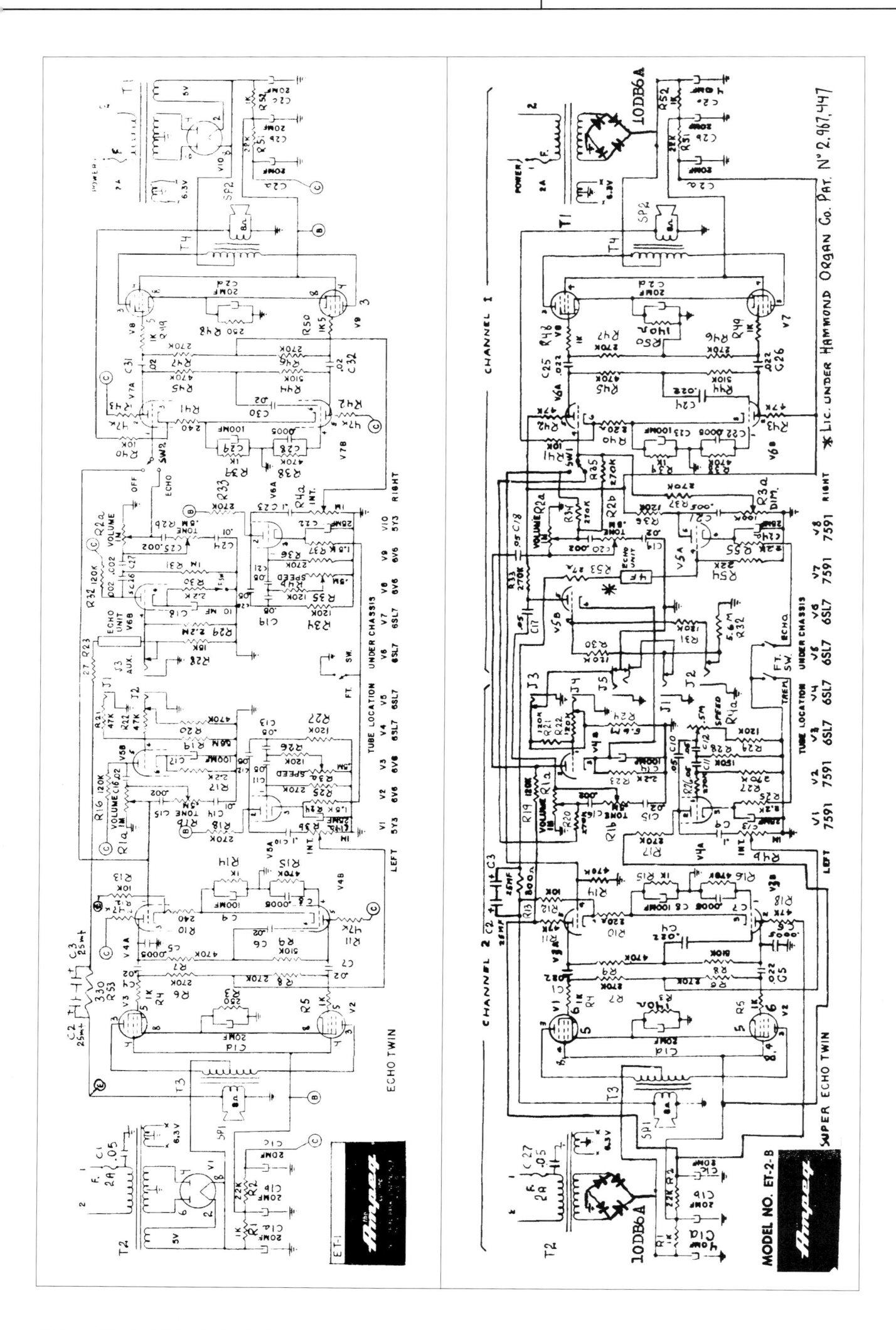

Ampeg

10 THE TUBE AMP BOOK

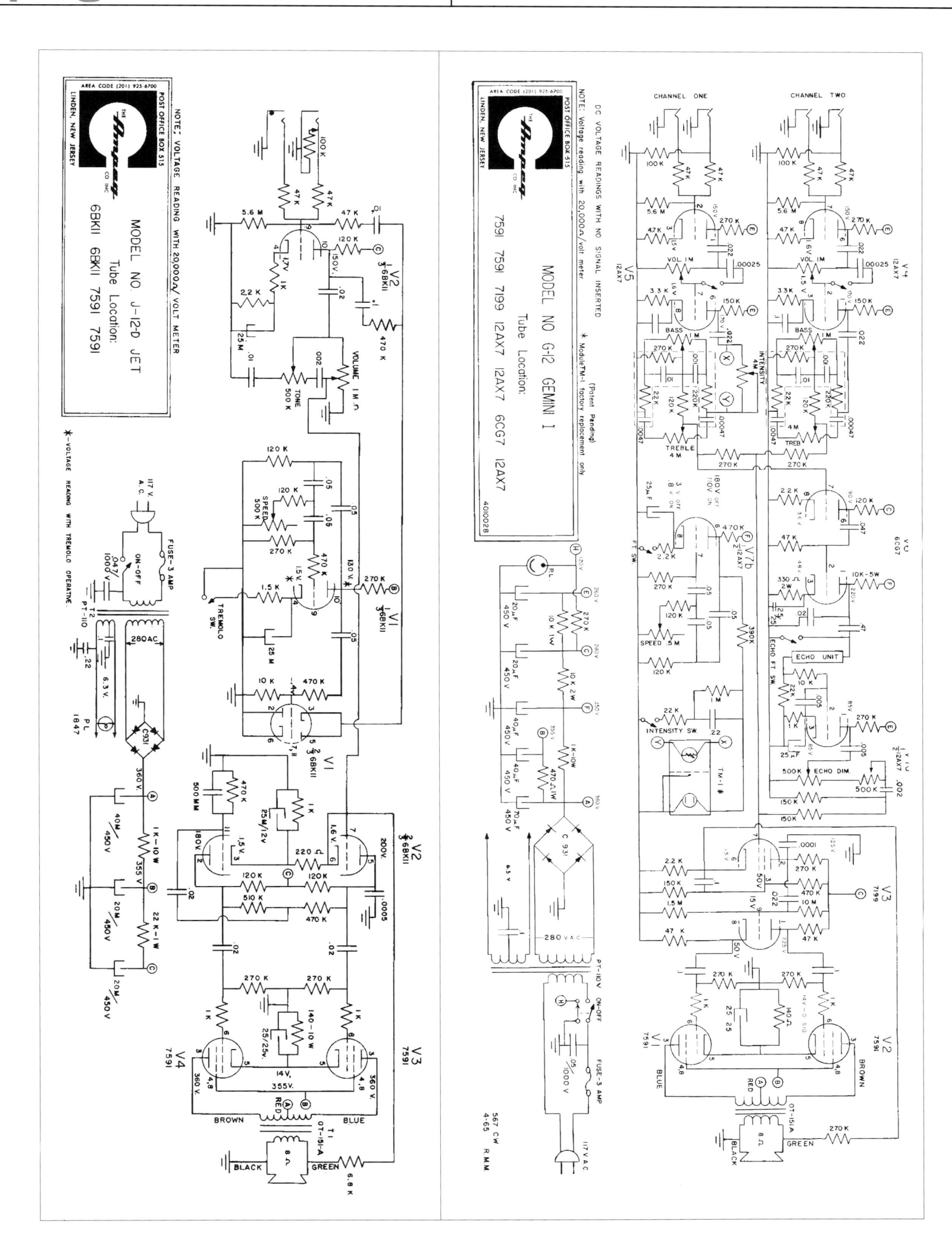

Ampeg

G-12
J-12D

SBT PREAMP
SB-12

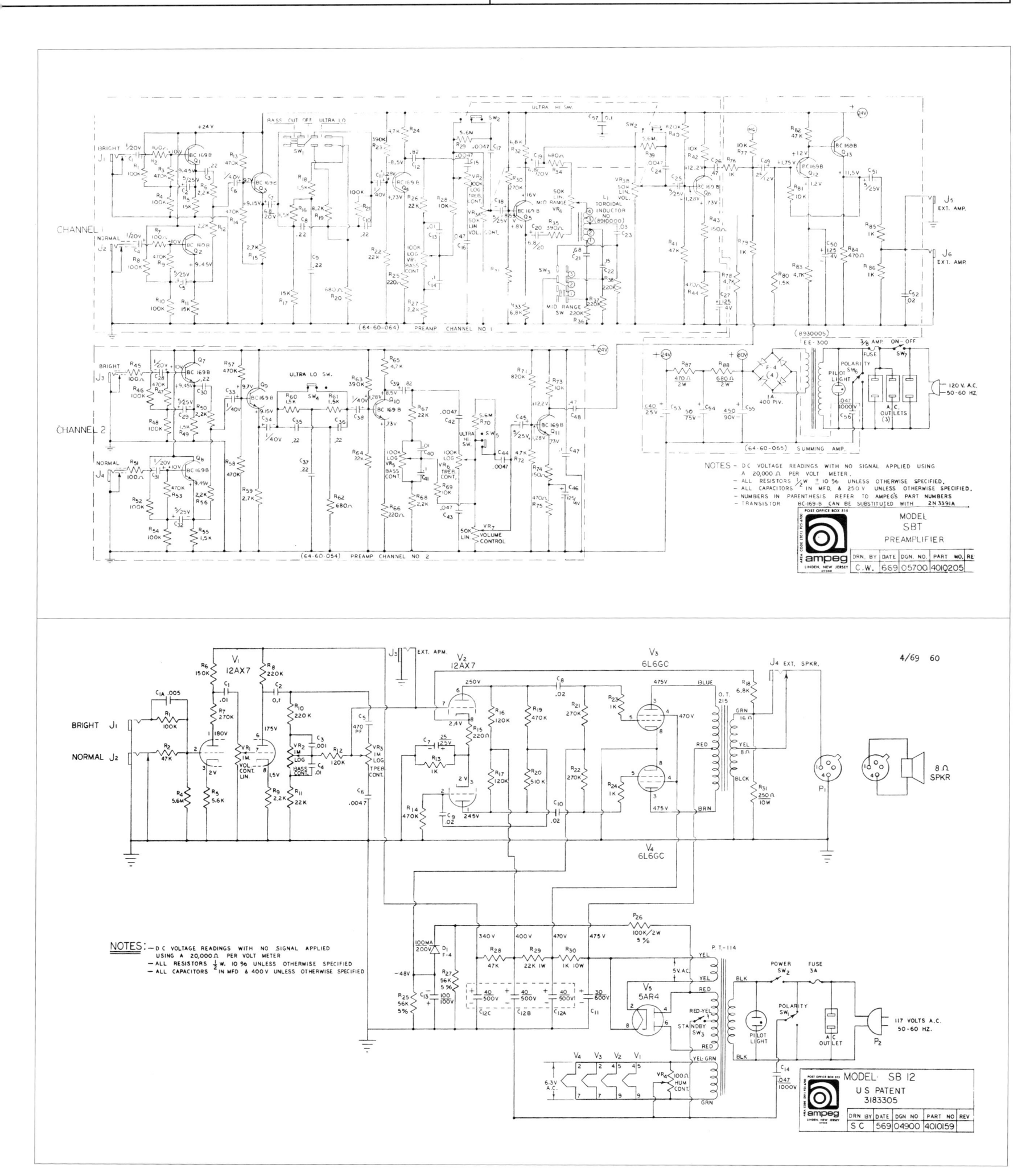

Ampeg

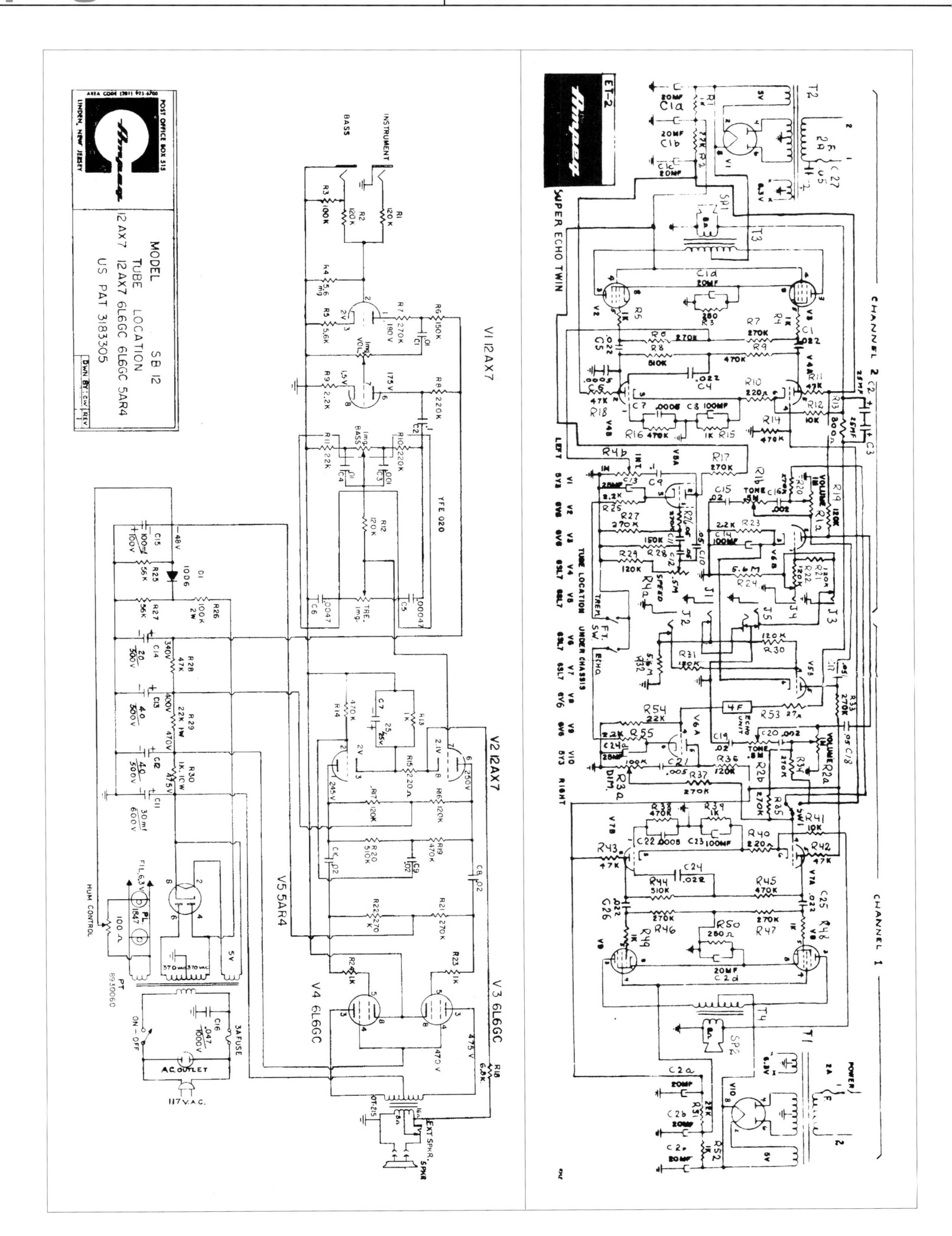

Ampeg

SBT-12 (REV I)
SUPER ECHO TWIN

THE TUBE AMP BOOK 13

SVT (REV A)
SVT PREAMP

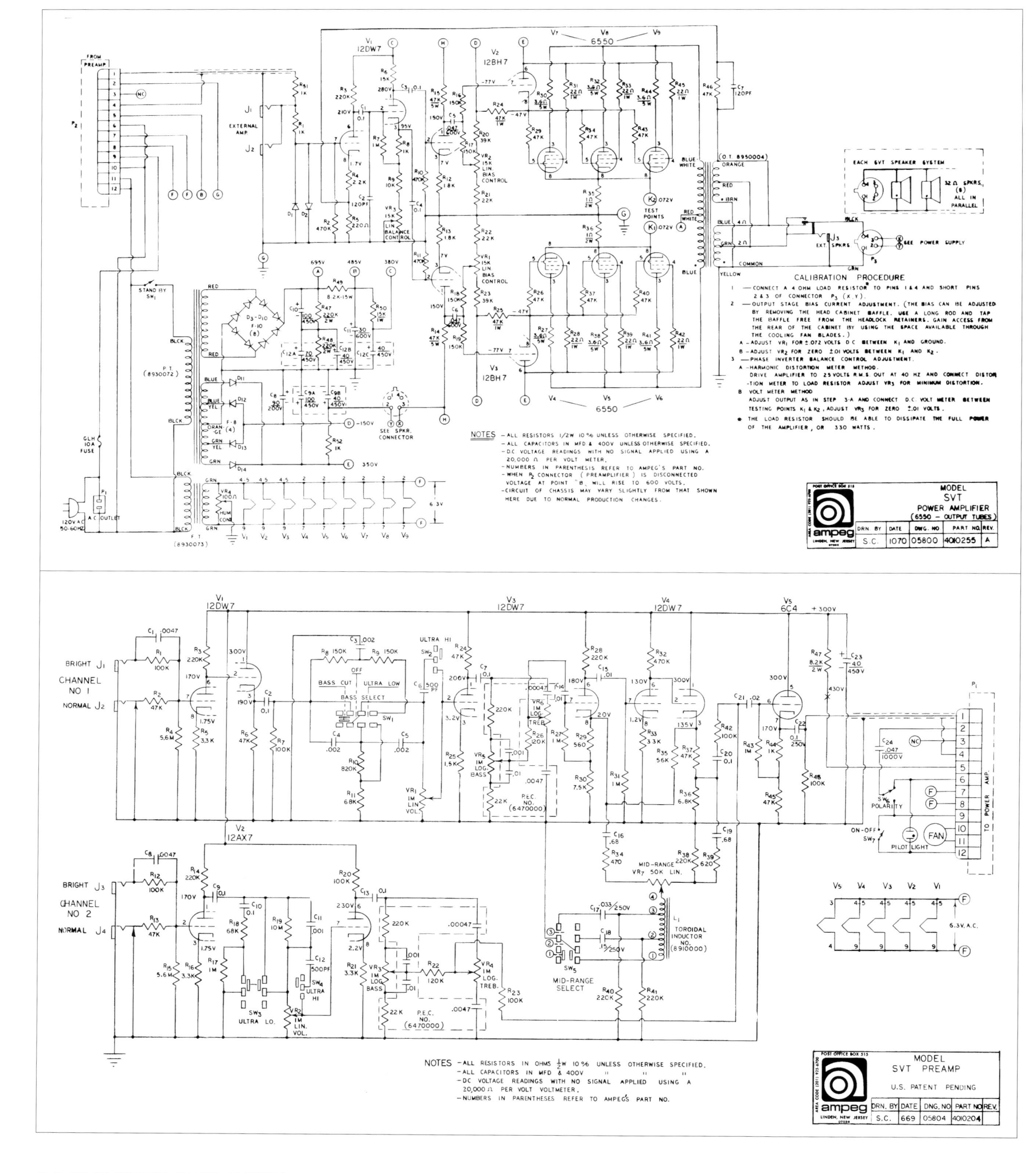

Ampeg

14 THE TUBE AMP BOOK

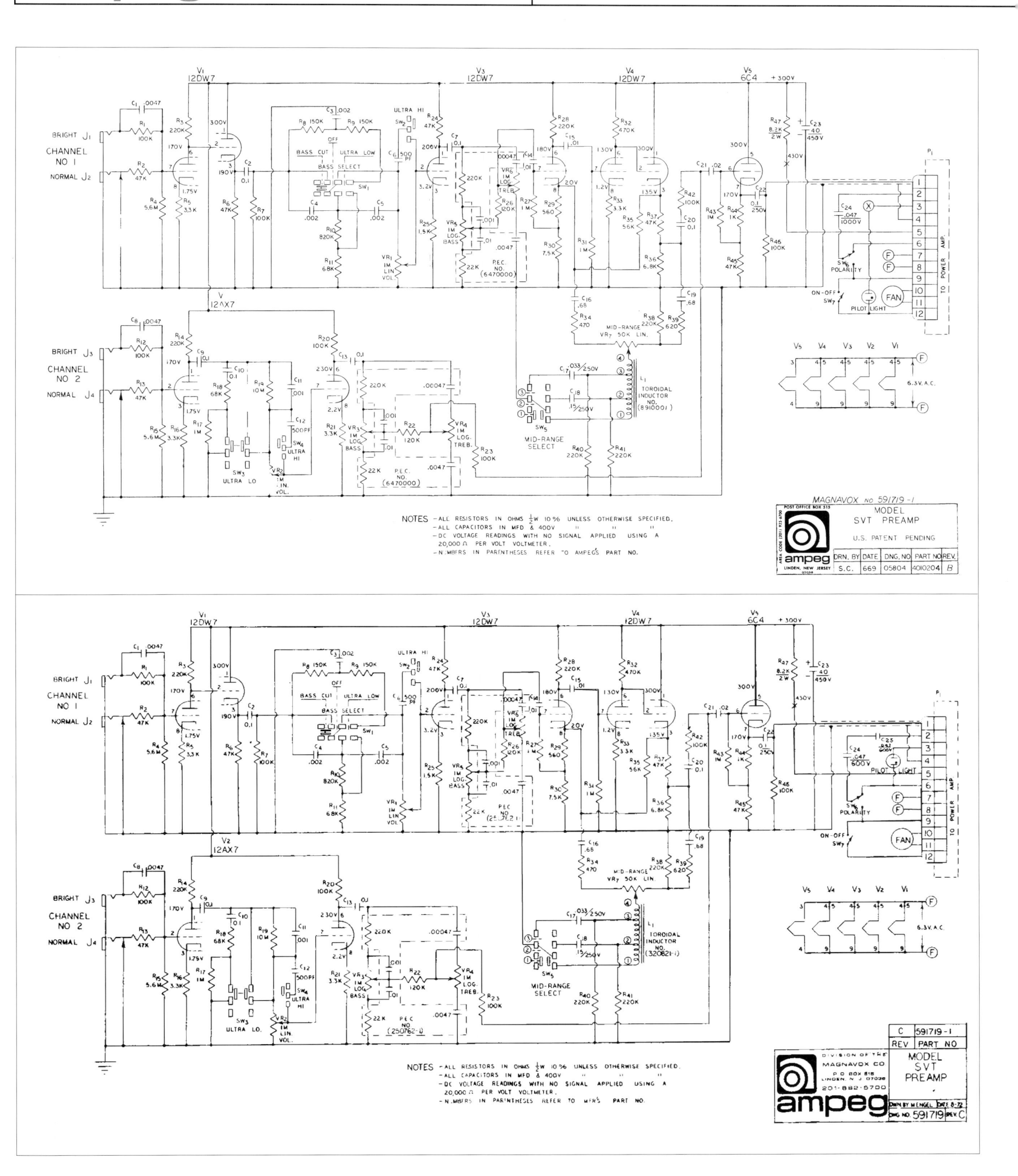

Ampeg

SVT PREAMP (REV B)
SVT PREAMP (REV C)

THE TUBE AMP BOOK 15

SVT POWER AMP (REV D)
SVT V9

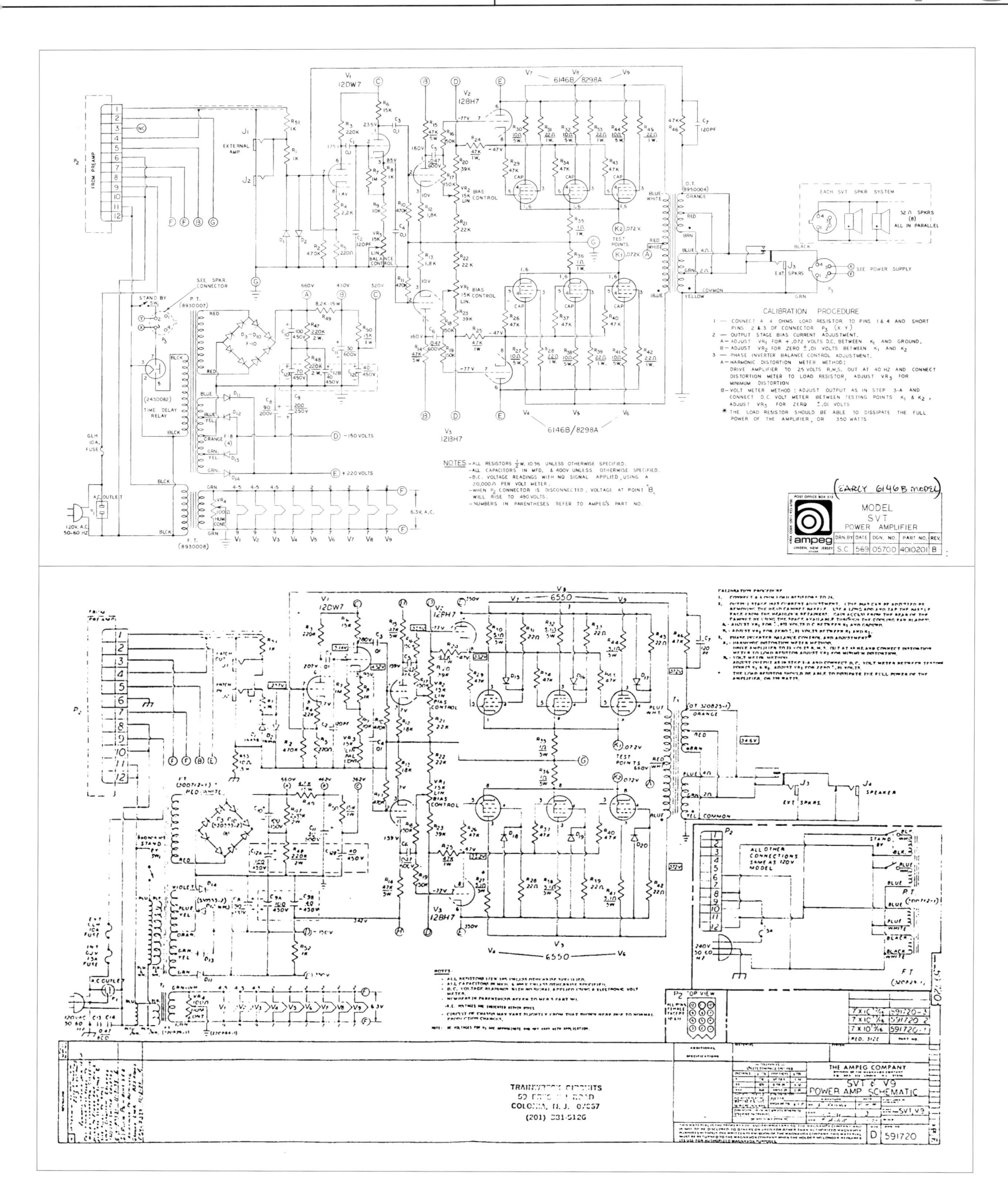

Ampeg

16 THE TUBE AMP BOOK

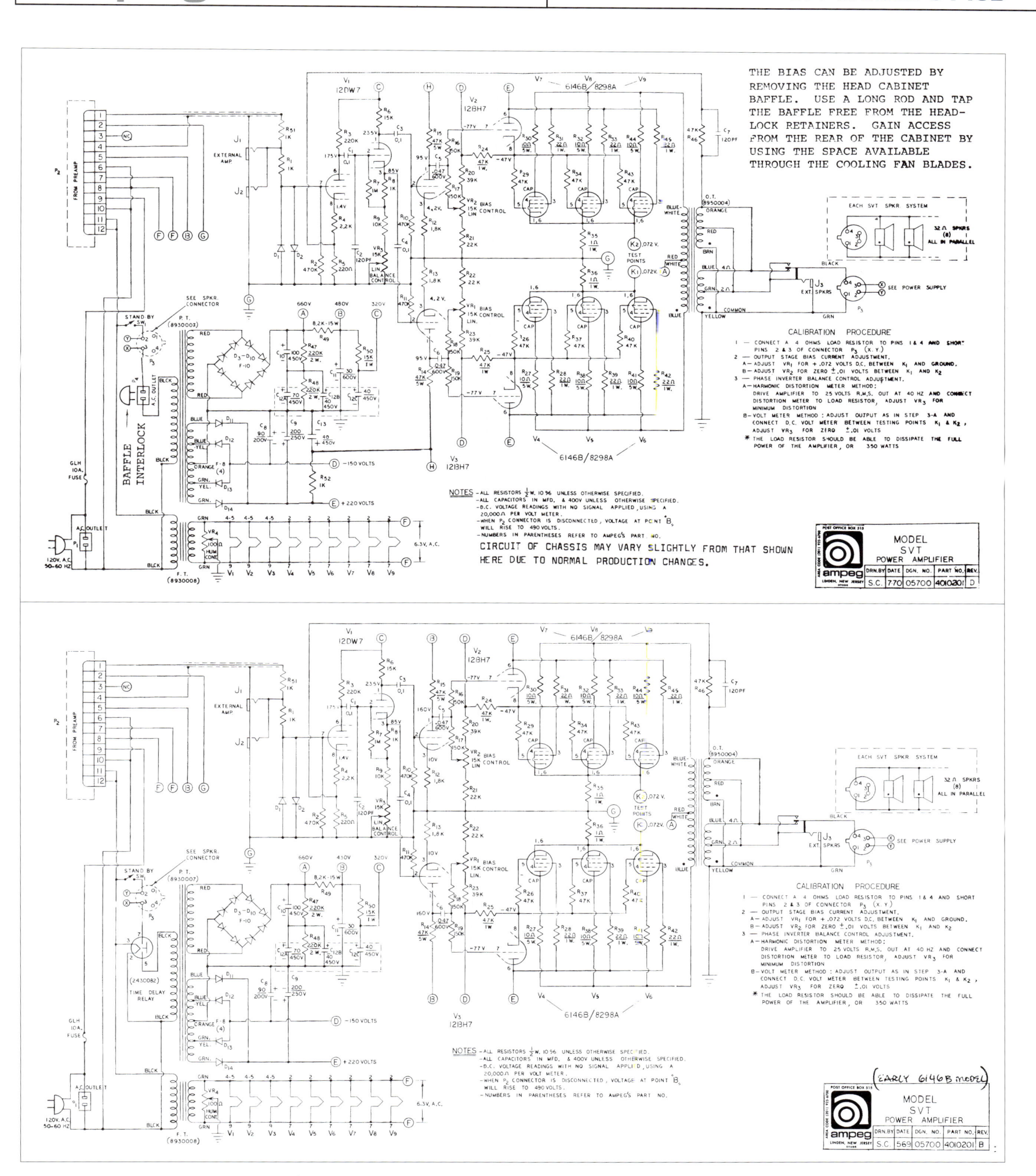

Ampeg

SVT MODEL 6146
SVT MODEL 6146B

CHB-20A
E30

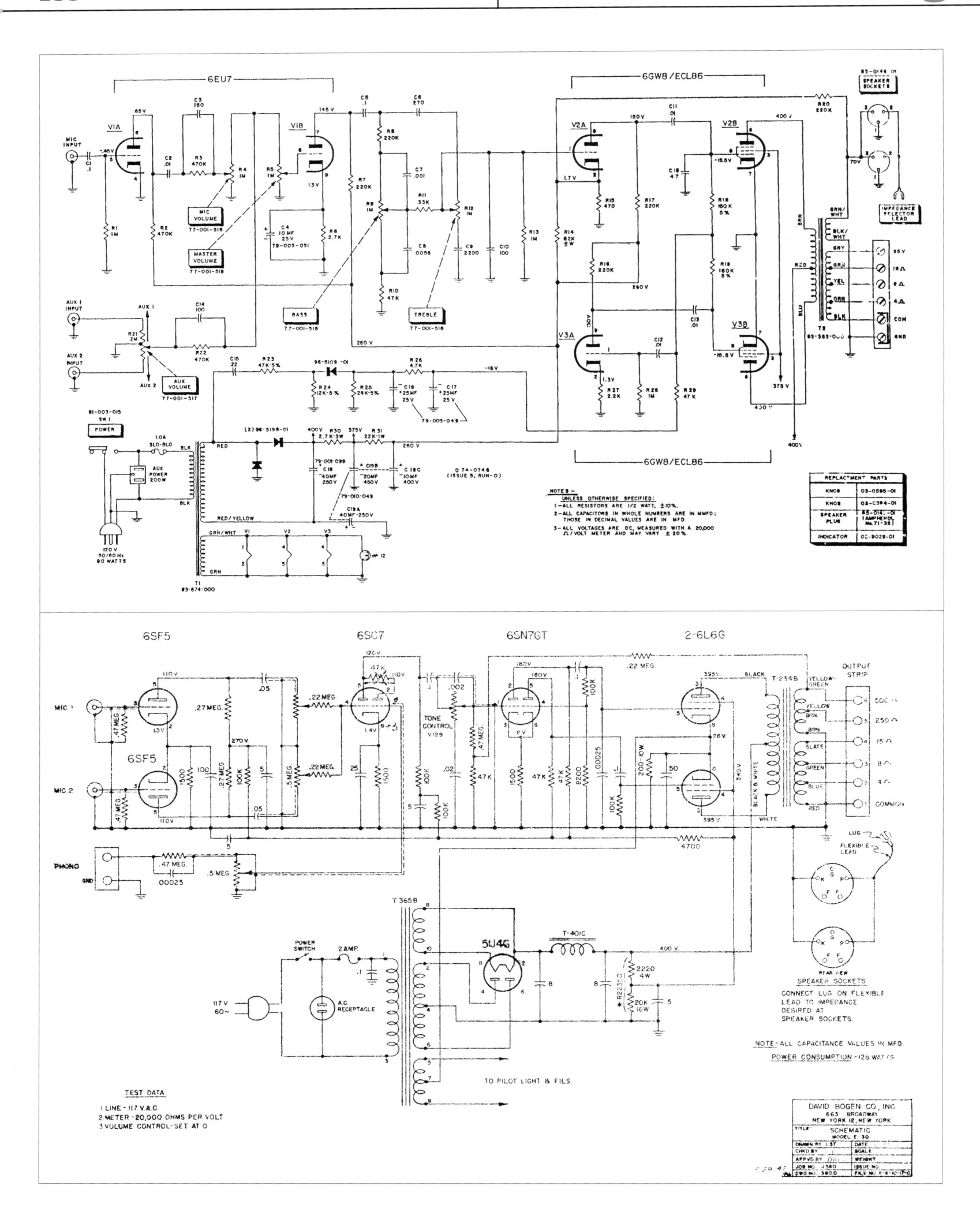

Bogen

18 THE TUBE AMP BOOK

Danelectro

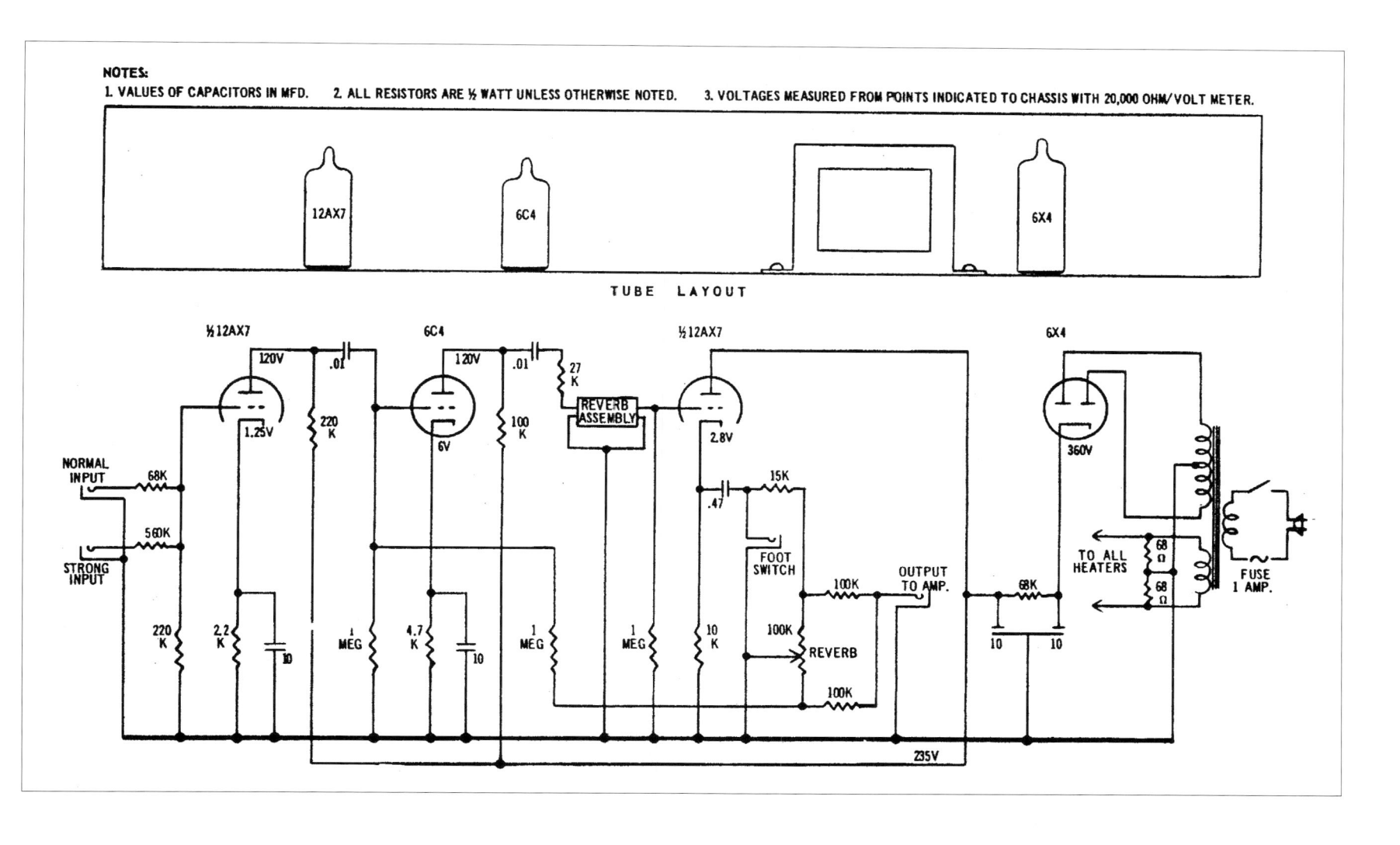

REVERB BOX 9100

NOTES:
1. VALUES OF CAPACITORS IN MFD.   2. ALL RESISTORS ARE ½ WATT UNLESS OTHERWISE NOTED.   3. VOLTAGES MEASURED FROM POINTS INDICATED TO CHASSIS WITH 20,000 OHM/VOLT METER.

TUBE LAYOUT

EA-10 DELUXE
EA-25 CENTURY

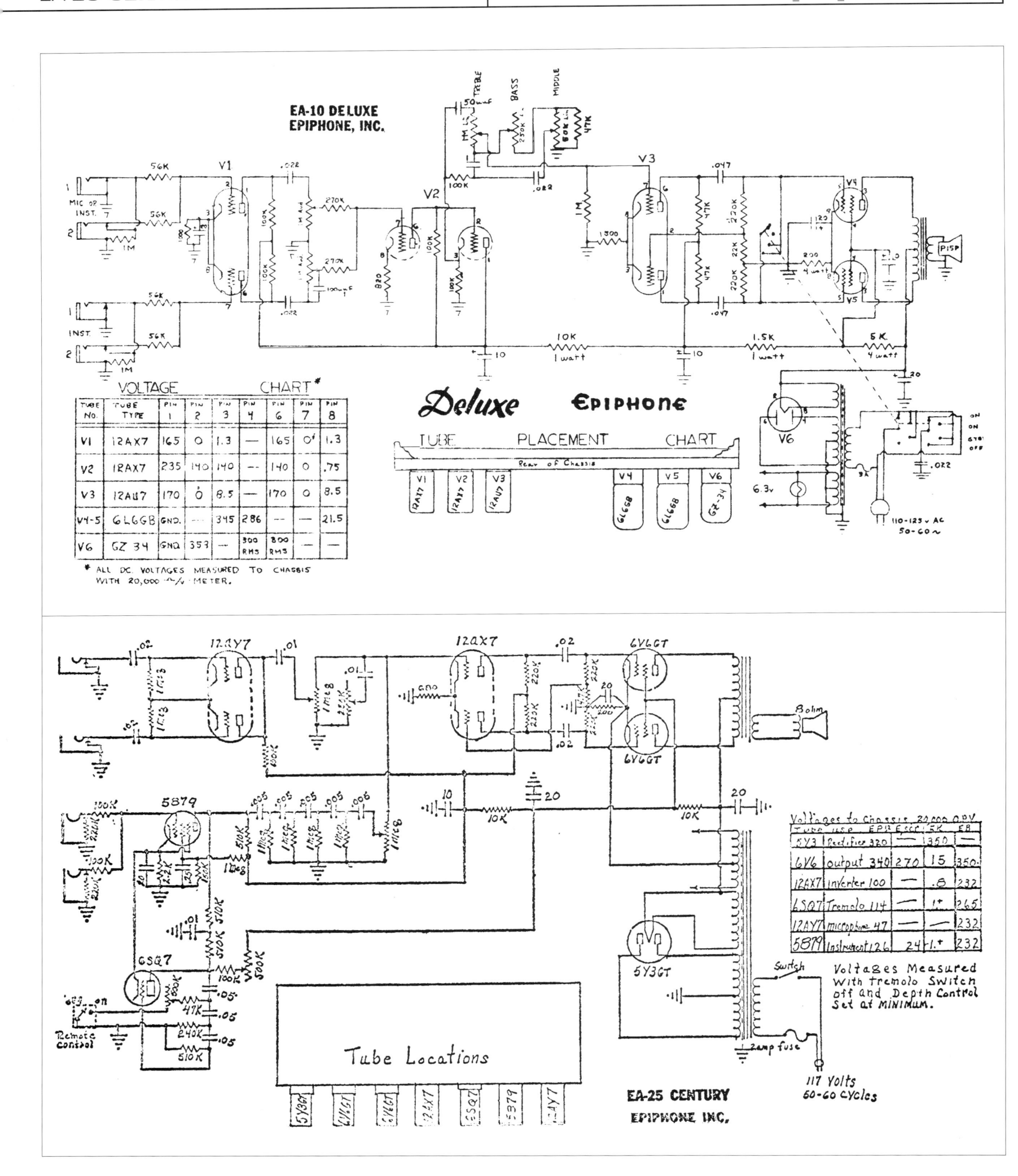

Epiphone

EA-10 DELUXE — EPIPHONE, INC.

Voltage Chart*

Tube No.	Tube Type	Pin 1	Pin 2	Pin 3	Pin 4	Pin 6	Pin 7	Pin 8
V1	12AX7	165	0	1.3	—	165	0	1.3
V2	12AX7	235	140	140	—	140	0	.75
V3	12AU7	170	0	8.5	—	170	0	8.5
V4-5	6L6GB	GND.	—	345	286	—	—	21.5
V6	GZ34	GND.	353	—	300 RMS	300 RMS	—	—

*ALL DC VOLTAGES MEASURED TO CHASSIS WITH 20,000 Ω/V METER.

Deluxe Epiphone — Tube Placement Chart

Rear of Chassis: V1 12AX7, V2 12AX7, V3 12AU7, V4 6L6GB, V5 6L6GB, V6 GZ-34

EA-25 CENTURY — EPIPHONE INC.

Voltages to Chassis 20,000 O.P.V.

Tube	Use	EP	ESCC	SK	EB
5Y3	Rectifier	320	—	1350	—
6V6	Output	340	270	15	350
12AX7	Inverter	100	—	.8	232
6SQ7	Tremolo	114	—	1*	265
12AY7	Microphone	47	—	—	232
5879	Instrument	126	—	24 1.*	232

Voltages Measured With Tremolo Switch off and Depth Control Set at MINIMUM.

117 Volts 60-60 cycles

Tube Locations: 5Y3GT, 6V6GT, 6V6GT, 12AX7, 6SQ7, 5879, 12AY7

20 THE TUBE AMP BOOK

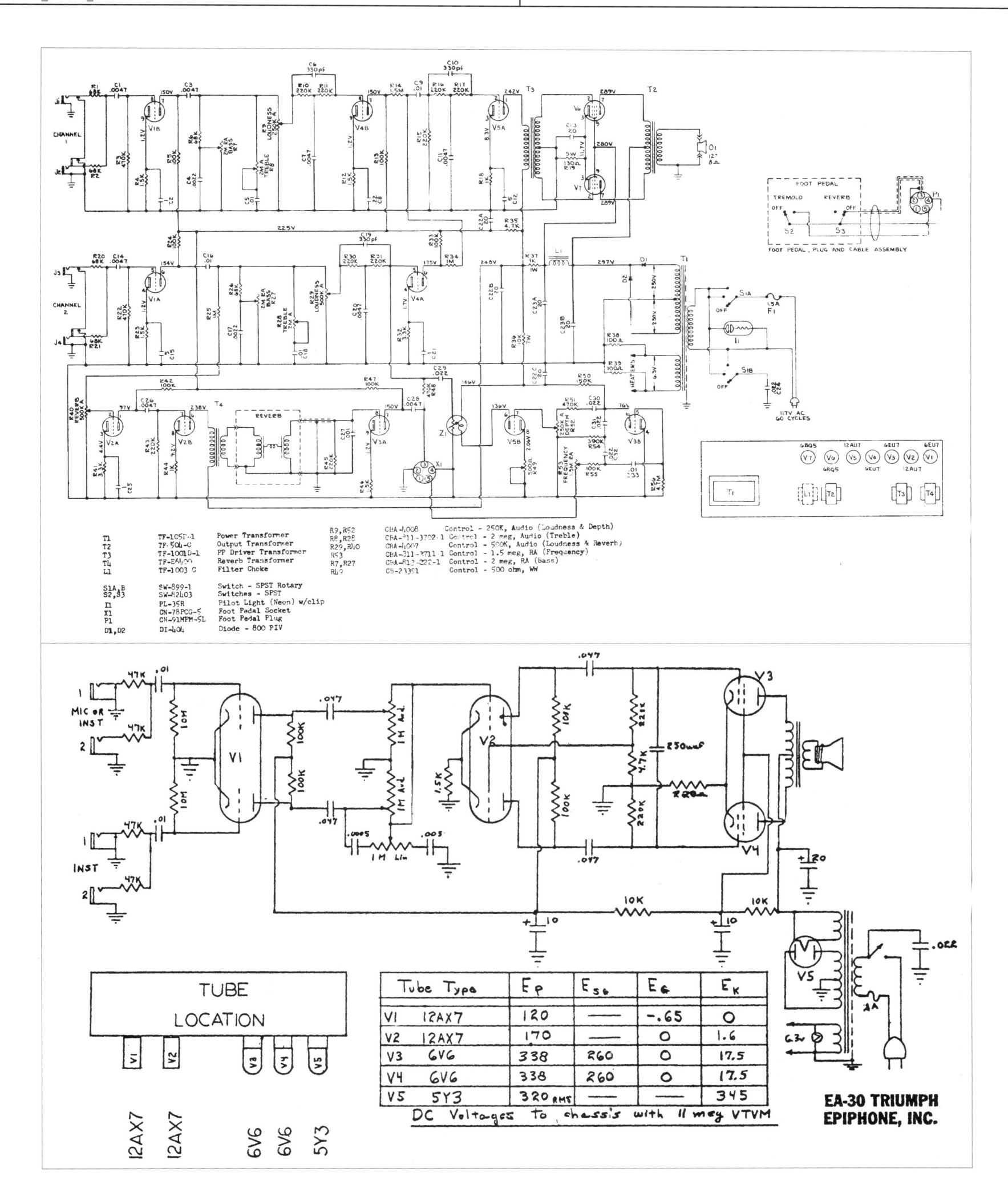

Epiphone

EA-26RVT
EA-30 TRIUMPH

T1	TF-105T-1	Power Transformer	R9,R52	CBA-4008	Control - 250K, Audio (Loudness & Depth)
T2	TP-504-C	Output Transformer	R8,R28	CBA-211-3702-1	Control - 2 meg, Audio (Treble)
T3	TF-1001D-1	PP Driver Transformer	R29,R40	CBA-4007	Control - 500K, Audio (Loudness & Reverb)
T4	TF-E4400	Reverb Transformer	R53	CBA-311-3711-1	Control - 1.5 meg, RA (Frequency)
L1	TF-1003 C	Filter Choke	R7,R27	CBA-812-222-1	Control - 2 meg, RA (bass)
			R49	CS-23391	Control - 500 ohm, WW
S1A,B	SW-899-1	Switch - SPST Rotary			
S2,S3	SW-82403	Switches - SPST			
I1	PL-35R	Pilot Light (Neon) w/clip			
X1	CN-78PCG-5	Foot Pedal Socket			
P1	CN-91MPM-5L	Foot Pedal Plug			
D1,D2	DI-404	Diode - 800 PIV			

Tube Type	Ep	Esc	Eg	Ek
V1	12AX7	120	—	-.65	0
V2	12AX7	170	—	0	1.6
V3	6V6	338	260	0	17.5
V4	6V6	338	260	0	17.5
V5	5Y3	320 RMS	—		345

DC Voltages to chassis with 11 meg VTVM

EA-30 TRIUMPH
EPIPHONE, INC.

THE TUBE AMP BOOK 21

Epiphone

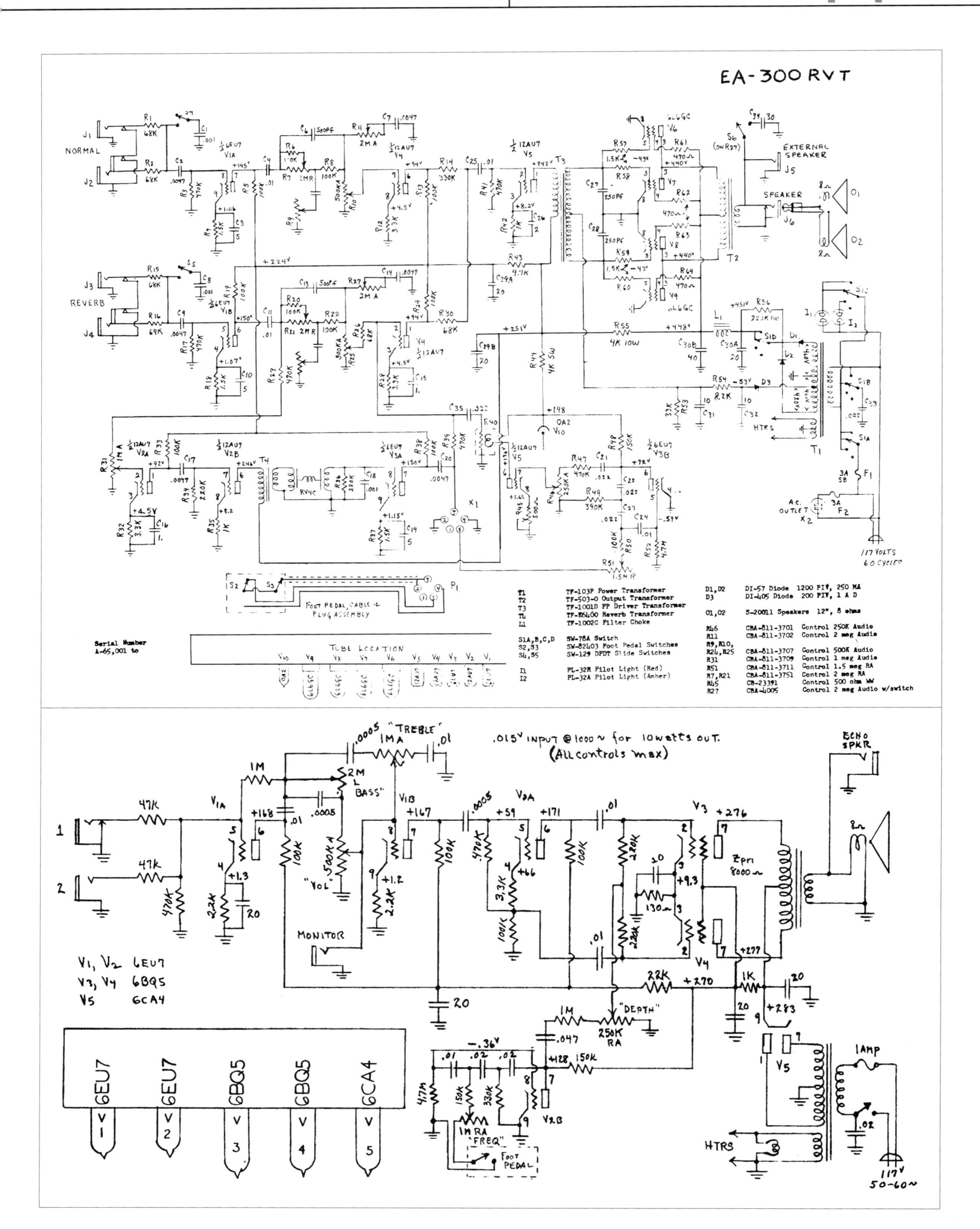

EA-300RVT
EA-35T

Epiphone

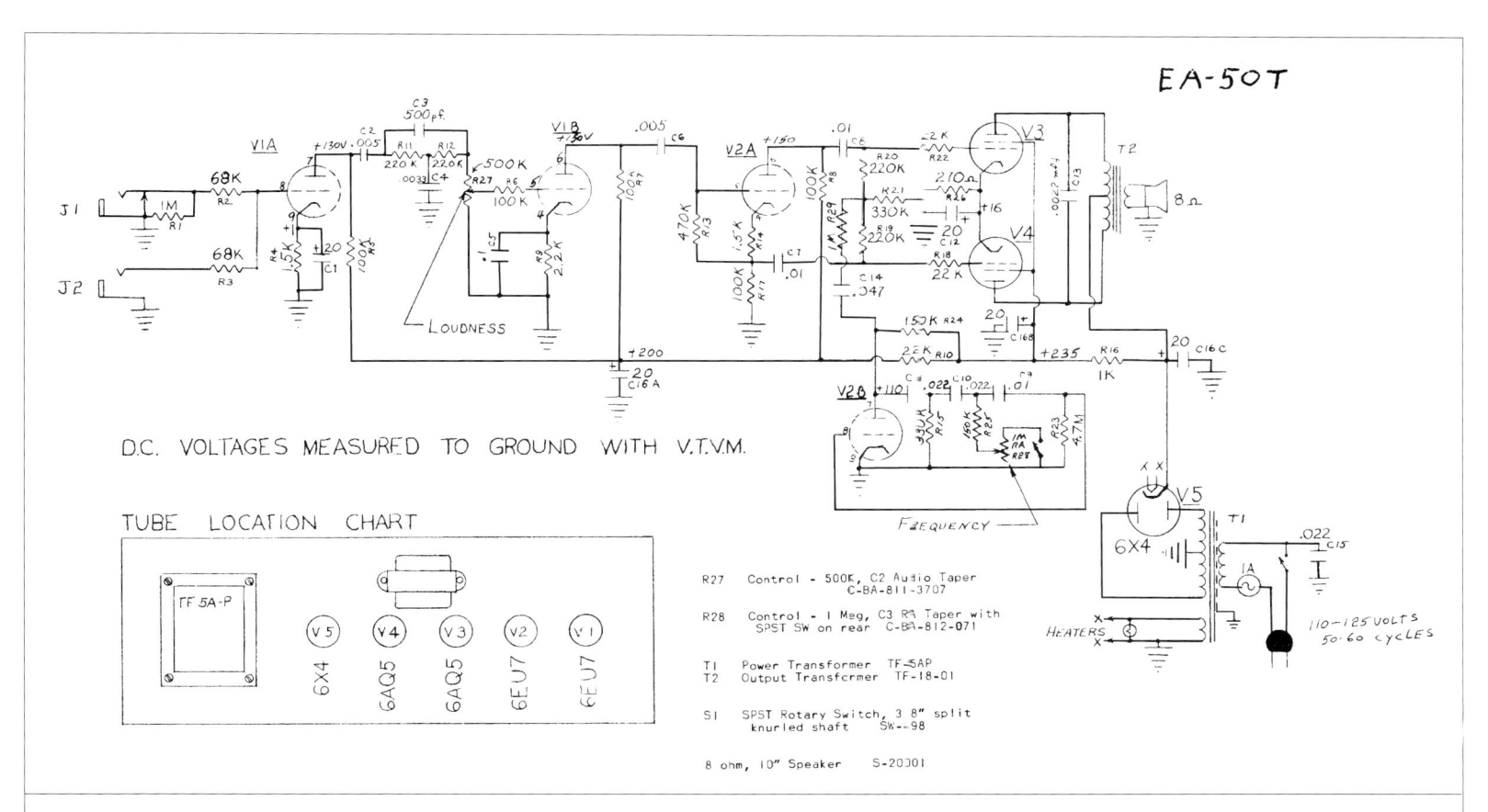

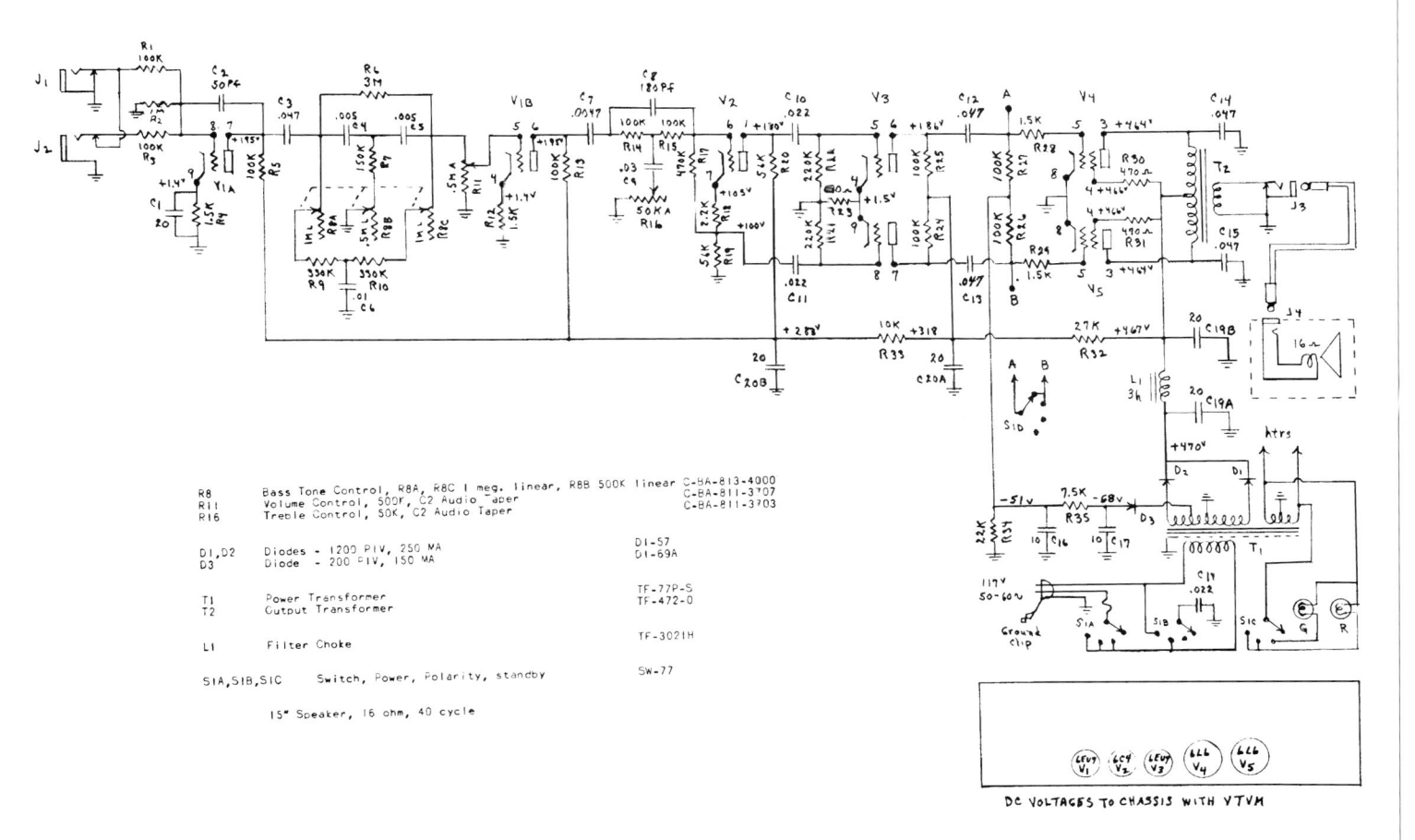

EA-50T
EA-72

EA-50T

TUBE LOCATION CHART

| TF 5A-P | V5 6X4 | V4 6AQ5 | V3 6AQ5 | V2 6EU7 | V1 6EU7 |

D.C. VOLTAGES MEASURED TO GROUND WITH V.T.V.M.

R27 Control - 500K, C2 Audio Taper   C-BA-811-3707
R28 Control - 1 Meg, C3 RA Taper with SPST SW on rear   C-BA-812-071

T1 Power Transformer   TF-5AP
T2 Output Transformer TF-18-01

S1 SPST Rotary Switch, 3/8" split knurled shaft   SW-98

8 ohm, 10" Speaker   S-20301

EA-72

R8   Bass Tone Control, R8A, R8C 1 meg. linear, R8B 500K linear   C-BA-813-4000
R11 Volume Control, 500K, C2 Audio Taper   C-BA-811-3707
R16 Treble Control, 50K, C2 Audio Taper   C-BA-811-3703

D1, D2   Diodes - 1200 PIV, 250 MA   D1-57
D3   Diode - 200 PIV, 150 MA   D1-69P

T1   Power Transformer   TF-77P-S
T2   Output Transformer   TF-472-0

L1   Filter Choke   TF-3021H

S1A, S1B, S1C   Switch, Power, Polarity, standby   SW-77

15" Speaker, 16 ohm, 40 cycle

DC VOLTAGES TO CHASSIS WITH VTVM

FUSE

The fuse used in this Amplifier is a type 3AG of three amperes rating.
DO NOT USE FUSES OF HIGHER RATING

SERVICE

If the amplifier is in need of servicing, it should be taken to a reliable radio man. The electrical diagram in this folder should be shown the repairman to assist him in servicing the amplifier.

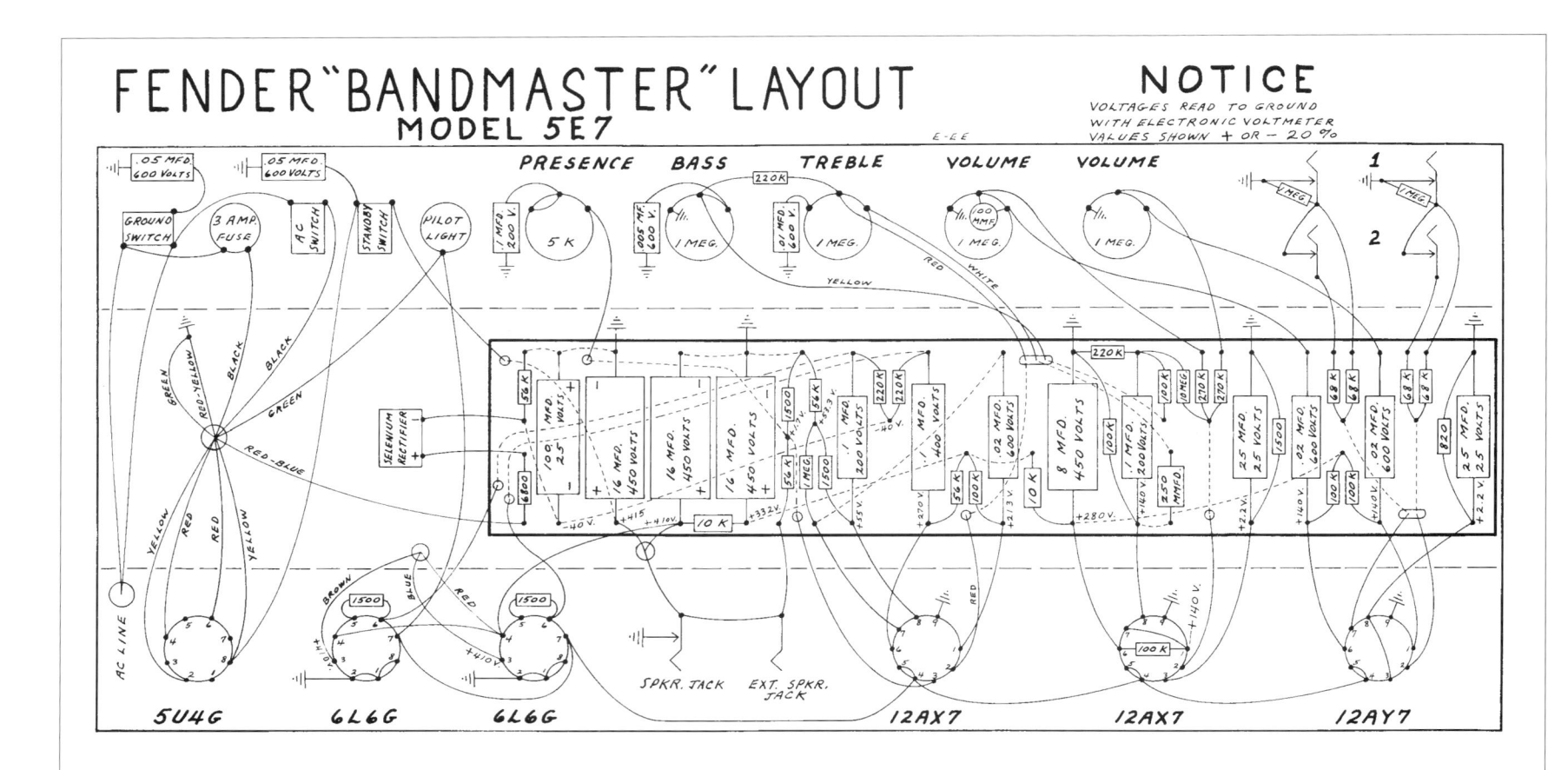

BANDMASTER 5E7
BANDMASTER 6G7

Fender

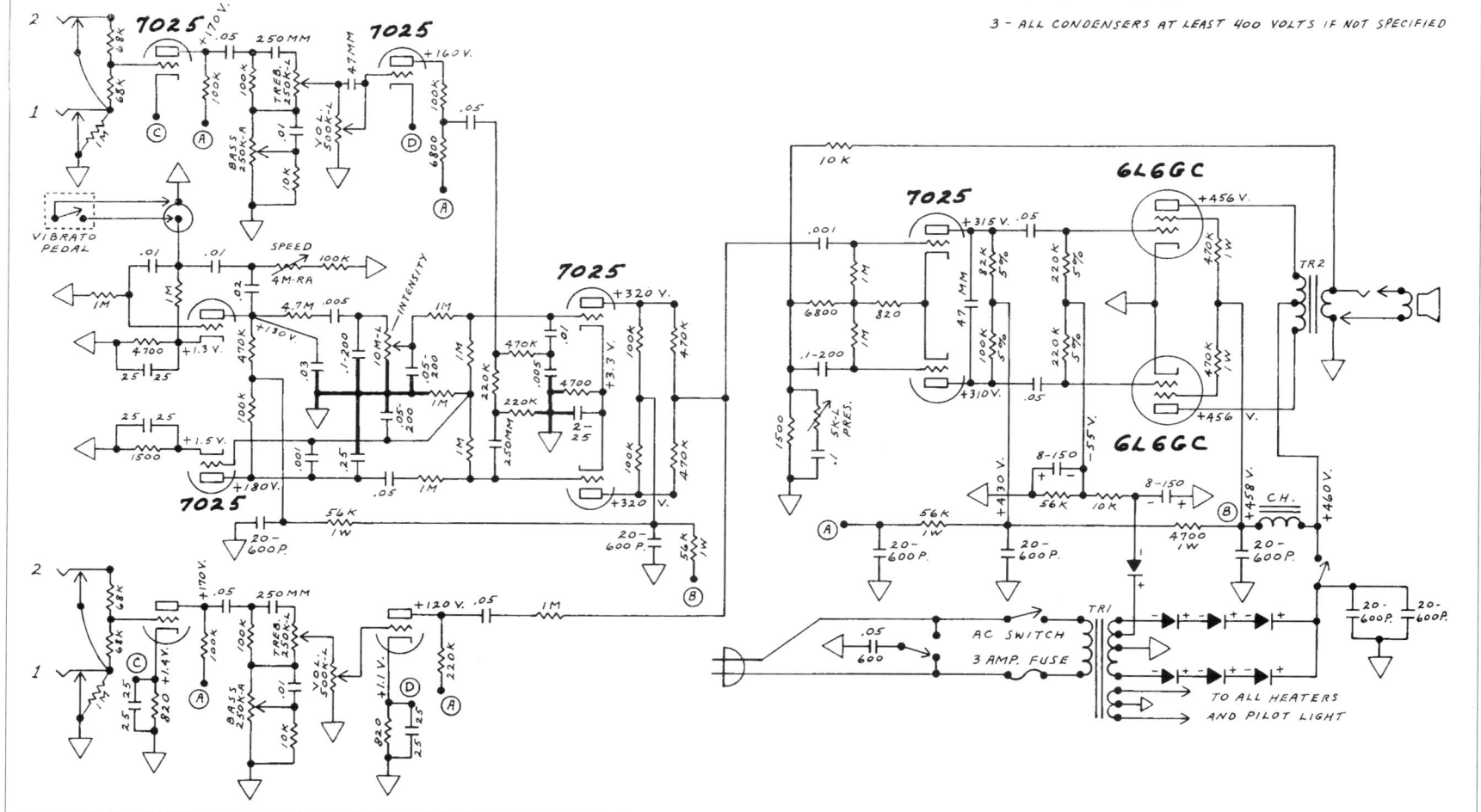

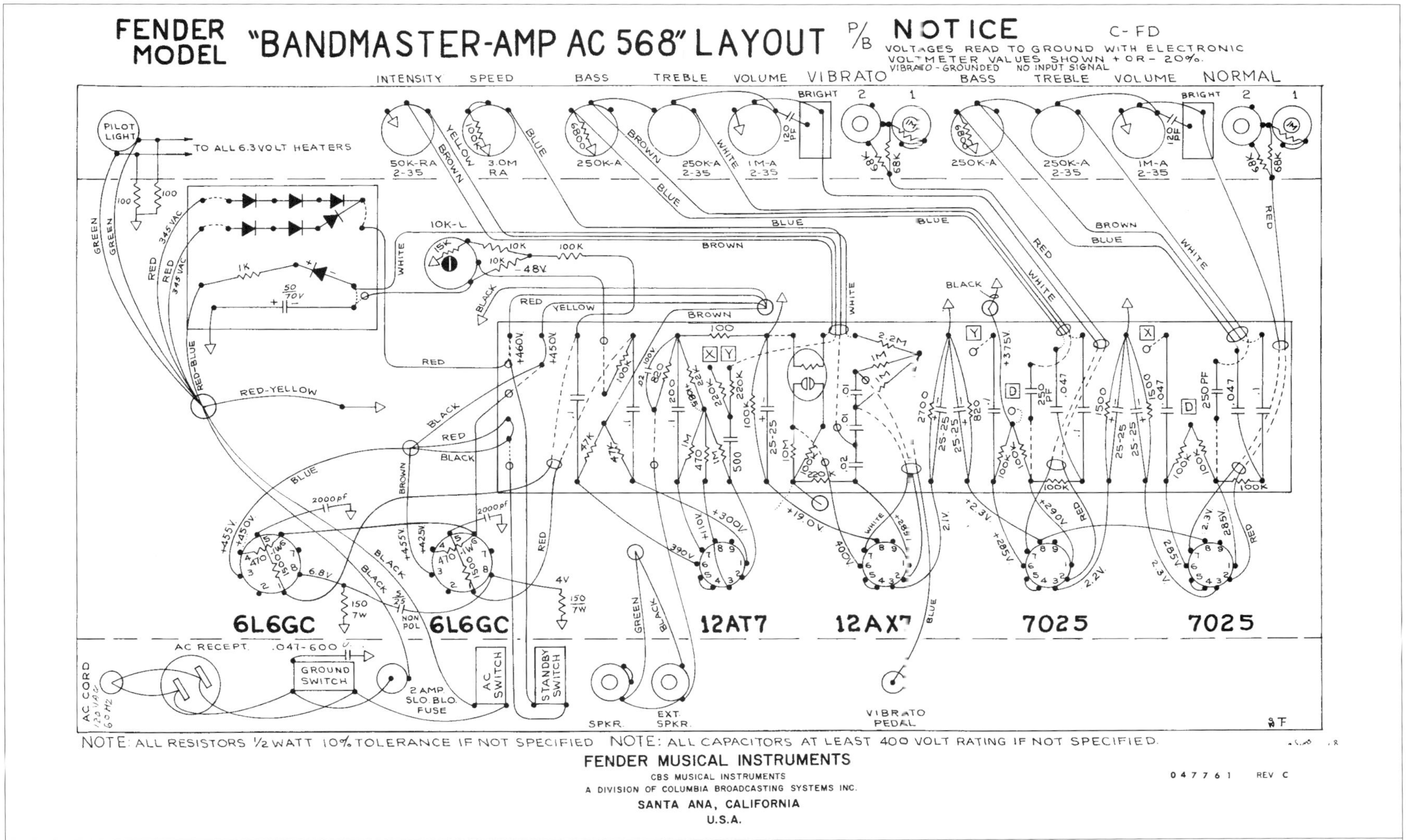
FENDER MODEL "BANDMASTER-AMP AC 568" LAYOUT
P/B
NOTICE
C-FD
VOLTAGES READ TO GROUND WITH ELECTRONIC VOLTMETER VALUES SHOWN + OR - 20%.
VIBRATO - GROUNDED NO INPUT SIGNAL
INTENSITY
SPEED
BASS
TREBLE
VOLUME
VIBRATO
NORMAL
BRIGHT
PILOT LIGHT
TO ALL 6.3 VOLT HEATERS
6L6GC
6L6GC
12AT7
12AX7
7025
7025
AC RECEPT.
GROUND SWITCH
AC CORD
2 AMP. SLO. BLO. FUSE
AC SWITCH
STANDBY SWITCH
SPKR.
EXT. SPKR.
VIBRATO PEDAL
NOTE: ALL RESISTORS 1/2 WATT 10% TOLERANCE IF NOT SPECIFIED
NOTE: ALL CAPACITORS AT LEAST 400 VOLT RATING IF NOT SPECIFIED.
FENDER MUSICAL INSTRUMENTS
CBS MUSICAL INSTRUMENTS
A DIVISION OF COLUMBIA BROADCASTING SYSTEMS INC.
SANTA ANA, CALIFORNIA
U.S.A.
047761
REV C

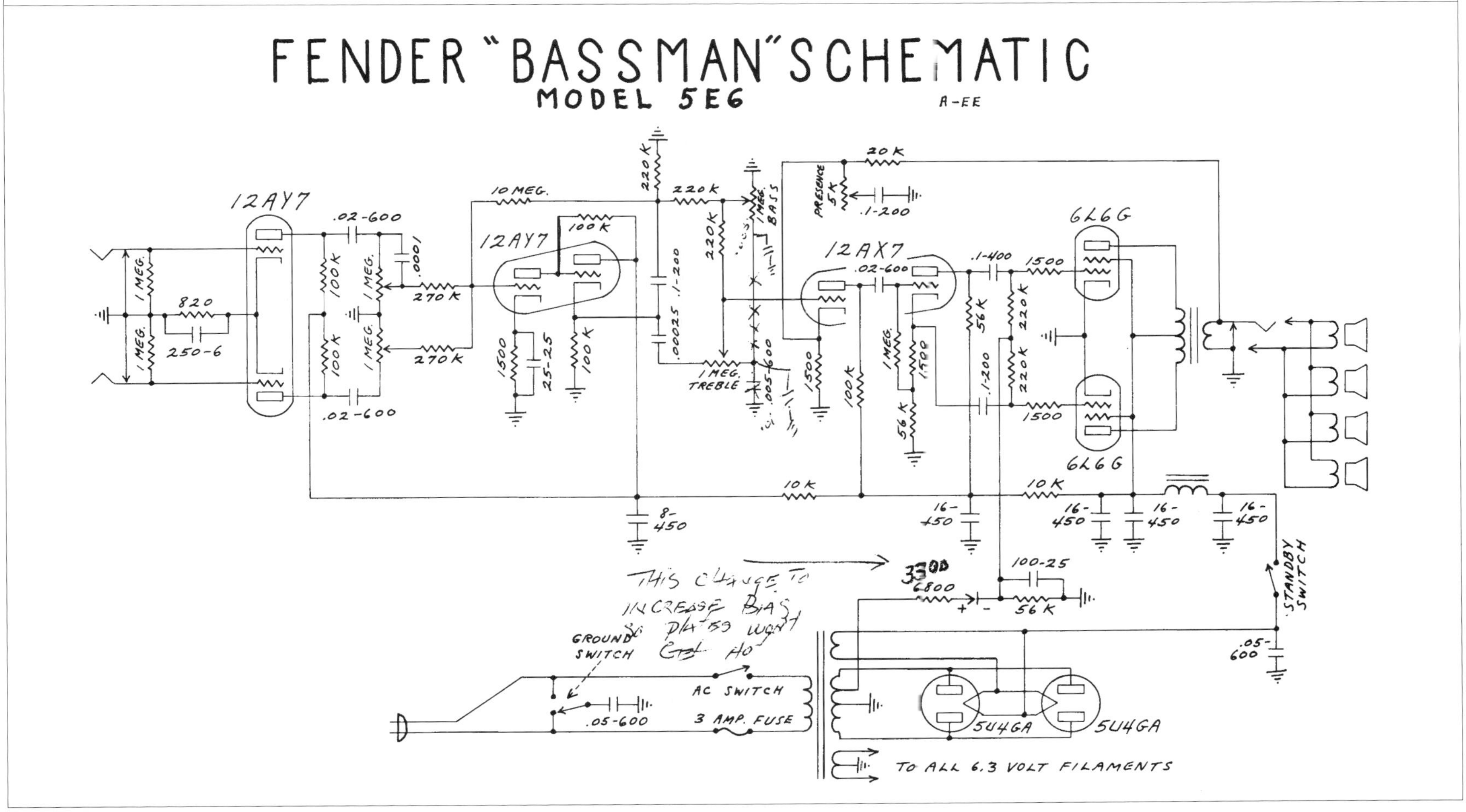
FENDER "BASSMAN" SCHEMATIC
MODEL 5E6
A-EE
12AY7
12AY7
12AX7
6L6G
6L6G
5U4GA
5U4GA
PRESENCE
BASS
TREBLE
GROUND SWITCH
AC SWITCH
3 AMP. FUSE
STANDBY SWITCH
TO ALL 6.3 VOLT FILAMENTS

BASSMAN **5E6-A**
BASSMAN **5F6**

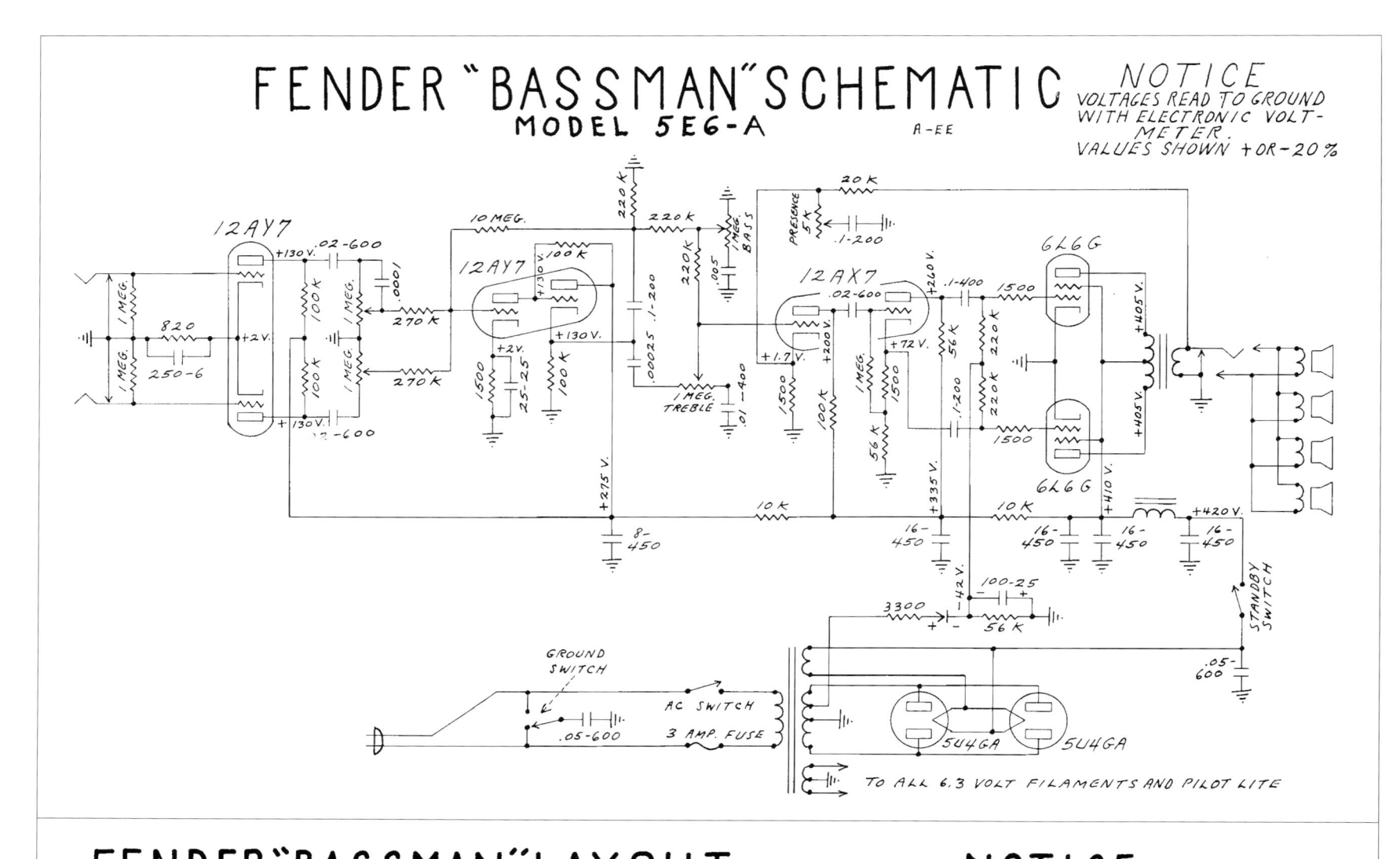

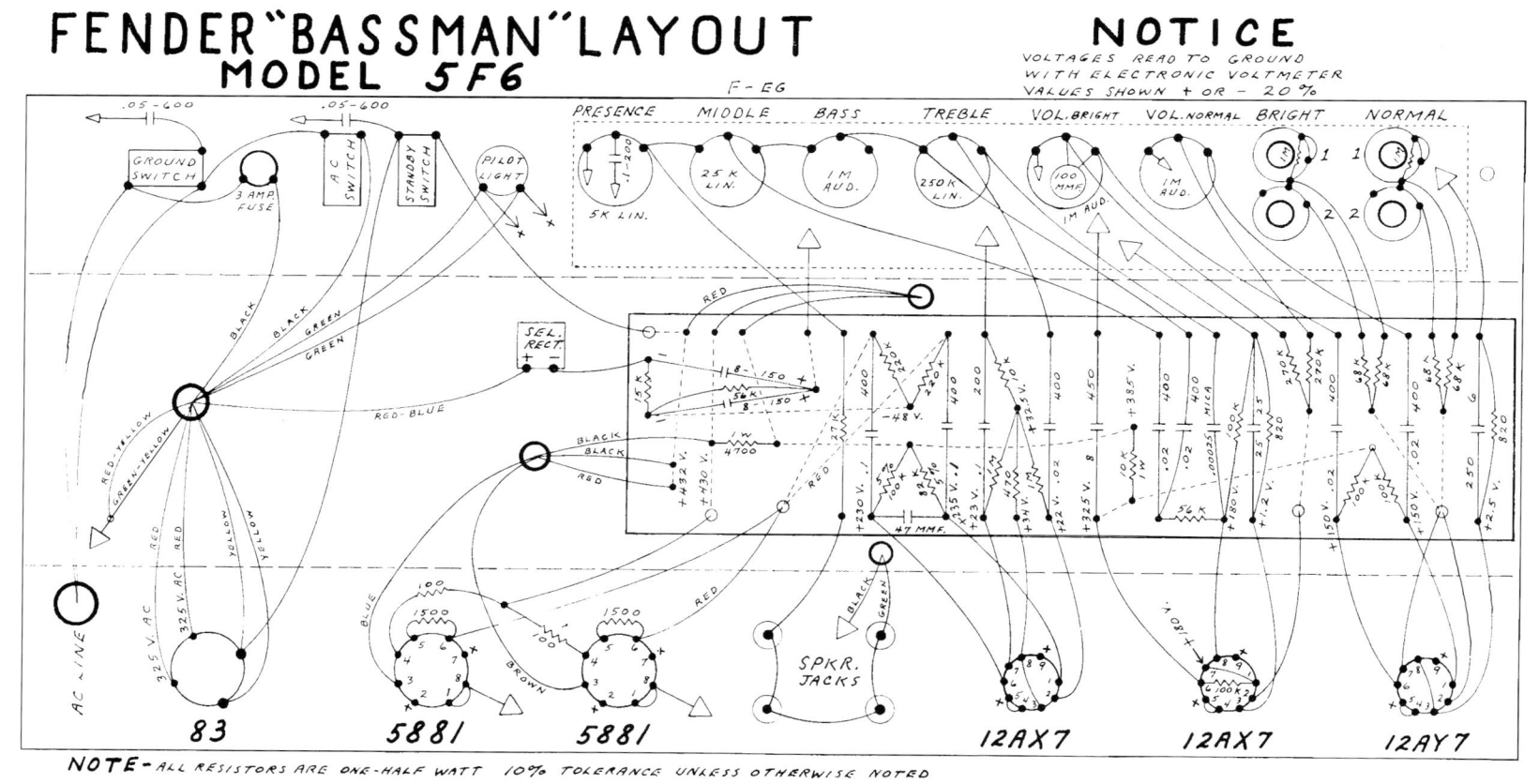

Fender

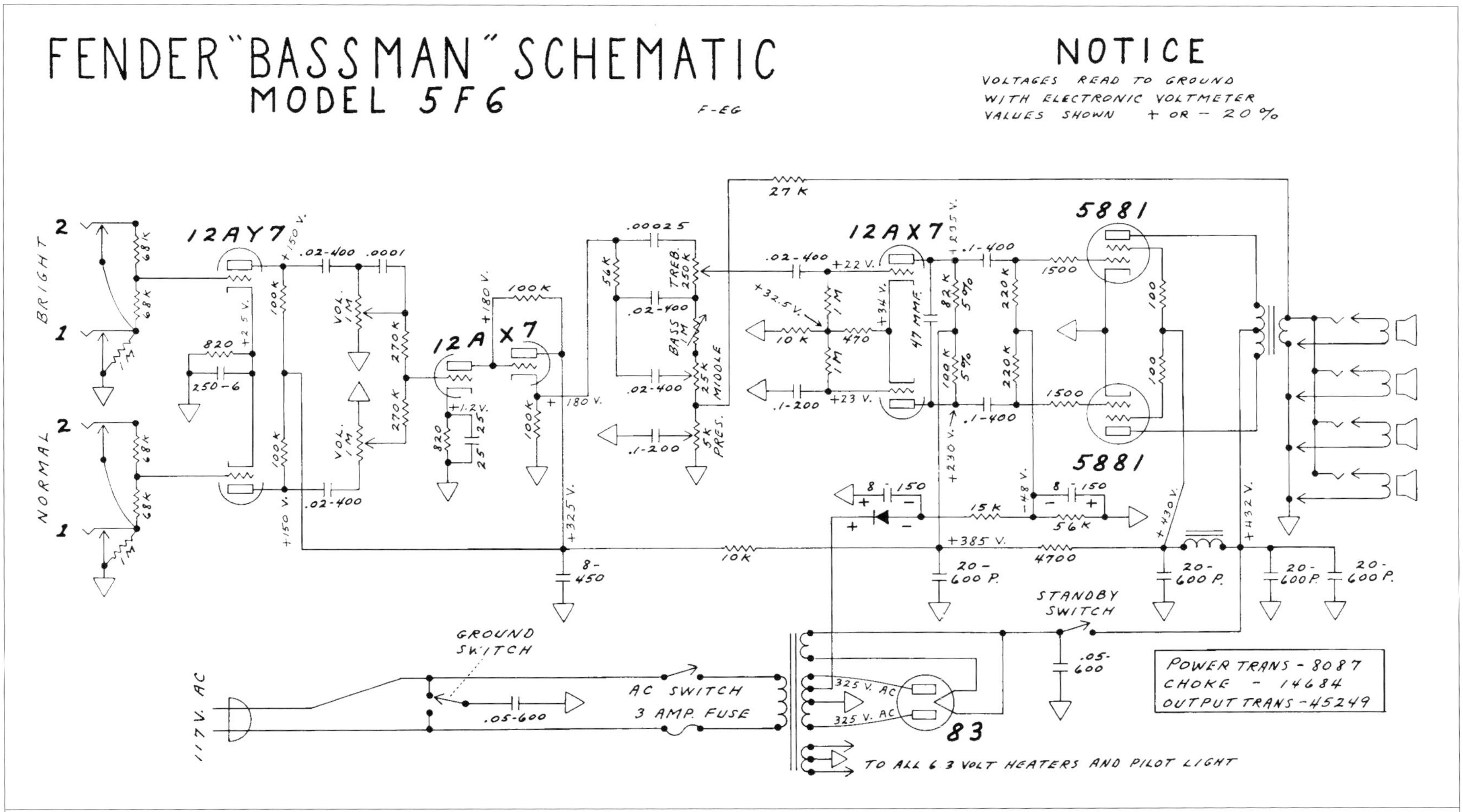
FENDER "BASSMAN" SCHEMATIC
MODEL 5F6
F-EG
NOTICE
VOLTAGES READ TO GROUND
WITH ELECTRONIC VOLTMETER
VALUES SHOWN + OR − 20 %
BRIGHT
NORMAL
12AY7
12AX7
12AX7
5881
5881
83
TREB. 250K
BASS 1M
25K MIDDLE
5K PRES.
STANDBY SWITCH
GROUND SWITCH
AC SWITCH
3 AMP. FUSE
117 V. AC
TO ALL 6.3 VOLT HEATERS AND PILOT LIGHT
POWER TRANS - 8087
CHOKE - 14684
OUTPUT TRANS - 45249

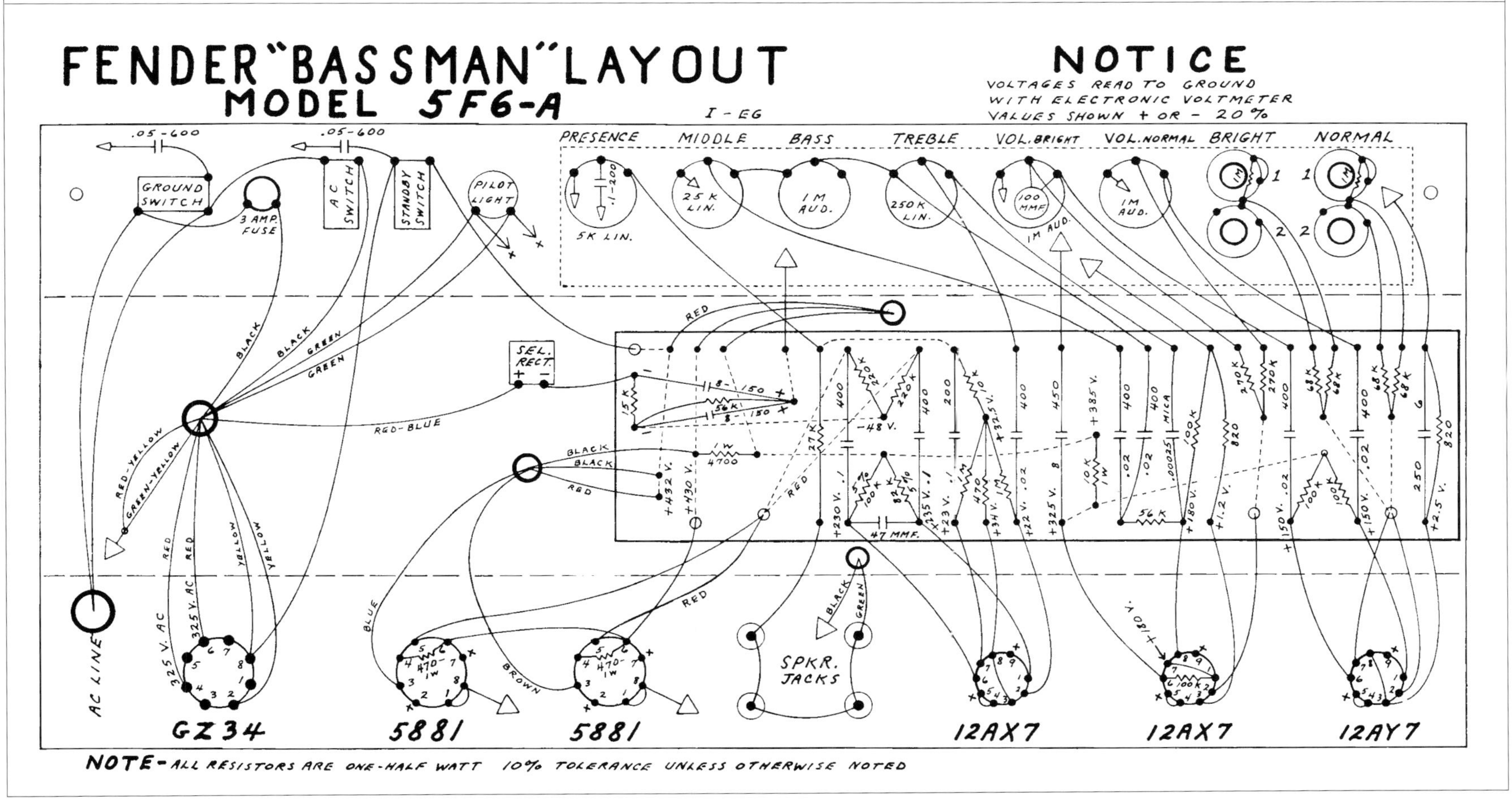
FENDER "BASSMAN" LAYOUT
MODEL 5F6-A
I-EG
NOTICE
VOLTAGES READ TO GROUND
WITH ELECTRONIC VOLTMETER
VALUES SHOWN + OR - 20%
PRESENCE
MIDDLE
BASS
TREBLE
VOL. BRIGHT
VOL. NORMAL
BRIGHT
NORMAL
GROUND SWITCH
3 AMP. FUSE
AC SWITCH
STANDBY SWITCH
PILOT LIGHT
SEL. RECT.
SPKR. JACKS
AC LINE
GZ34
5881
5881
12AX7
12AX7
12AY7
NOTE - ALL RESISTORS ARE ONE-HALF WATT 10% TOLERANCE UNLESS OTHERWISE NOTED

BASSMAN 5F6-A
BASSMAN 6G6-A

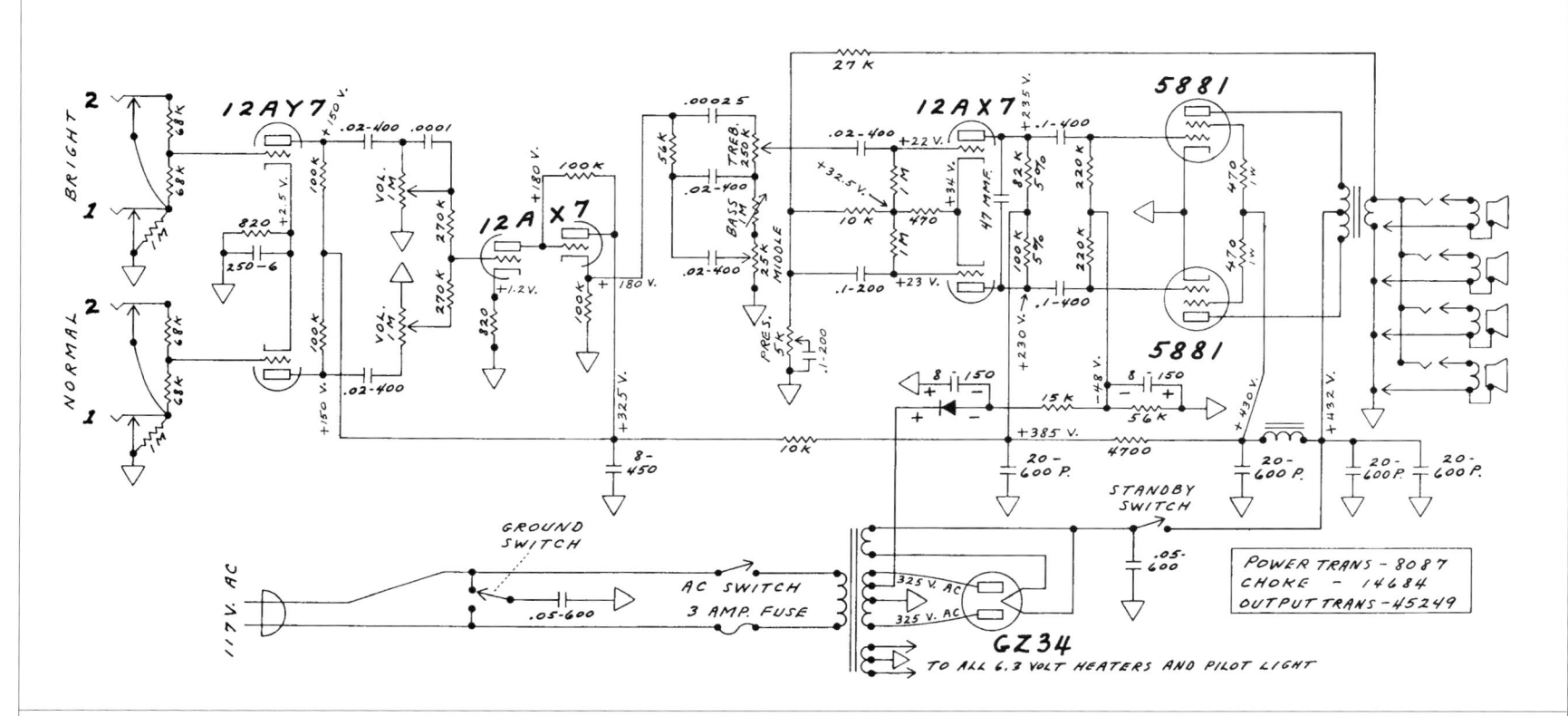

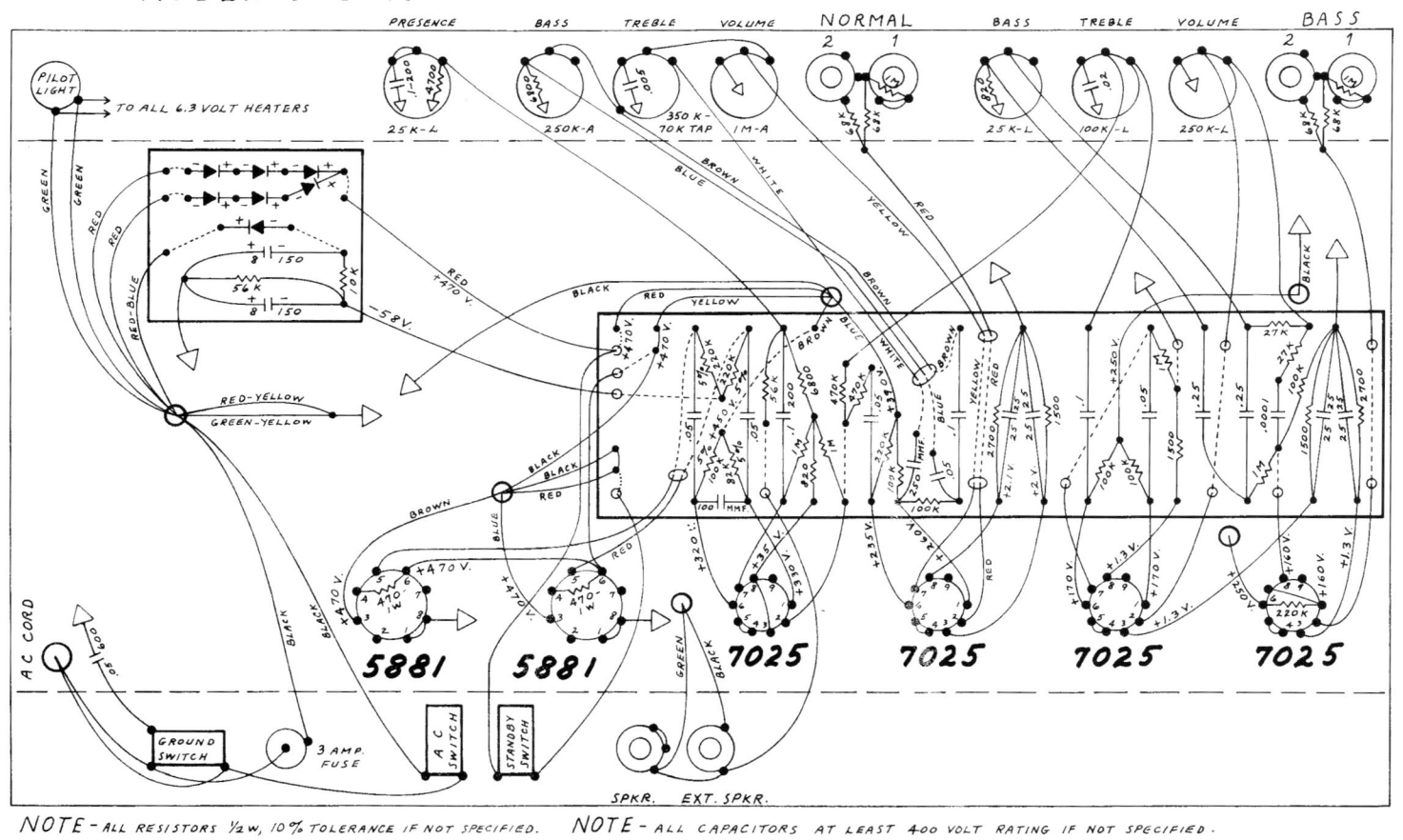

Fender

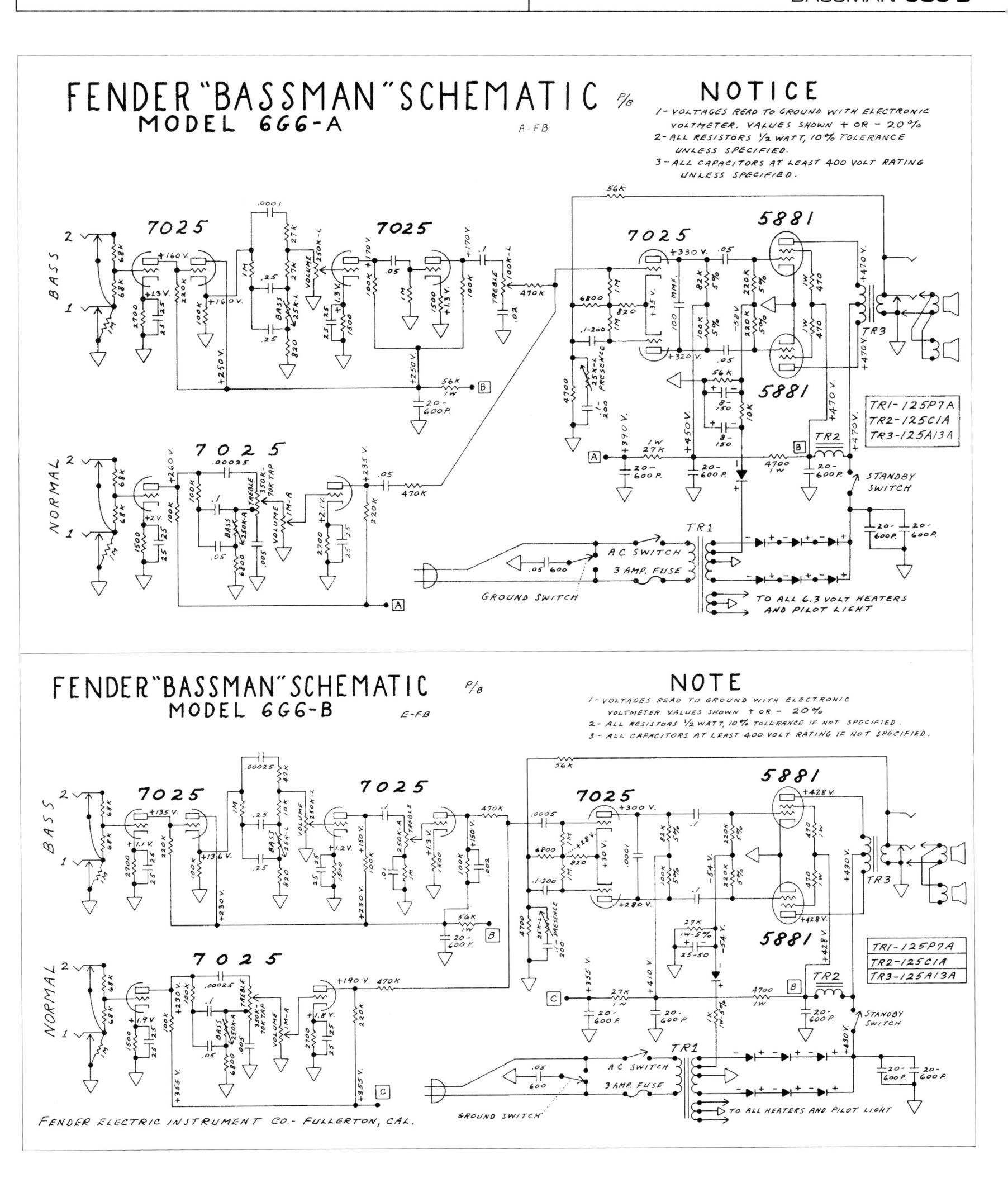

Fender

BASSMAN 6G6-A
BASSMAN 6G6-B

59 BASSMAN **REV A**
59 BASSMAN **REV E**

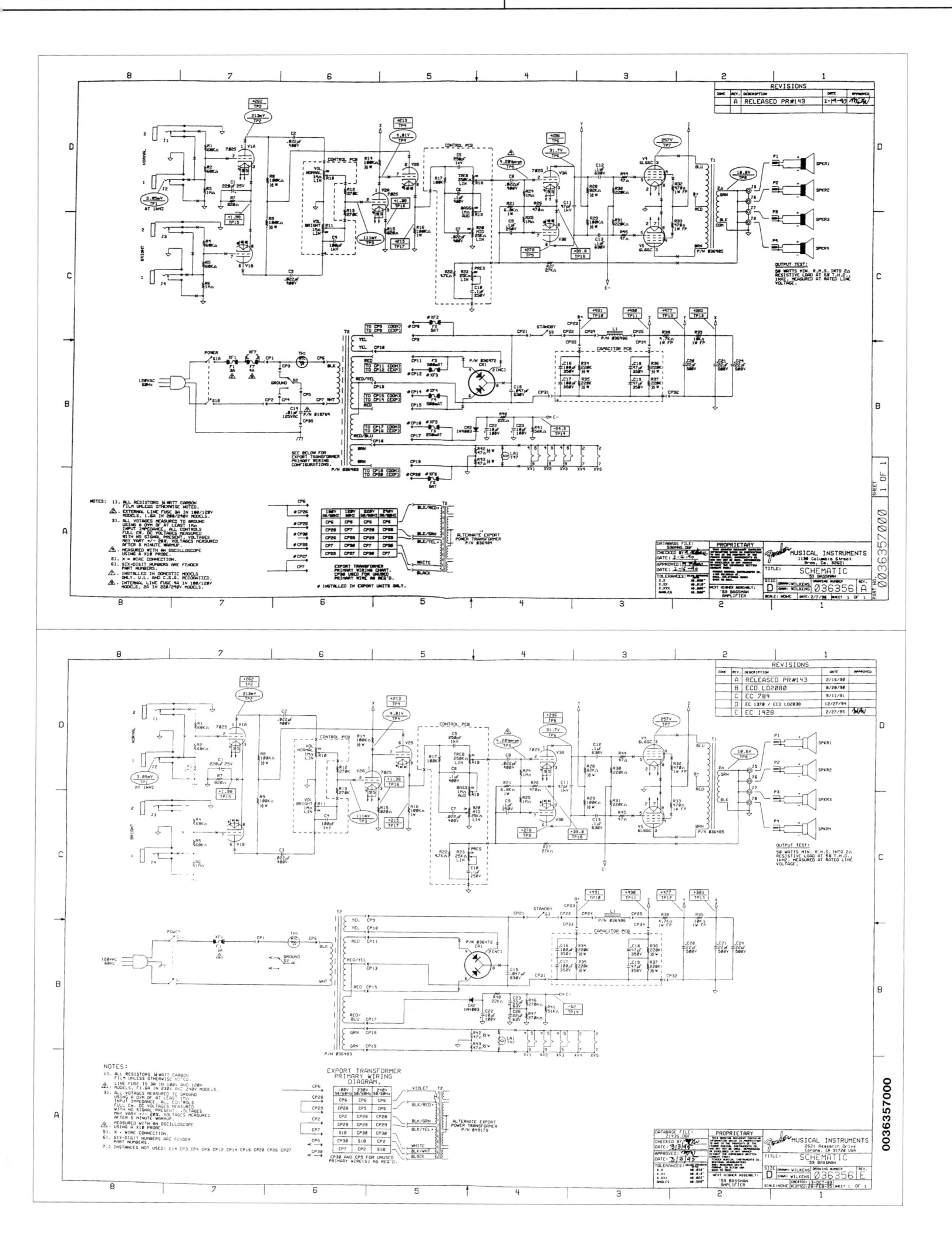

Fender

Fender

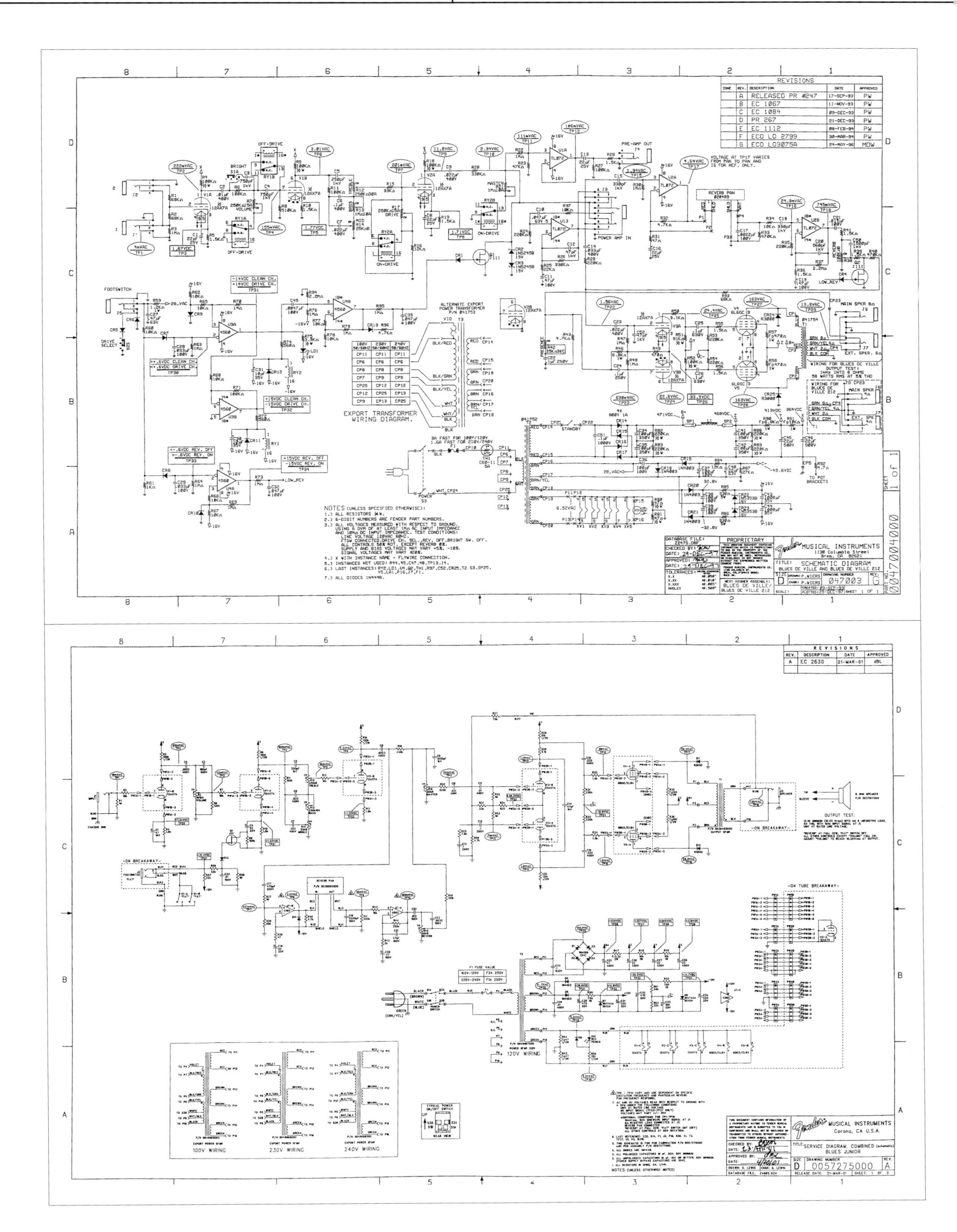

BLUES DE VILLE
BLUES JUNIOR

THE TUBE AMP BOOK

BASSMAN **AC568**
CHAMP-AMP **5E1**

Fender

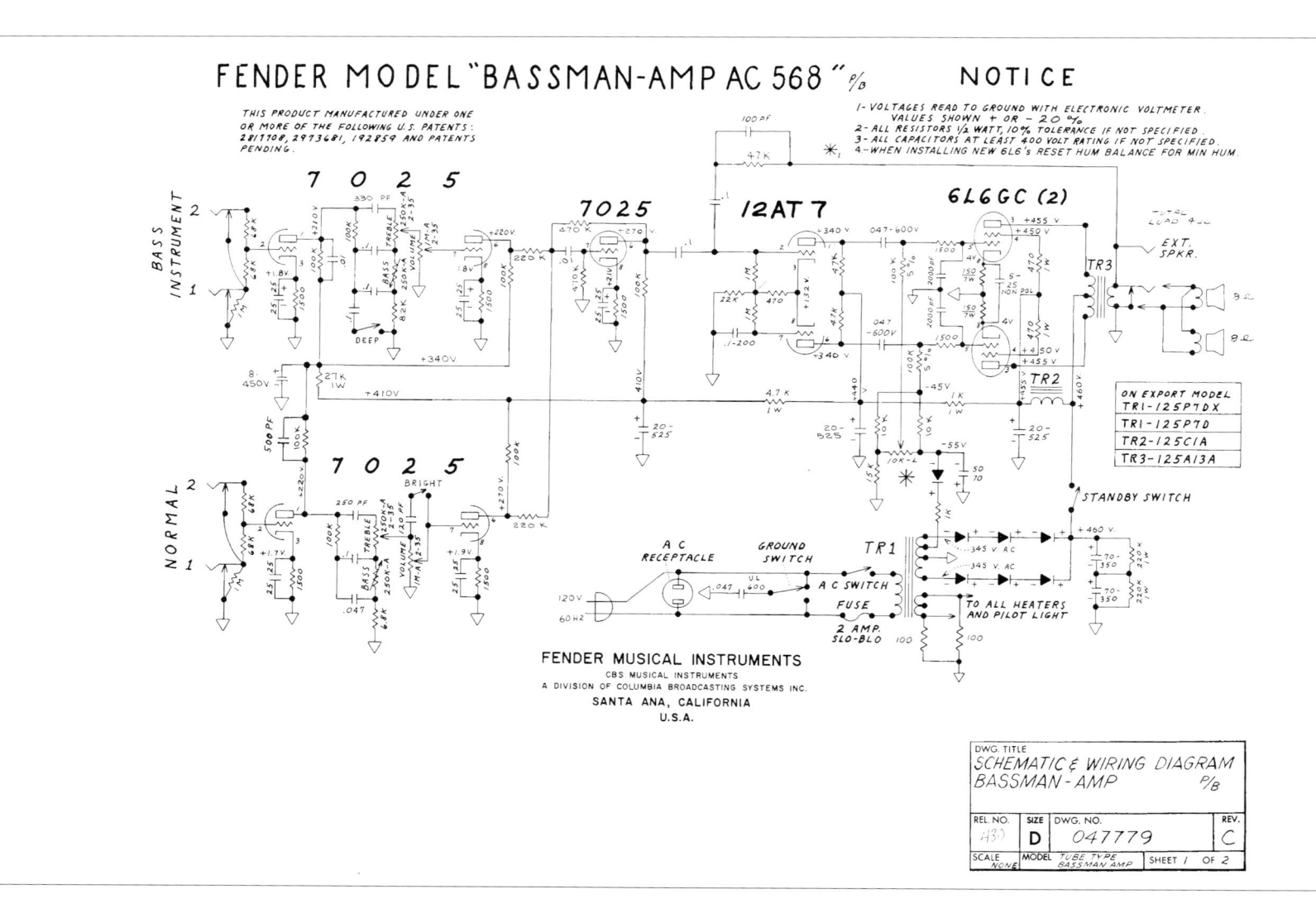

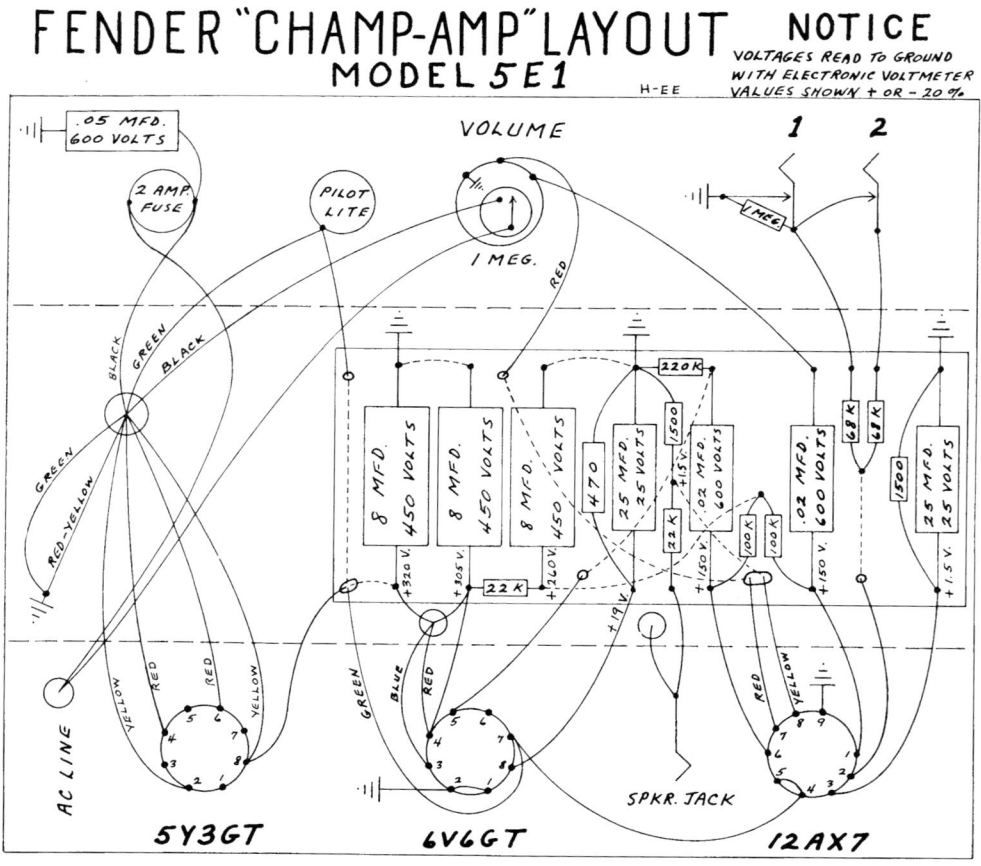

Fender

CHAMP-AMP **5E1**
VIBRO-CHAMP **AA764**

FENDER "CHAMP-AMP" SCHEMATIC
MODEL 5E1

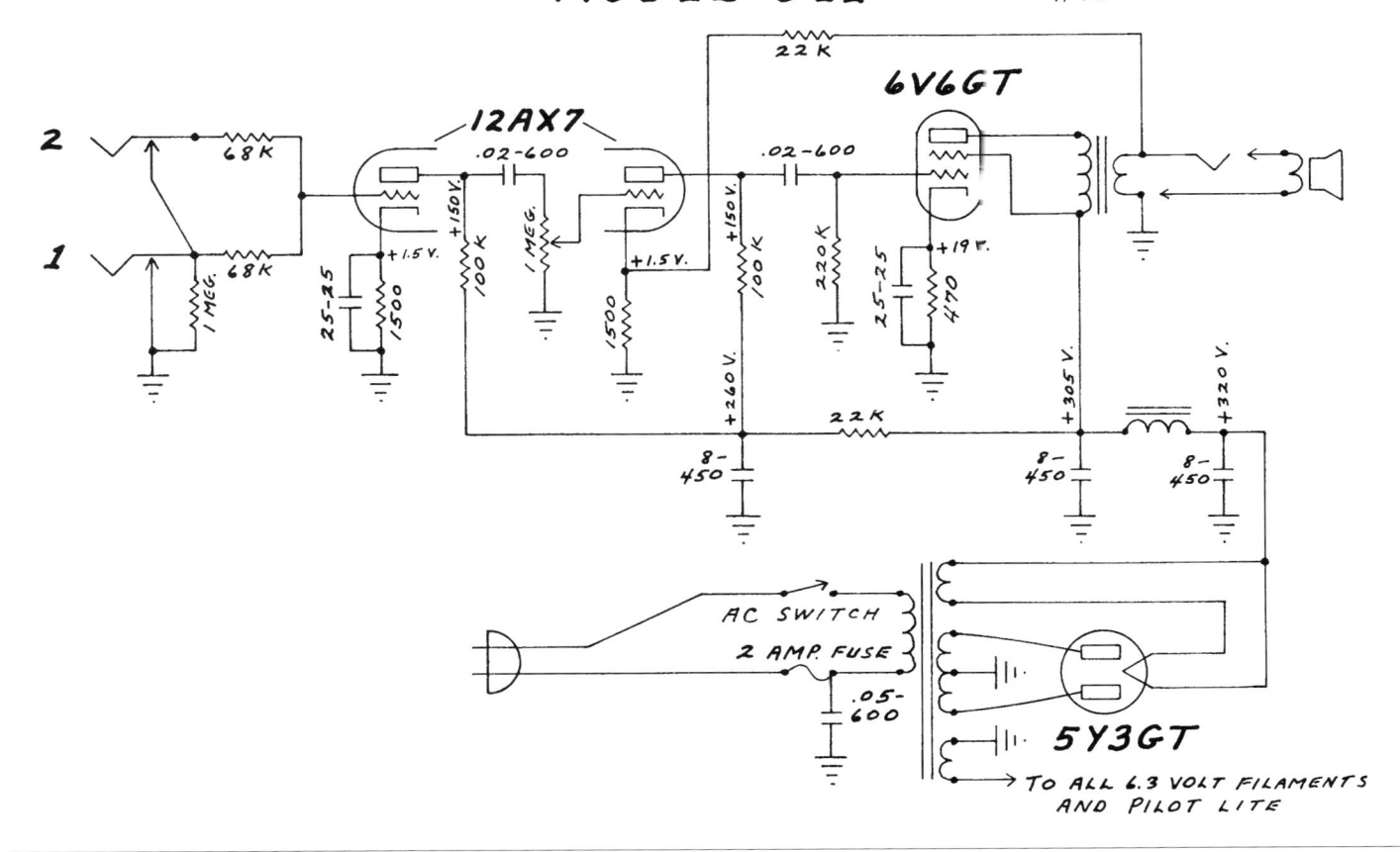

FENDER MODEL "VIBRO-CHAMP-AMP AA764" LAYOUT

NOTICE
1 - VOLTAGES READ TO GROUND WITH ELECTRONIC VOLTMETER VALUES SHOWN + OR - 20%
2 - ALL RESISTORS 1/2 WATT 10% TOLERANCE IF NOT SPECIFIED.
3 - ALL CAPACITORS AT LEAST 400 VOLT RATING IF NOT SPECIFIED.

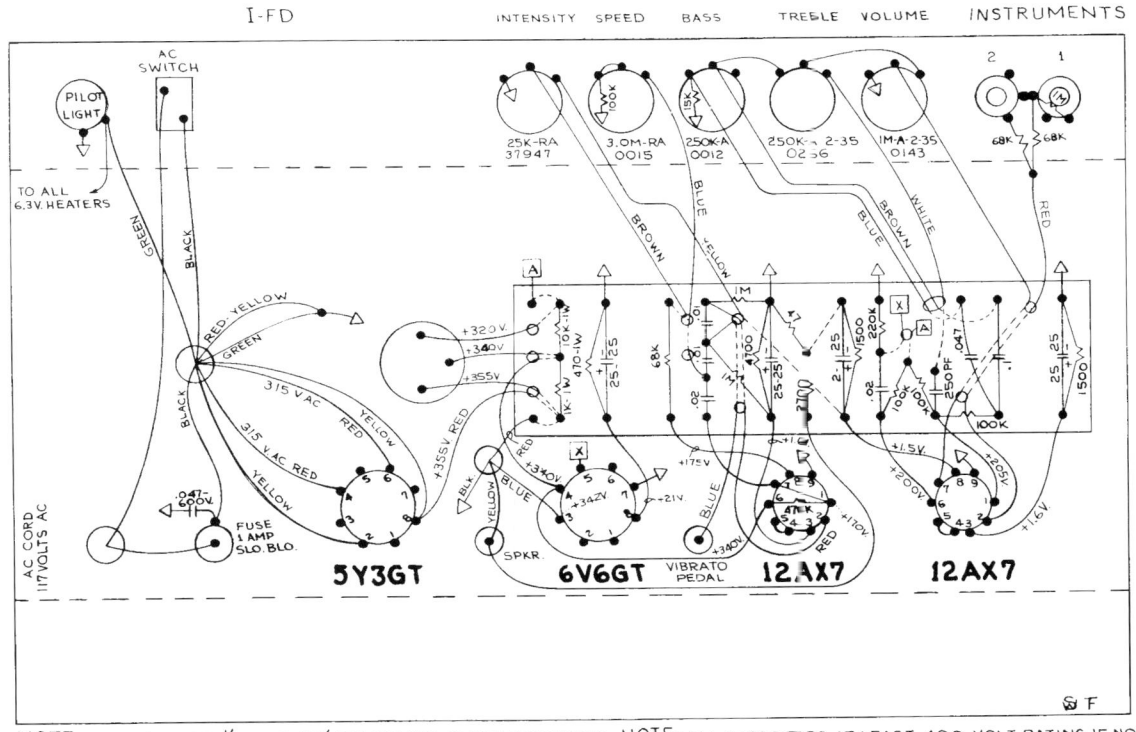

NOTE - ALL RESISTORS 1/2 WATT 10% TOLERANCE, IF NOT SPECIFIED.   NOTE - ALL CAPACITORS AT LEAST 400 VOLT RATING IF NOT SPECIFIED

FENDER ELECTRIC INSTRUMENT COMPANY
FULLERTON, CALIFORNIA
U.S.A.

THE TUBE AMP BOOK 33

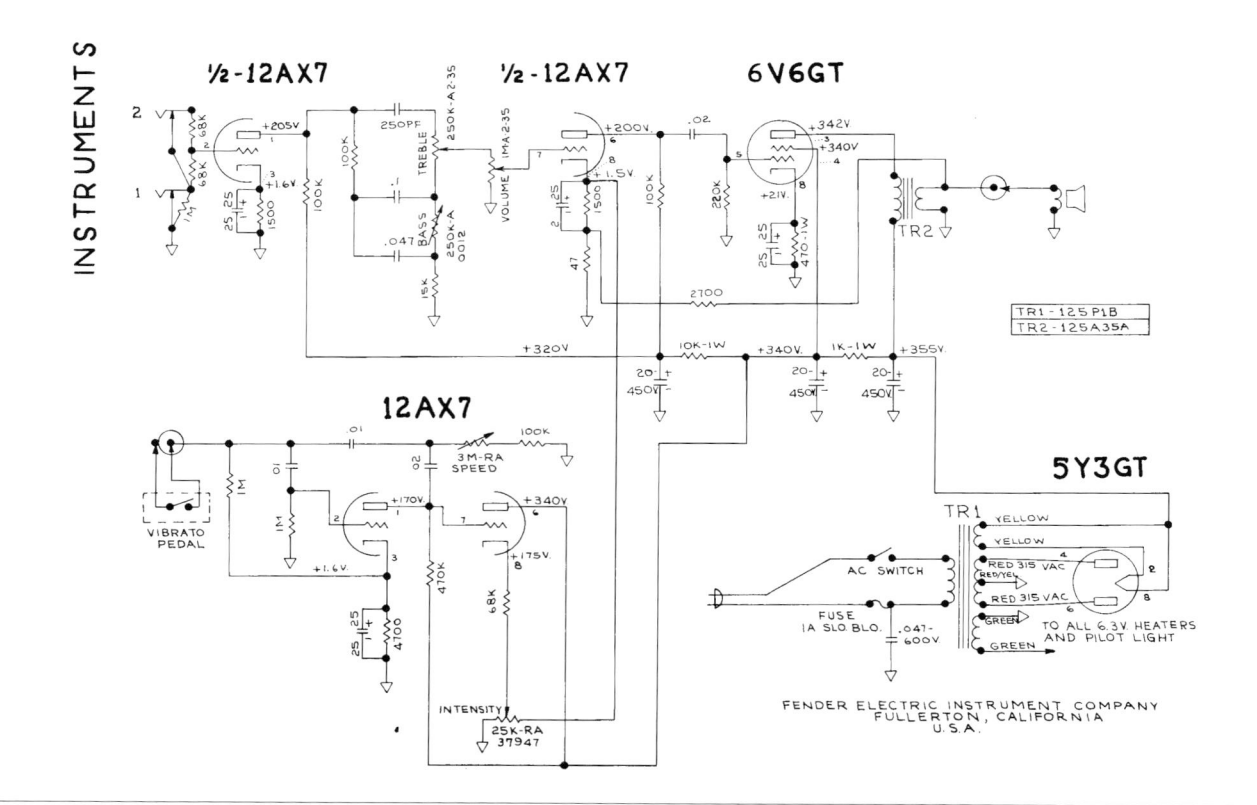

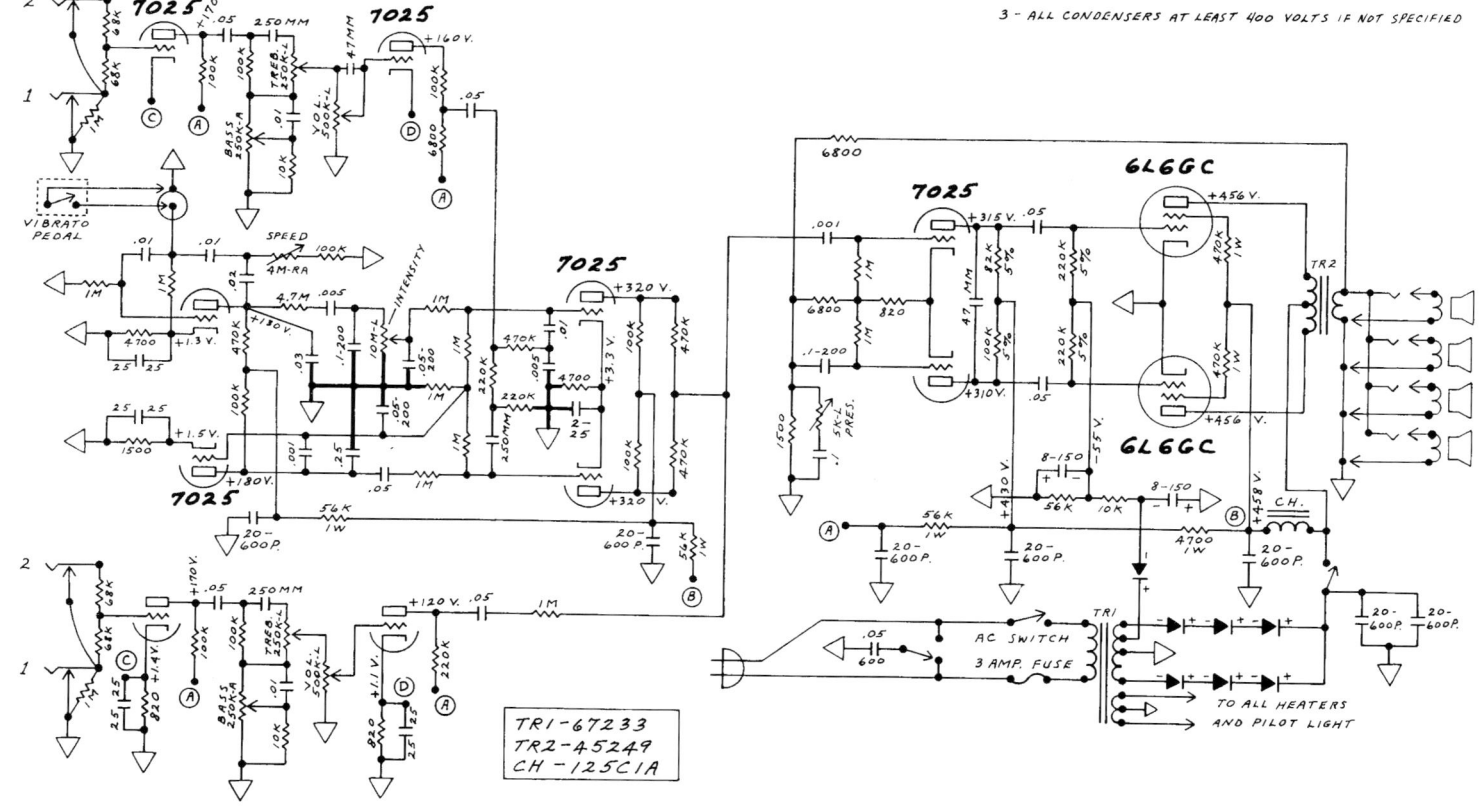

VIBRO-CHAMP AA764
CONCERT 6G12

Fender

Fender

DELUXE 5C3
DELUXE 5D3

FENDER "DELUXE" SCHEMATIC
MODEL 5C3

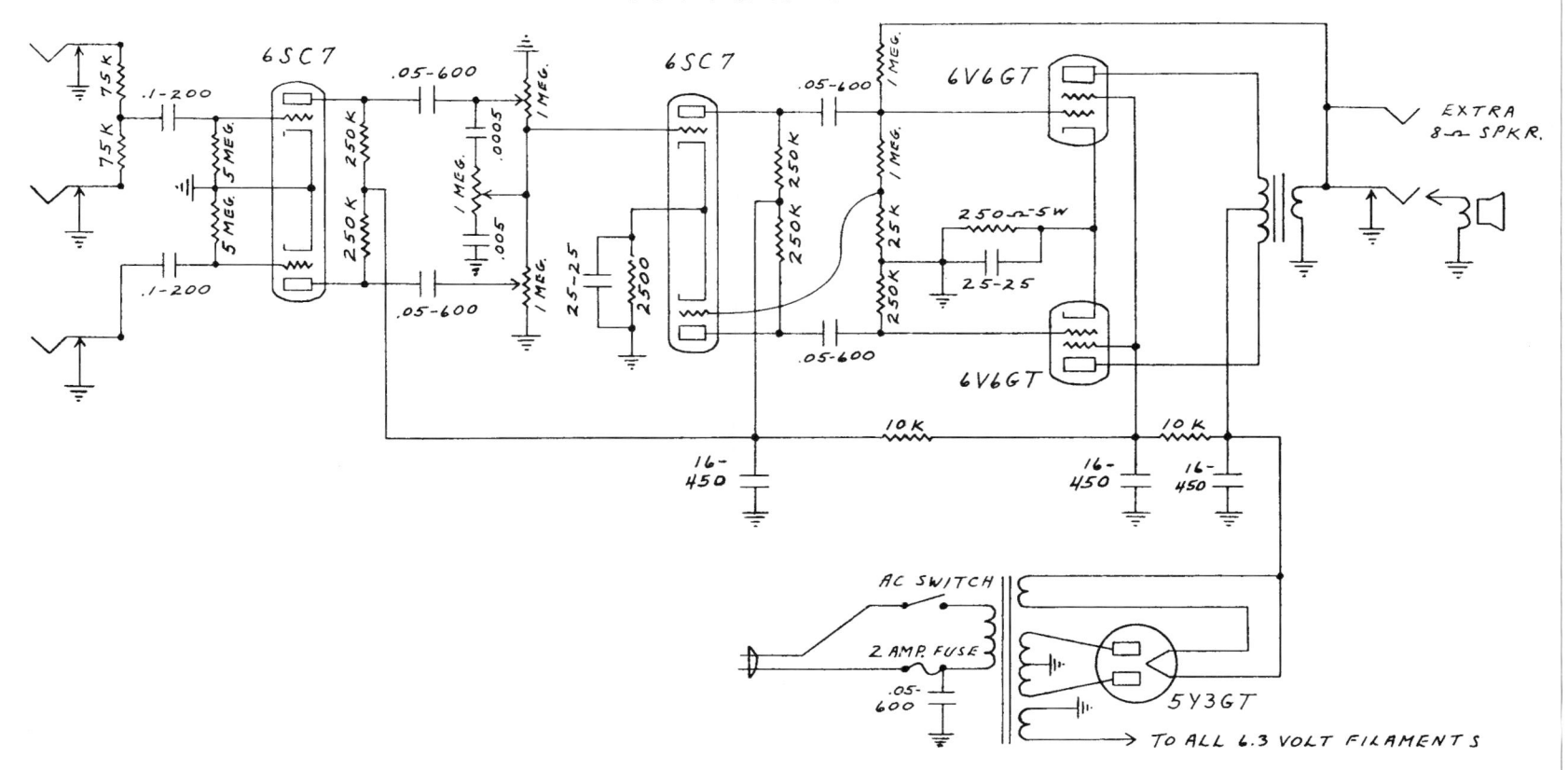

FENDER DELUXE SCHEMATIC
MODEL 5D3

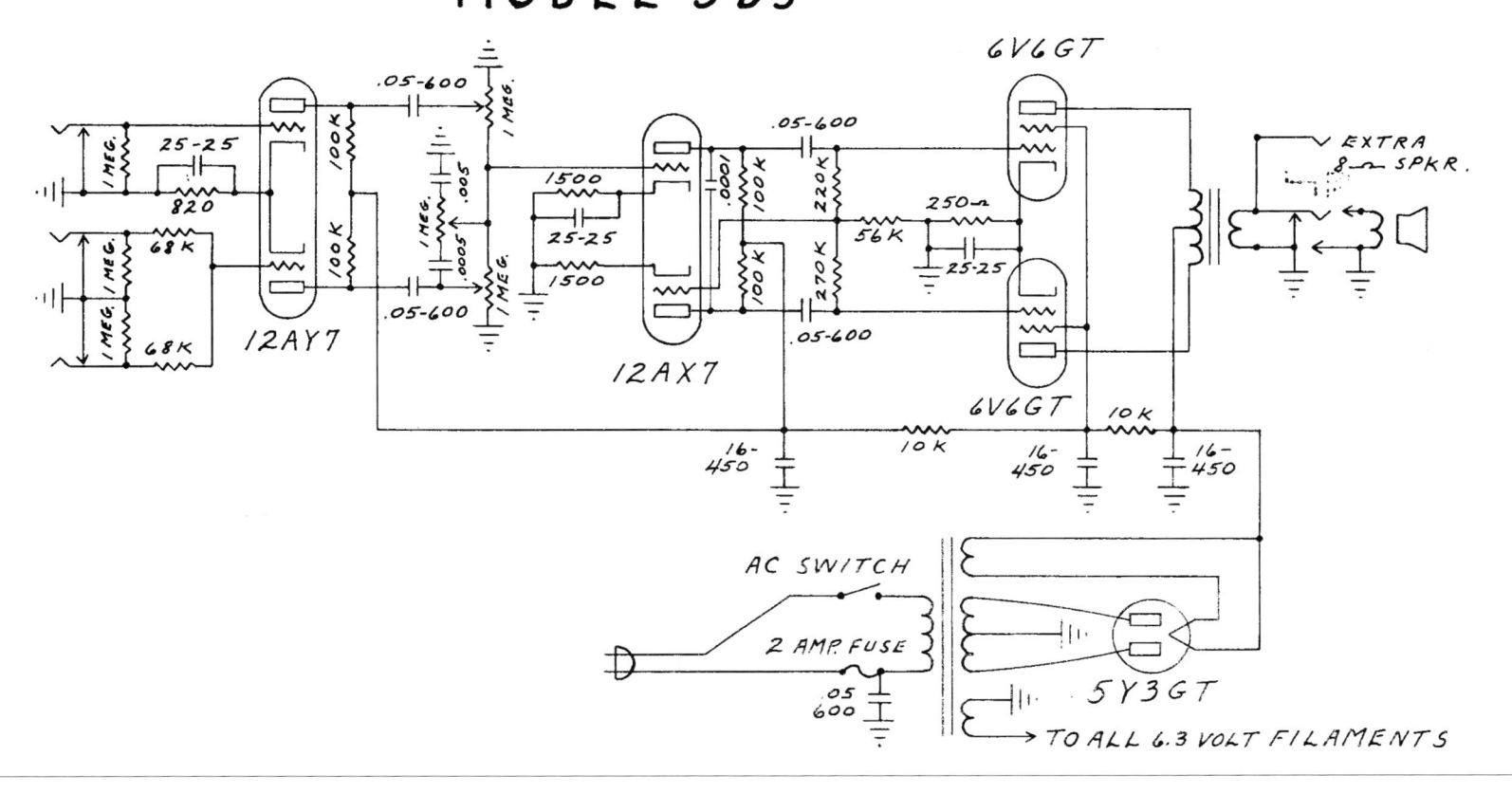

THE TUBE AMP BOOK

DELUXE **5D4**
DELUXE **5E3**

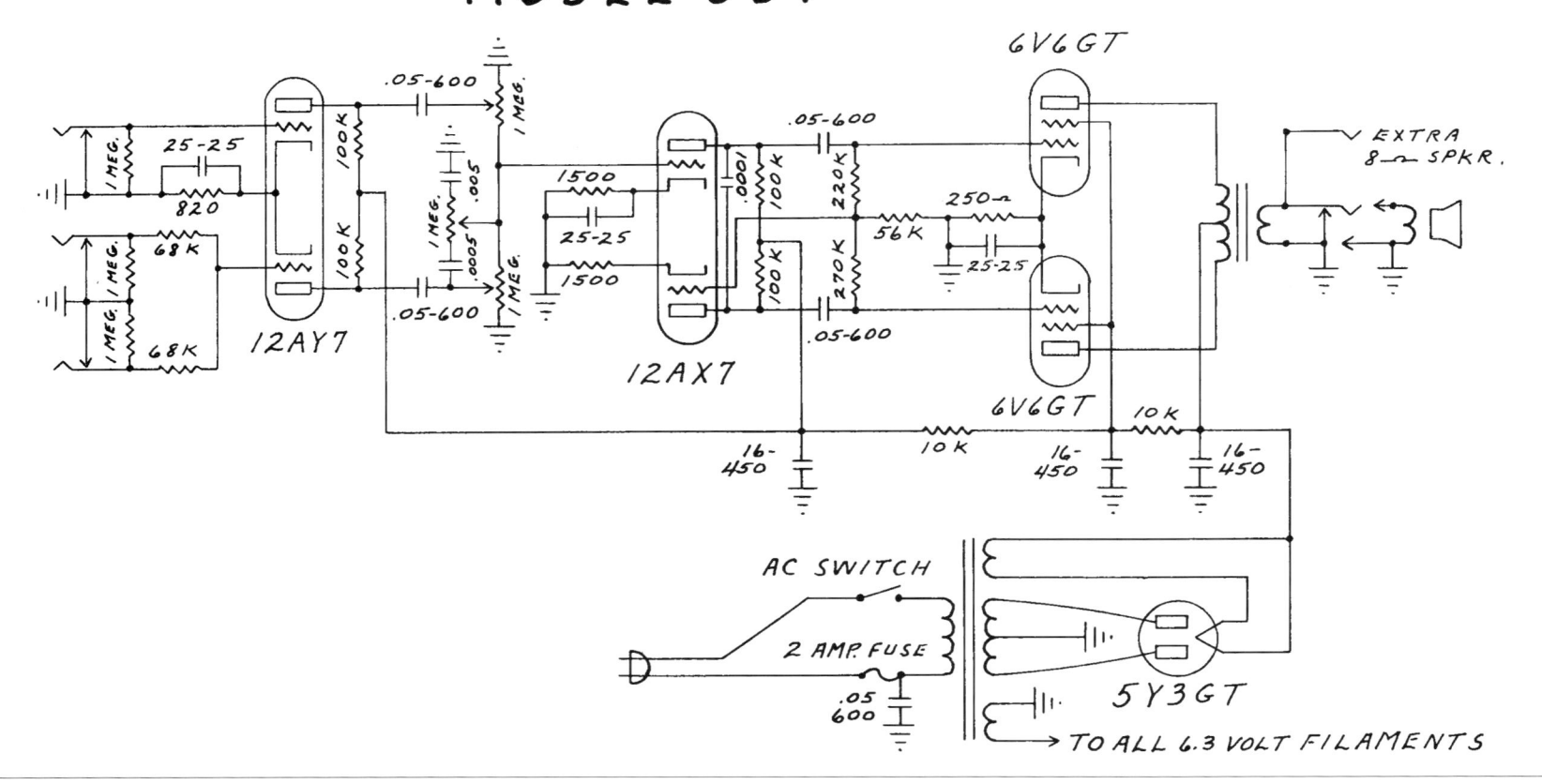

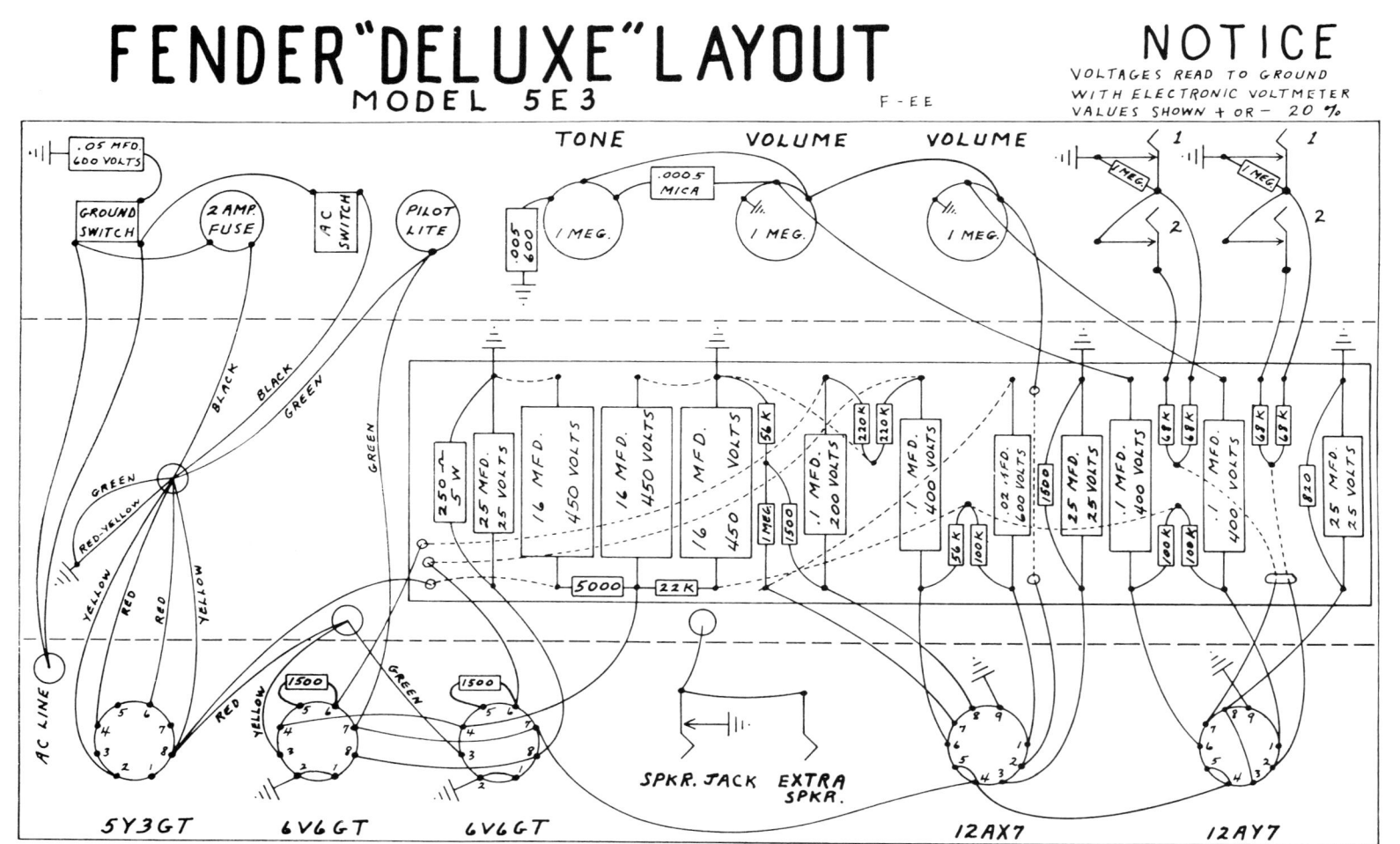

Fender

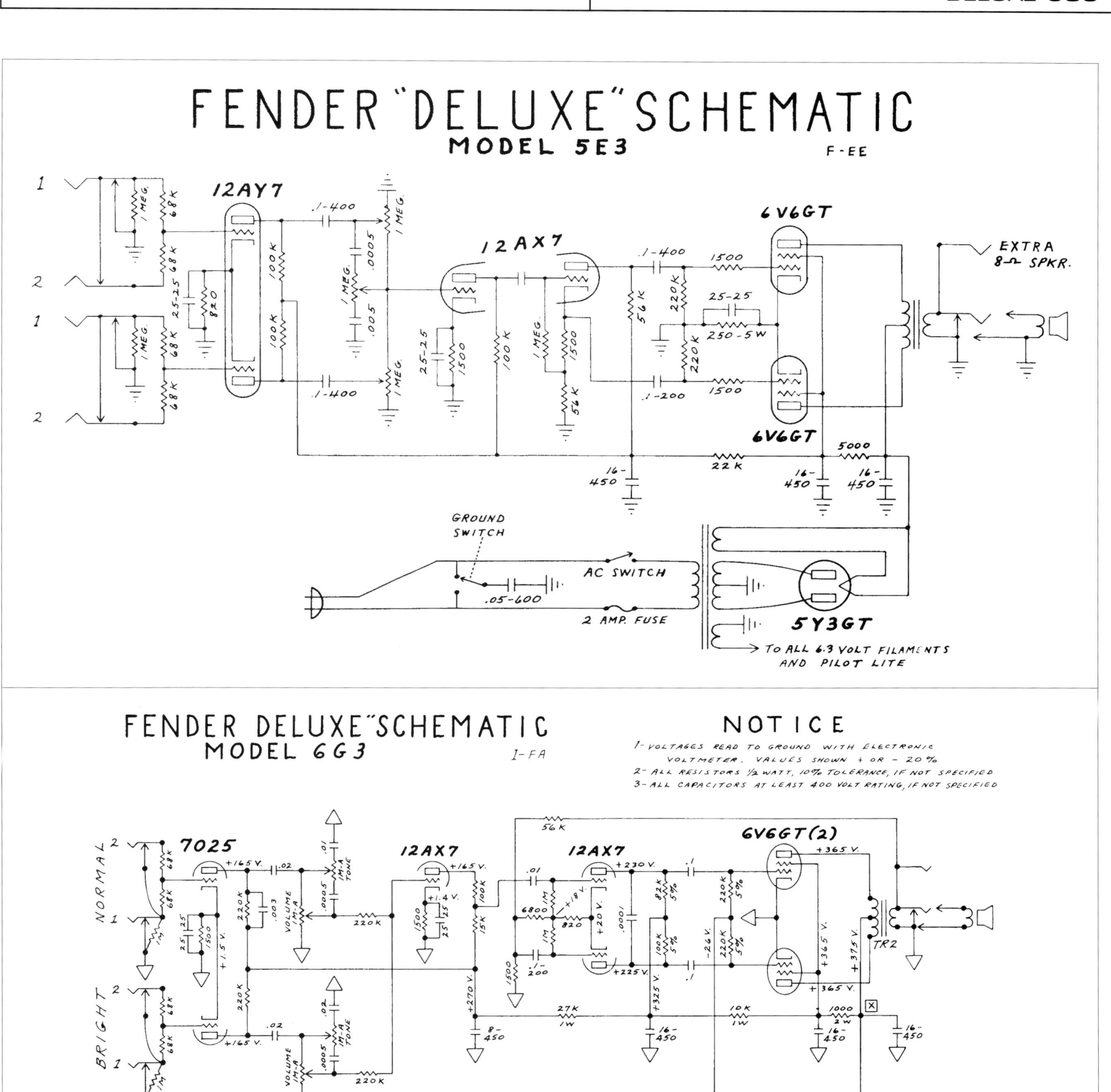

Fender

DELUXE **5E3**
DELUXE **6G3**

FENDER "DELUXE" SCHEMATIC
MODEL 5E3 — F-EE

FENDER DELUXE SCHEMATIC
MODEL 6G3 — 1-FA

NOTICE
1 - VOLTAGES READ TO GROUND WITH ELECTRONIC VOLTMETER. VALUES SHOWN + OR - 20%
2 - ALL RESISTORS ½ WATT, 10% TOLERANCE, IF NOT SPECIFIED
3 - ALL CAPACITORS AT LEAST 400 VOLT RATING, IF NOT SPECIFIED

TR1-125P2A
TR2-125A1A

THE TUBE AMP BOOK

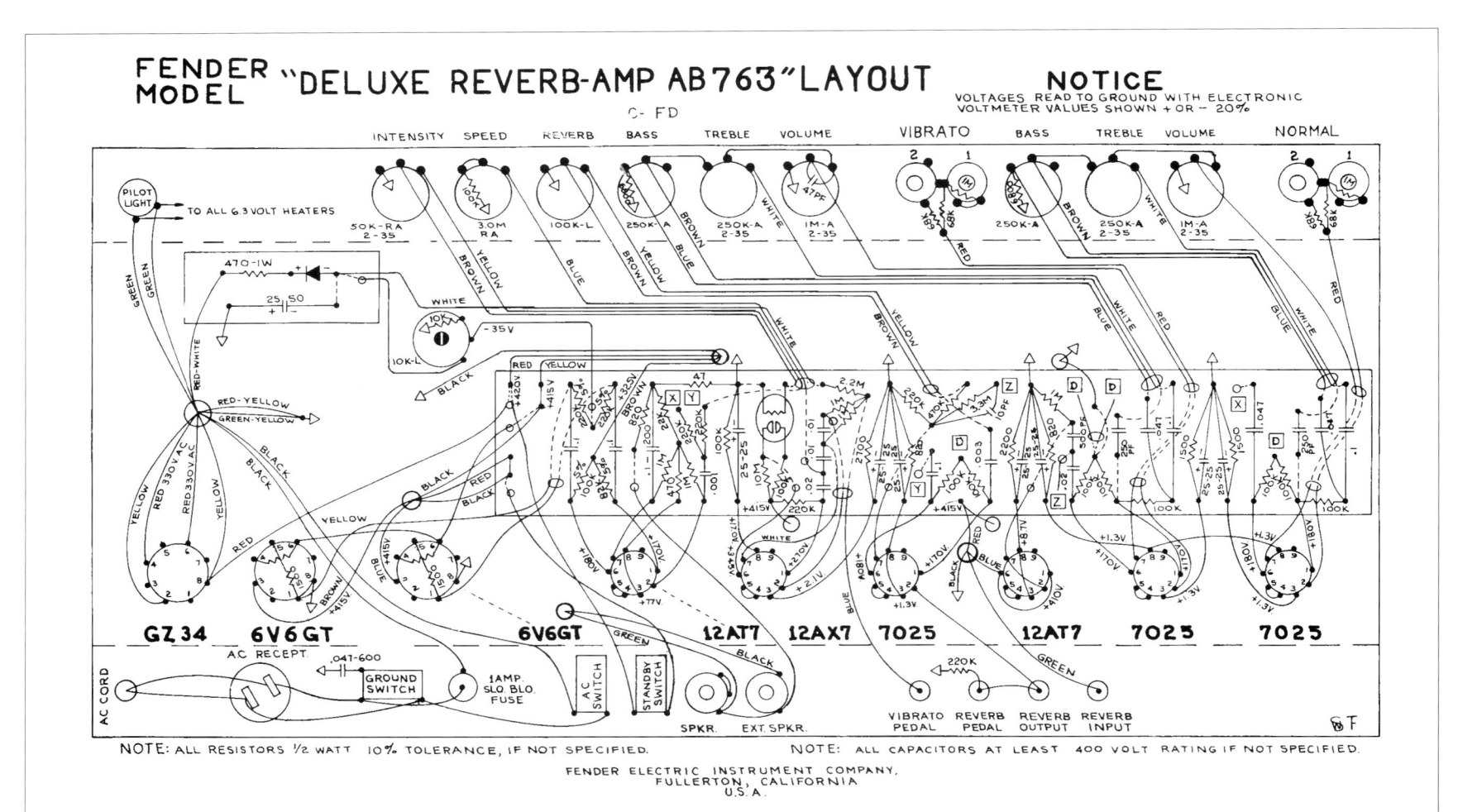

DELUXE REVERB-AMP **AB763**
DELUXE REVERB-AMP **AB763**

Fender

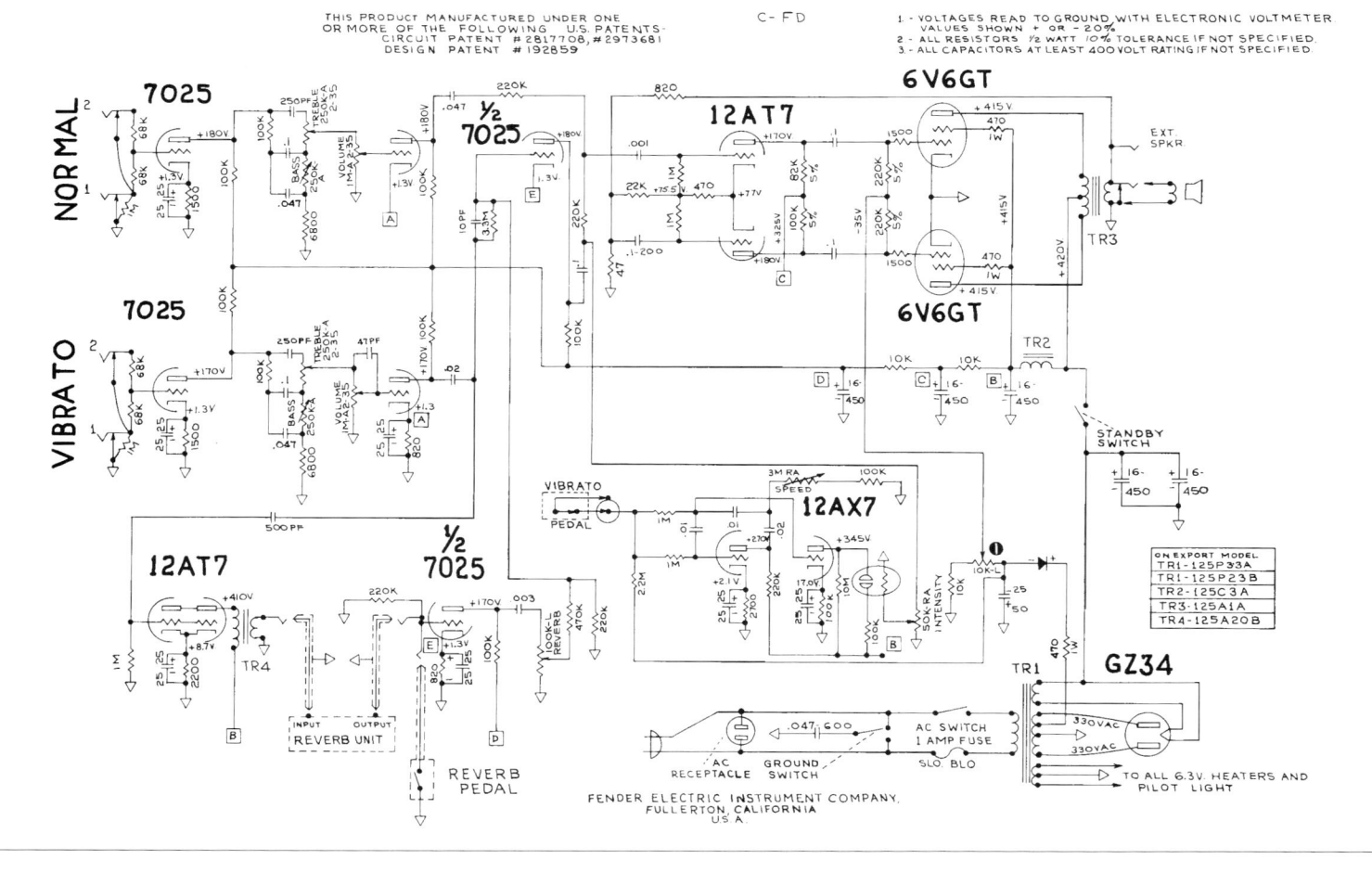

Fender

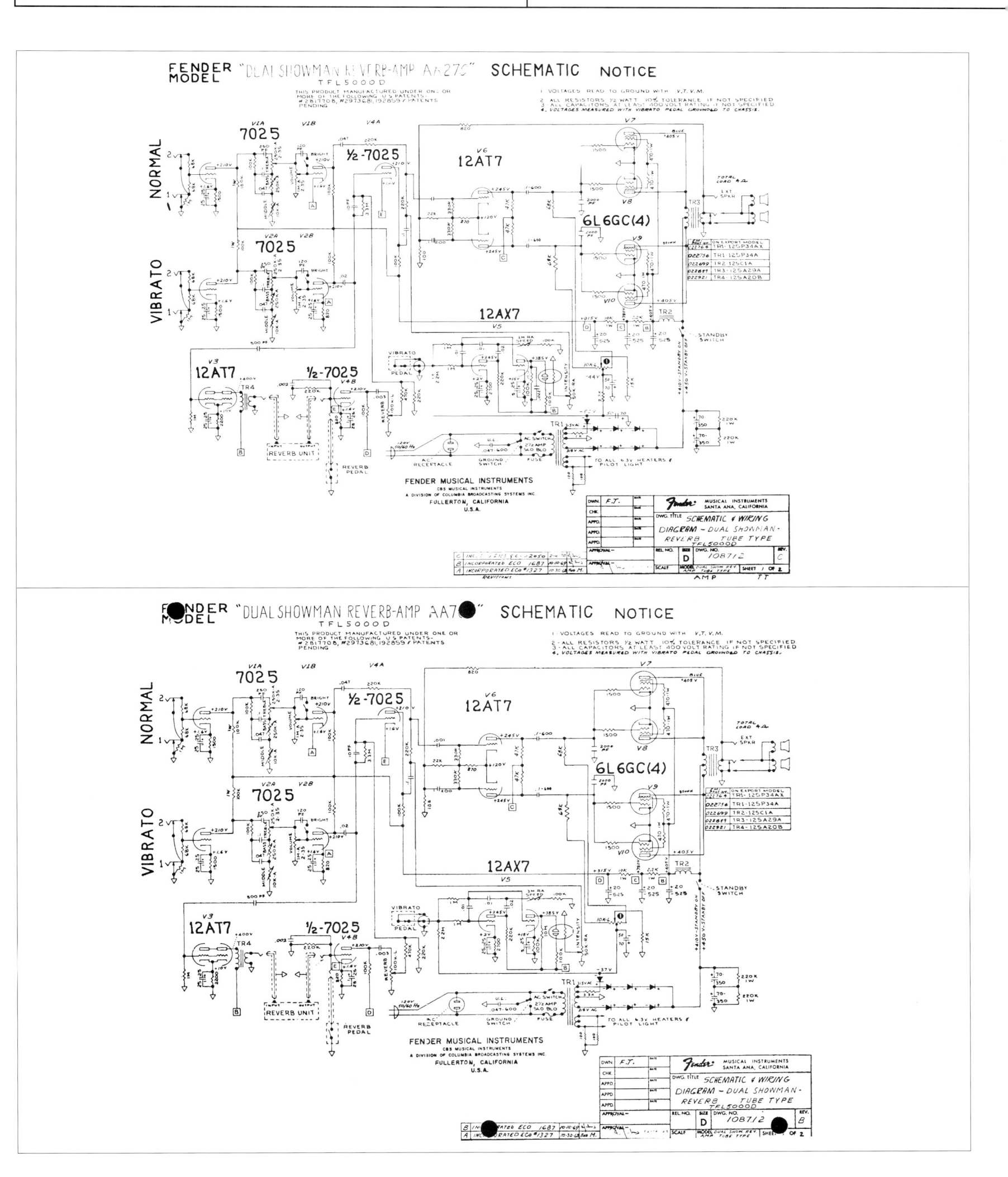

DUAL SHOWMAN REVERB **AA270**
DUAL SHOWMAN REVERB **AA769**

HARVARD 5F10
HARVARD 6G10

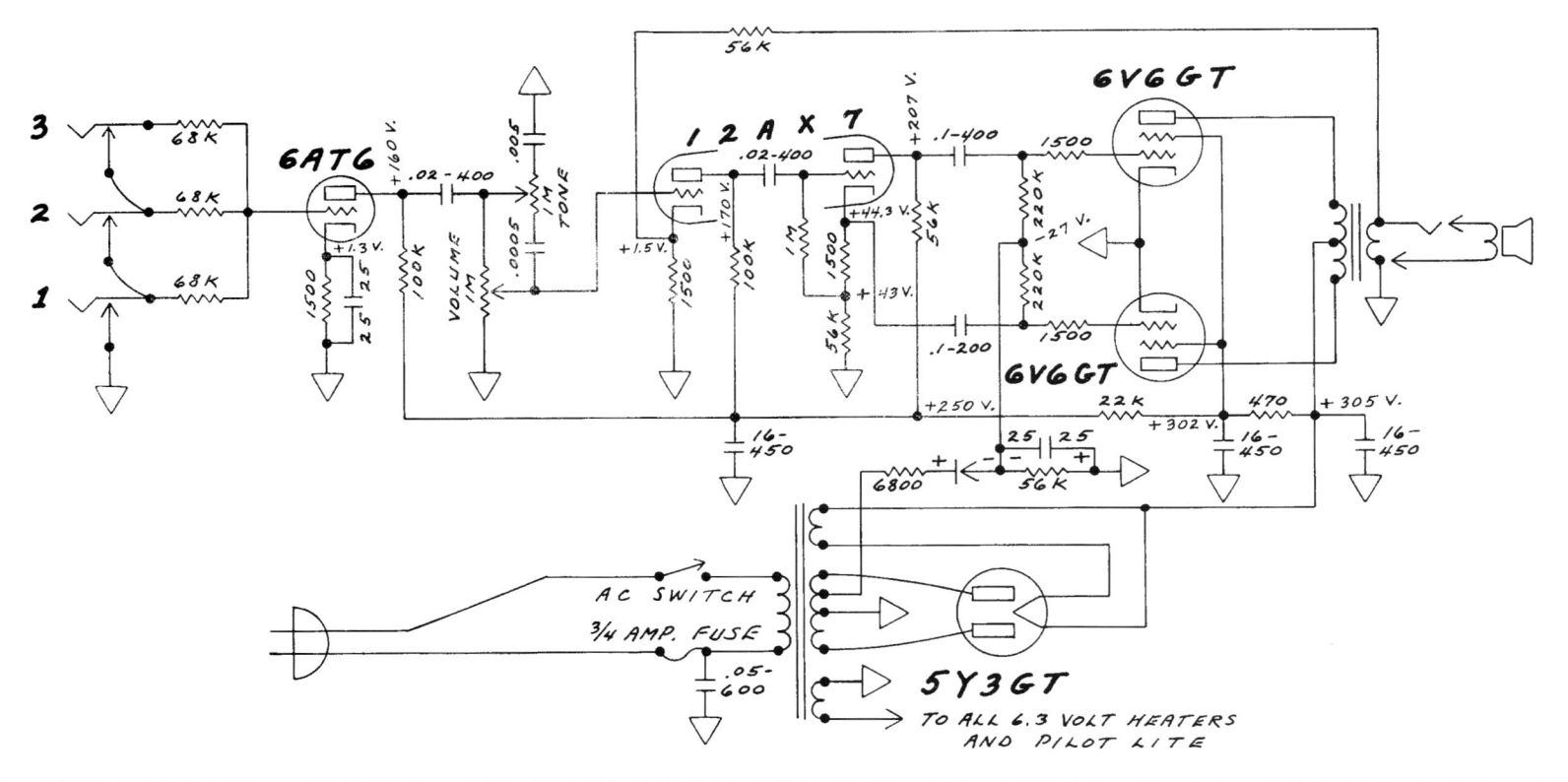

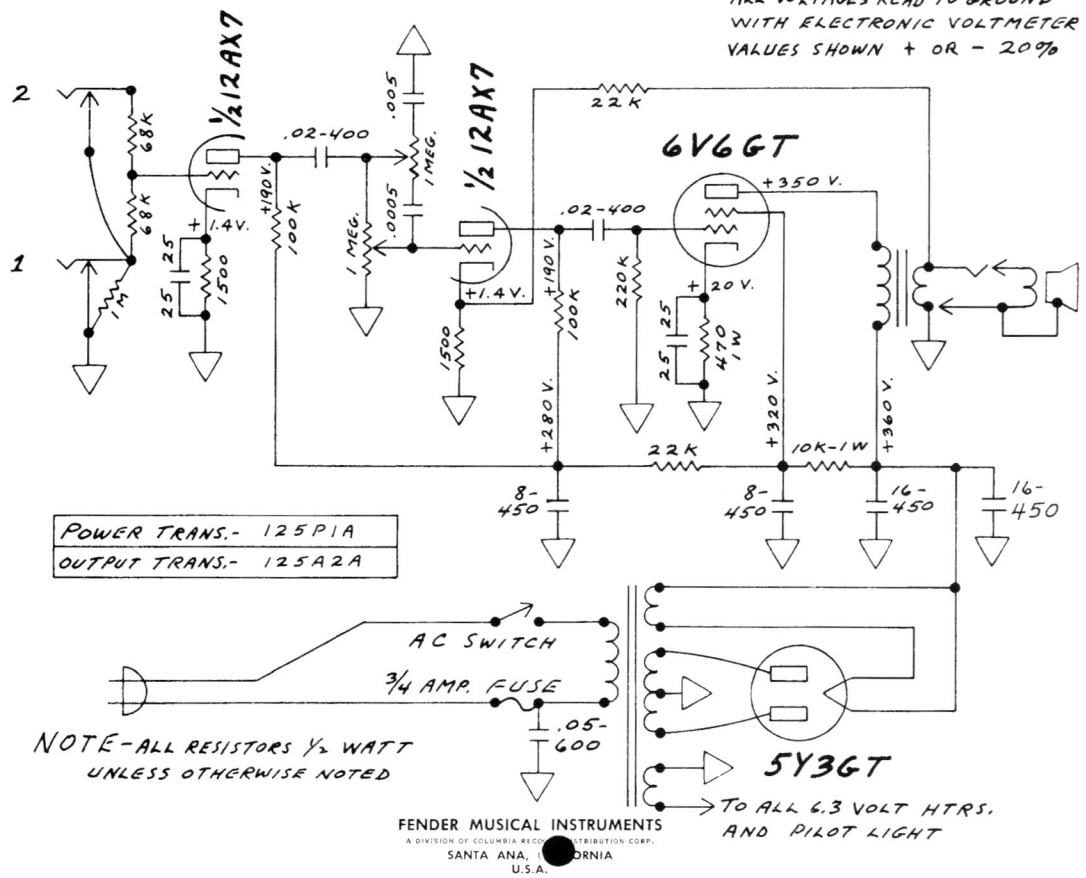

Fender

Fender

PRINCETON **5C2**
PRINCETON **5E2**

FENDER "PRINCETON"
MODEL 5C2 G-DH

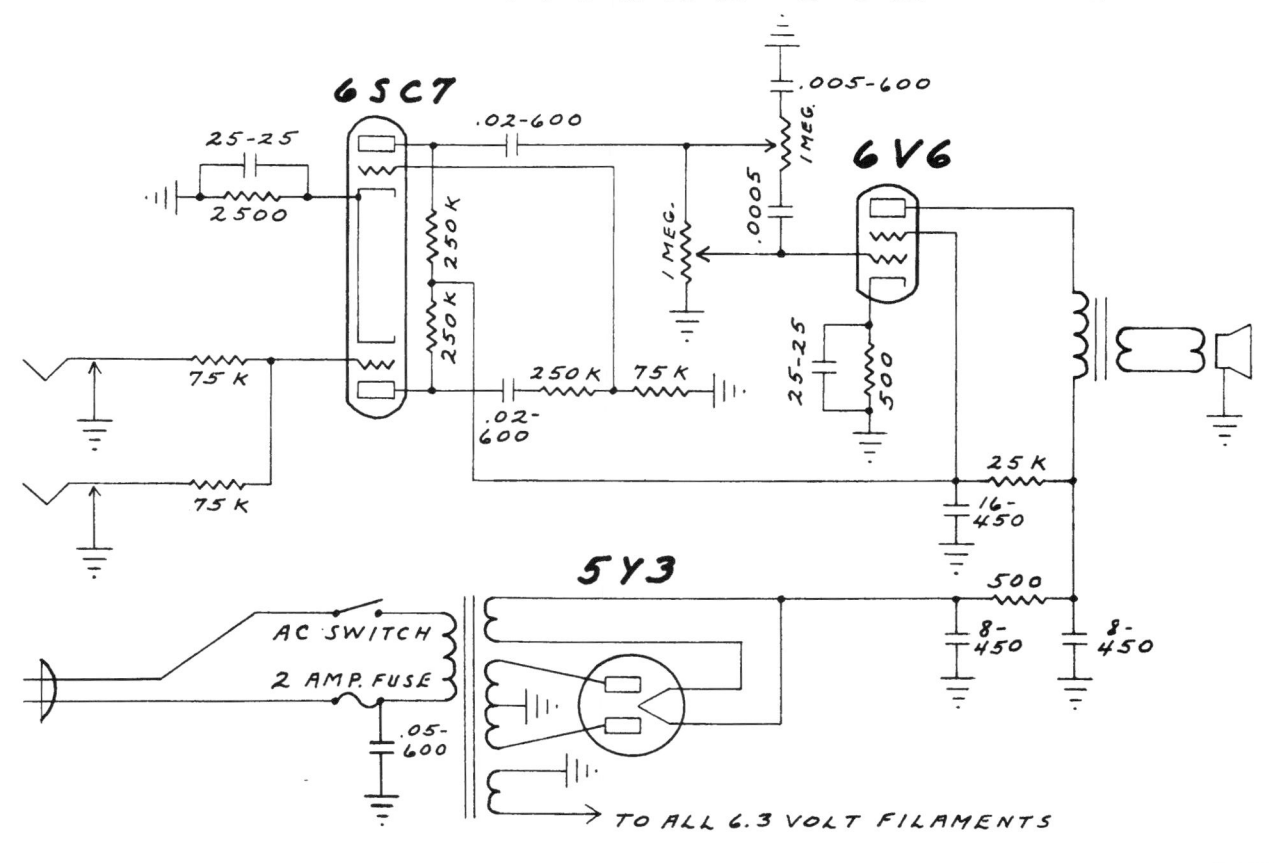

FENDER "PRINCETON" SCHEMATIC
MODEL 5E2 E-EE

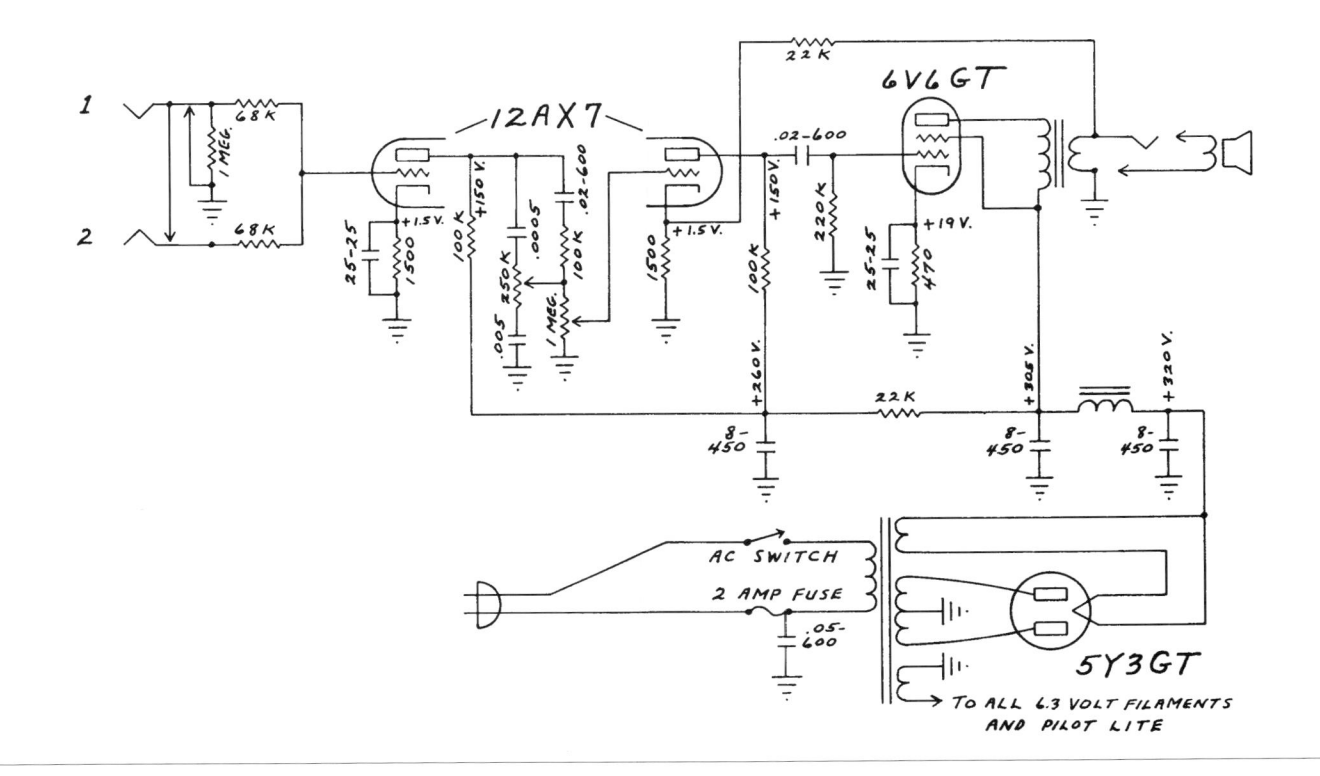

THE TUBE AMP BOOK 41

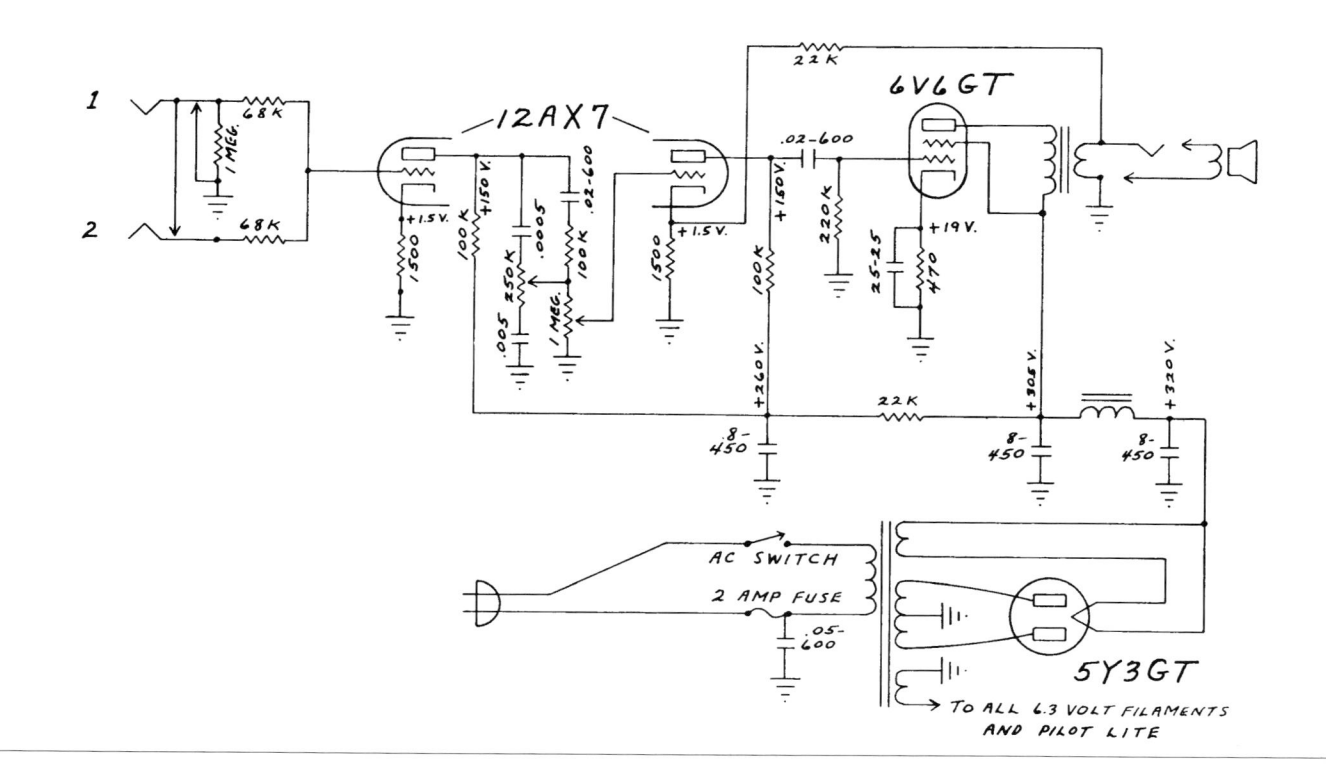

PRINCETON 5F2
PRINCETON 5F2

Fender

Fender

PRINCETON **5F2-A**
PRINCETON **5F2-A**

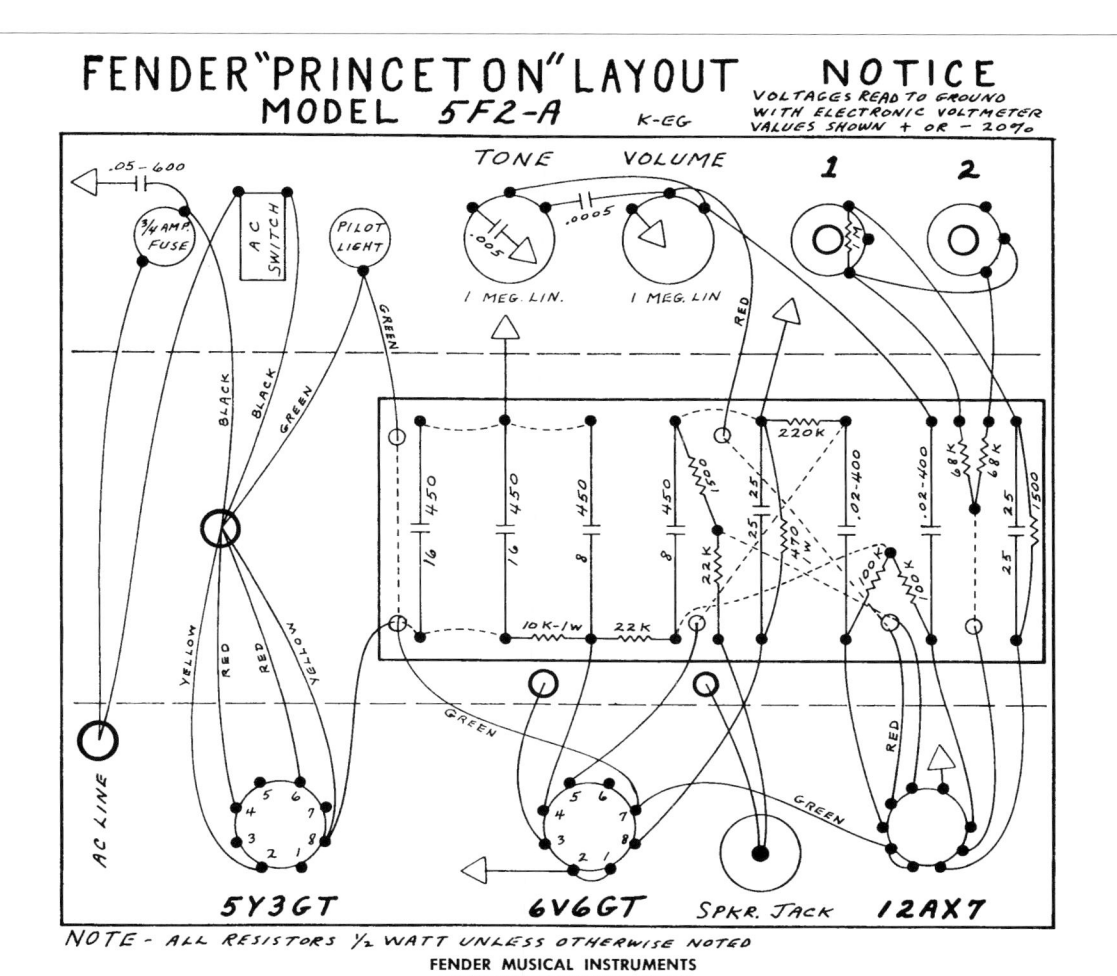

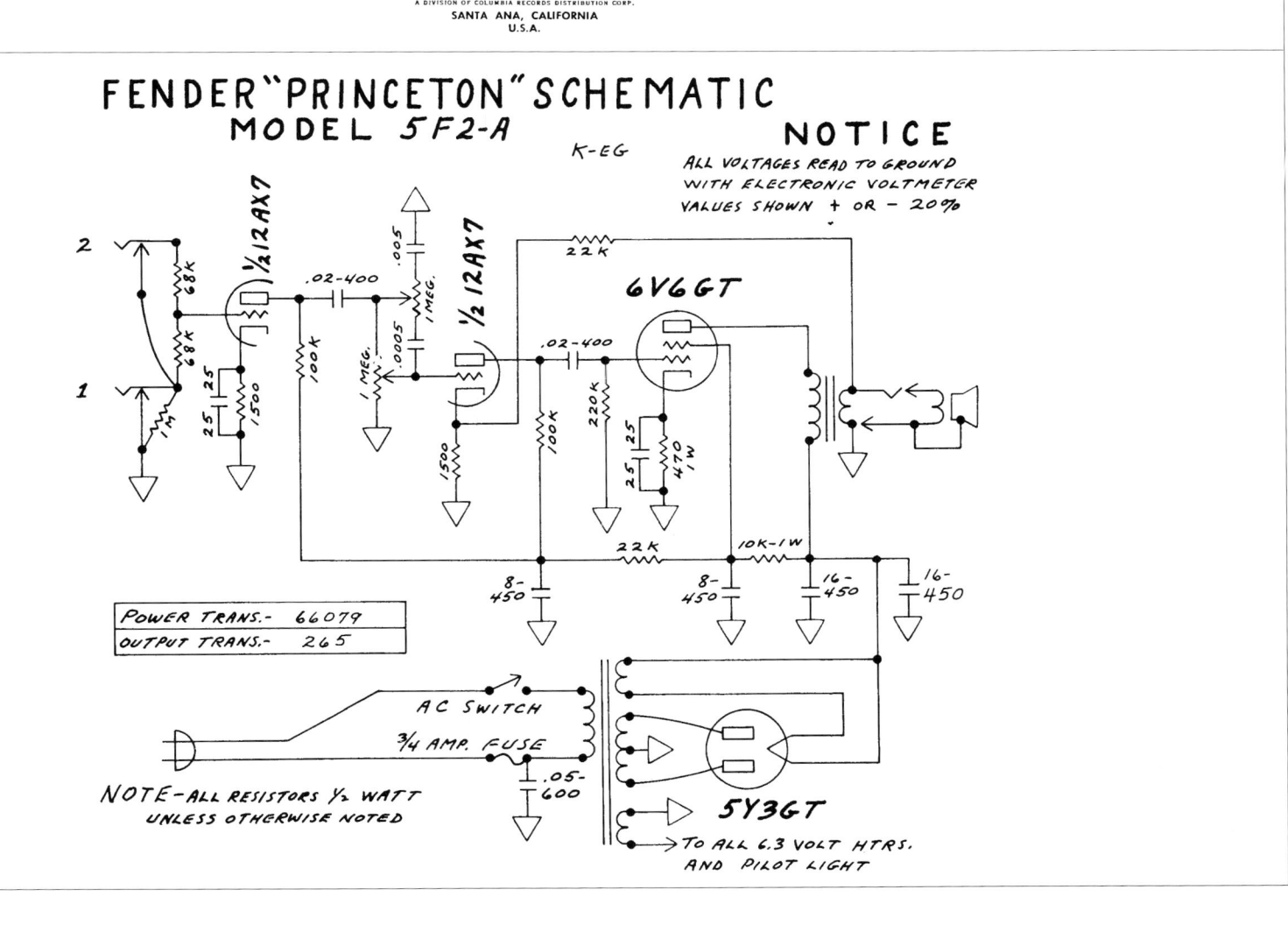

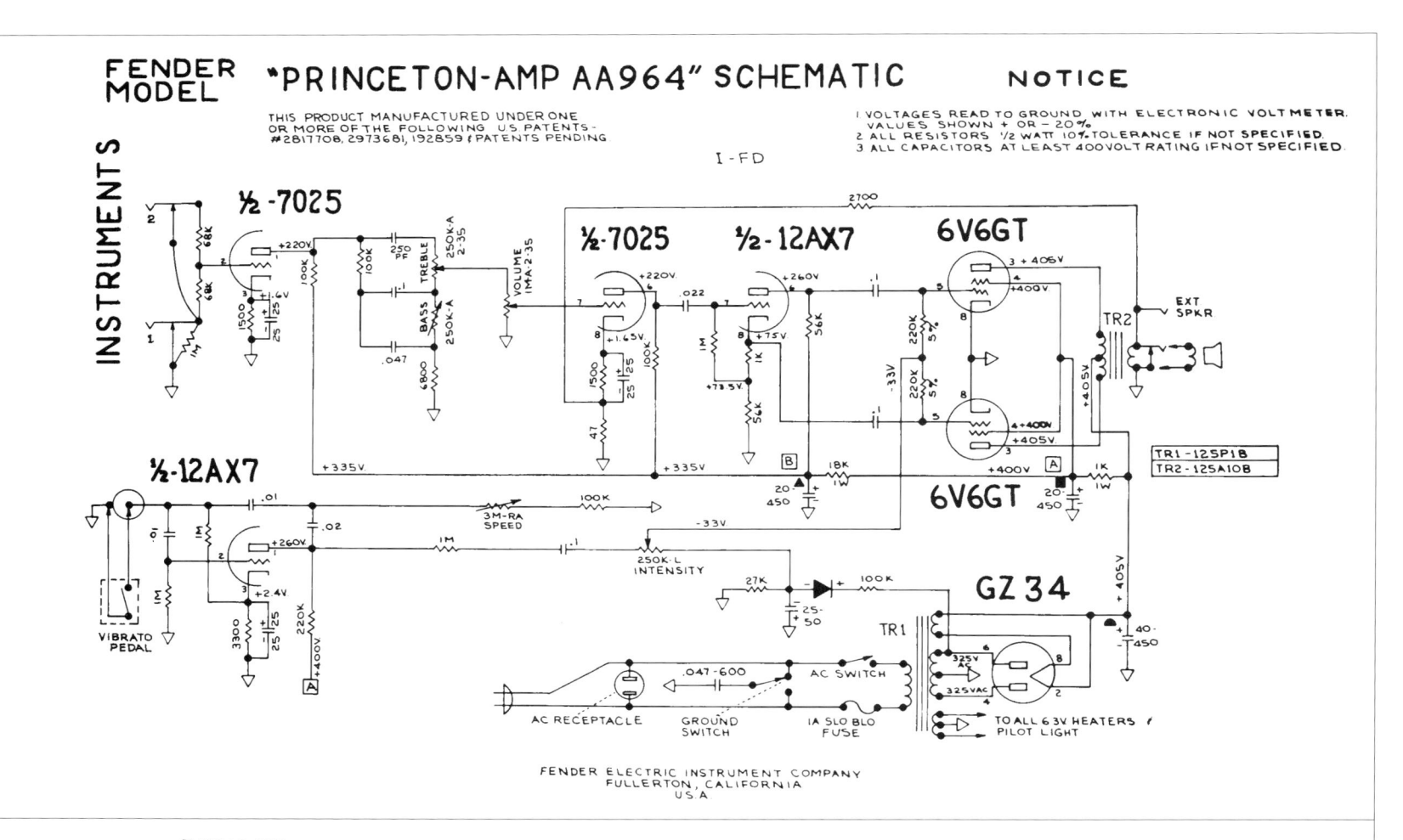

PRINCETON AA964
PRINCETON REVERB AA1164

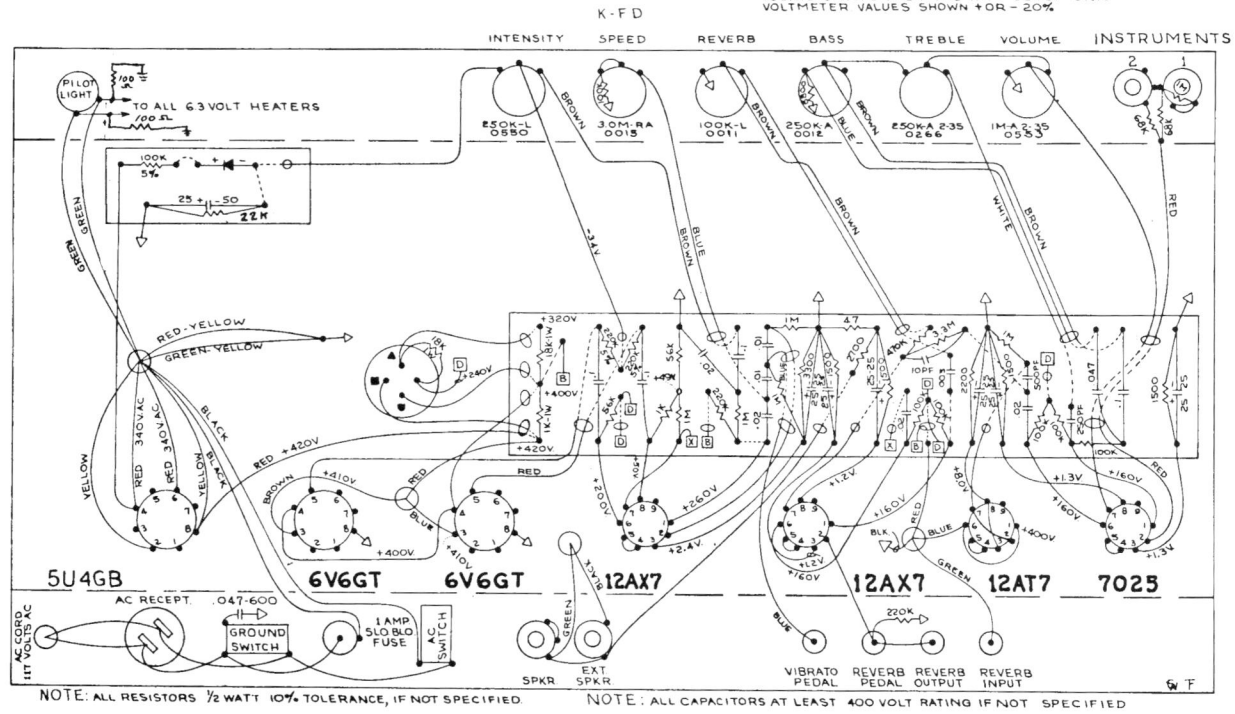

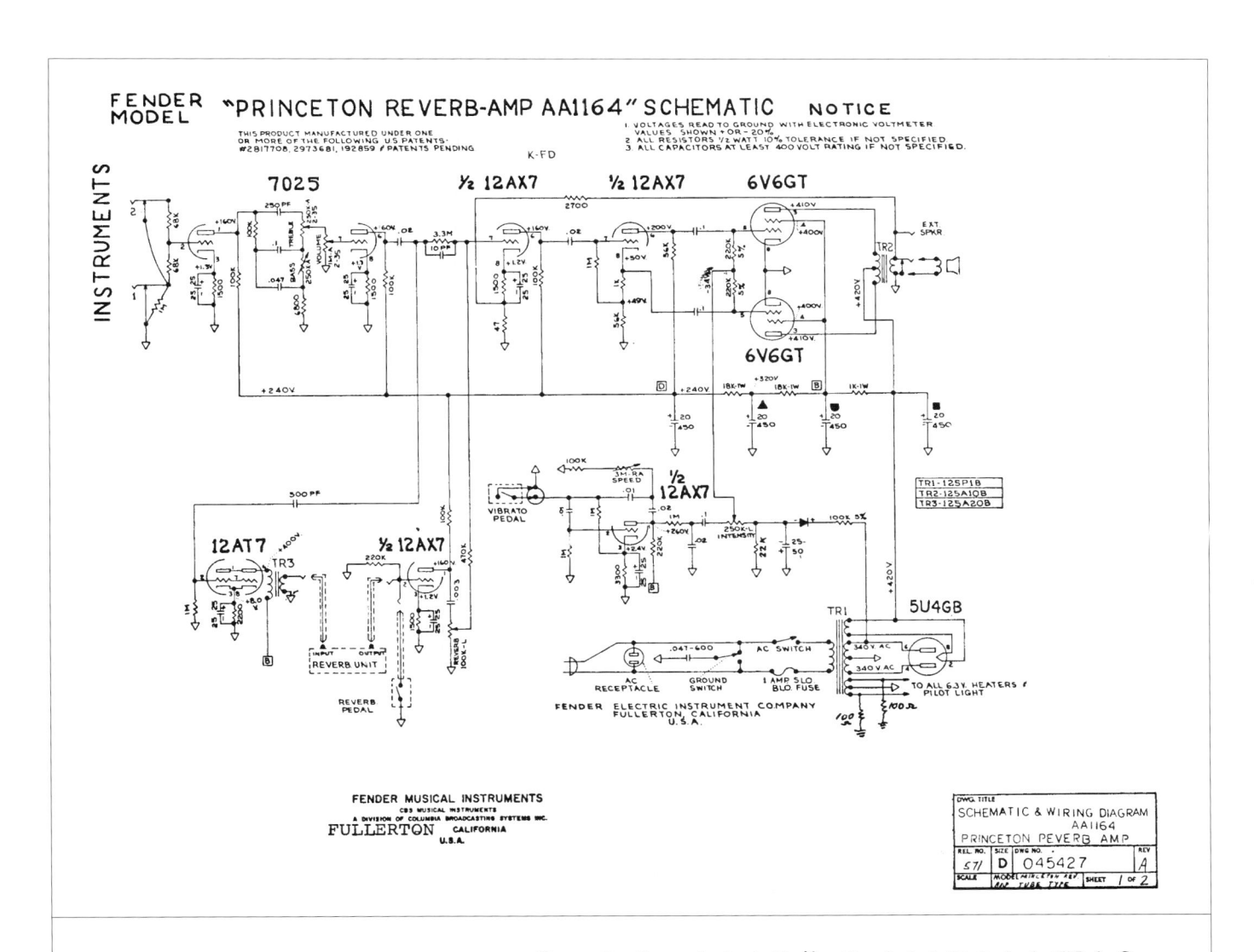

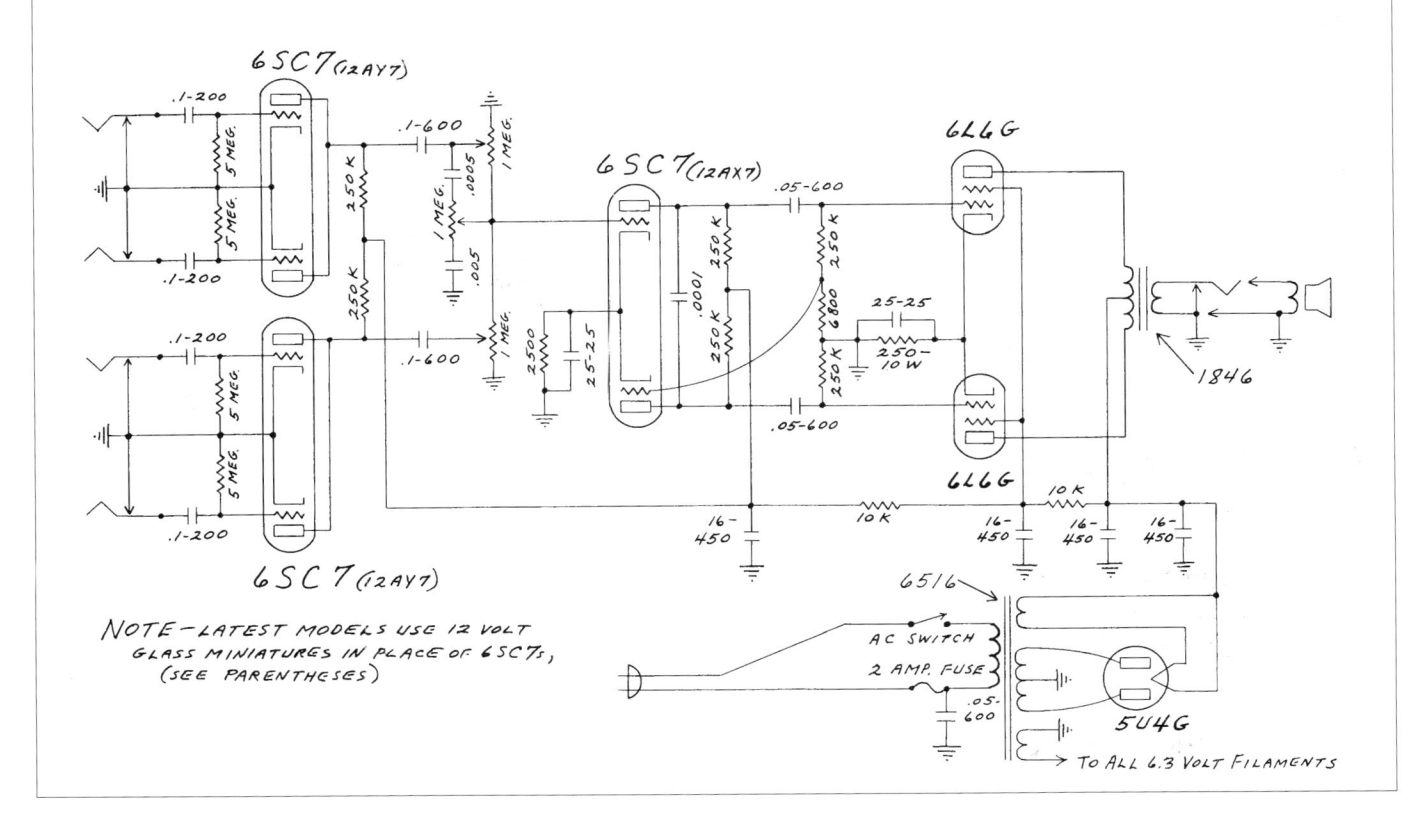

Fender

PRINCETON REVERB AA1164
PRO-AMP 5C5

PRO-AMP 5D5
PRO-AMP 5E5

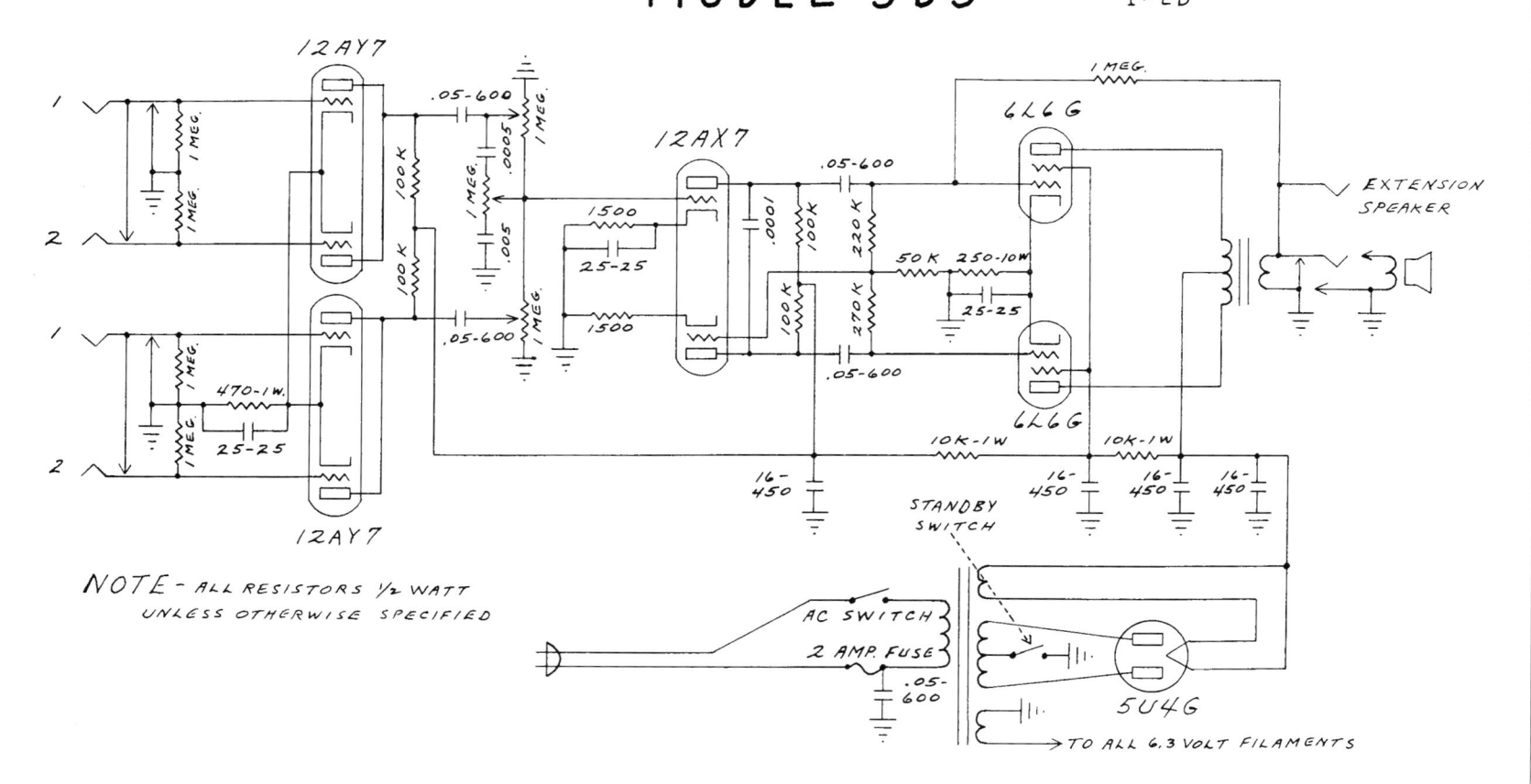

Fender

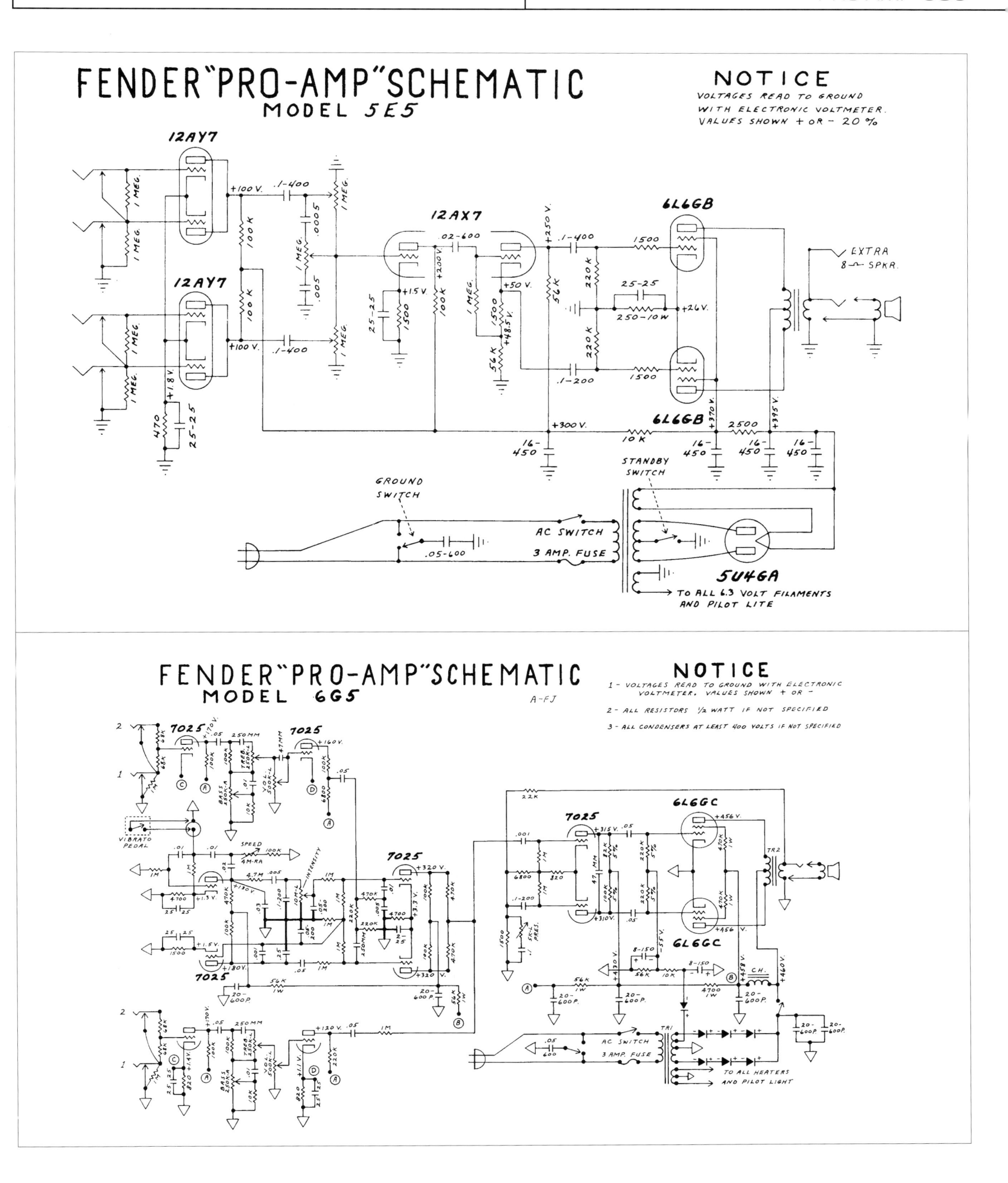

Fender

PRO-AMP **5E5**
PRO-AMP **6G5**

FENDER "PRO-AMP" SCHEMATIC MODEL 5E5

NOTICE
VOLTAGES READ TO GROUND WITH ELECTRONIC VOLTMETER. VALUES SHOWN + OR − 20%

FENDER "PRO-AMP" SCHEMATIC MODEL 6G5
A-FJ

NOTICE
1 - VOLTAGES READ TO GROUND WITH ELECTRONIC VOLTMETER. VALUES SHOWN + OR −
2 - ALL RESISTORS ½ WATT IF NOT SPECIFIED
3 - ALL CONDENSERS AT LEAST 400 VOLTS IF NOT SPECIFIED

THE TUBE AMP BOOK

PRO-AMP **6G5-A**
PRO-AMP **6G5-A**

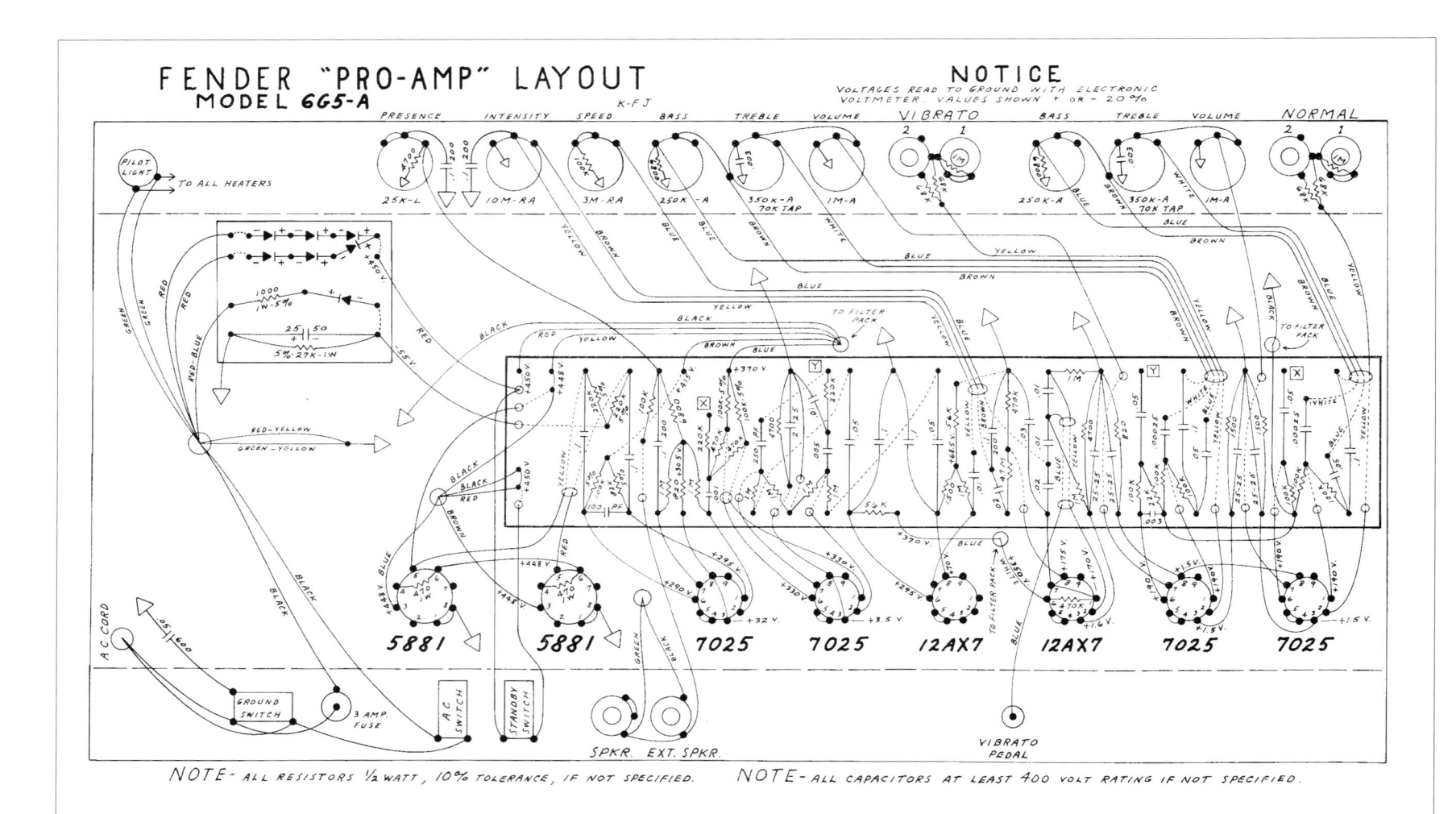

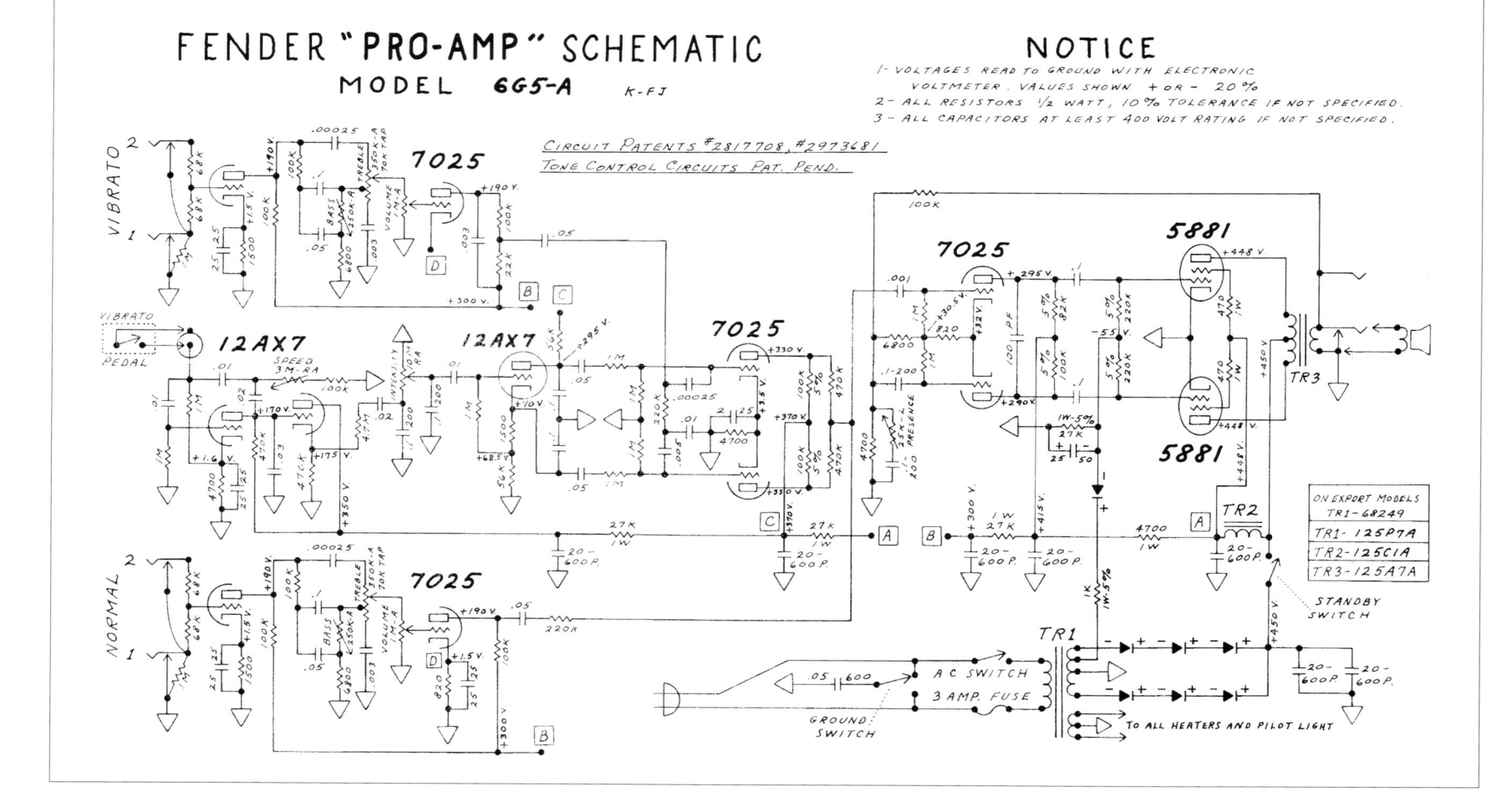

Fender

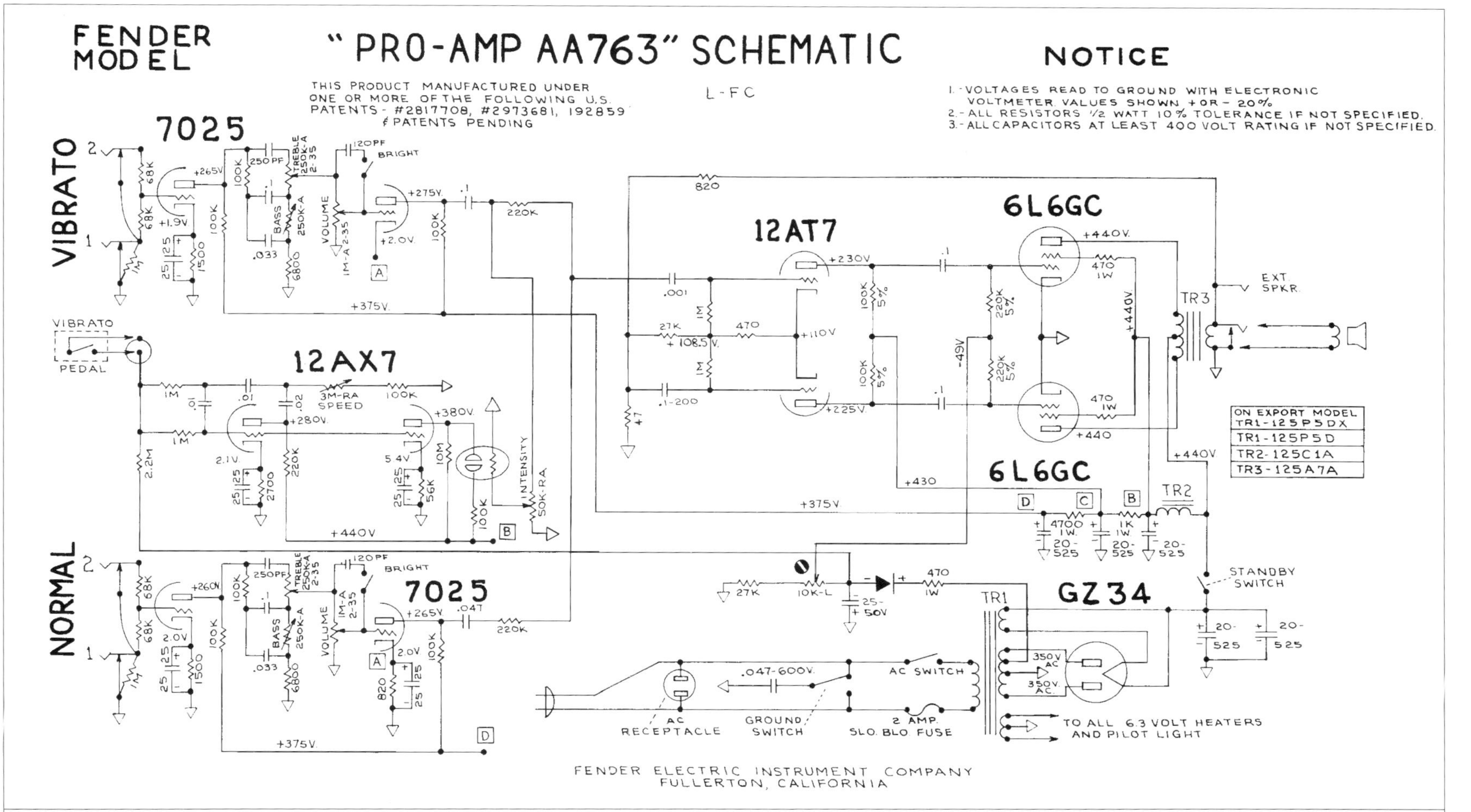
FENDER MODEL "PRO-AMP AA763" SCHEMATIC
THIS PRODUCT MANUFACTURED UNDER ONE OR MORE OF THE FOLLOWING U.S. PATENTS - #2817708, #2973681, 192859 & PATENTS PENDING
L-FC
NOTICE
1.-VOLTAGES READ TO GROUND WITH ELECTRONIC VOLTMETER VALUES SHOWN + OR - 20%
2.- ALL RESISTORS 1/2 WATT 10% TOLERANCE IF NOT SPECIFIED.
3.- ALL CAPACITORS AT LEAST 400 VOLT RATING IF NOT SPECIFIED.
VIBRATO
NORMAL
7025
12AX7
12AT7
6L6GC
GZ34
VIBRATO PEDAL
SPEED
INTENSITY
EXT. SPKR.
ON EXPORT MODEL
TR1-125P5DX
TR1-125P5D
TR2-125C1A
TR3-125A7A
STANDBY SWITCH
AC SWITCH
AC RECEPTACLE
GROUND SWITCH
2 AMP SLO. BLO FUSE
TO ALL 6.3 VOLT HEATERS AND PILOT LIGHT
FENDER ELECTRIC INSTRUMENT COMPANY
FULLERTON, CALIFORNIA

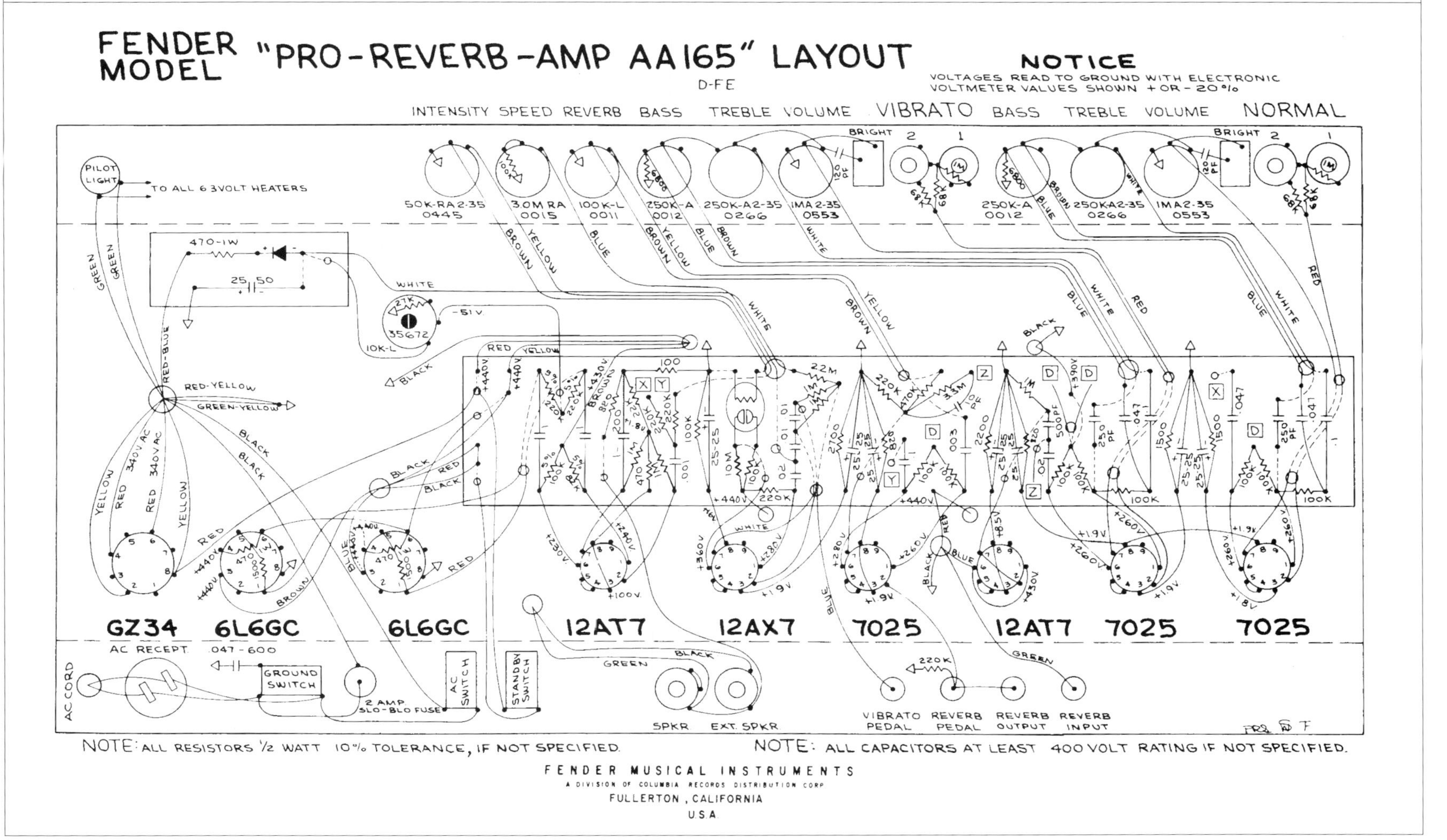
FENDER MODEL "PRO-REVERB-AMP AA165" LAYOUT
D-FE
NOTICE
VOLTAGES READ TO GROUND WITH ELECTRONIC VOLTMETER VALUES SHOWN + OR - 20%
INTENSITY SPEED REVERB BASS TREBLE VOLUME VIBRATO BASS TREBLE VOLUME NORMAL
PILOT LIGHT
TO ALL 6.3VOLT HEATERS
GZ34
6L6GC
6L6GC
12AT7
12AX7
7025
12AT7
7025
7025
AC RECEPT
ACCORD
GROUND SWITCH
2 AMP SLO-BLO FUSE
AC SWITCH
STANDBY SWITCH
SPKR
EXT. SPKR
VIBRATO PEDAL
REVERB PEDAL
REVERB OUTPUT
REVERB INPUT
NOTE: ALL RESISTORS 1/2 WATT 10% TOLERANCE, IF NOT SPECIFIED.
NOTE: ALL CAPACITORS AT LEAST 400 VOLT RATING IF NOT SPECIFIED.
FENDER MUSICAL INSTRUMENTS
A DIVISION OF COLUMBIA RECORDS DISTRIBUTION CORP.
FULLERTON, CALIFORNIA
U.S.A.

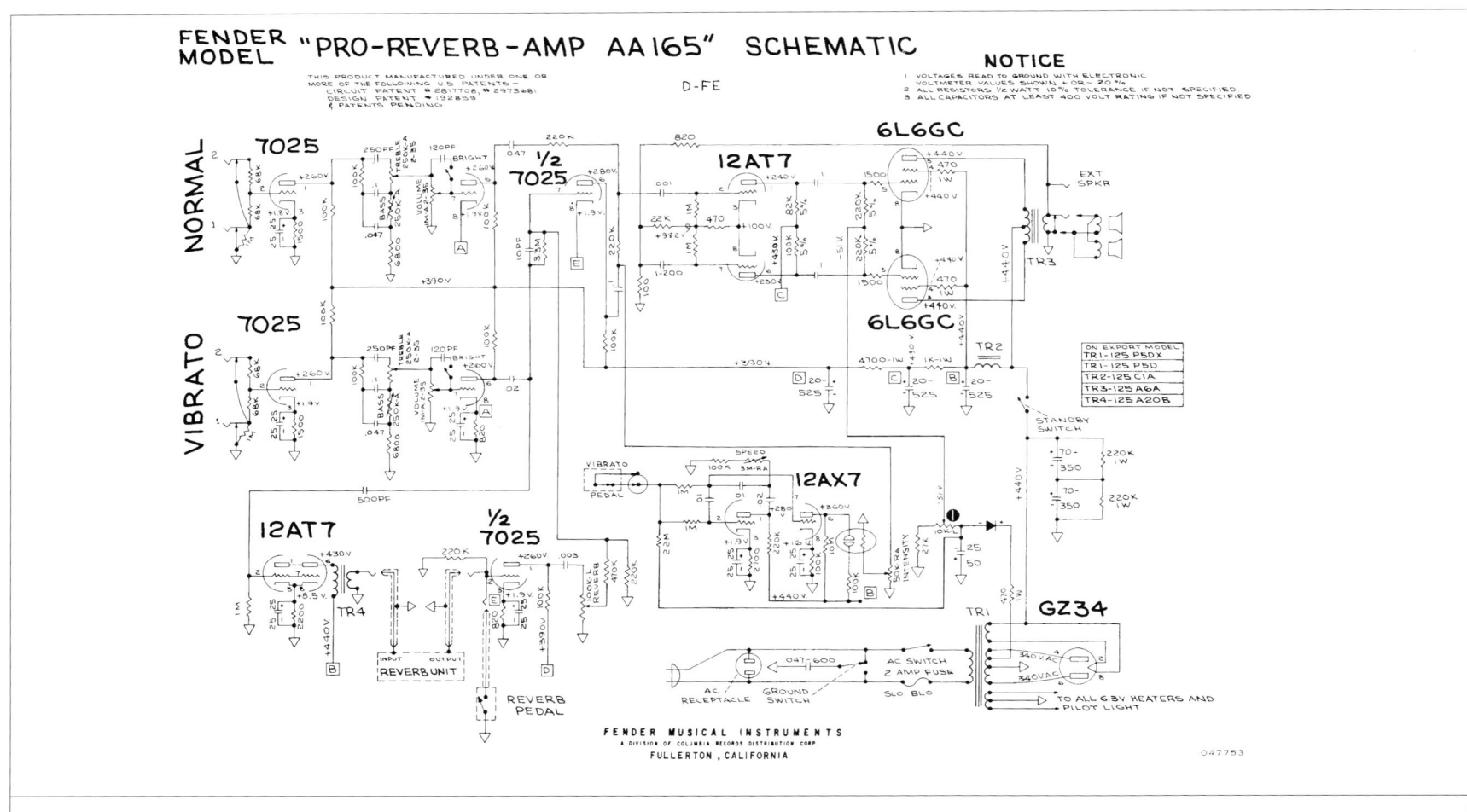

Pro Reverb AA165
Pro Reverb AA1069

Fender

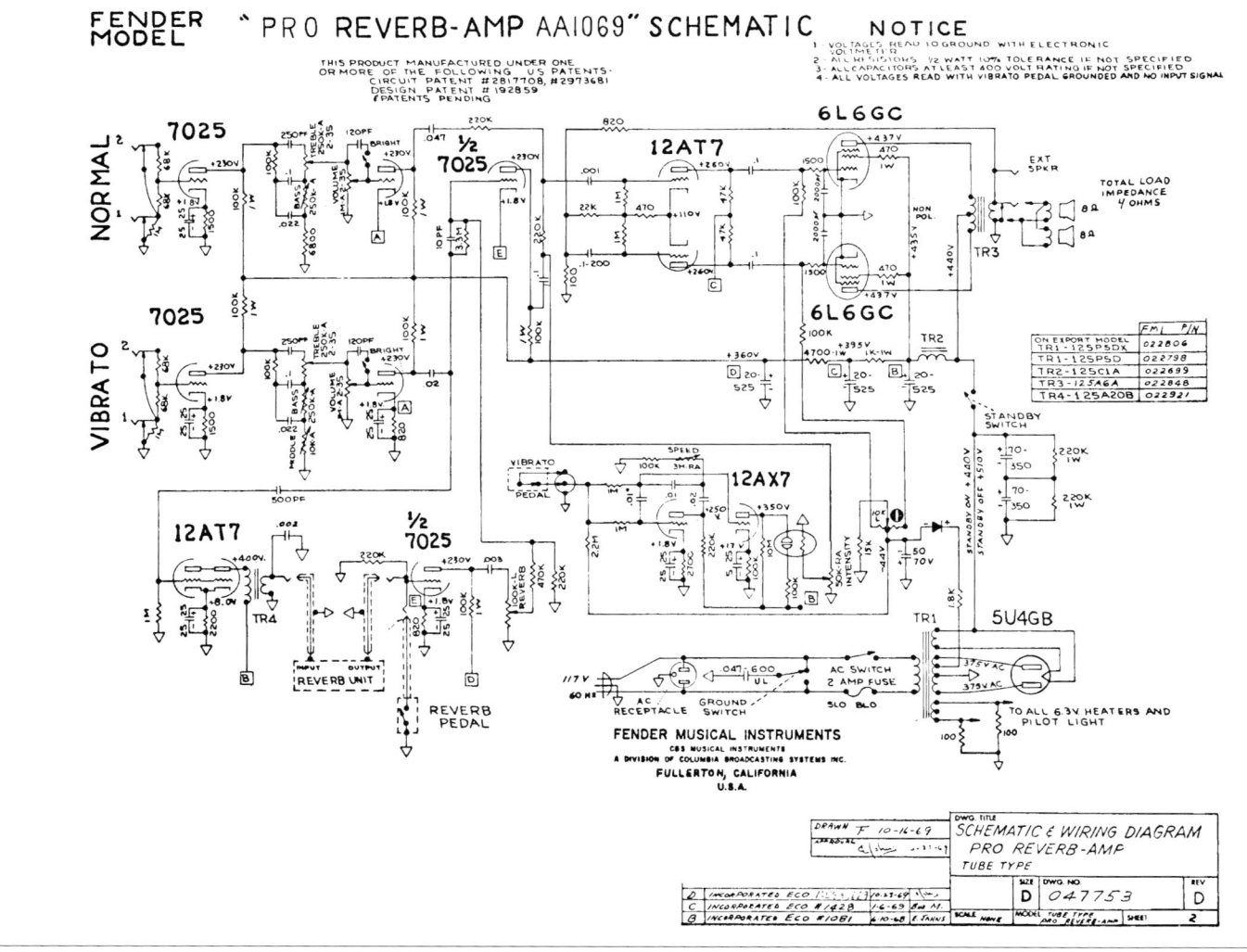

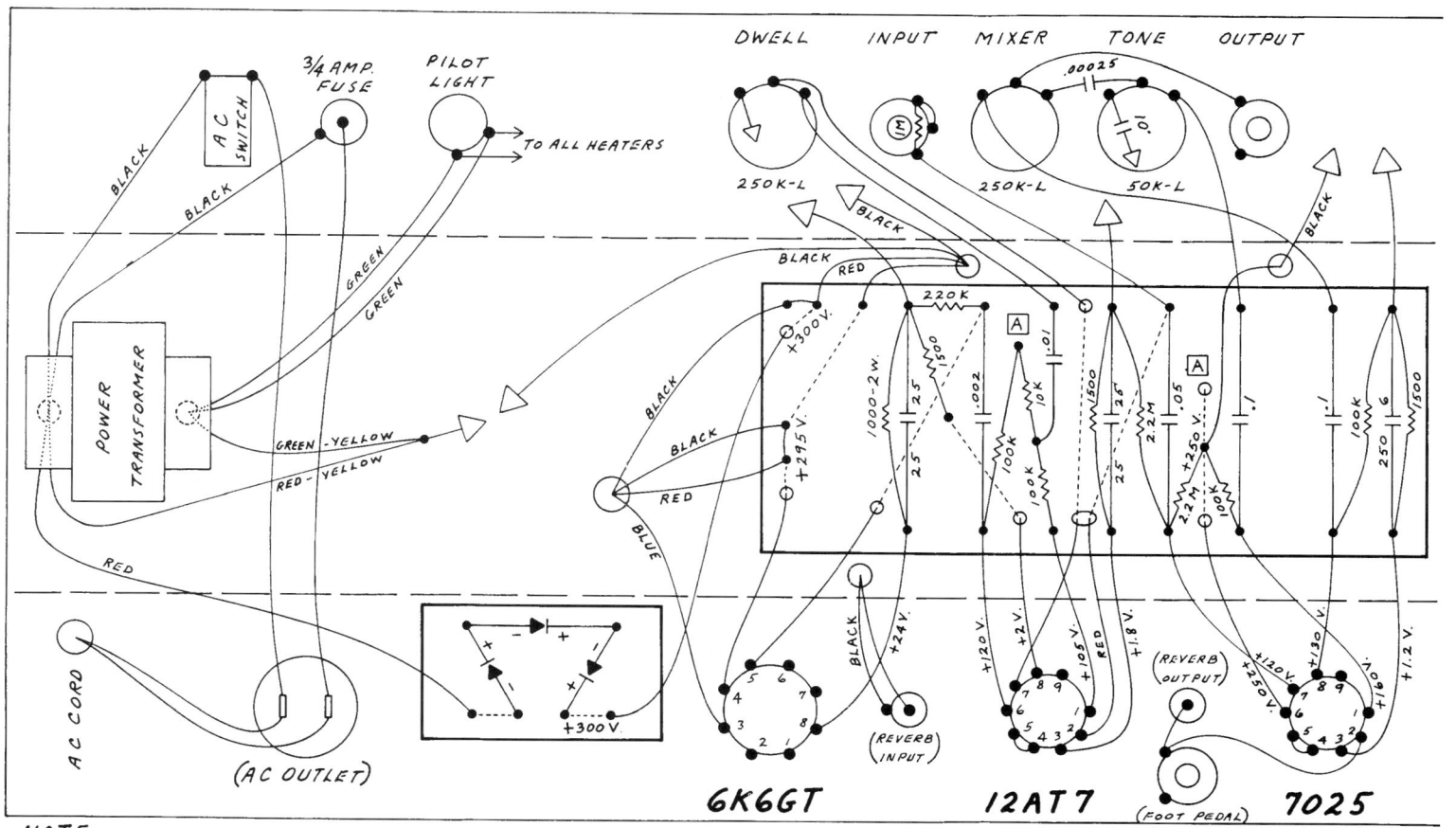

Fender

REVERB **6G15**
REVERB **6G15**

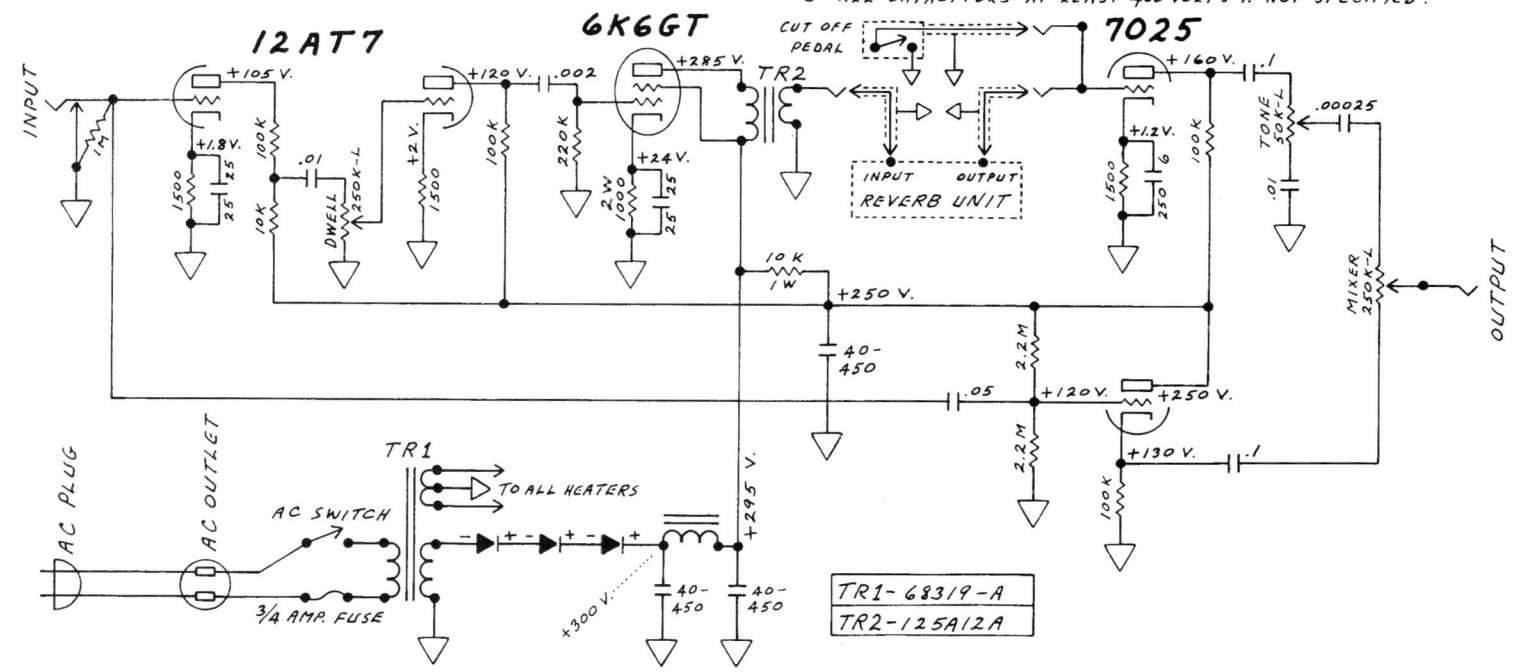

SHOWMAN 6G14-A
Fender

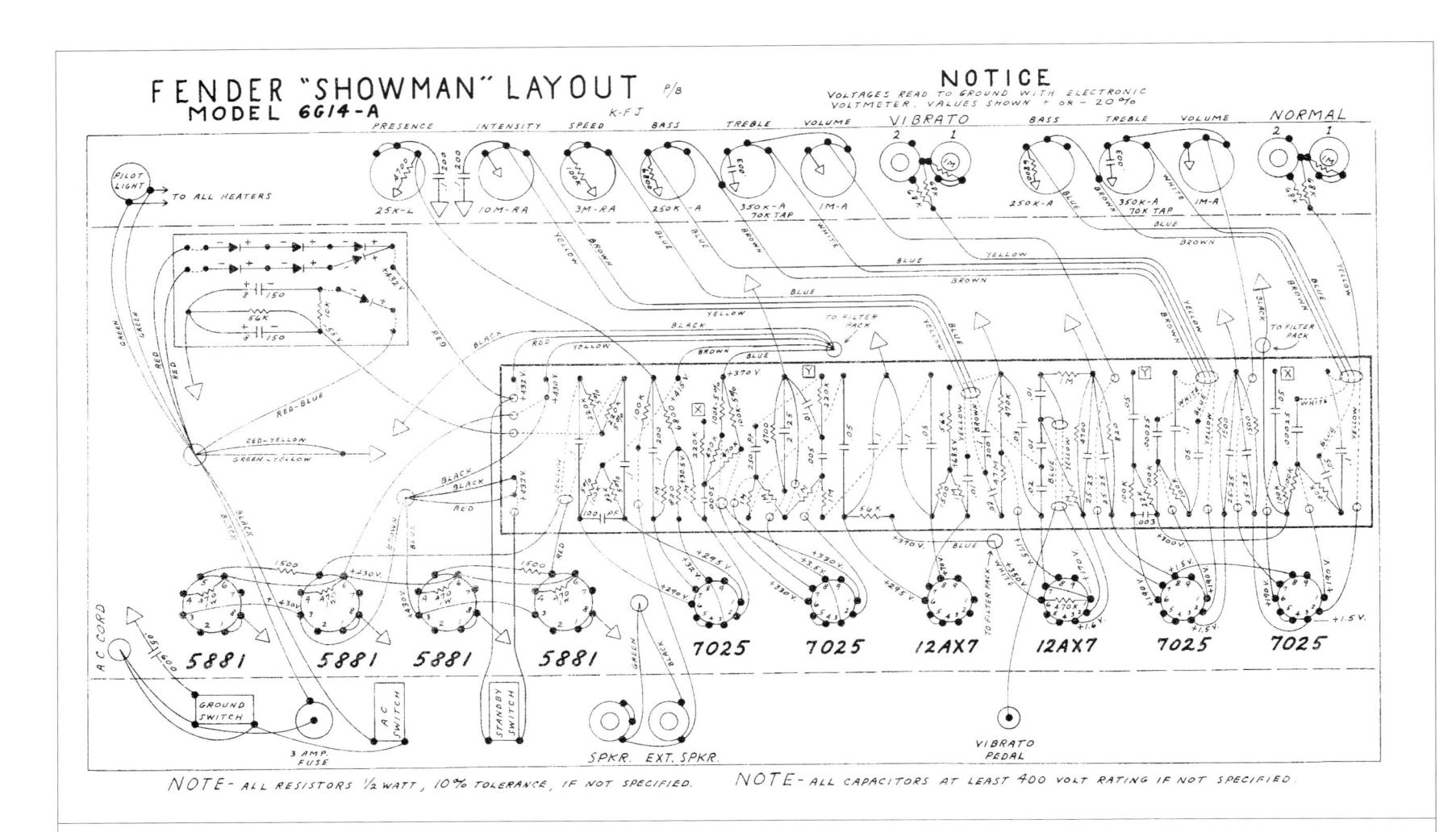

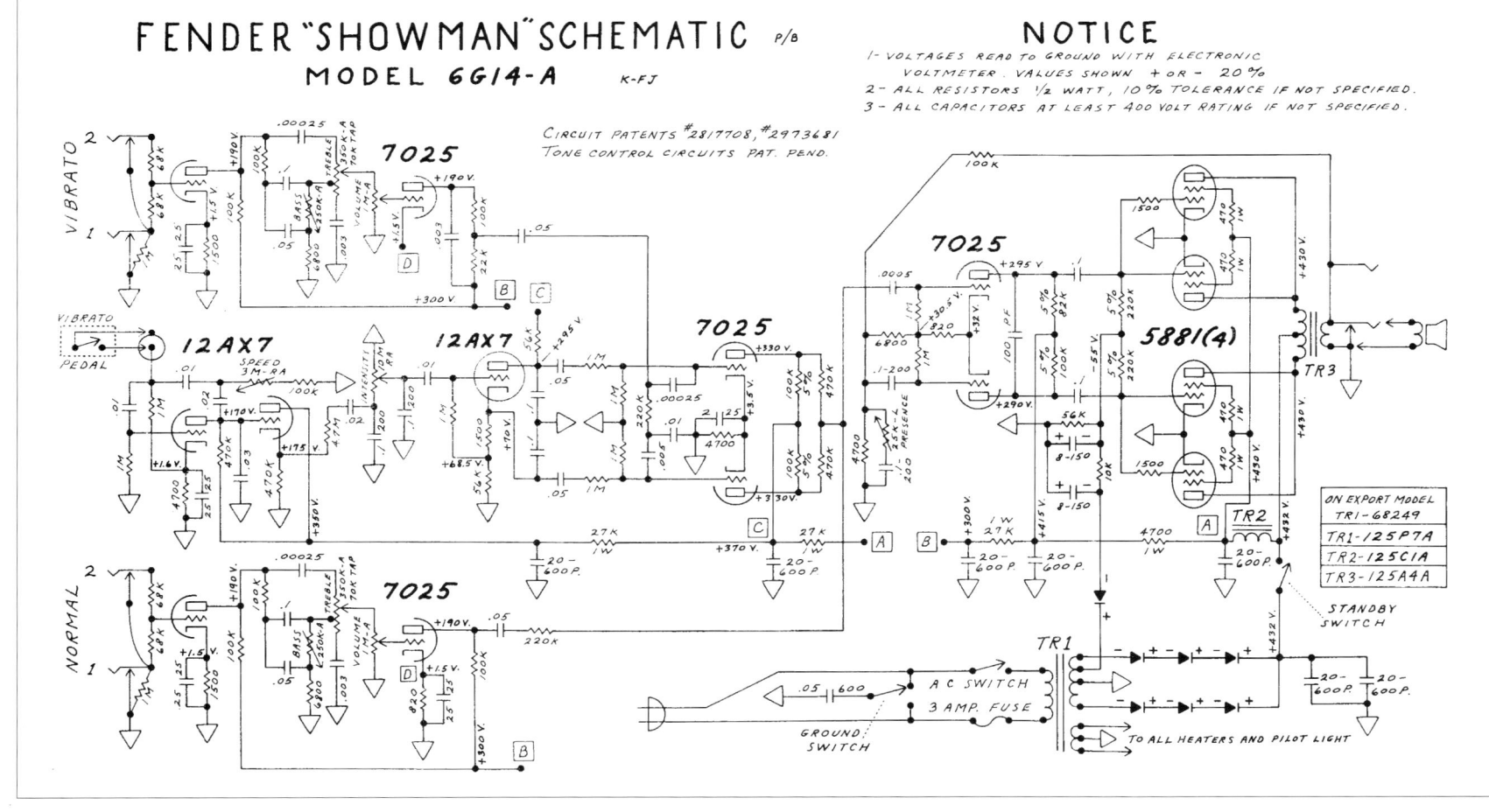

Fender

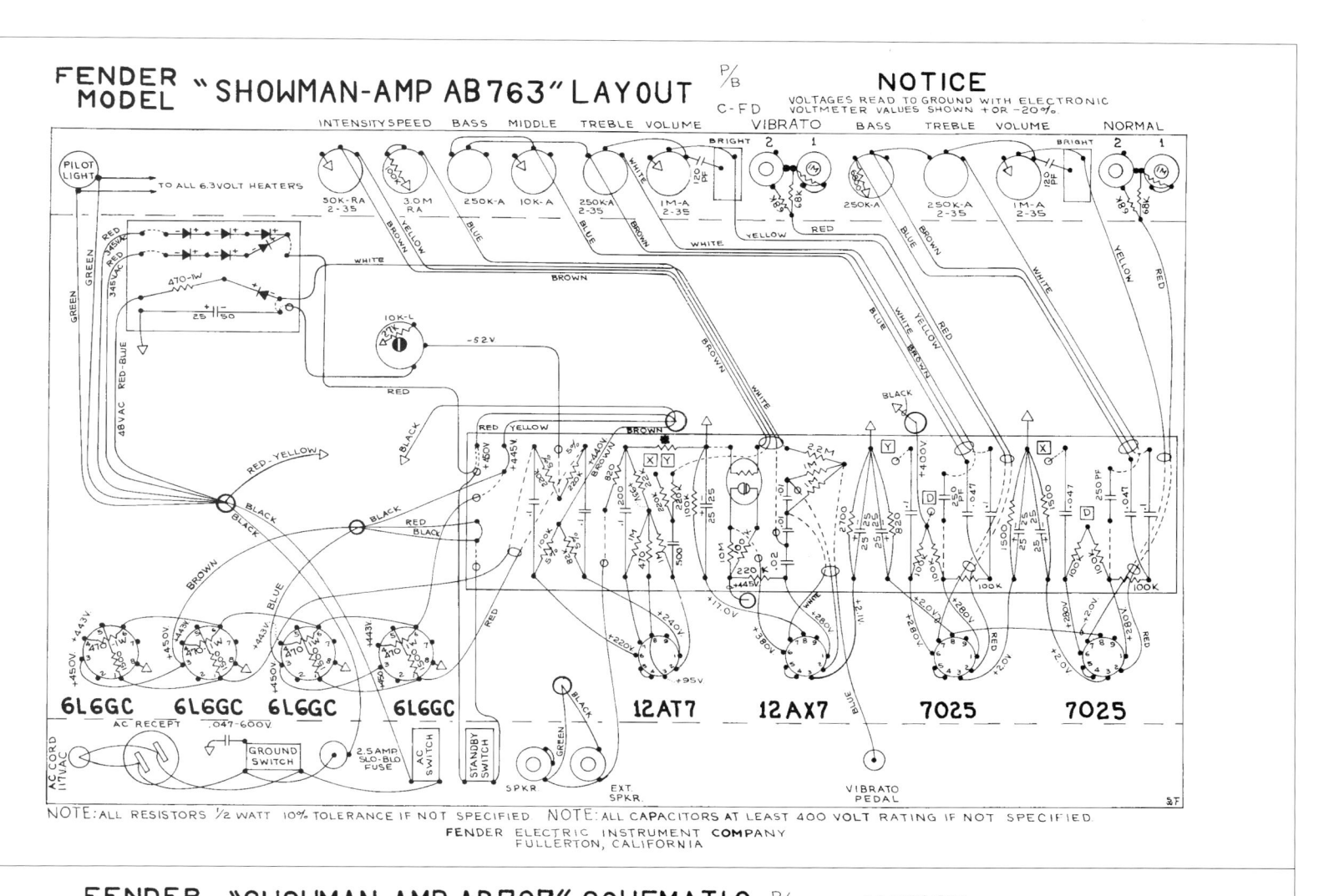

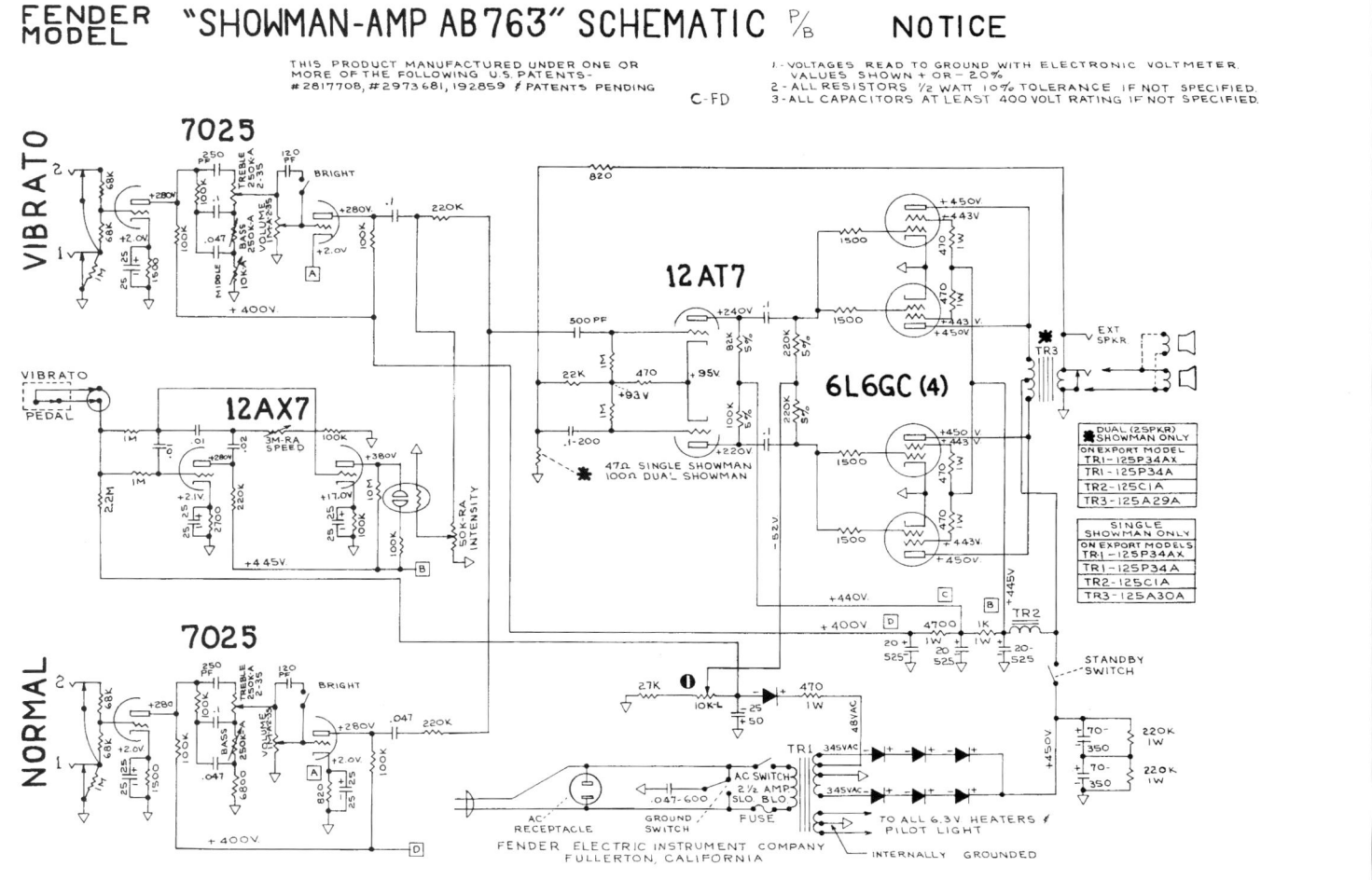

SHOWMAN **AB763**
SHOWMAN **AB763**

THE TUBE AMP BOOK

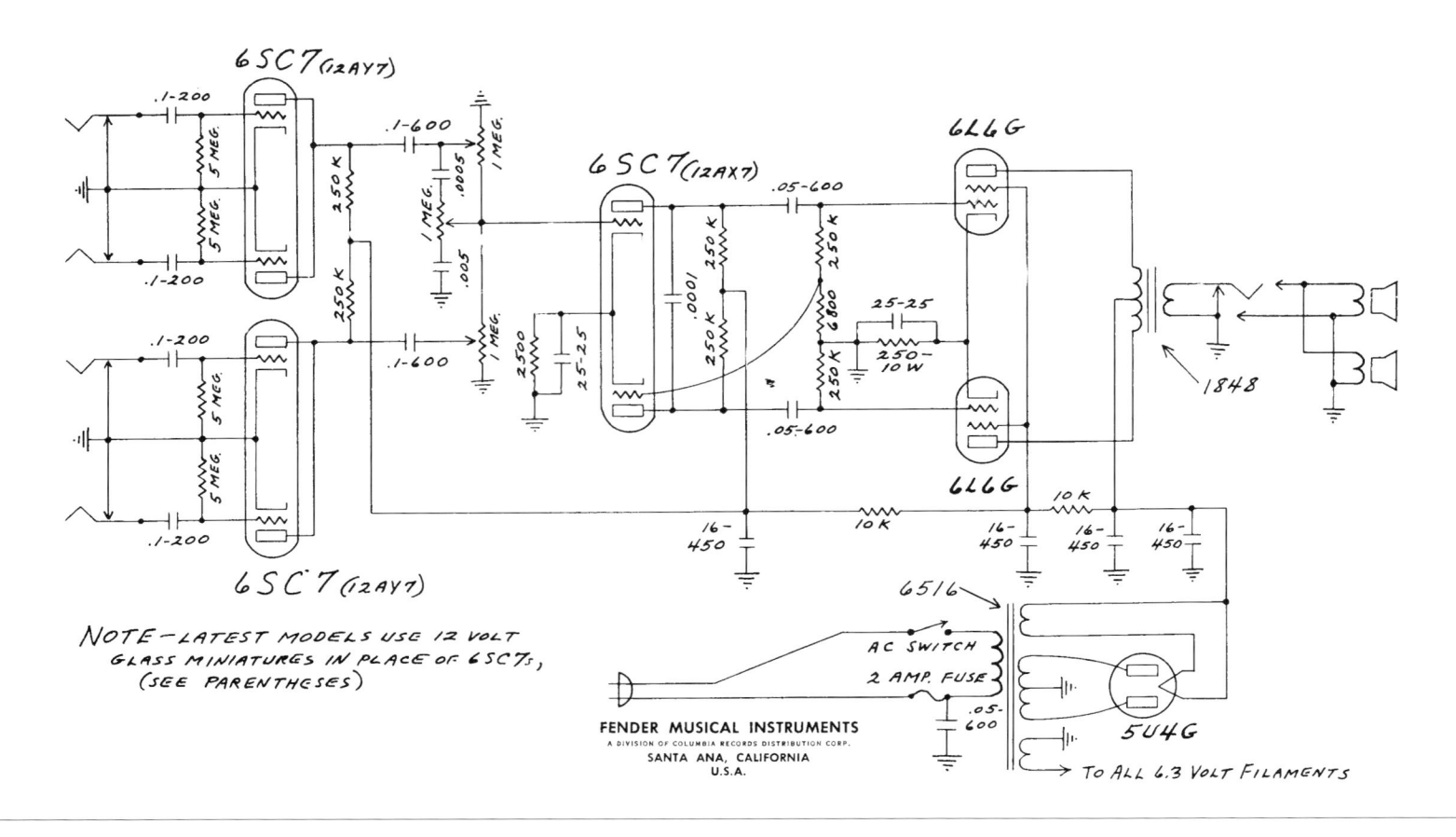

Fender

SUPER-AMP **5C4**
SUPER-AMP **5D4**

FENDER "SUPER-AMP" SCHEMATIC
MODEL 5C4

NOTE — LATEST MODELS USE 12 VOLT GLASS MINIATURES IN PLACE OF 6SC7s, (SEE PARENTHESES)

FENDER MUSICAL INSTRUMENTS
A DIVISION OF COLUMBIA RECORDS DISTRIBUTION CORP.
SANTA ANA, CALIFORNIA
U.S.A.

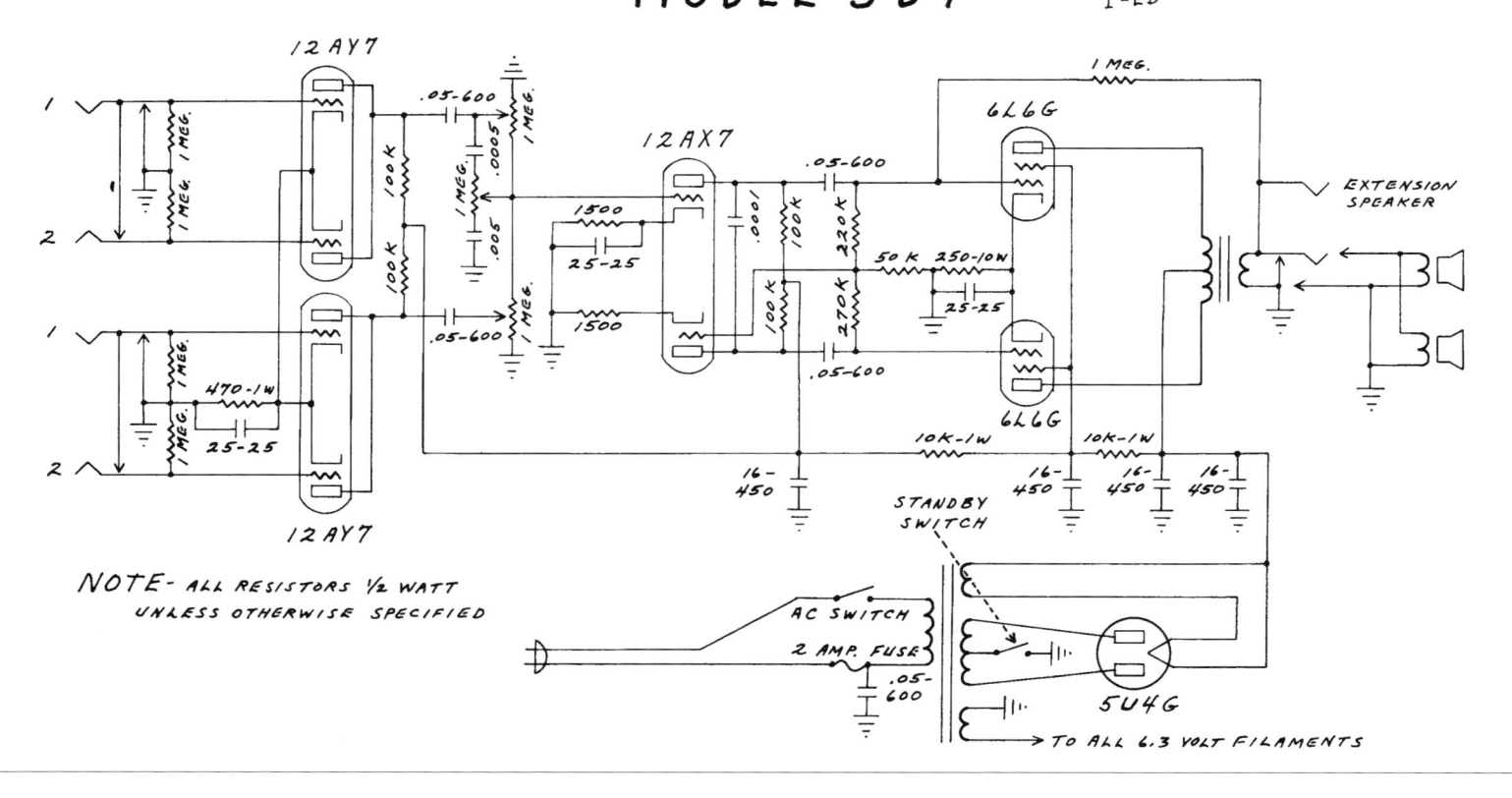

FENDER "SUPER-AMP" SCHEMATIC
MODEL 5D4   I-ED

NOTE — ALL RESISTORS ½ WATT UNLESS OTHERWISE SPECIFIED

54 THE TUBE AMP BOOK

Fender

SUPER-AMP 5E4-A
SUPER-AMP 5F4

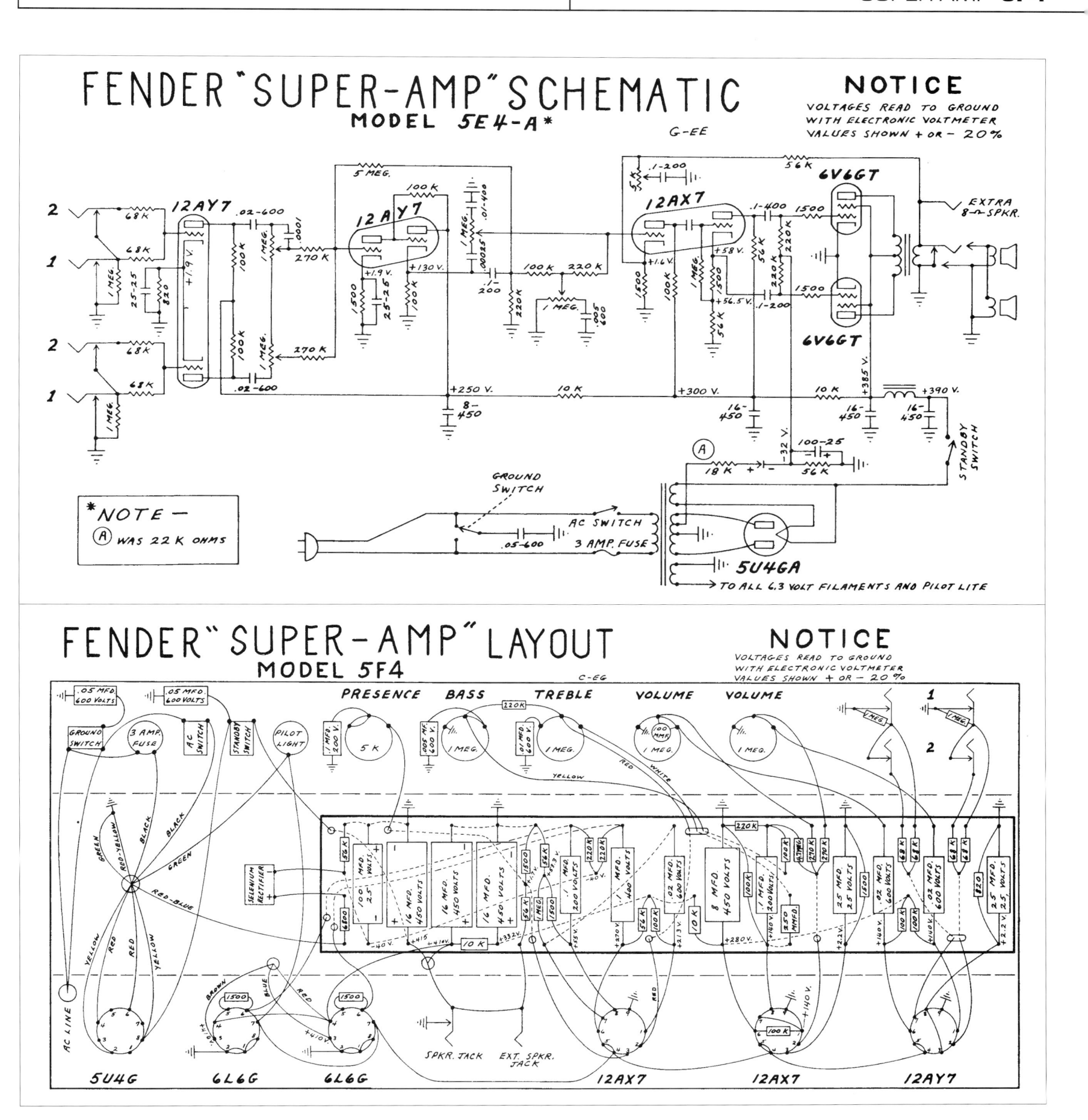

SUPER-AMP 5F4
SUPER REVERB AA1069

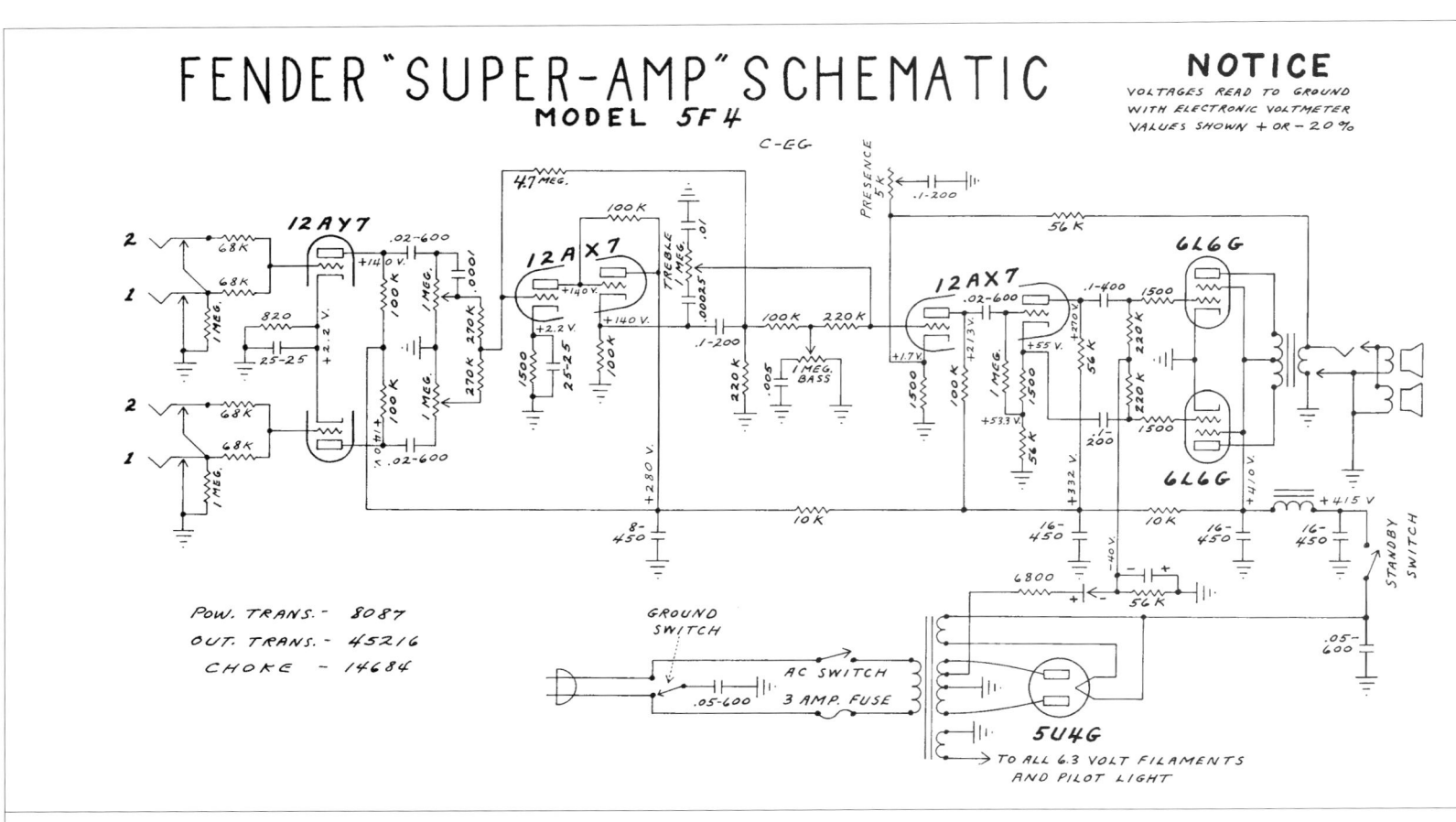

Fender

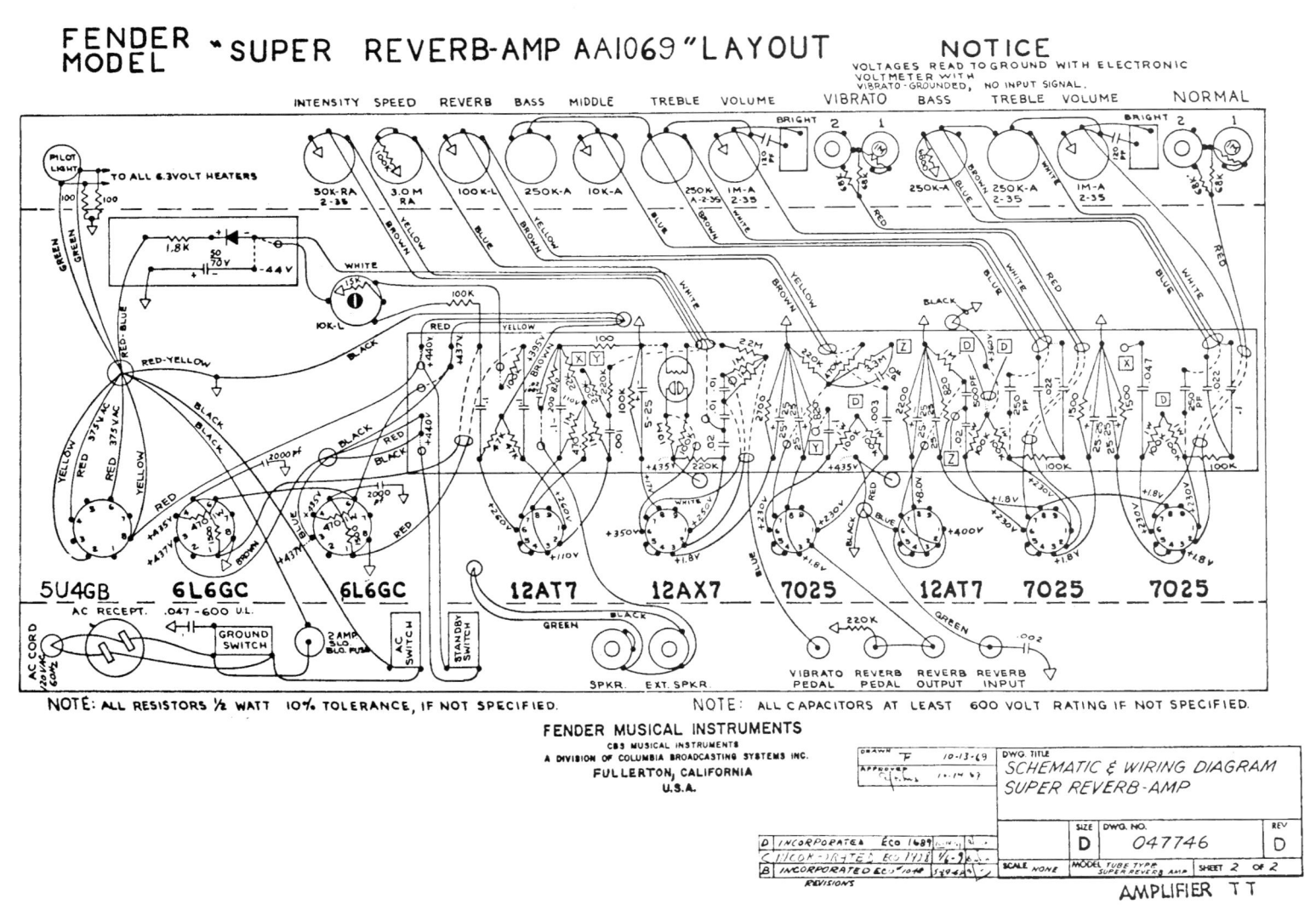

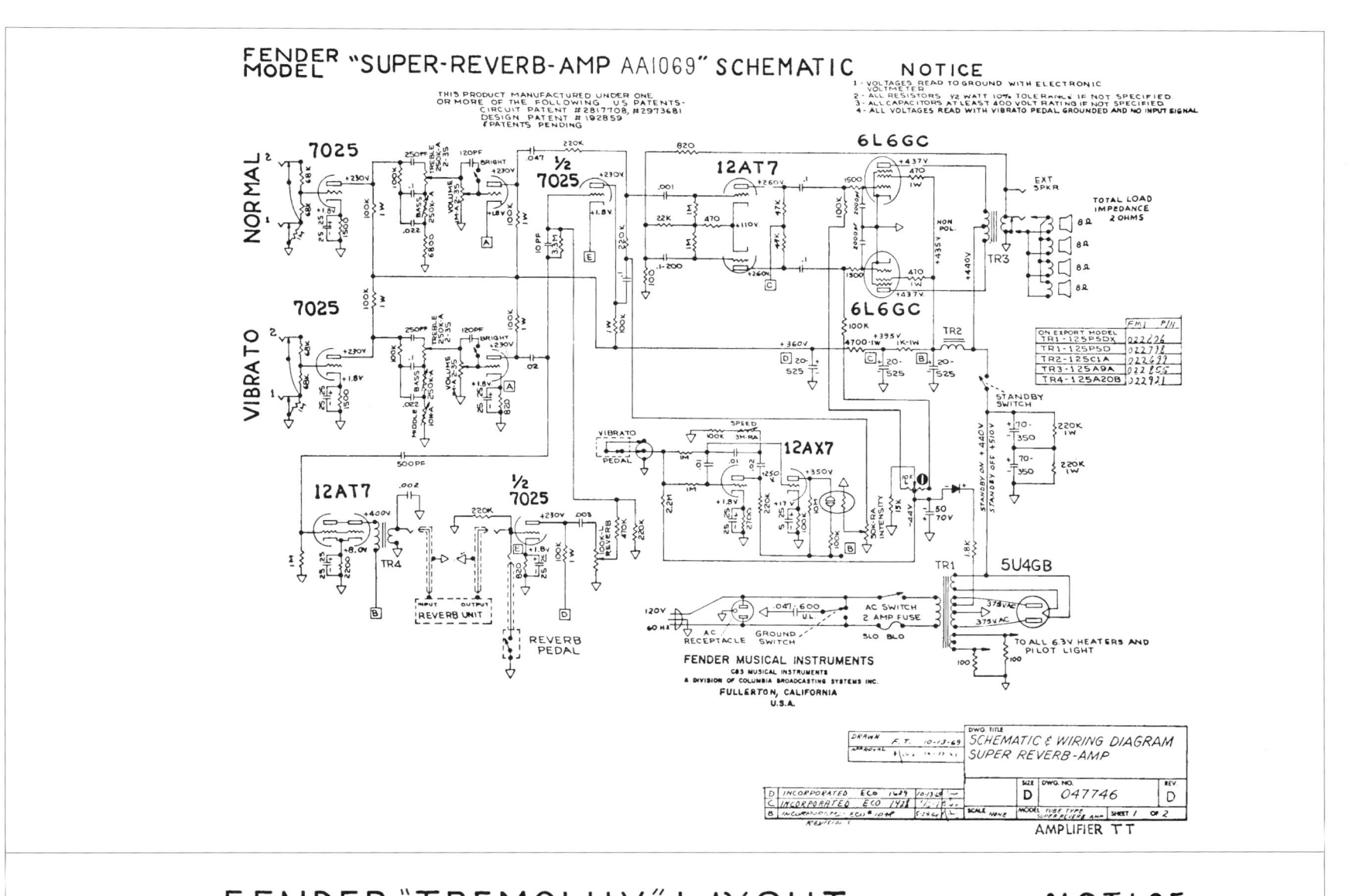

Fender

SUPER REVERB AA1069
TREMOLUX 5E9-A

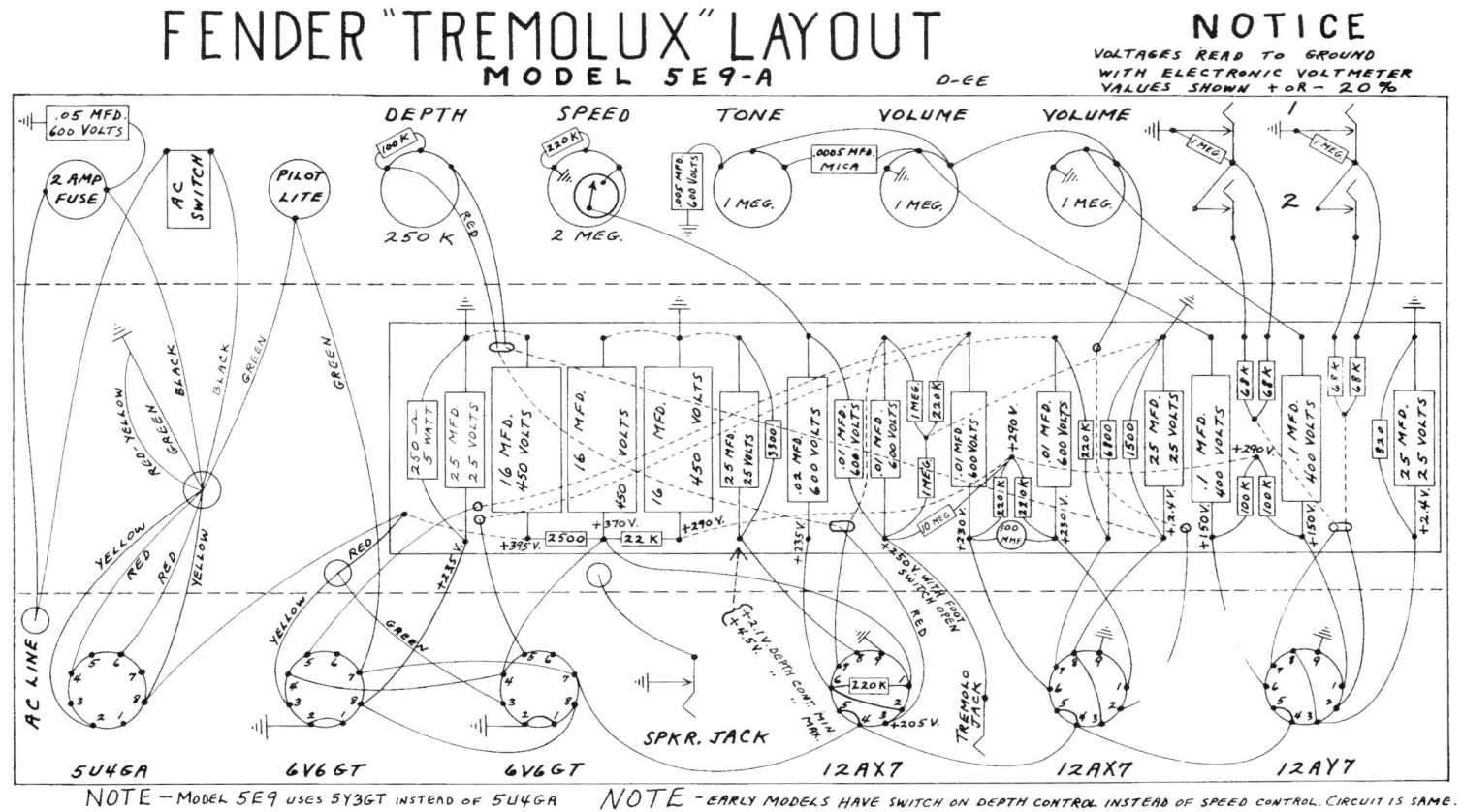

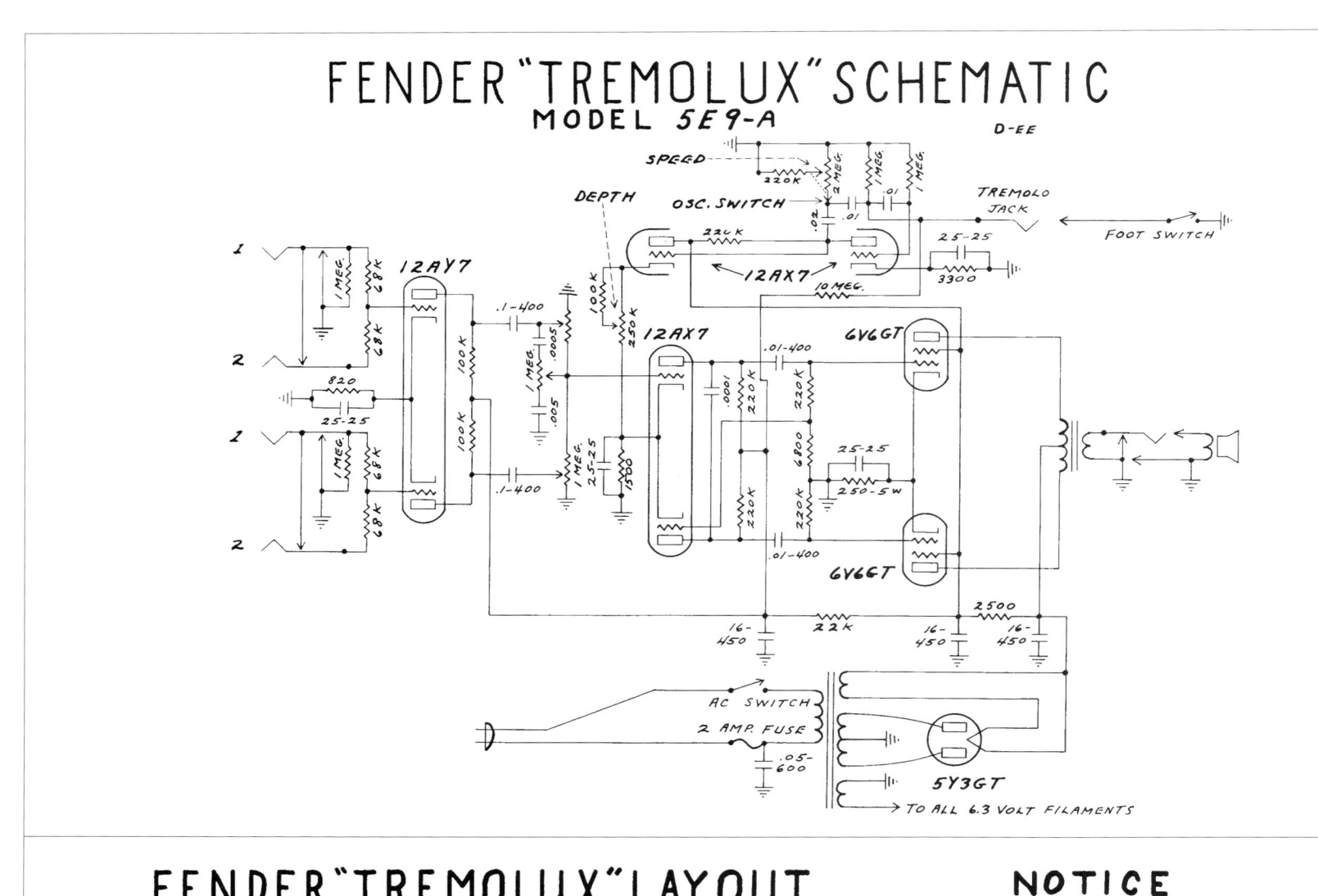

Fender

TREMOLUX 5E9-A
TREMOLUX 5G9

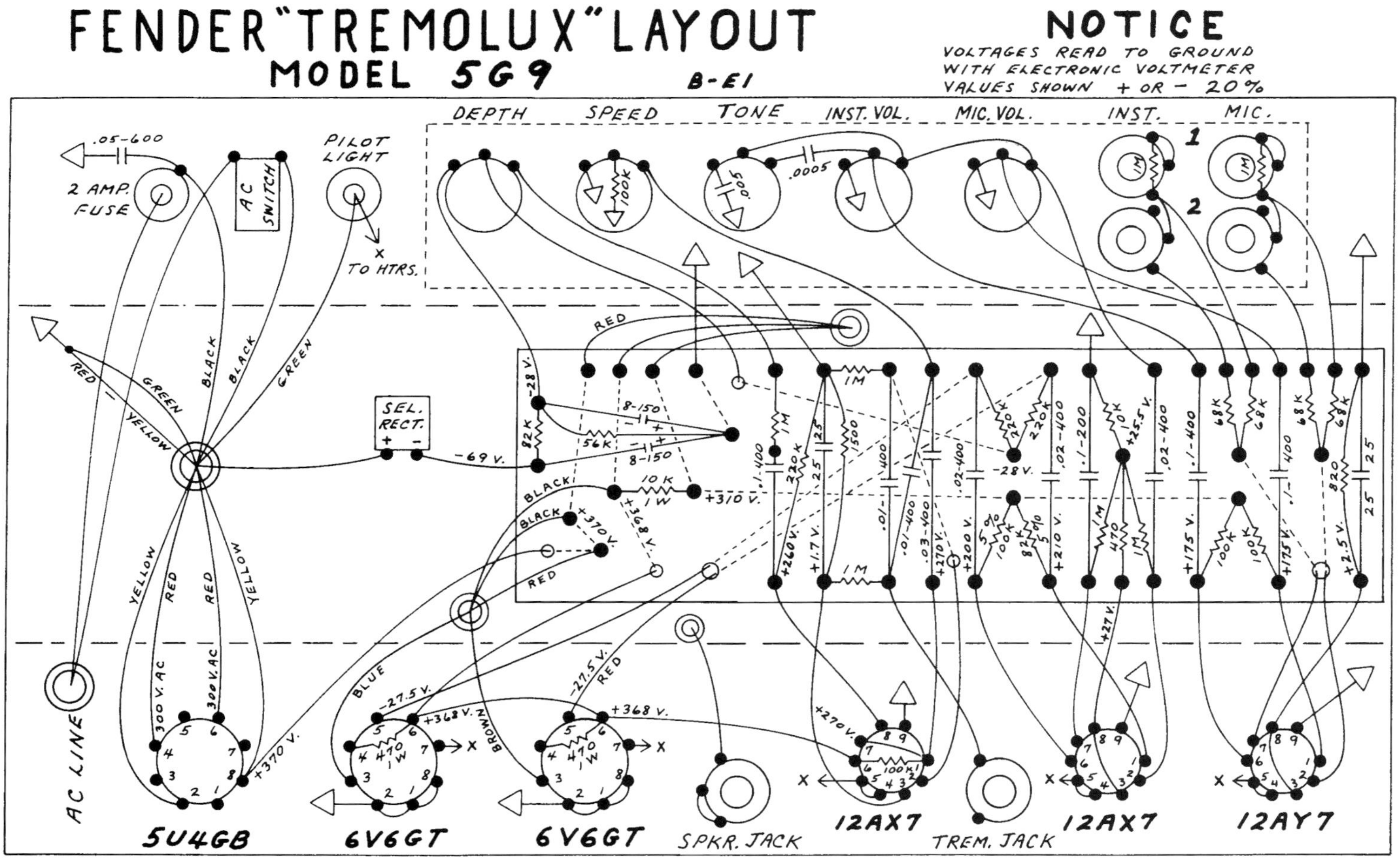

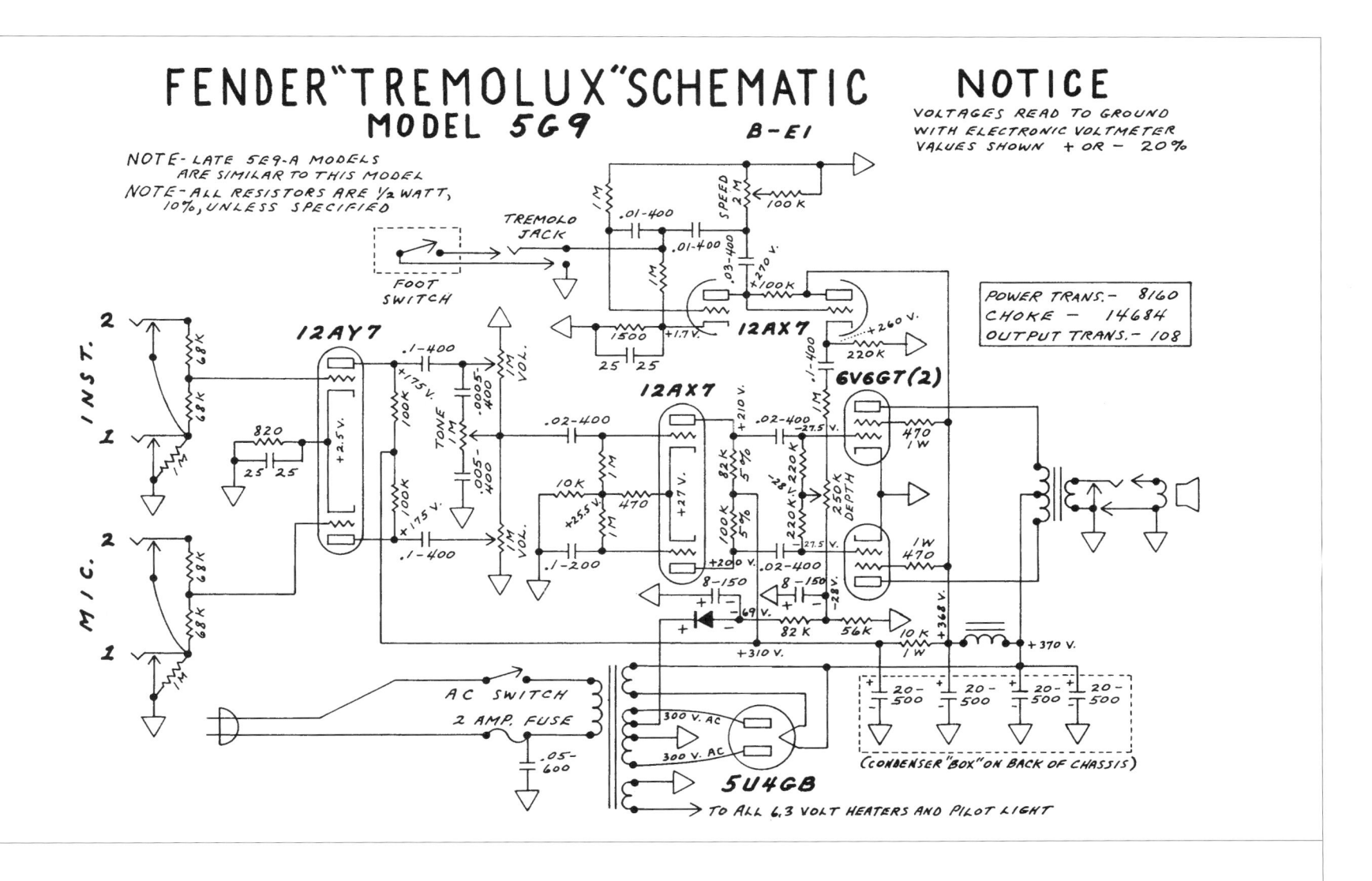

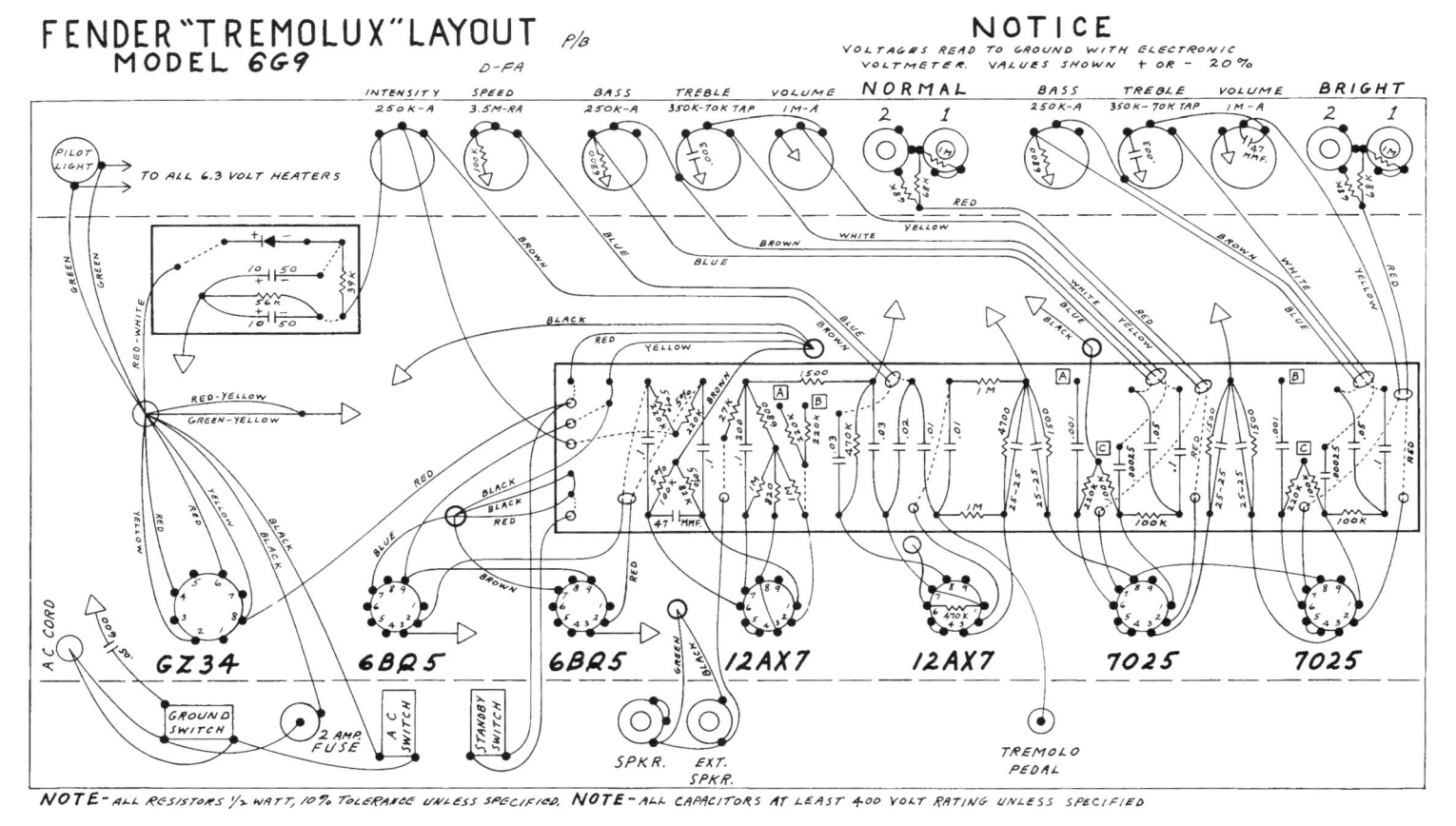

Fender

TREMOLUX 5G9
TREMOLUX 6G9

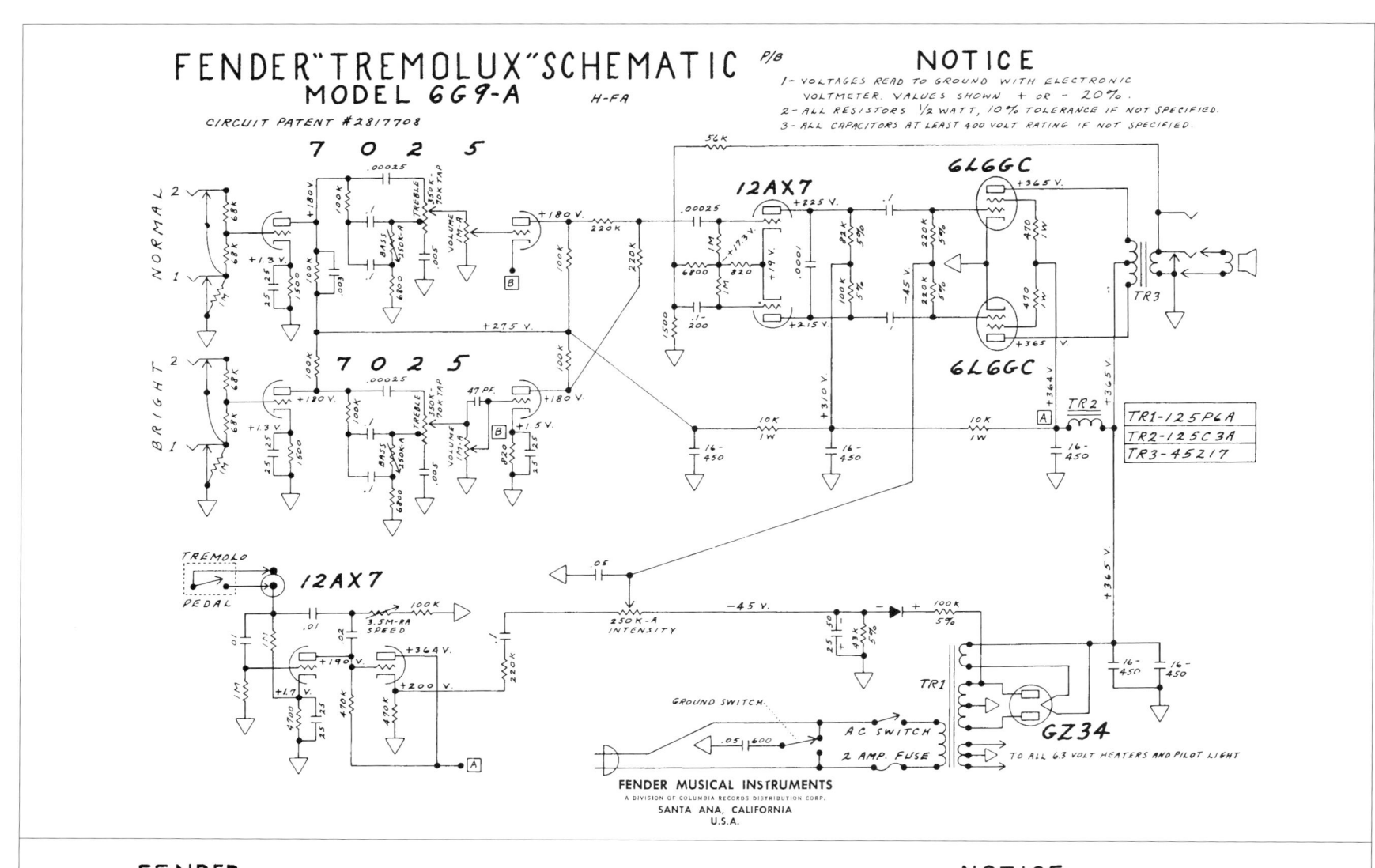

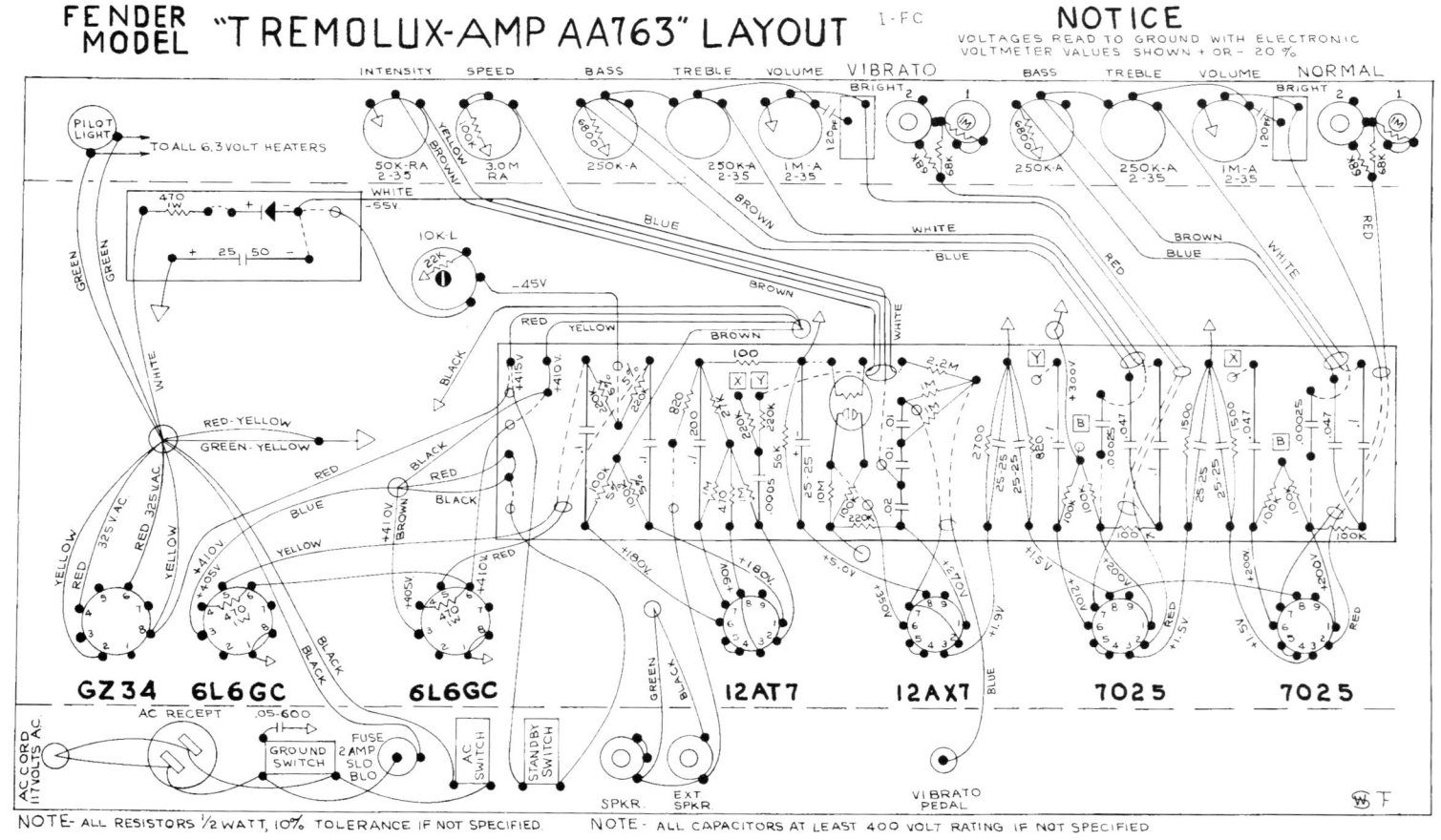

Fender

TREMOLUX 6G9-A
TREMOLUX AA763

Fender

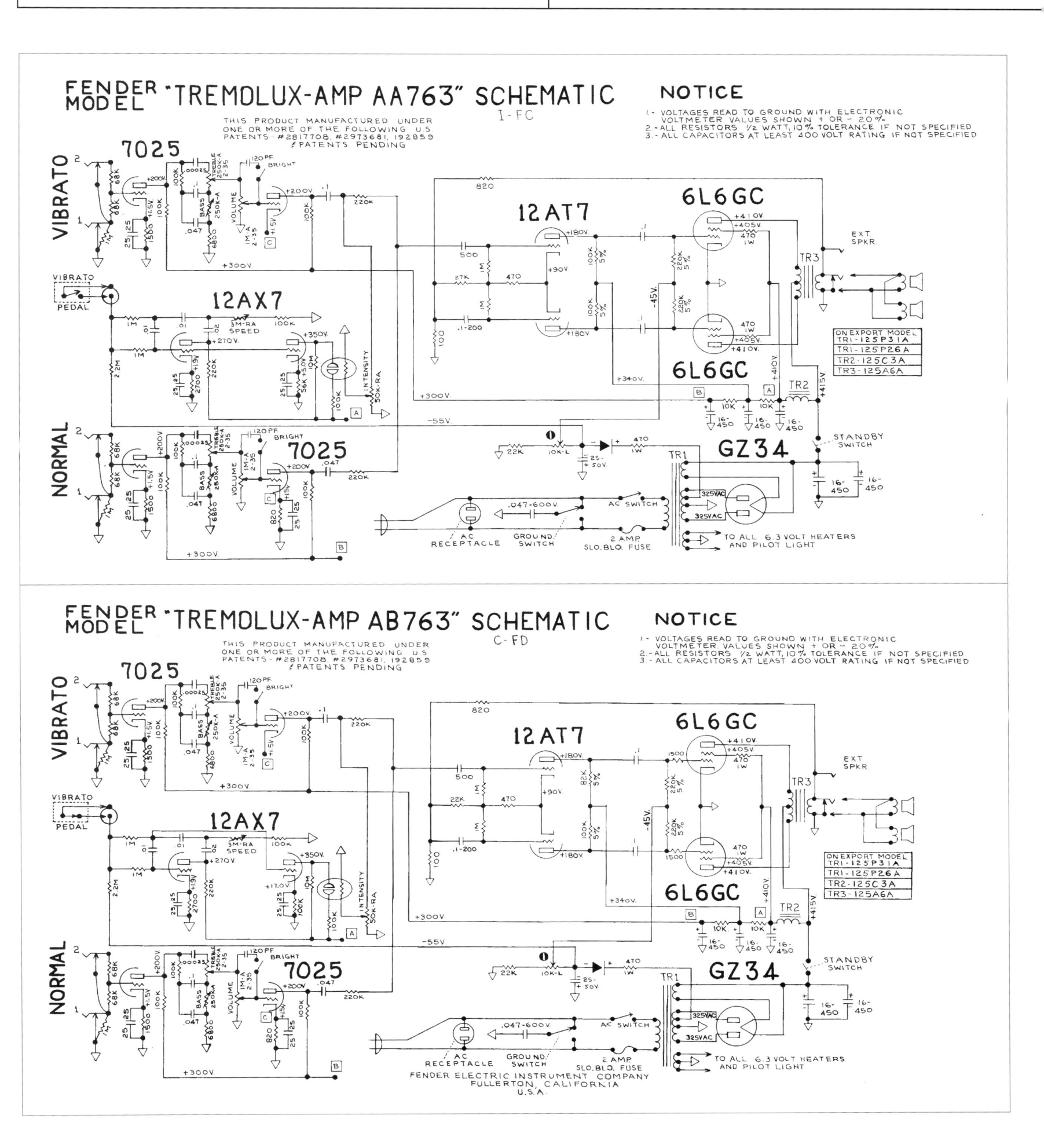

TREMOLUX **AA763**
TREMOLUX **AB763**

TWIN-AMP **5C8**
TWIN-AMP **5C8**

Fender

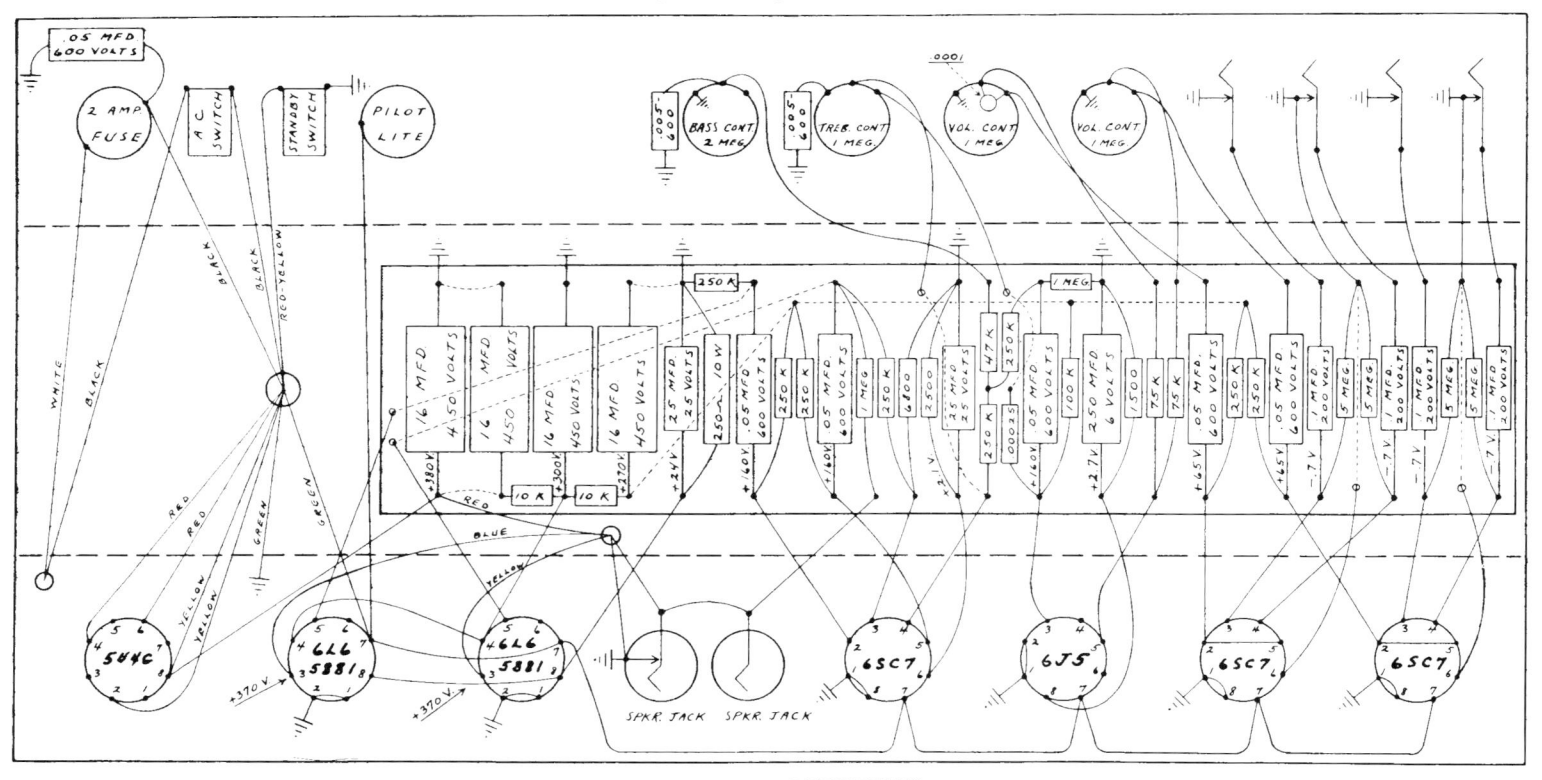

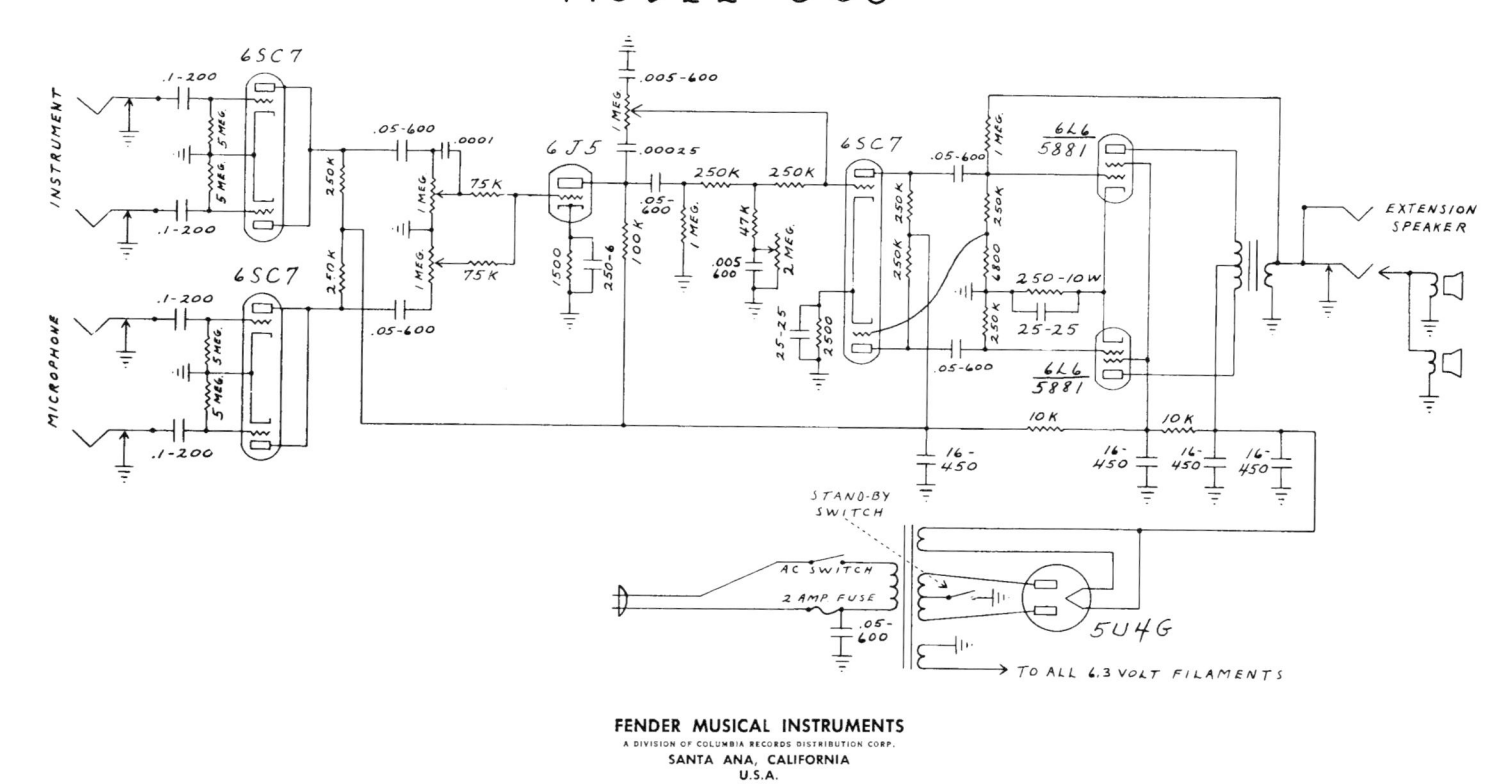

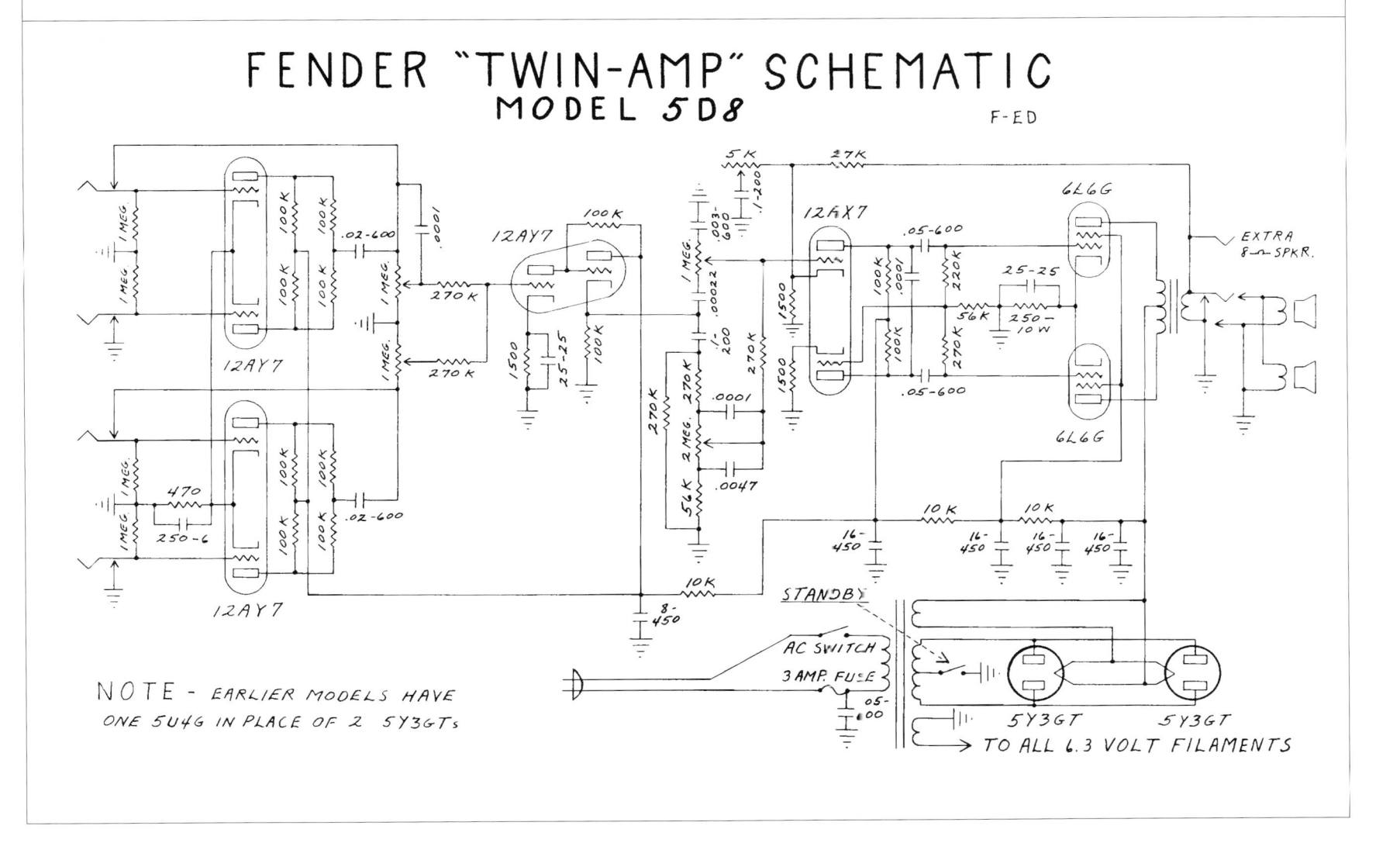

Fender

TWIN-AMP **5D8**
TWIN-AMP **5D8**

FENDER "TWIN-AMP" LAYOUT
MODEL 5D8 F-ED

NOTICE: ALL VOLTAGES READ TO GROUND WITH ELECTRONIC VOLTMETER. VALUES SHOWN + OR -20%

FENDER "TWIN-AMP" SCHEMATIC
MODEL 5D8 F-ED

NOTE — EARLIER MODELS HAVE ONE 5U4G IN PLACE OF 2 5Y3GTs

THE TUBE AMP BOOK 63

TWIN-AMP **5E8-A**
TWIN-AMP **5F8**

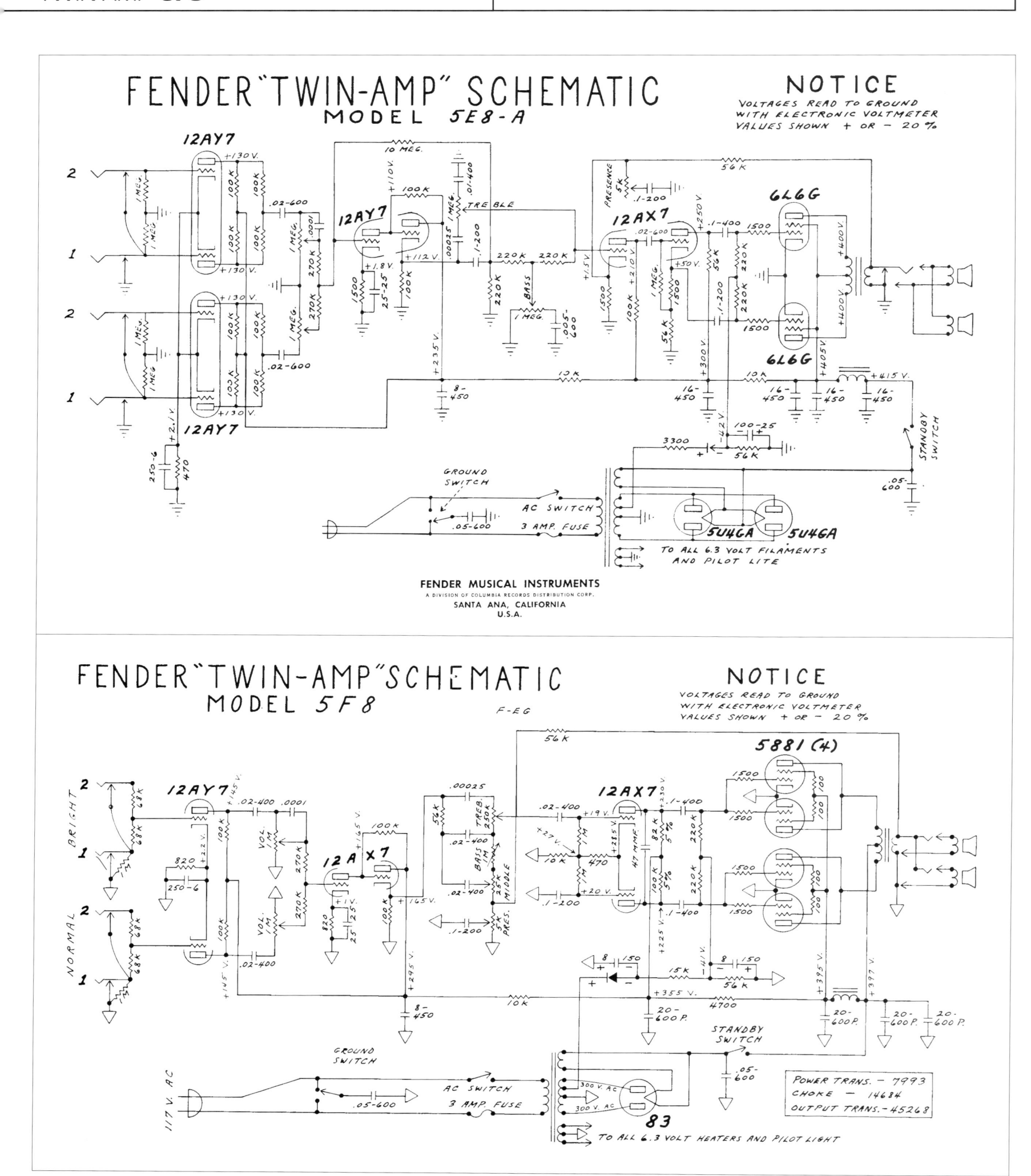

Fender

Fender

TWIN-AMP **5F8-A**
TWIN-AMP **5F8-A**

FENDER "TWIN-AMP" LAYOUT MODEL 5F8-A

NOTICE — VOLTAGES READ TO GROUND WITH ELECTRONIC VOLTMETER VALUES SHOWN + OR − 20%

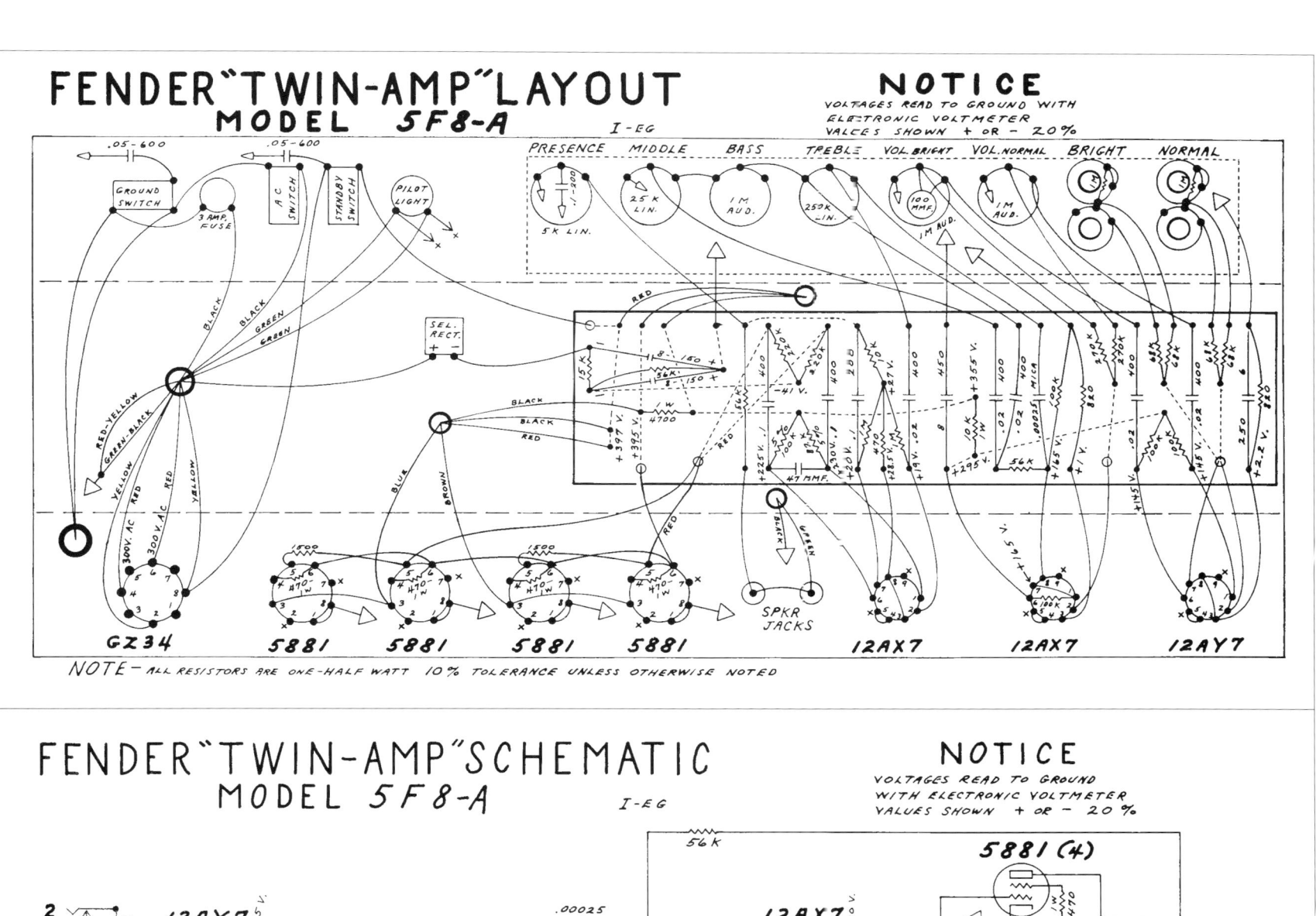

NOTE — ALL RESISTORS ARE ONE-HALF WATT 10% TOLERANCE UNLESS OTHERWISE NOTED

FENDER "TWIN-AMP" SCHEMATIC MODEL 5F8-A

NOTICE — VOLTAGES READ TO GROUND WITH ELECTRONIC VOLTMETER VALUES SHOWN + OR − 20%

POWER TRANS. — 7993
CHOKE — 14634
OUTPUT TRANS. — 45268

THE TUBE AMP BOOK 65

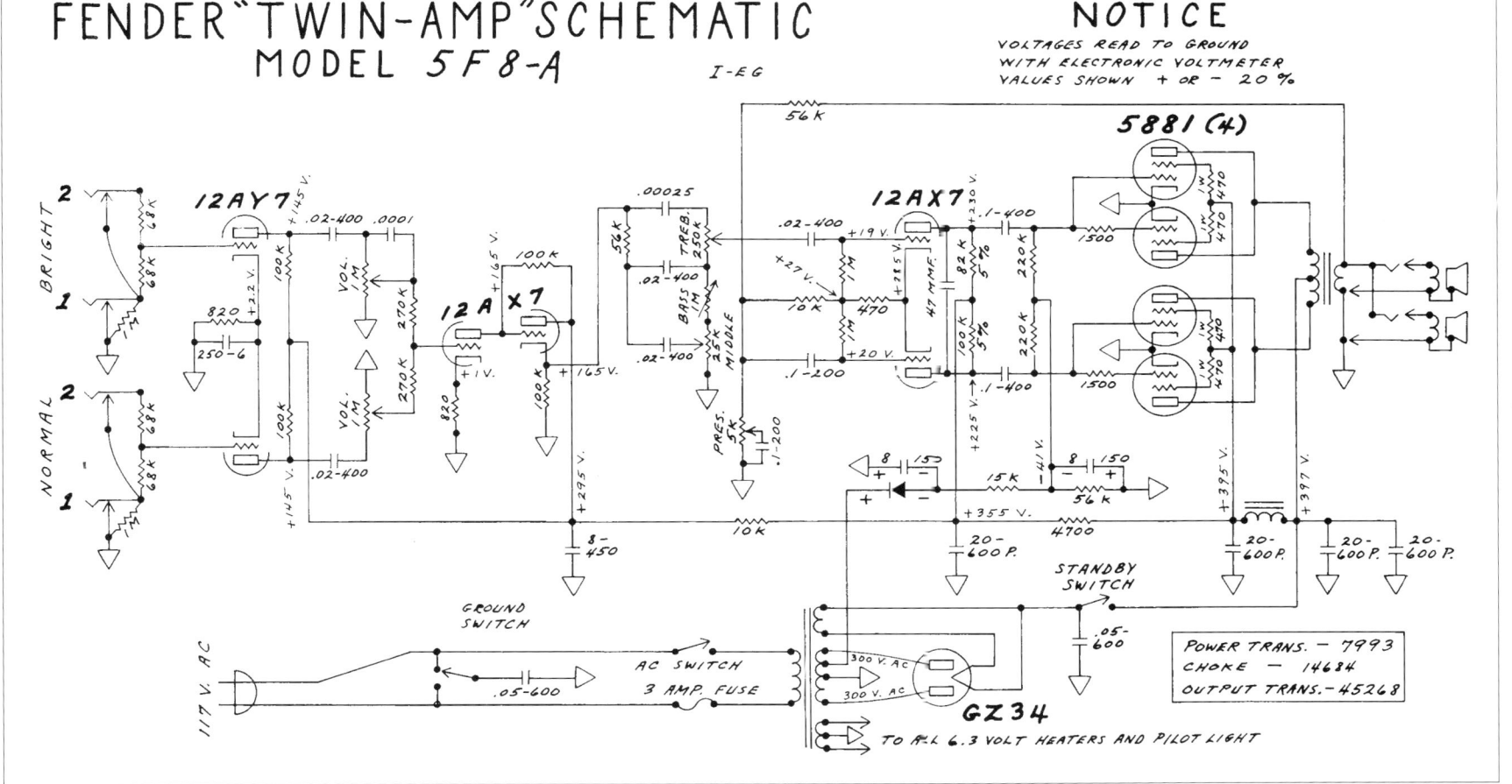

Fender

TWIN-AMP **6G8**
TWIN-AMP **6G8**

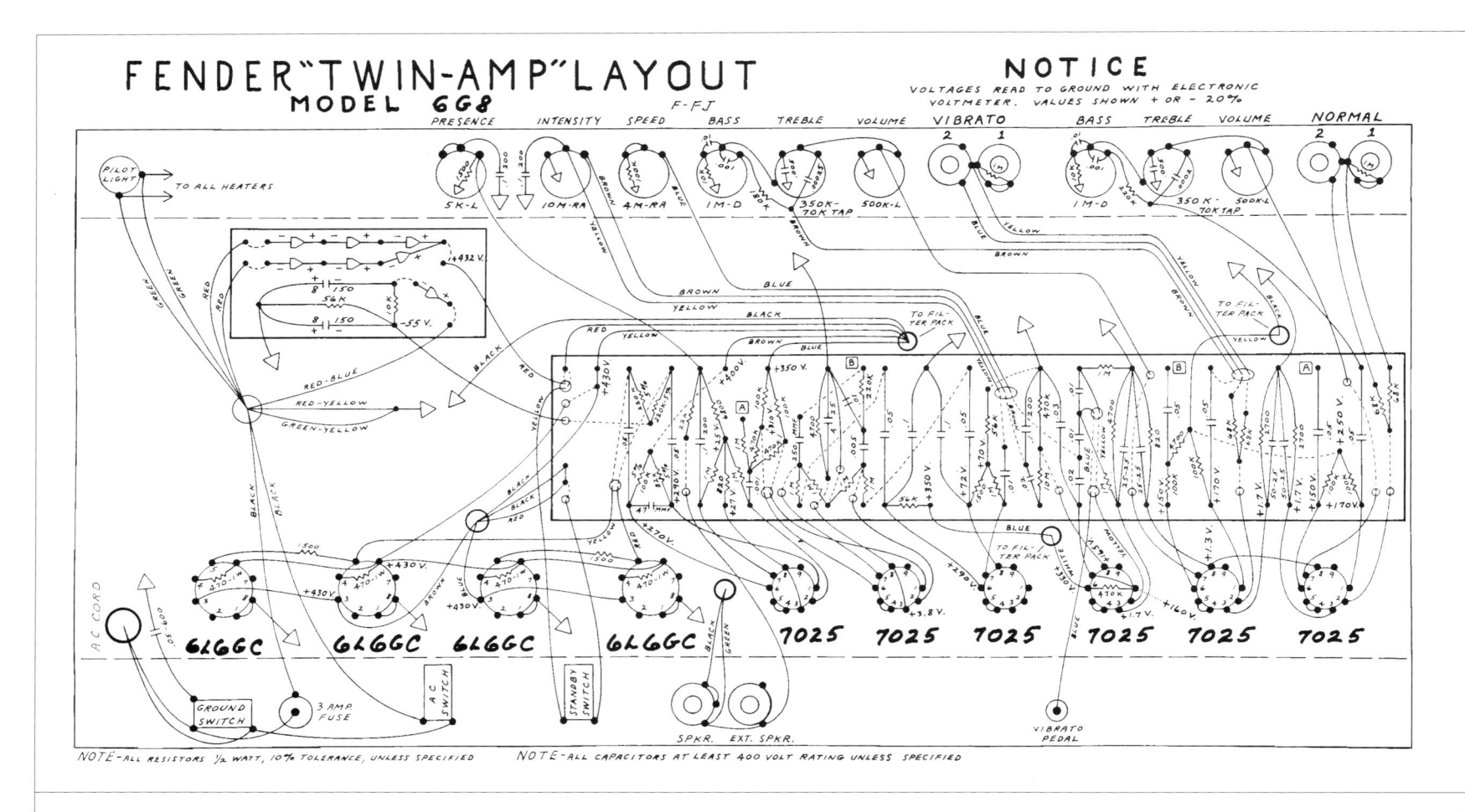

FENDER "TWIN-AMP" LAYOUT — MODEL 6G8

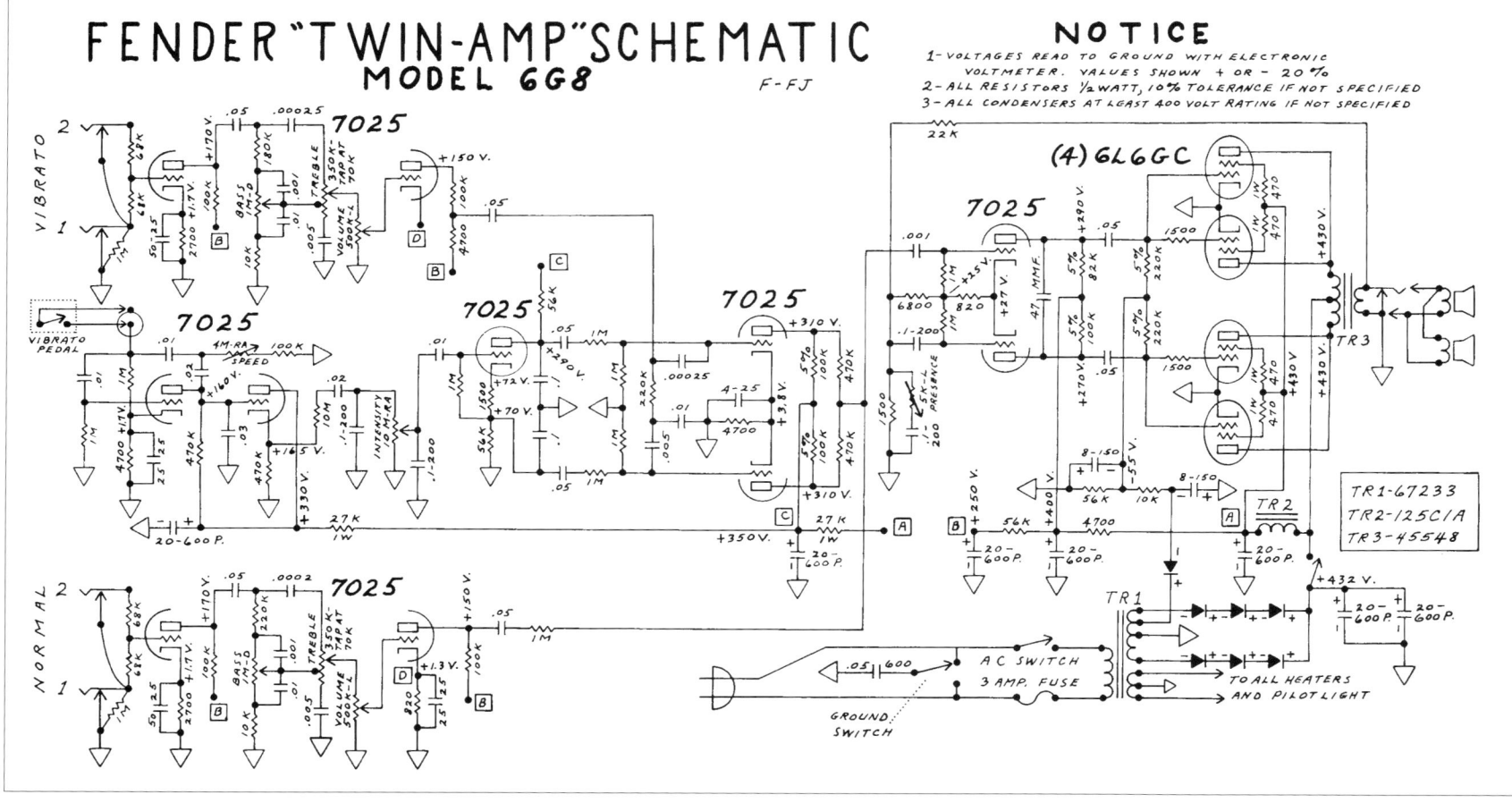

FENDER "TWIN-AMP" SCHEMATIC — MODEL 6G8

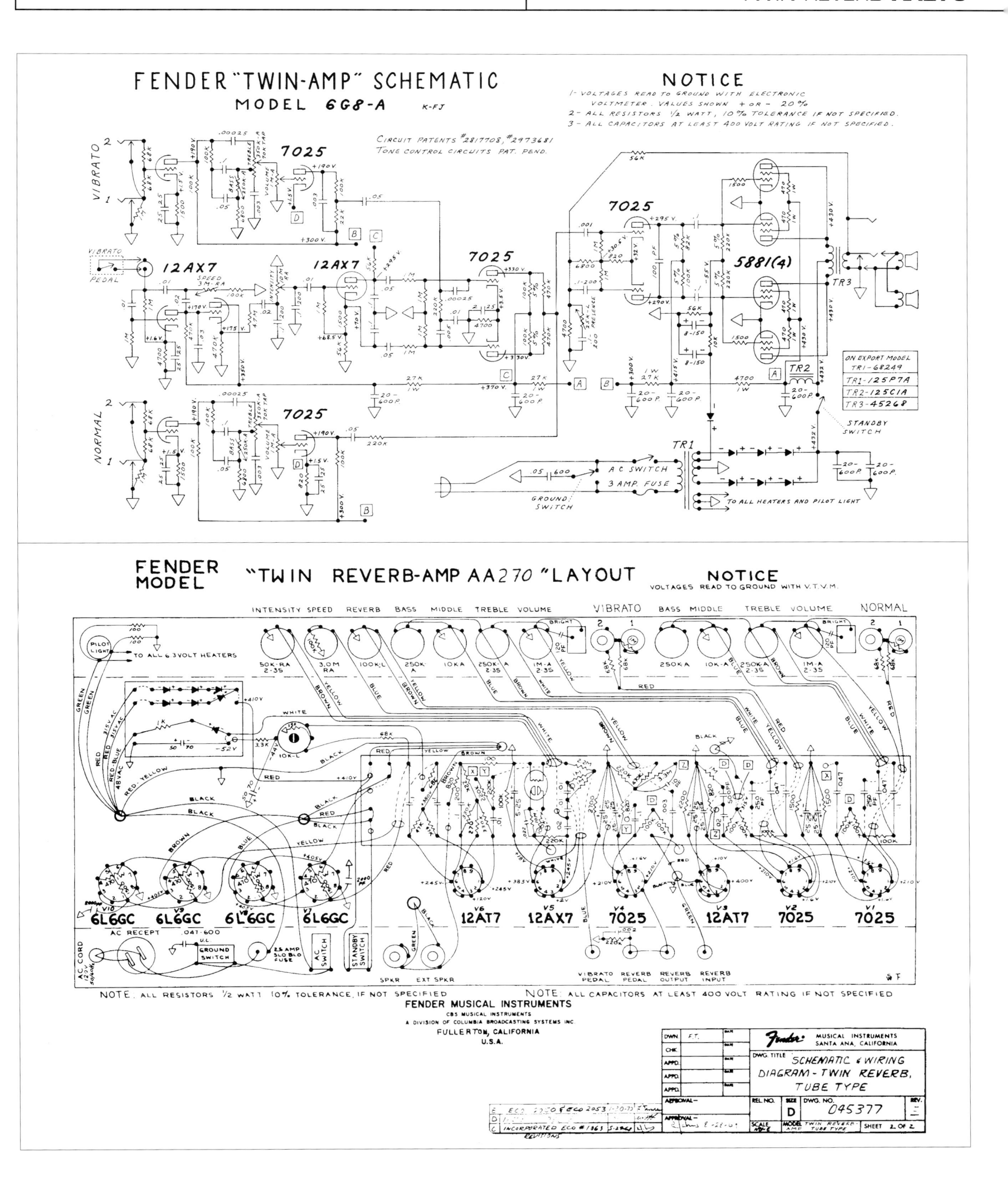

Fender

TWIN-AMP 6G8-A
TWIN REVERB AA270

TWIN REVERB AA270
TWIN REVERB AA769

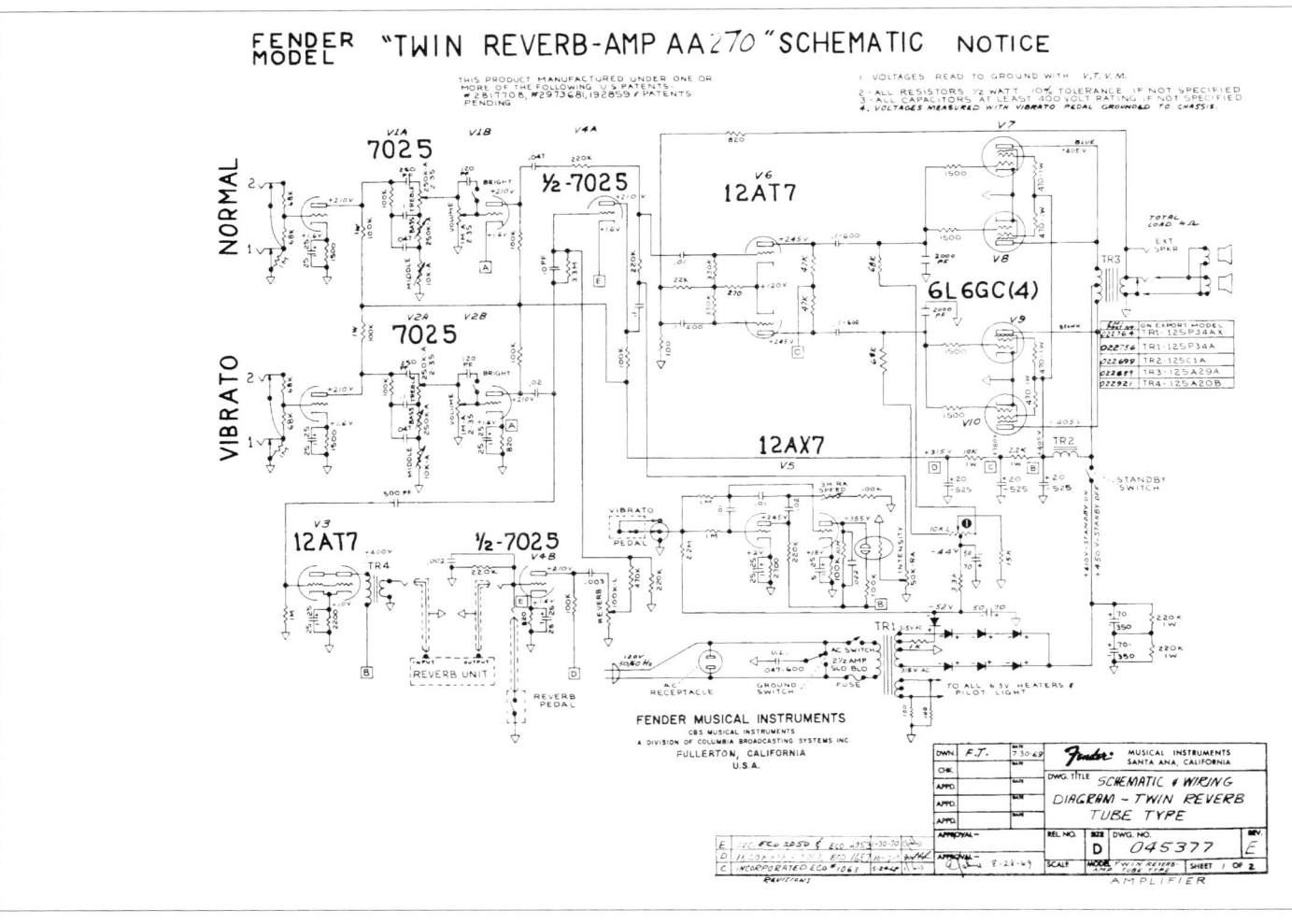

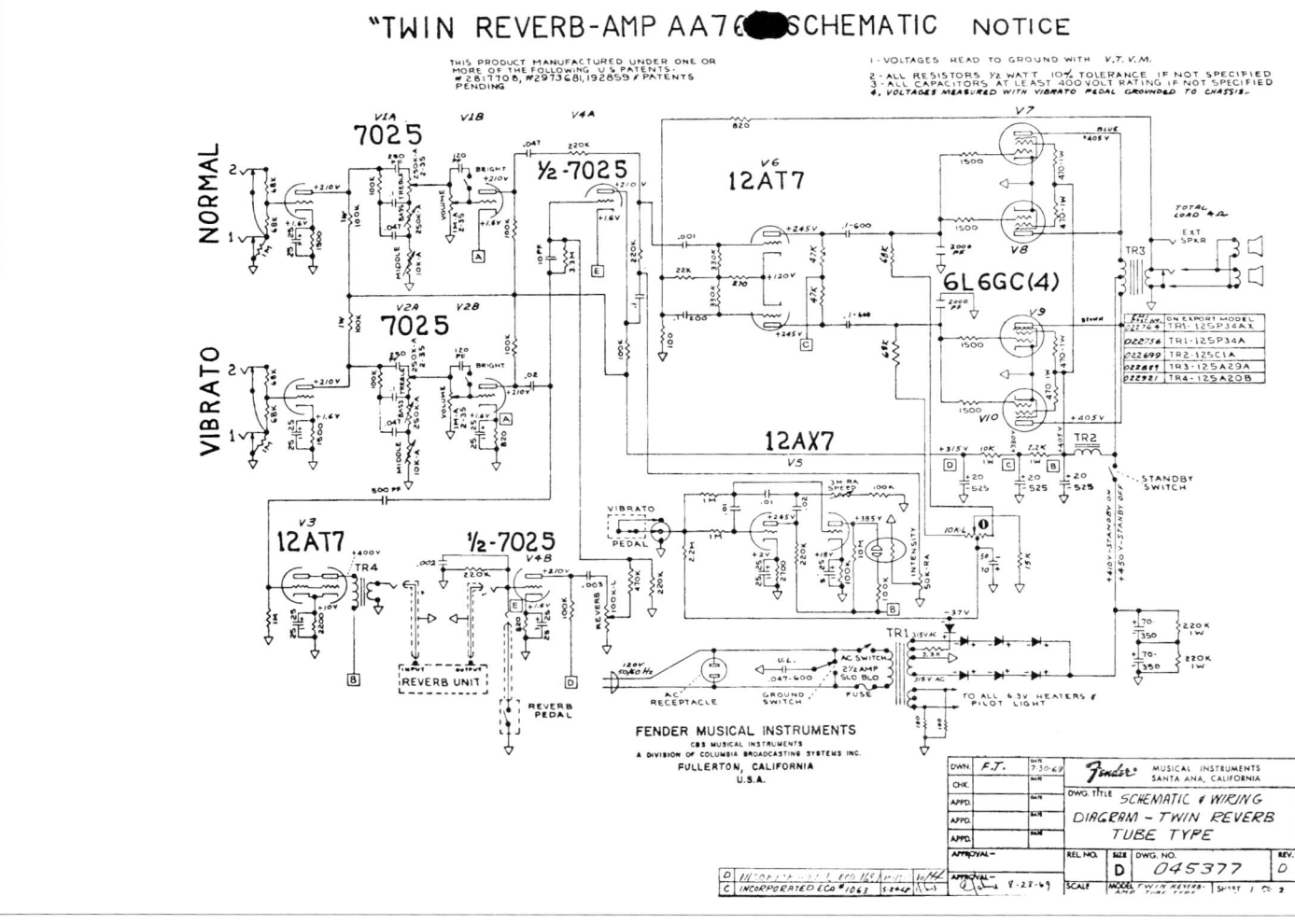

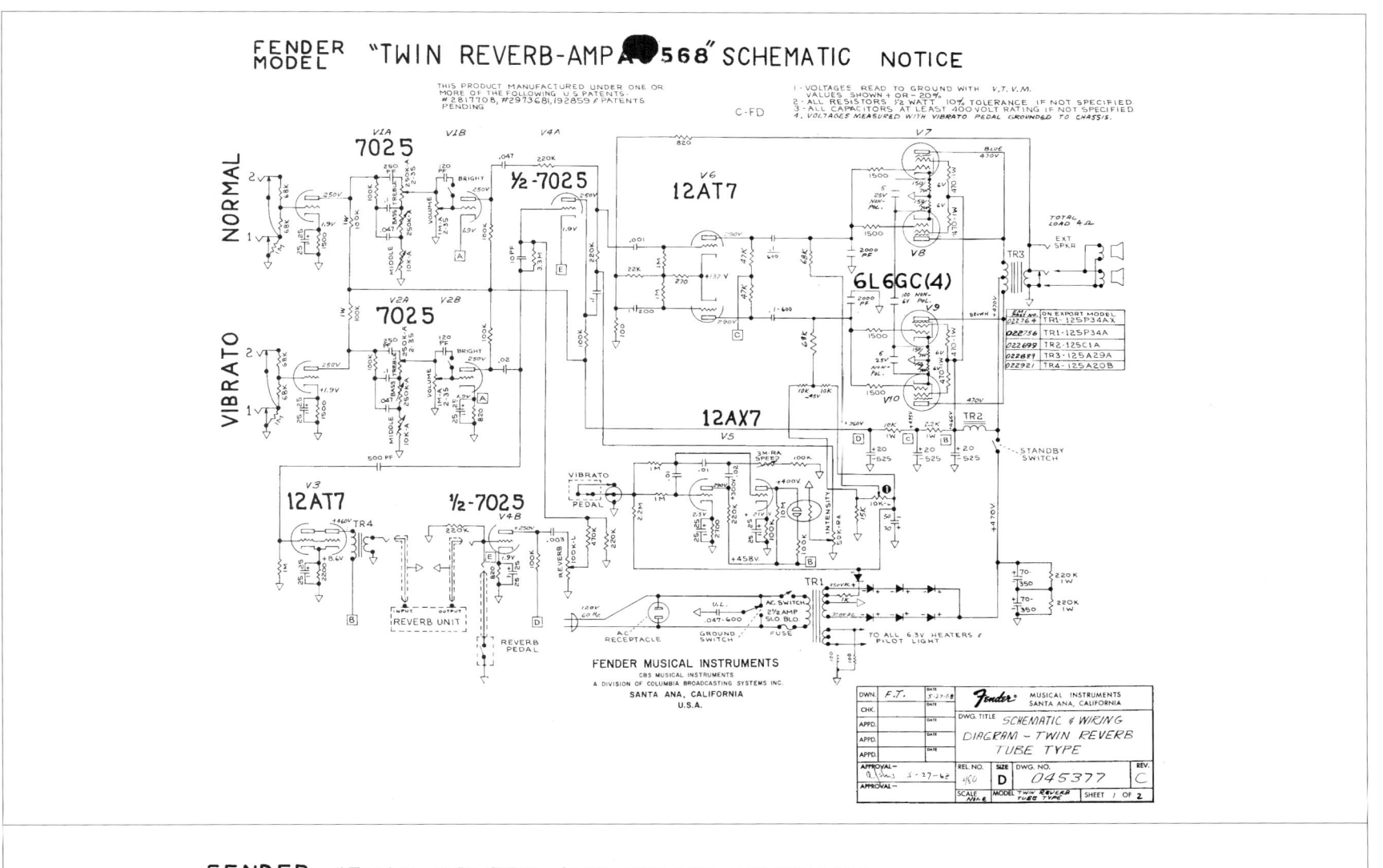

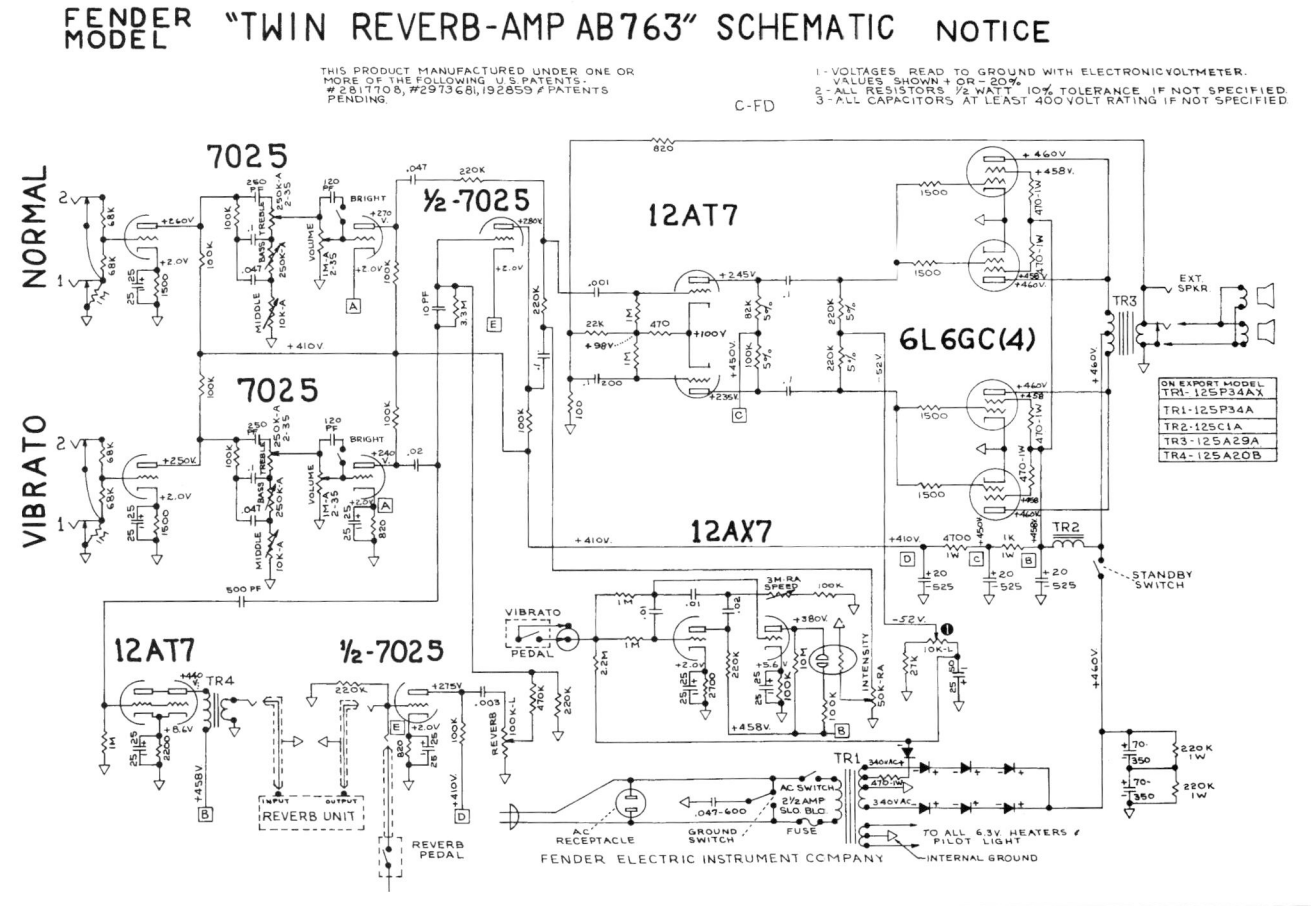

Fender

TWIN REVERB AB568
TWIN REVERB AB763

VIBRASONIC 5G13
VIBRASONIC 5G13

Fender

FENDER "VIBRASONIC" LAYOUT
MODEL 5G13 A-FJ

NOTICE
VOLTAGES READ TO GROUND WITH ELECTRONIC VOLTMETER. VALUES SHOWN + OR - 20%

NOTE - ALL RESISTORS ½ WATT, 10% TOLERANCE UNLESS SPECIFIED   NOTE - ALL CAPACITORS AT LEAST 400 VOLT RATING UNLESS SPECIFIED

FENDER "VIBRASONIC" SCHEMATIC
MODEL 5G13 A-FJ

NOTICE
1 - VOLTAGES READ TO GROUND WITH ELECTRONIC VOLTMETER. VALUES SHOWN + OR -
2 - ALL RESISTORS ½ WATT IF NOT SPECIFIED
3 - ALL CONDENSERS AT LEAST 400 VOLTS IF NOT SPECIFIED

70 THE TUBE AMP BOOK

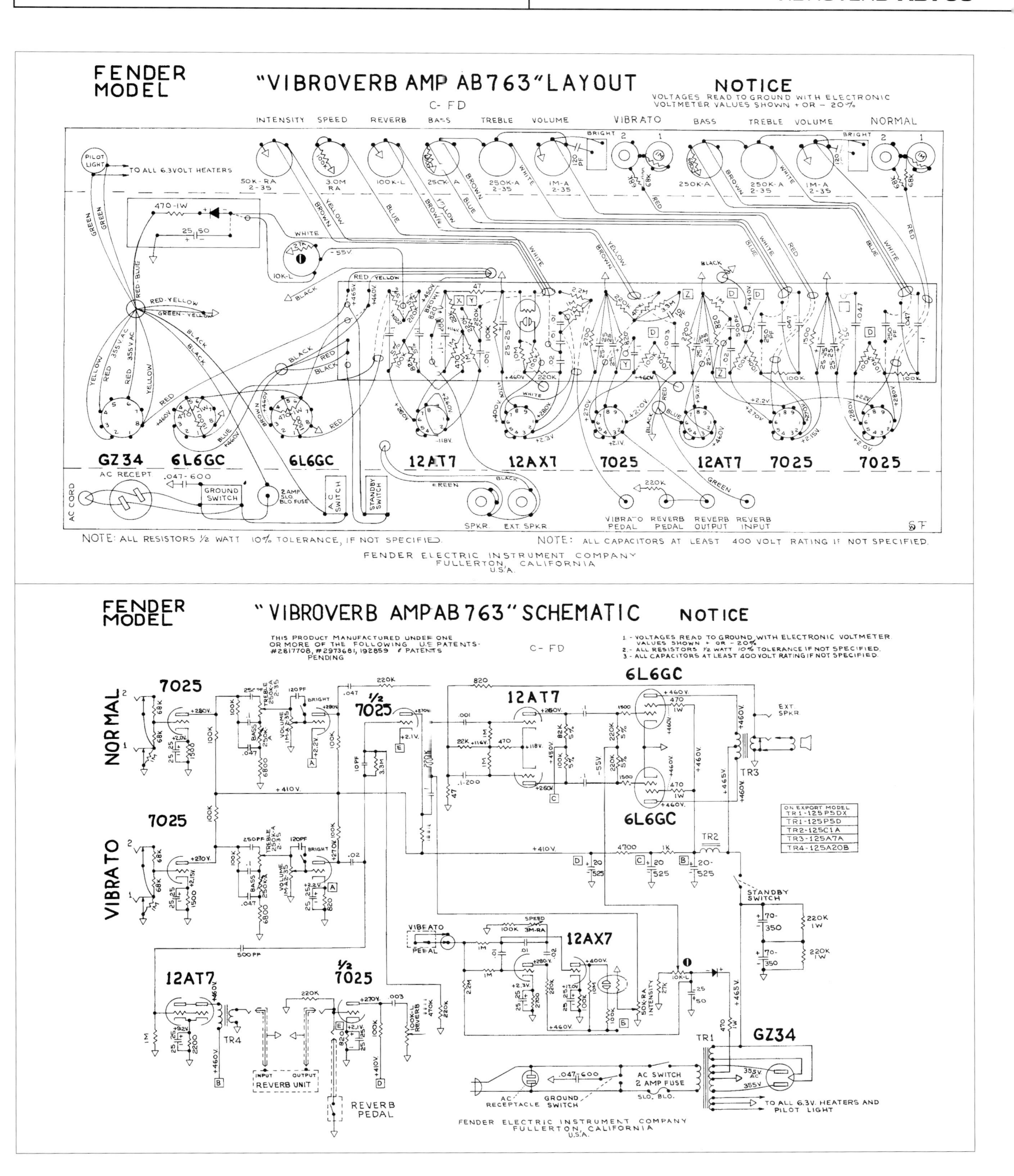

Fender

VIBROVERB AB763

Fender

VIBROLUX 5E11
VIBROLUX 5F11

FENDER "VIBROLUX" LAYOUT — MODEL 5E11

NOTICE: VOLTAGES READ TO GROUND WITH ELECTRONIC VOLTMETER. VALUES SHOWN + OR − 20%

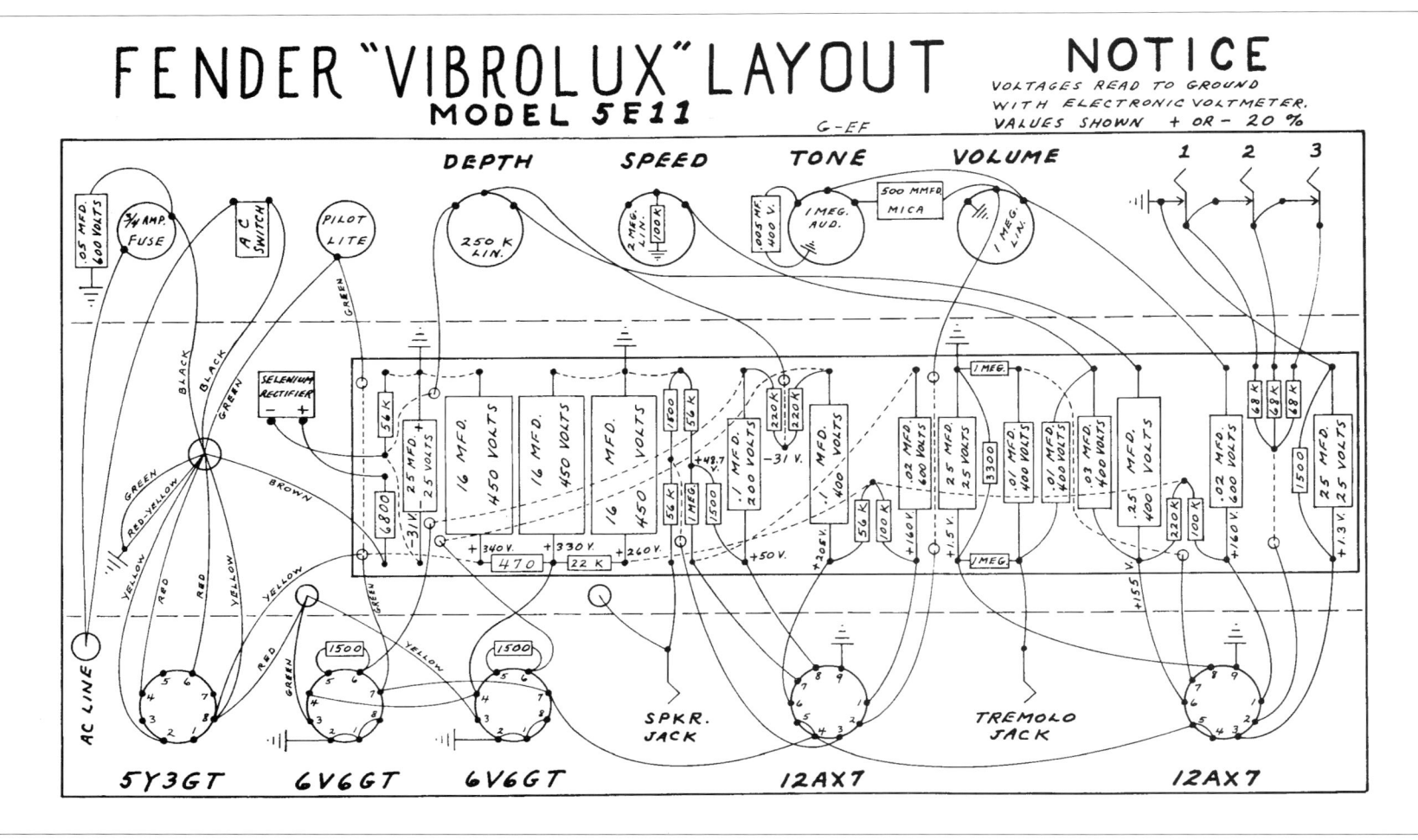

Tubes: 5Y3GT, 6V6GT, 6V6GT, 12AX7, 12AX7

FENDER "VIBROLUX" SCHEMATIC — MODEL 5F11

NOTICE: VOLTAGES READ TO GROUND WITH V.T.V.M. VALUES SHOWN + OR − 20%

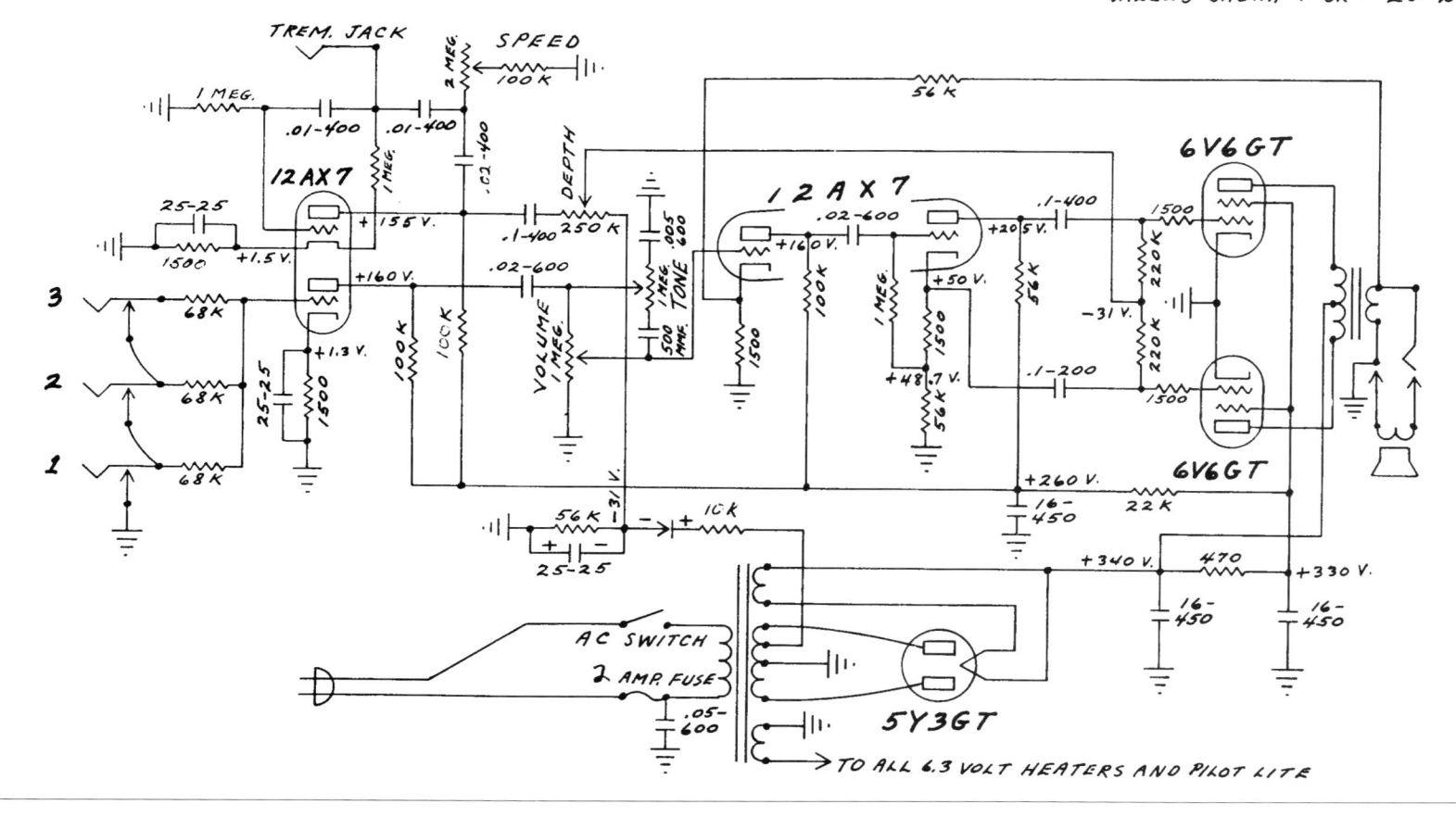

72 THE TUBE AMP BOOK

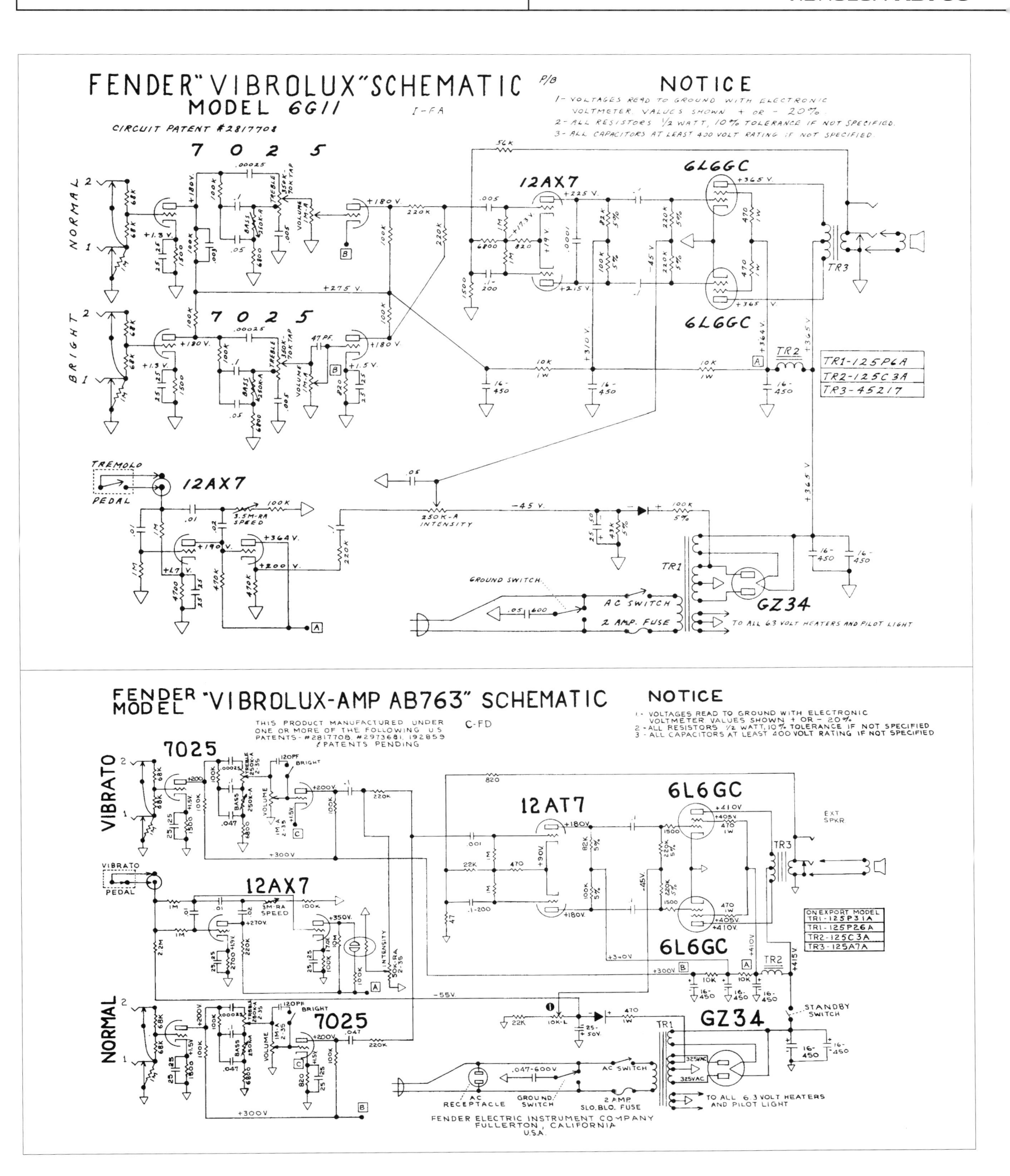

Fender

VIBROLUX **6G11**
VIBROLUX **AB763**

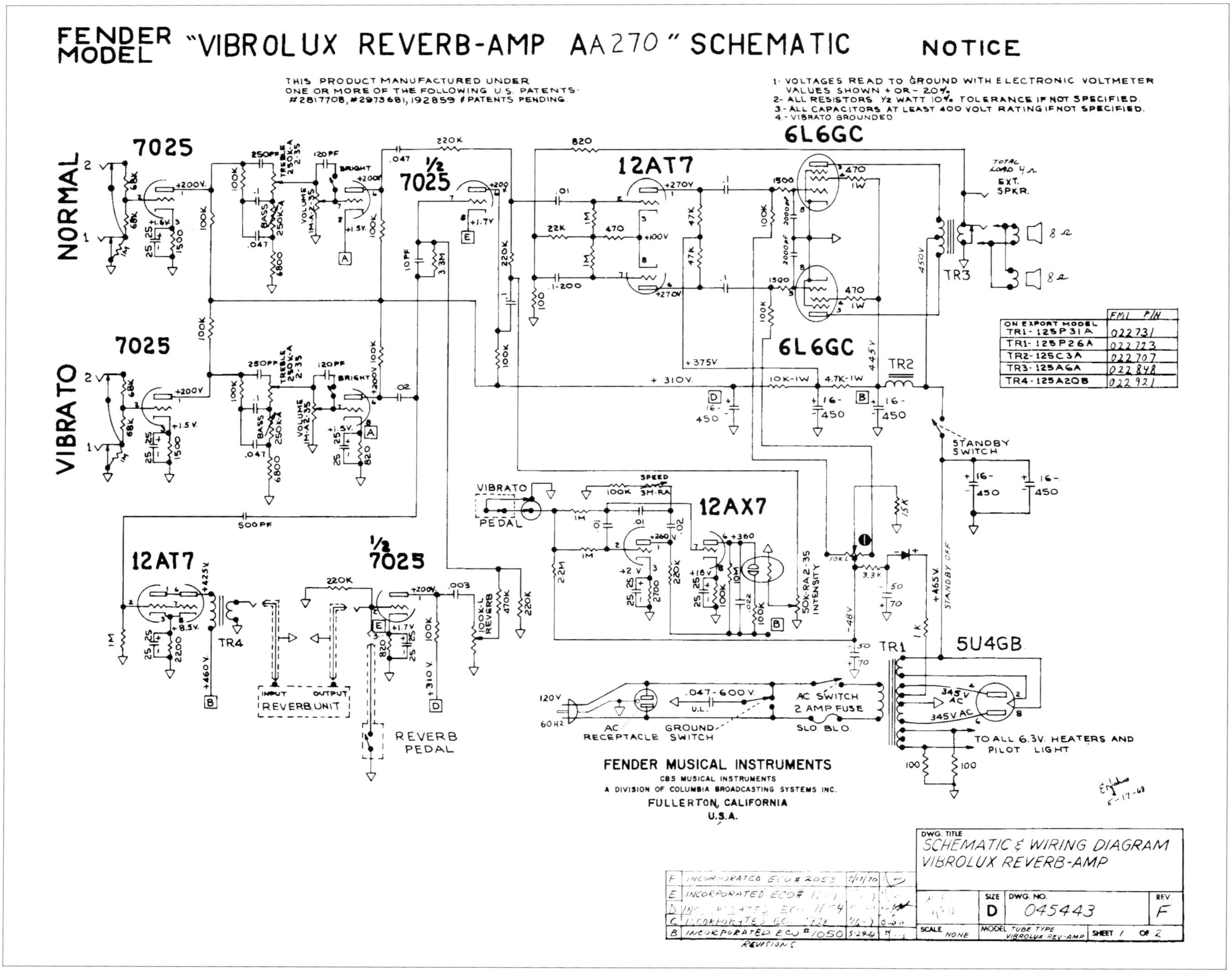
FENDER MODEL "VIBROLUX REVERB-AMP AA270" SCHEMATIC
THIS PRODUCT MANUFACTURED UNDER ONE OR MORE OF THE FOLLOWING U.S. PATENTS- #2817708, #2973681, 192859 & PATENTS PENDING.
NOTICE
1- VOLTAGES READ TO GROUND WITH ELECTRONIC VOLTMETER VALUES SHOWN + OR - 20%.
2- ALL RESISTORS ½ WATT 10% TOLERANCE IF NOT SPECIFIED.
3- ALL CAPACITORS AT LEAST 400 VOLT RATING IF NOT SPECIFIED.
4- VIBRATO GROUNDED
NORMAL
VIBRATO
7025
½ 7025
12AT7
6L6GC
12AX7
5U4GB
TR1
TR2
TR3
TR4
VIBRATO PEDAL
REVERB UNIT
REVERB PEDAL
STANDBY SWITCH
AC RECEPTACLE
GROUND-SWITCH
AC SWITCH 2 AMP FUSE SLO BLO
TO ALL 6.3V. HEATERS AND PILOT LIGHT
EXT. SPKR.
FENDER MUSICAL INSTRUMENTS
CBS MUSICAL INSTRUMENTS
A DIVISION OF COLUMBIA BROADCASTING SYSTEMS INC.
FULLERTON, CALIFORNIA
U.S.A.
ON EXPORT MODEL | FMI P/N
TR1-125P31A | 022731
TR1-125P26A | 022723
TR2-125C3A | 022707
TR3-125A6A | 022848
TR4-125A20B | 022921
DWG. TITLE SCHEMATIC & WIRING DIAGRAM VIBROLUX REVERB-AMP
SIZE D
DWG. NO. 045443
REV F
SCALE NONE
MODEL TUBE TYPE VIBROLUX REV-AMP
SHEET 1 OF 2
REVISIONS

Fender

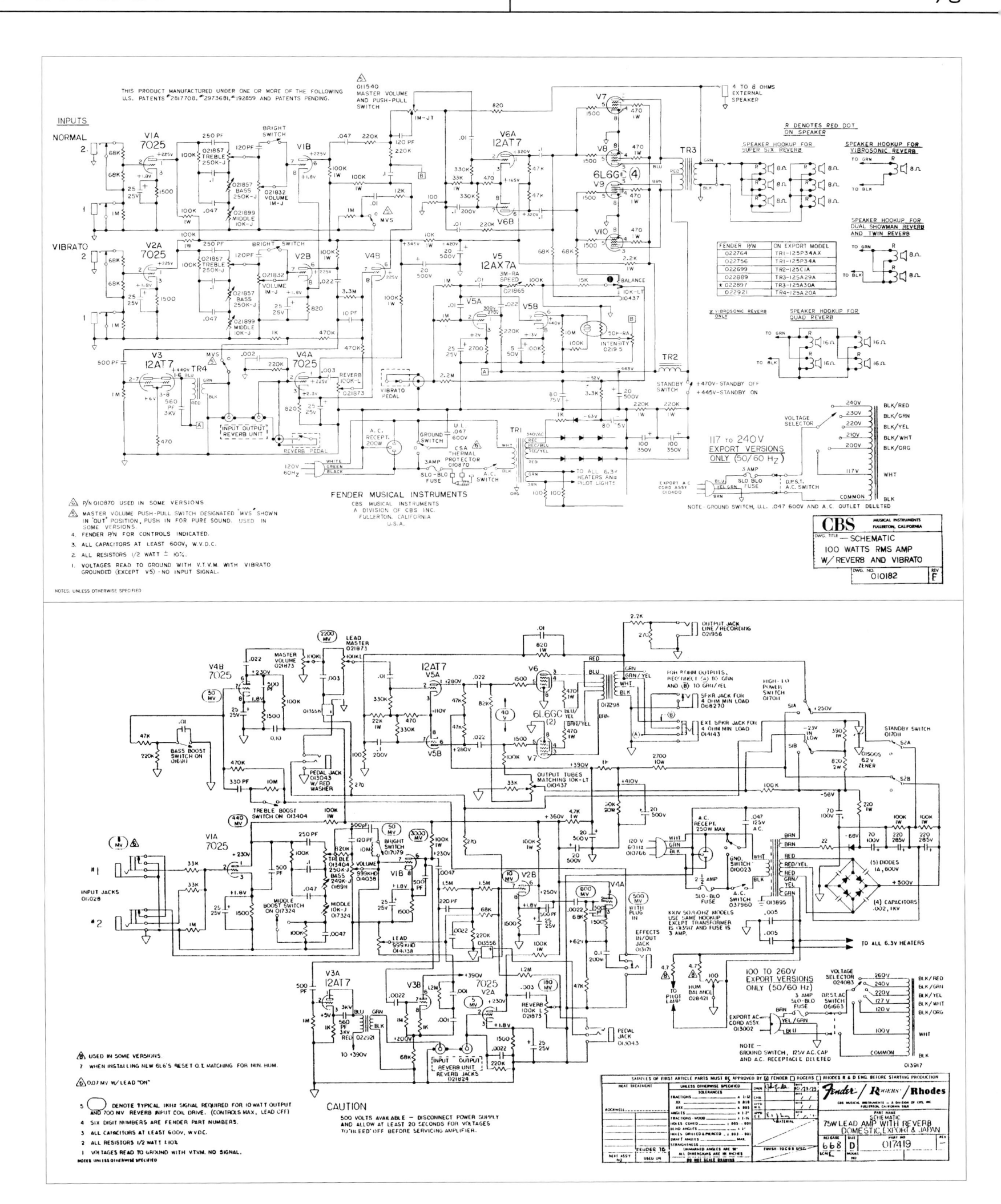

CBS 100W / 75

THE TUBE AMP BOOK 75

BANSHEE
BTO OVERDRIVE CIRCUIT

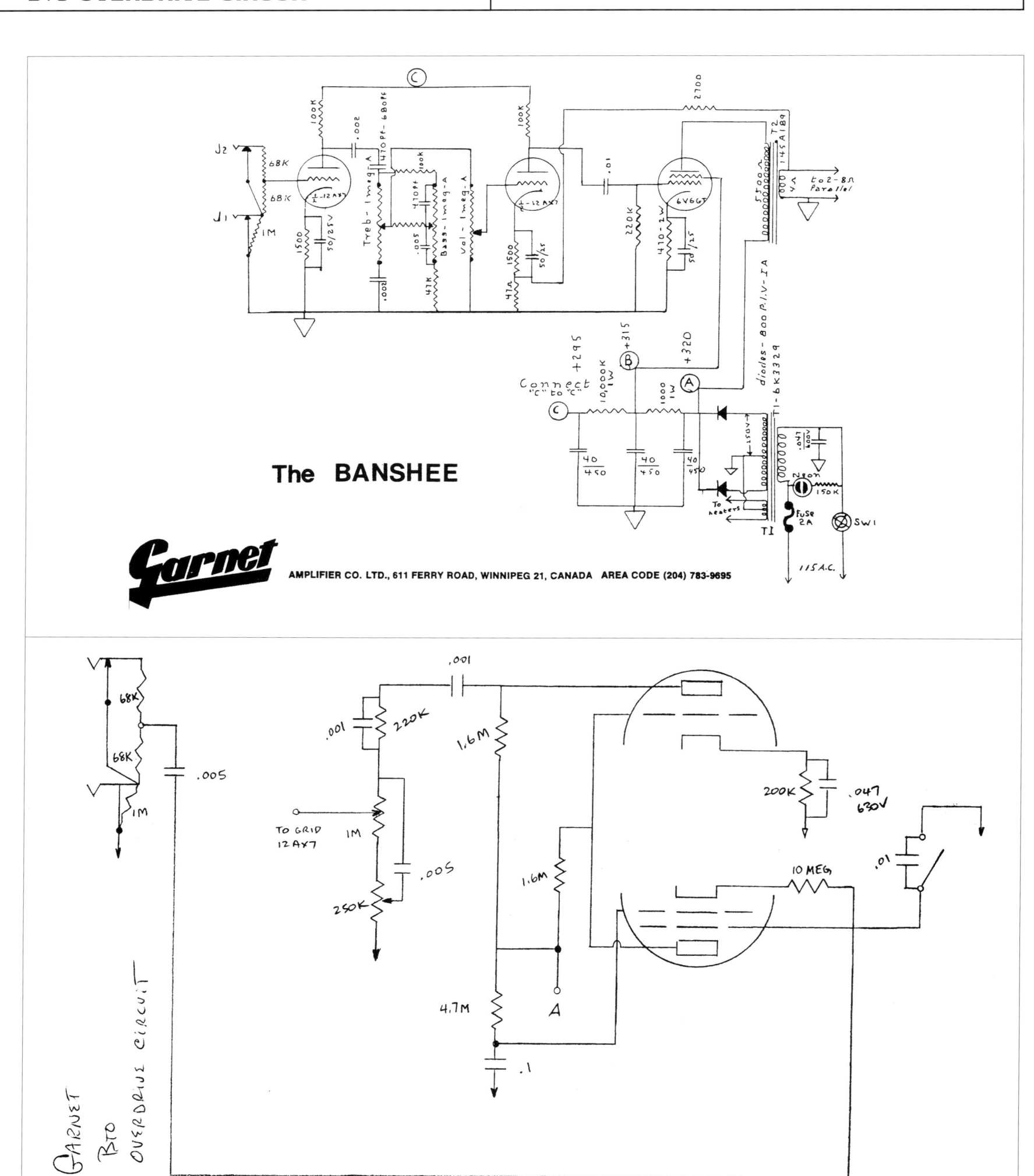

Garnet

The BANSHEE

Garnet AMPLIFIER CO. LTD., 611 FERRY ROAD, WINNIPEG 21, CANADA AREA CODE (204) 783-9695

GARNET BTO OVERDRIVE CIRCUIT

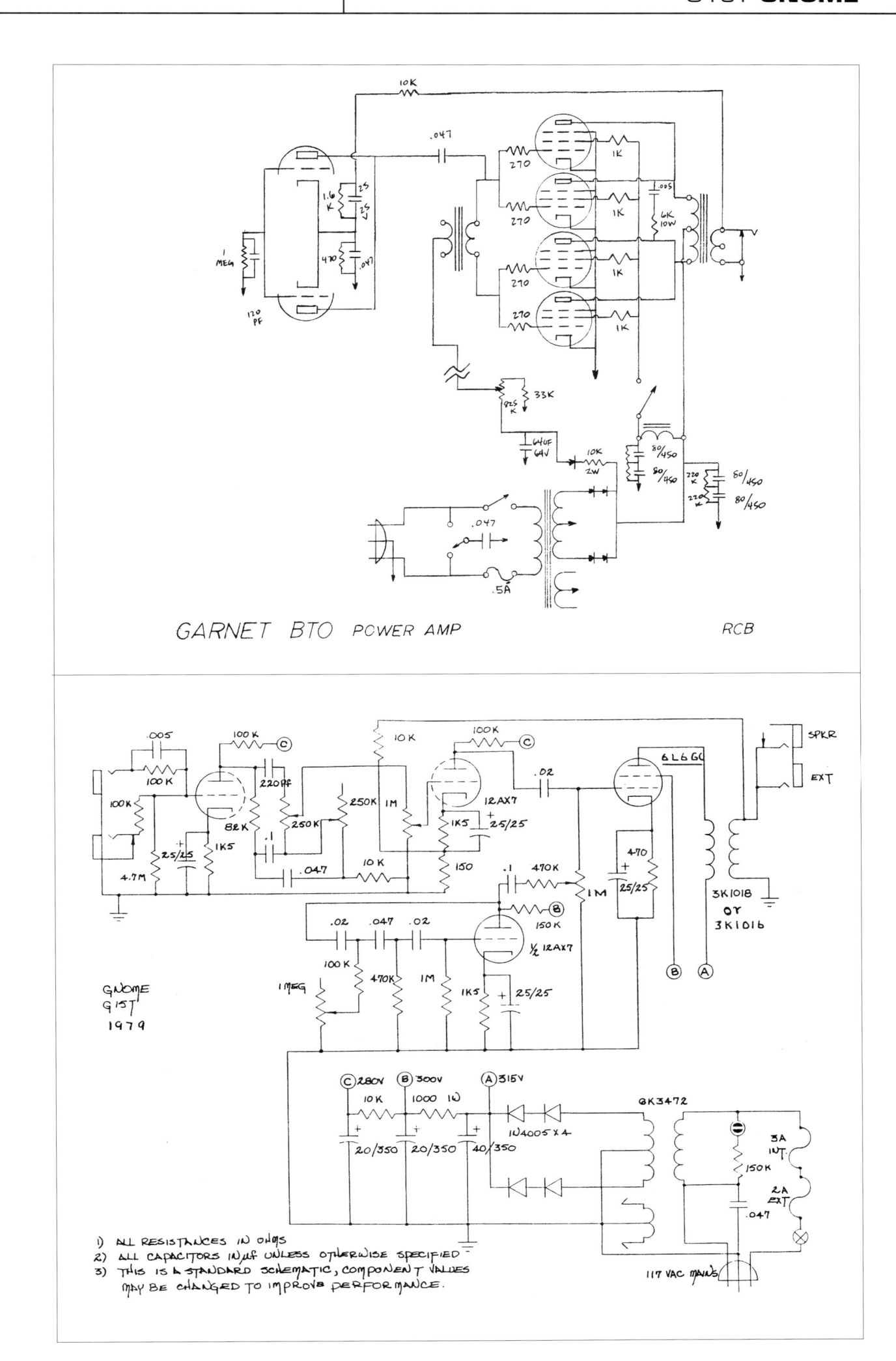

Garnet

BTO POWER AMP
G15T GNOME

GARNET BTO POWER AMP — RCB

GNOME G15T 1979

1) ALL RESISTANCES IN OHMS
2) ALL CAPACITORS IN µF UNLESS OTHERWISE SPECIFIED
3) THIS IS A STANDARD SCHEMATIC, COMPONENT VALUES MAY BE CHANGED TO IMPROVE PERFORMANCE.

G45-90-100 **PRE**
G90TR **POWER SUPPLY**

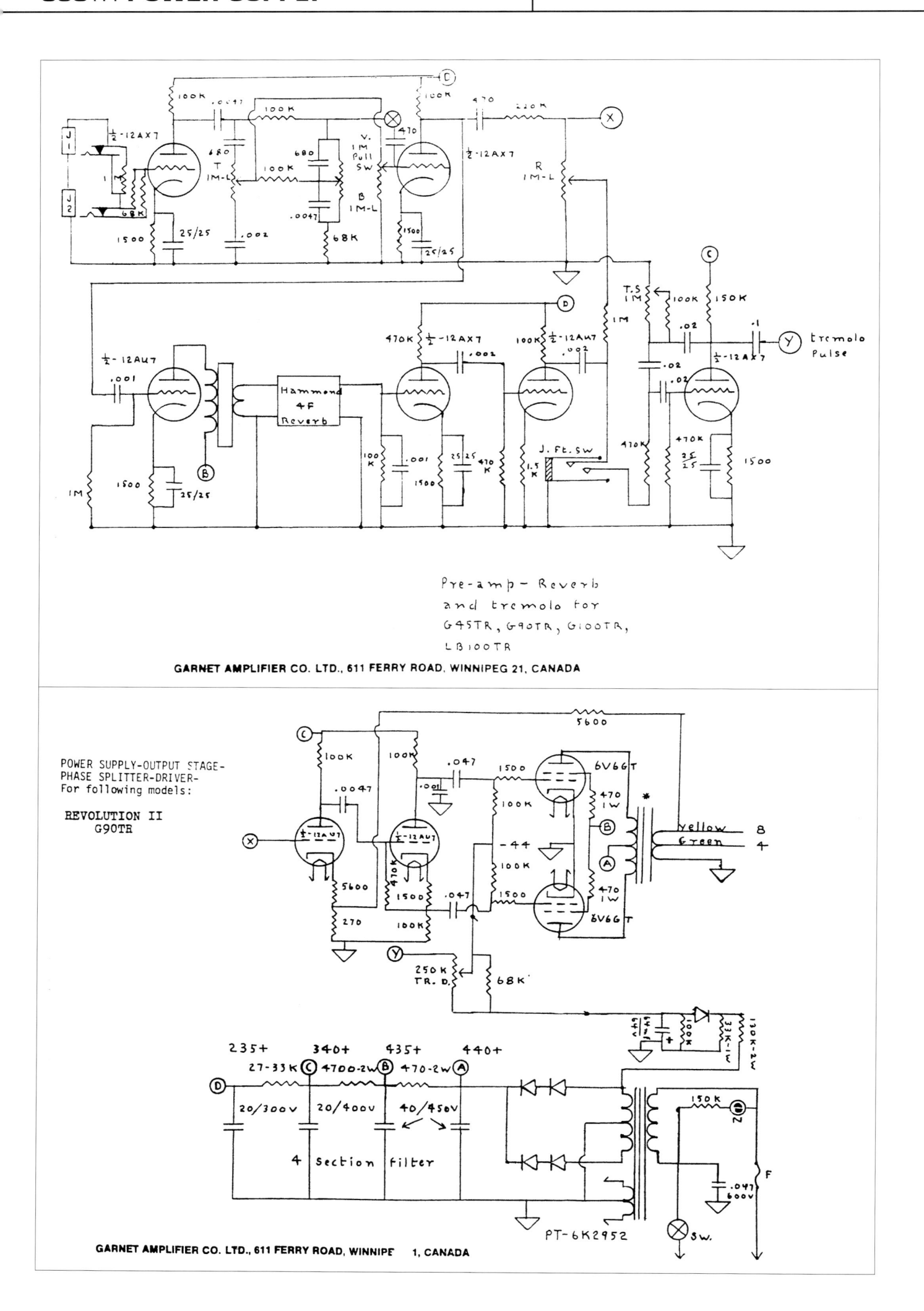

Garnet

Pre-amp - Reverb and tremolo for G45TR, G90TR, G100TR, LB100TR

GARNET AMPLIFIER CO. LTD., 611 FERRY ROAD, WINNIPEG 21, CANADA

POWER SUPPLY-OUTPUT STAGE-
PHASE SPLITTER-DRIVER-
For following models:

REVOLUTION II
G90TR

GARNET AMPLIFIER CO. LTD., 611 FERRY ROAD, WINNIPEG 21, CANADA

78 THE TUBE AMP BOOK

Garnet

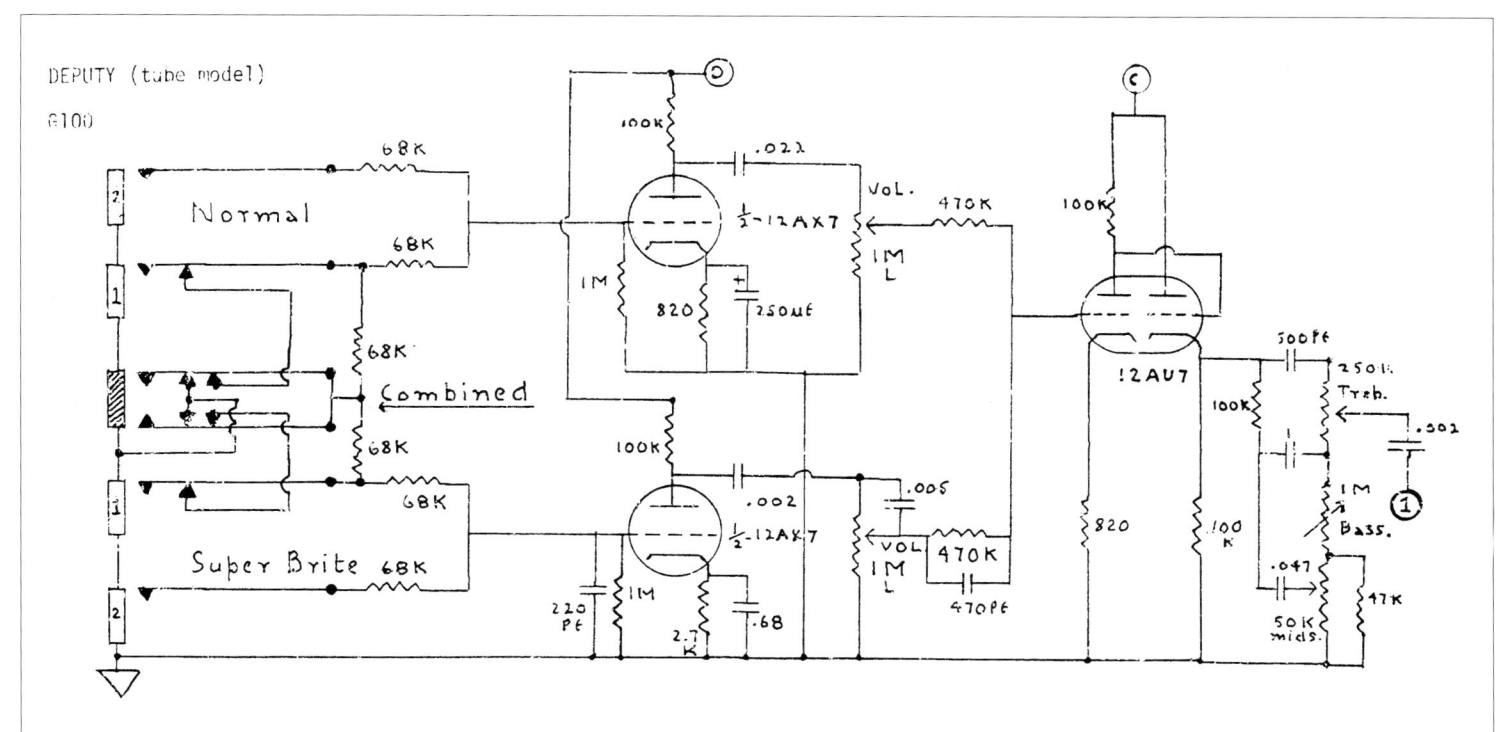

G100 PREAMP
G200R ENFORCER

DEPUTY (tube model)
G100

pre-amp, mixer, 2nd pre-amp and tone circuits.

GARNET AMPLIFIER LTD. 1360 Sargent Ave., Winnipeg, Manitoba, Canada. R3E 0G5.

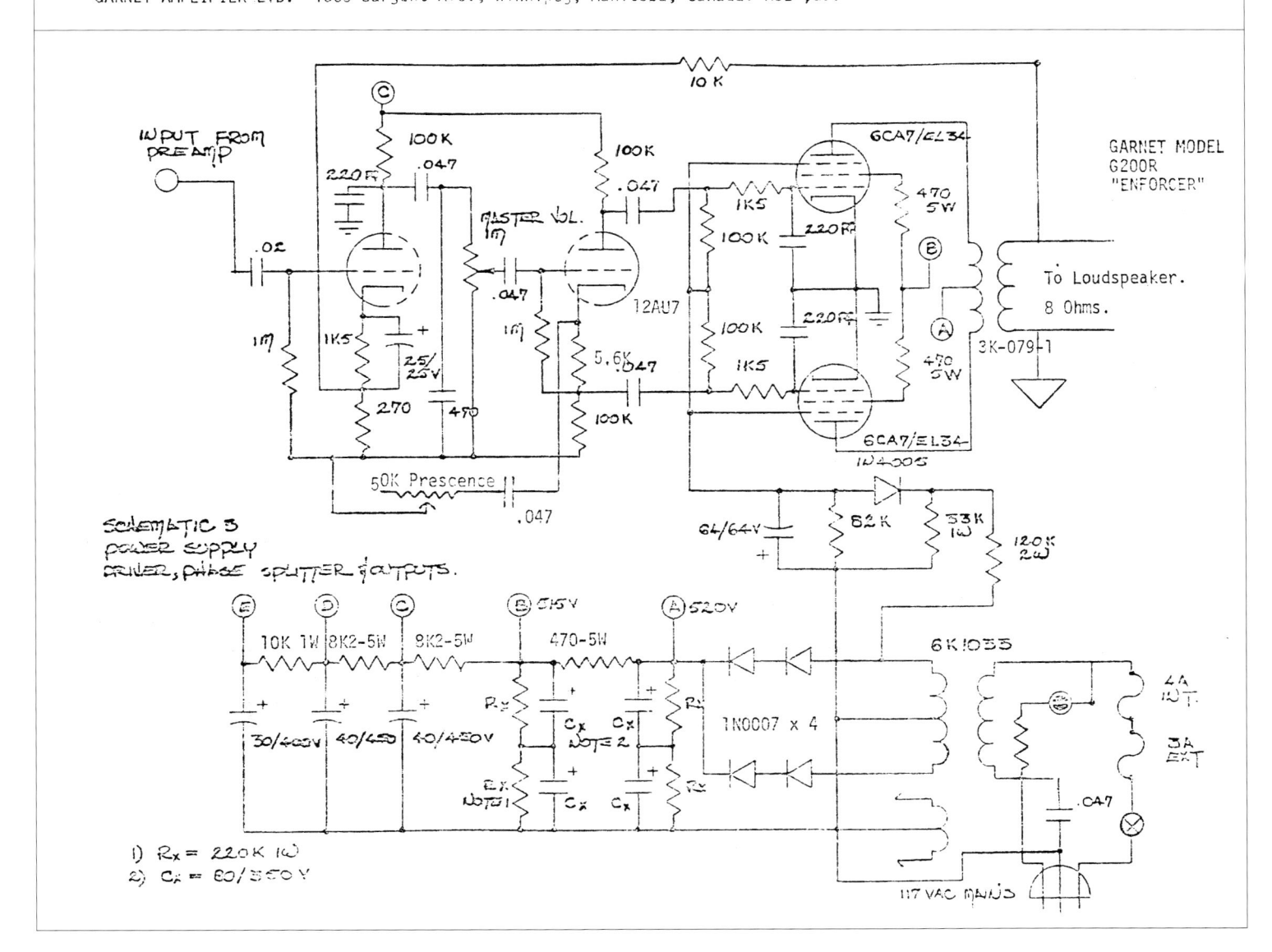

GARNET MODEL G200R "ENFORCER"

SCHEMATIC 3
POWER SUPPLY
DRIVER, PHASE SPLITTER & OUTPUTS.

1) R_x = 220K 1W
2) C_x = 80/350V

G250TR & G250D **PREAMPS**
G250TR & G250D **POWER SUPPLY**

Garnet

Modification June 1974
R4 to 2.7K
C1 to .1

Points F and F go to "pull sw." on rear MASTER VOLUME

Part 2- Two channel pre-amps for SESSIONMAN G250TR and G250D

JUNE 1974

GARNET AMPLIFIER CO. LTD
1360 SARGENT AVE, WINNIPEG, CANADA

Power Supply - Output - phase - splitter - driver for following -
G250TR - G250D
250PA - LB200F
all 190 models

R1 - 10K - models 200F, 400F, 190
6K - models 250FTR - 250PA

Some models have
+ and 8Ω
V.C. taps

tremolo inserted 1 and 2 -
some models.

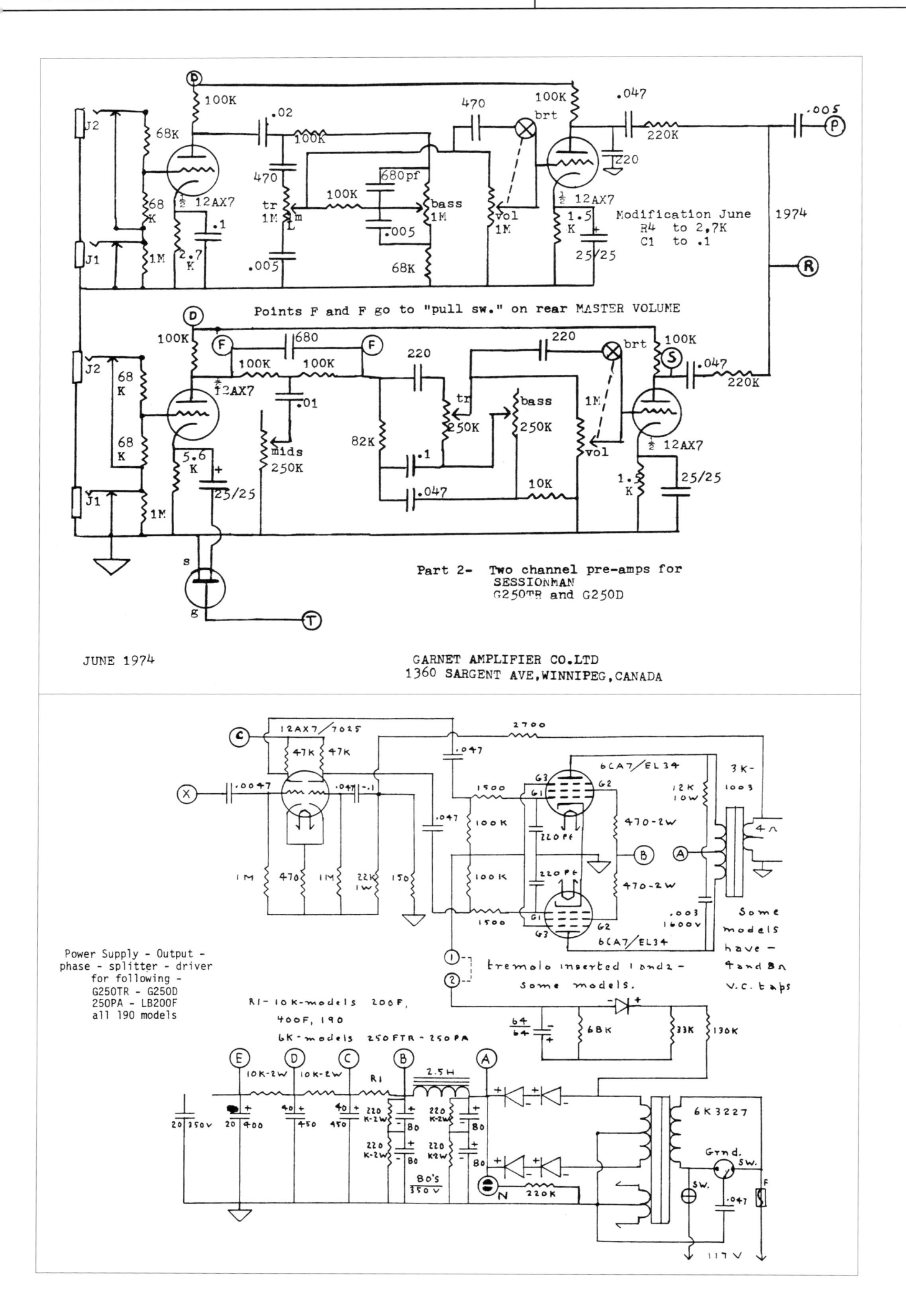

80 THE TUBE AMP BOOK

Garnet

G250TR & G250D REVERB & PRES
G250TR, G250D, G200S & G200

GARNET AMPLIFIER CO. LTD
Winnipeg, Canada

#3- JUNE 1974/75

REVERBERATION DRIVER AND PRE-AMPS plus TREMOLO SYSTEM.

FET "clamp" is shown on part 2 (Pre-amps)

SESSIONMAN
G250TR
G250D

Pull "SLOW SPEED" switch on rear of TREMOLO speed control

G250TR G250D (production May '74)
G200S G200C

*R1 is 6K/10W in G250TR and D
*R1 is 30K/2W in G200S and C
Filter network at "E" is not used in G200S

Voltages
	PRO 200S	SESSIONMAN
A)	520	520
B)	515	515
C)	400	390
D)	380	315
C)	290	

R2,R3,R4,R5- 220K/1W   C1,2,3,4- 80/350V

GARNET AMPLIFIER CO. LTD.   1360 SARGENT AVE, WINNIPEG CANADA

THE TUBE AMP BOOK 81

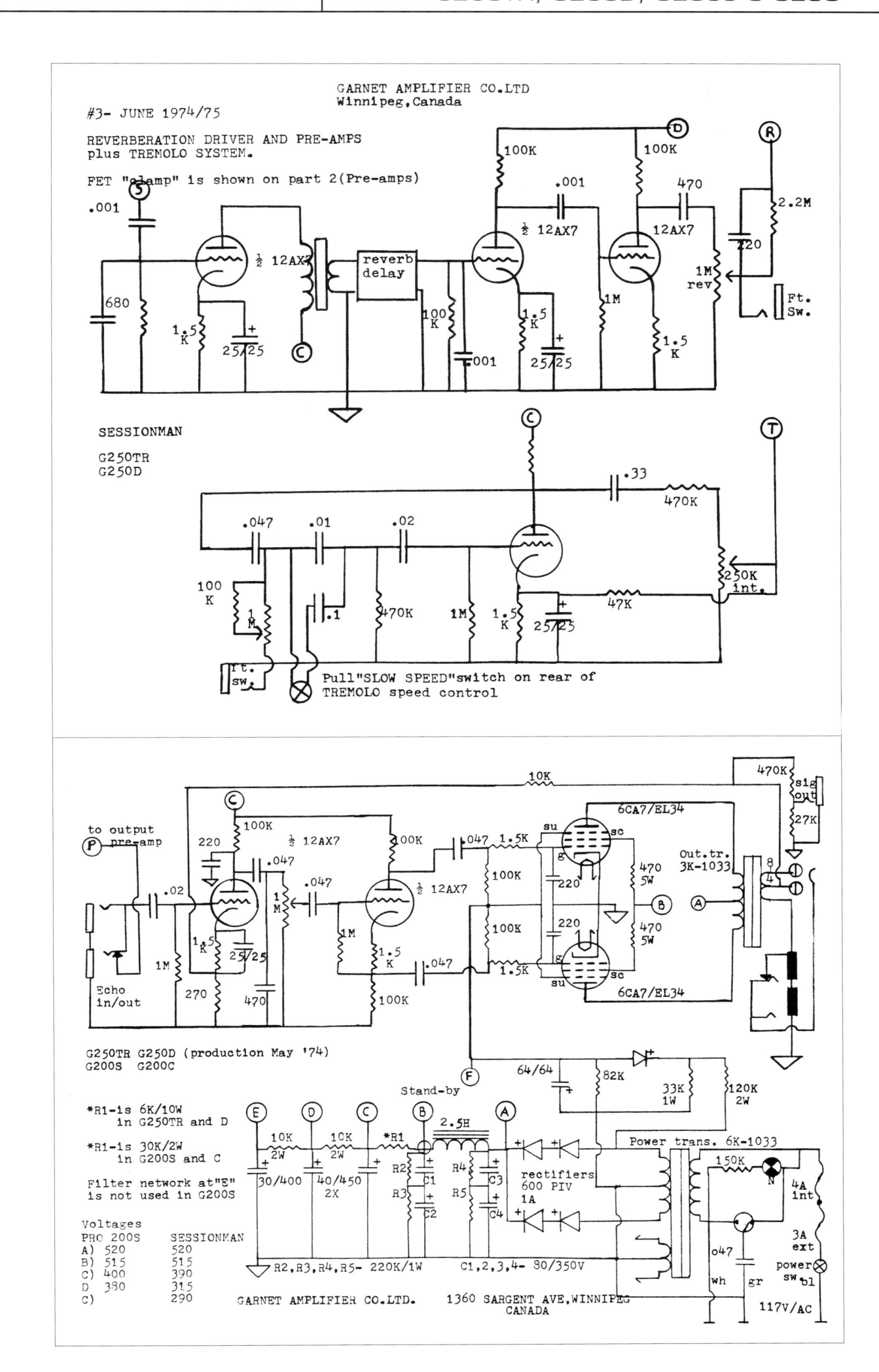

HERZOG **RANDY BACHMAN**
H-ZOG

Garnet

Randy Bachman's "Herzog"

AMPLIFIER CO. LTD., 611 FERRY ROAD, WINNIPEG 21, CANADA

"H-ZOG"

GARNET AMPLIFIER LTD.
1360 SARGENT AVE, WINNIPEG, CANADA.
R3E 0G5

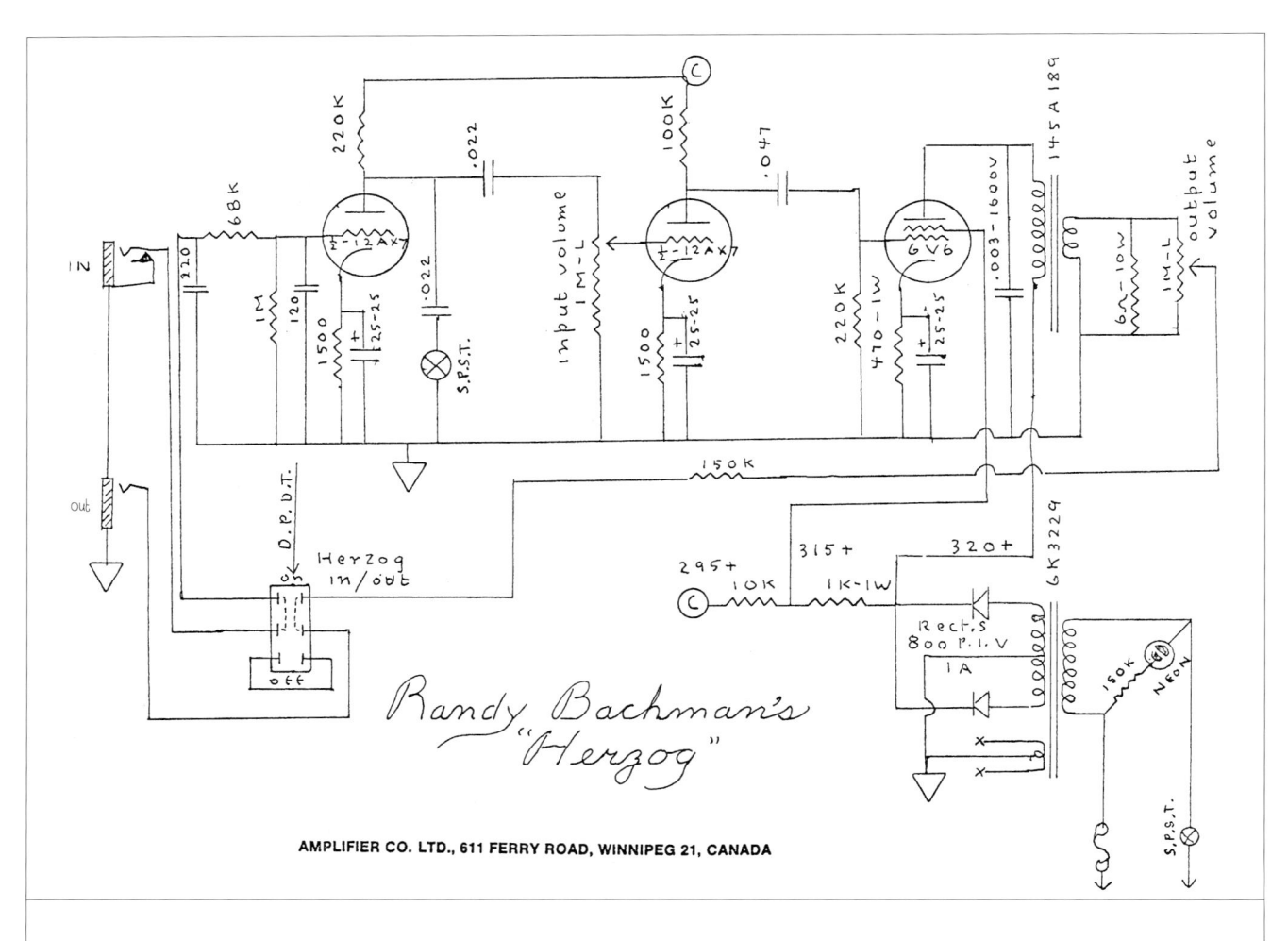

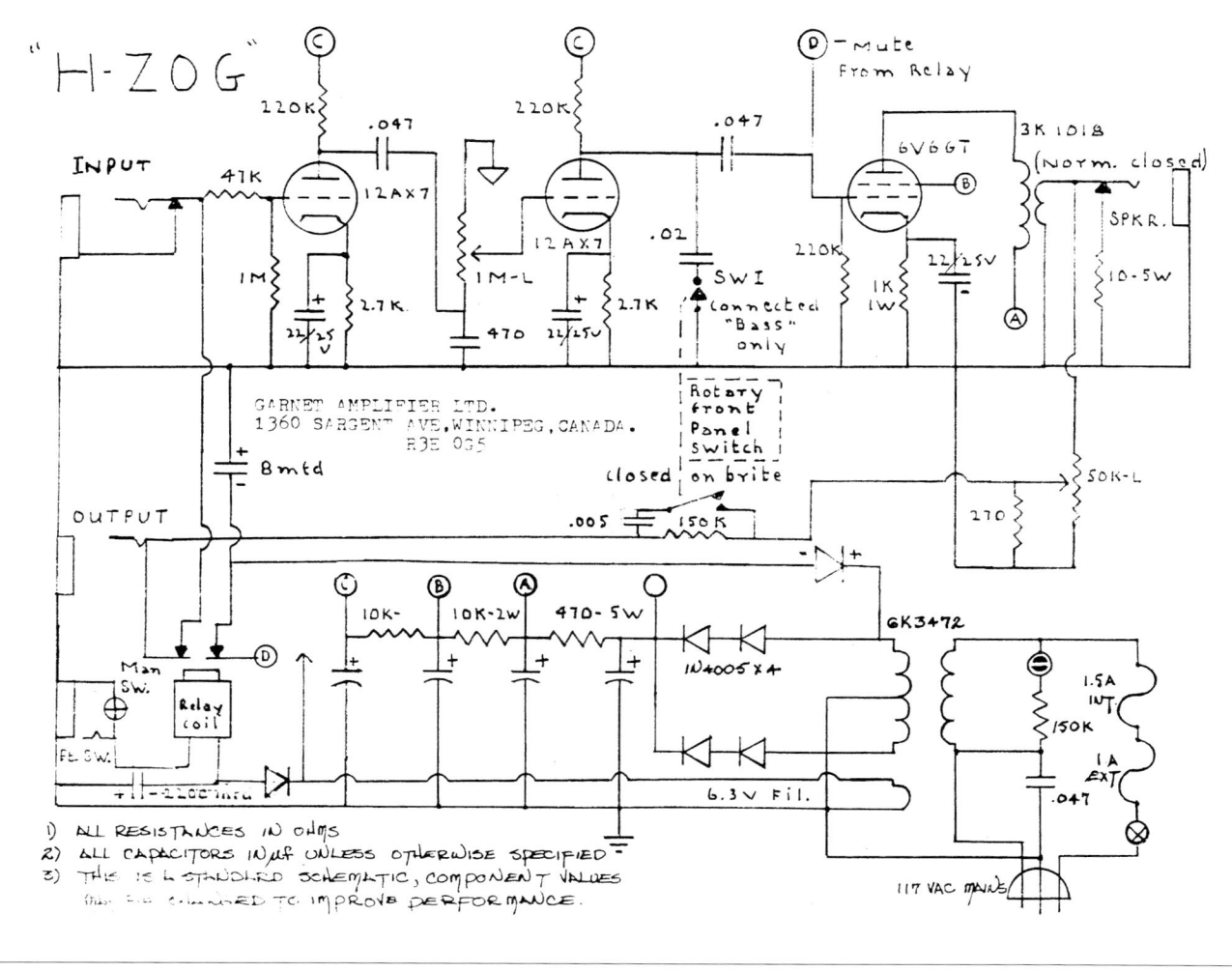

1) ALL RESISTANCES IN OHMS
2) ALL CAPACITORS IN µF UNLESS OTHERWISE SPECIFIED
3) THIS IS A STANDARD SCHEMATIC, COMPONENT VALUES MAY BE CHANGED TO IMPROVE PERFORMANCE.

82 THE TUBE AMP BOOK

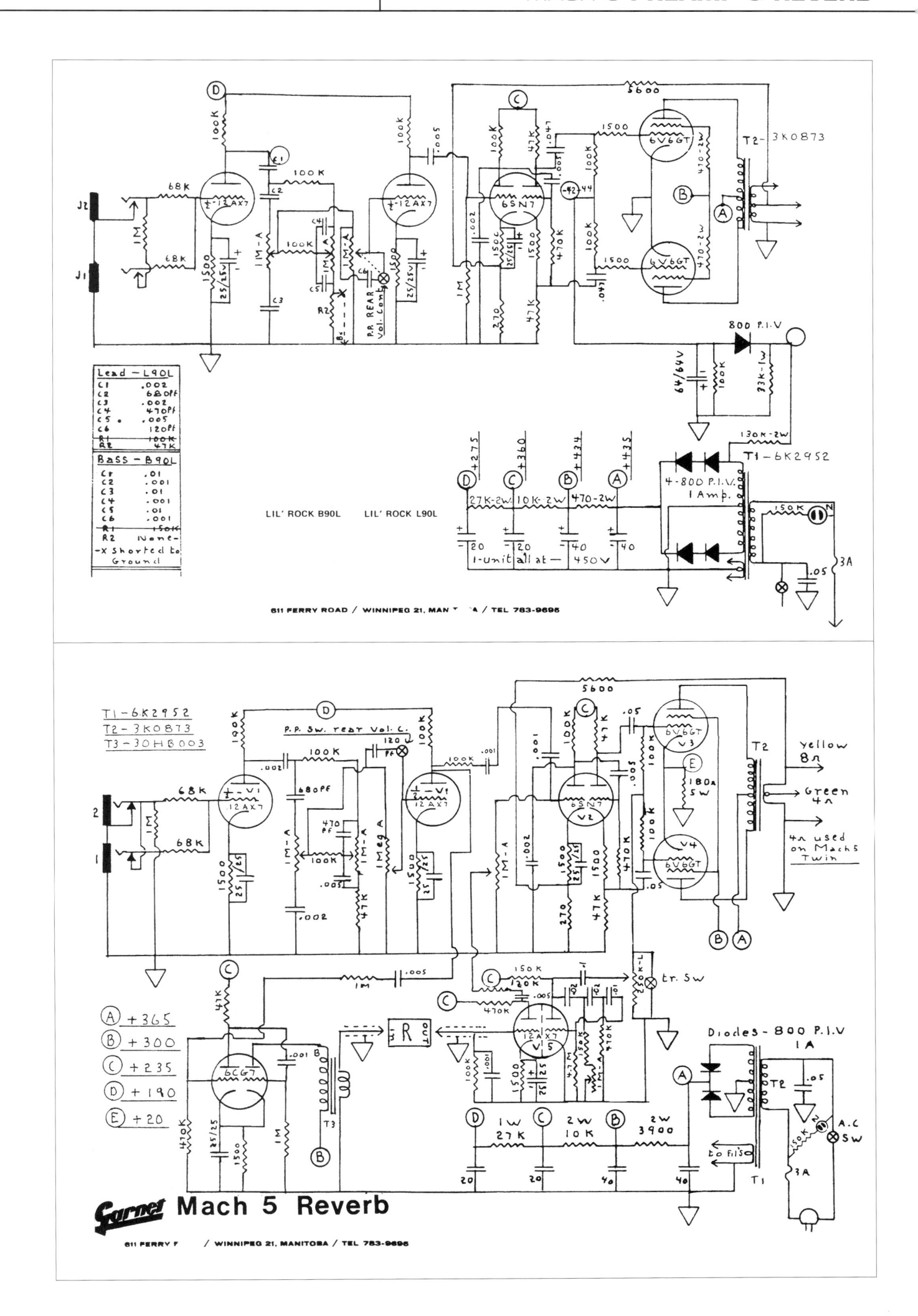

Garnet

LIL ROCK B90L
MACH 5 PREAMP & REVERB

PA90
PA90 MIXER

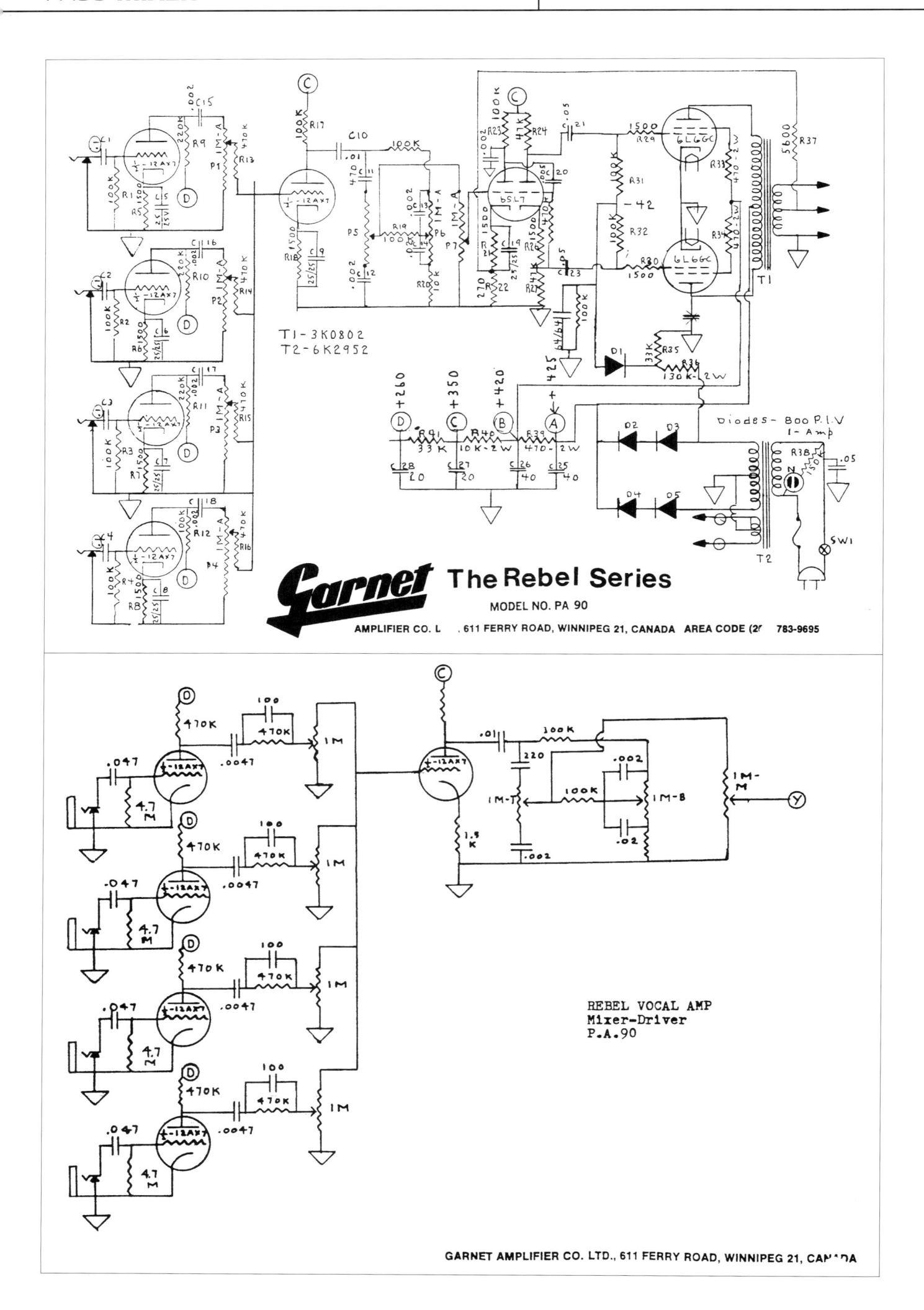

Garnet

Garnet The Rebel Series
MODEL NO. PA 90
AMPLIFIER CO. L , 611 FERRY ROAD, WINNIPEG 21, CANADA AREA CODE (2) 783-9695

REBEL VOCAL AMP
Mixer-Driver
P.A.90

GARNET AMPLIFIER CO. LTD., 611 FERRY ROAD, WINNIPEG 21, CANADA

84 THE TUBE AMP BOOK

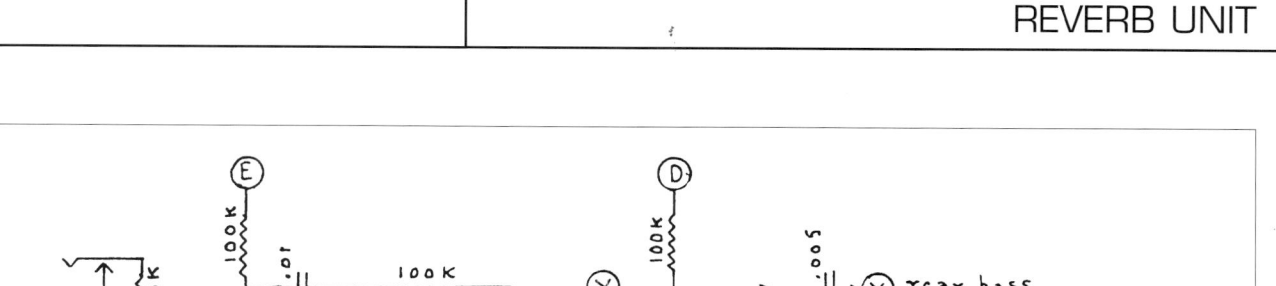

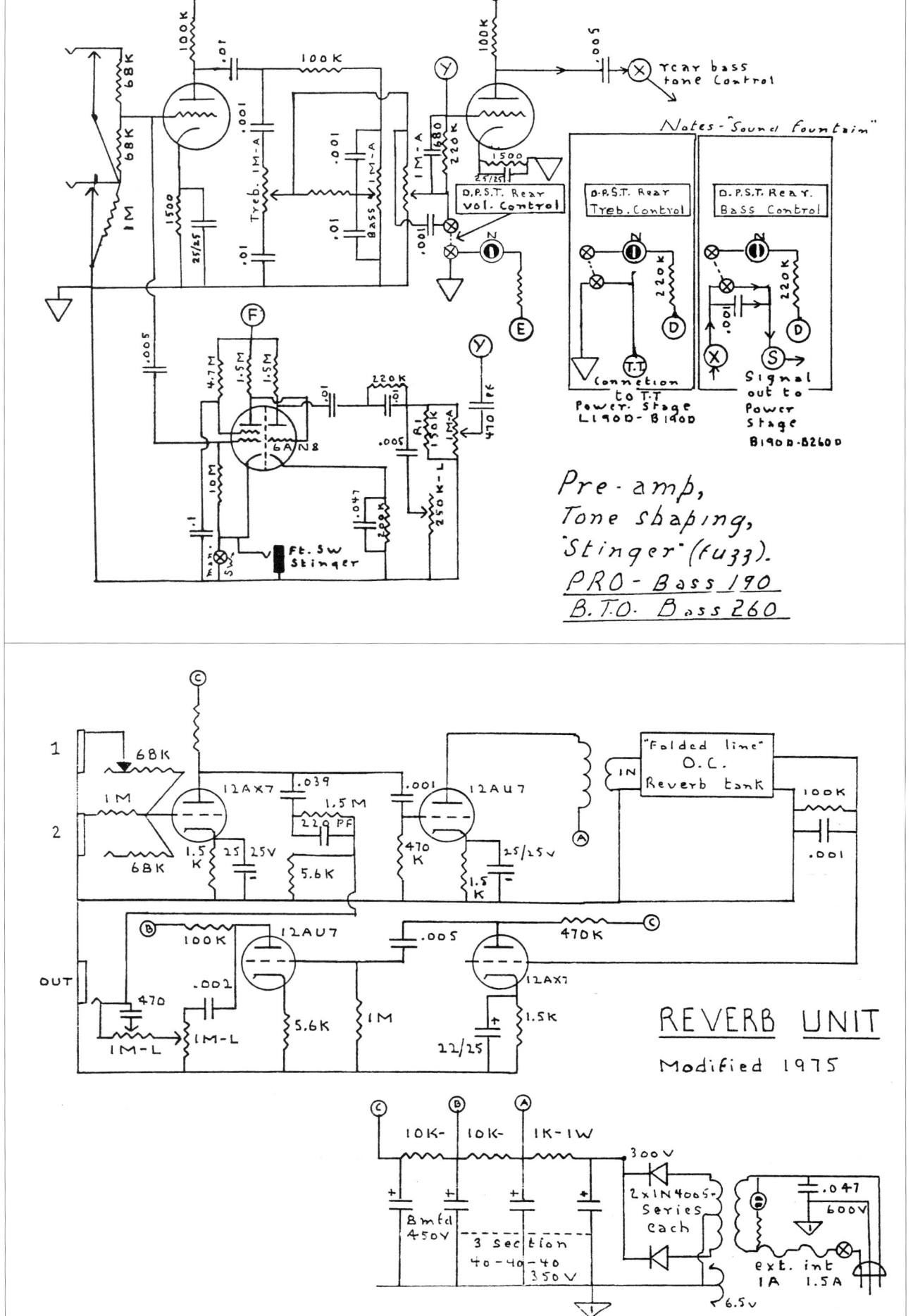

Garnet

PRO BASS 190 STINGER FUZZ & PRE
REVERB UNIT

Pre-amp, Tone shaping, "Stinger" (fuzz).
PRO-Bass 190
B.T.O. Bass 260

REVERB UNIT
Modified 1975

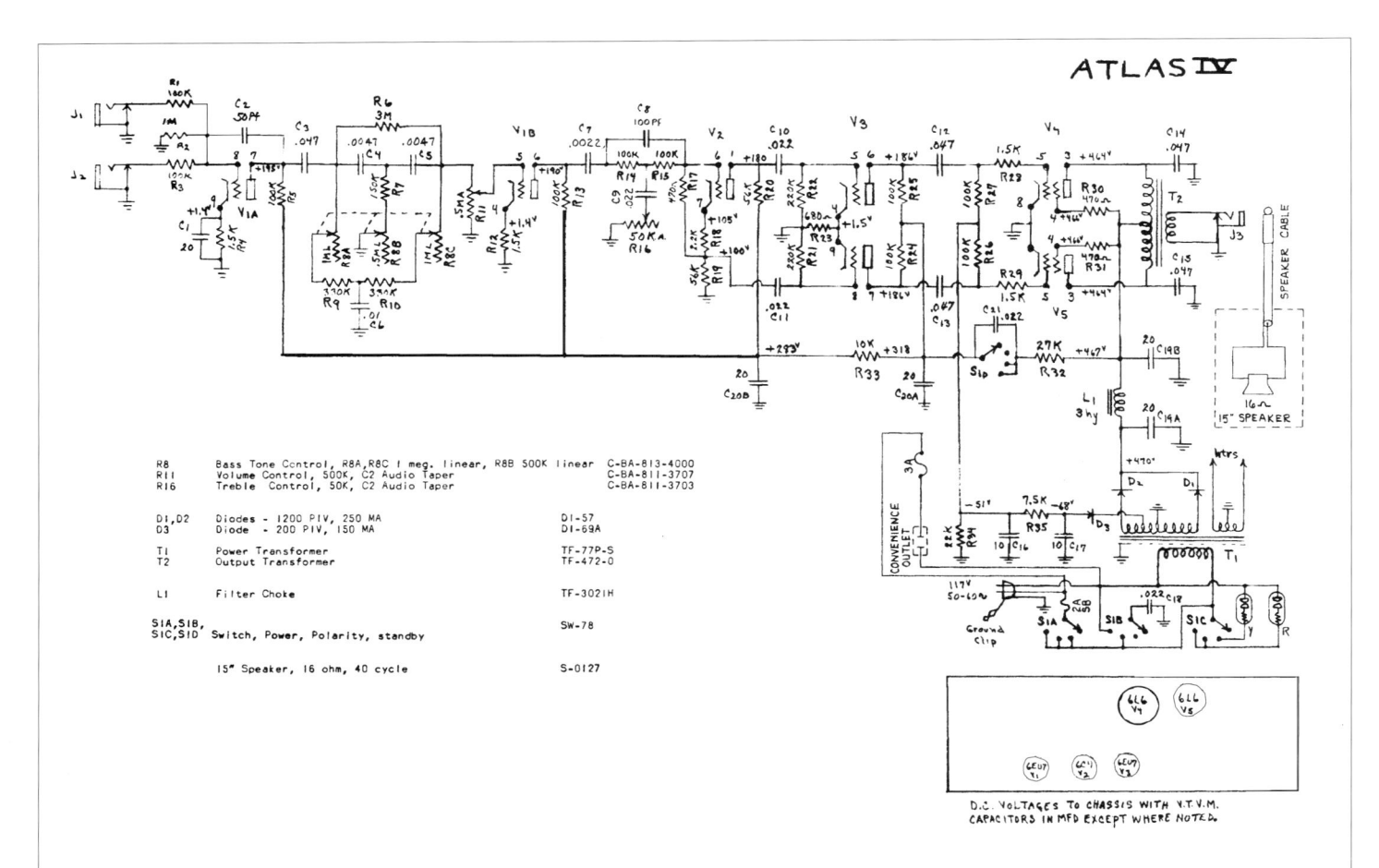

ATLAS IV
ATLAS MEDALIST

Gibson

ATLAS IV

R8	Bass Tone Control, R8A,R8C 1 meg. linear, R8B 500K linear		C-BA-813-4000
R11	Volume Control, 500K, C2 Audio Taper		C-BA-811-3707
R16	Treble Control, 50K, C2 Audio Taper		C-BA-811-3703
D1,D2	Diodes - 1200 PIV, 250 MA		D1-57
D3	Diode - 200 PIV, 150 MA		D1-69A
T1	Power Transformer		TF-77P-S
T2	Output Transformer		TF-472-0
L1	Filter Choke		TF-3021H
S1A,S1B, S1C,S1D	Switch, Power, Polarity, standby		SW-78
	15" Speaker, 16 ohm, 40 cycle		S-0127

D.C. VOLTAGES TO CHASSIS WITH V.T.V.M.
CAPACITORS IN MFD EXCEPT WHERE NOTED.

FUSE

The fuse used in this Amplifier is a type 3AG of three amperes rating
DO NOT USE FUSES OF HIGHER RATING

SERVICE

If the amplifier is in need of servicing, it should be taken to a reliable radio man. The electrical diagram in this folder should be shown the repairman to assist him in servicing the amplifier.

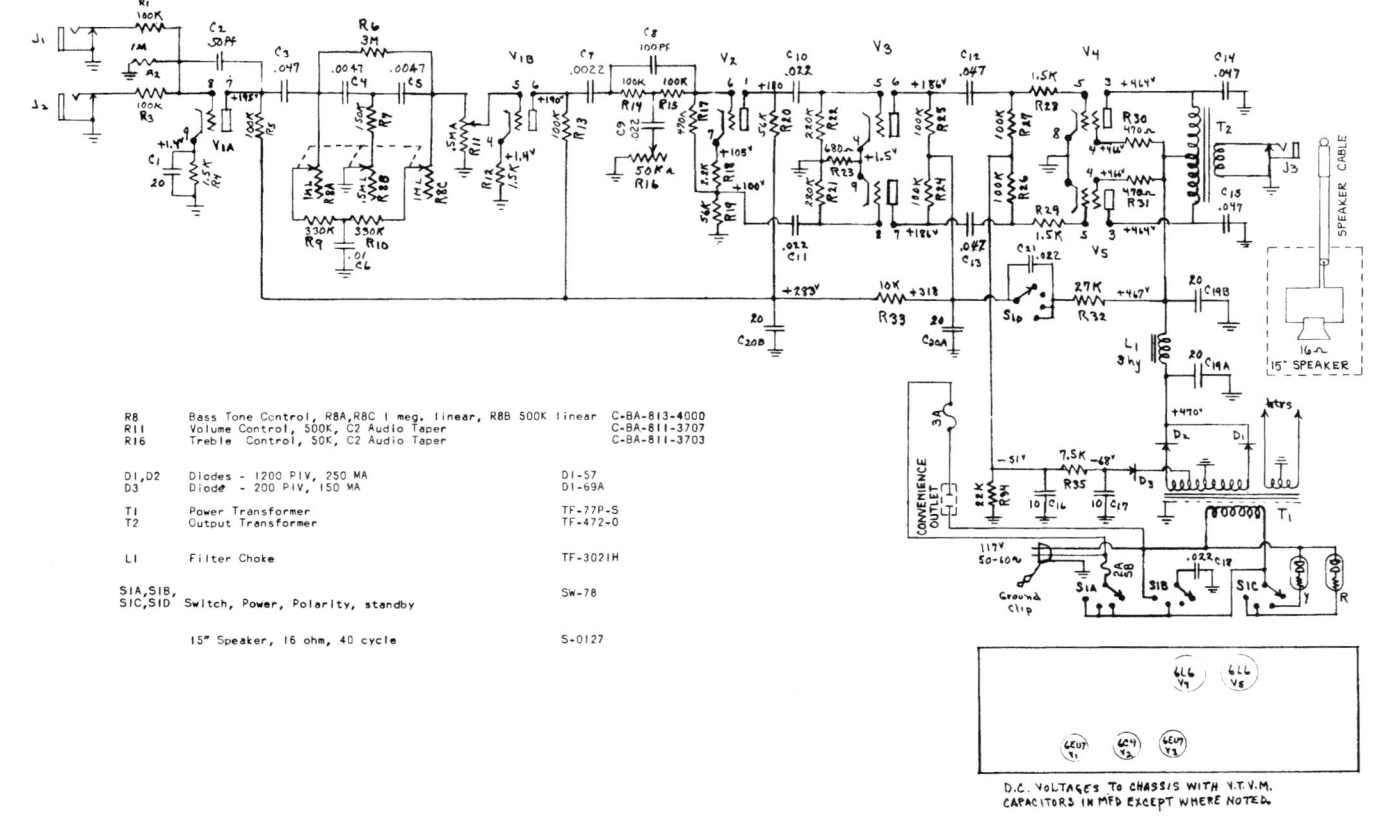

ATLAS MEDALIST

R8	Bass Tone Control, R8A,R8C 1 meg. linear, R8B 500K linear		C-BA-813-4000
R11	Volume Control, 500K, C2 Audio Taper		C-BA-811-3707
R16	Treble Control, 50K, C2 Audio Taper		C-BA-811-3703
D1,D2	Diodes - 1200 PIV, 250 MA		D1-57
D3	Diode - 200 PIV, 150 MA		D1-69A
T1	Power Transformer		TF-77P-S
T2	Output Transformer		TF-472-0
L1	Filter Choke		TF-3021H
S1A,S1B, S1C,S1D	Switch, Power, Polarity, standby		SW-78
	15" Speaker, 16 ohm, 40 cycle		S-0127

D.C. VOLTAGES TO CHASSIS WITH V.T.V.M.
CAPACITORS IN MFD EXCEPT WHERE NOTED.

86 THE TUBE AMP BOOK

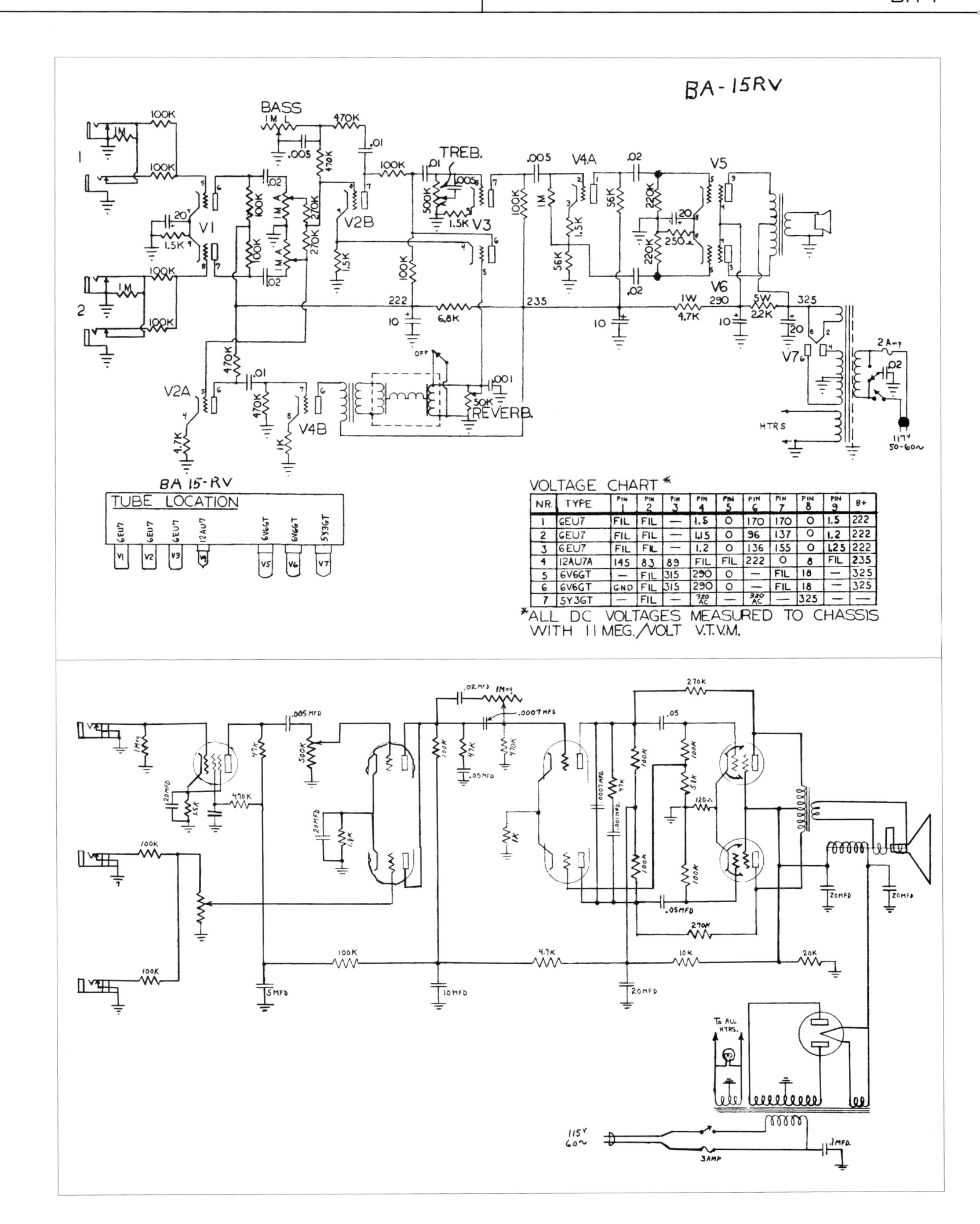

Gibson

BA-15RV
BR-1

BA-15RV

NR.	TYPE	PIN 1	PIN 2	PIN 3	PIN 4	PIN 5	PIN 6	PIN 7	PIN 8	PIN 9	B+
1	6EU7	FIL	FIL	—	1.5	0	170	170	0	1.5	222
2	6EU7	FIL	FIL	—	1.5	0	96	137	0	1.2	222
3	6EU7	FIL	FIL	—	1.2	0	136	155	0	1.25	222
4	12AU7A	145	83	89	FIL	FIL	222	0	8	FIL	235
5	6V6GT	—	FIL	315	290	0	—	FIL	18	—	325
6	6V6GT	GND	FIL	315	290	0	—	FIL	18	—	325
7	5Y3GT	—	FIL	—	320 AC	—	320 AC	—	325	—	—

*ALL DC VOLTAGES MEASURED TO CHASSIS WITH 11 MEG./VOLT V.T.V.M.

BA 15-RV TUBE LOCATION

6EU7	6EU7	6EU7	12AU7	6V6GT	6V6GT	5Y3GT
V1	V2	V3	V4	V5	V6	V7

THE TUBE AMP BOOK

BR-3
BR-4

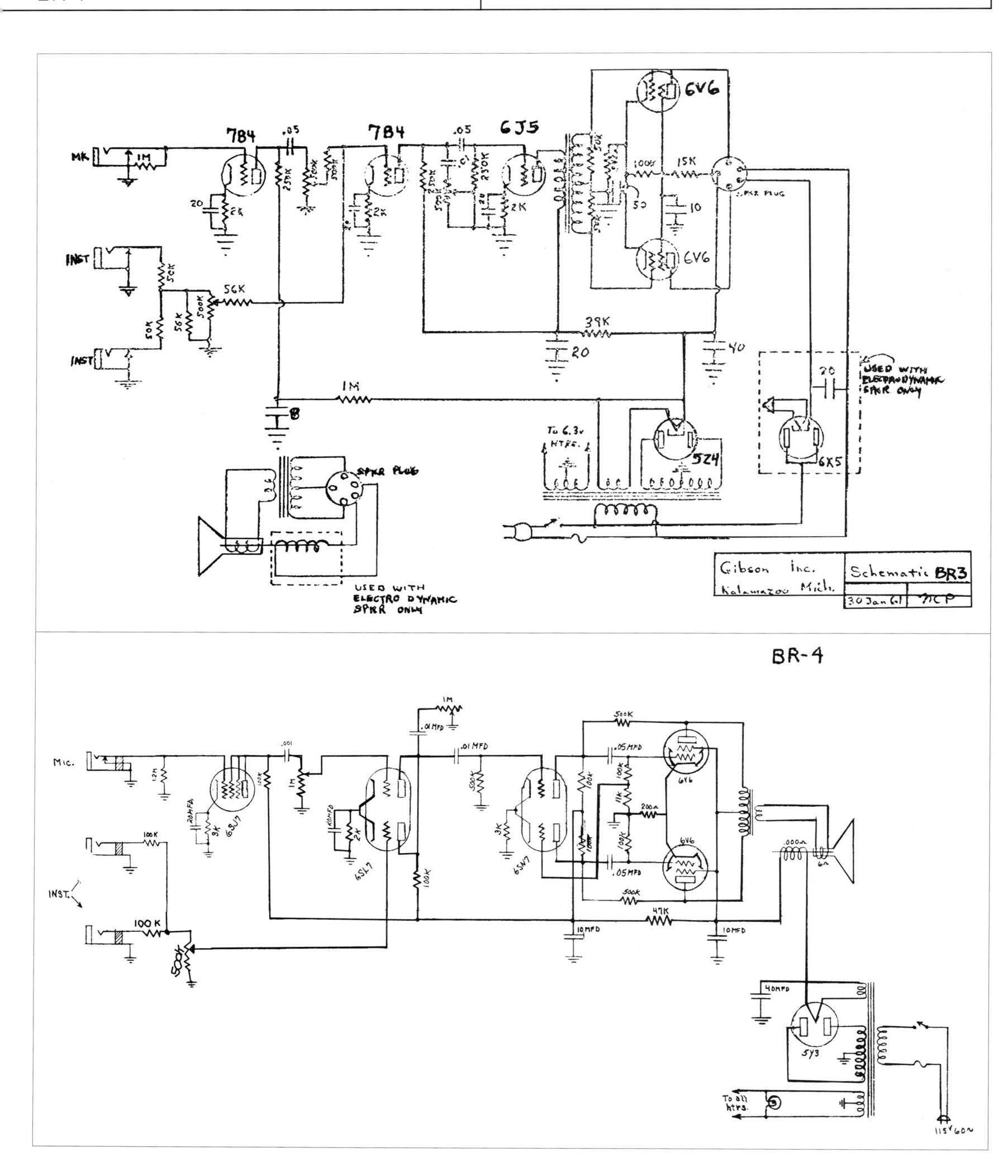

Gibson

BR-4

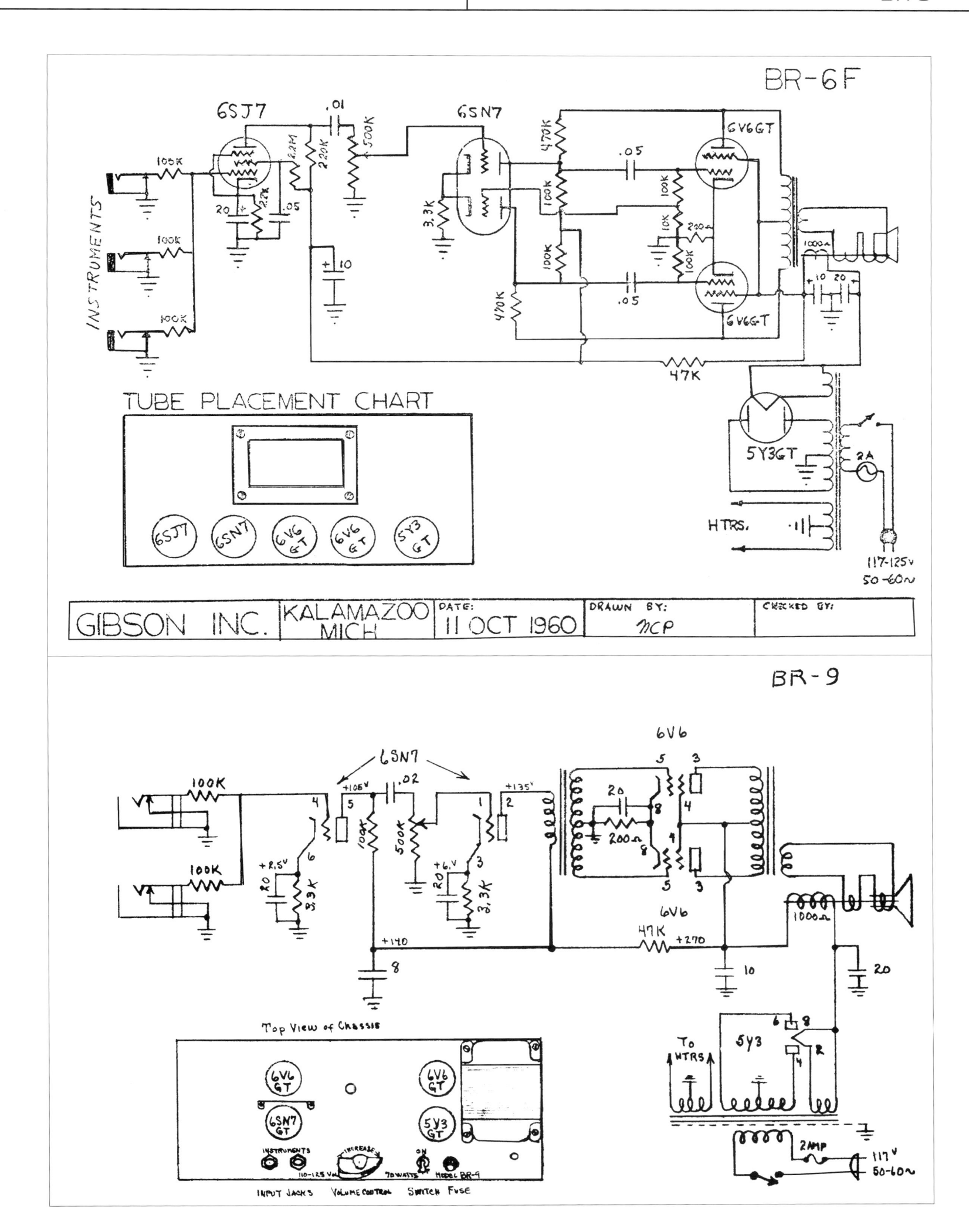

Gibson

BR-6F
BR-9

EH-100
EH-100
Gibson

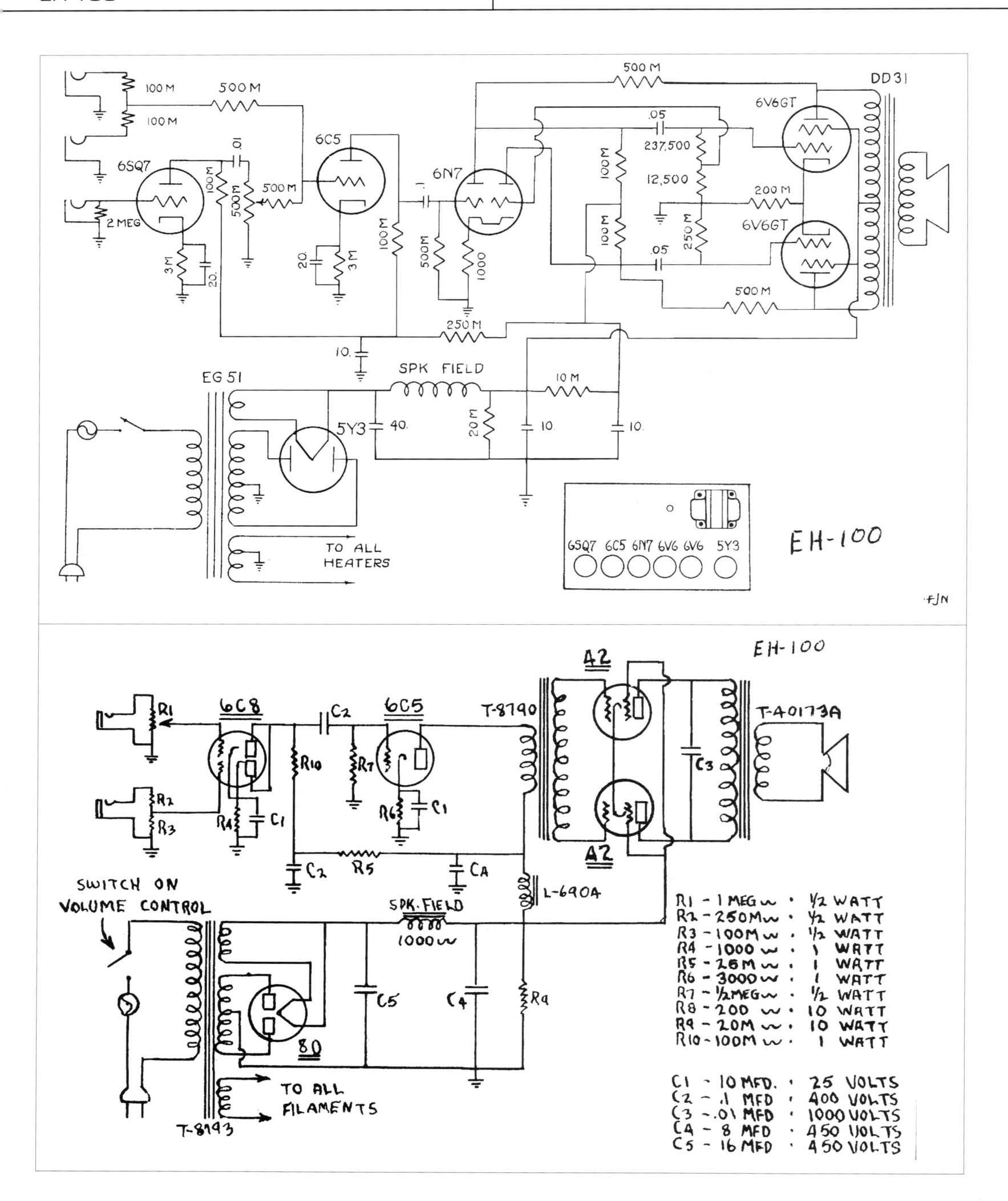

EH-100 (top schematic)

Tubes: 6SQ7, 6C5, 6N7, 6V6GT, 6V6GT, 5Y3
Transformer: EG 51
Speaker: DD 31

EH-100 (bottom schematic)

Tubes: 6C8, 6C5, A2, A2, 80
Transformers: T-8790, T-40173A, T-8193
Choke: L-690A

Switch on Volume Control
To All Filaments

R1	1 MEG Ω	½ WATT
R2	250 MΩ	½ WATT
R3	100 MΩ	½ WATT
R4	1000 Ω	1 WATT
R5	25 MΩ	1 WATT
R6	3000 Ω	1 WATT
R7	½ MEG Ω	½ WATT
R8	200 Ω	10 WATT
R9	20 MΩ	10 WATT
R10	100 MΩ	1 WATT
C1	10 MFD	25 VOLTS
C2	.1 MFD	400 VOLTS
C3	.01 MFD	1000 VOLTS
C4	8 MFD	450 VOLTS
C5	16 MFD	450 VOLTS

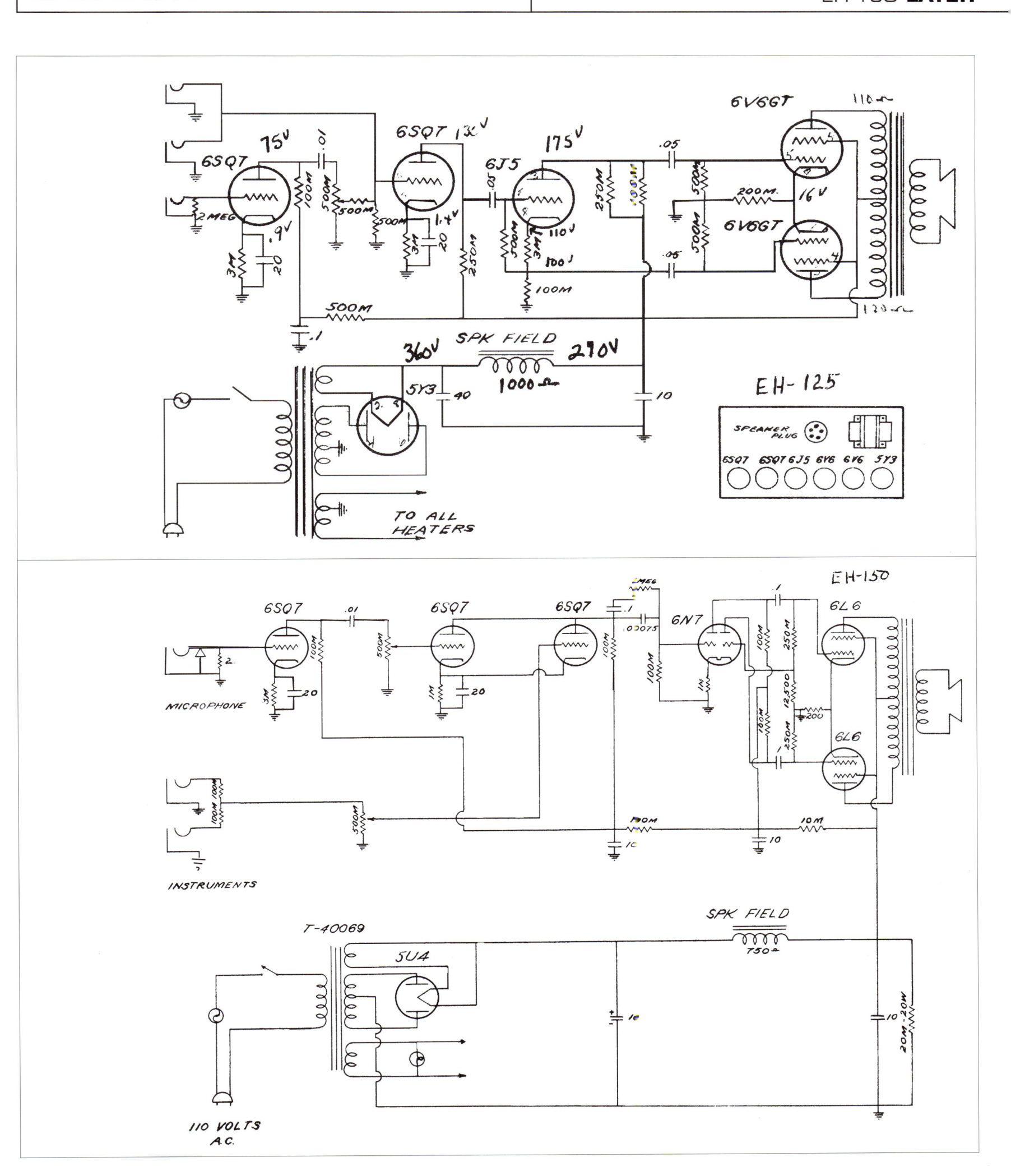

Gibson

EH-125
EH-150 **LATER**

EH-160
EH-185 (6J7 PRE)

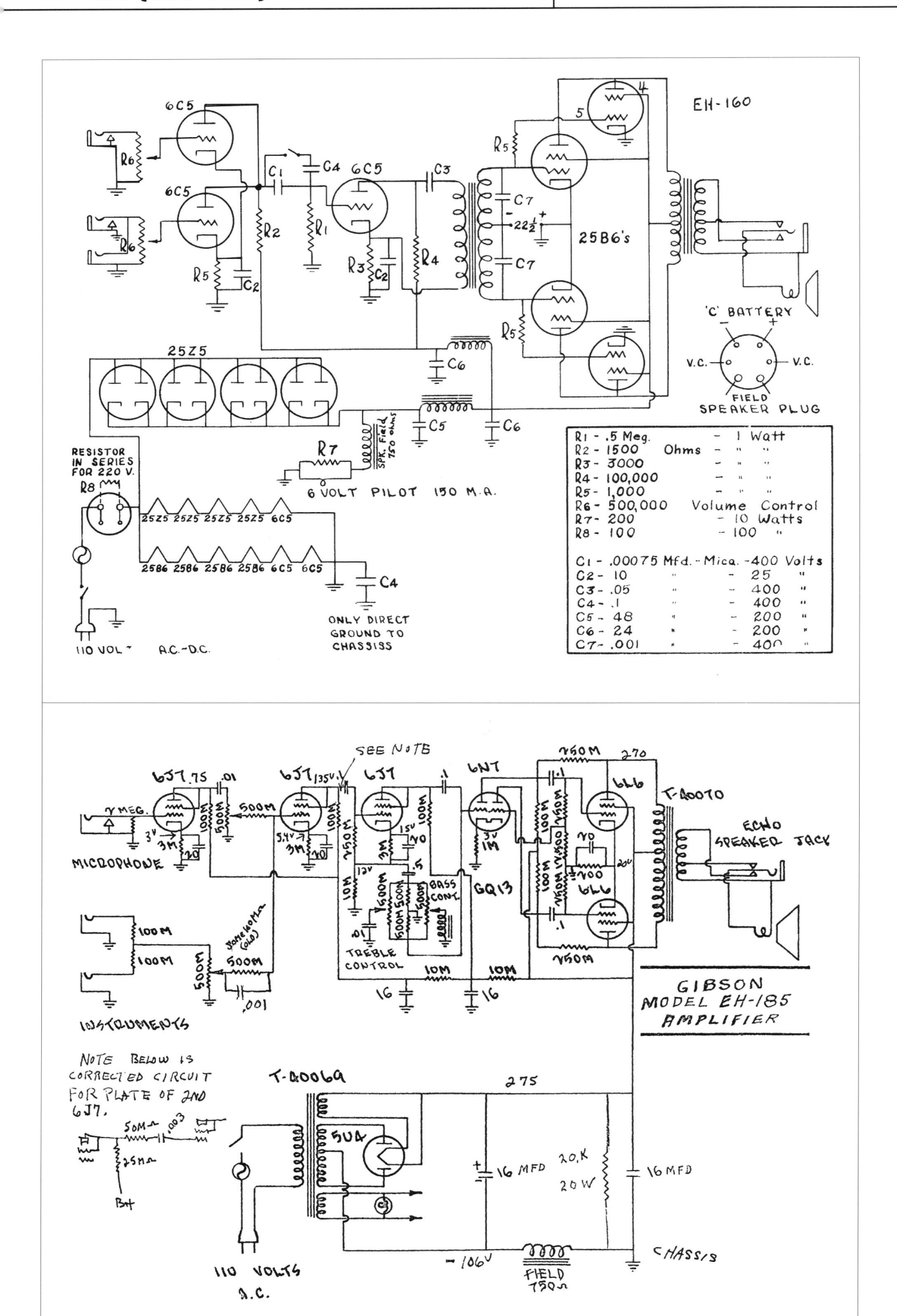

Gibson

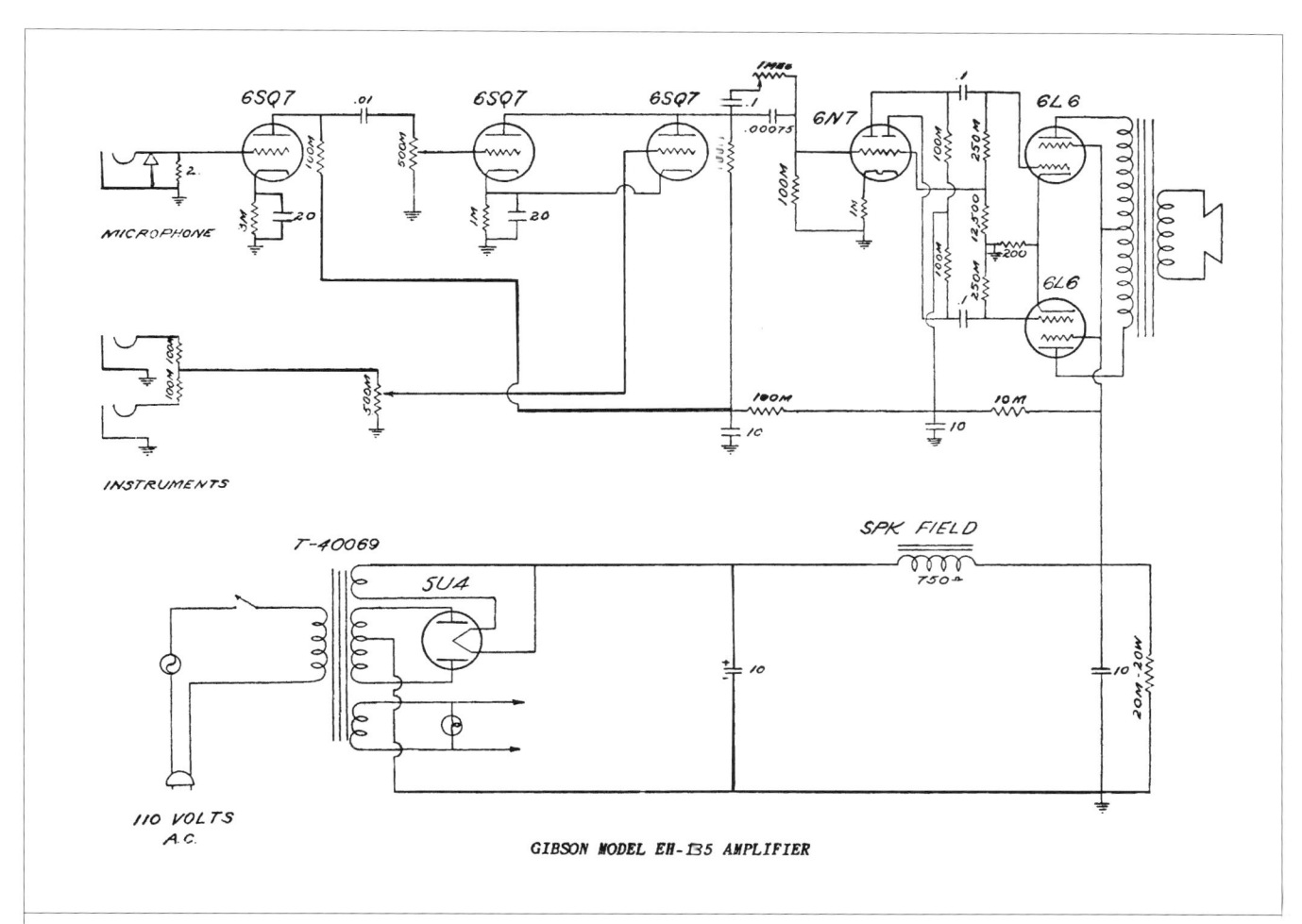

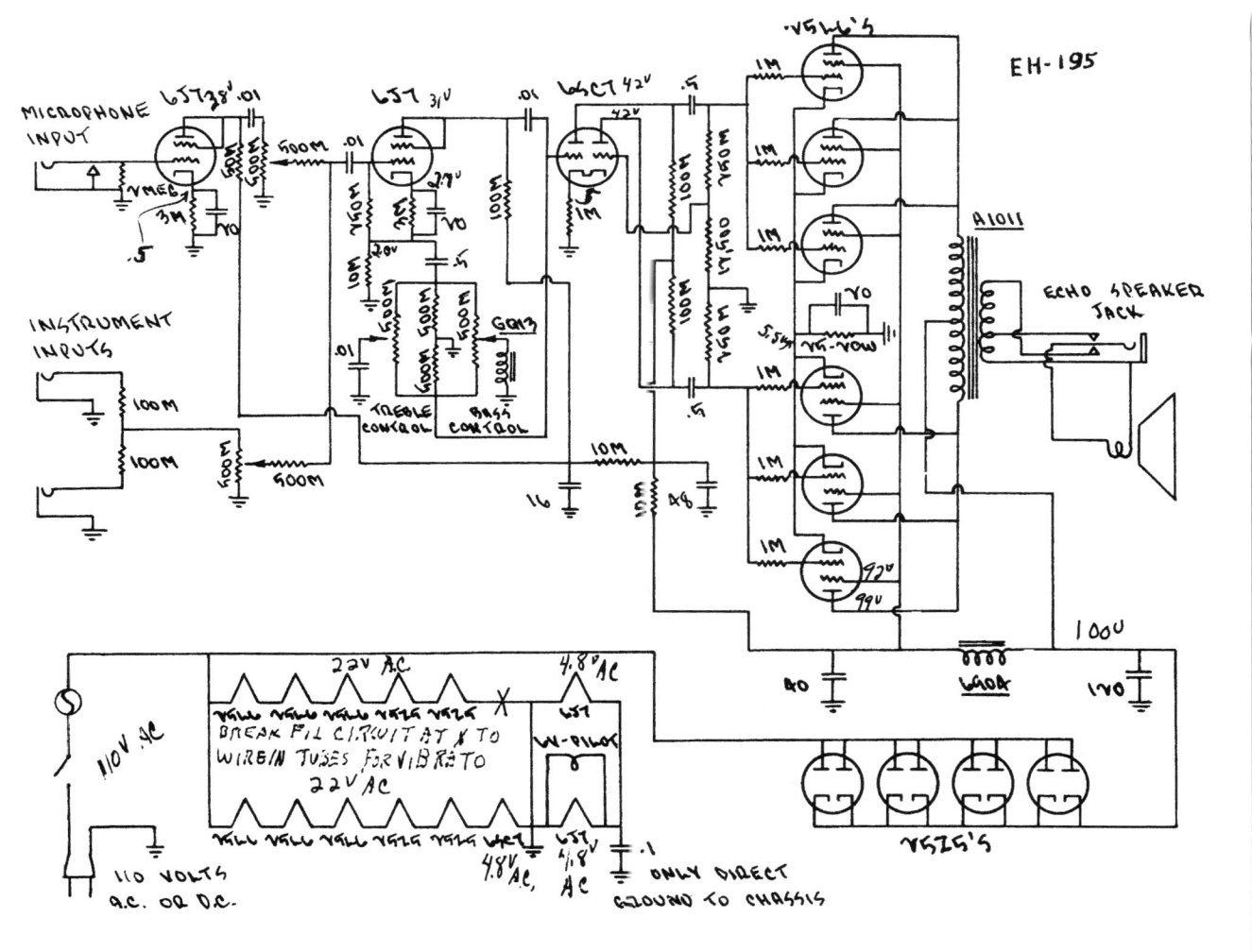

Gibson

EH-185 (6SQ7 PRE)
EH-195

Falcon GA2-RVT — Gibson

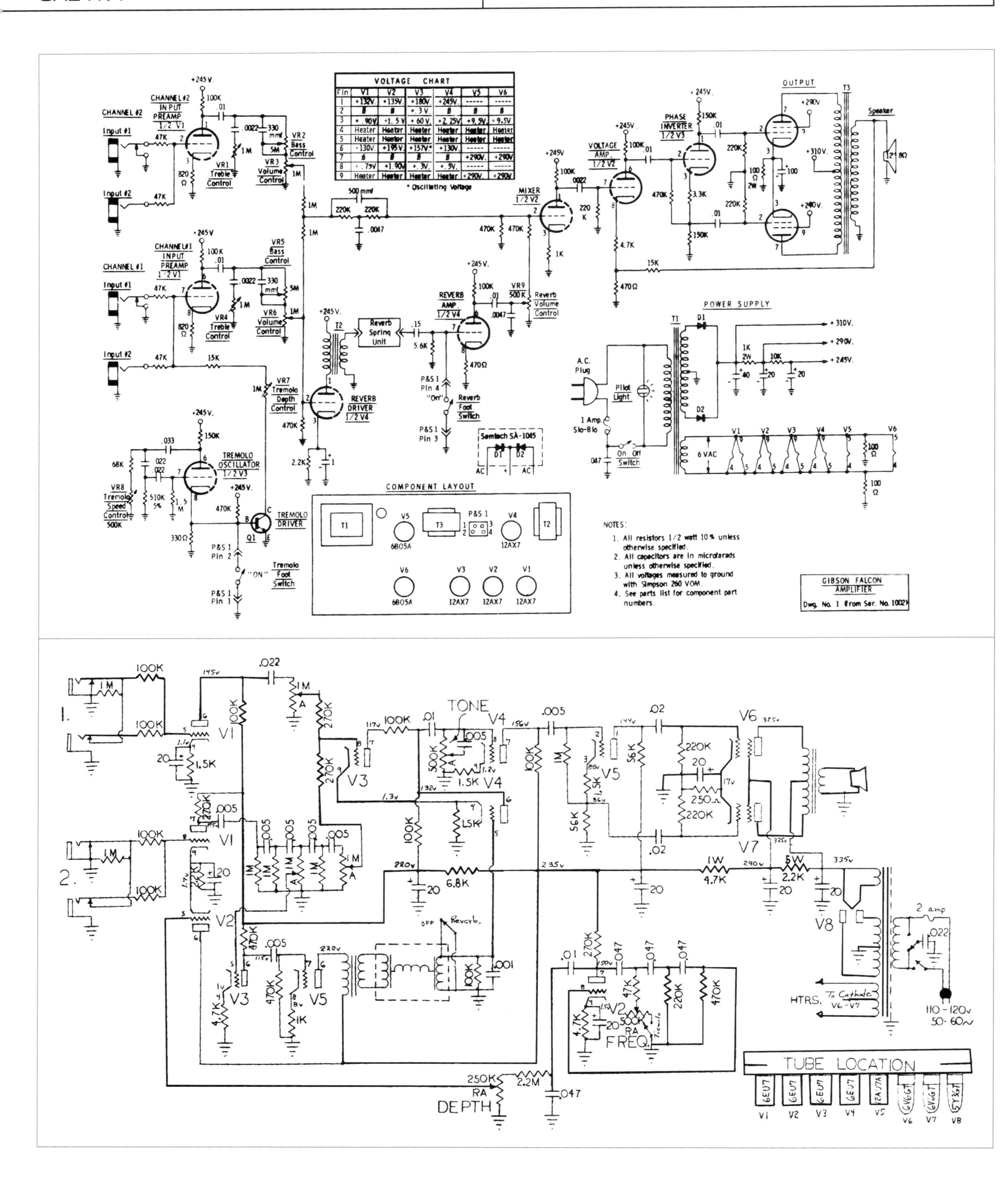

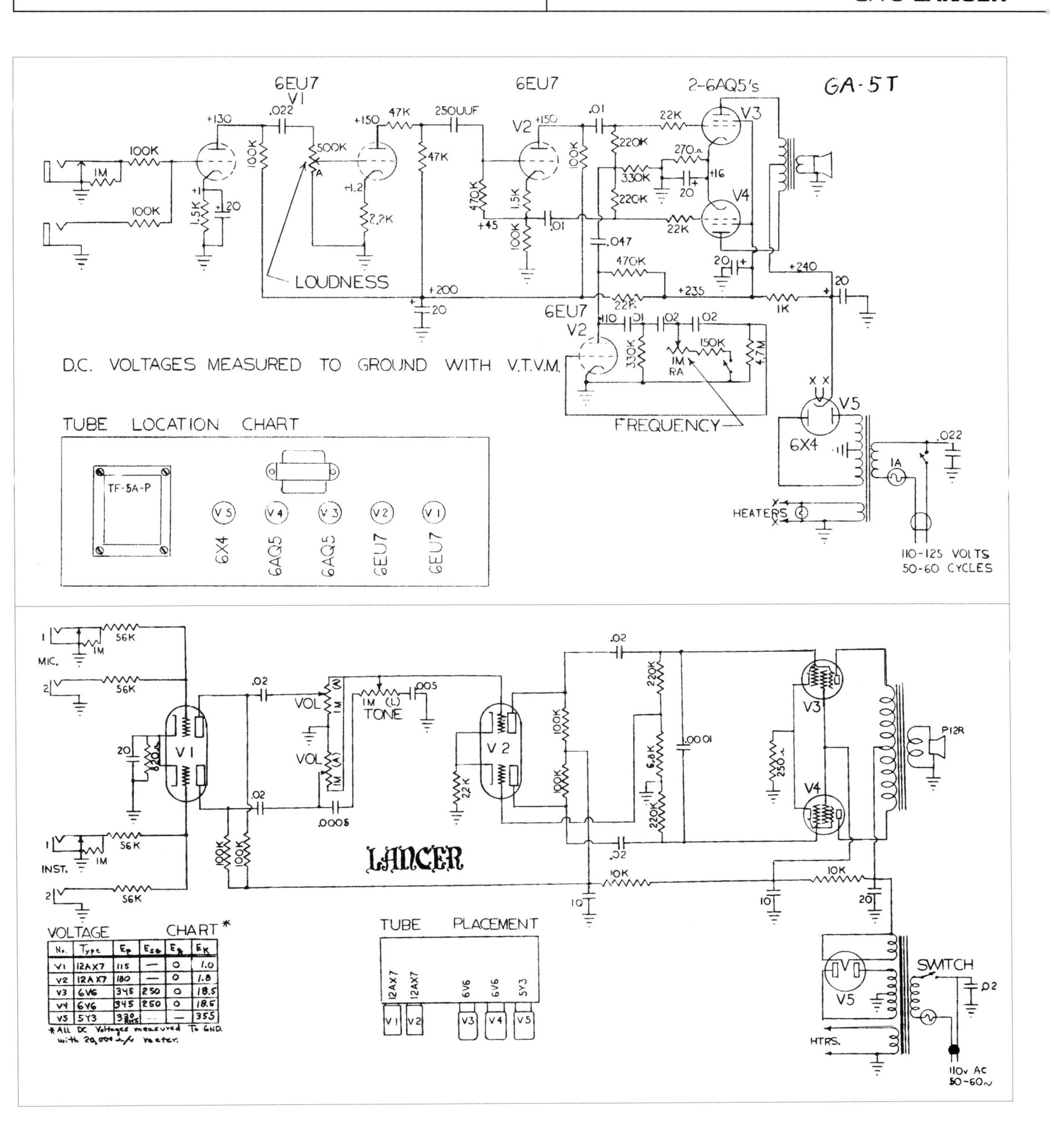

Gibson

GA-5T
GA-6 LANCER

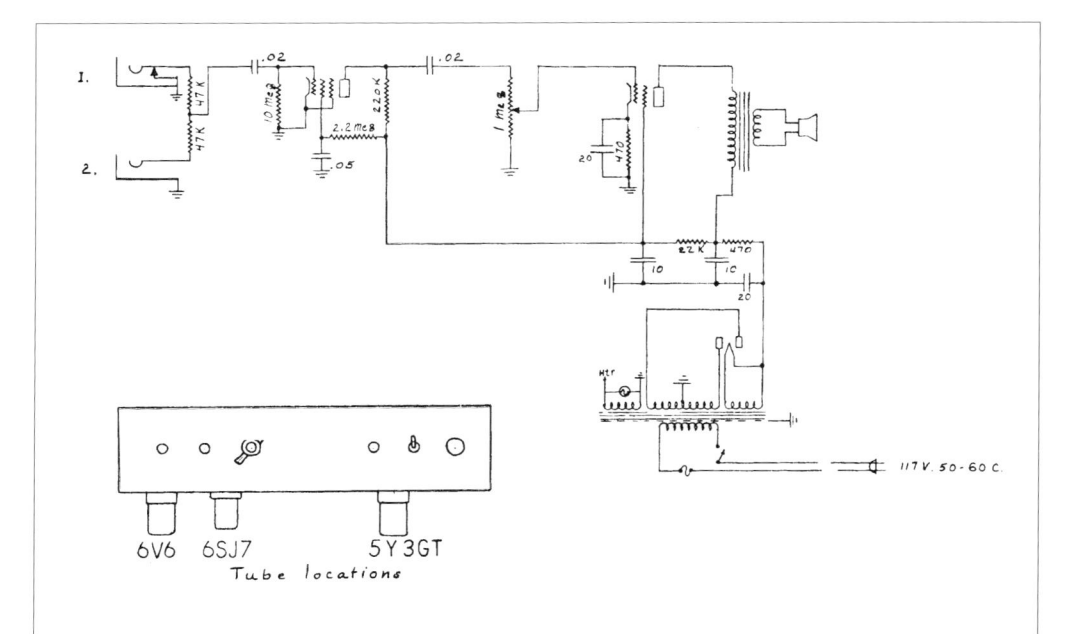

LES PAUL GA-5 EARLY
LES PAUL GA-5 LATER

Gibson

Les Paul

When only one instrument is used plug into #1 input jack.

This amplifier designed for 105-125 volt, 50-60 cycle current. Damage will result if connected to improper power source.

Use the above schematic to facilitate service by a reliable radio man.

Do not use higher rating fuse than one ampere, type 3 A.G.

This amplifier was carefully checked and in good playing condition when shipped. If damaged when received call transportation company immediately and place claim.

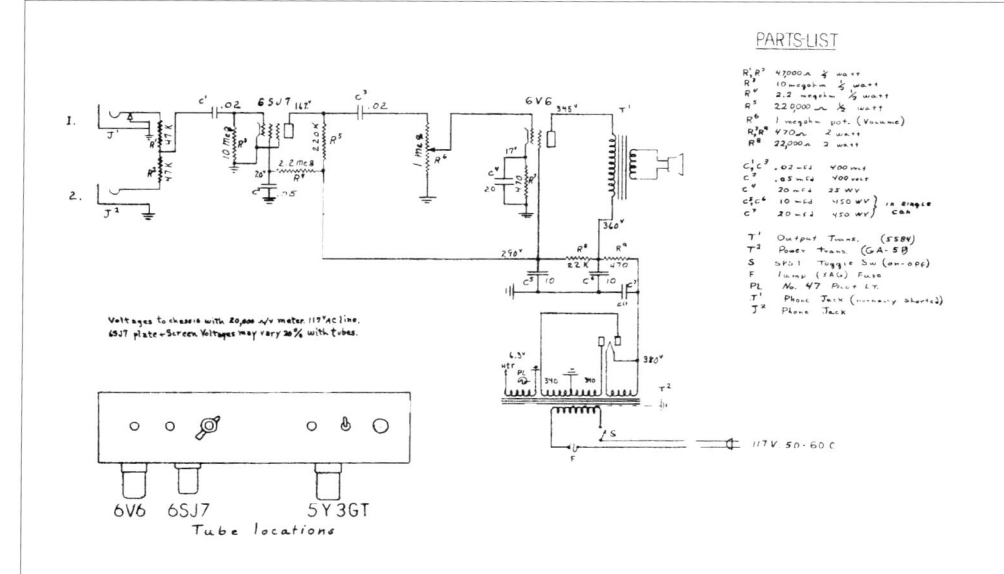

When only one instrument is used plug into #1 input jack.

This amplifier designed for 105-125 volt, 50-60 cycle current. Damage will result if connected to improper power source.

Use the above schematic to facilitate service by a reliable radio man.

Do not use higher rating fuse than one ampere, type 3 A.G.

This amplifier was carefully checked and in good playing condition when shipped. If damaged when received call transportation company immediately and place claim.

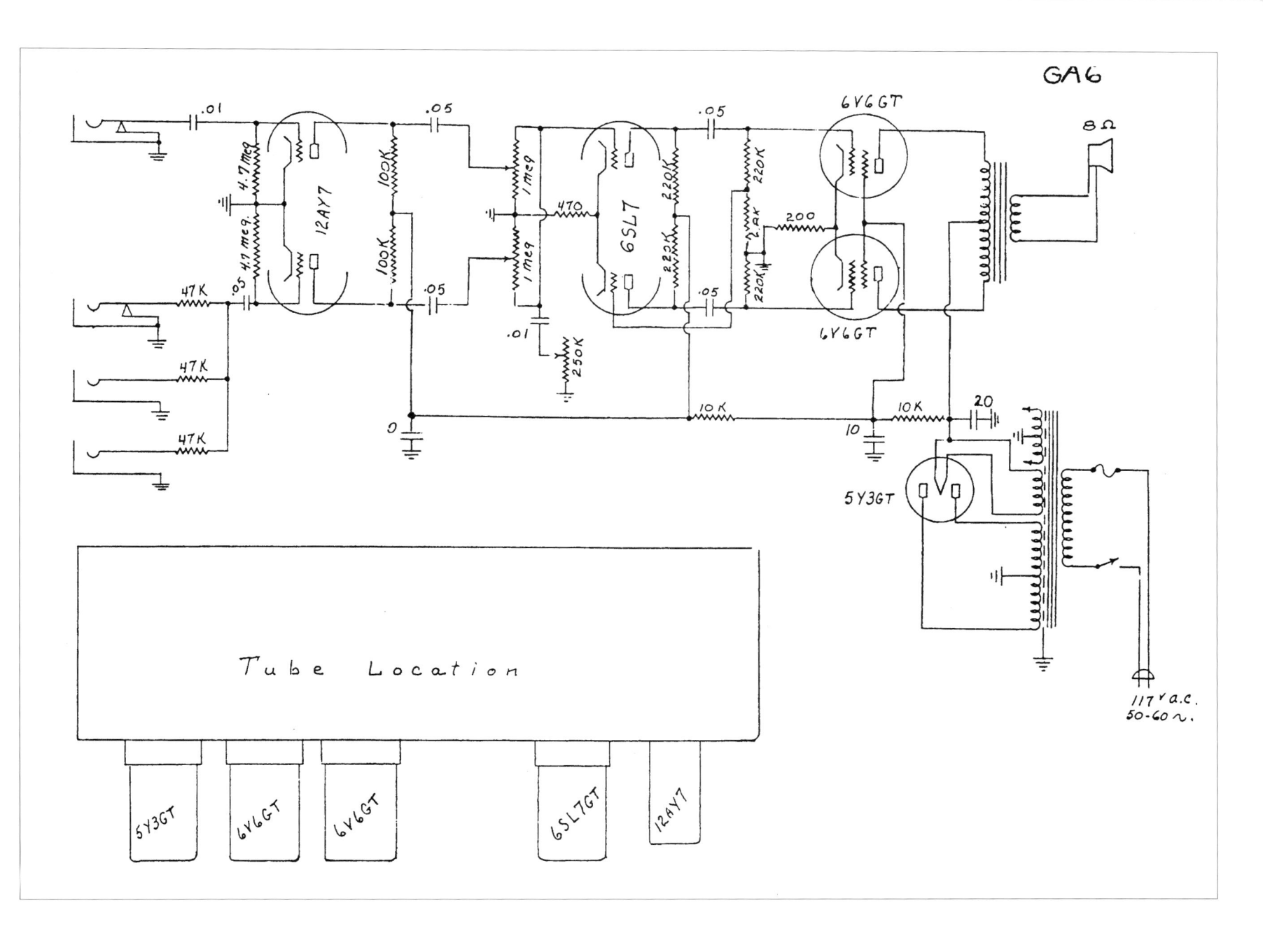

Gibson

GA-6

GA-8 6V6
GA-15RVT

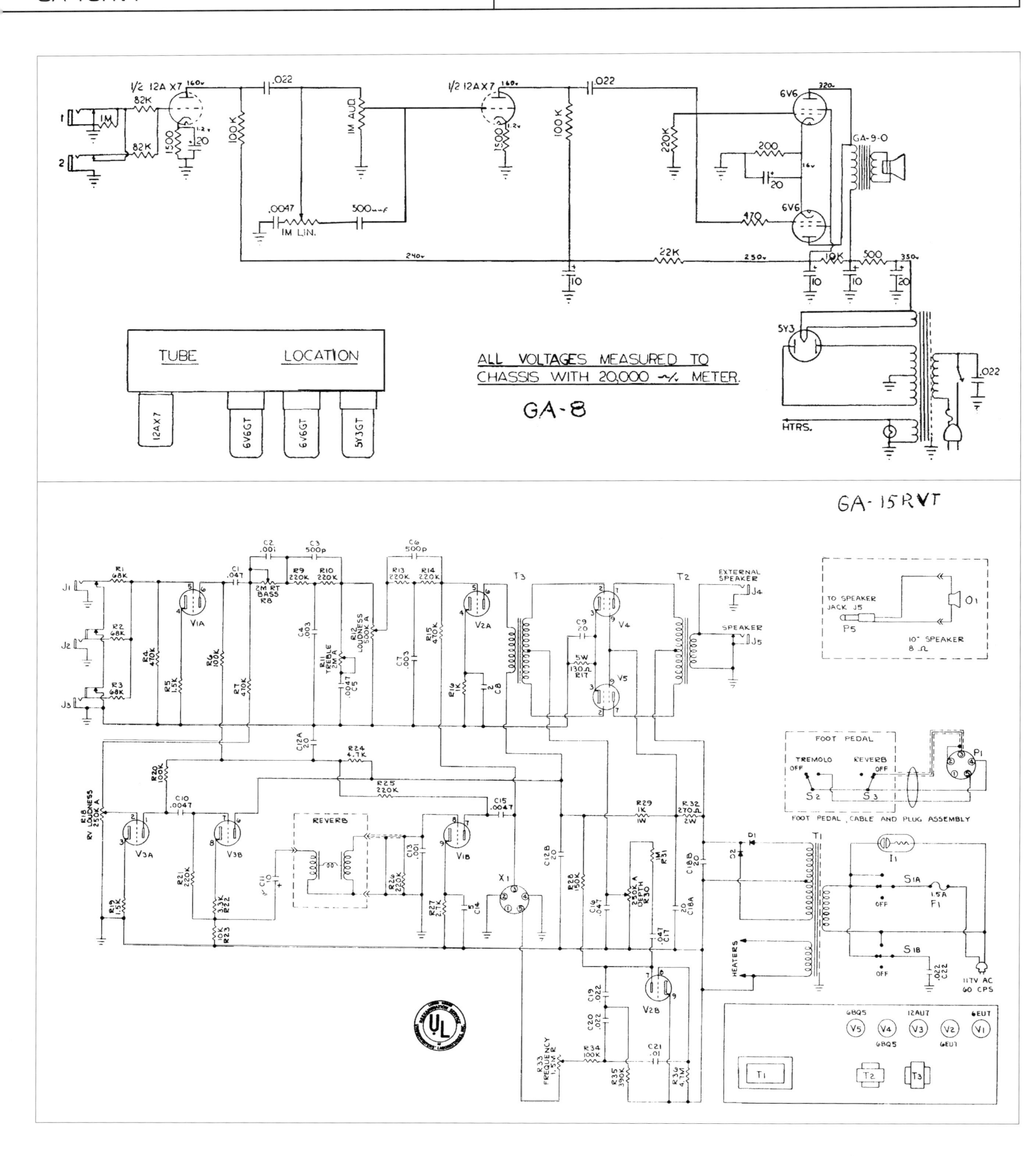

Gibson

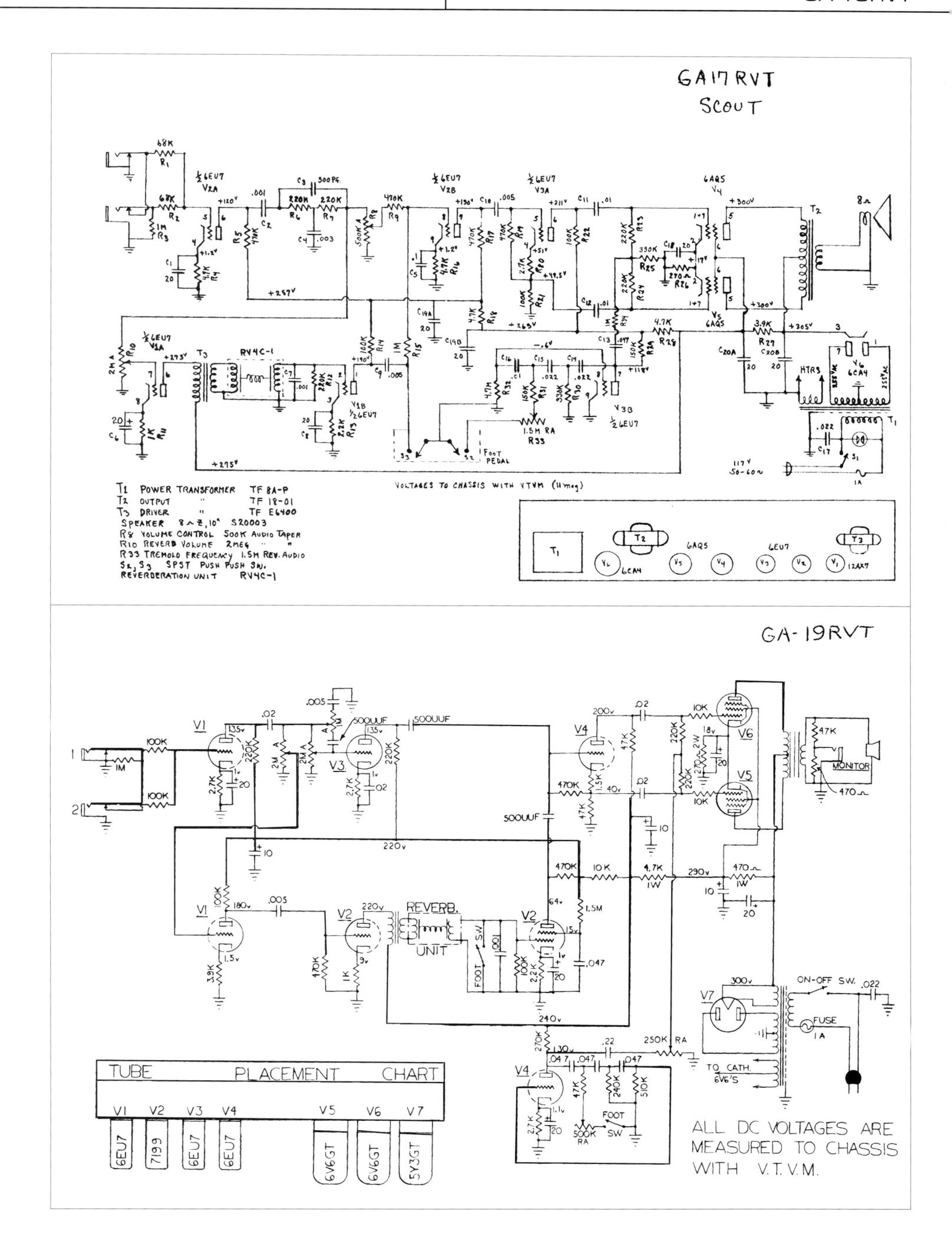

Gibson

GA-17RVT SCOUT
GA-19RVT

GA-20
GA-20T

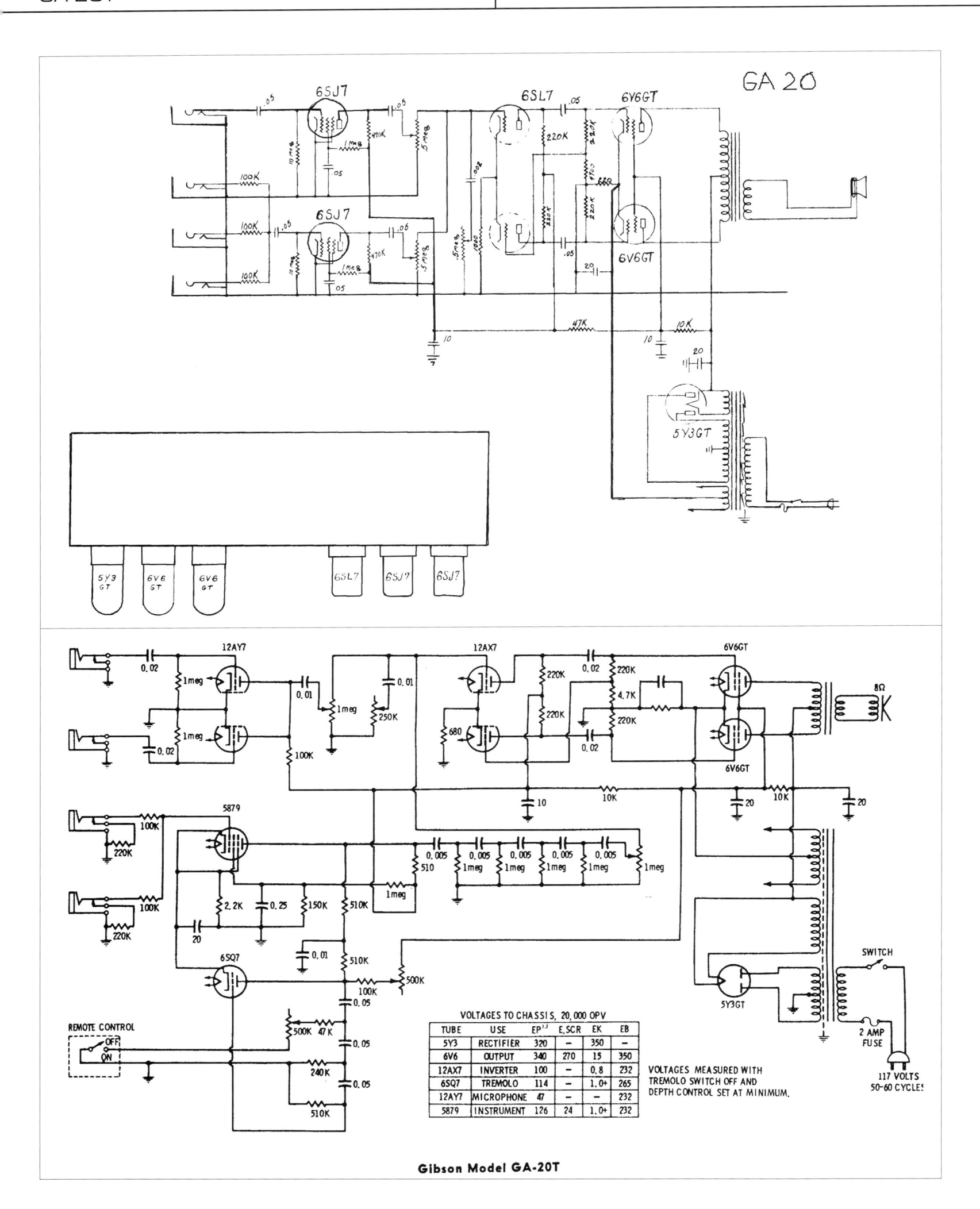

Gibson

Gibson Model GA-20T

100 THE TUBE AMP BOOK

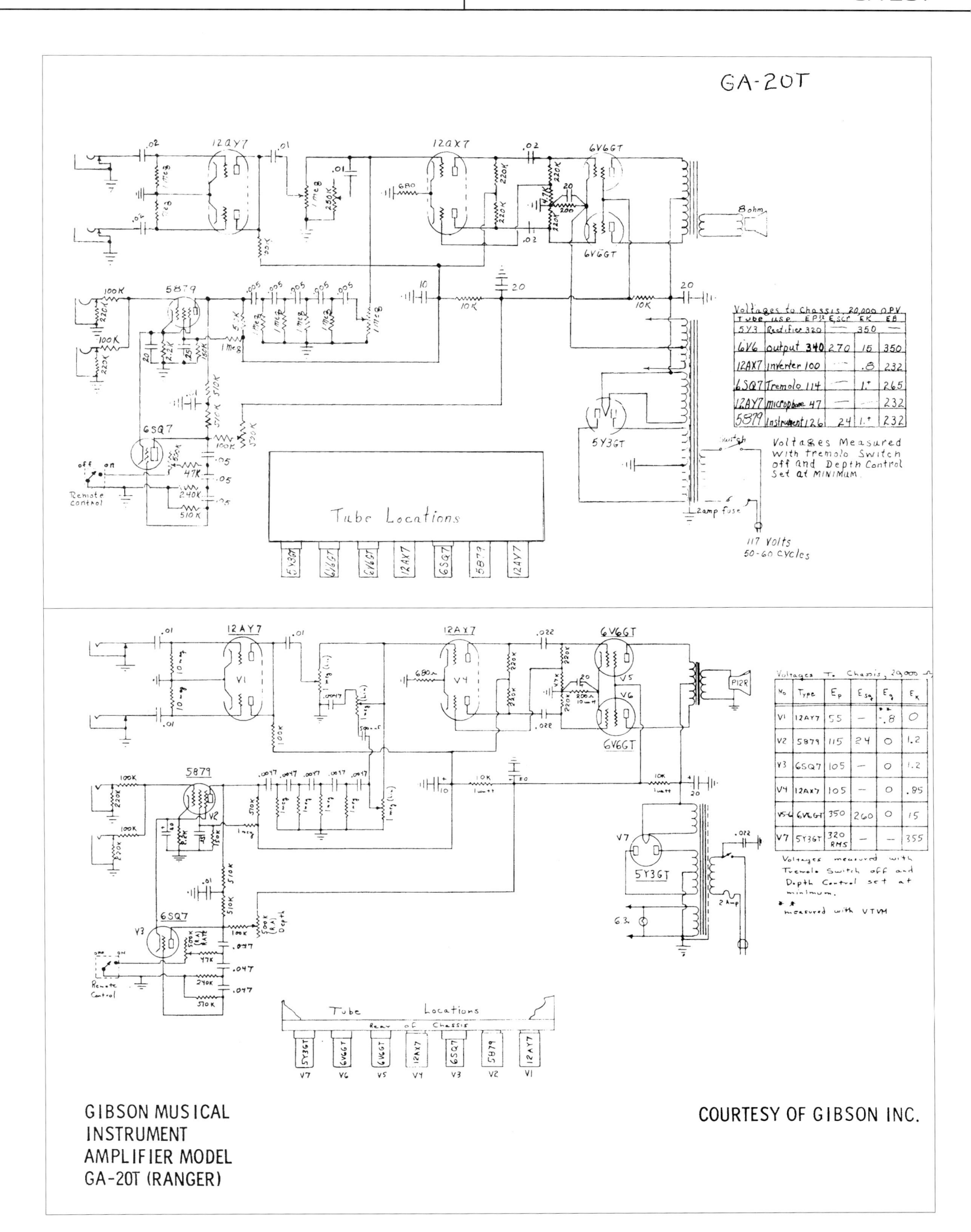

Gibson

GA-20T
GA-20T

GIBSON MUSICAL
INSTRUMENT
AMPLIFIER MODEL
GA-20T (RANGER)

COURTESY OF GIBSON INC.

GA-25RVT
GA-30 (6SJ7 PREAMP)

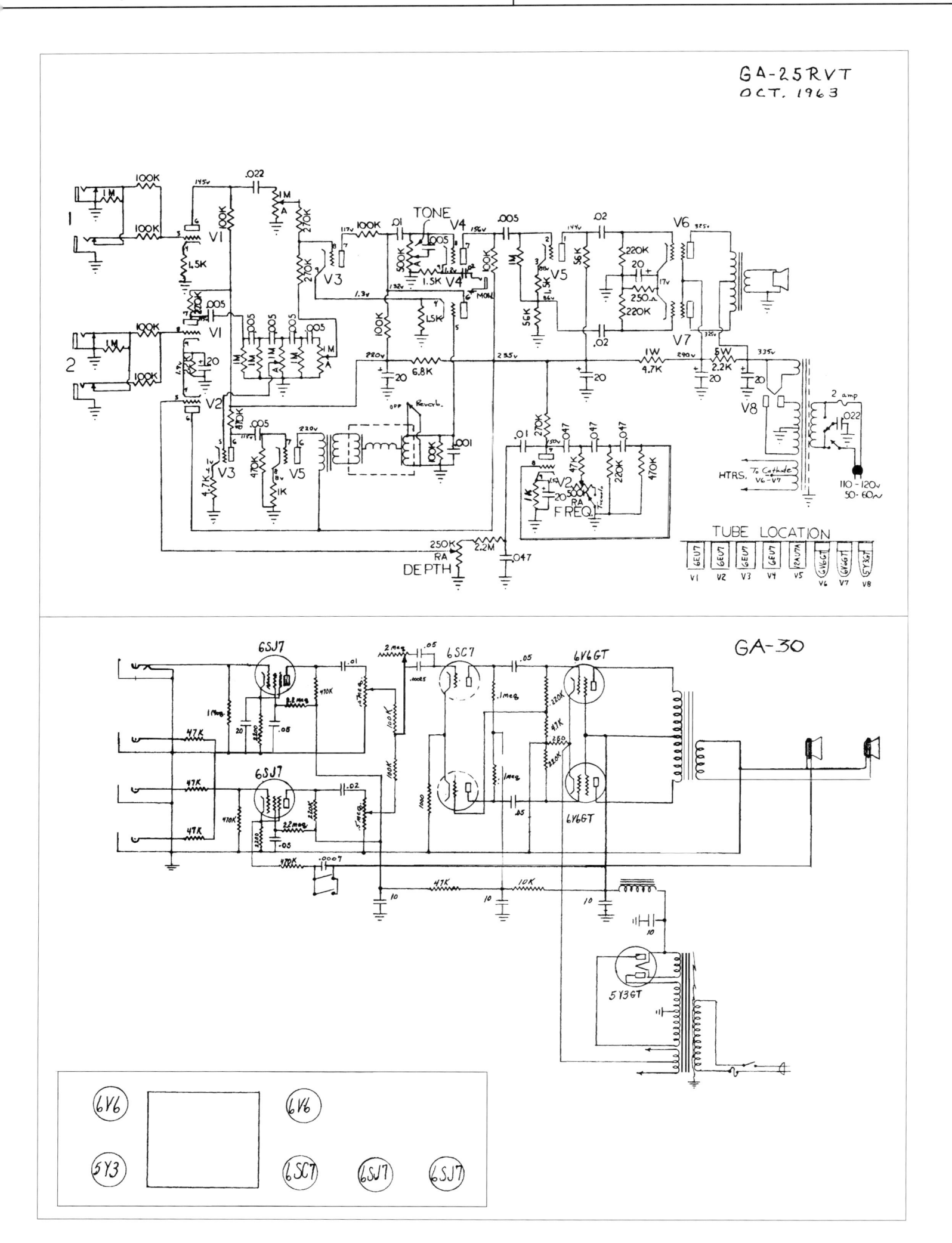

Gibson

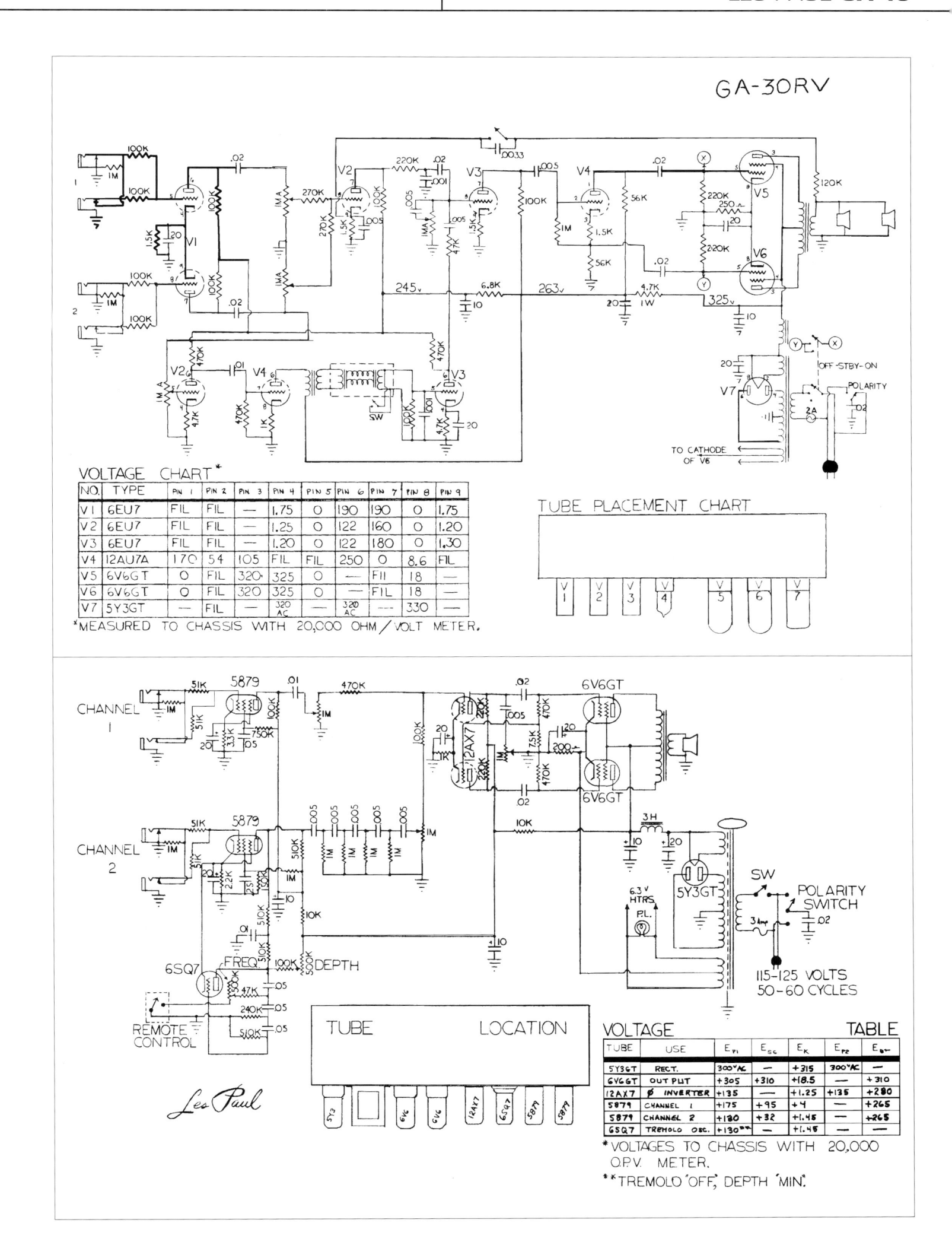

Gibson

GA-30RV
LES PAUL **GA-40**

GA-30RV

VOLTAGE CHART*

NO.	TYPE	PIN 1	PIN 2	PIN 3	PIN 4	PIN 5	PIN 6	PIN 7	PIN 8	PIN 9
V1	6EU7	FIL	FIL	—	1.75	0	190	190	0	1.75
V2	6EU7	FIL	FIL	—	1.25	0	122	160	0	1.20
V3	6EU7	FIL	FIL	—	1.20	0	122	180	0	1.30
V4	12AU7A	170	54	105	FIL	FIL	250	0	8.6	FIL
V5	6V6GT	0	FIL	320	325	0	—	FIL	18	
V6	6V6GT	0	FIL	320	325	0	—	FIL	18	
V7	5Y3GT	—	FIL	—	320 AC	—	320 AC	—	330	

*MEASURED TO CHASSIS WITH 20,000 OHM/VOLT METER.

TUBE PLACEMENT CHART

V1 V2 V3 V4 V5 V6 V7

Les Paul

TUBE LOCATION

5Y3 6V6 6V6 12AX7 6SQ7 5879 5879

VOLTAGE TABLE

TUBE	USE	E_{P1}	E_{SC}	E_K	E_{P2}	$E_{B^{++}}$
5Y3GT	RECT.	300vAC	—	+315	300vAC	—
6V6GT	OUTPUT	+305	+310	+18.5	—	+310
12AX7	∅ INVERTER	+135	—	+1.25	+135	+280
5879	CHANNEL 1	+175	+95	+4	—	+265
5879	CHANNEL 2	+180	+32	+1.45	—	+265
6SQ7	TREMOLO OSC.	+130**	—	+1.45		

* VOLTAGES TO CHASSIS WITH 20,000 O.P.V. METER.
** TREMOLO 'OFF', DEPTH 'MIN'.

THE TUBE AMP BOOK 103

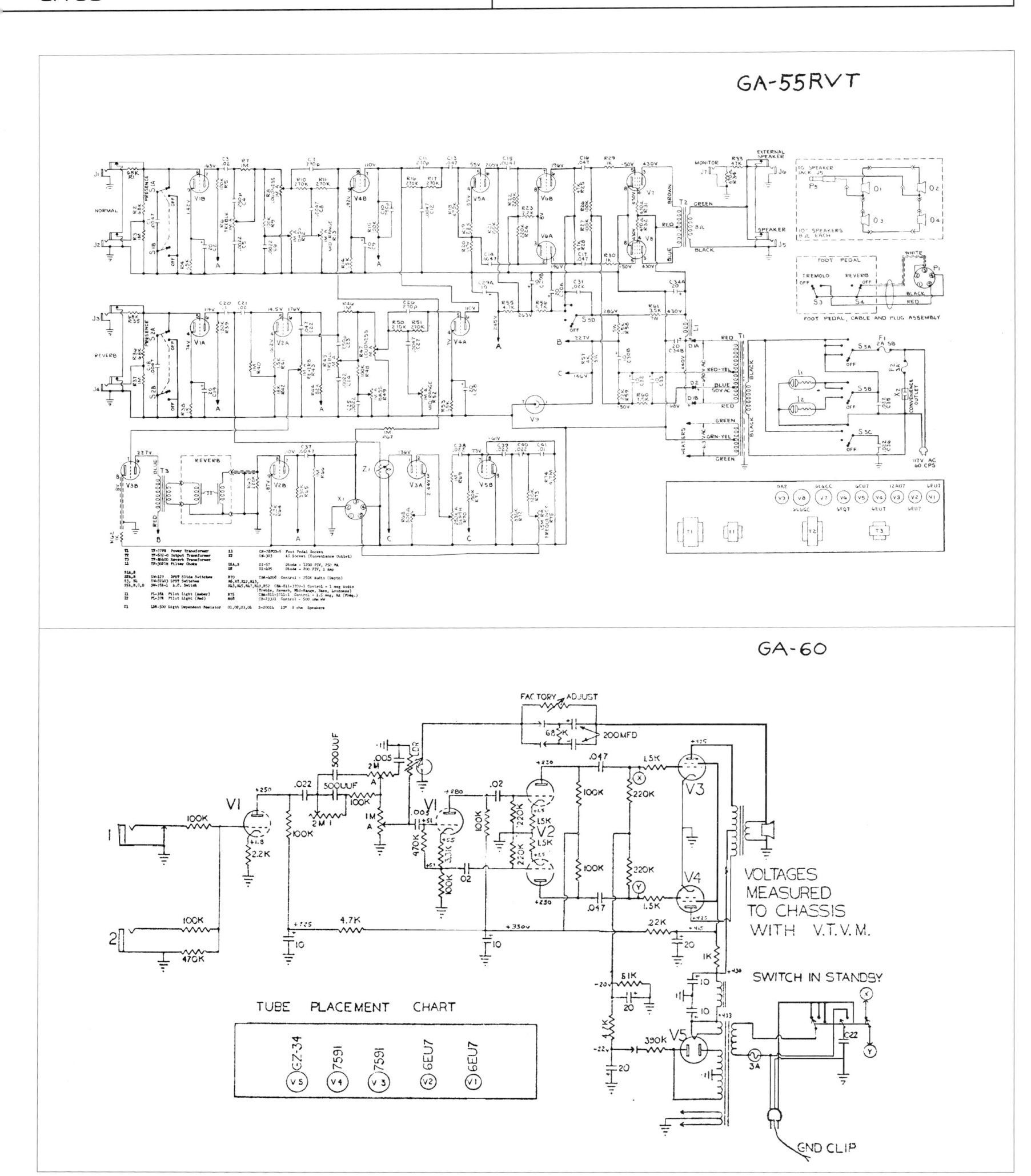

Gibson

GA-55RVT
GA-60

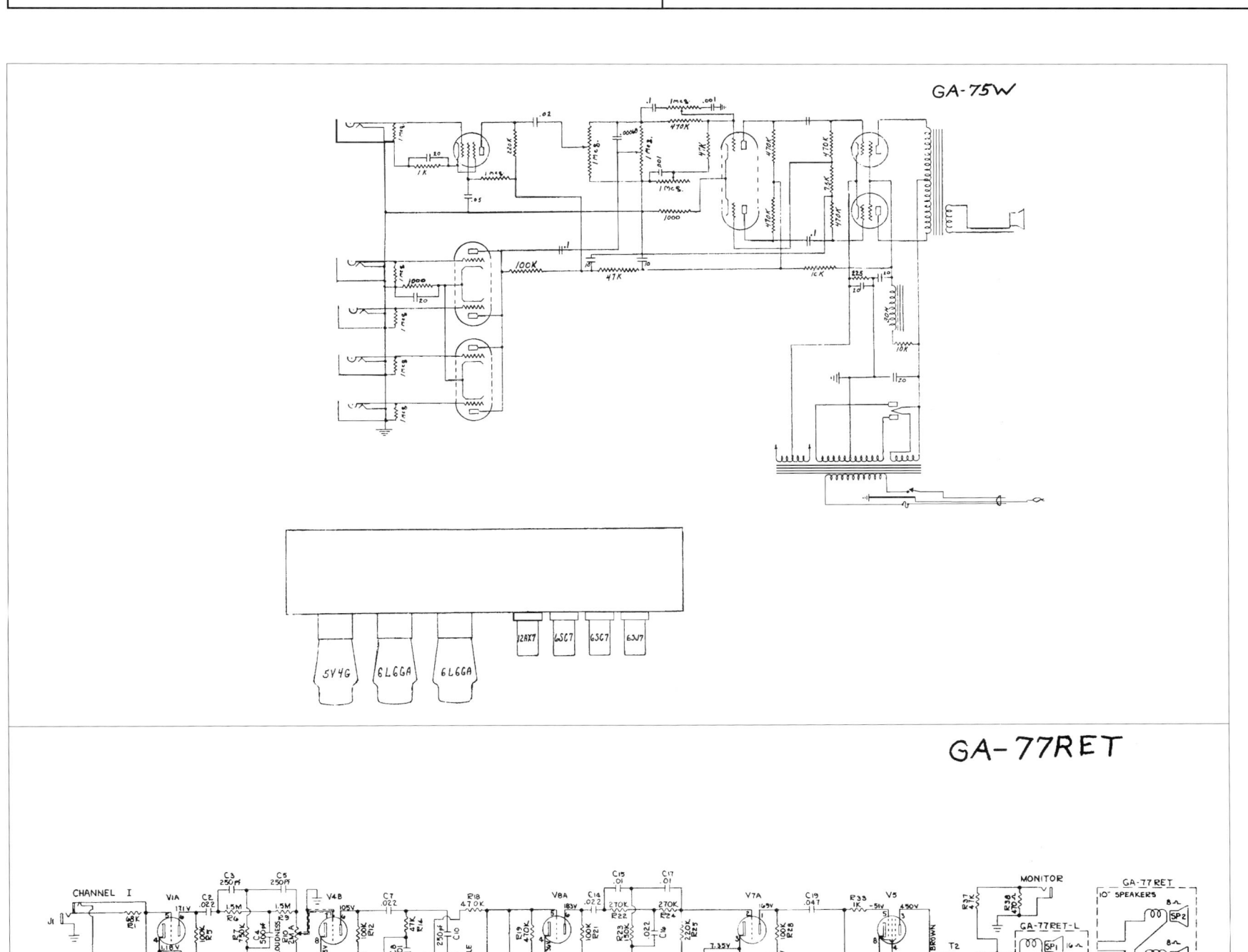

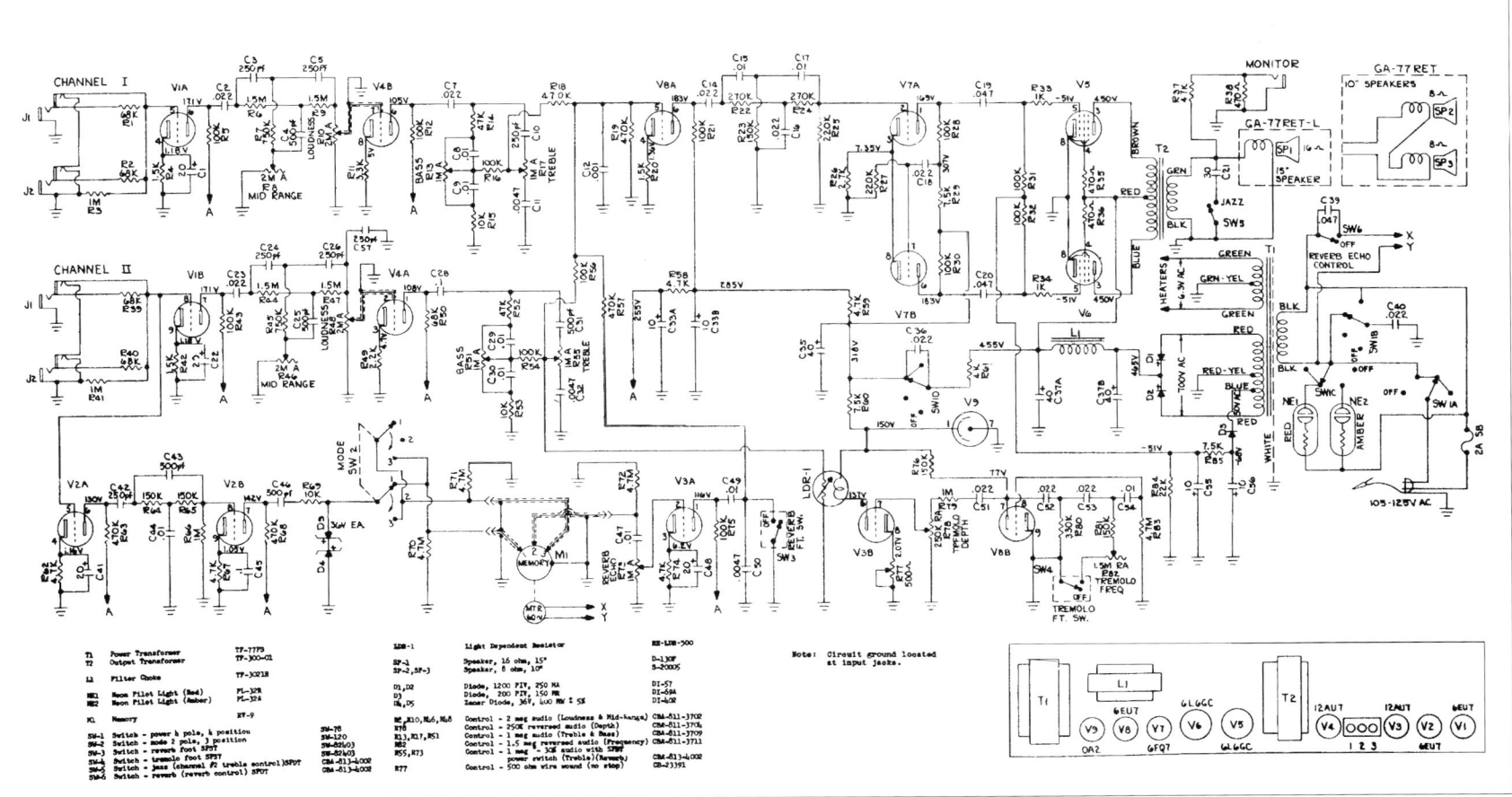

Gibson

GA-75W
GA-77RET

GA-83S STEREO-VIB (PREAMP) — Gibson

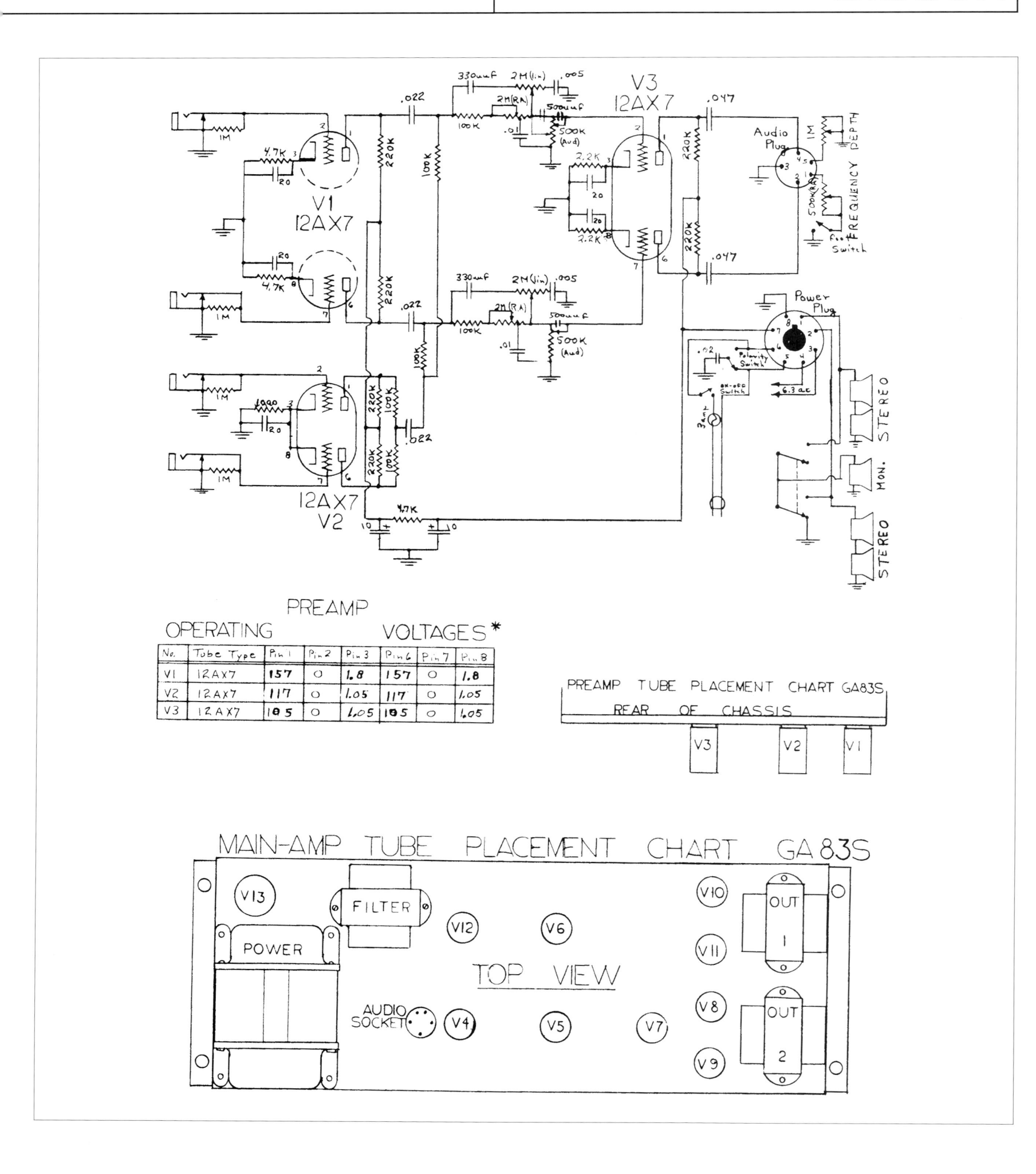

PREAMP OPERATING VOLTAGES*

No.	Tube Type	Pin 1	Pin 2	Pin 3	Pin 6	Pin 7	Pin 8
V1	12AX7	157	0	1.8	157	0	1.8
V2	12AX7	117	0	1.05	117	0	1.05
V3	12AX7	105	0	1.05	105	0	1.05

PREAMP TUBE PLACEMENT CHART GA83S — REAR OF CHASSIS: V3, V2, V1

MAIN-AMP TUBE PLACEMENT CHART GA83S (TOP VIEW)

Gibson

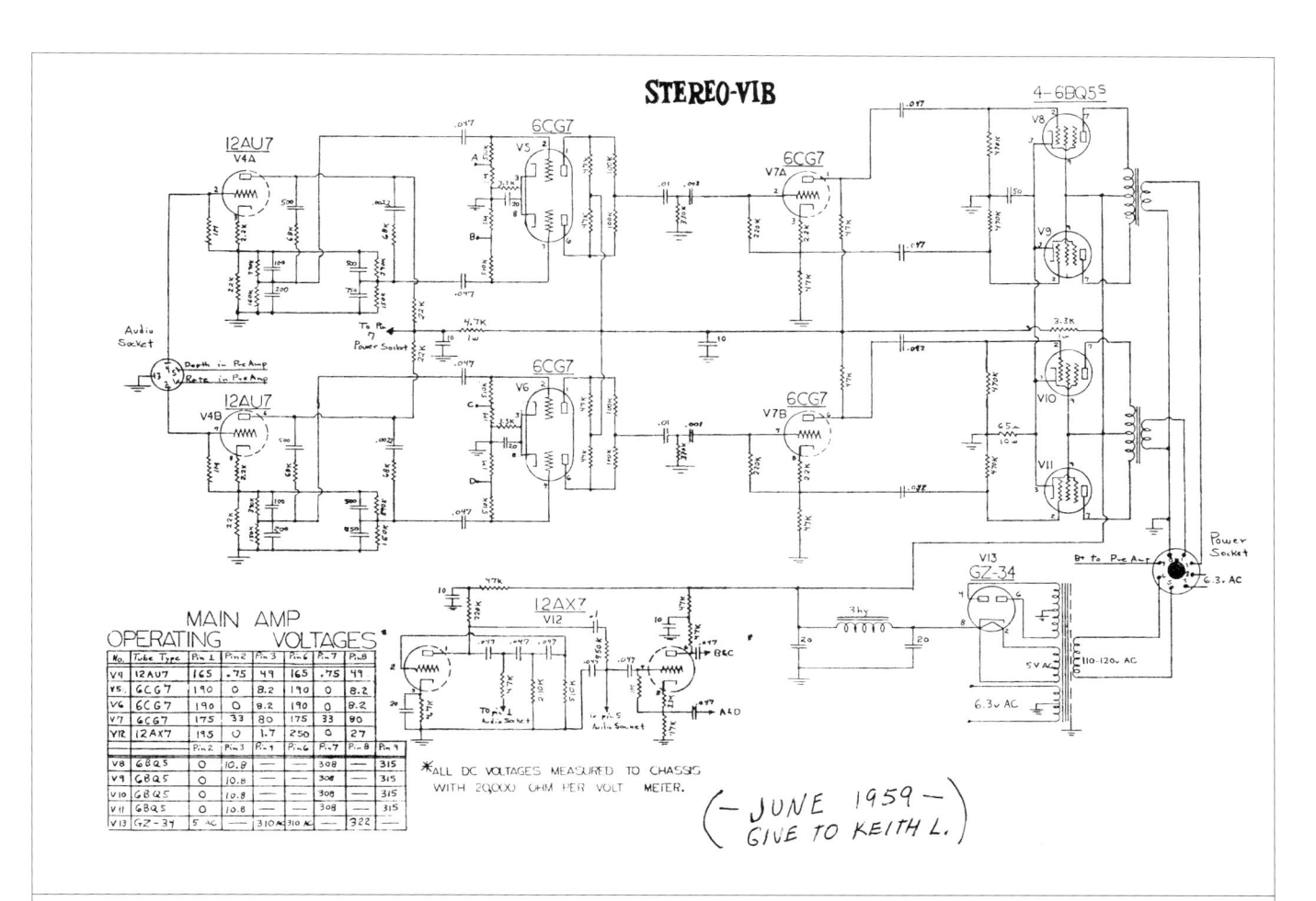

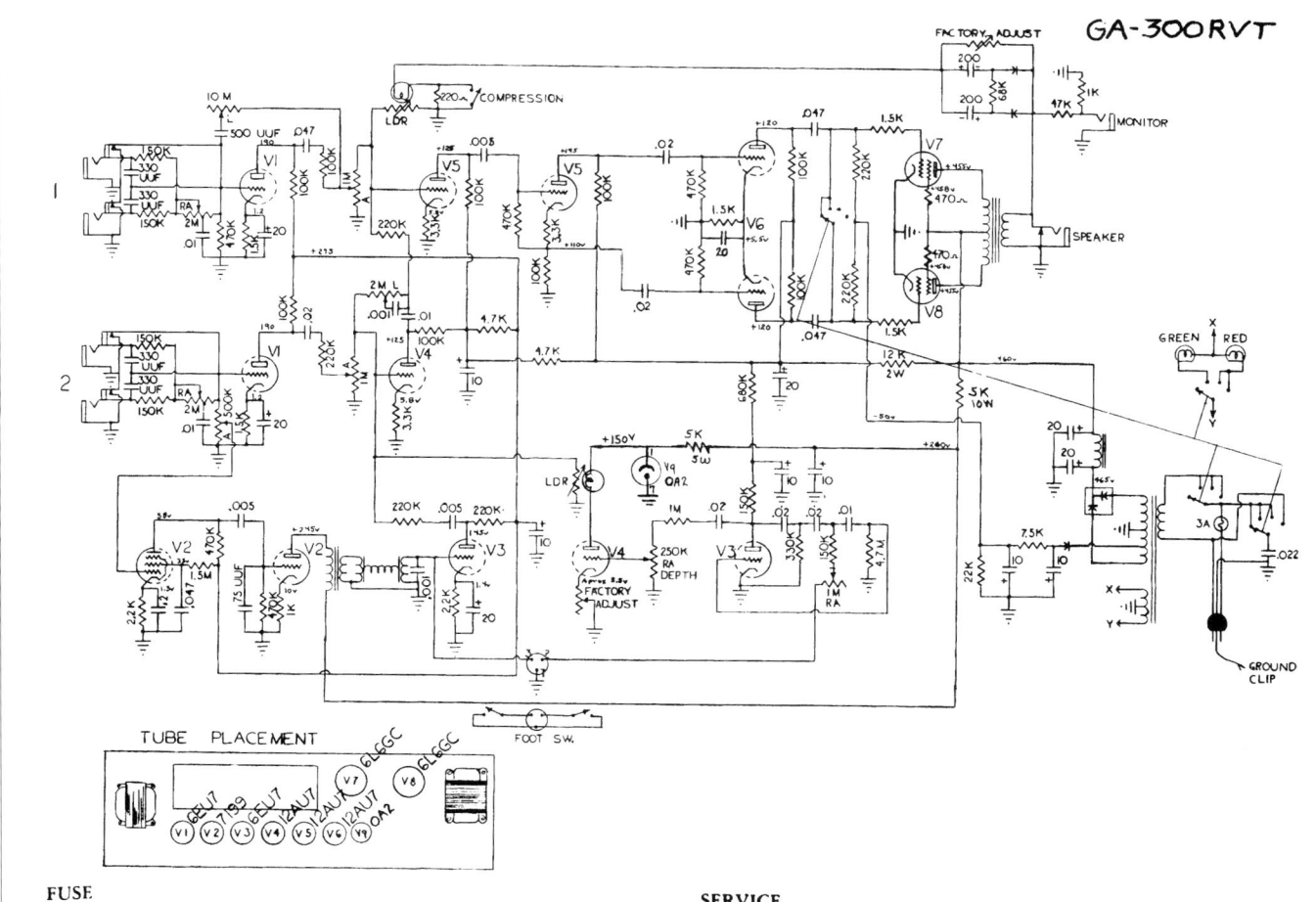

GA83S STEREO-VIB
GA-300RVT

FUSE
The fuse used in this Amplifier is a type 3AG of three amperes rating.
DO NOT USE FUSES OF HIGHER RATING

SERVICE
If the amplifier is in need of servicing, it should be taken to a reliable radio man. The electrical diagram in this folder should be shown the repairman to assist him in servicing the amplifier.

HAWK
REVERB-12

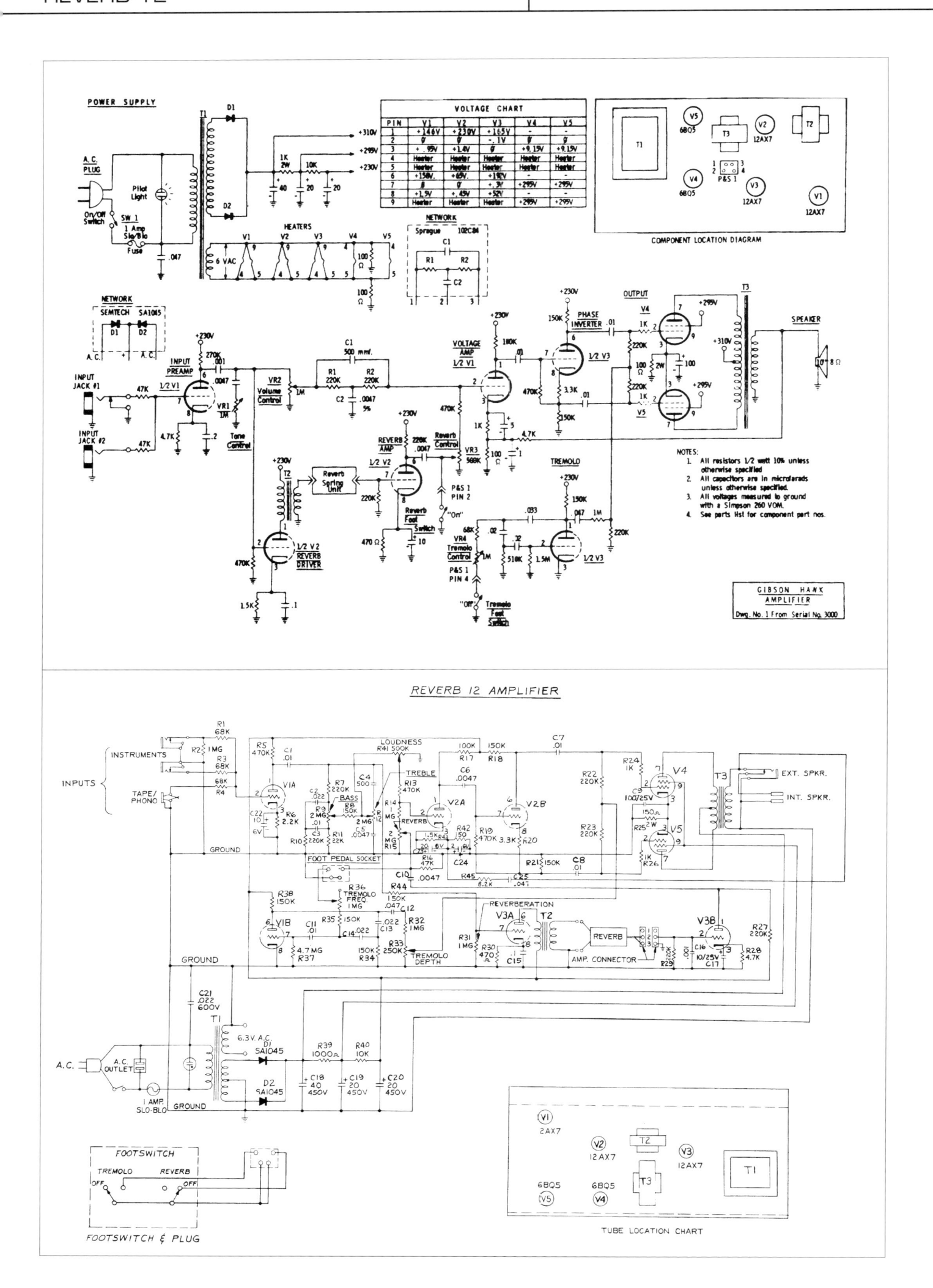

Gibson

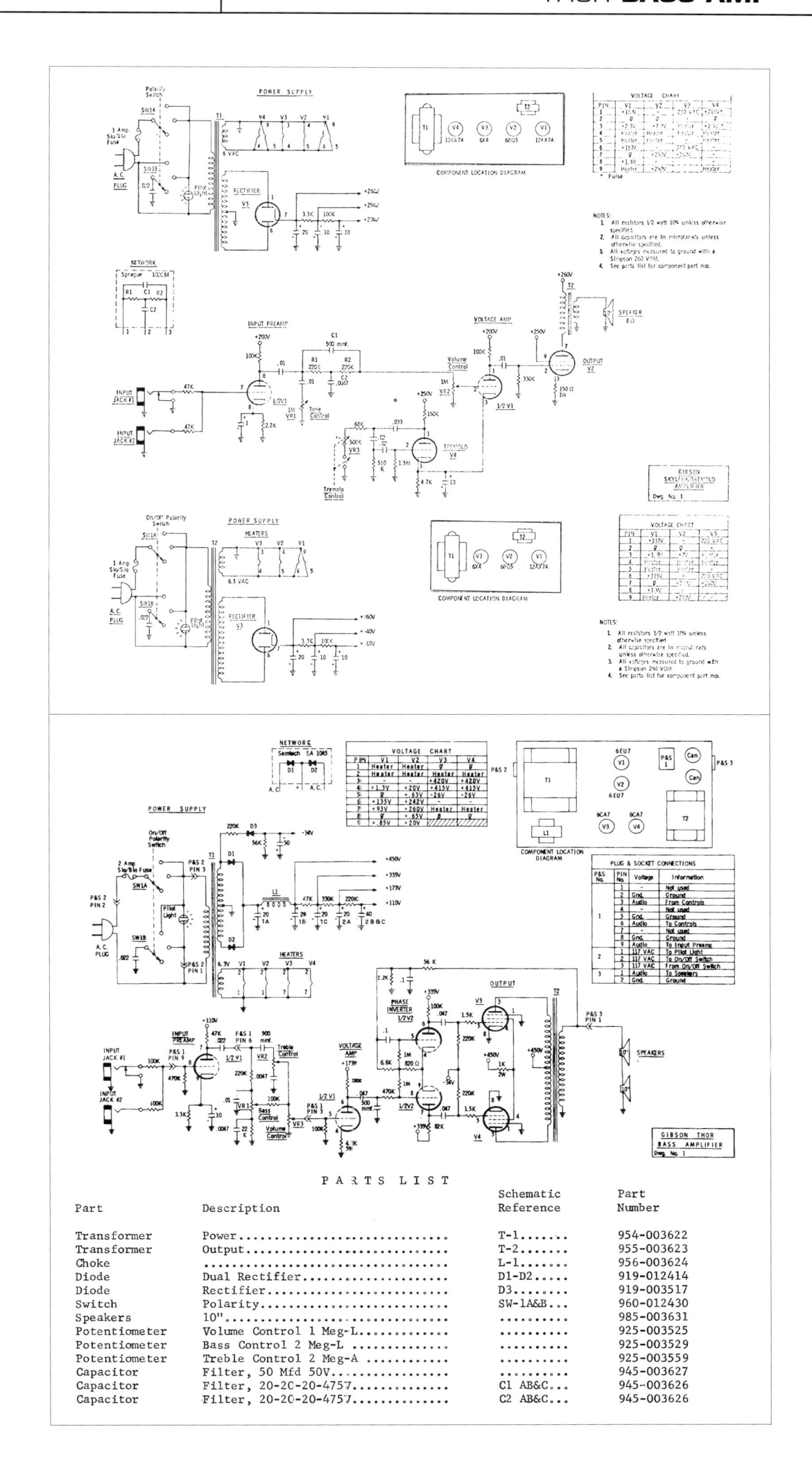

Gibson

SKYLARK/TREMOLO
THOR BASS AMP

PARTS LIST

Part	Description	Schematic Reference	Part Number
Transformer	Power	T-1	954-003622
Transformer	Output	T-2	955-003623
Choke		L-1	956-003624
Diode	Dual Rectifier	D1-D2	919-012414
Diode	Rectifier	D3	919-003517
Switch	Polarity	SW-1A&B	960-012430
Speakers	10"		985-003631
Potentiometer	Volume Control 1 Meg-L		925-003525
Potentiometer	Bass Control 2 Meg-L		925-003529
Potentiometer	Treble Control 2 Meg-A		925-003559
Capacitor	Filter, 50 Mfd 50V		945-003627
Capacitor	Filter, 20-20-20-4757	C1 AB&C	945-003626
Capacitor	Filter, 20-20-20-4757	C2 AB&C	945-003626

THE TUBE AMP BOOK 109

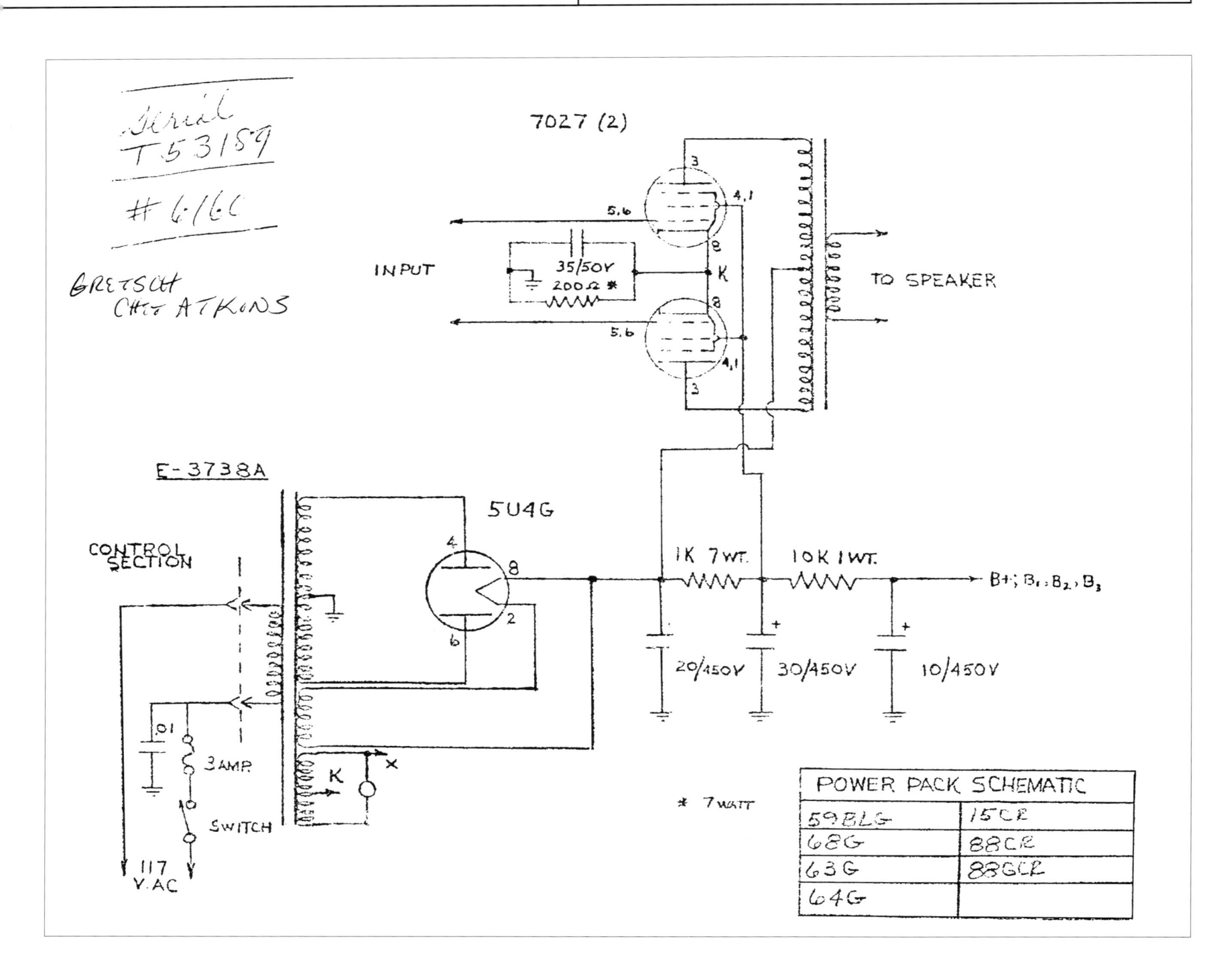

Gretsch

CHET ATKINS

GT Electronics

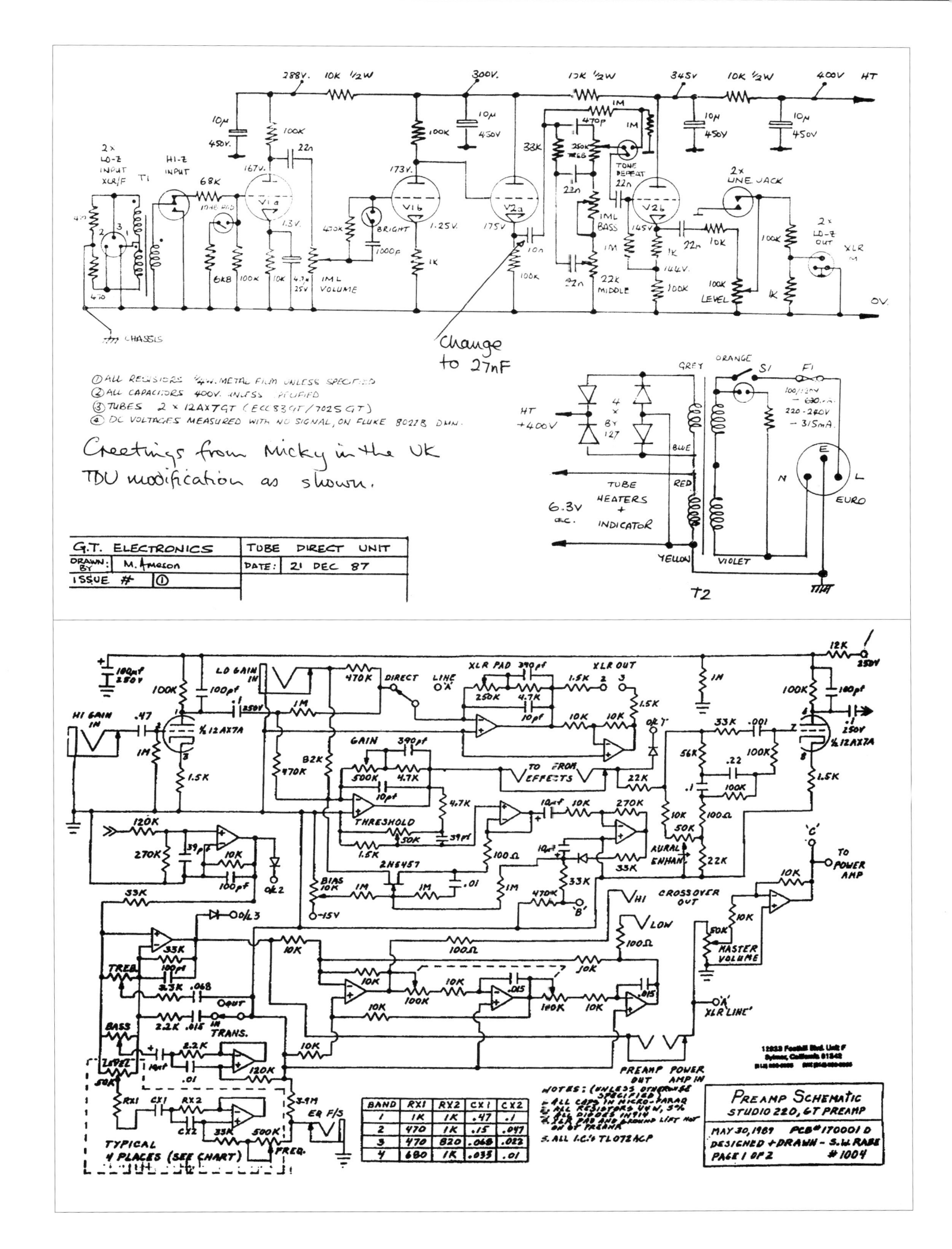

**TUBE DIRECT UNIT
STUDIO 220 PREAMP**

ORGAN REVERB AO-35 (EARLY A-100)
ORGAN AO-39

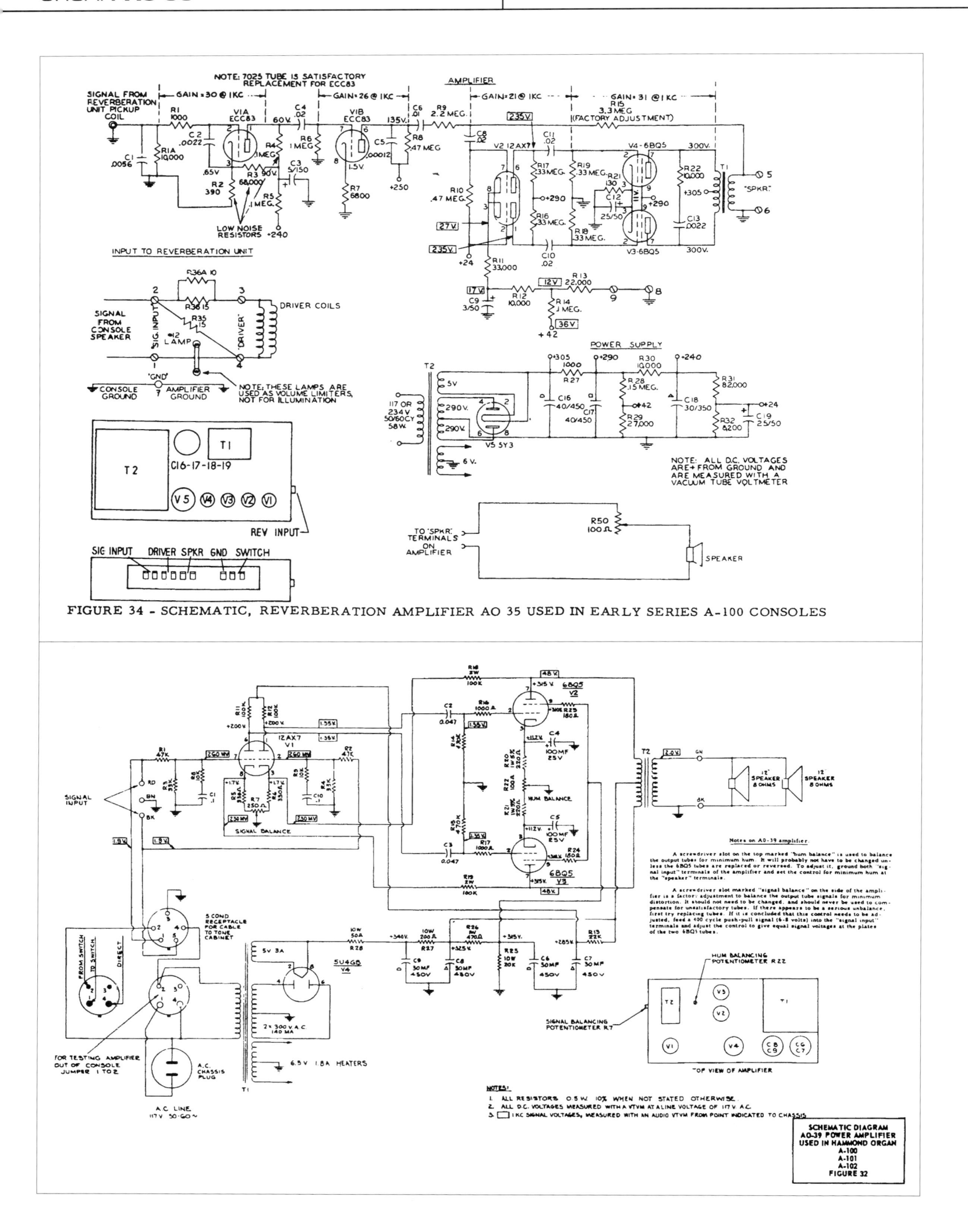

Hammond

FIGURE 34 - SCHEMATIC, REVERBERATION AMPLIFIER AO 35 USED IN EARLY SERIES A-100 CONSOLES

SCHEMATIC DIAGRAM
AO-39 POWER AMPLIFIER
USED IN HAMMOND ORGAN
A-100
A-101
A-102
FIGURE 32

Hammond

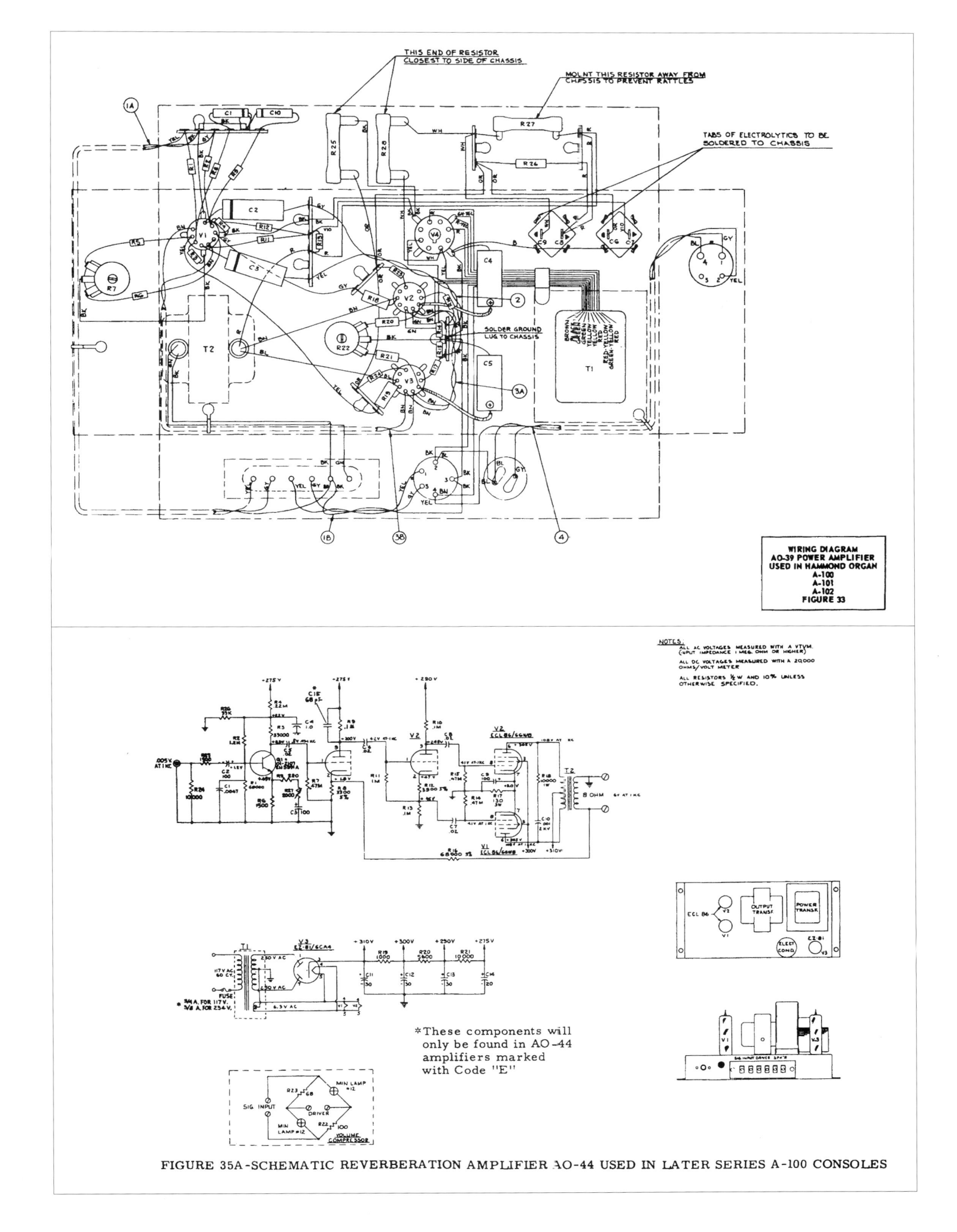

PWR-AMP AO-39 WIRING
ORGAN REVERB AO-44 EXP (A-100)

WIRING DIAGRAM
AO-39 POWER AMPLIFIER
USED IN HAMMOND ORGAN
A-100
A-101
A-102
FIGURE 33

*These components will only be found in AO-44 amplifiers marked with Code "E"

FIGURE 35A - SCHEMATIC REVERBERATION AMPLIFIER AO-44 USED IN LATER SERIES A-100 CONSOLES

THE TUBE AMP BOOK 113

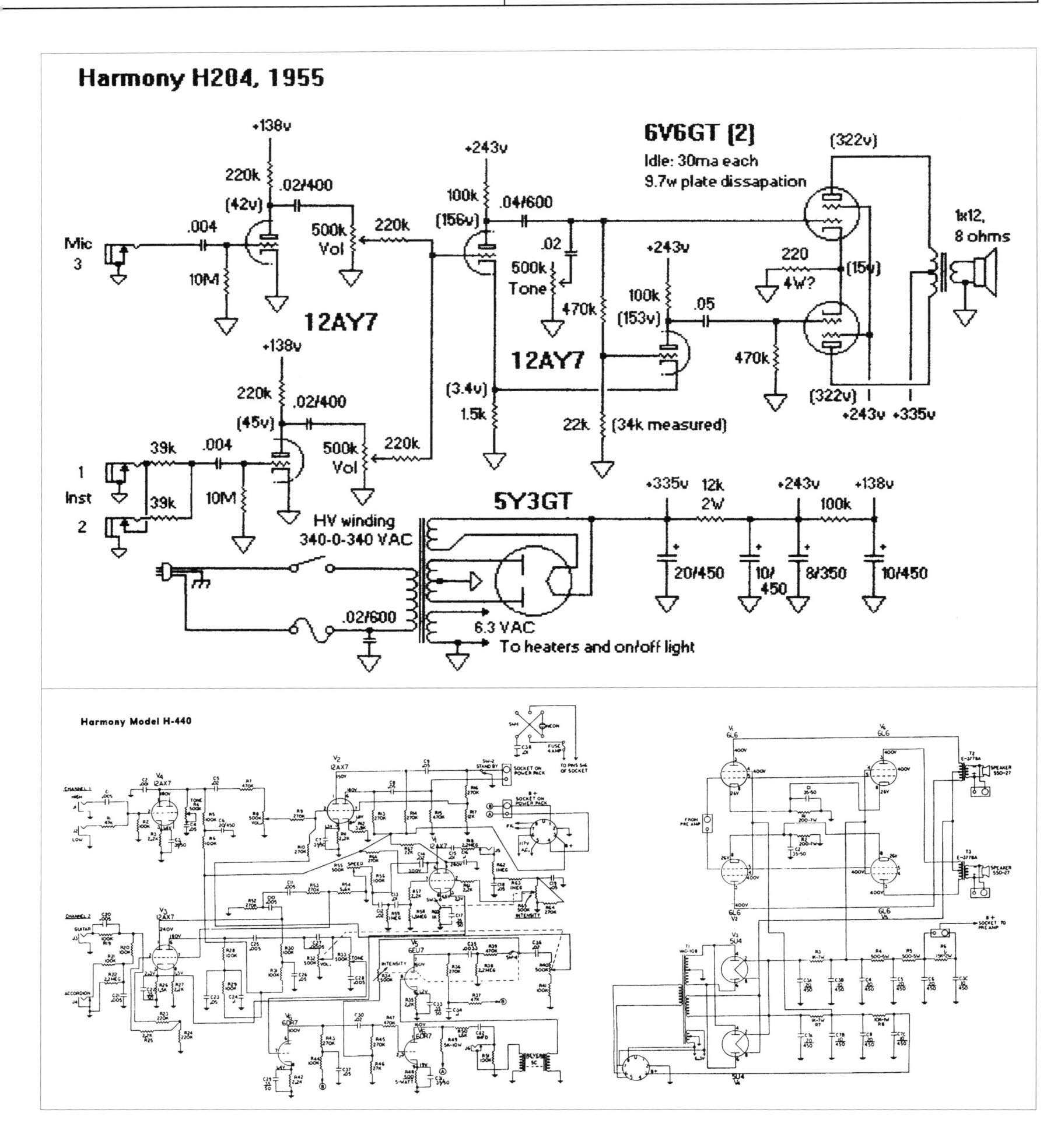

Harmony

H-204 (1995)
H-440

Harmony H204, 1955

Harmony Model H-440

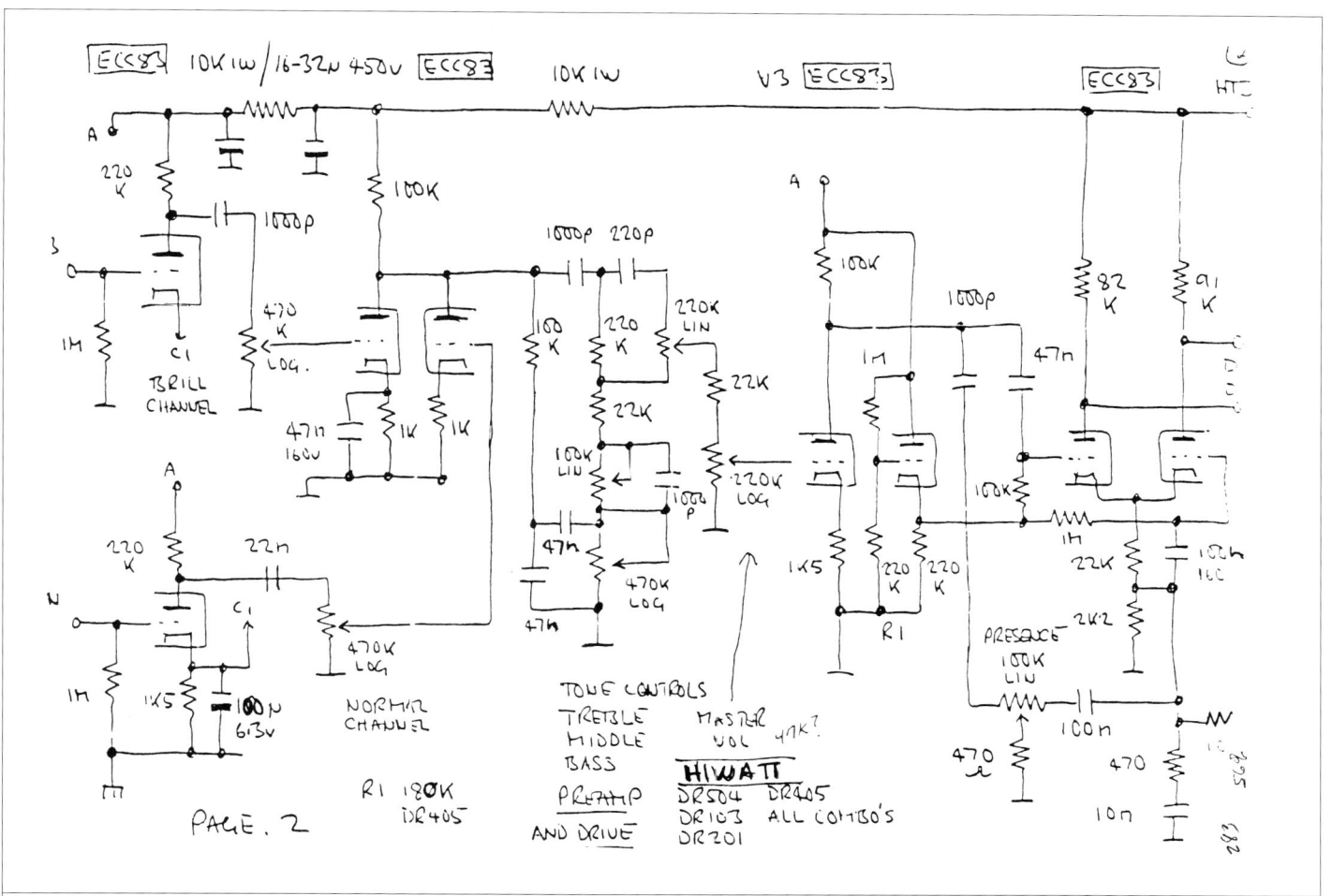

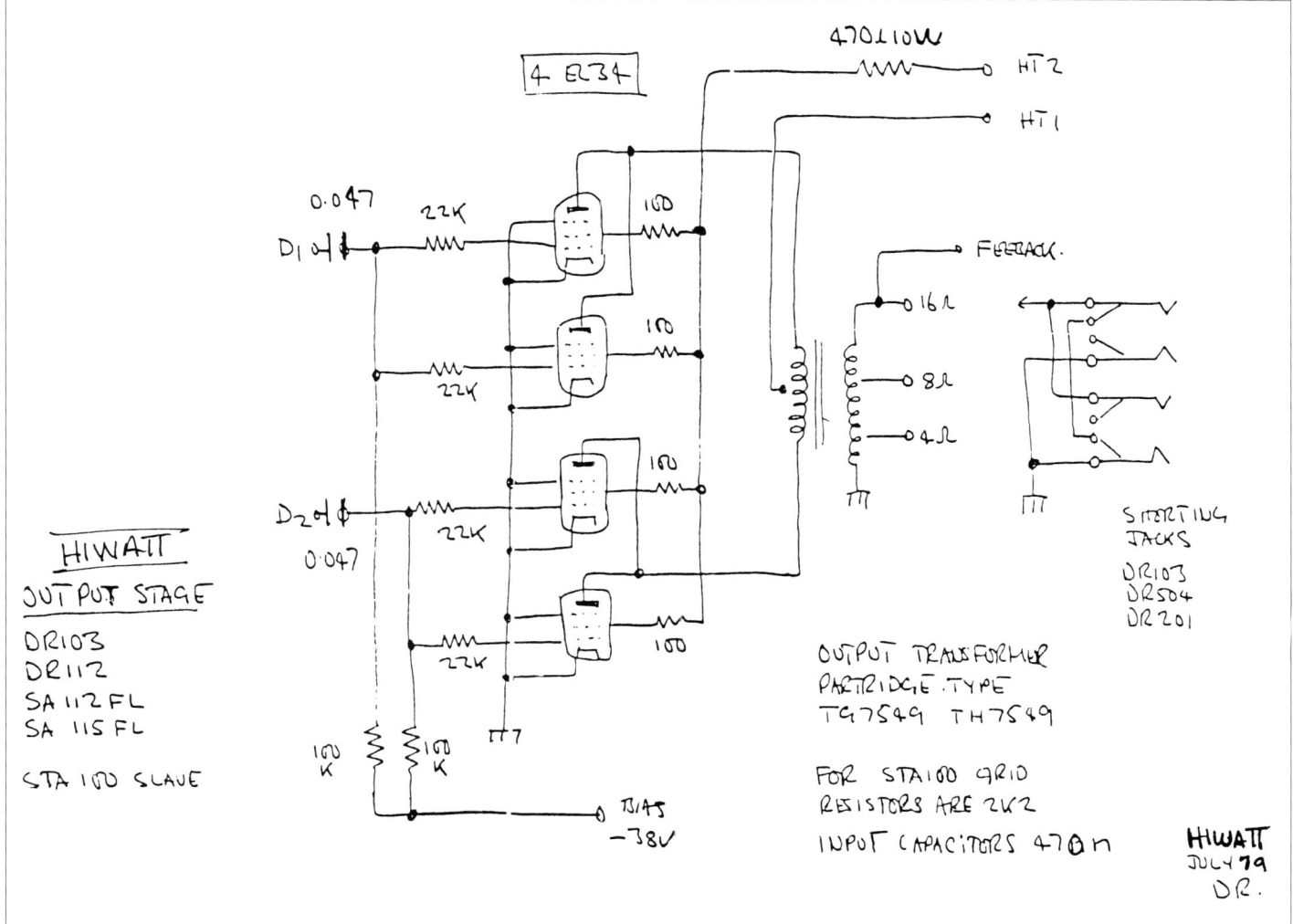

Hiwatt

DR PREAMP STAGE
DR 103-112 **OUTPUT**

DR201-203 OUTPUT
POWER SUPPLY 50W

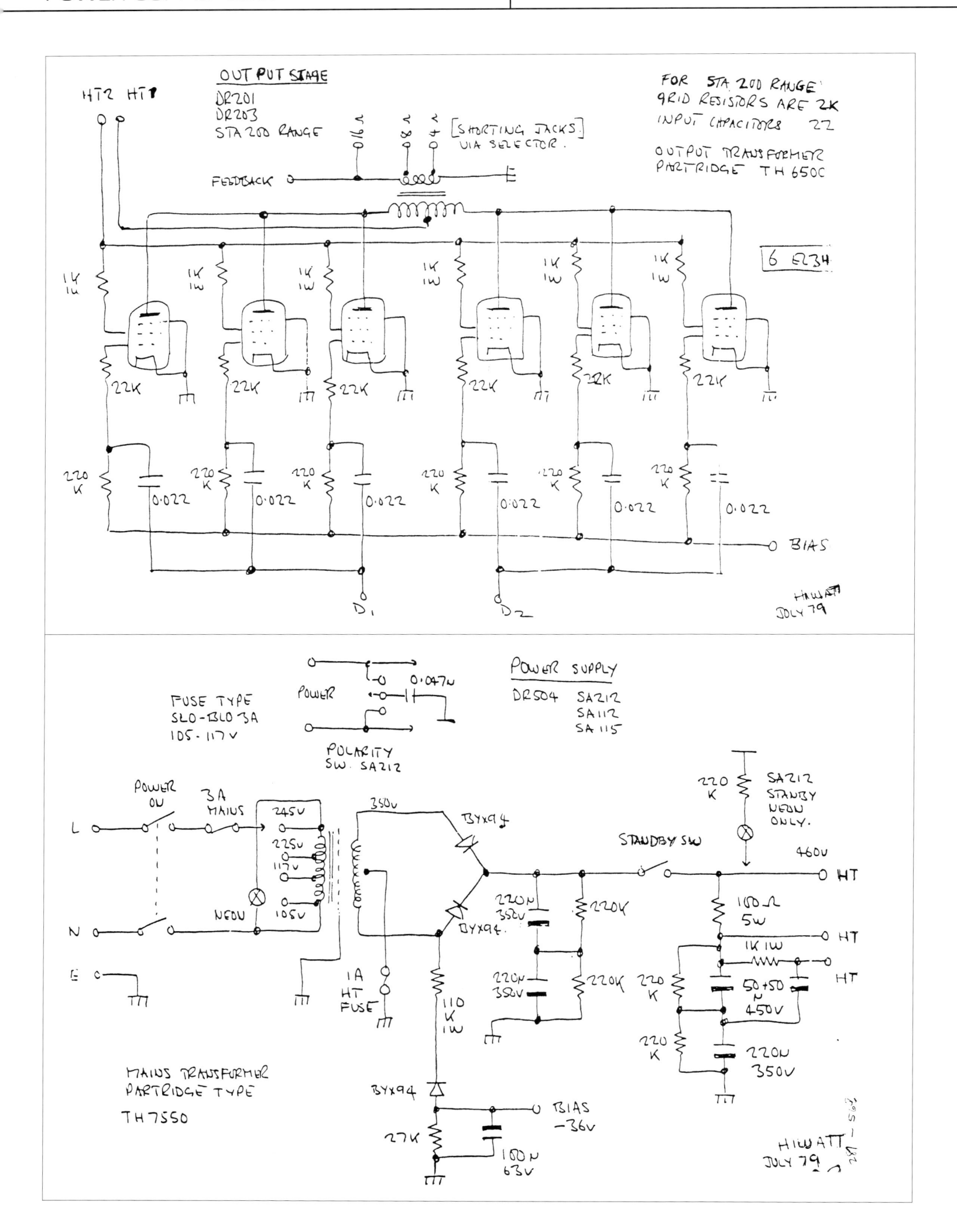

Hiwatt

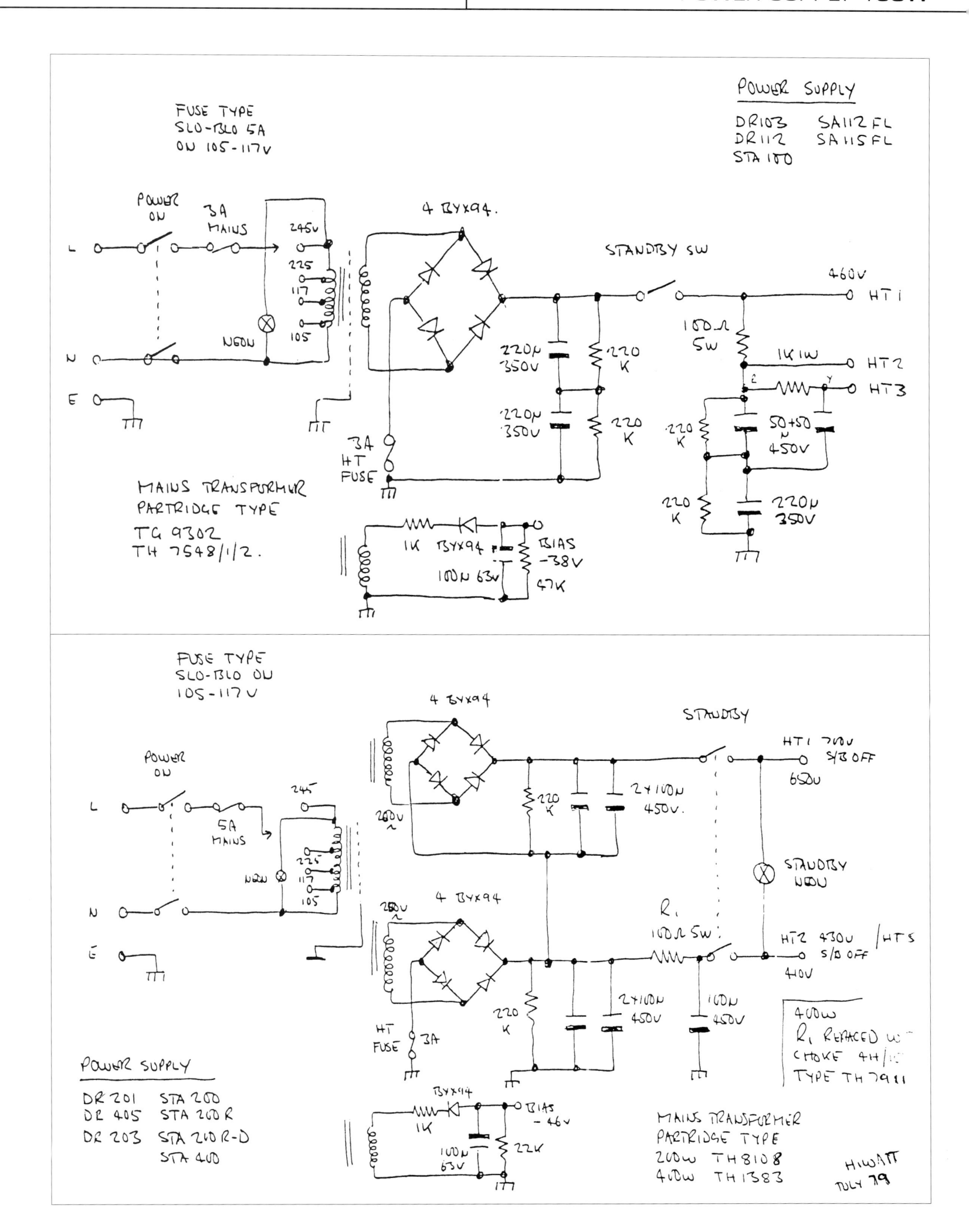

Hiwatt

POWER SUPPLY 100W
POWER SUPPLY 150W

STA 100-200 PREAMP
STA 400 SLAVE

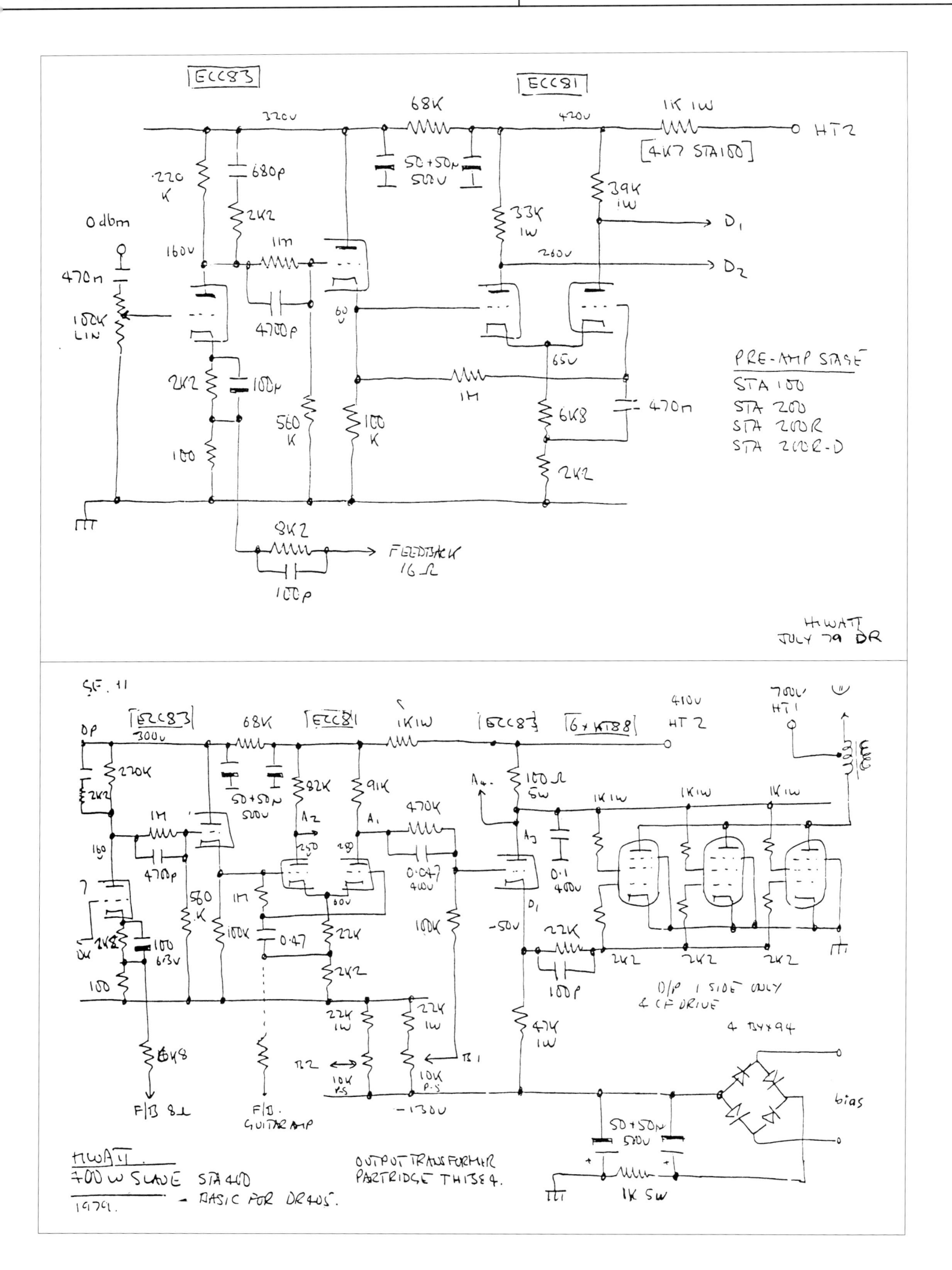

Hiwatt

Kay

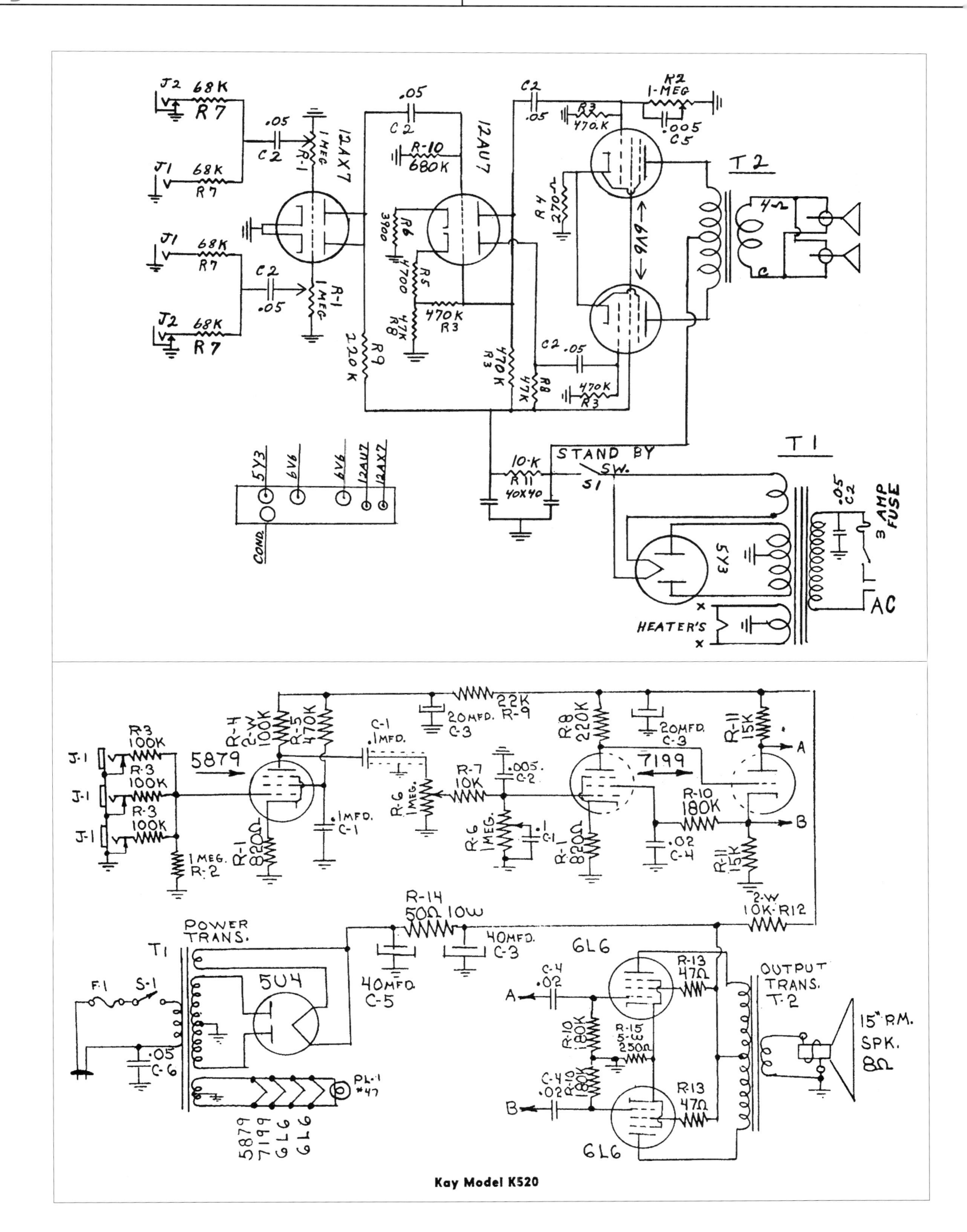

K-505
K-520

Kay Model K520

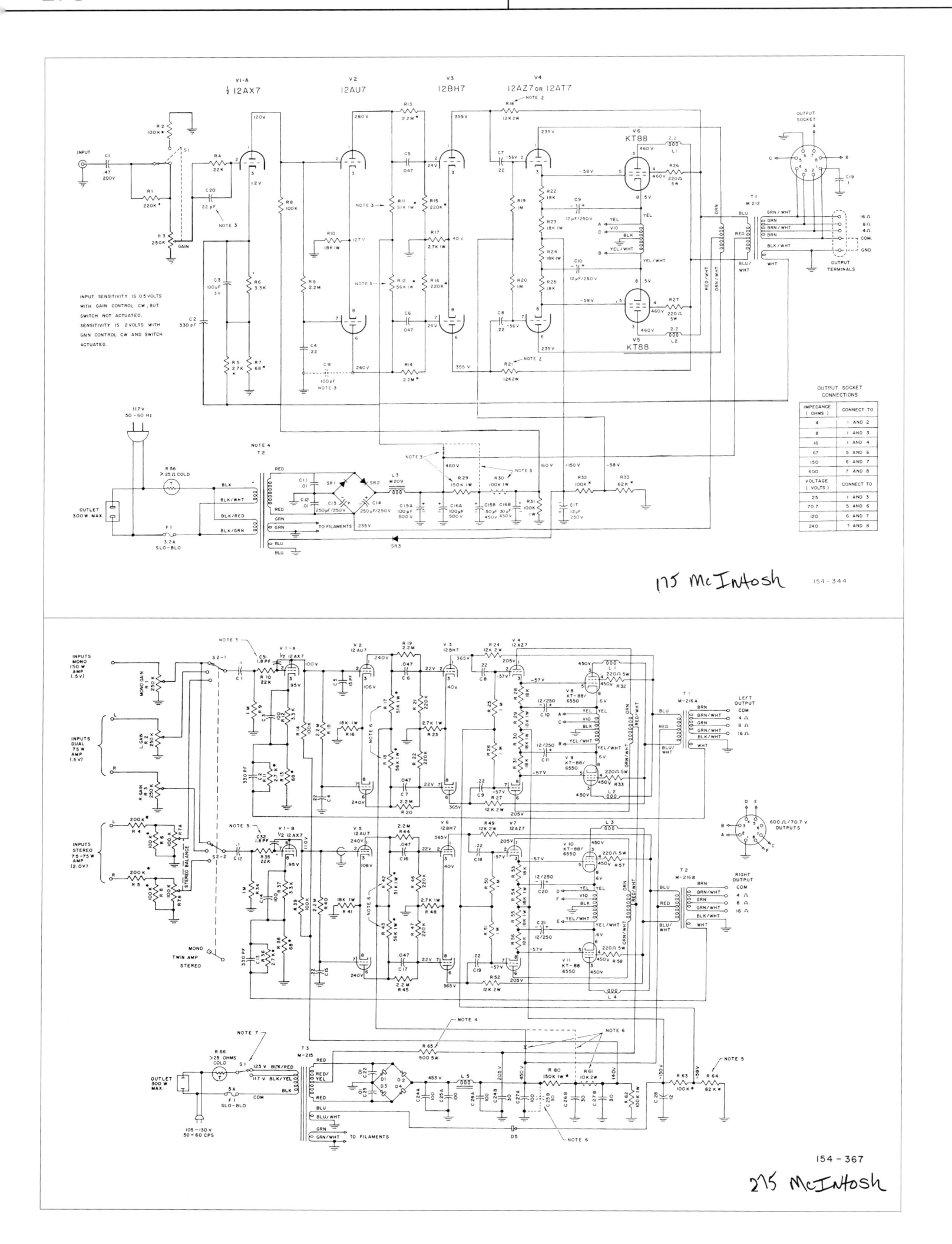

McIntosh

175
275

120 THE TUBE AMP BOOK

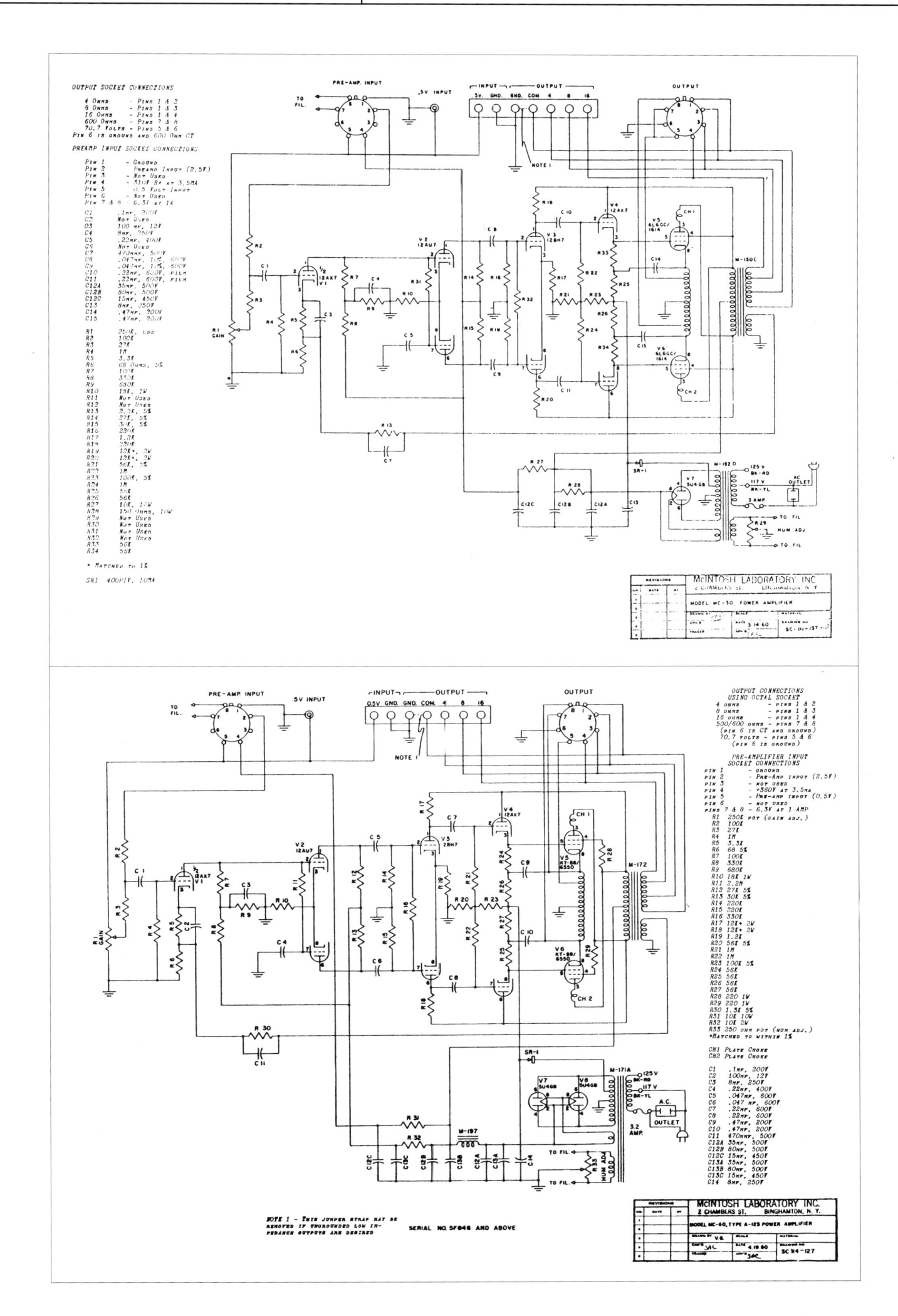

McIntosh

MC-30 POWER AMP
MC-60 TYPE A

MC-225
MC-2100 POWER SUPPLY

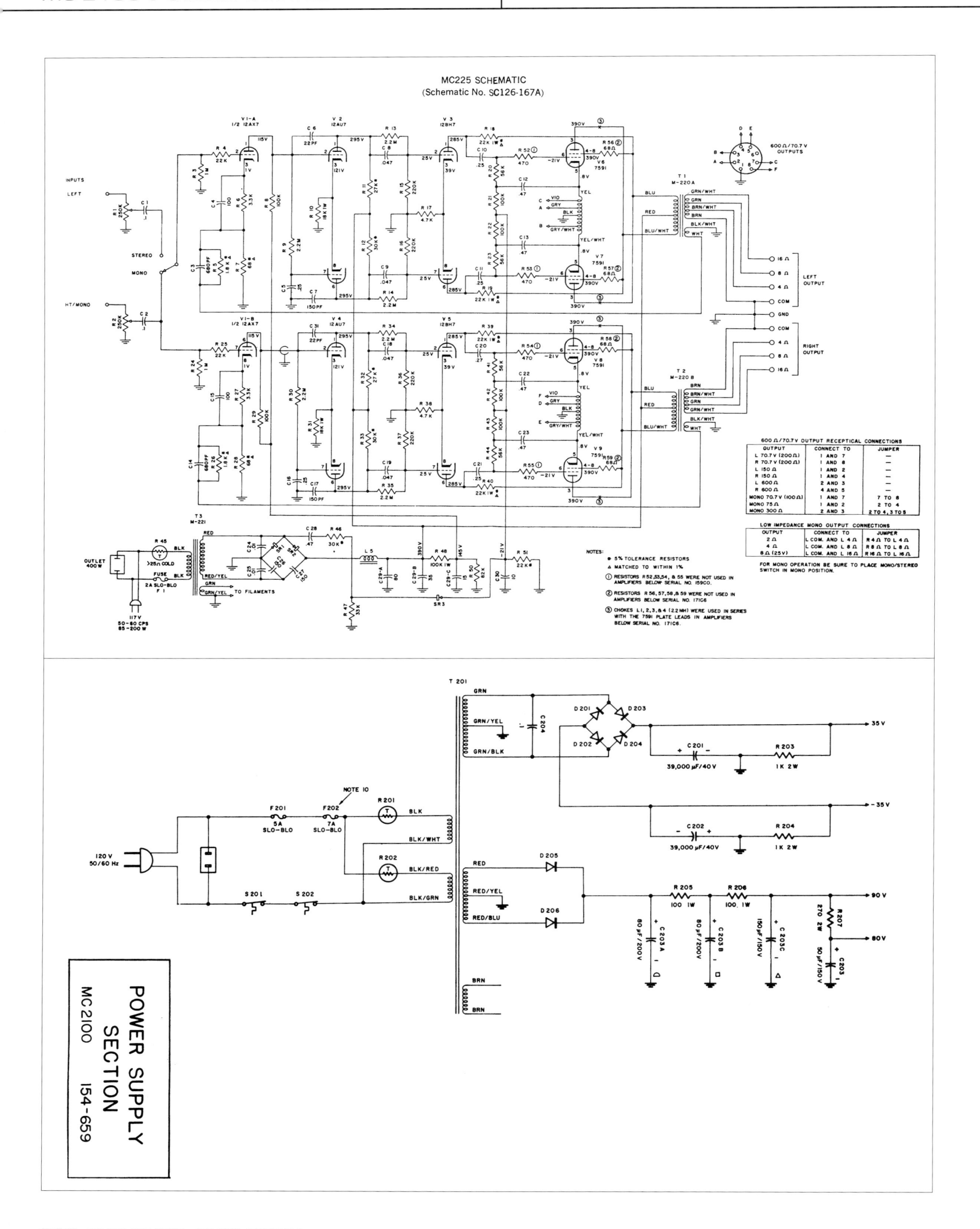

McIntosh

MC225 SCHEMATIC
(Schematic No. SC126-167A)

POWER SUPPLY SECTION
MC 2100
154-659

122 THE TUBE AMP BOOK

Maestro

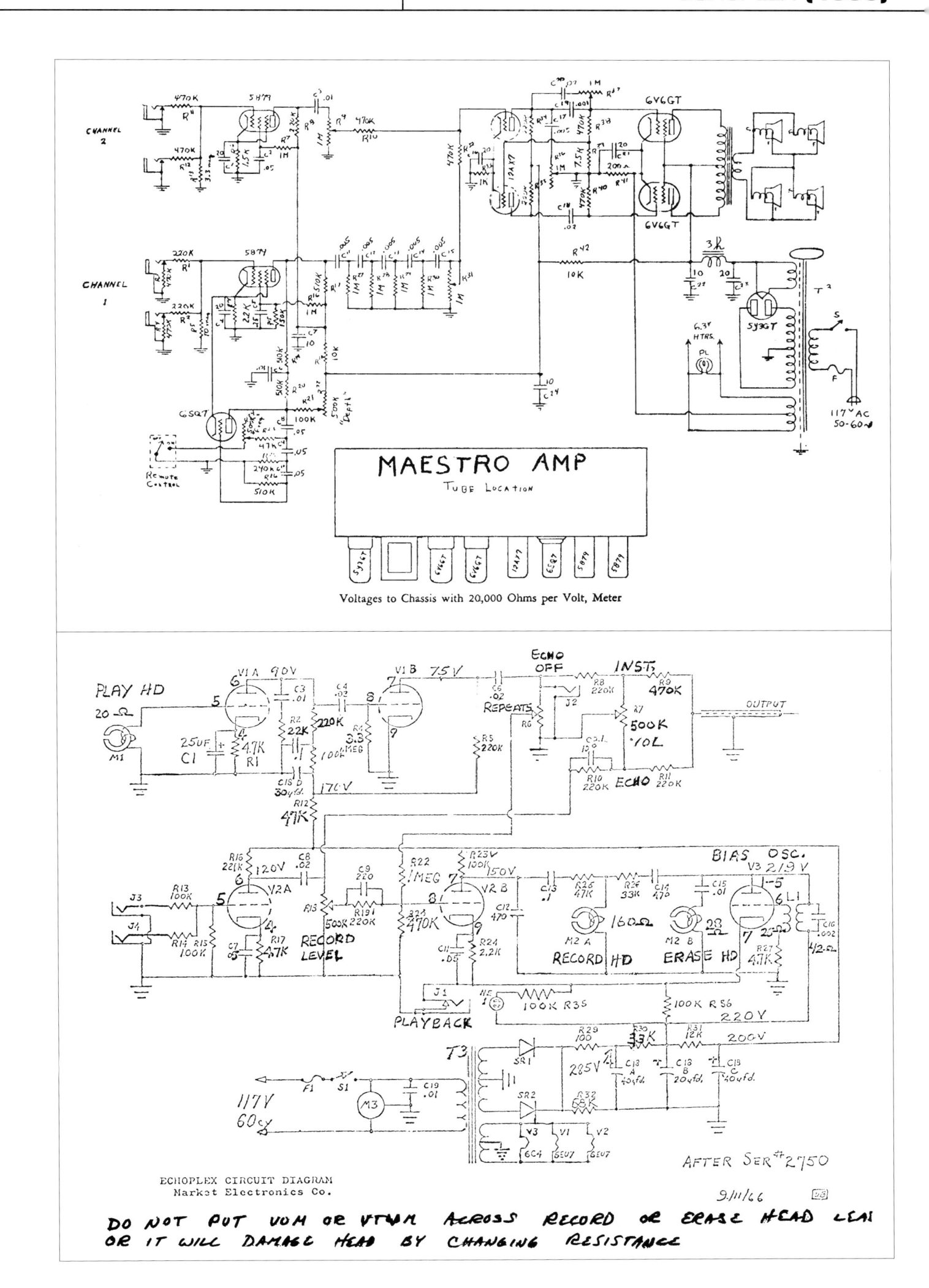

6V6 TREMOLO AMP
ECHOPLEX (1966)

ECHOPLEX EP-2
GA-78RV

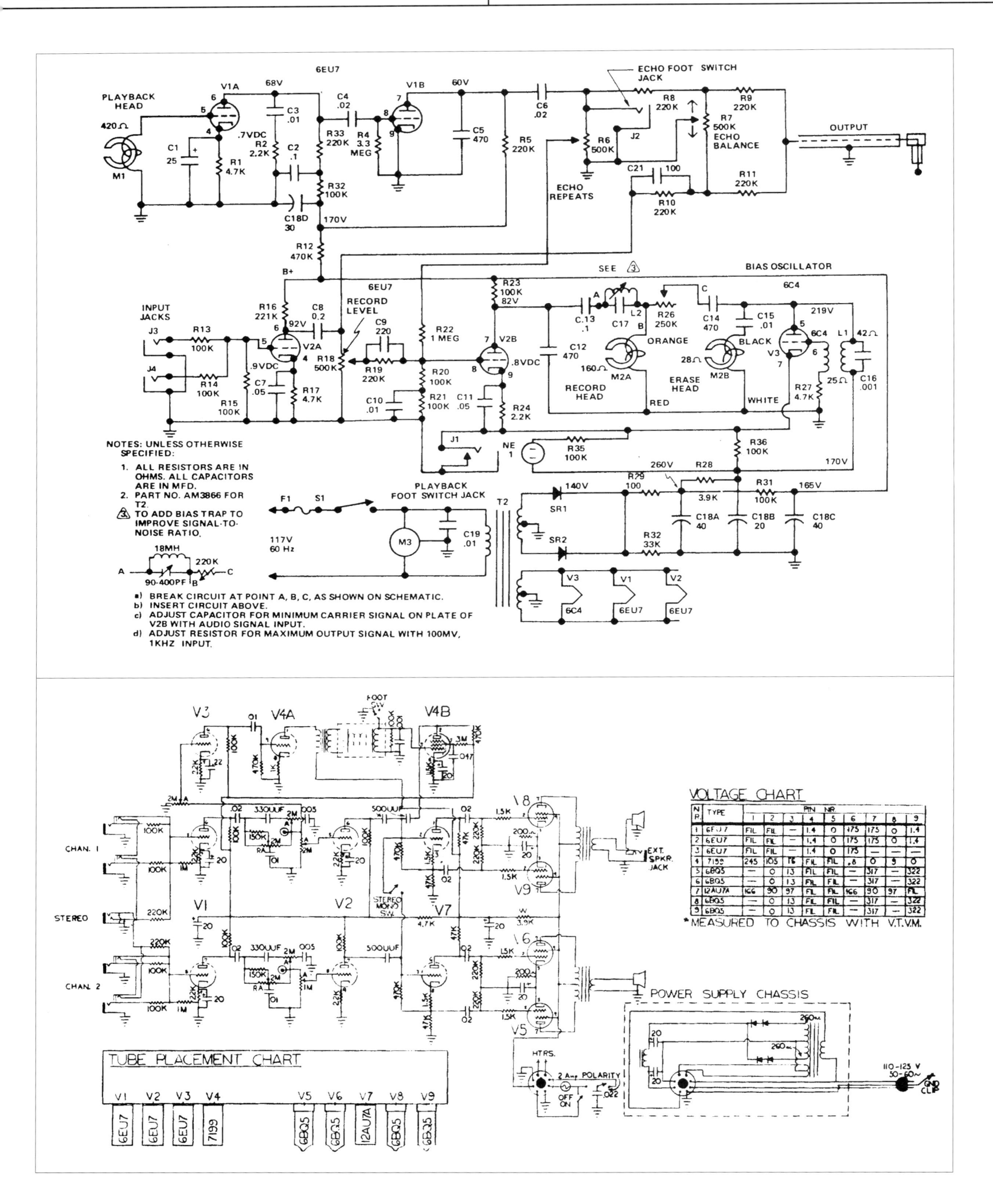

Maestro

Maestro

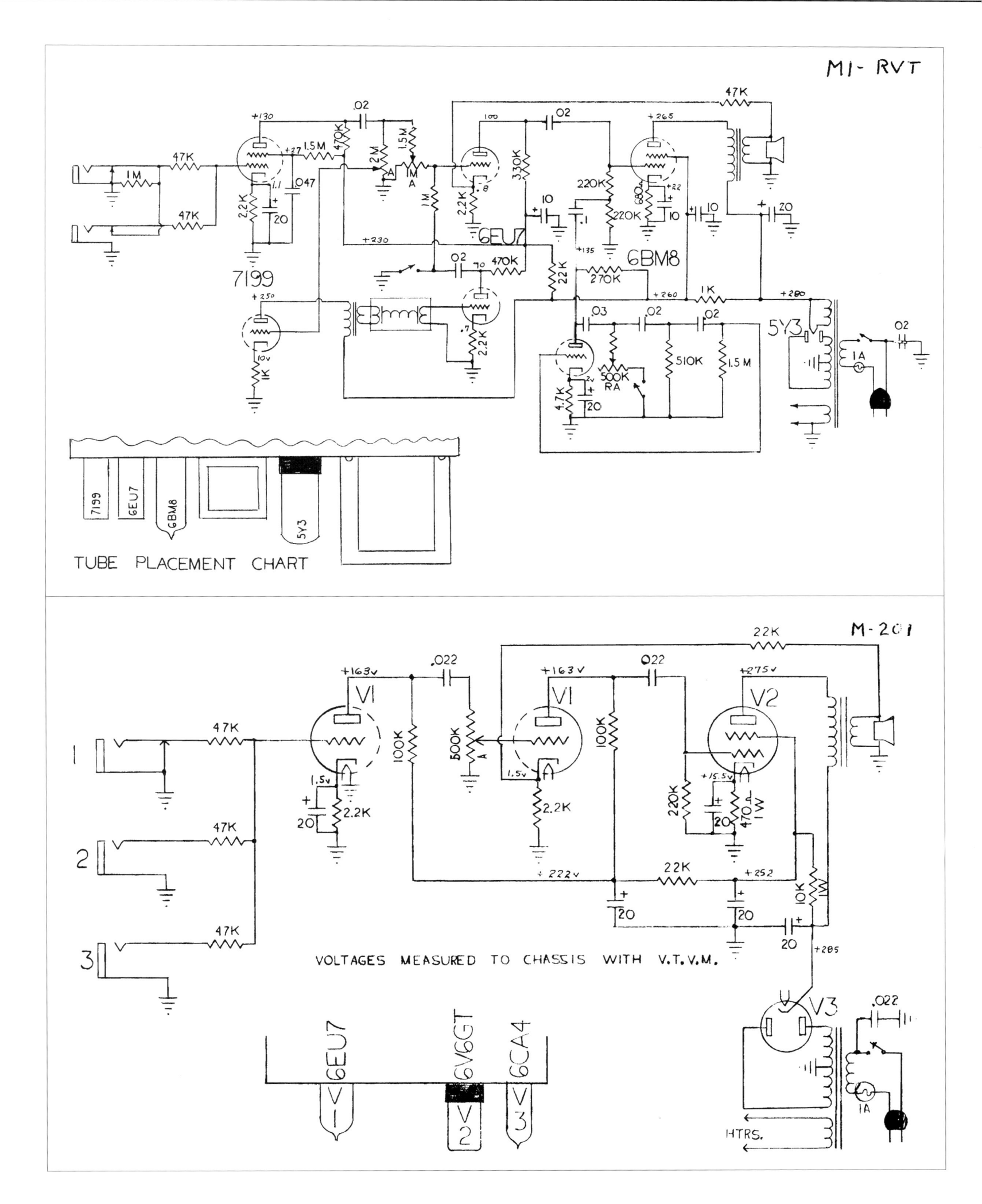

M1-RVT
M-201

M-216RVT
MA-40RVT

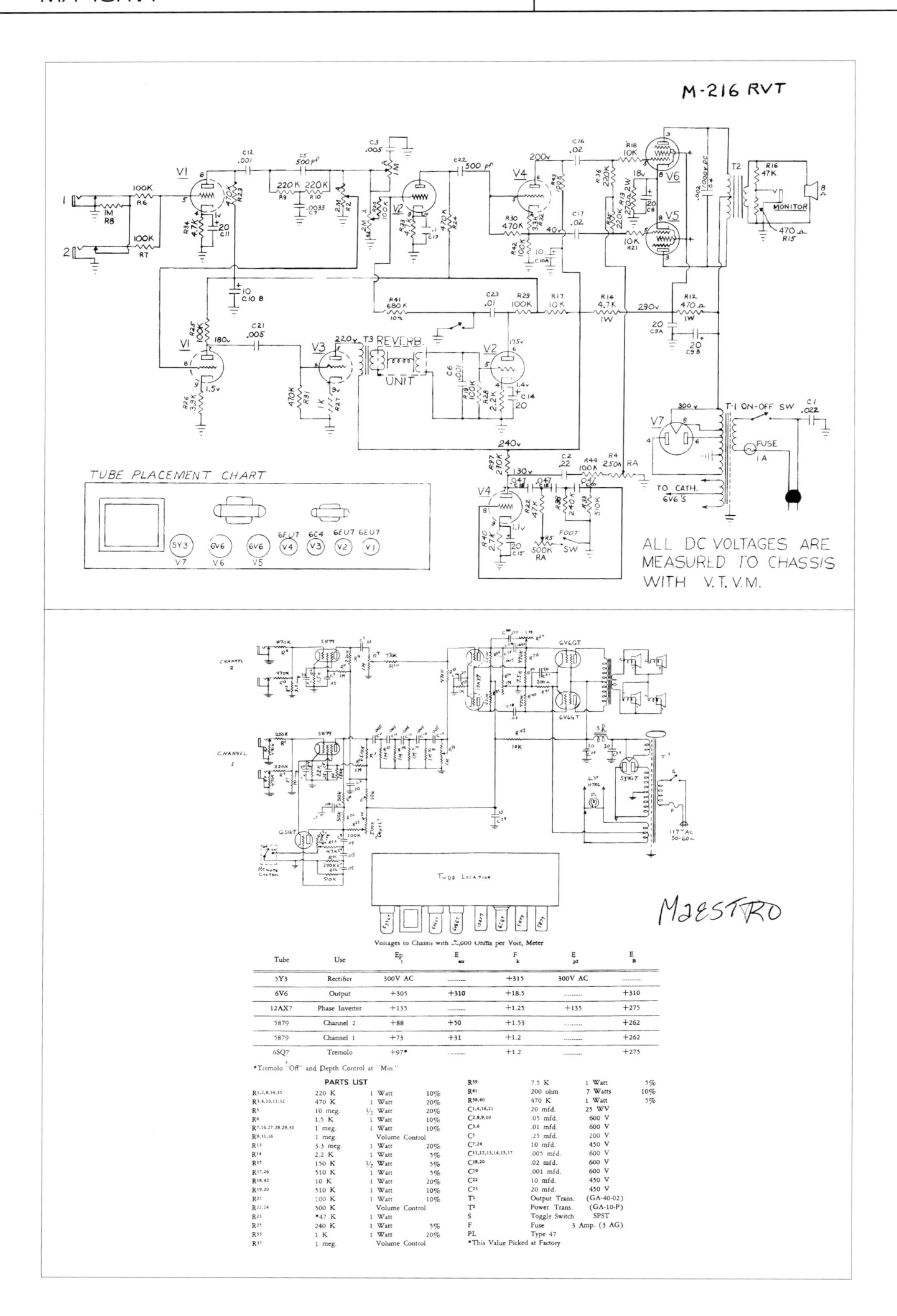

Maestro

M-216 RVT

ALL DC VOLTAGES ARE MEASURED TO CHASSIS WITH V.T.V.M.

TUBE PLACEMENT CHART: 5Y3 (V7), 6V6 (V6), 6V6 (V5), 6EU7 (V4), 6C4 (V3), 6EU7 (V2), 6EU7 (V1)

Maestro

Voltages to Chassis with 20,000 Ohms per Volt Meter

Tube	Use	E_{p_1}	E_{sc_1}	E_k	E_{p_2}	E_B
5Y3	Rectifier	300V AC		+315		300V AC
6V6	Output	+305	+310	+18.5		+310
12AX7	Phase Inverter	+135		+1.25	+135	+275
5879	Channel 2	+88	+50	+1.53		+262
5879	Channel 1	+73	+31	+1.2		+262
6SQ7	Tremolo	+97*			+1.2	+275

*Tremolo "Off" and Depth Control at "Min."

PARTS LIST

R1,2,8,34,35	220 K	1 Watt	10%
R3,4,10,11,32	470 K	1 Watt	20%
R5	10 meg.	½ Watt	20%
R6	1.5 K	1 Watt	10%
R7,16,27,28,29,30	1 meg.	1 Watt	10%
R9,31,36	1 meg.	Volume Control	
R12	3.3 meg.	1 Watt	20%
R14	2.2 K	1 Watt	5%
R15	150 K	½ Watt	5%
R17,26	510 K	1 Watt	5%
R18,42	10 K	1 Watt	20%
R19,20	510 K	1 Watt	20%
R21	100 K	1 Watt	10%
R22,24	500 K	Volume Control	
R23	*47 K	1 Watt	
R25	240 K	1 Watt	5%
R33	1 K	1 Watt	20%
R37	1 meg.	Volume Control	
R39	7.5 K	1 Watt	5%
R41	200 ohm	7 Watts	10%
R38,40	470 K	1 Watt	5%
C1,4,16,21	20 mfd.	25 WV	
C2,8,9,10	.05 mfd.	600 V	
C3,6	.01 mfd.	600 V	
C5	.25 mfd.	200 V	
C7,24	10 mfd.	450 V	
C11,12,13,14,15,17	.005 mfd.	600 V	
C18,20	.02 mfd.	600 V	
C19	.001 mfd.	600 V	
C22	10 mfd.	450 V	
C23	20 mfd.	450 V	
T1	Output Trans.	(GA-40-02)	
T2	Power Trans.	(GA-10-P)	
S	Toggle Switch	SPST	
F	Fuse	3 Amp. (3 AG)	
PL	Type 47		

*This Value Picked at Factory

126 THE TUBE AMP BOOK

Magnatone

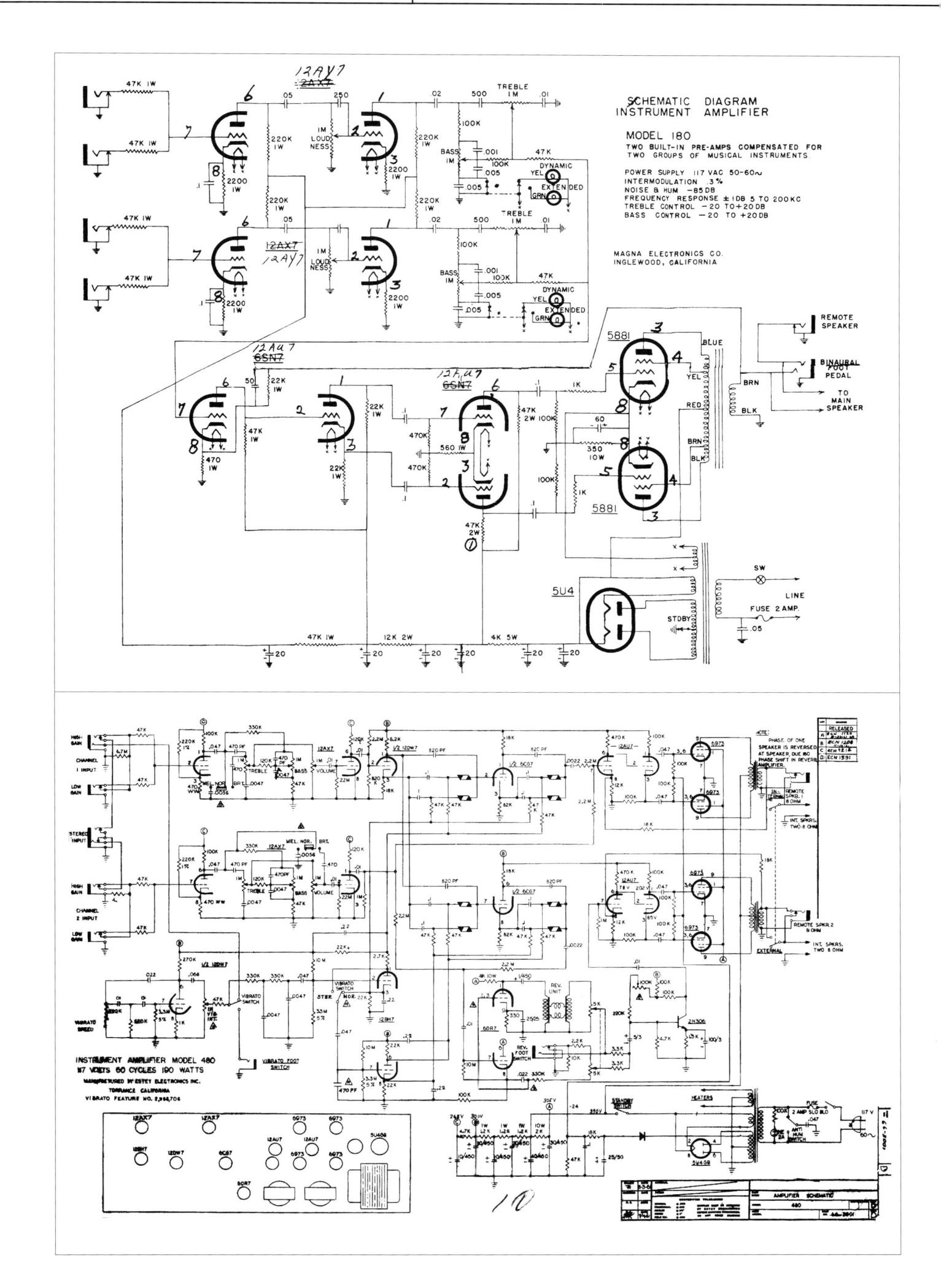

180 / 480

50W
50W MASTER VOL (MODEL 2204)

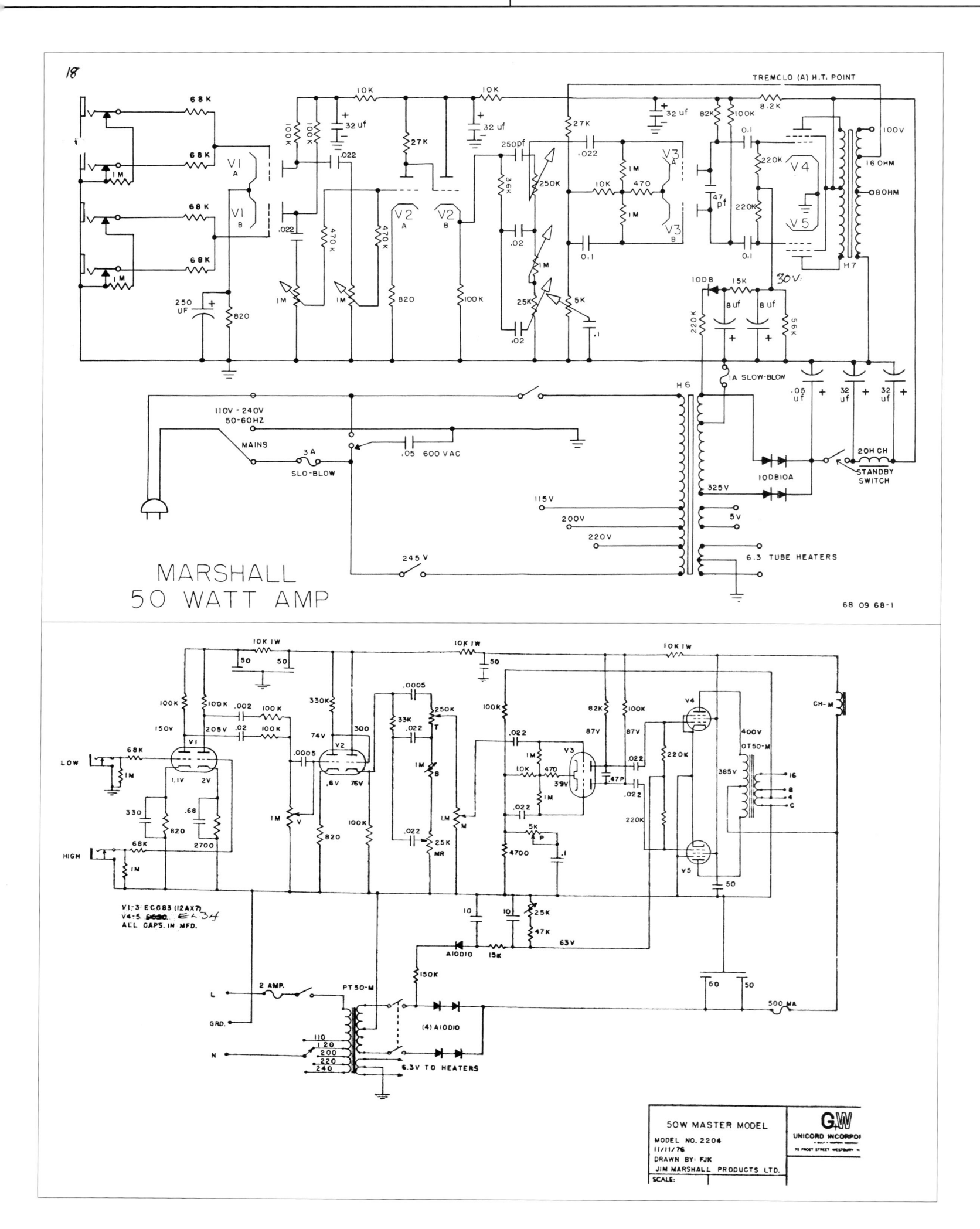

Marshall

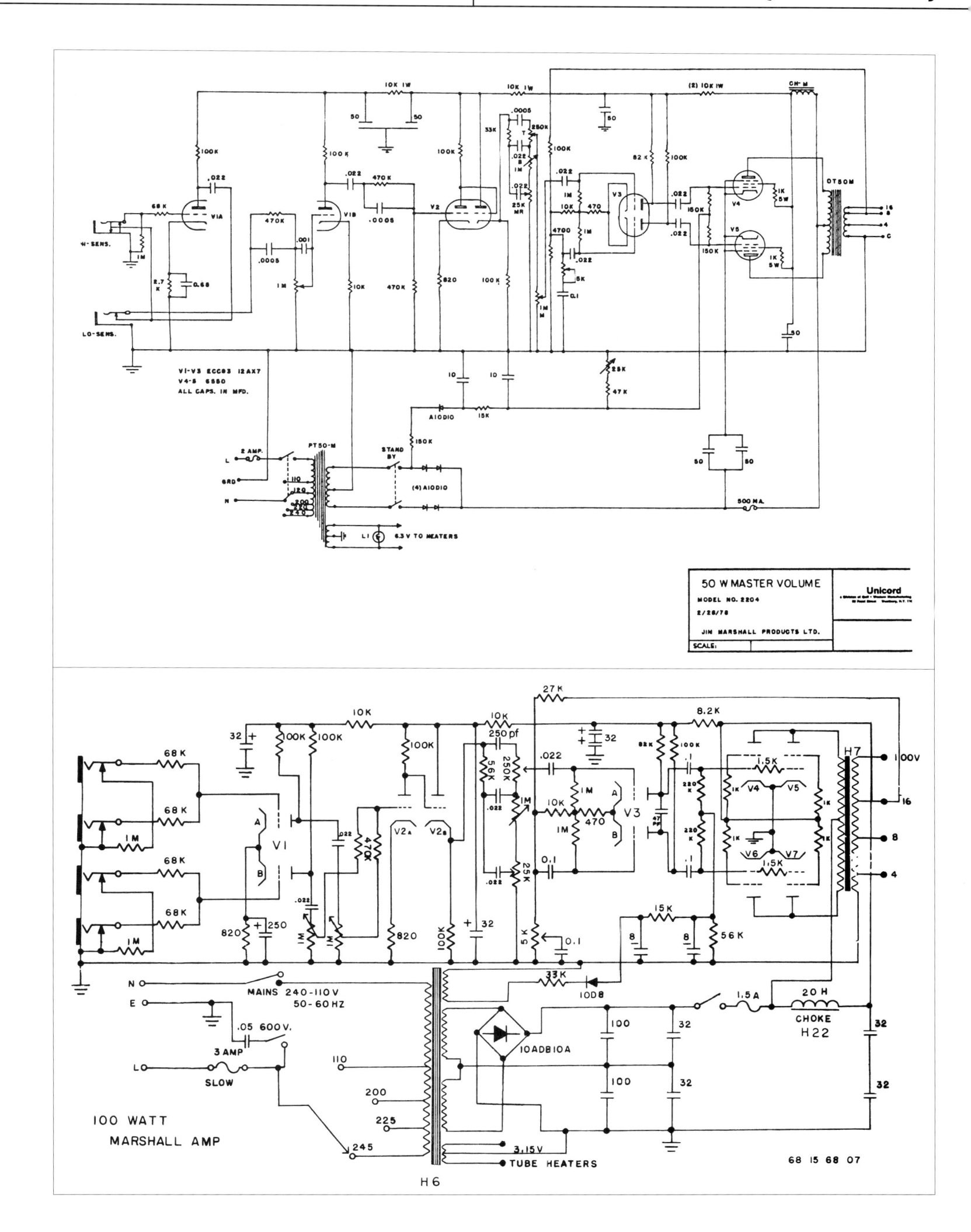

Marshall

50W MASTER VOL (MODEL 2204 REV)
BASS AMP (MODEL 1992)

100W PA (MODEL 1968)
200W PA (MODEL 1966)

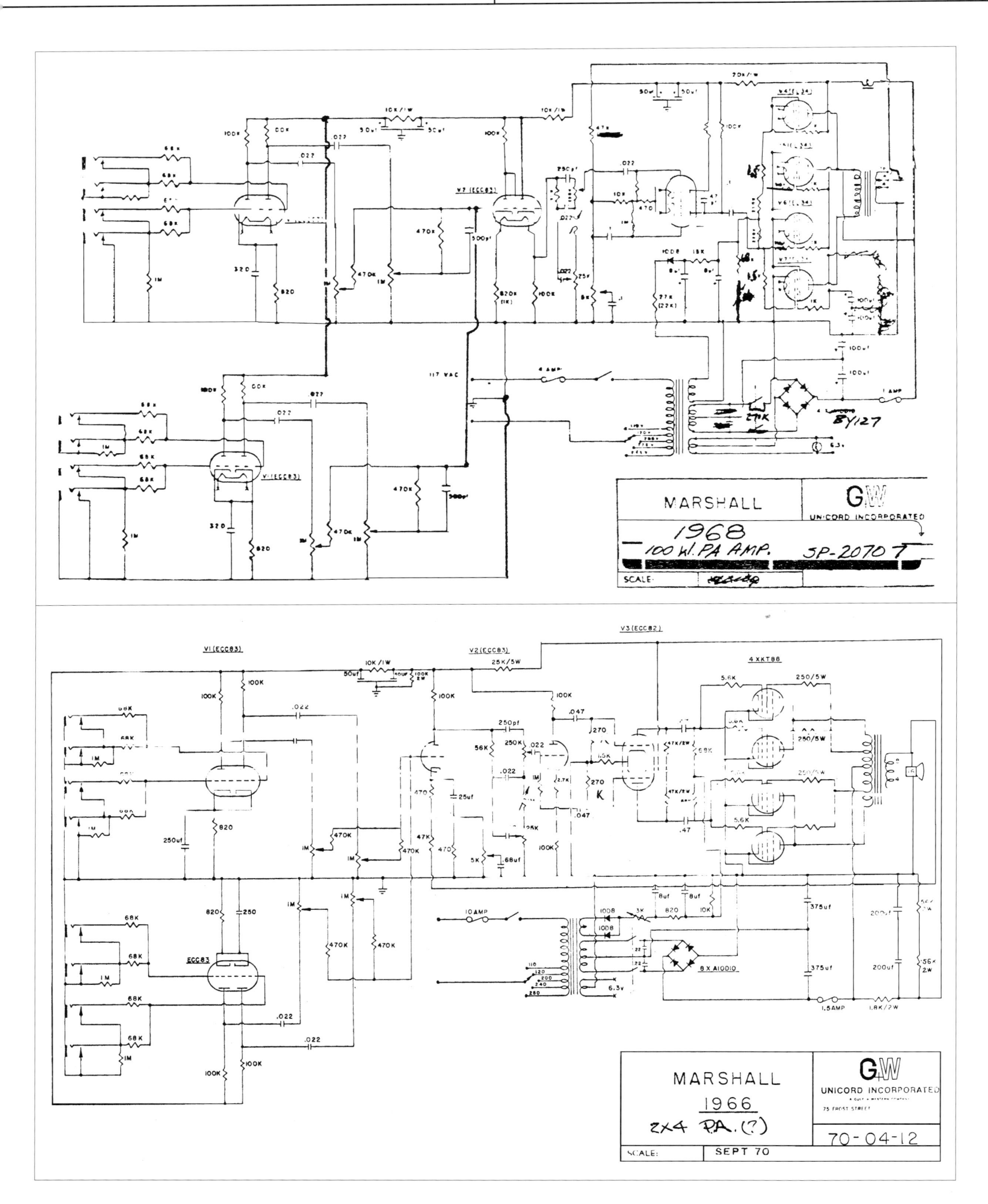

Marshall

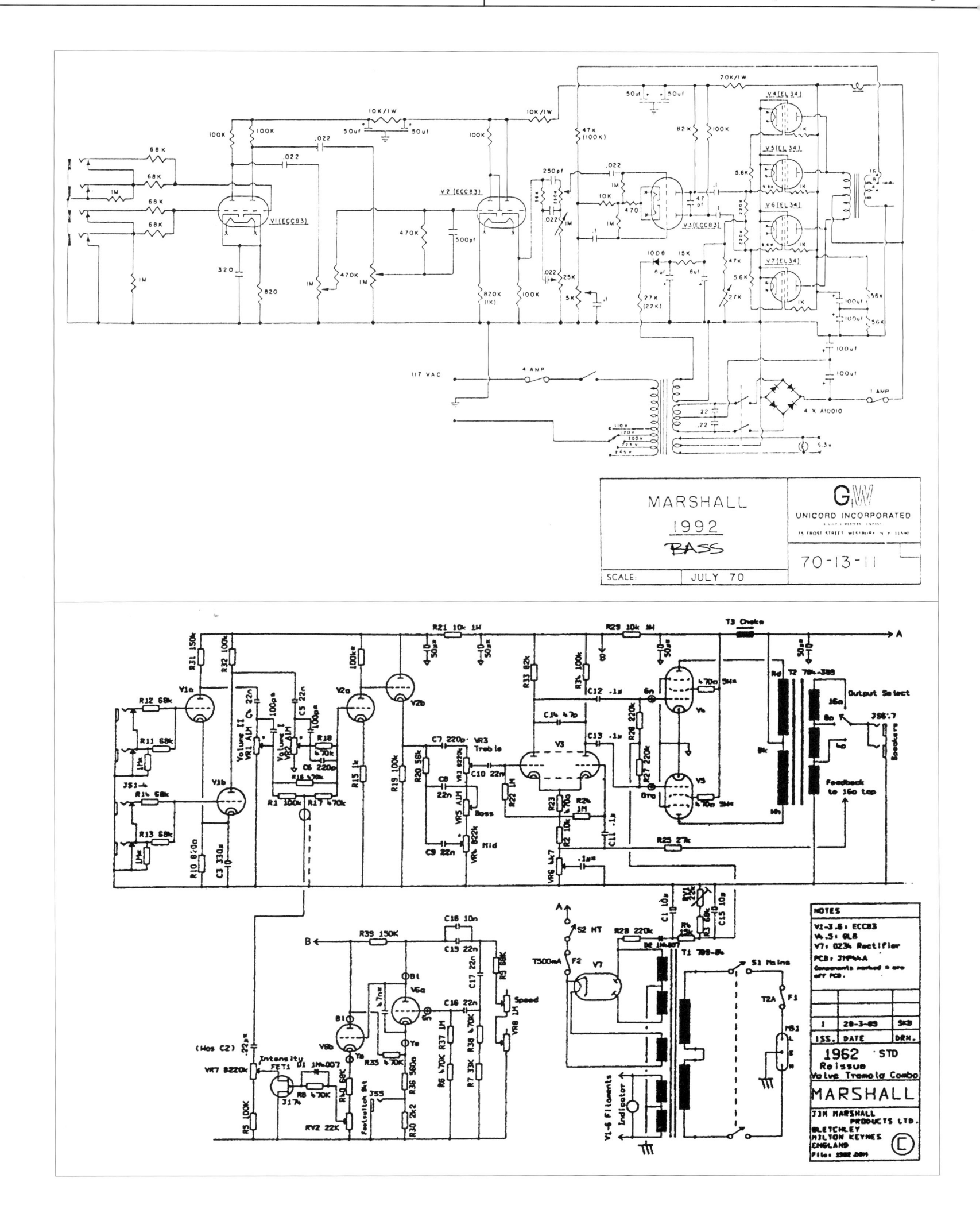

Marshall

BASS AMP (MODEL 1992)
BLUESBREAKER REISSUE (MODEL 1962)

JCM 800 BASS (MODEL 1992)
JCM 800 LEAD (50 & 100W PWR)

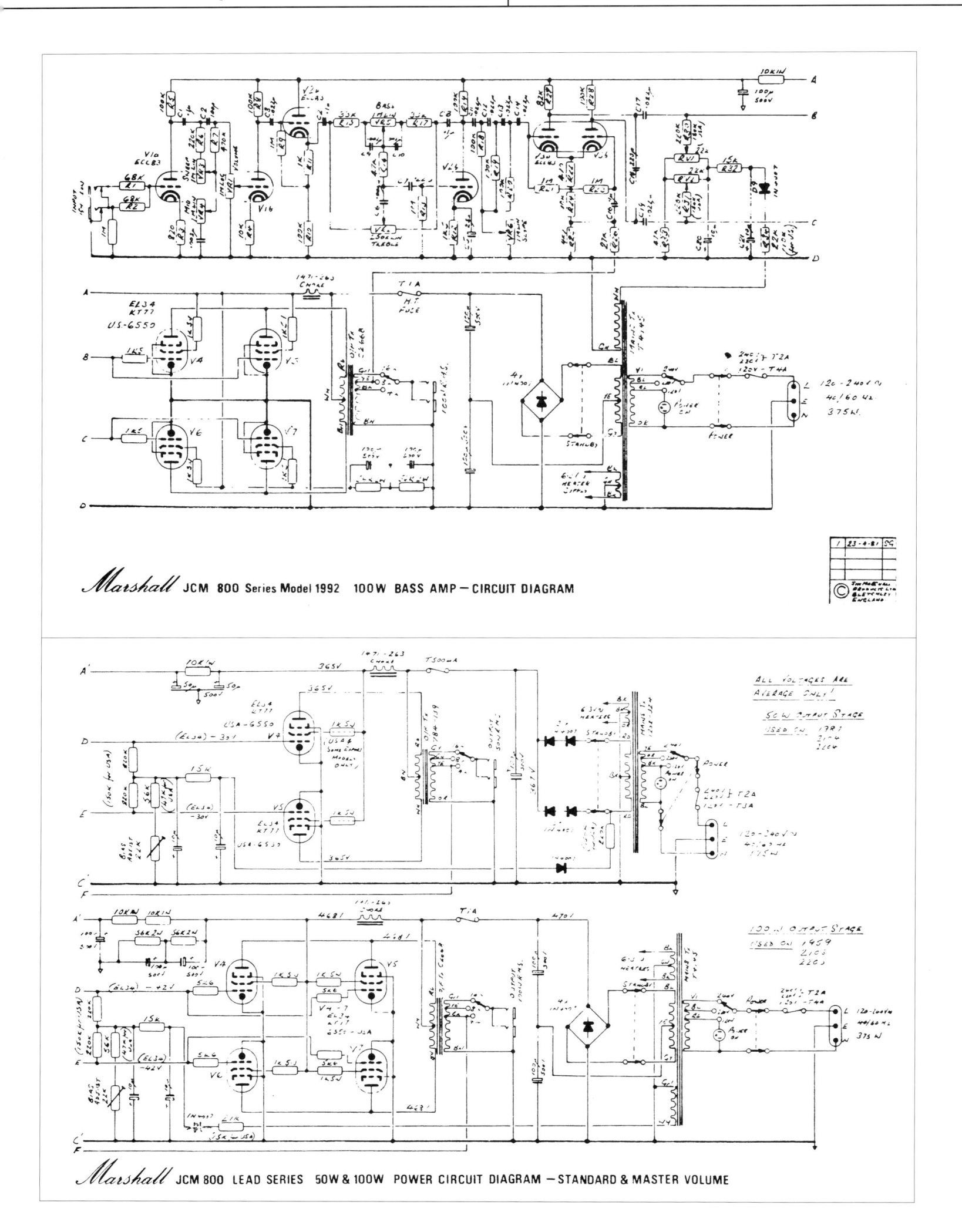

Marshall

Marshall JCM 800 Series Model 1992 100W BASS AMP — CIRCUIT DIAGRAM

Marshall JCM 800 LEAD SERIES 50W & 100W POWER CIRCUIT DIAGRAM — STANDARD & MASTER VOLUME

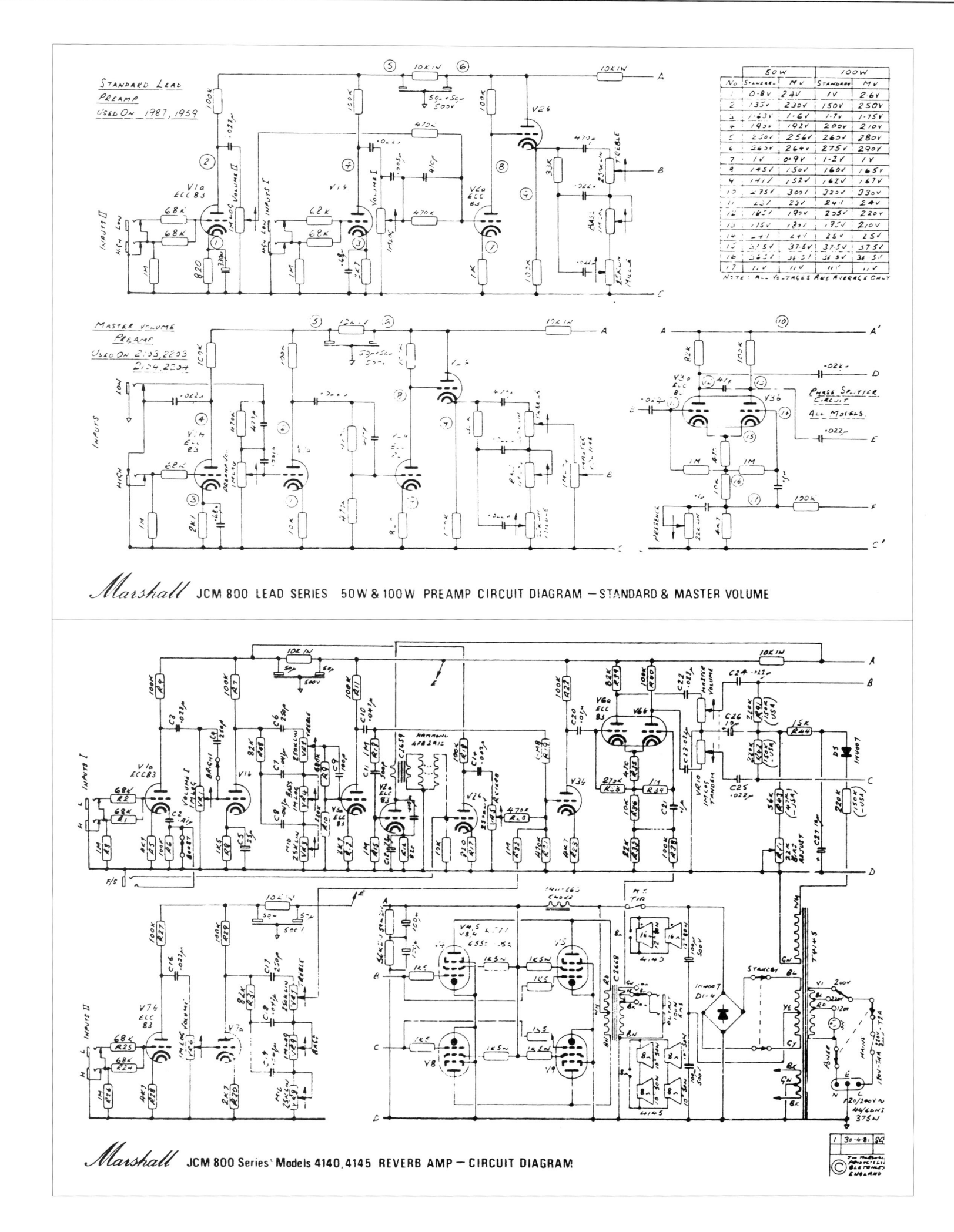

Marshall

JCM 800 LEAD (PREAMP)
JCM 800 REV (MODEL 4140/4145)

Marshall JCM 800 LEAD SERIES 50W & 100W PREAMP CIRCUIT DIAGRAM — STANDARD & MASTER VOLUME

Marshall JCM 800 Series Models 4140, 4145 REVERB AMP — CIRCUIT DIAGRAM

THE TUBE AMP BOOK 133

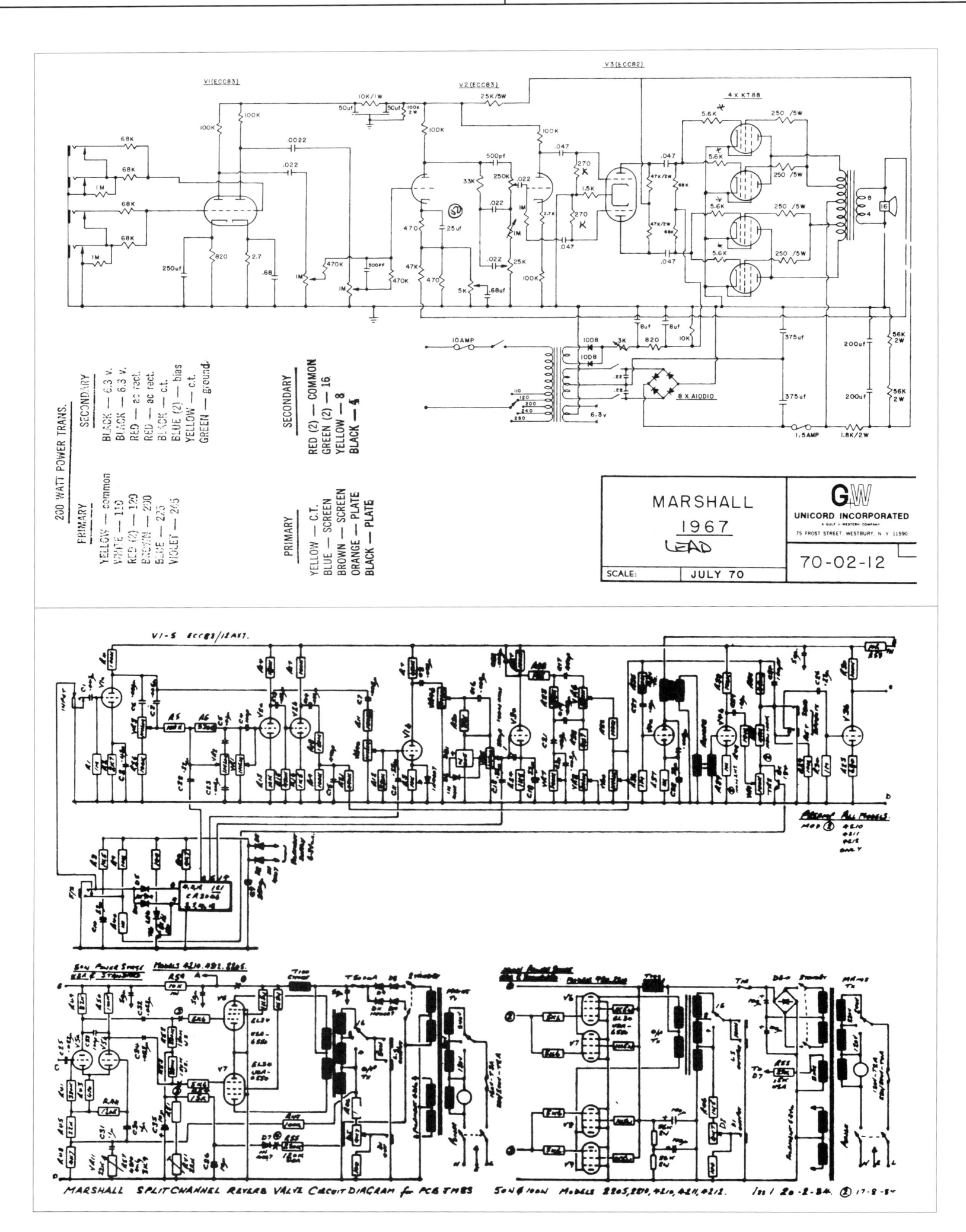

MARSHALL

MAJOR 200W (MODEL 1963)
REVERB (2205/2810/4210 REV)

Marshall

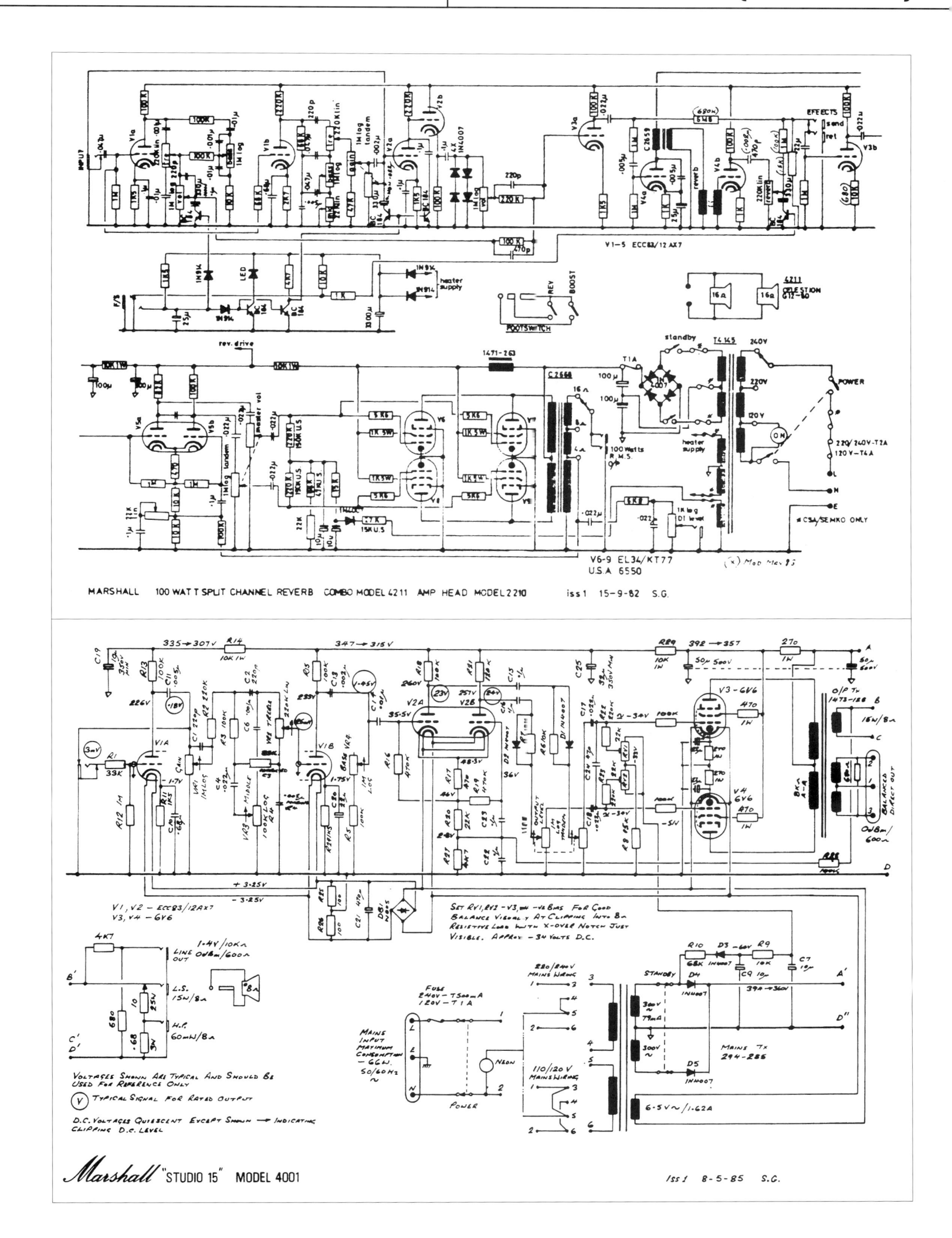

REVERB (2210 HEAD/4211 COMBO)
STUDIO 15 (MODEL 4001)

SUPER PA (MODEL 1963)

Marshall

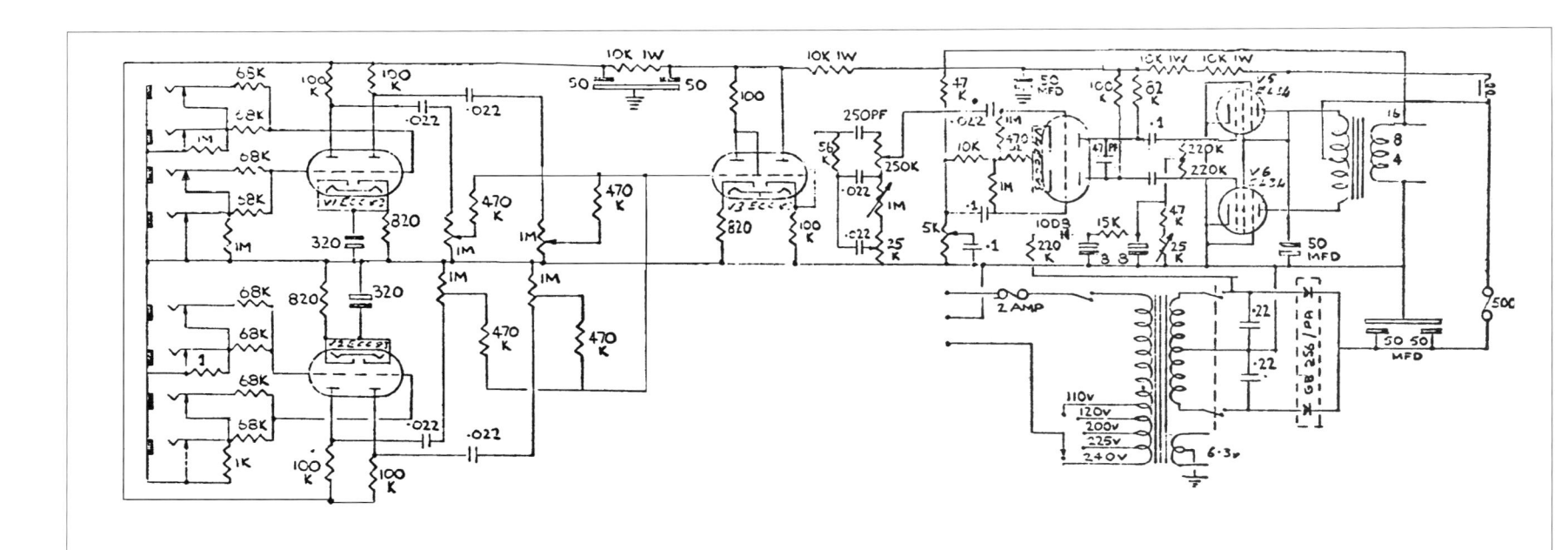

PIN N°	V1	V2	V3	V4	V5	V6	7
1	155v	155v	150v	210v	-	-	
2	-	-	-	+	HEATER	HEATER	
3	1.4v	1.4v	1v	37v	430v	430v	
4	HEATERS	HEATERS	HEATERS	HEATERS	435v	435v	
5	"	"	"	"	-31	-31	
6	155v	155v	270v	200v	N.C.	N.C.	
7	-	-	150v	+	HEATER	HEATER	
8	1.4v	1.4v	150v	37v	-	-	
9	HEATER	HEATERS	HEATERS	HEATERS	N.P.	N.P.	

1963 SUPER PA — July '70'

JIM MARSHALL PRODUCTS LTD.
ALL MARSHALL AMPLIFIERS ARE SUBJECT TO CONTINUOUS DEVELOPMENT AND IMPROVMENT CONSIQUENTLY THE UNITS MAY INCORPORATE MINOR CHANGES IN DETAIL FROM THE INFORMATION CONTAINED ABOVE.

Mesa/Boogie

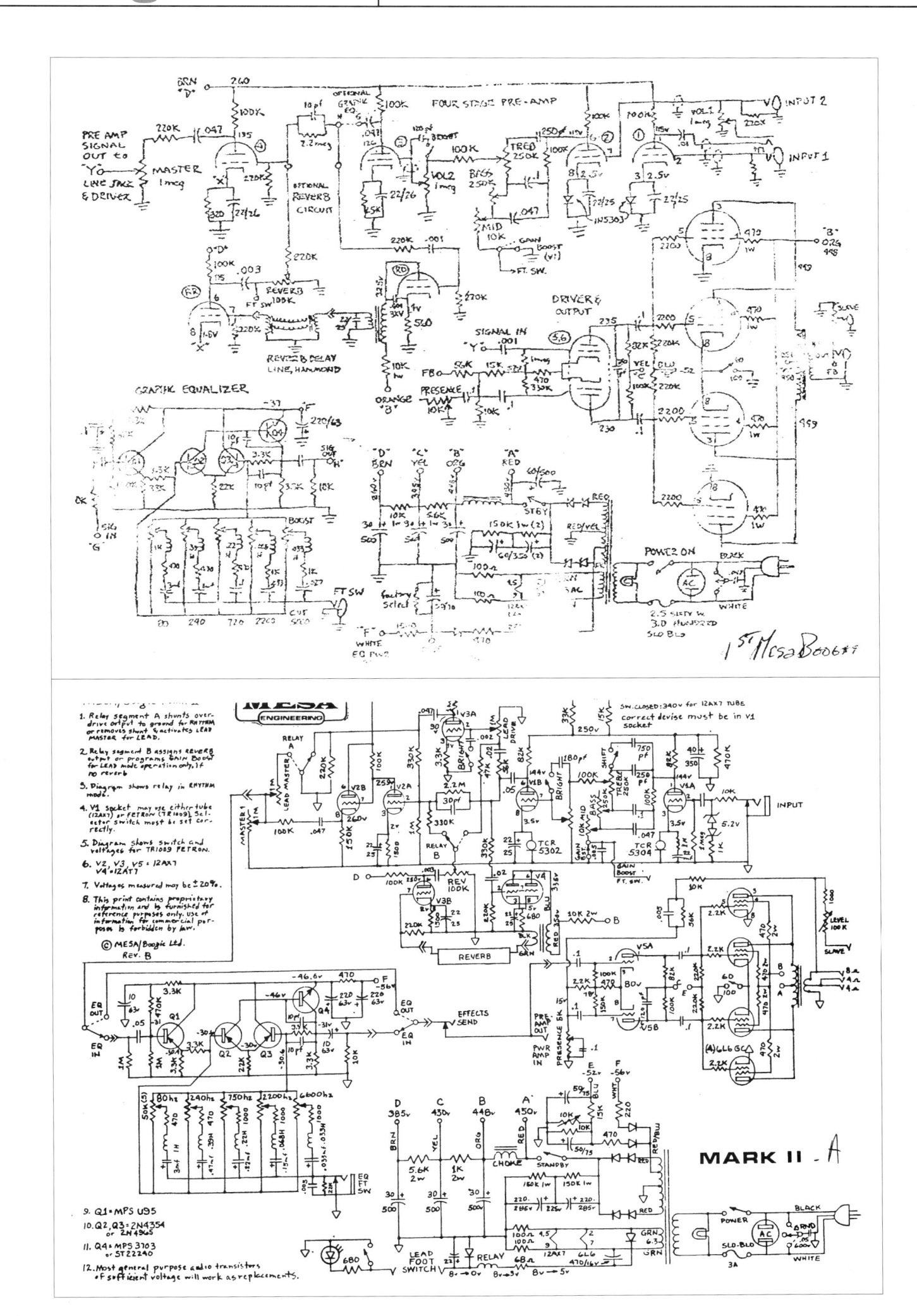

BOOGIE MARK I
BOOGIE MARK IIA

BOOGIE MARK IIB
BOOGIE MARK IIC & PREAMP

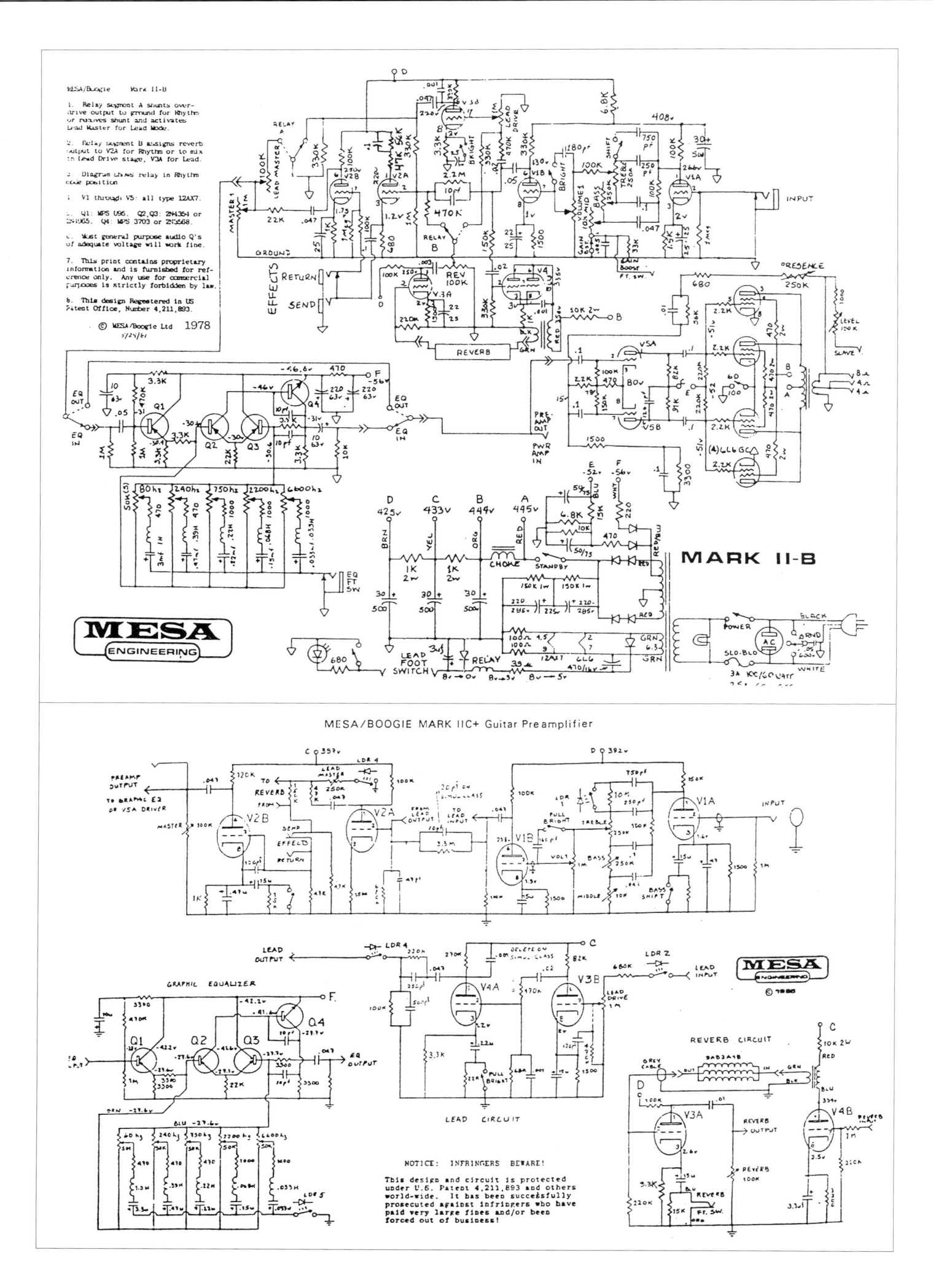

Mesa/Boogie

Mesa/Boogie

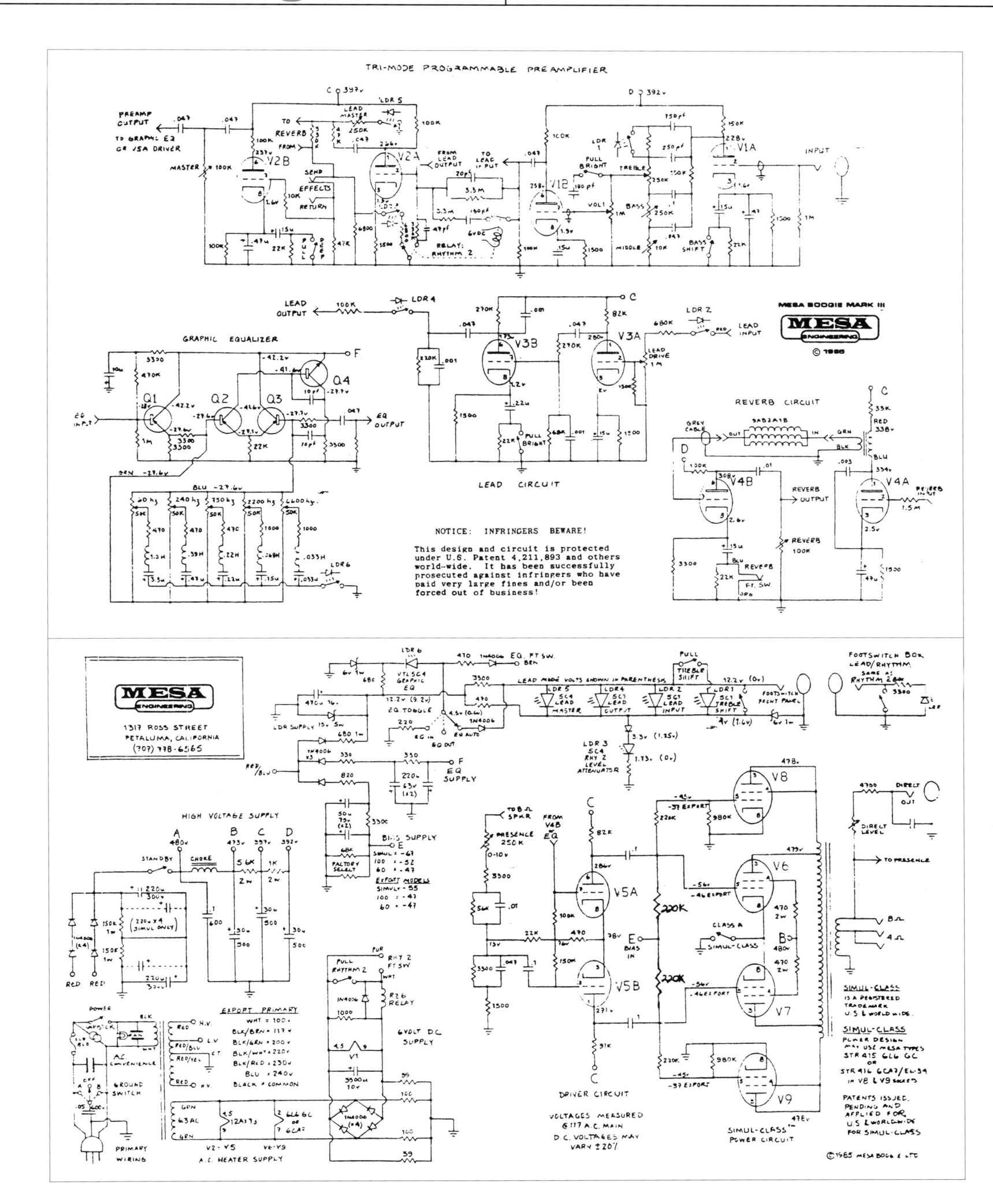

BOOGIE MARK III PREAMP
BOOGIE MARK III PWR/OUTPUT

BOOGIE BASS 400

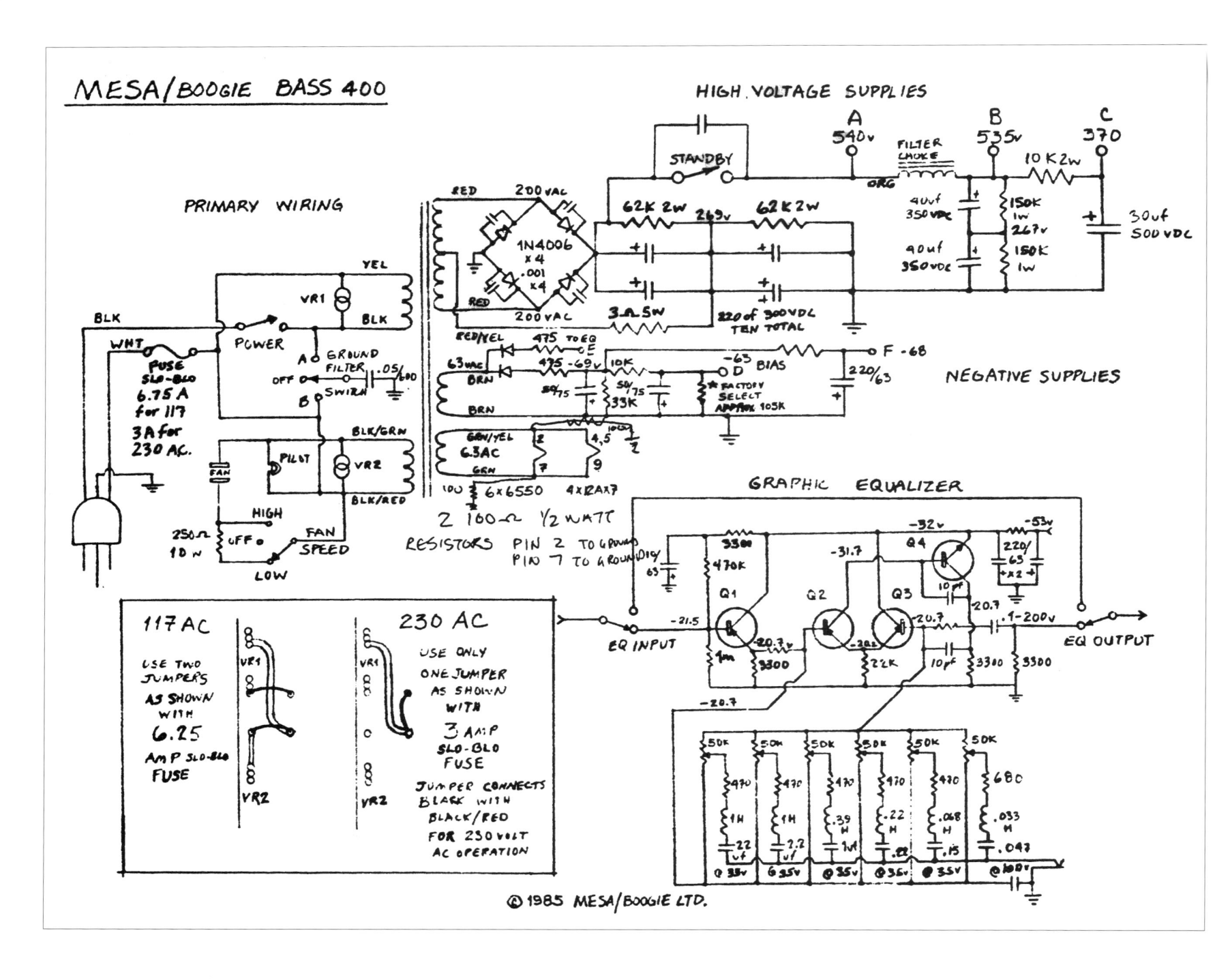

Mesa/Boogie

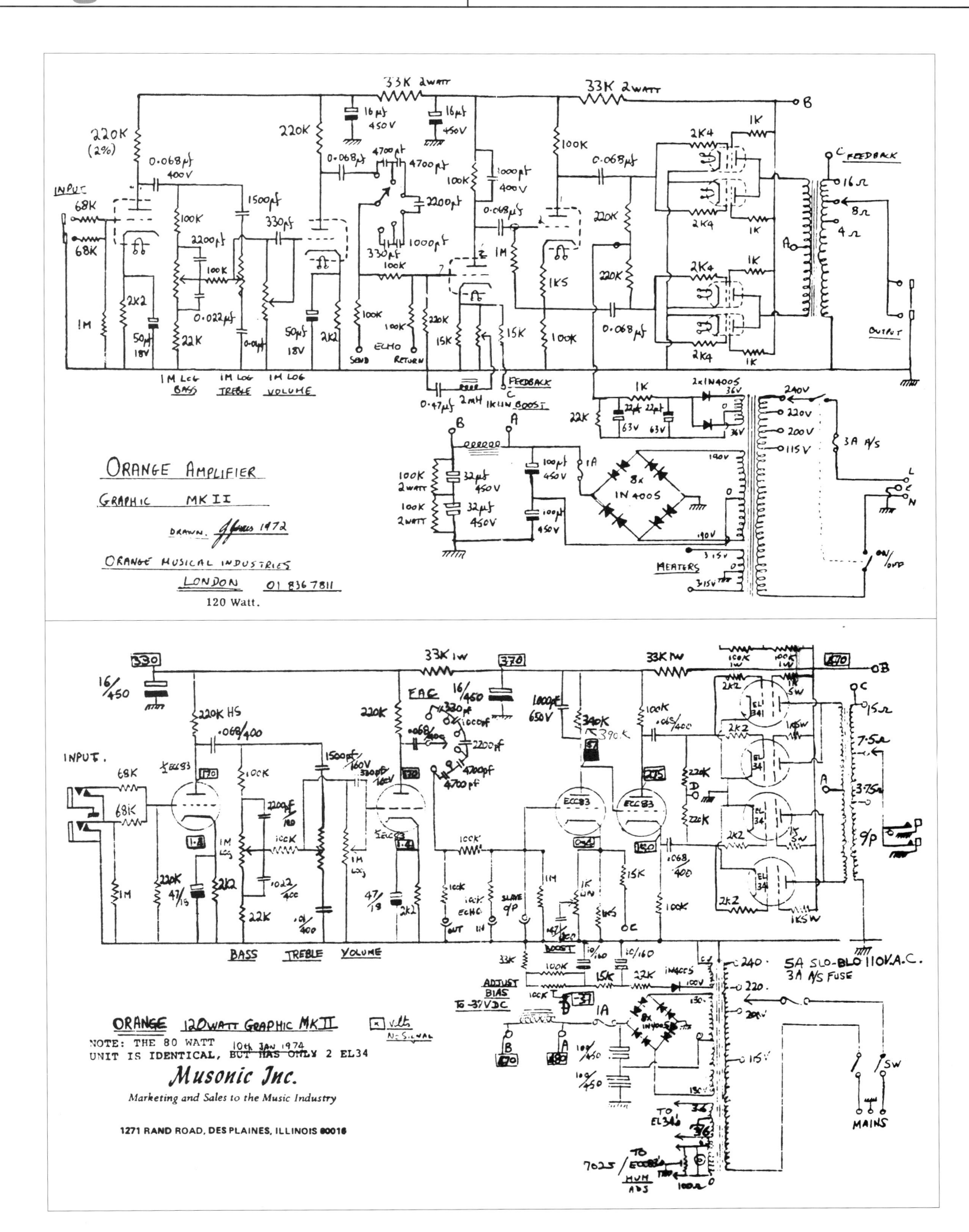

Orange

120W GRAPHIC (EARLY MODEL)
120W GRAPHIC MKII

MODEL 125

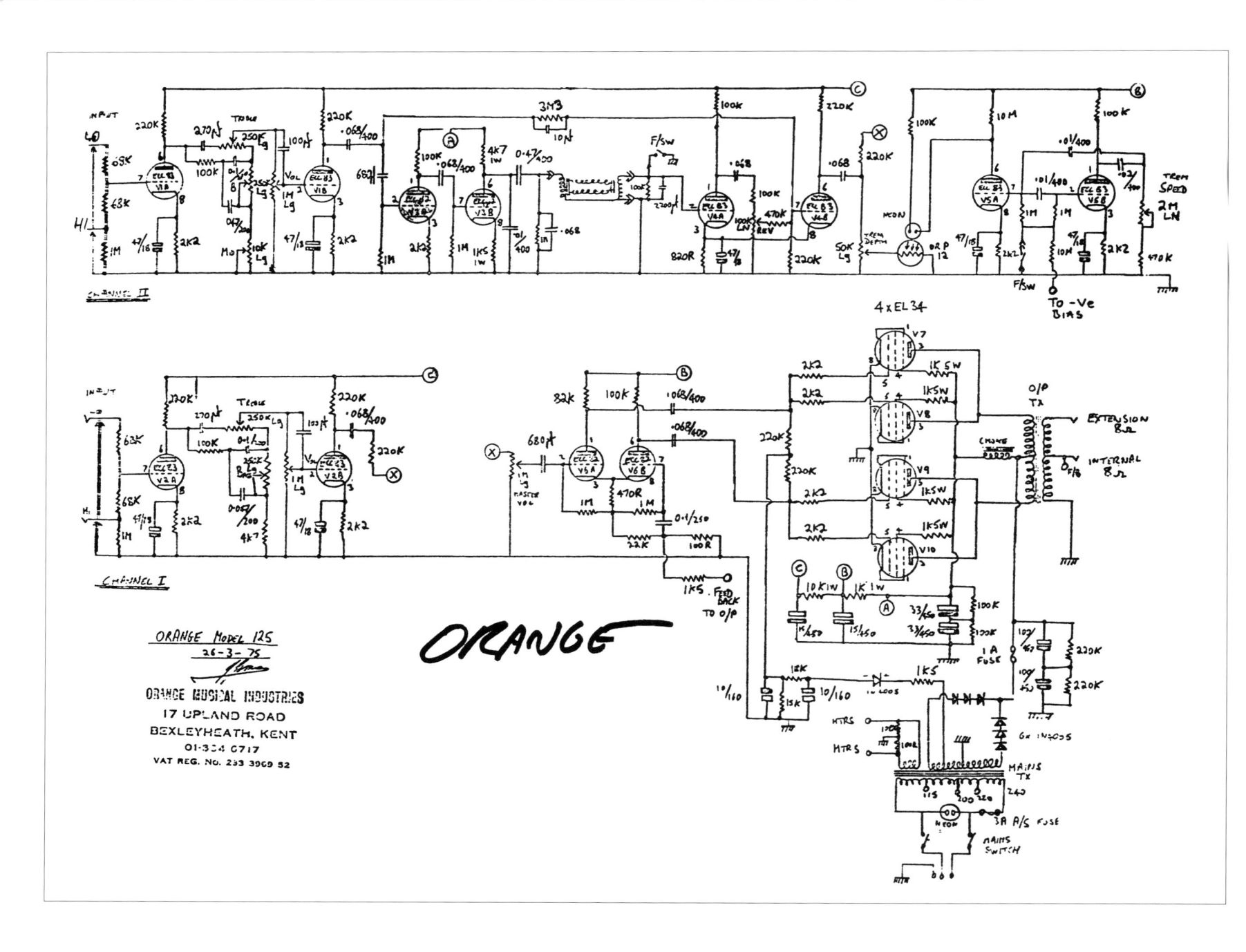

Orange

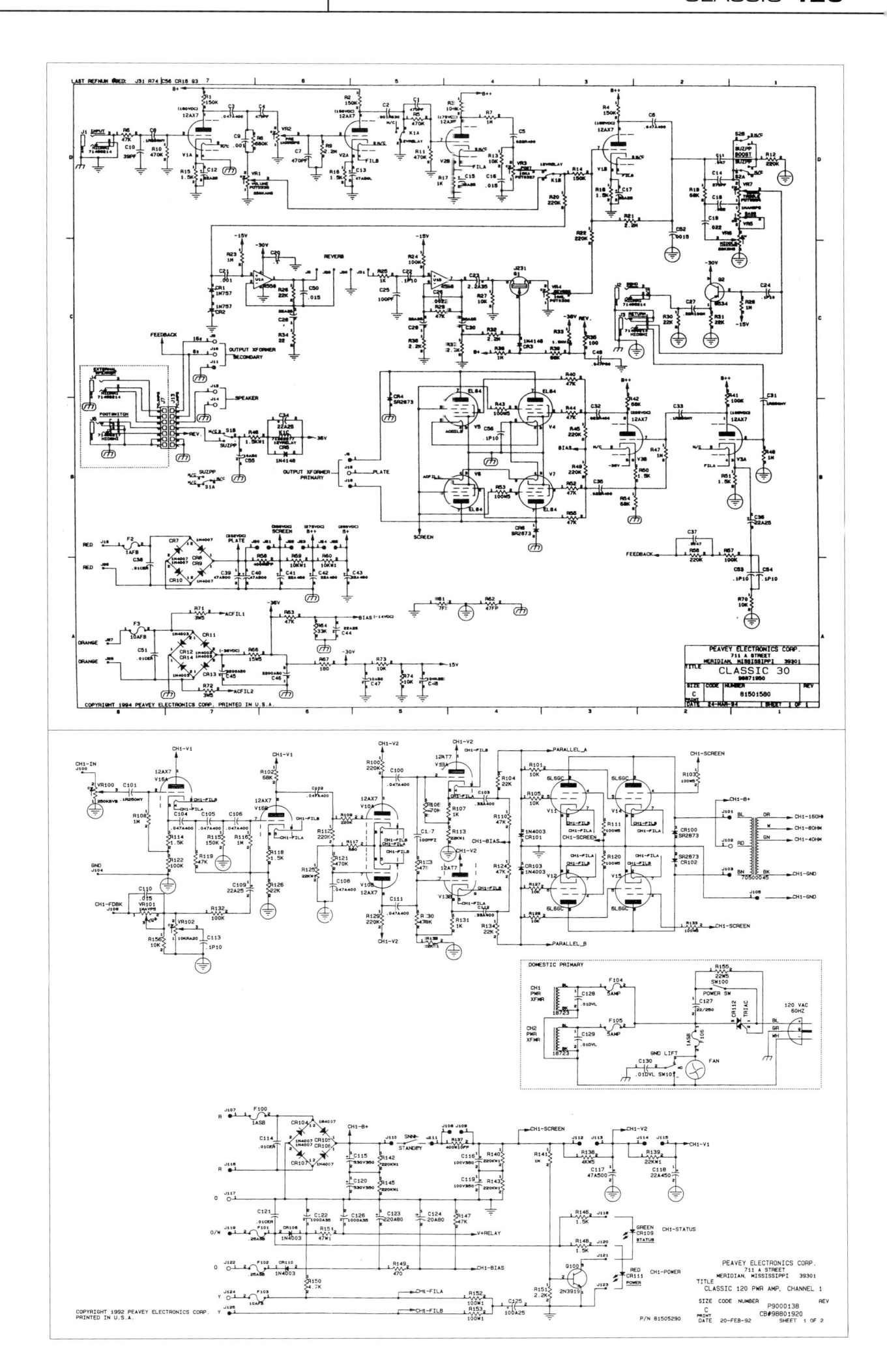

Peavey

CLASSIC 30
CLASSIC 120

THE TUBE AMP BOOK

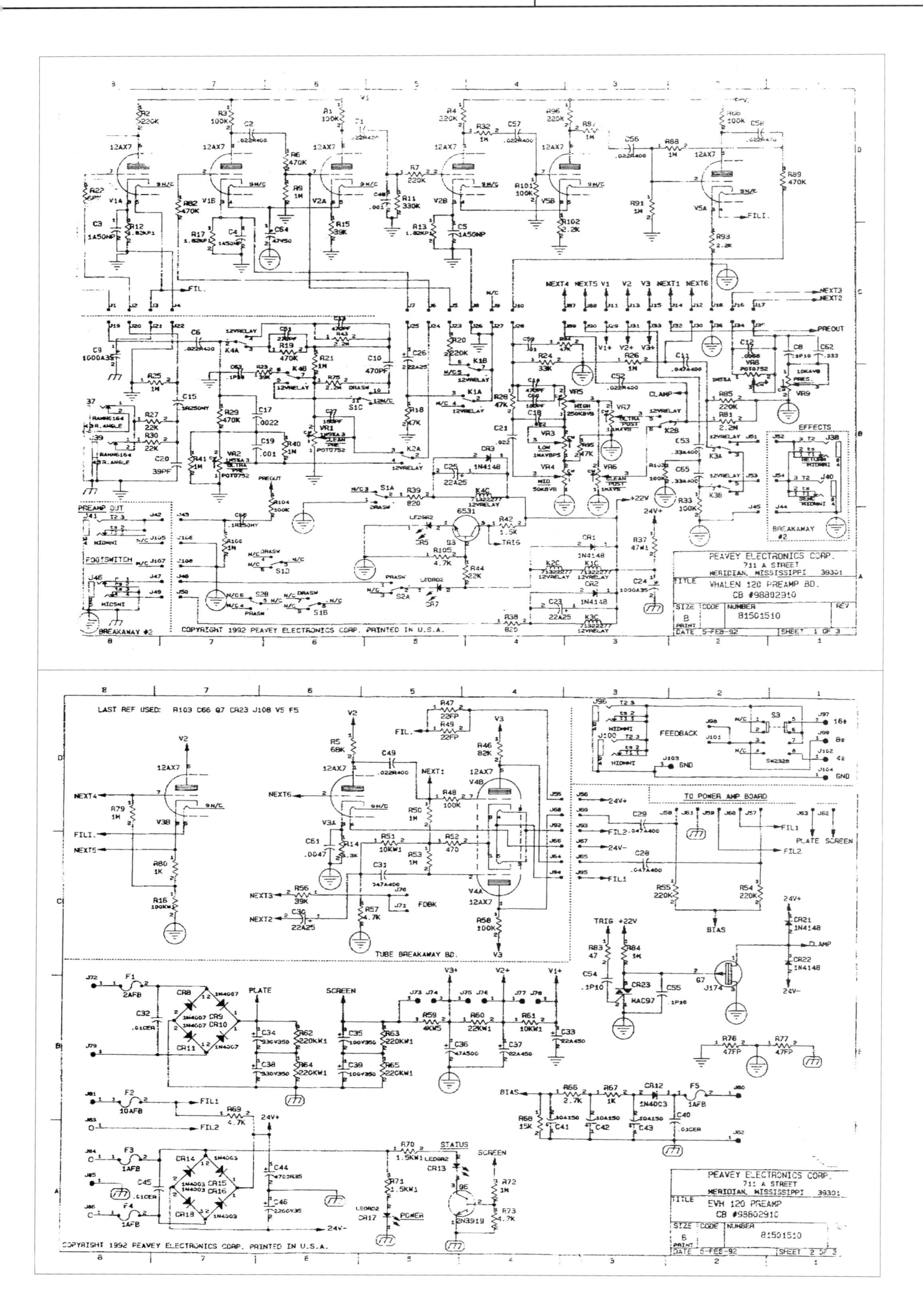

VAN HALEN 120 PREAMP (DIAG 1)
VAN HALEN 120 PREAMP (DIAG 2)

Peavey

144 THE TUBE AMP BOOK

RCA Theremin

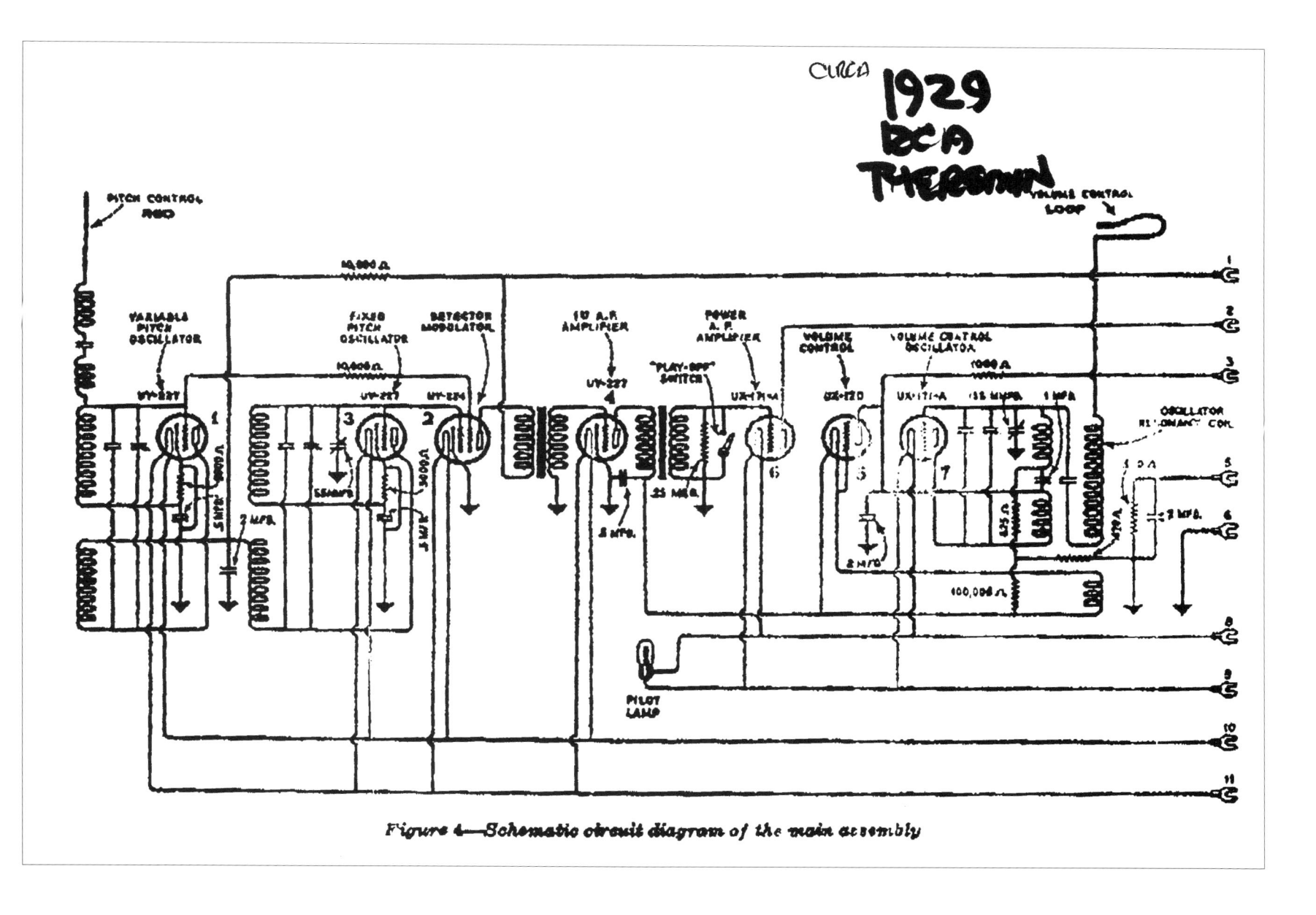

1929 **THEREMIN**

Figure 4—Schematic circuit diagram of the main assembly

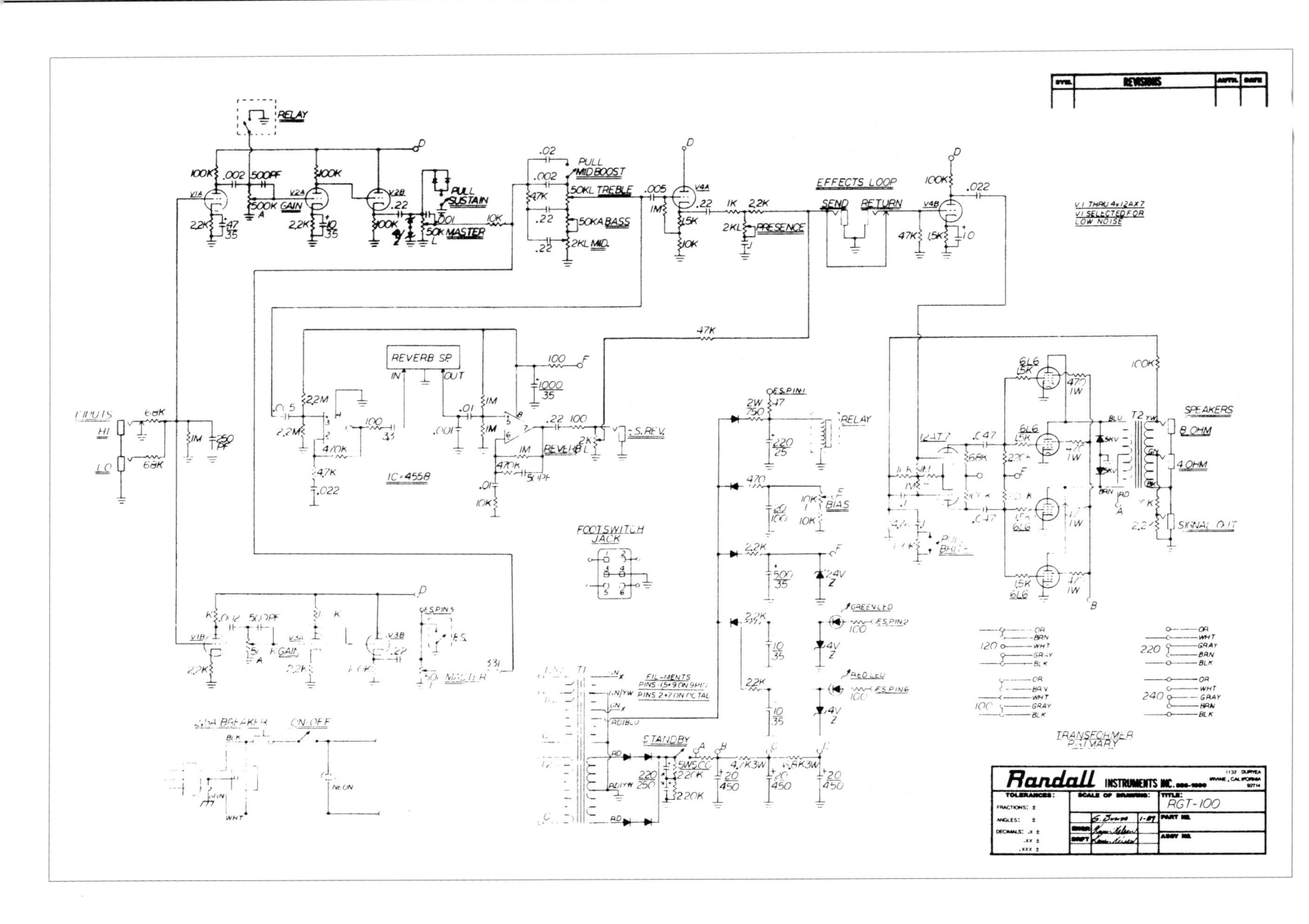

Revox

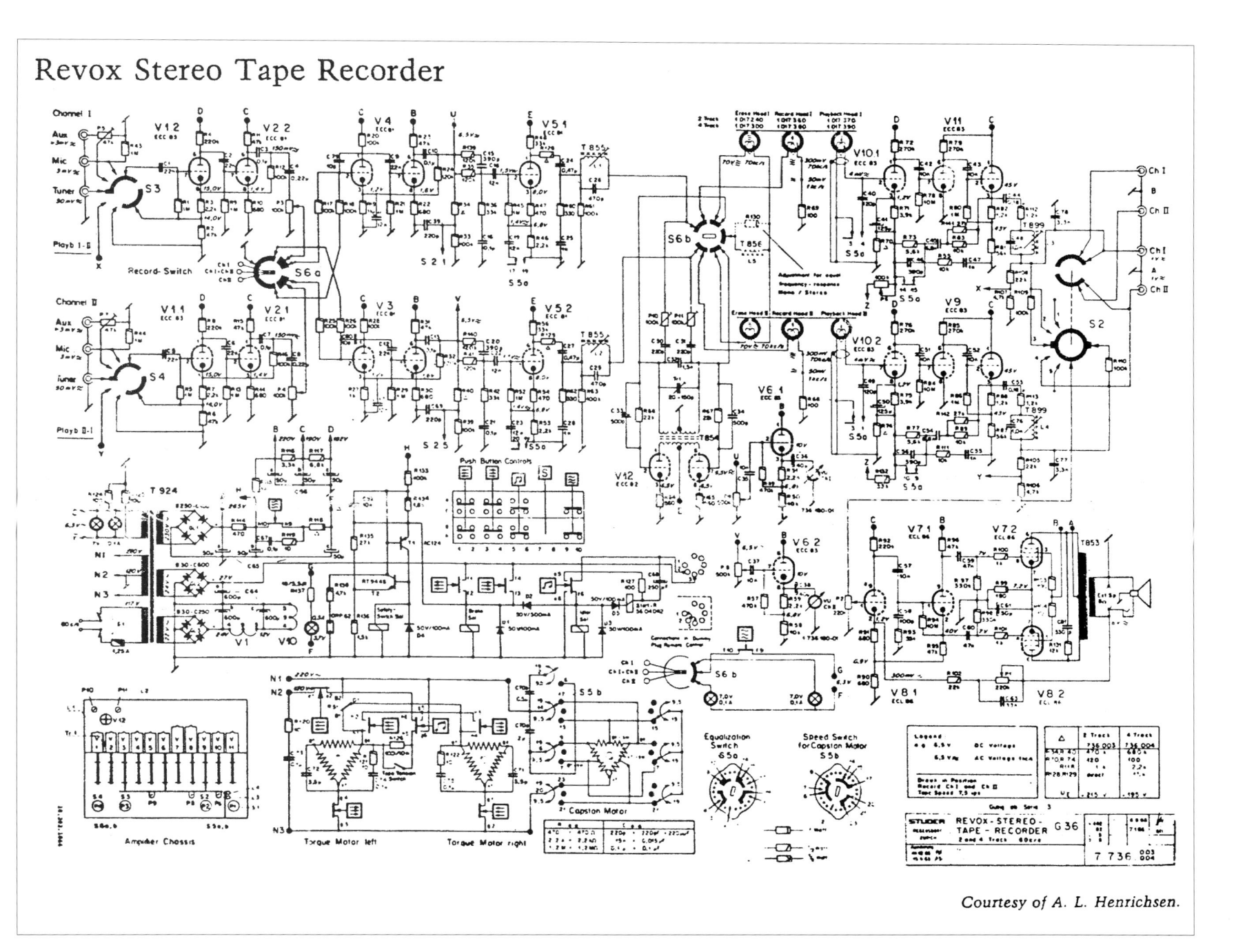

STEREO TAPE RECORDER

Revox Stereo Tape Recorder

Courtesy of A. L. Henrichsen.

B16/16D
B SERIES GUITAR

Rickenbacker

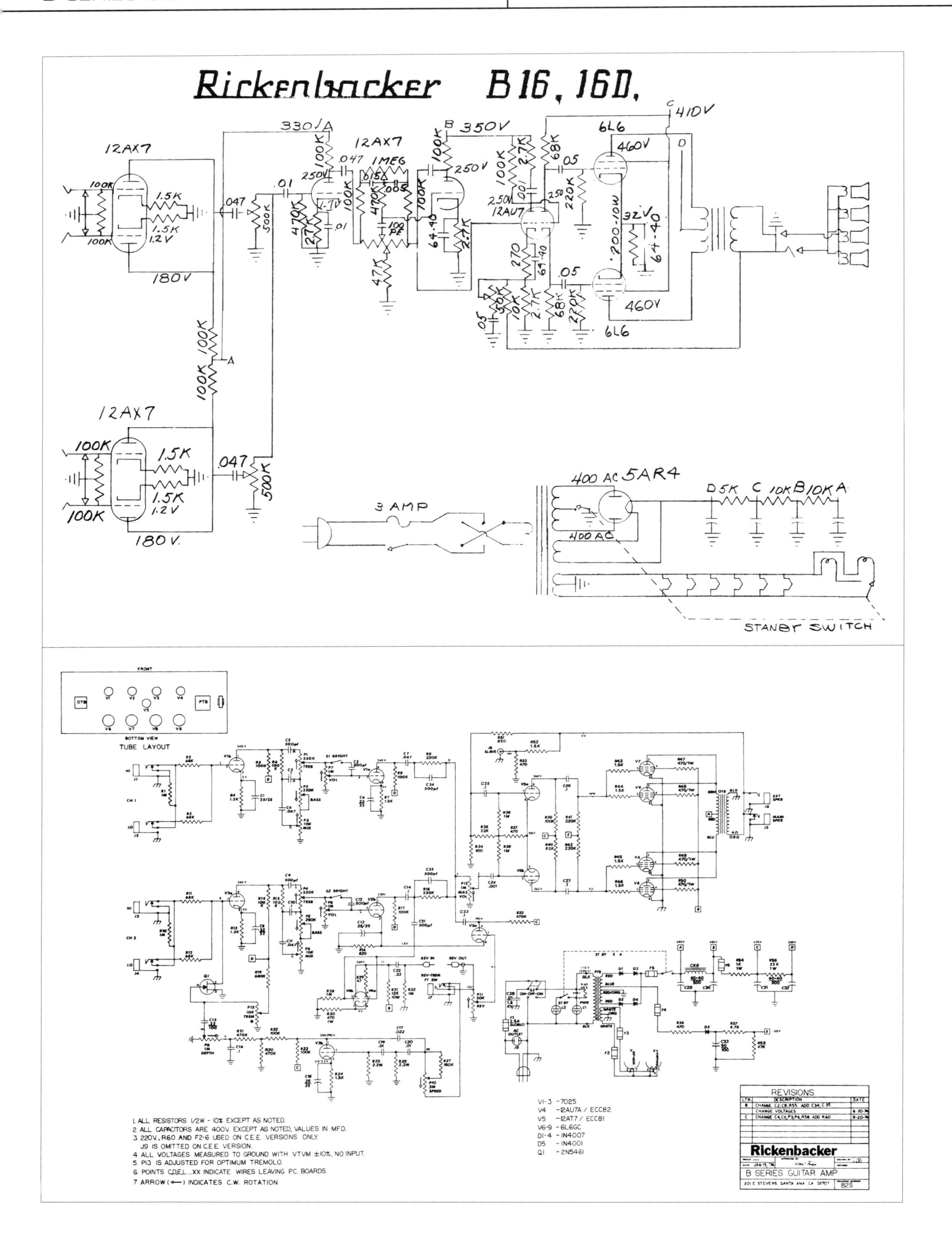

Selmer

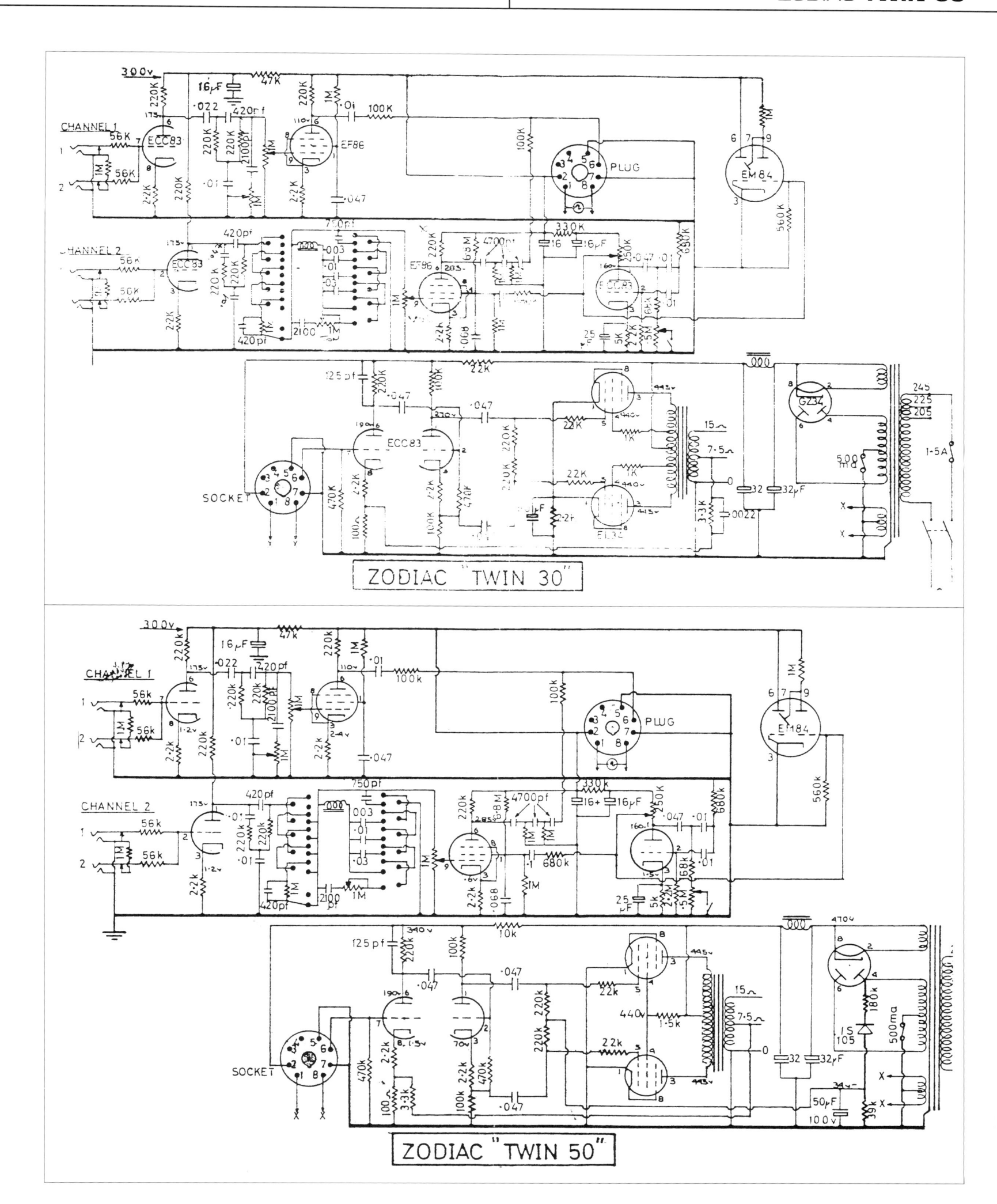

ZODIAC **TWIN 30**
ZODIAC **TWIN 50**

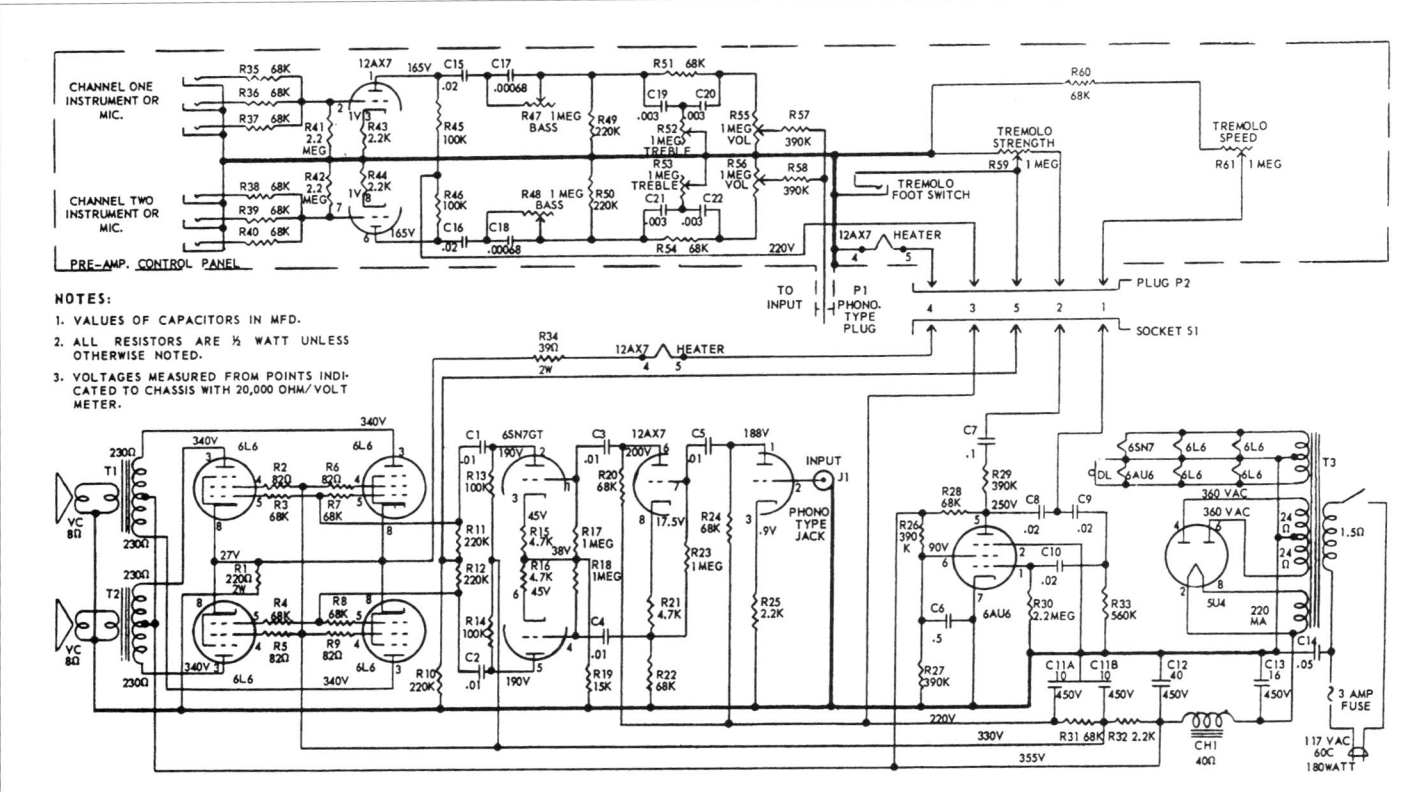

Silvertone

1396
1474

SILVERTONE MUSICAL INSTRUMENT AMPLIFIER MODEL 1396, Ch. 185.10500

SCHEMATIC DIAGRAM COURTESY SEARS, ROEBUCK & CO.

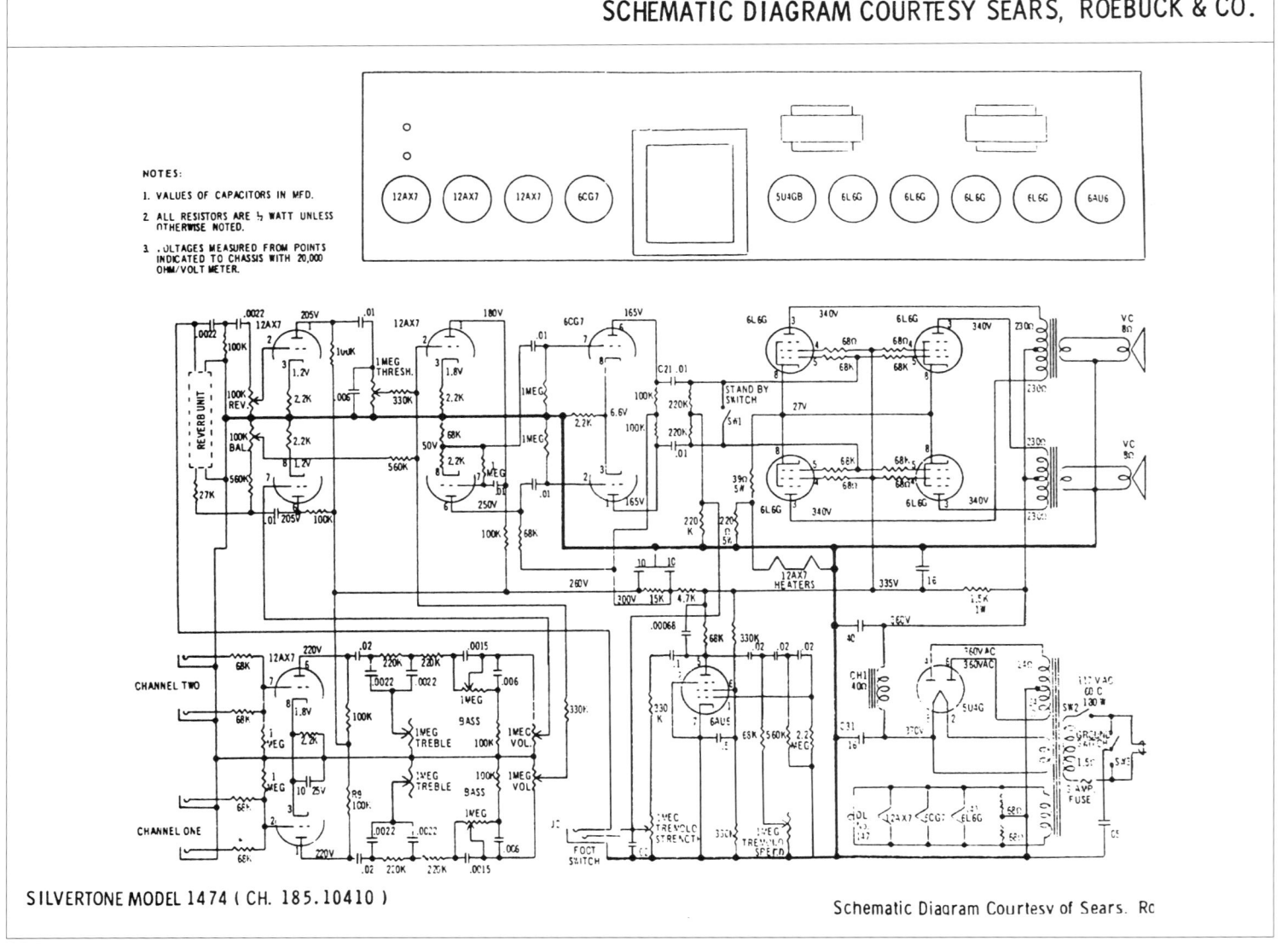

SILVERTONE MODEL 1474 (CH. 185.10410)

Schematic Diagram Courtesy of Sears, Roebuck & Co.

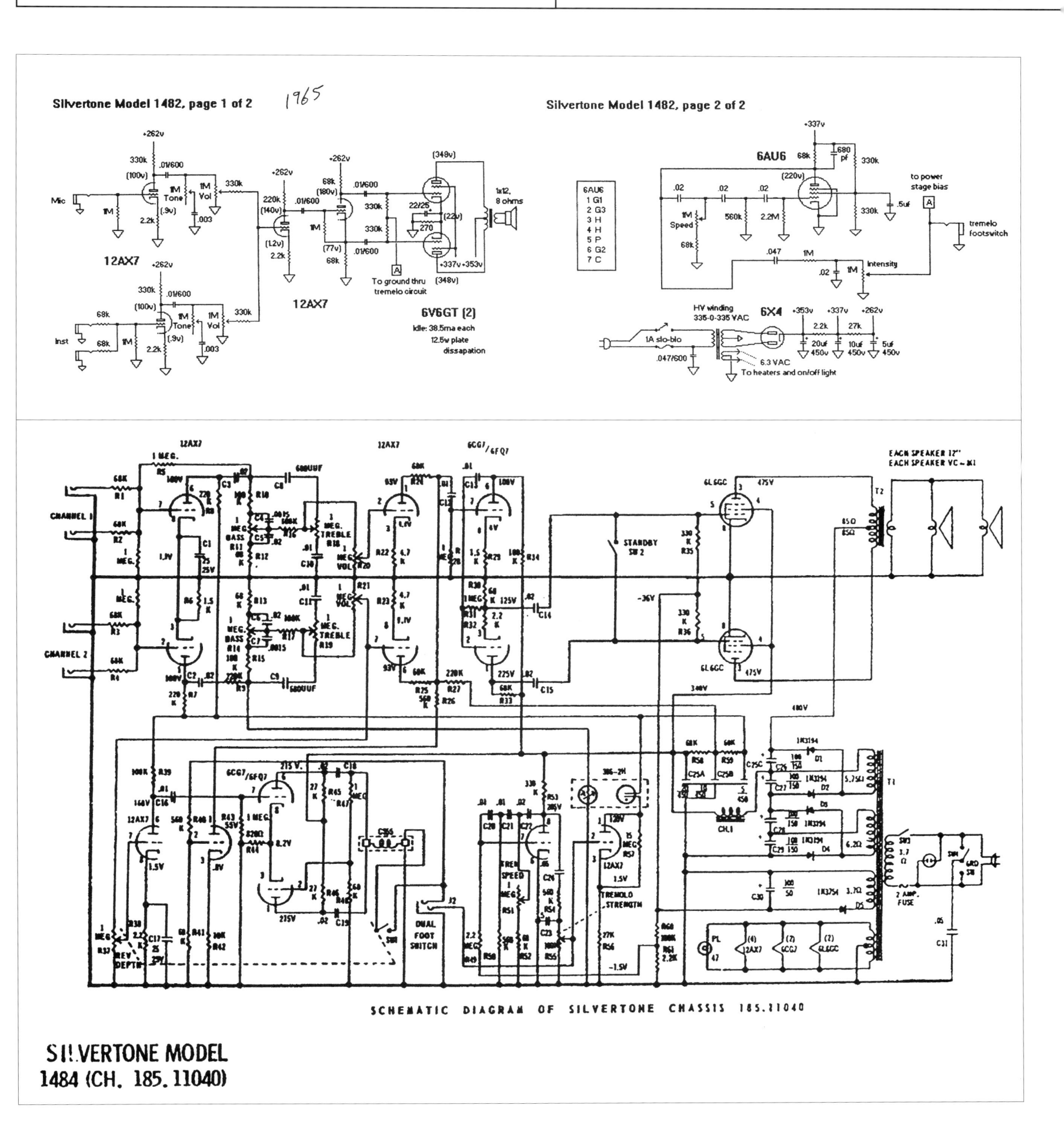

Silvertone

1482
1484

1485
2 X 6V6 AMP

Silvertone

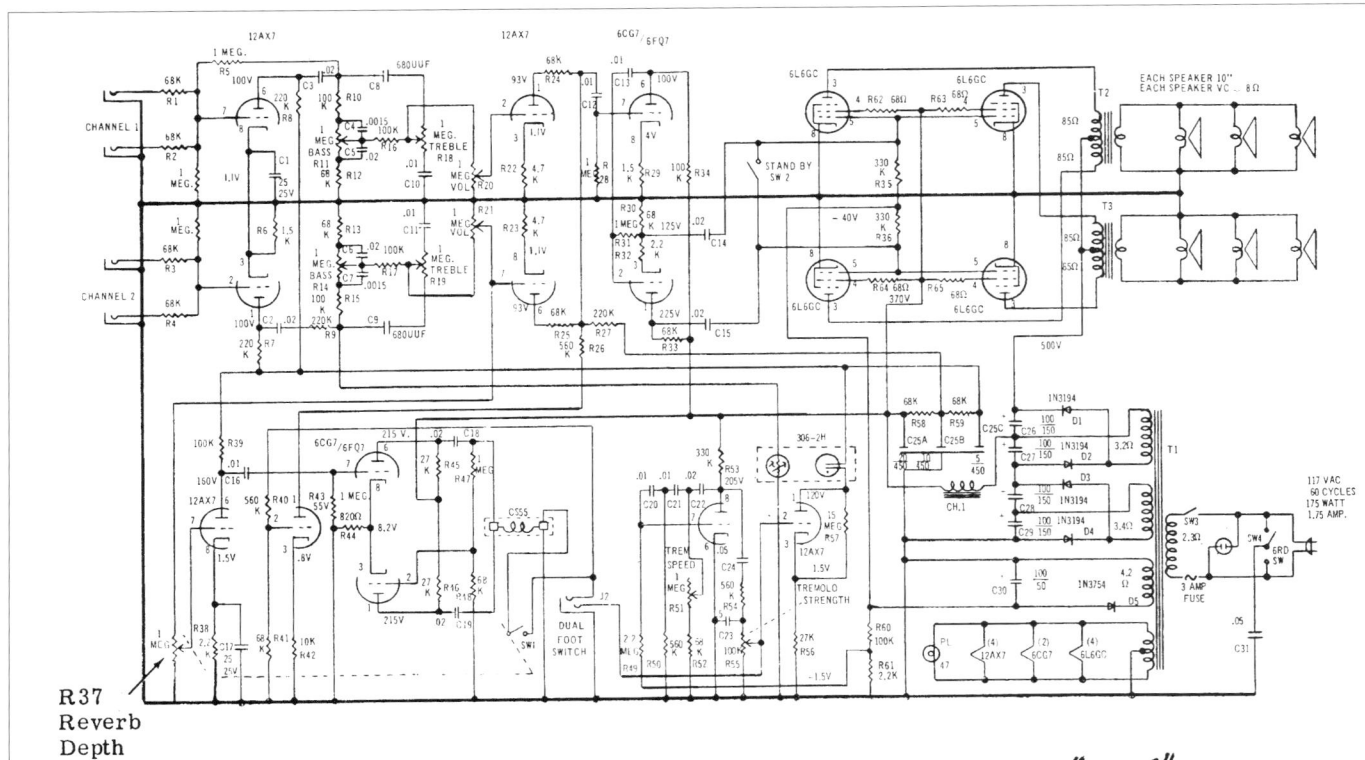

SCHEMATIC DIAGRAM COURTESY OF SEARS ROEBUCK & CO. *April 1964*

Less two "6L6s" similar model 14

SILVERTONE MODEL 1485 (CH. 185.11050)

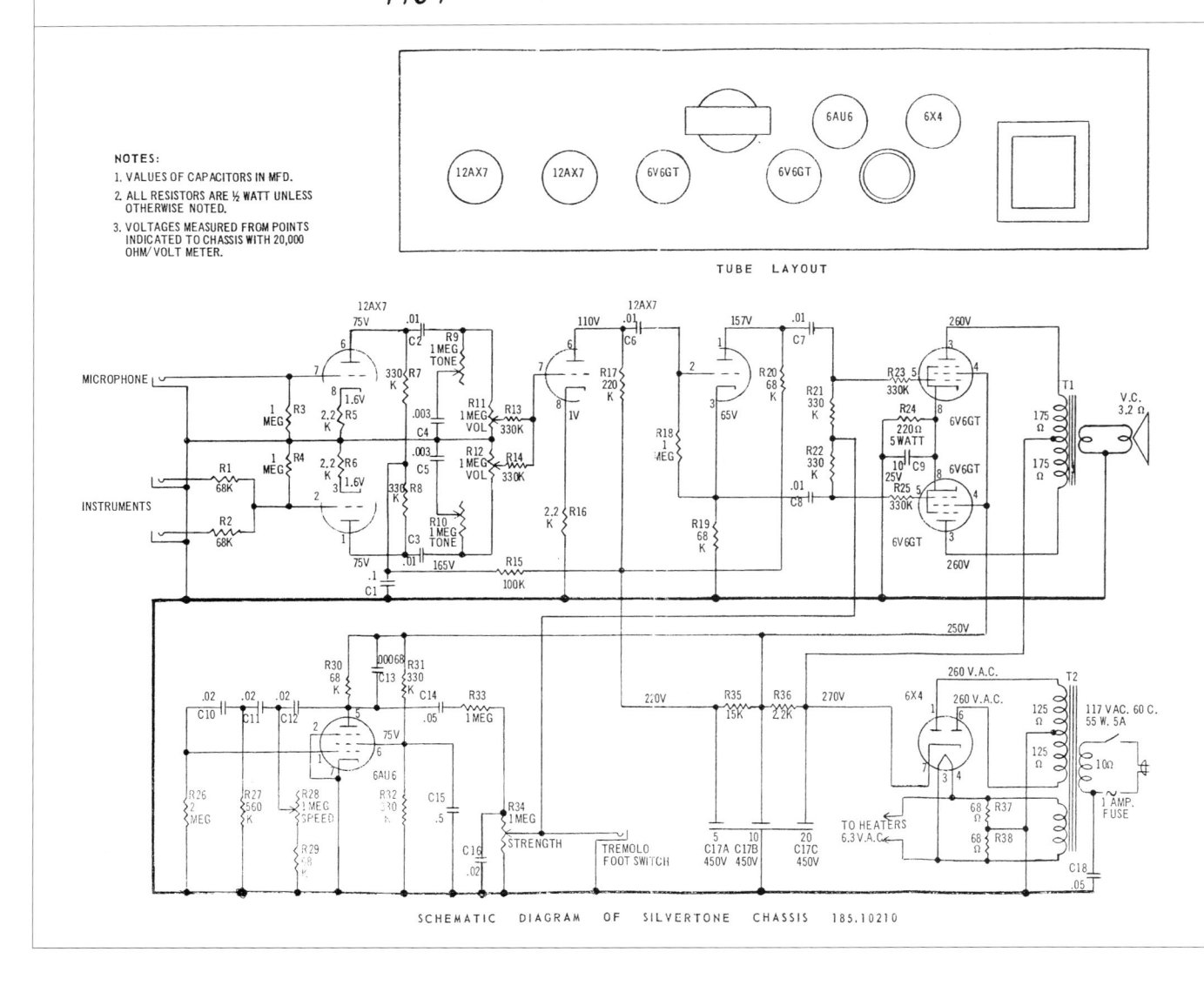

NOTES:
1. VALUES OF CAPACITORS IN MFD.
2. ALL RESISTORS ARE ½ WATT UNLESS OTHERWISE NOTED.
3. VOLTAGES MEASURED FROM POINTS INDICATED TO CHASSIS WITH 20,000 OHM/VOLT METER.

TUBE LAYOUT

R37 Reverb Depth

SCHEMATIC DIAGRAM OF SILVERTONE CHASSIS 185.10210

152 THE TUBE AMP BOOK

Silvertone

2 X 6V6 AMP REVISED
OLD 2 X 6L6 TREMOLO AMP

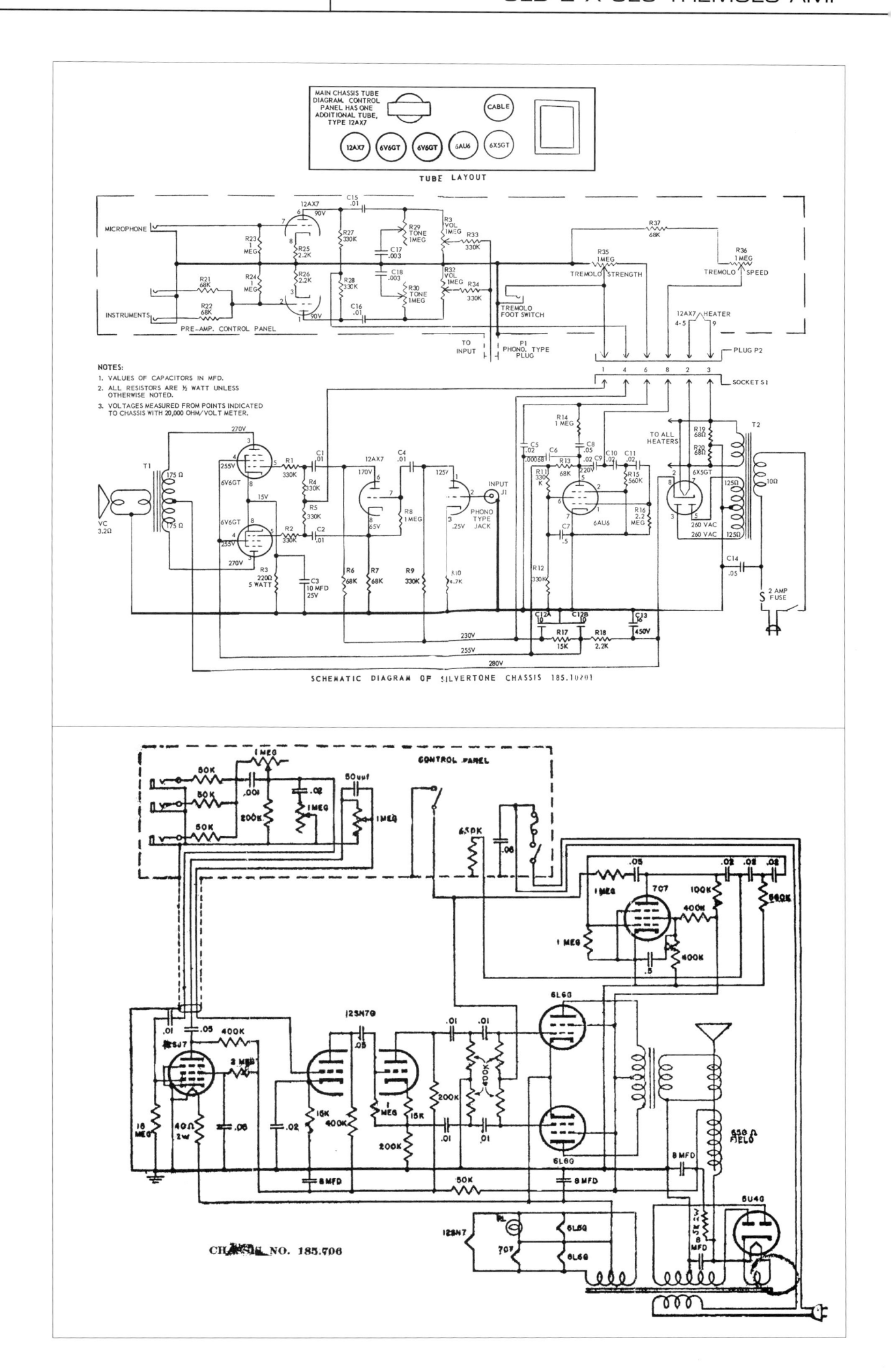

THE TUBE AMP BOOK 153

BIG 12E & 25RT
STUDIO-20

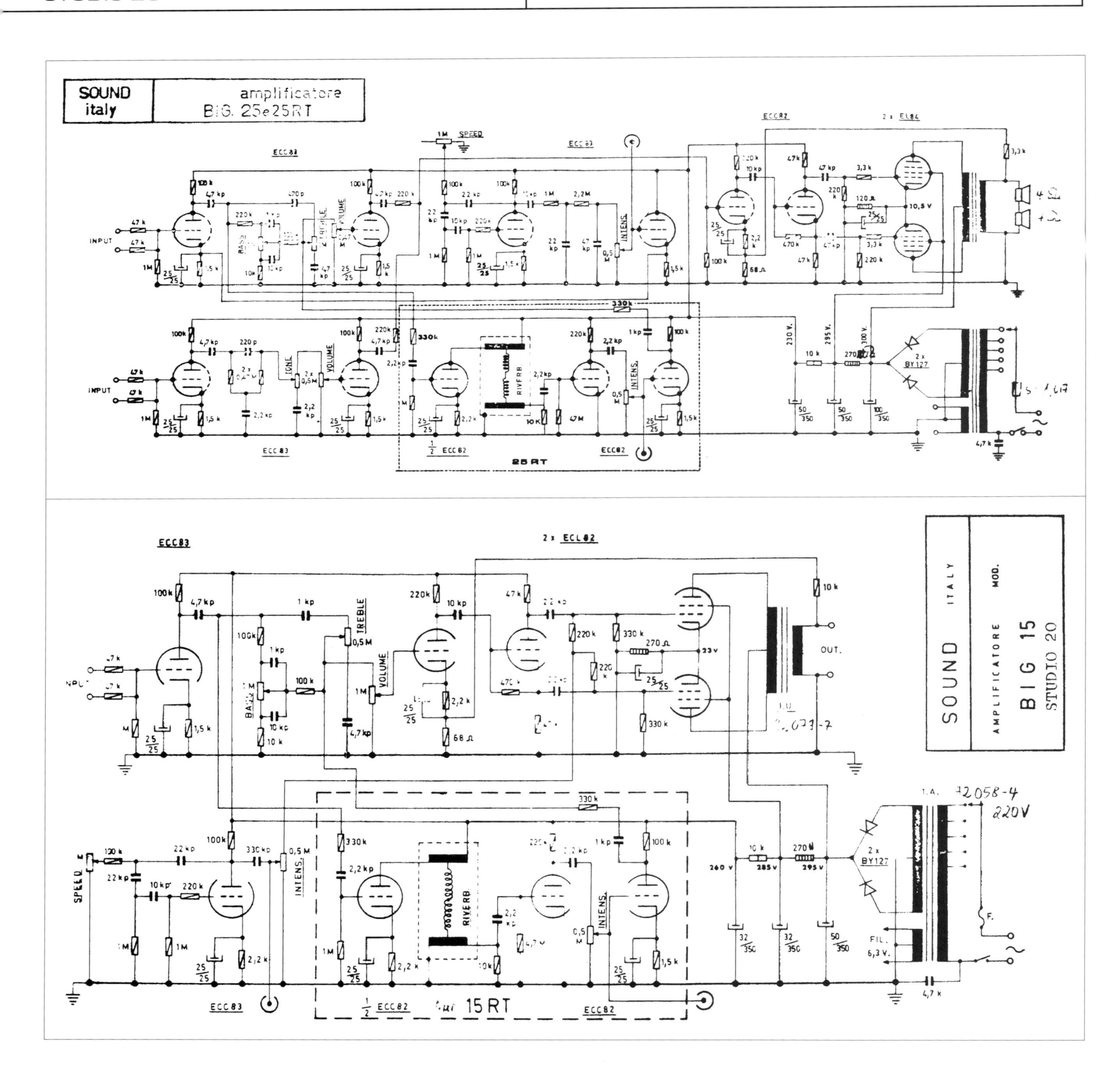

Sound

Sound City

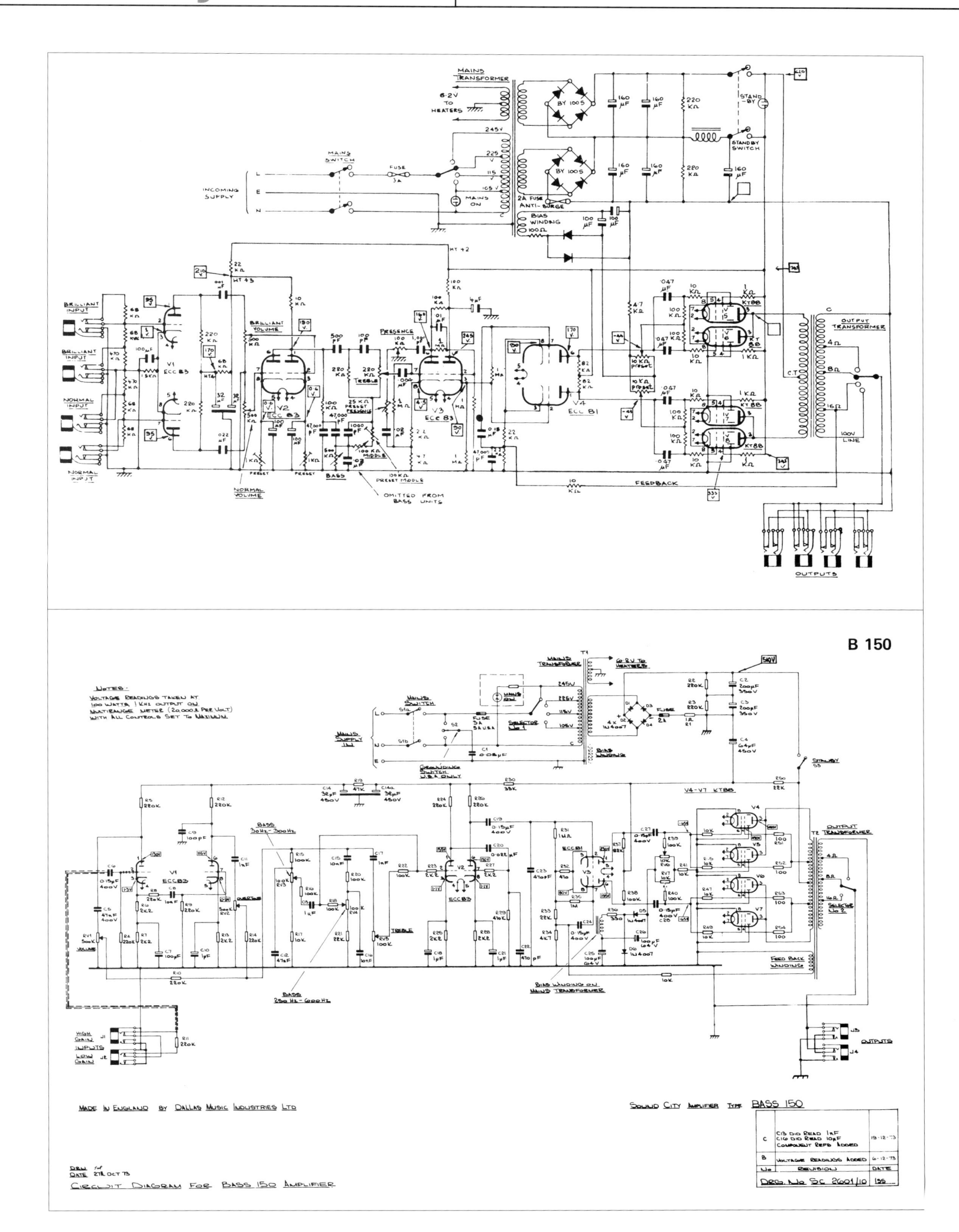

200W LE
BASS 150W

CONCORD
L-B 120 MARK IV

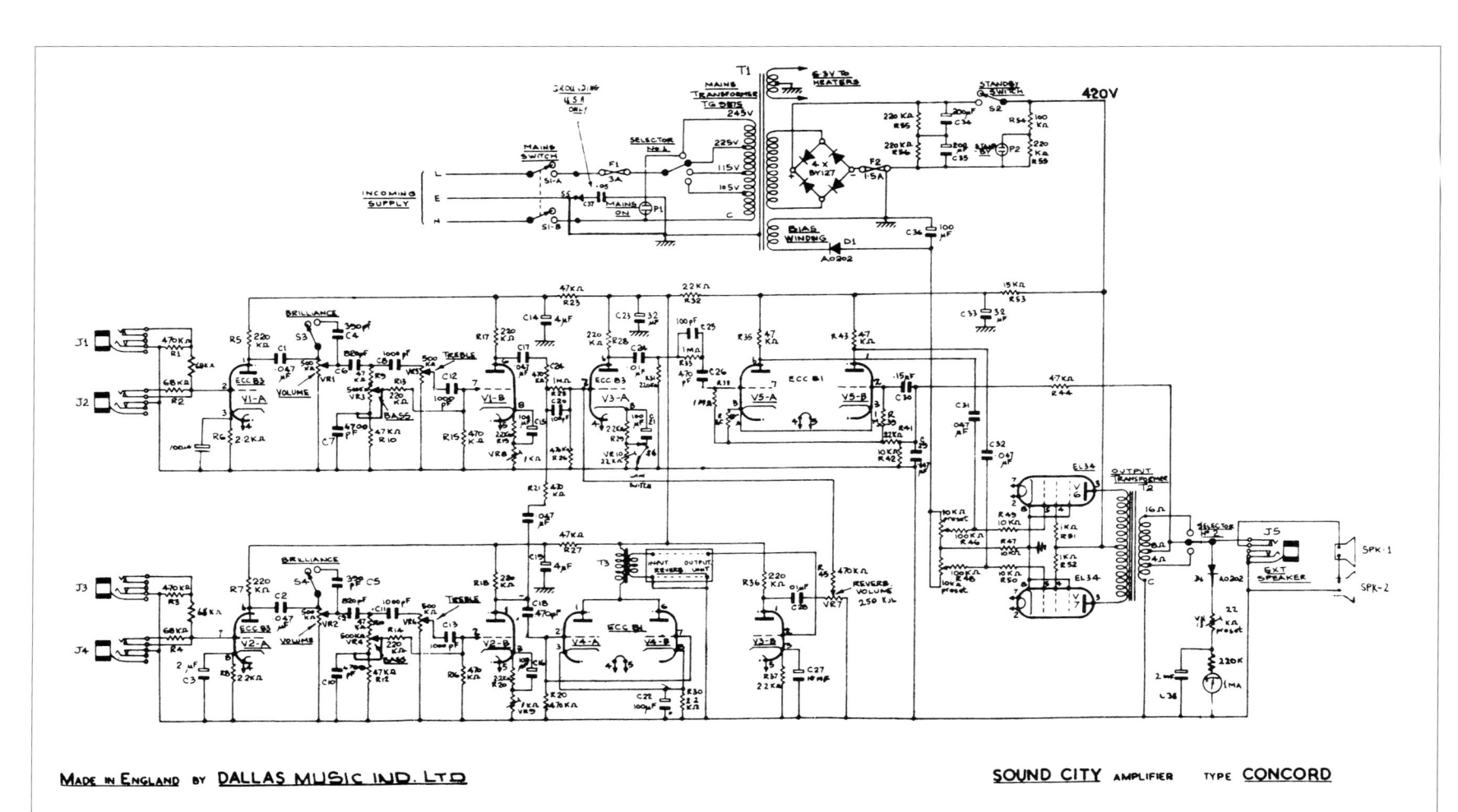

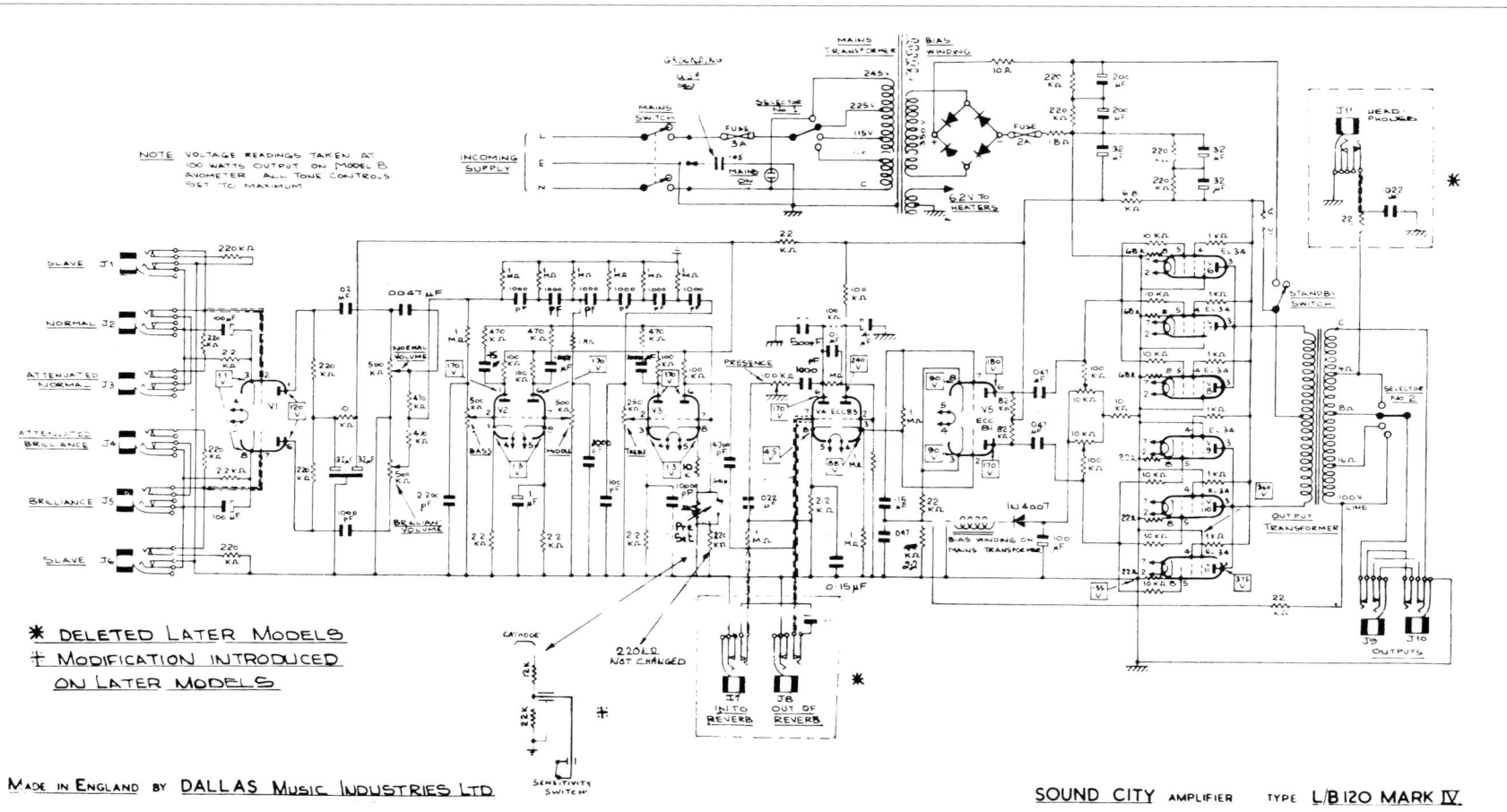

Sound City

Sound City

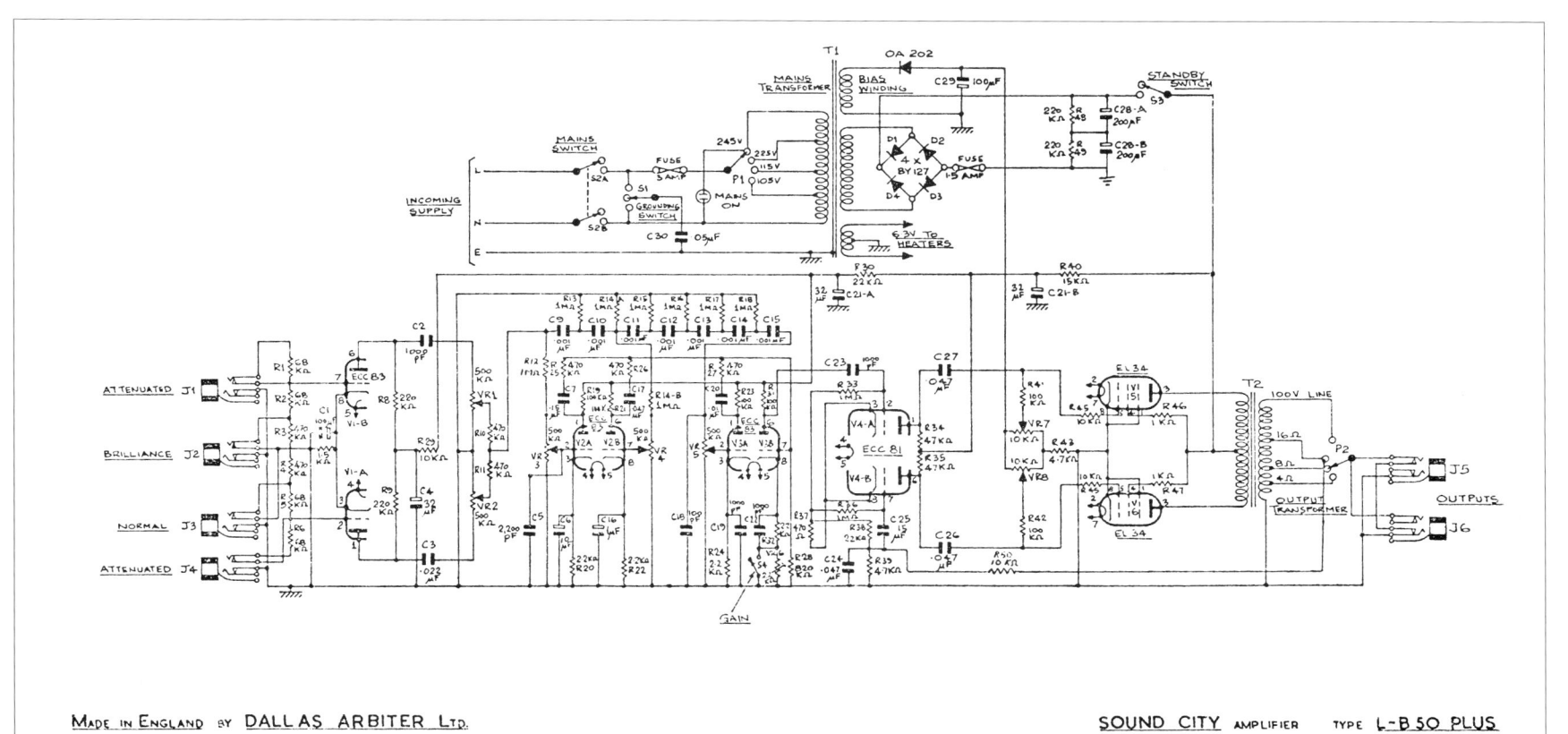

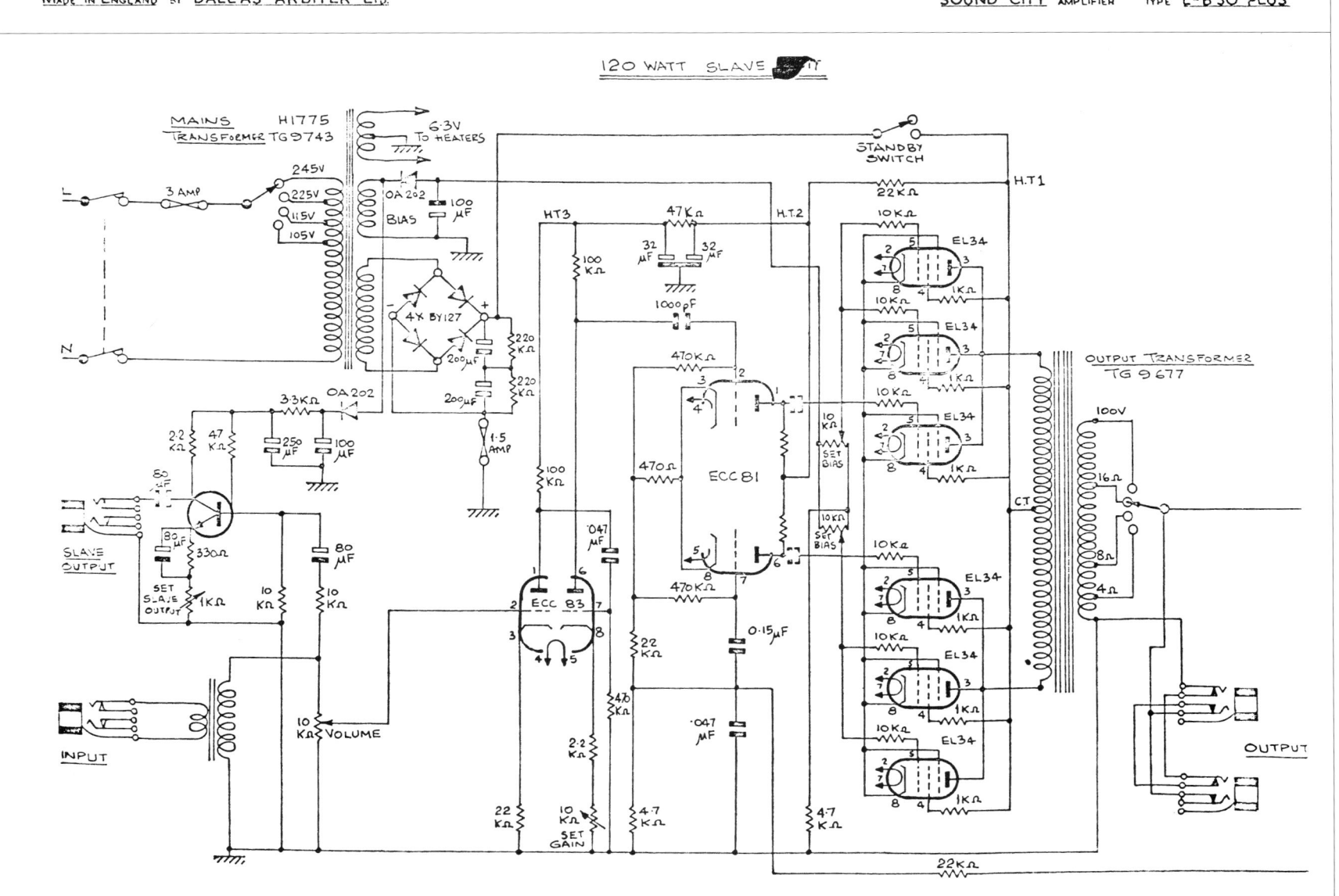

L-B PLUS 50
120W SLAVE

MIG 50 MASTER VOLUME
MIG 60 LEAD

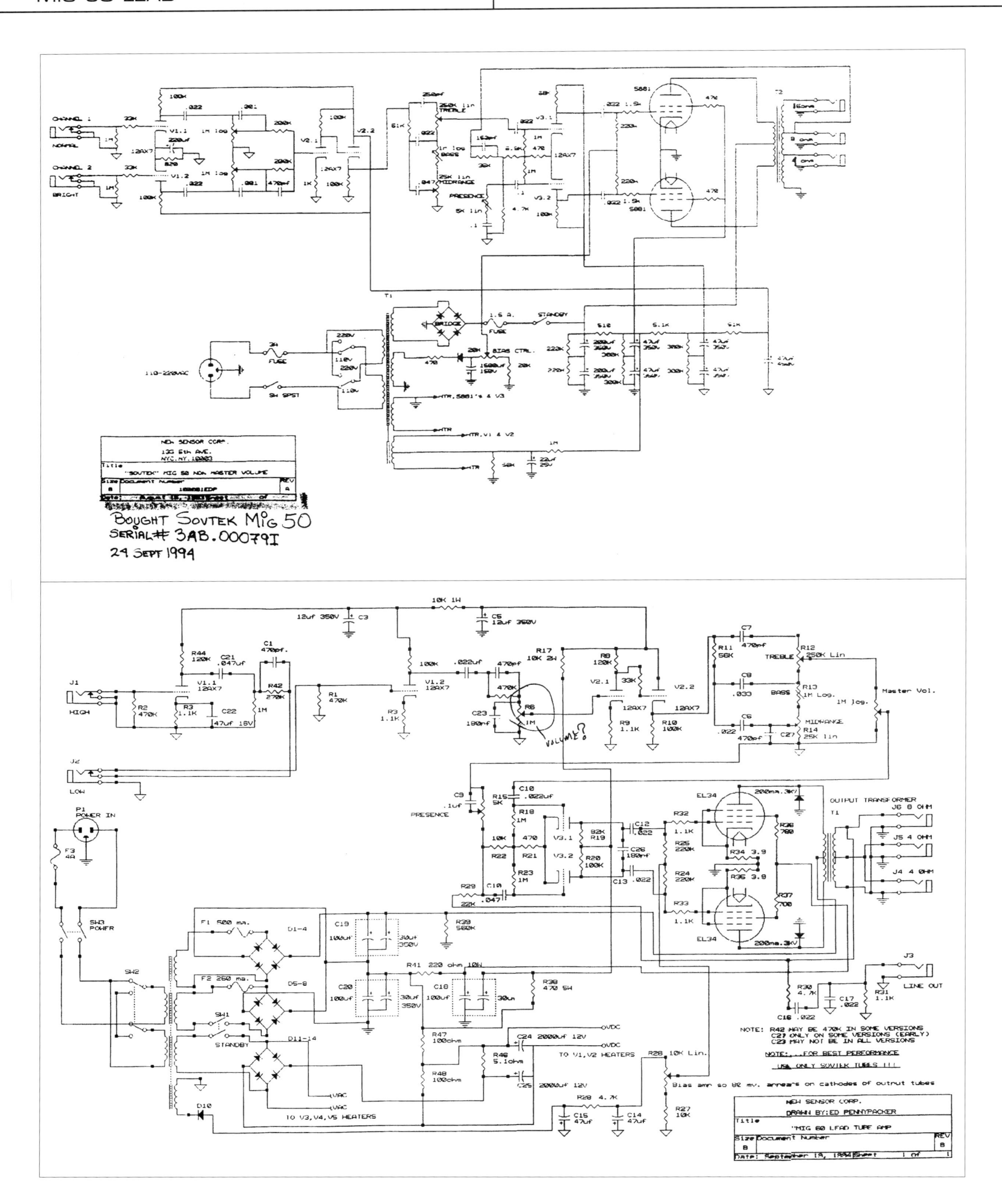

Sovtek

158 THE TUBE AMP BOOK

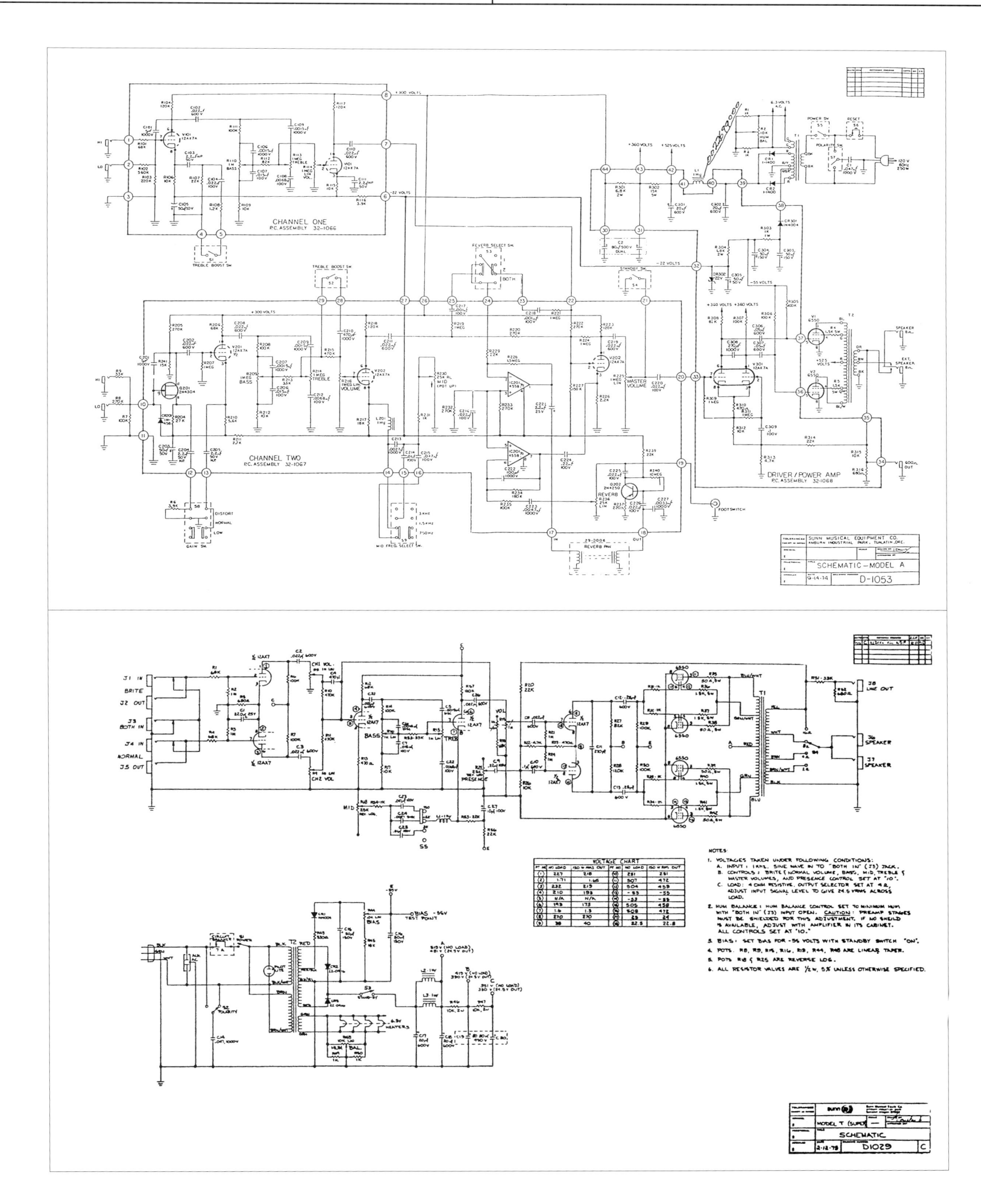

Sunn

MODEL A
MODEL T SUPER

THE TUBE AMP BOOK

1200 / 2000S — Sunn

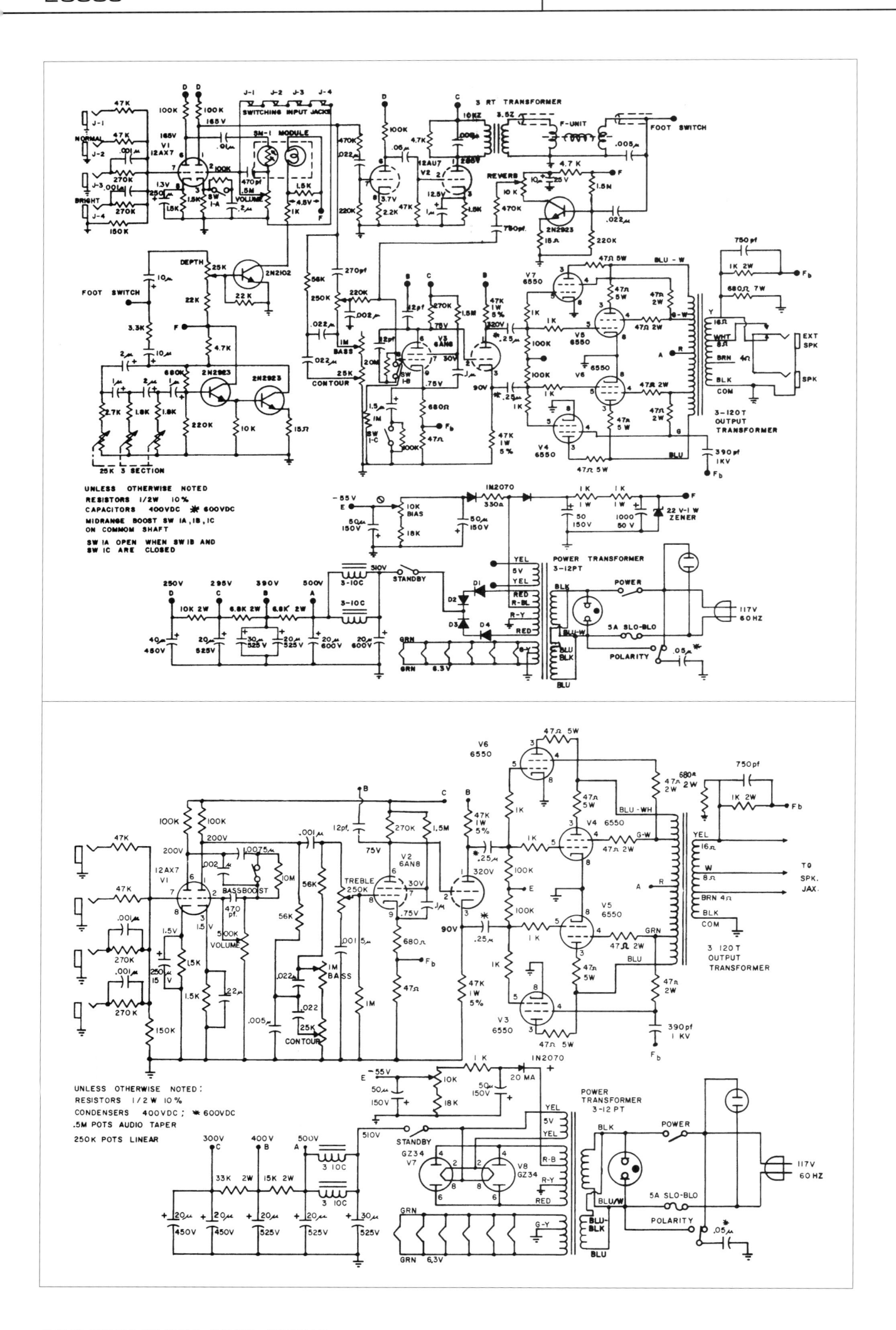

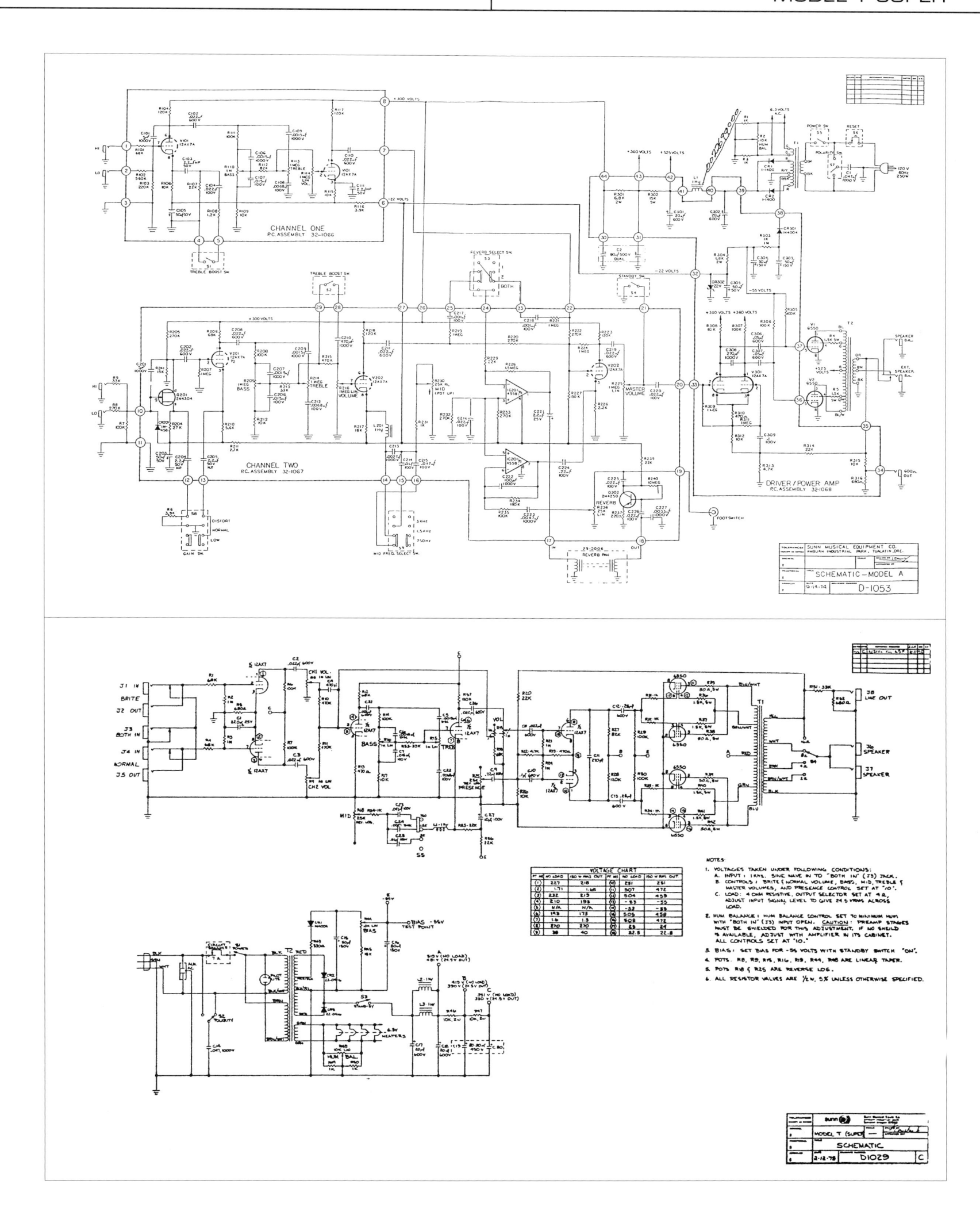

Sunn

MODEL A
MODEL T SUPER

THE TUBE AMP BOOK 161

SENTURA 1
SONIC II / 1-40 / 200S / SORADO

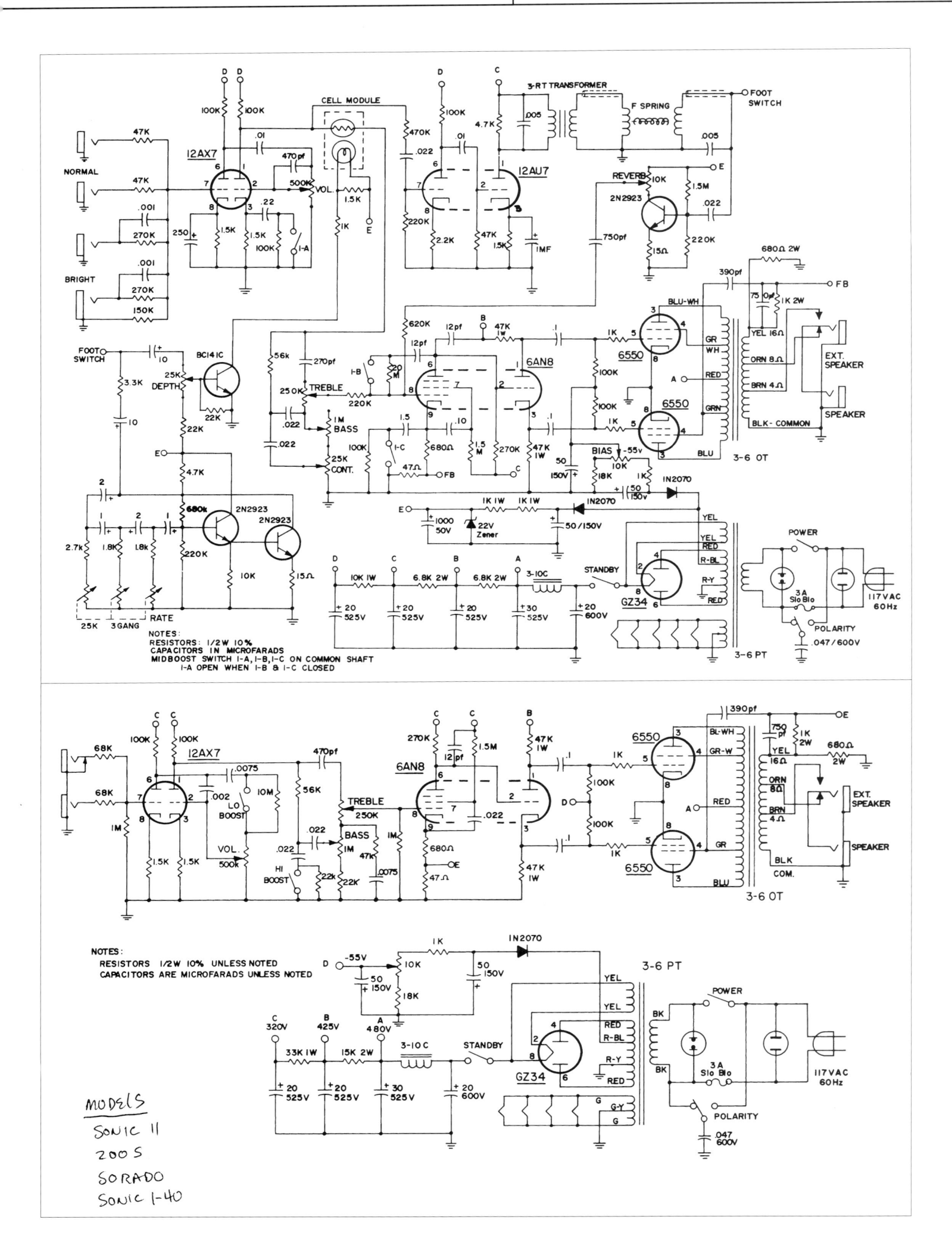

SUNN

162 THE TUBE AMP BOOK

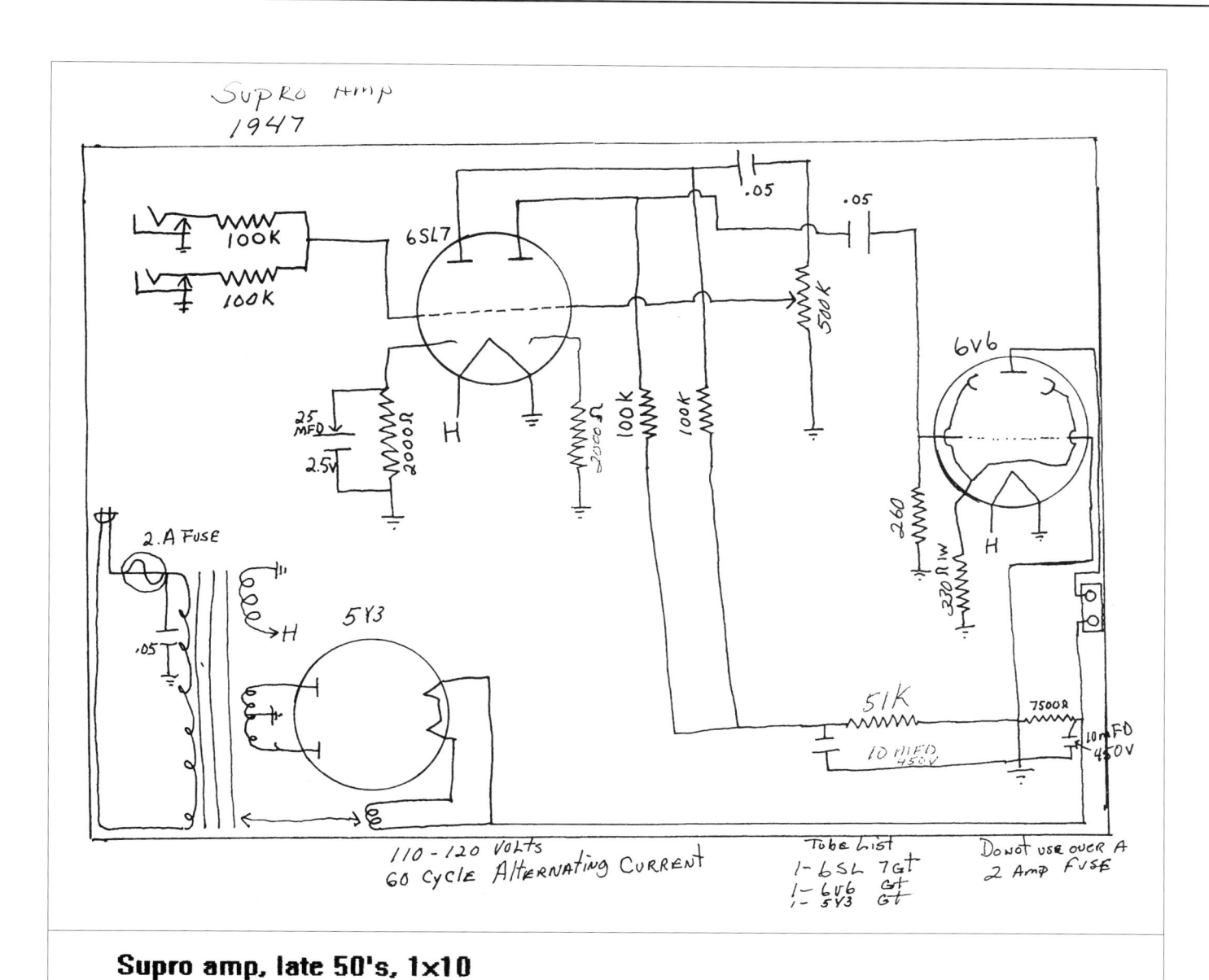

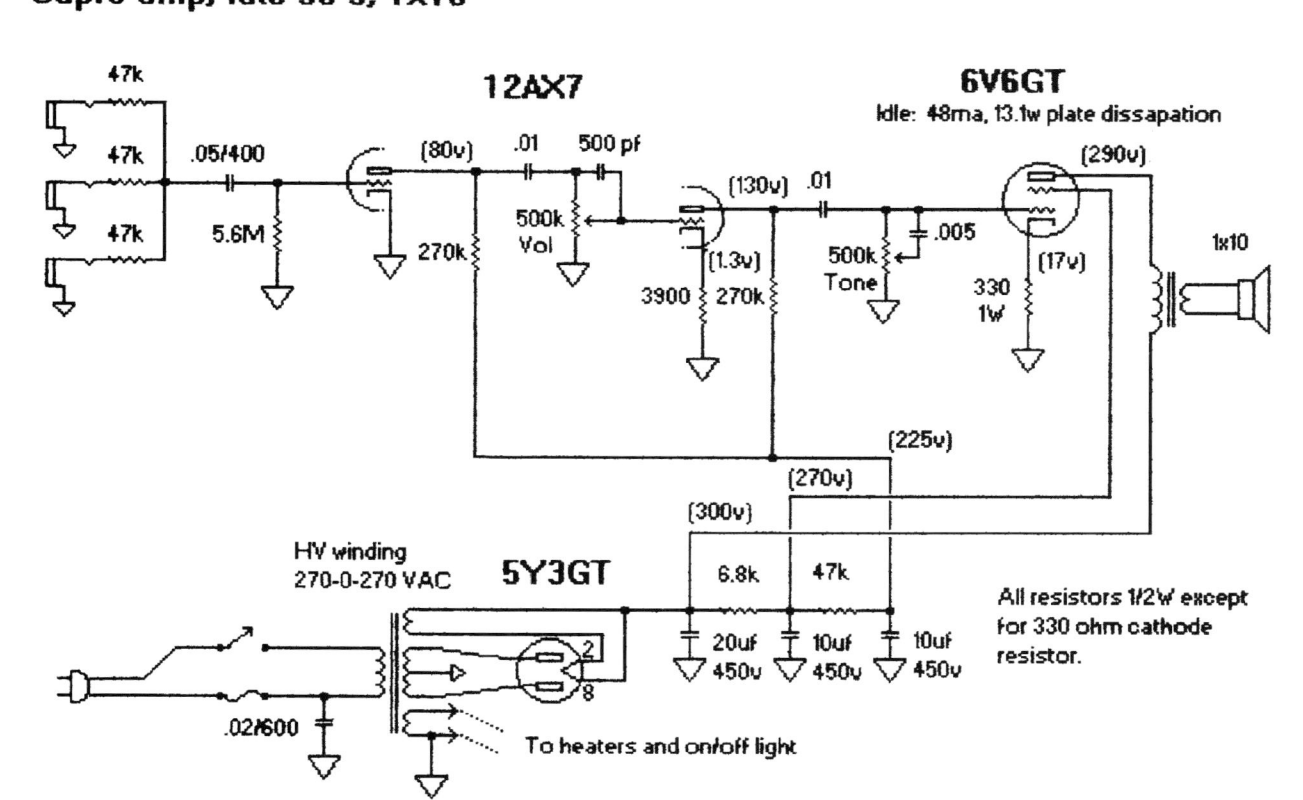

Supro

AMP **1947**
LATE '50S **1 X 6V6 AMP**

Supro amp, late 50's, 1x10

THE TUBE AMP BOOK 163

REVERB
S6625

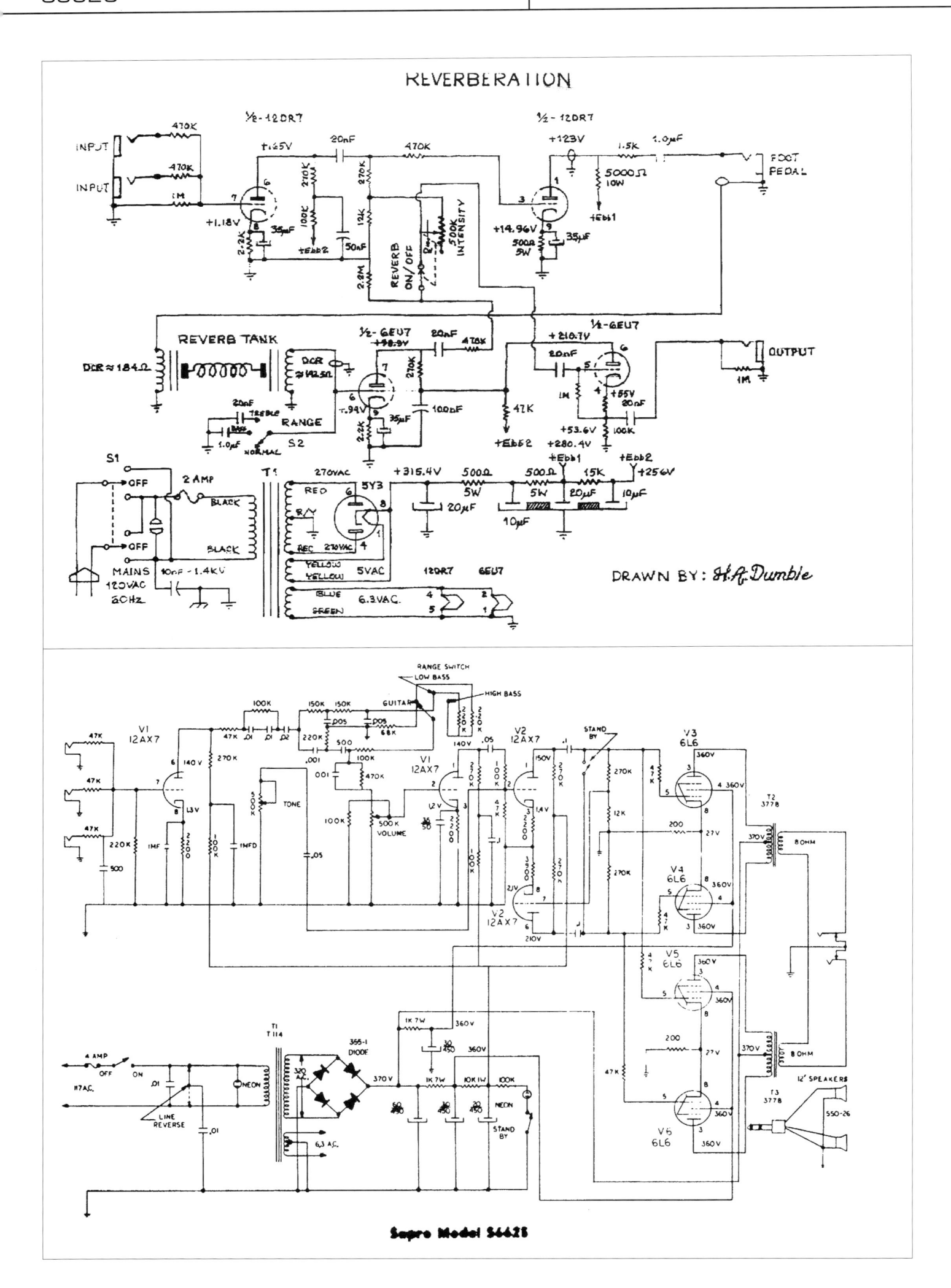

Supro

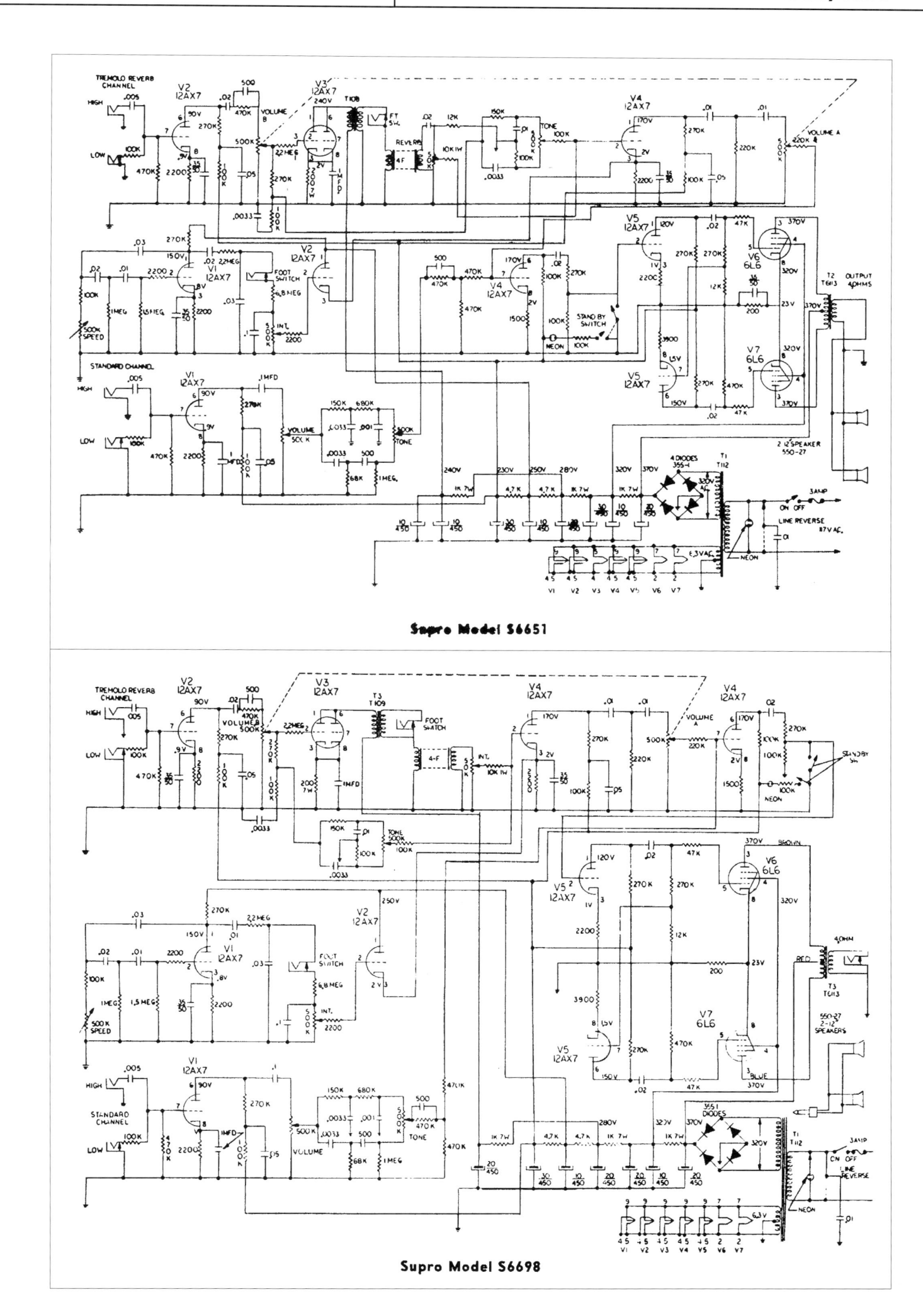

Supro

S6651
S6698 TREM/REV

Supro Model S6651

Supro Model S6698

THE TUBE AMP BOOK

STEEL GUITAR AMP **1946** | **Supro**

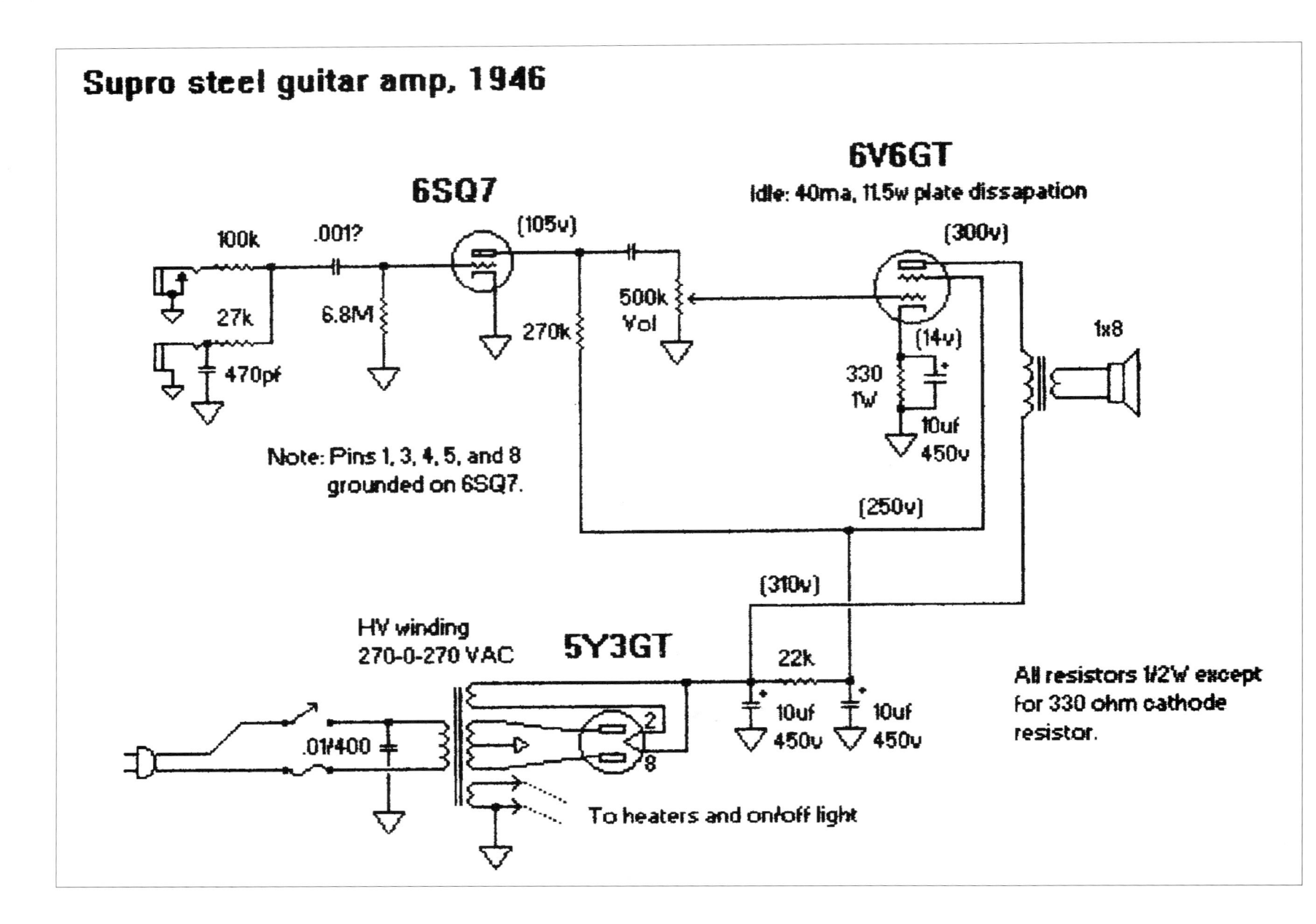

Supro steel guitar amp, 1946

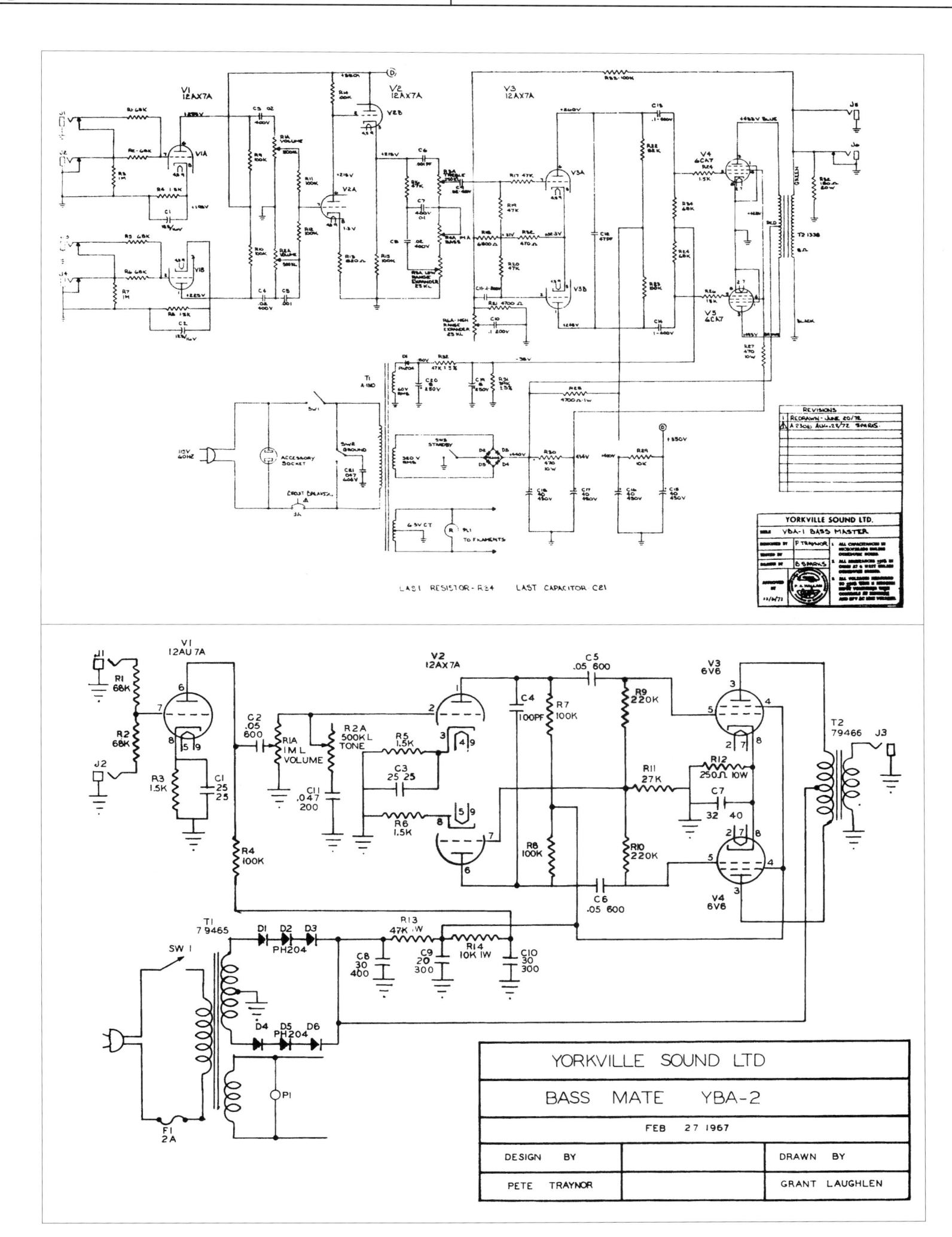

Traynor

YBA-1 BASS MASTER
YBA-2 BASS MATE

YBA-3 POWER AMP
YBA-3 CUSTOM SPECIAL PREAMP

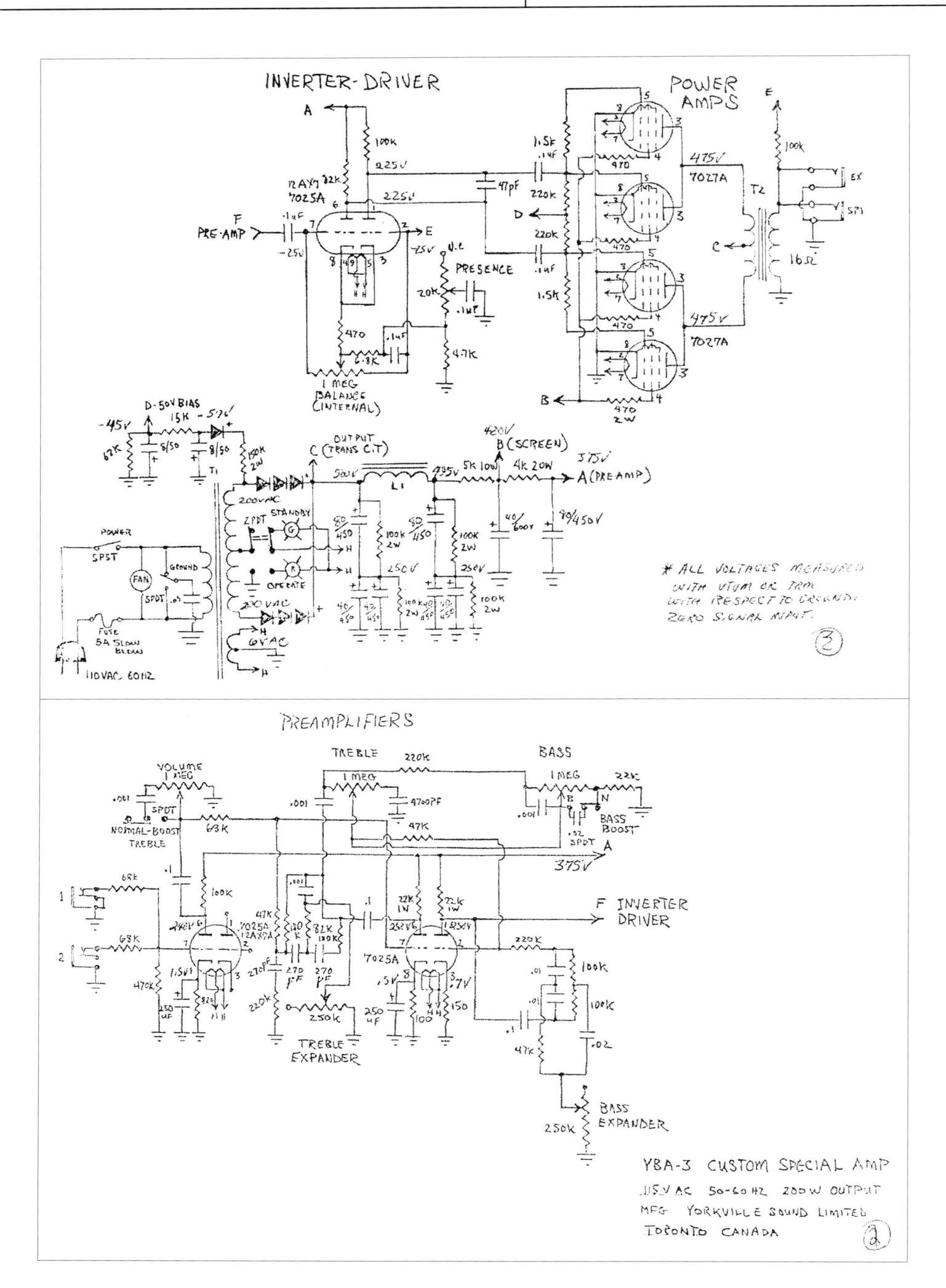

Traynor

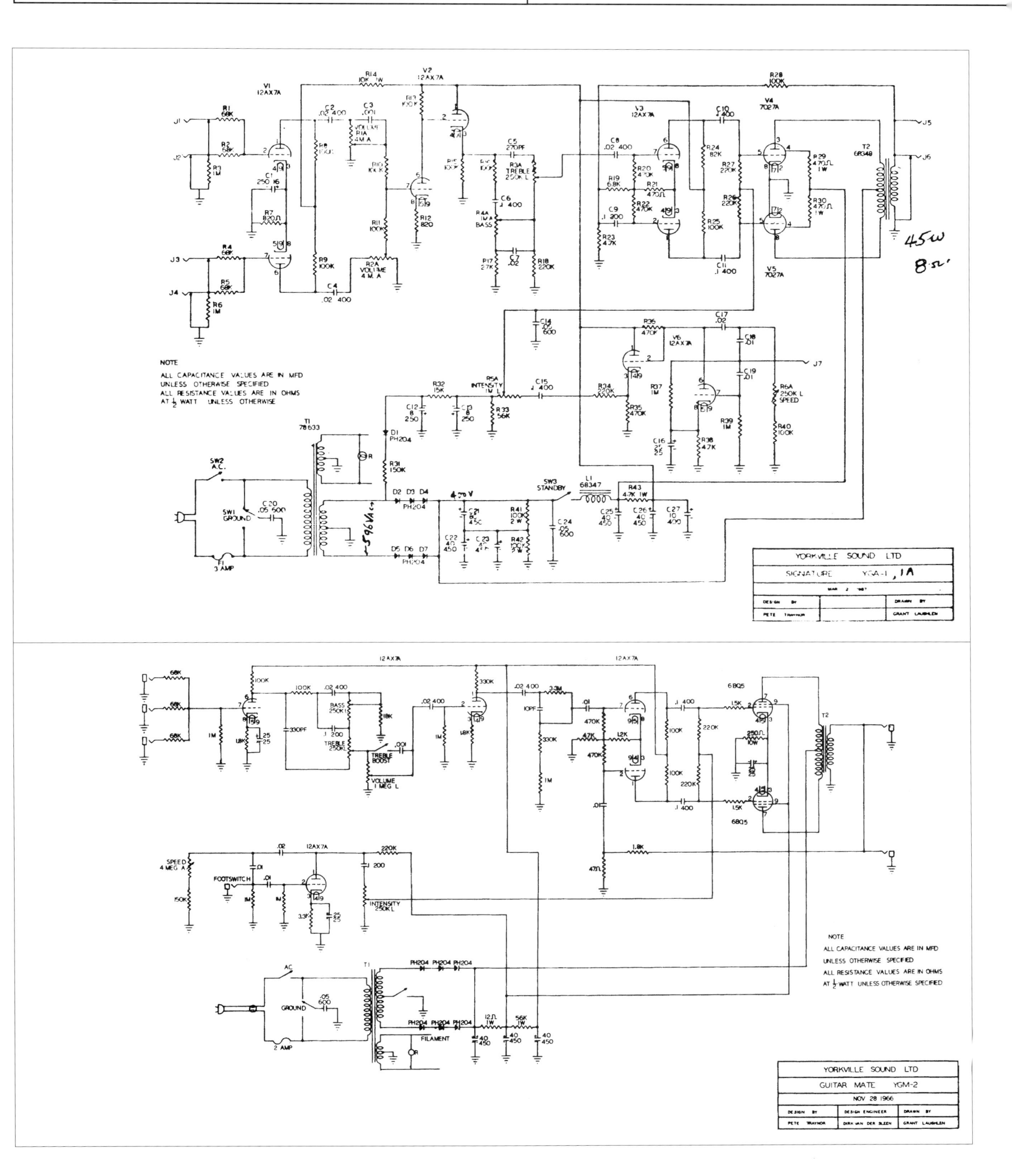

Traynor

YGA-1 1A
YGM-2 GUITAR MATE

U-1011 LEAD AMP
U-1226 LEAD AMP

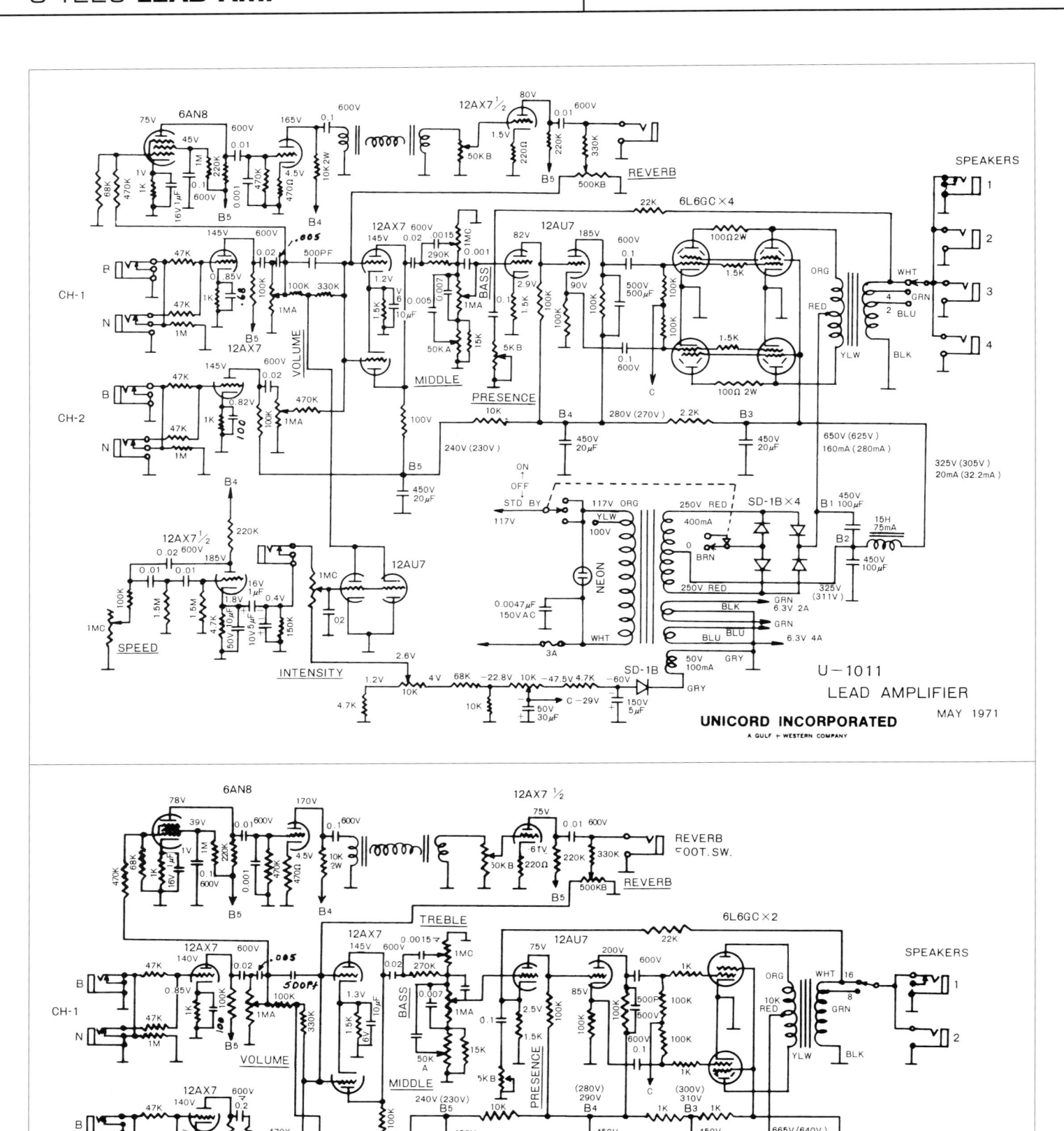

UNIVOX

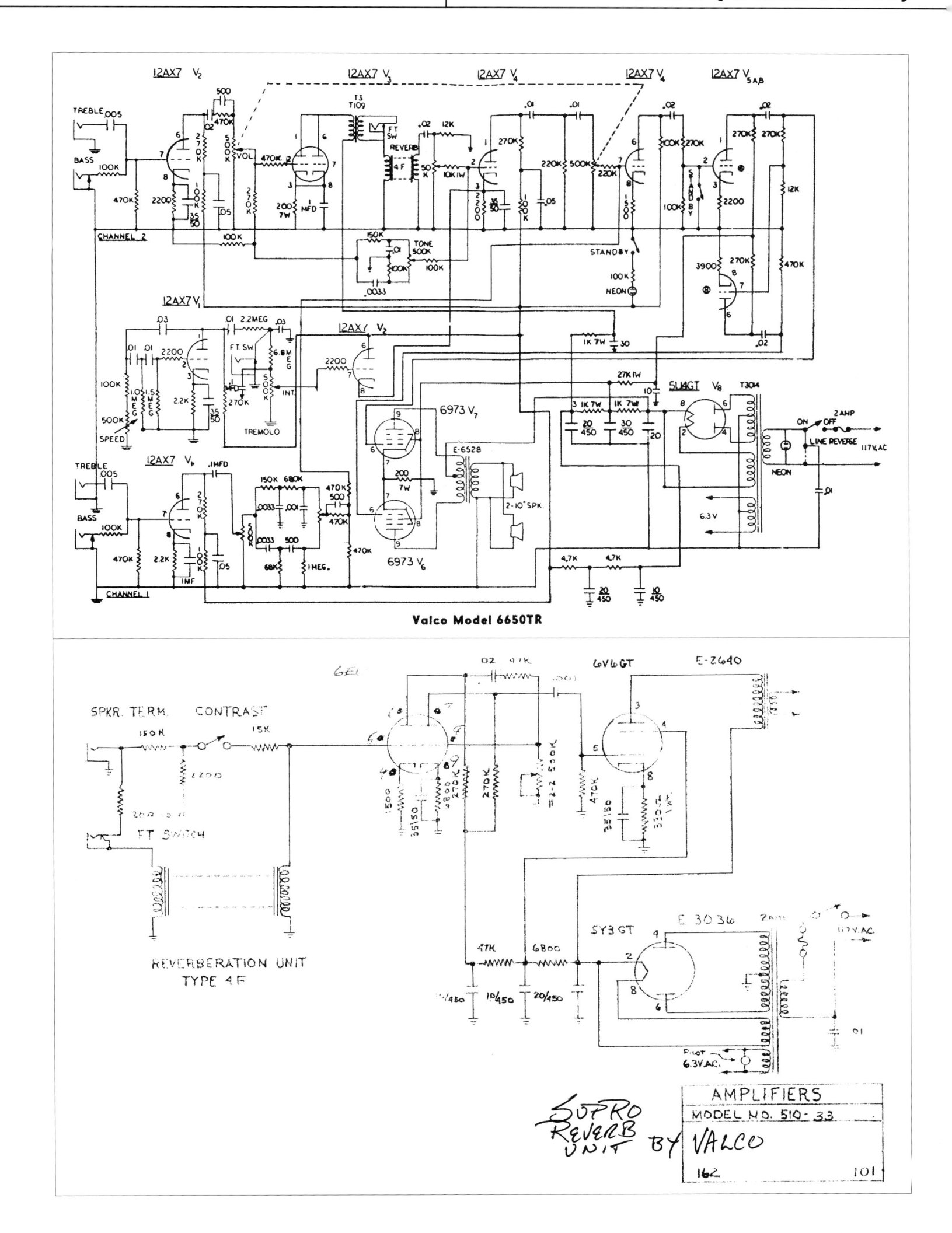

Valco

6650 TR
SUPER REVERB (MODEL 510-33)

AC10
AC15 MK II (1959)

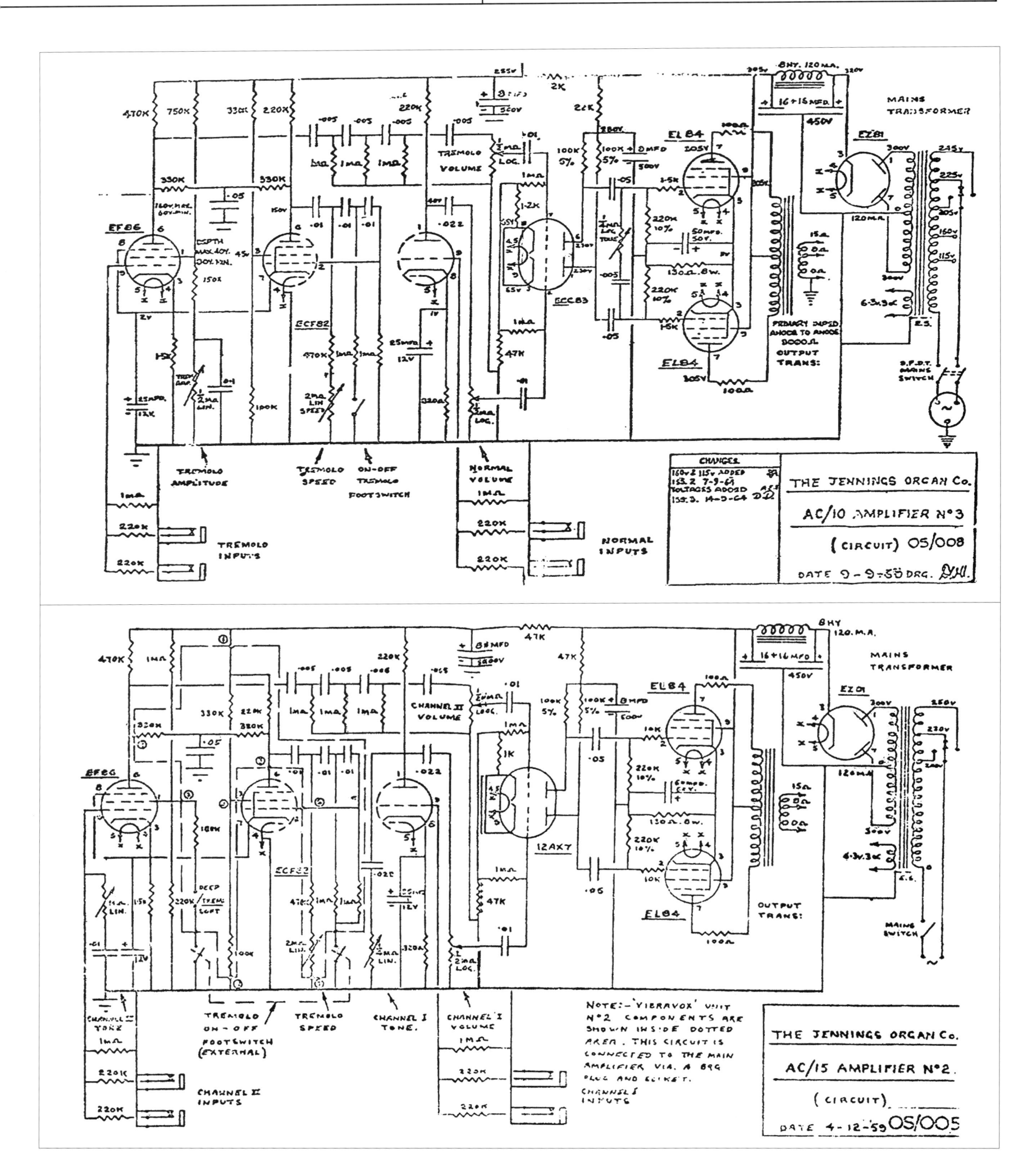

Vox

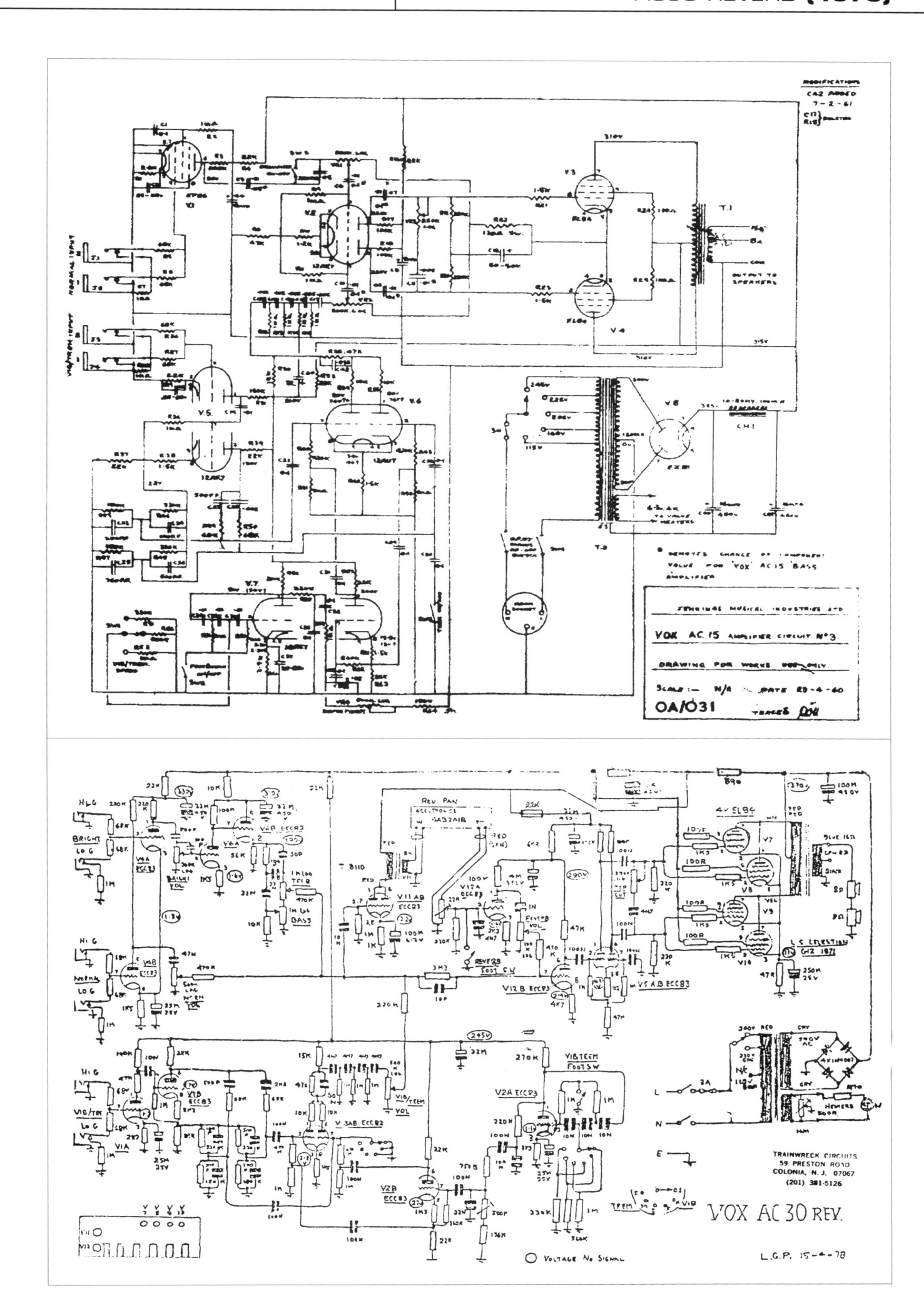

Vox

AC15 MK III (1960)
AC30 REVERB (1978)

AC30 SILVER JUBILEE (VOX LTD)
AC30 (S.S. RECT)

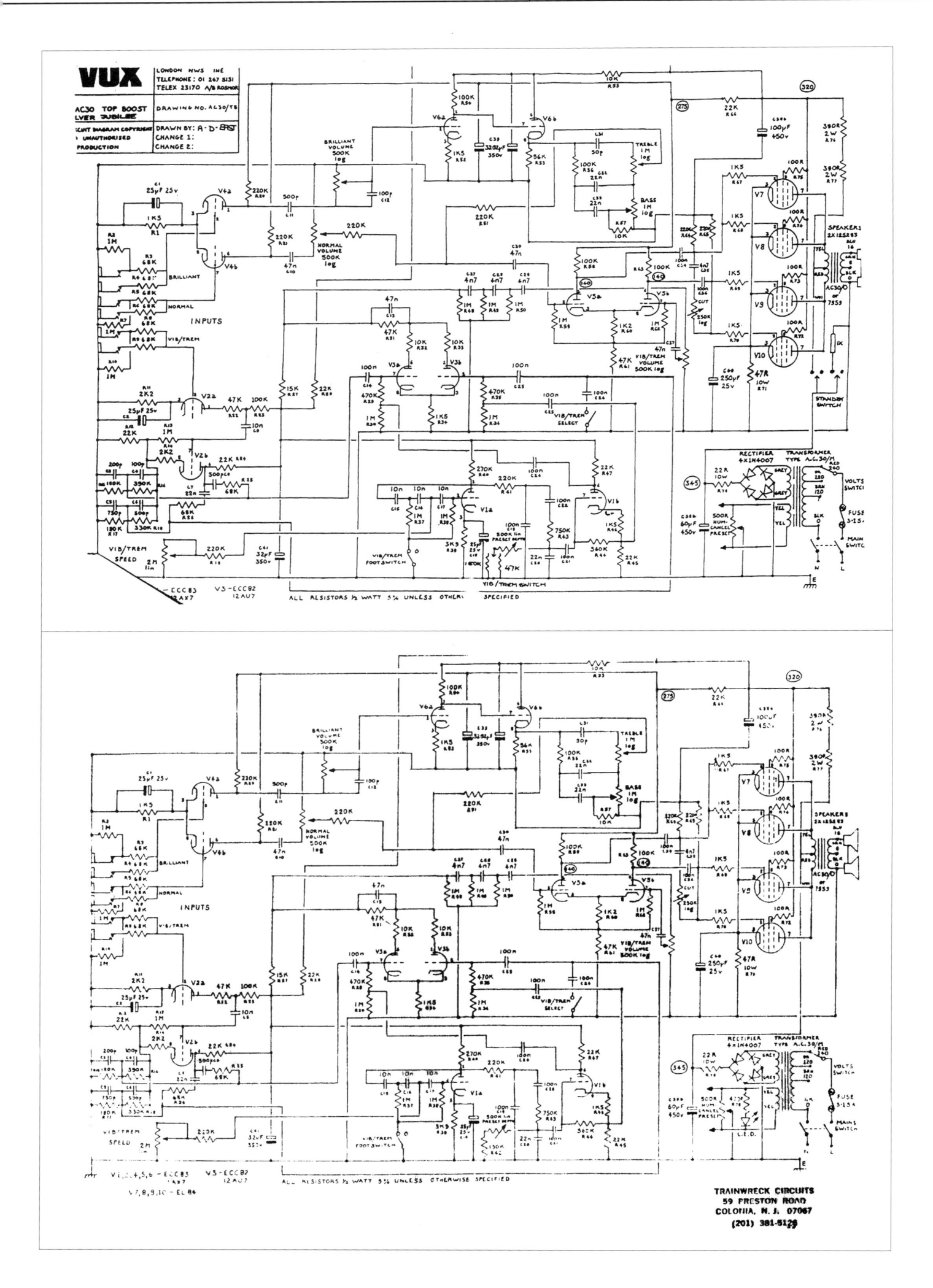

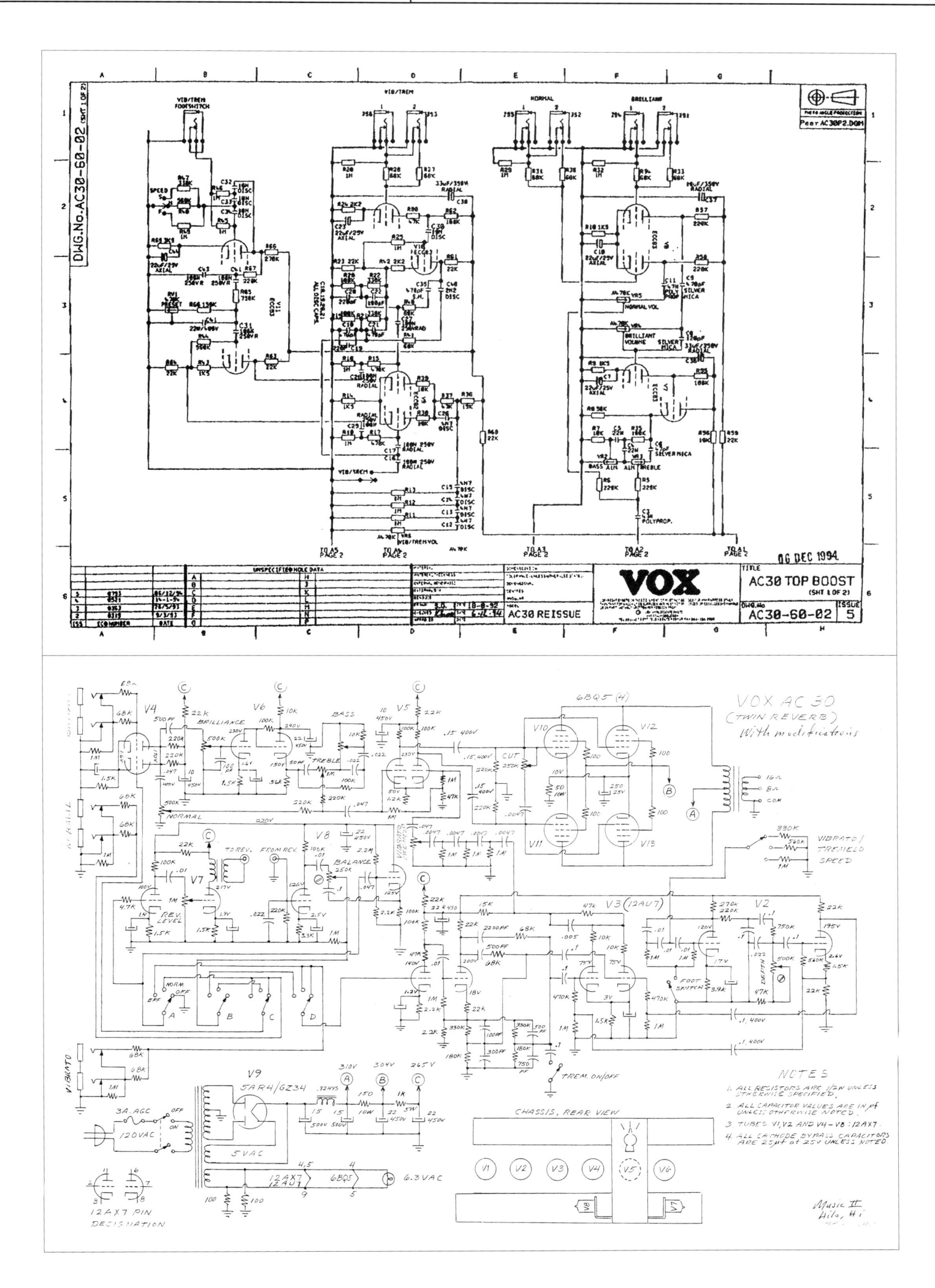

VOX

AC30 TOP BOOST REISSUE (1994)
AC30 TWIN REVERB (W/MOD)

AC50/2 (GZ34 RECT)
AC50/4 MK 11 (S.S. RECT)

Vox

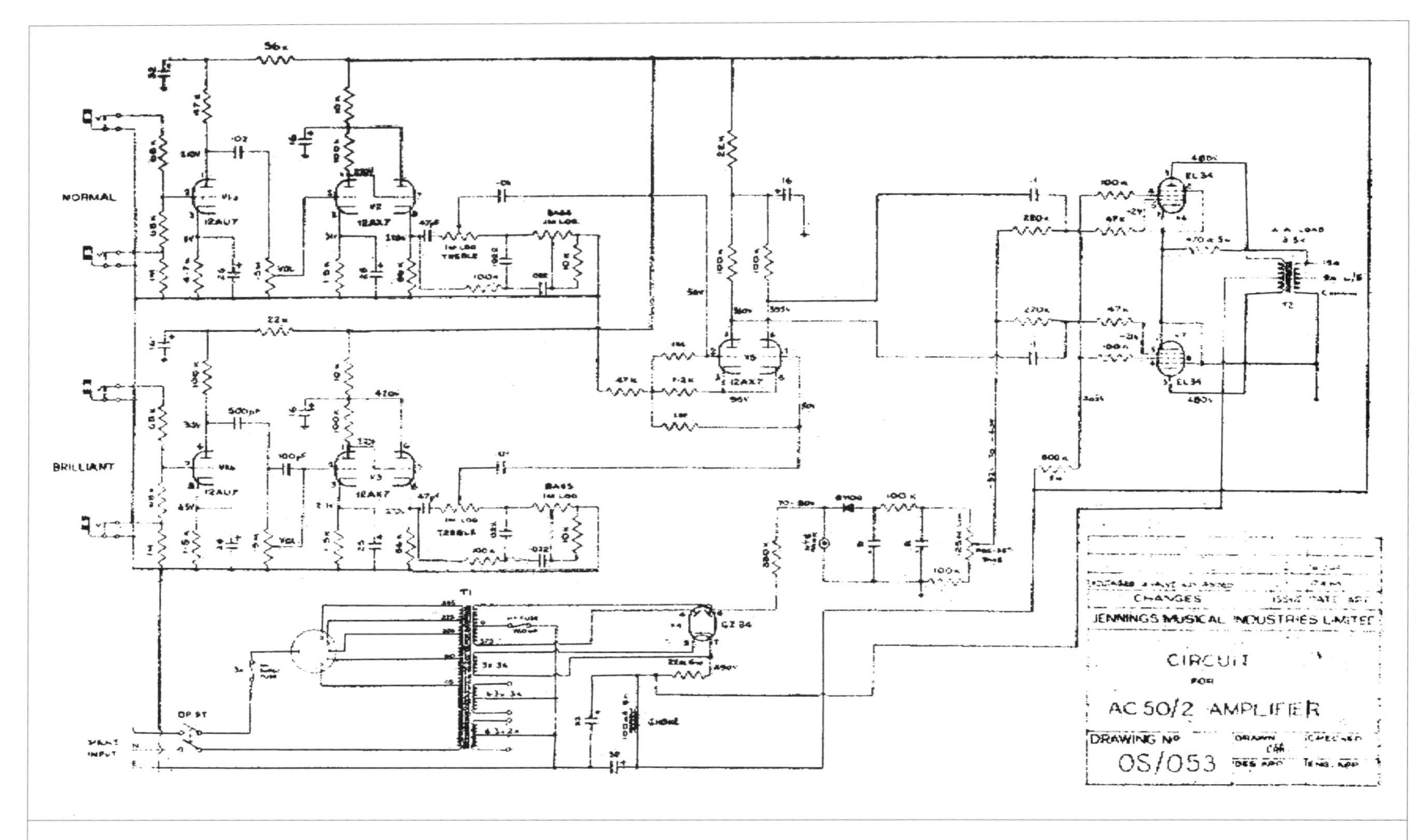

AC 50

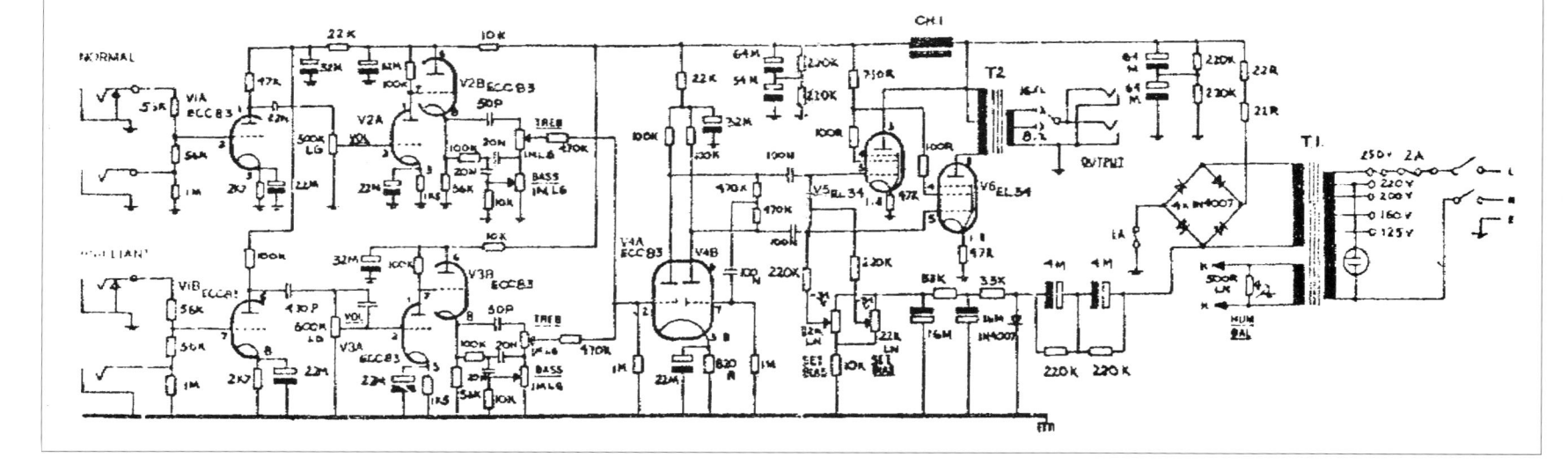

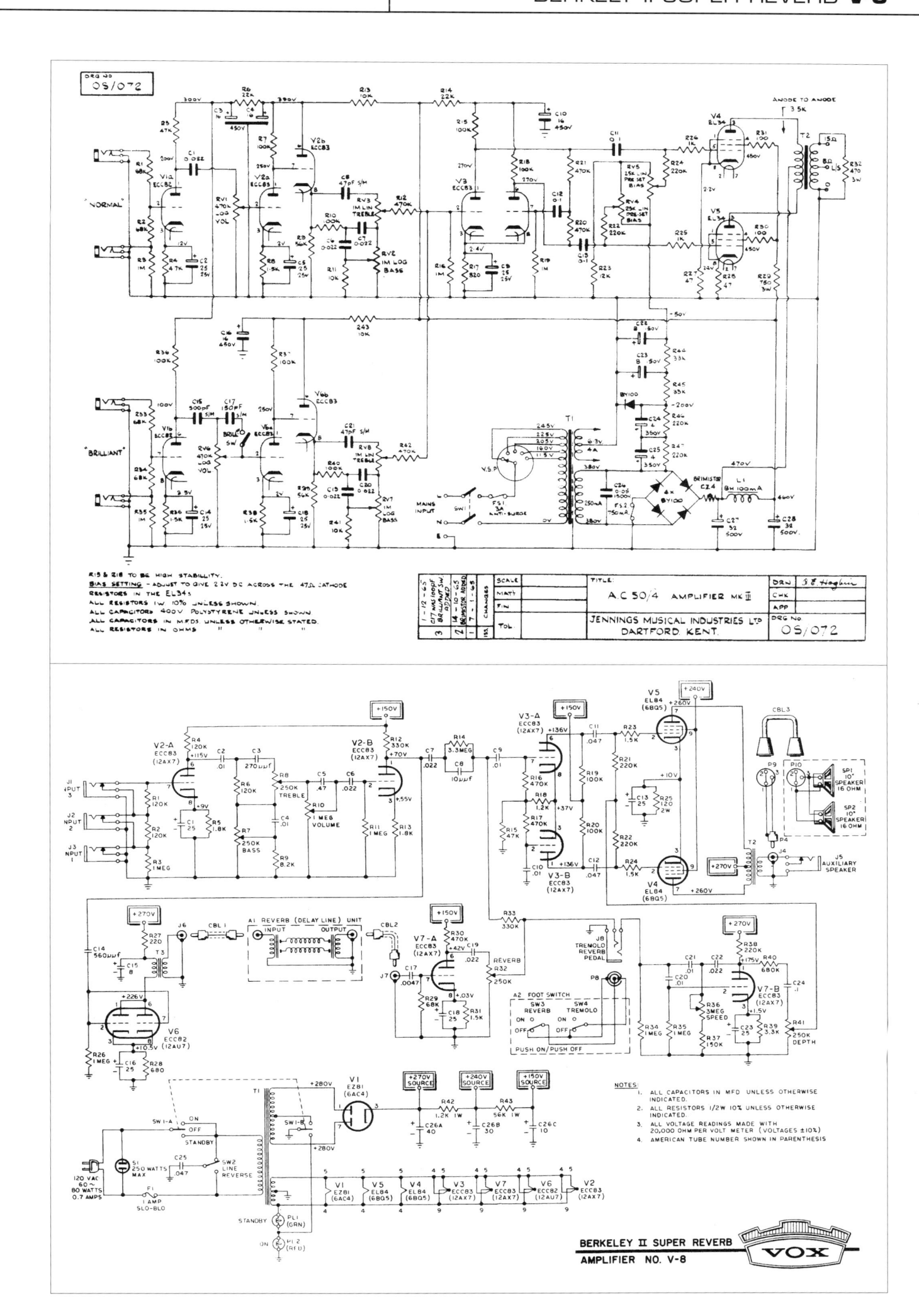

Vox

AC50/4 MK III
BERKELY II SUPER REVERB V-8

CAMBRIDGE REVERB V-3
V-15 (VOX LTD)

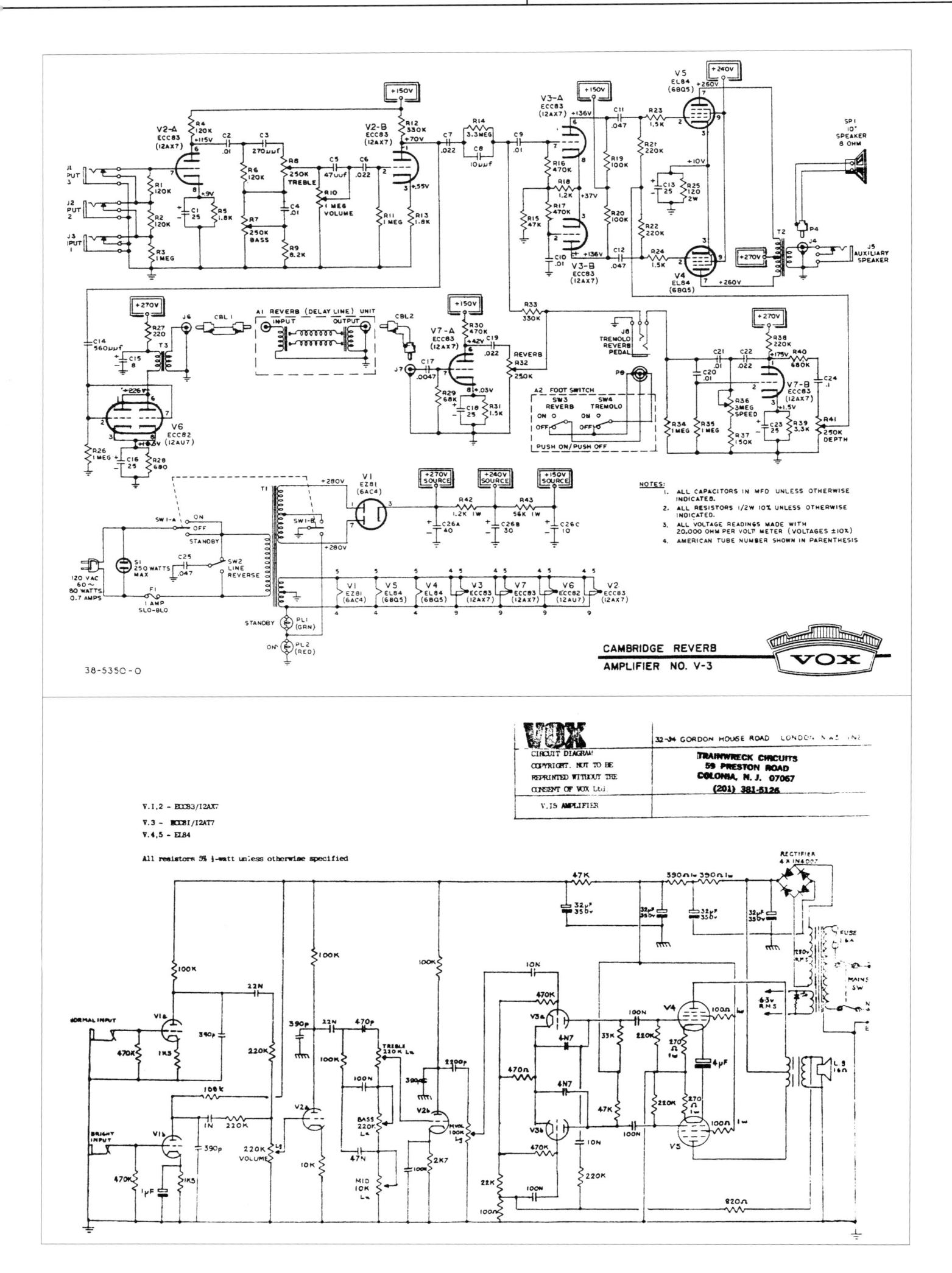

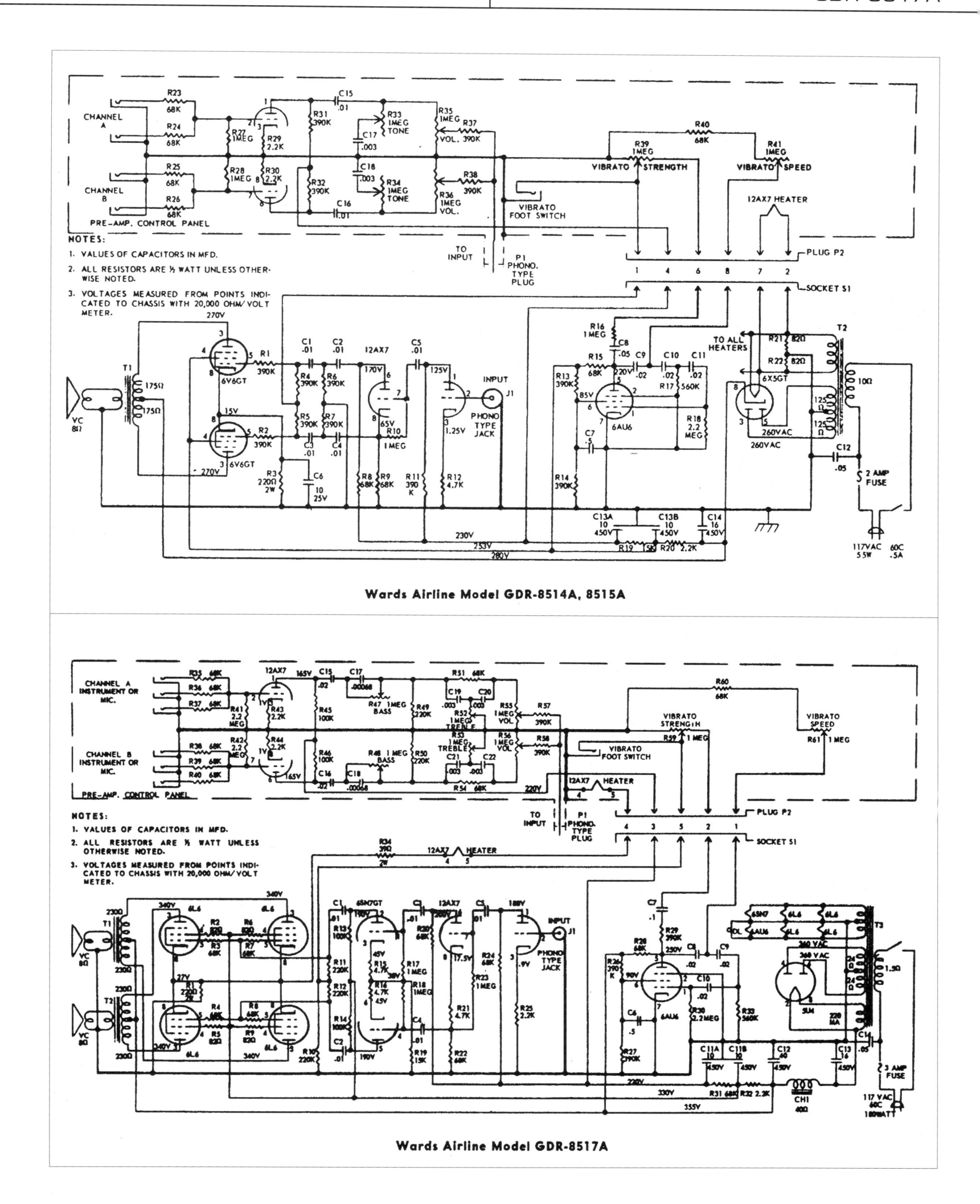

Wards Airline

GDR 8514A/8515/A
GDR 8517A

Wards Airline Model GDR-8514A, 8515A

Wards Airline Model GDR-8517A

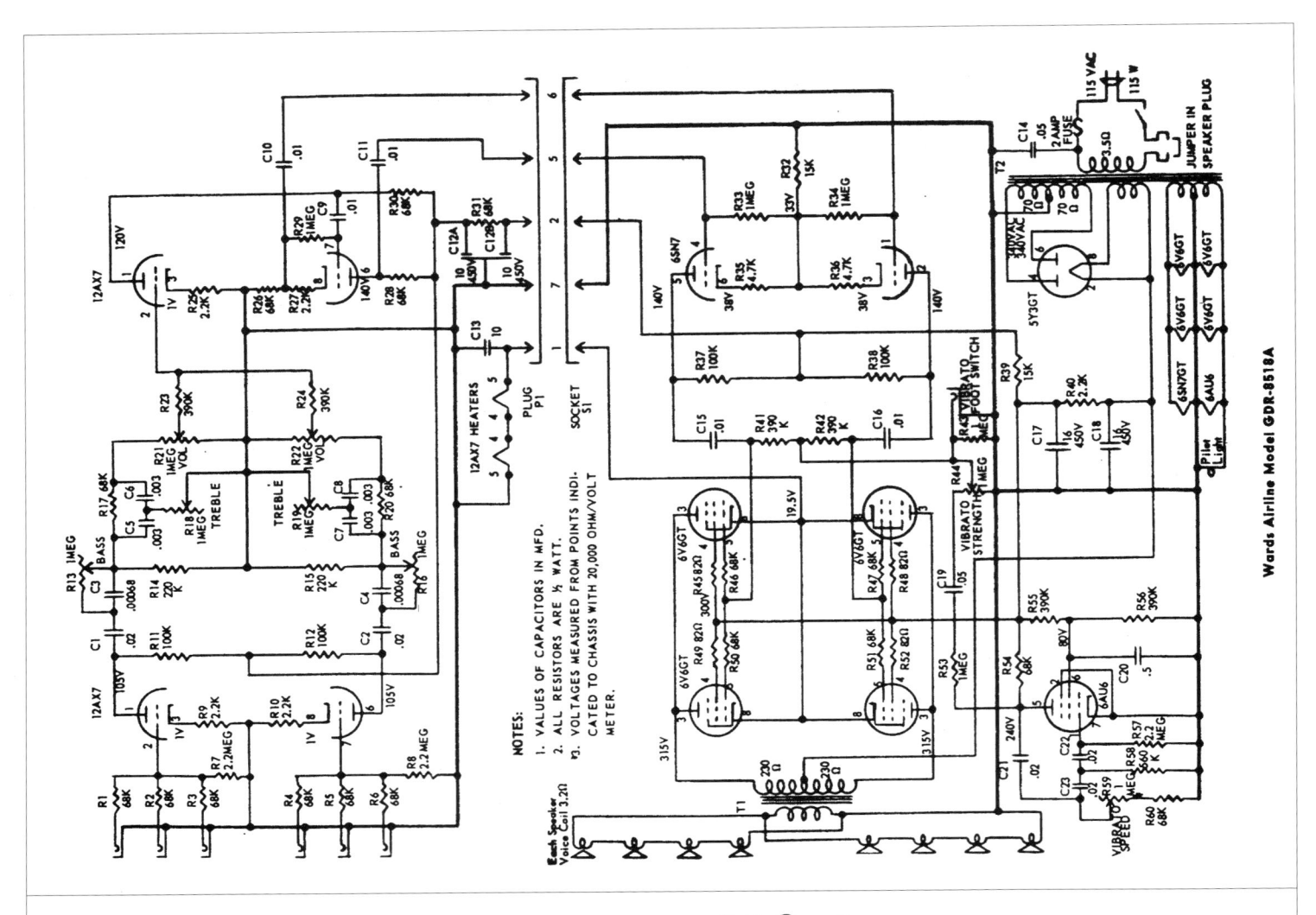

Wards Airline

GDR-8518A
GDR 9012A

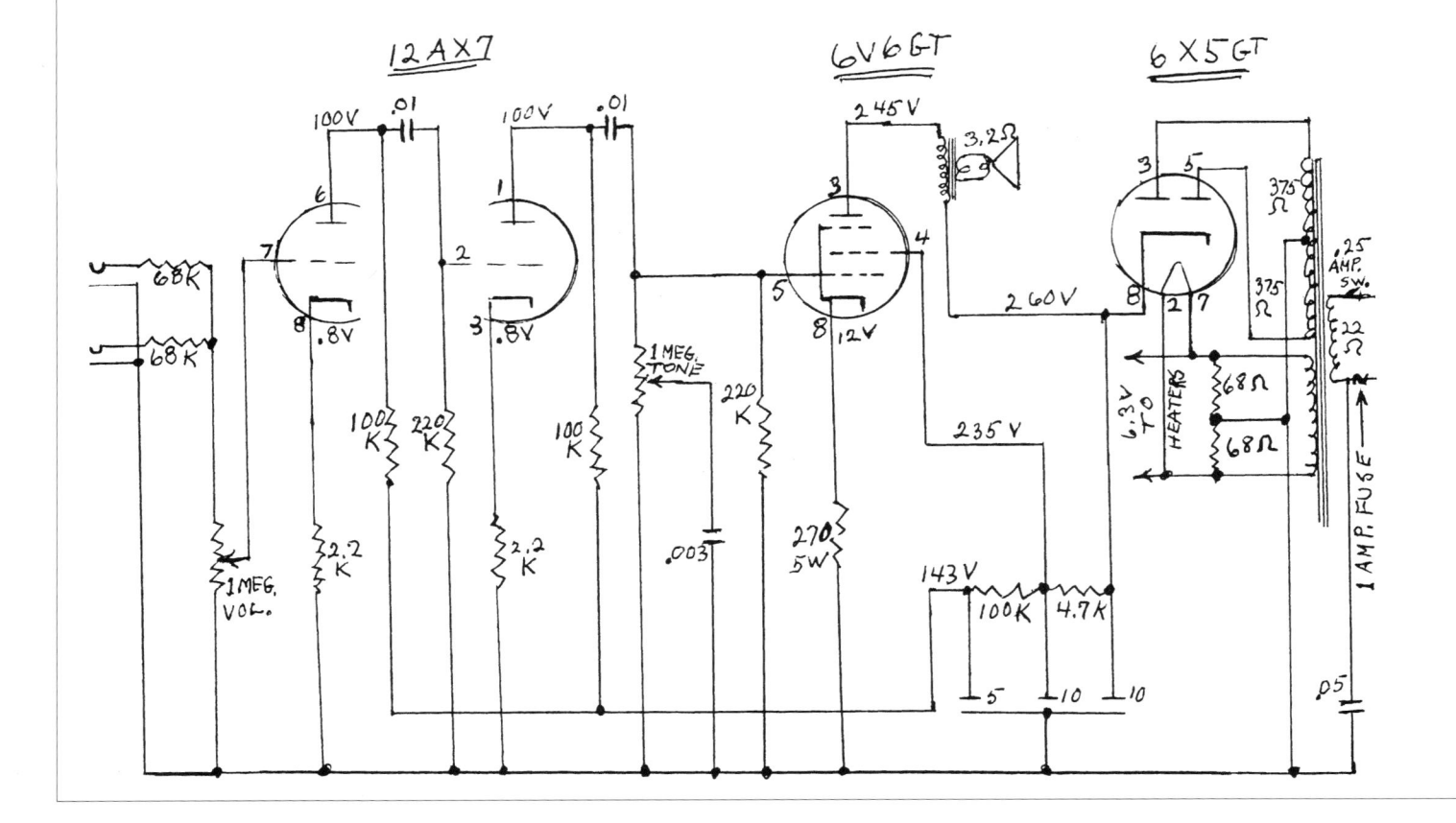

Wards Airline

GIM 9151A
GVC-9058A

NOTE:
1. All resistors ½ watt 10% unless otherwise noted.
2. All capacitor values are in MFD unless otherwise noted.
3. D.C. Voltages measured with 20,000 Ohm/volt meter, with all controls set at minimum.

Wards Airline Model GVC-9058A

NOTE:
ALL COND. SHOWN IN MFD UNLESS NOTED
ALL RESISTORS 1/2 W. UNLESS NOTED

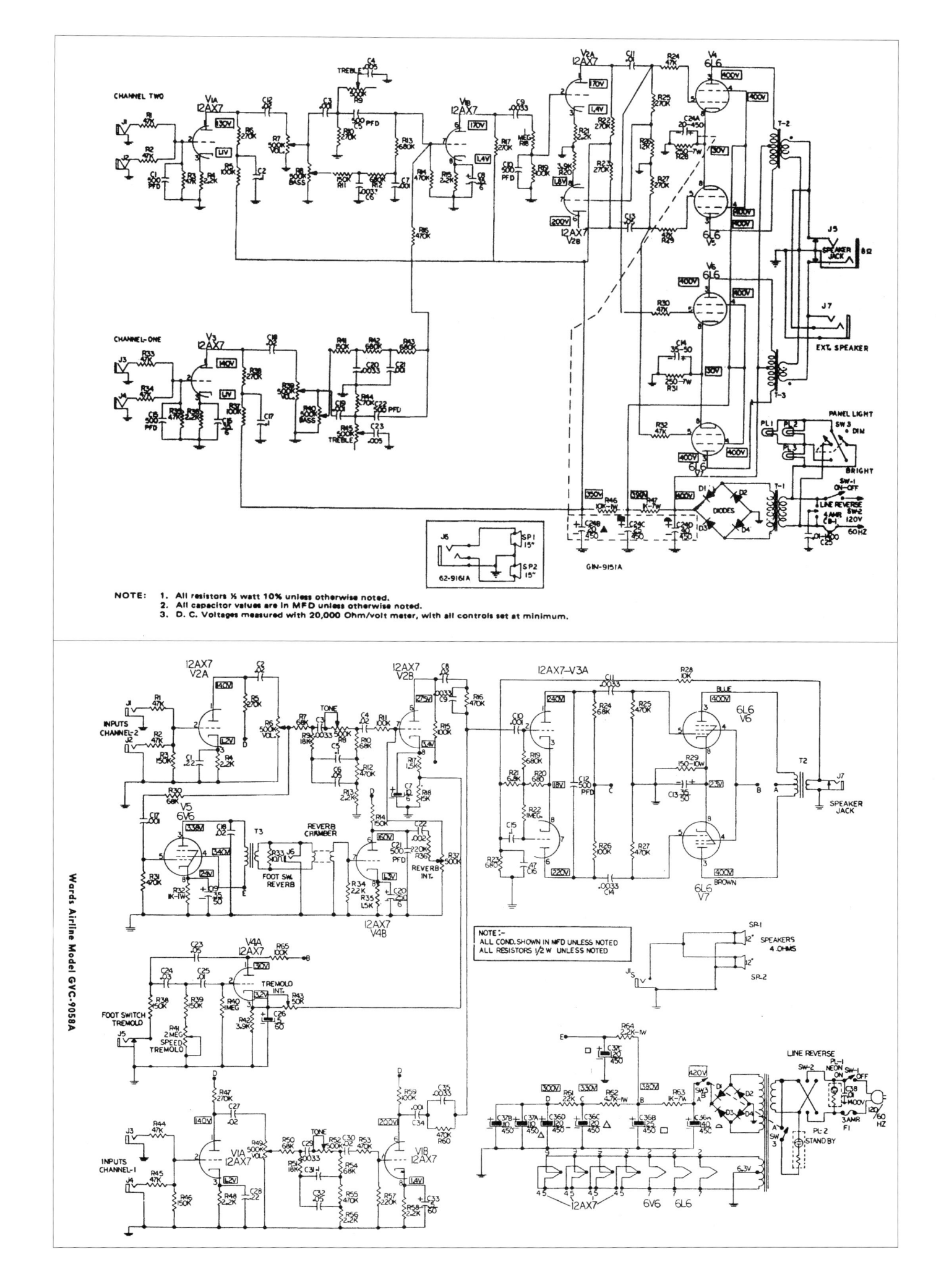

THE TUBE AMP BOOK 181

Watkins

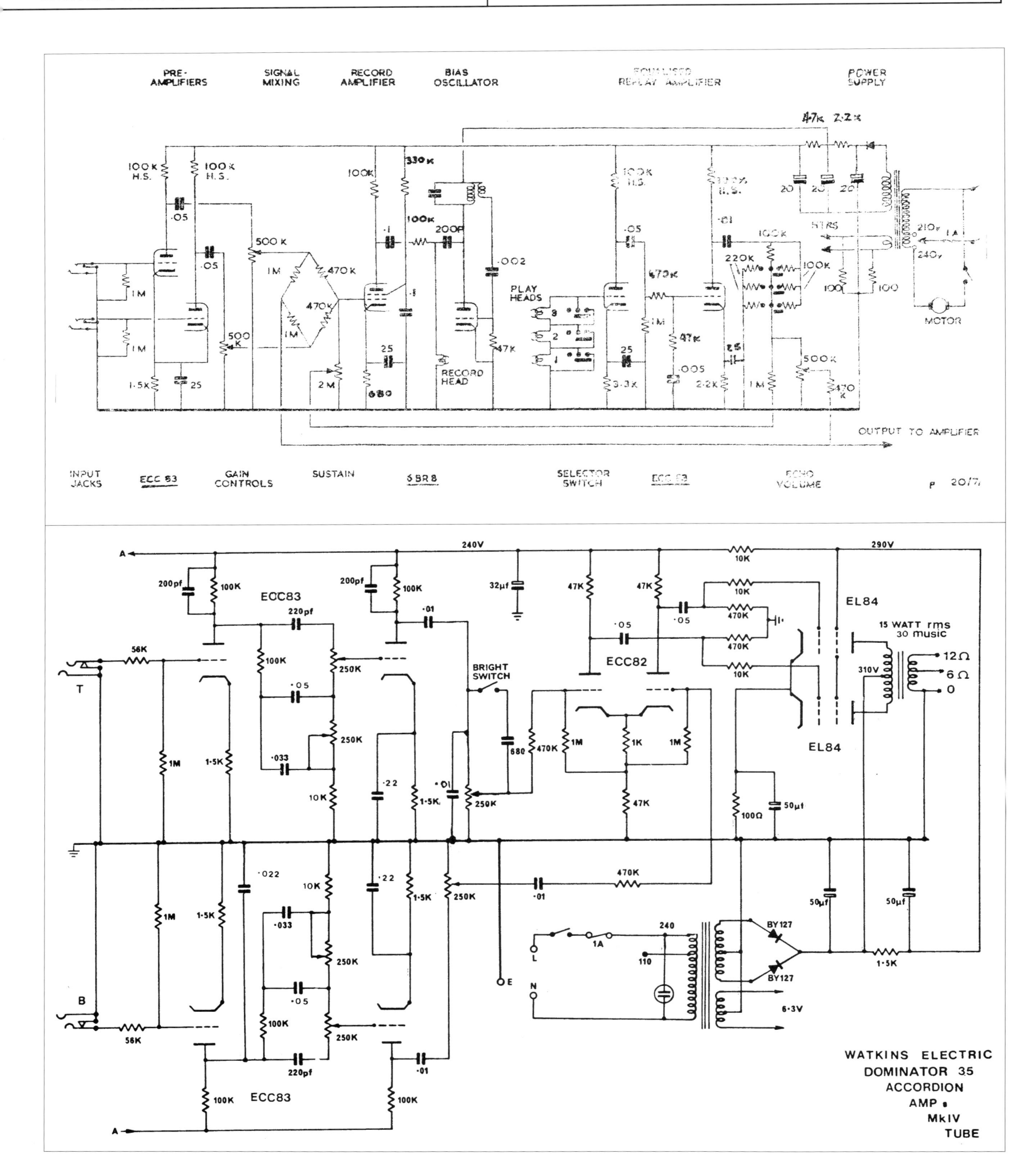

1959 COPICAT MK II
DOMINATOR 35 MK IV

Schematic: Copicat MK II — Pre-Amplifiers, Signal Mixing, Record Amplifier, Bias Oscillator, Equalised Replay Amplifier, Power Supply. Labels: Input Jacks, ECC83, Gain Controls, Sustain, 6BR8, Selector Switch, ECC83, Echo Volume. Output to Amplifier.

Schematic: Watkins Electric Dominator 35 Accordion Amp MkIV Tube — ECC83, ECC82, EL84 × 2, 15 Watt rms / 30 music, 12Ω / 6Ω / 0, Bright Switch, BY127 rectifiers, 240V / 290V / 310V, 6.3V heaters.

182 THE TUBE AMP BOOK

Watkins

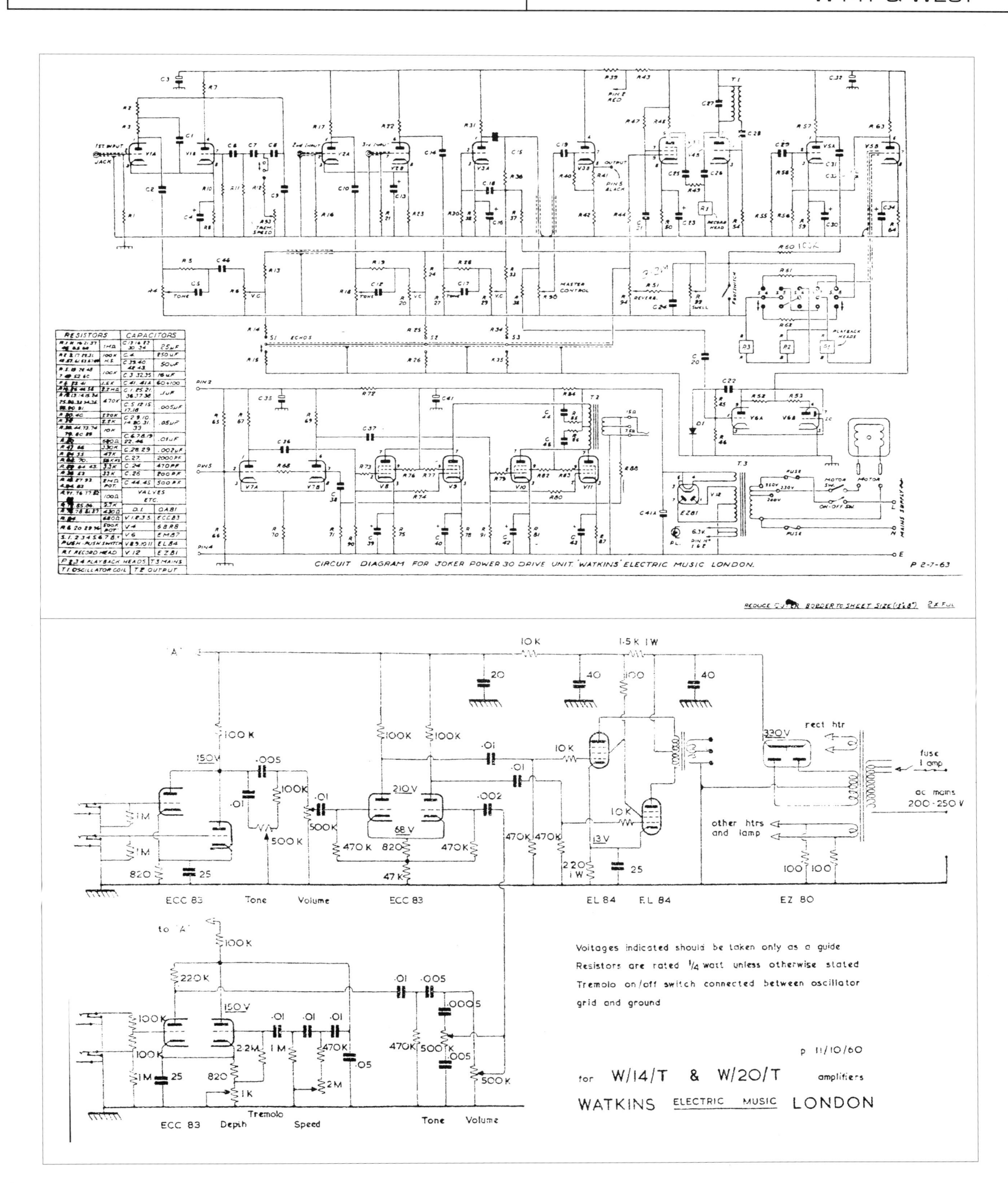

JOKER 30
W14T & W20T

CLUBMAN MK 8
DOMINATOR 50

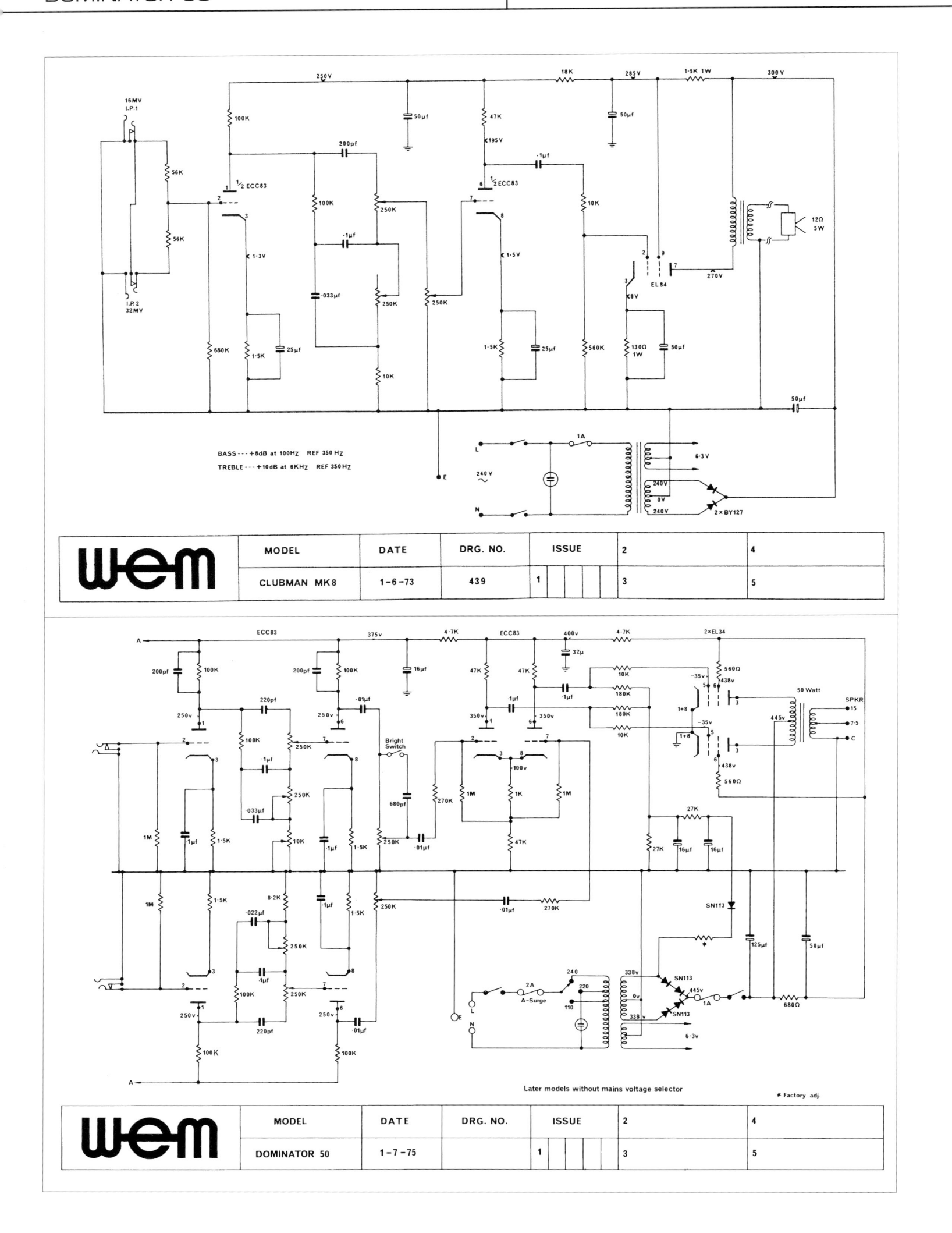

Watkins/WEM

Watkins/WEM

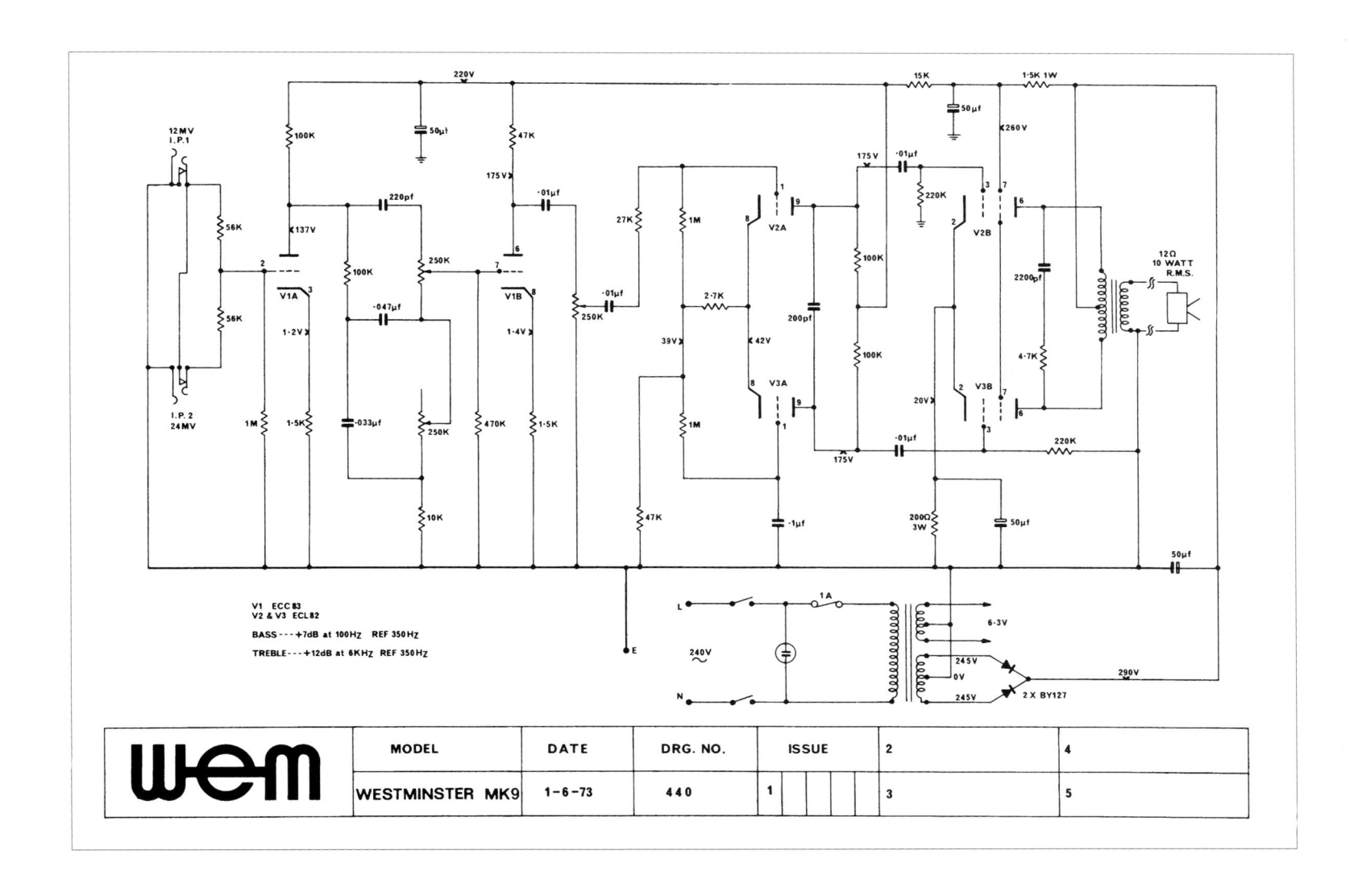

WESTMINSTER MK9

Western Electric

140A AMP / 142A AMP

FIG. 45 (MFG. DISC.)
140A AMPLIFIER

NOTES:
1. VOLTAGES SHALL BE MEASURED BETWEEN POINT INDICATED & B-(NOT CHASSIS). VALUES SHOWN ARE FOR 120V LINE VOLTAGE MEASUREMENTS WITH VOLTMETER WHICH HAS RESISTANCE OF AT LEAST 20,000 OHMS PER VOLT. MEASURED VALUES MAY DEPART ±10% FROM VALUES SHOWN.

OUTPUT CONNECTIONS

NOMINAL LOAD	LOAD RANGE	STRAP	OUTPUT TERMINALS
A.C. SUPPLY			
4 Ω	2 TO 6		1,2
8 Ω	4 TO 12		2,3
24 Ω	12 TO 36		1,3
250 Ω	125 TO 375	4-6, 5-7	4,7
1000 Ω	500 TO 1500	5-6	4,7
70 VOLTS		3-5-6, 5-7	1,7
D.C. SUPPLY			
1.5 Ω	1.0 TO 2.3		1,2
3 Ω	2.0 TO 4.5		2,3
8 Ω	6 TO 14		1,3
100 Ω	60 TO 150	4-6, 5-7	4,7
400 Ω	250 TO 600	5-6	4,7
70 VOLTS		3-4, 5-6	1,7

INPUT CONNECTIONS

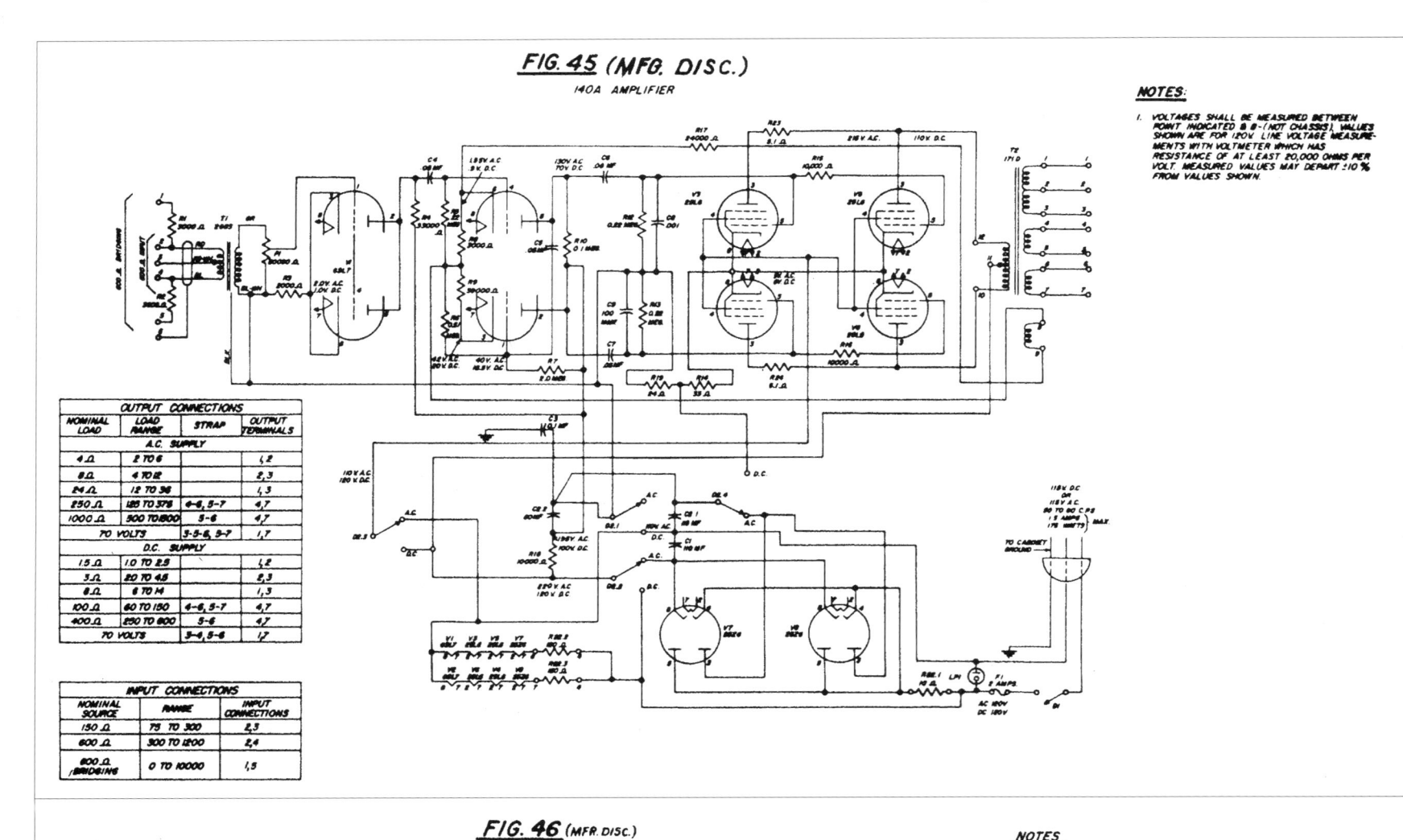

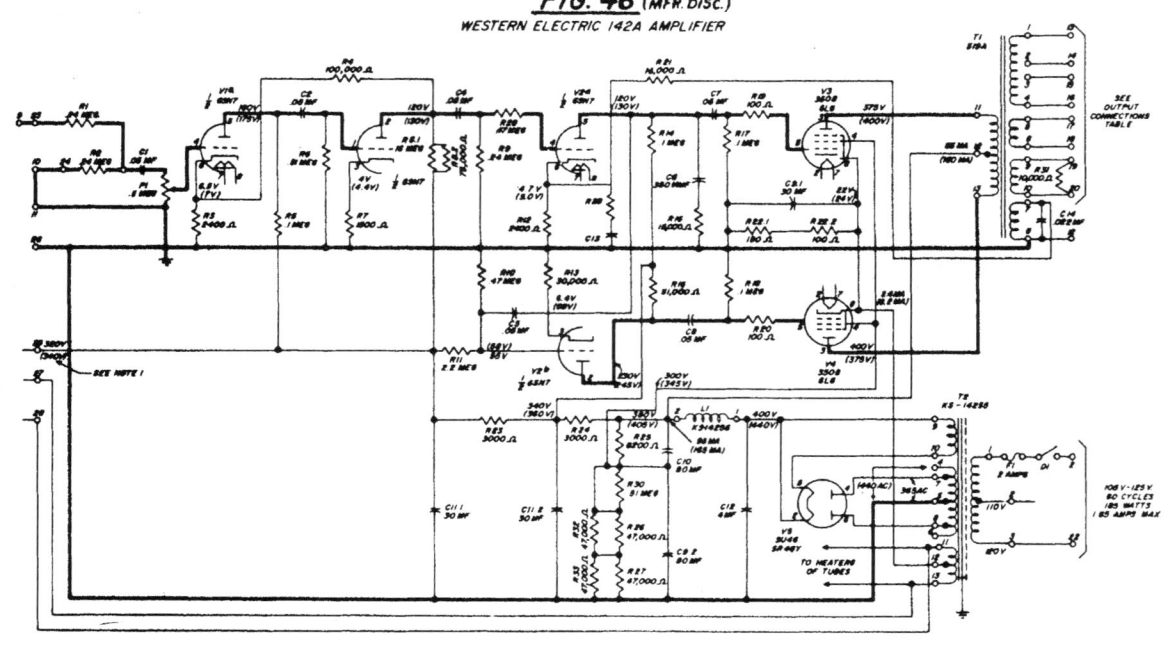

NOMINAL SOURCE	RANGE	INPUT CONNECTIONS
150 Ω	75 TO 300	2,3
600 Ω	300 TO 1200	2,4
600 Ω /BRIDGING	0 TO 10000	1,5

FIG. 46 (MFR. DISC.)
WESTERN ELECTRIC 142A AMPLIFIER

NOTES:
1. CIRCUIT SHOWN FOR 12 WATTS POWER SUPPLY OUTPUT. THE NUMBERS IN PARENTHESES ARE THE VALUES FOR THE 25 WATT CONDITION.
2. WHEN FIG. 35 IS ADDED IT BECOMES A 142B AMPLIFIER.
3. WHEN FIG. 36 IS ADDED IT BECOMES A 142C AMPLIFIER.
4. WHEN FIG. 37 IS ADDED IT BECOMES A 142D AMPLIFIER.

OUTPUT CONNECTIONS TABLE

NOMINAL LOAD IMPEDANCE	WORKING RANGE OF LOAD IMPEDANCE	STRAP TERMINALS	OUTPUT CONNECTIONS
200 Ω	150 Ω TO 300 Ω		19 & 20
24 Ω	18 Ω TO 36 Ω	13-15	13 & 18
12 Ω	9 Ω TO 18 Ω	13-15, 14-16-17	15 & 18
8 Ω	6 Ω TO 12 Ω	14-15	13 & 16
4 Ω	3 Ω TO 6 Ω		17 & 18
2 Ω	1.5 Ω TO 3 Ω	13-15, 14-16	13 & 16
400 Ω	300 Ω TO 600 Ω	14-15, 16-17, 18-19	13 & 20

70 V. LOUDSPEAKER DISTRIBUTION LINE CONNECTIONS

POWER OUTPUT CONDITION	STRAP TERMINALS	OUTPUT CONNECTIONS
12 WATTS	14-15, 16-17, 18-19	13 & 20
25 WATTS		19 & 20

Western Electric

142A AMP (CONVERTED FOR 25W)
143A AMP

FIG. 47 (MFR. DISC.)
WESTERN ELECTRIC 142A AMPLIFIER
CONVERTED FOR 25 WATT OUTPUT

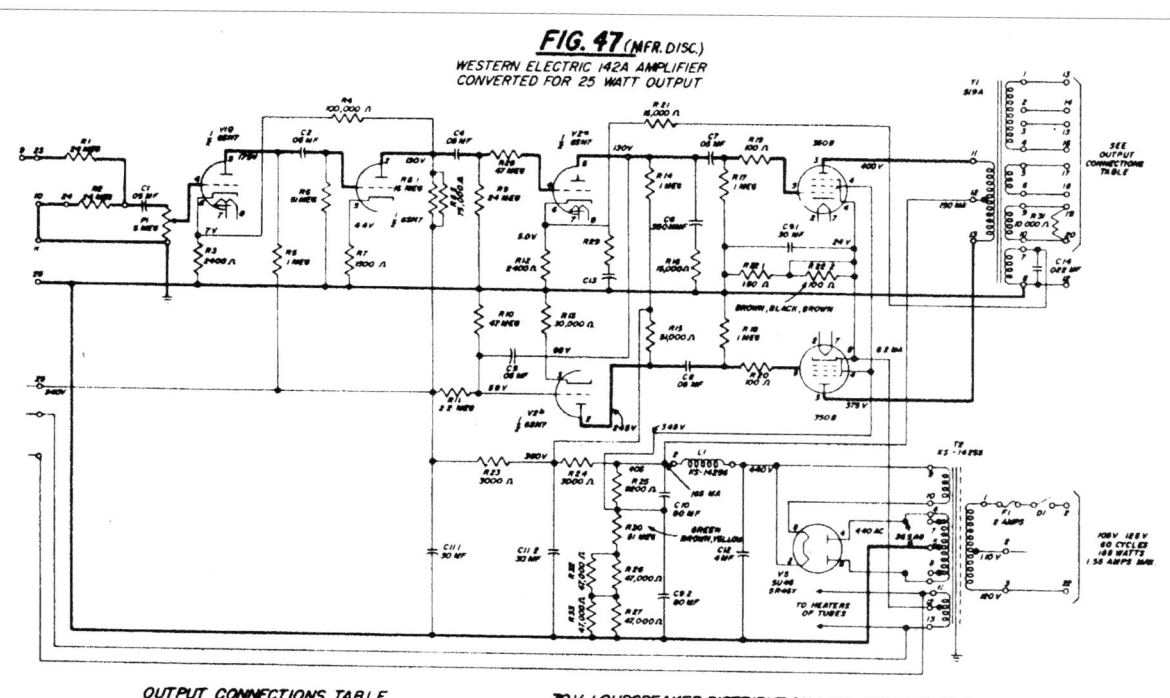

NOTES

1. FOR 25 WATTS OUTPUT THE FOLLOWING CHANGES ARE NECESSARY:
   A. USE 350B TUBES.
   B. SHORT R22.
   C. AT TRANSFORMER T2 TRANSFER LEAD FROM TERMINAL 7 TO TERMINAL 4 AND LEAD FROM TERMINAL 8 TO TERMINAL 6.
   D. REMOVE SHORT ACROSS R30.

2. LINE INPUT CONNECTIONS 142C & 142D

SOURCE OHMS	STRAP TERMINALS	CONNECT TO
600 Ω	6-7	4 AND 8
150 Ω	4-7, 6-8	4 AND 8

3.

AMPLIFIER	REMOVE STRAP BETWEEN TERMS.	STRAP TERMINALS
142 B	9 - 23	10 TO 11
142 C	9	10 TO 11
142 D	—	10 TO 11

COLOR CODE FOR FIXED RESISTORS

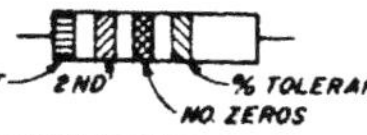

COLOR	1ST BAND	2ND BAND	3RD BAND	END BAND
BLACK	0	0	NONE	GOLD 5%
BROWN	1	1	0	SILVER 10%
RED	2	2	00	NONE 20%
ORANGE	3	3	000	
YELLOW	4	4	0000	
GREEN	5	5	00000	
BLUE	6	6	000000	
VIOLET	7	7	0000000	
GRAY	8	8	00000000	
WHITE	9	9	000000000	

OUTPUT CONNECTIONS TABLE

NOMINAL LOAD IMPEDANCE	WORKING RANGE OF LOAD IMPEDANCE	STRAP TERMINALS	OUTPUT CONNECTIONS
200 Ω	150 Ω TO 300 Ω	—	19 & 20
24 Ω	18 Ω TO 36 Ω	14-15, 16-17	13 & 18
12 Ω	9 Ω TO 18 Ω	13-15, 14-16-17	13 & 18
8 Ω	6 Ω TO 12 Ω	14-15	13 & 16
4 Ω	3 Ω TO 6 Ω	—	17 & 20
2 Ω	1.5 Ω TO 3 Ω	13-15, 14-16	13 & 16
400 Ω	300 Ω TO 600 Ω	14-15, 16-17, 18-19	13 & 20

70 V. LOUDSPEAKER DISTRIBUTION LINE CONNECTIONS

POWER OUTPUT CONDITION	STRAP TERMINALS	OUTPUT CONNECTIONS
12 WATTS	14-15, 16-17, 18-19	13 & 20
25 WATTS	—	19 & 20

FIG. 52
WE 143A AMPLIFIER

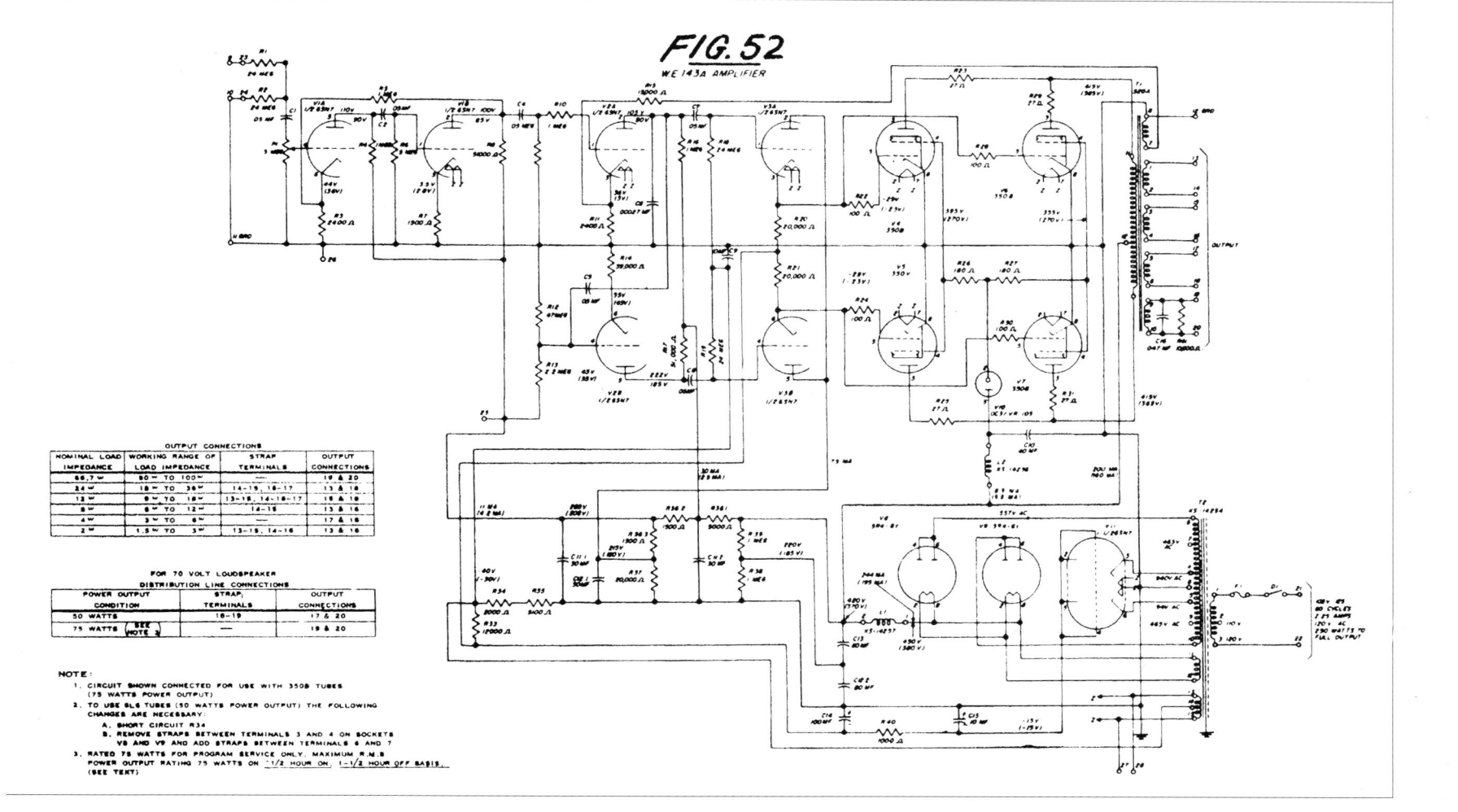

OUTPUT CONNECTIONS

NOMINAL LOAD IMPEDANCE	WORKING RANGE OF LOAD IMPEDANCE	STRAP TERMINALS	OUTPUT CONNECTIONS
66.6 W	50 W TO 100 W	—	19 & 20
24 W	18 W TO 36 W	14-15, 16-17	13 & 18
12 W	9 W TO 18 W	13-15, 14-16-17	13 & 18
8 W	6 W TO 12 W	14-15	13 & 16
4 W	3 W TO 6 W	—	17 & 20
2 W	1.5 W TO 3 W	13-15, 14-16	13 & 16

FOR 70 VOLT LOUDSPEAKER DISTRIBUTION LINE CONNECTIONS

POWER OUTPUT CONDITION	STRAP TERMINALS	OUTPUT CONNECTIONS
50 WATTS	18-19	17 & 20
75 WATTS (SEE NOTE 3)	—	19 & 20

NOTE:
1. CIRCUIT SHOWN CONNECTED FOR USE WITH 350B TUBES (75 WATTS POWER OUTPUT)
2. TO USE 6L6 TUBES (50 WATTS POWER OUTPUT) THE FOLLOWING CHANGES ARE NECESSARY:
   A. SHORT CIRCUIT R34.
   B. REMOVE STRAPS BETWEEN TERMINALS 3 AND 4 ON SOCKETS V8 AND V9 AND ADD STRAPS BETWEEN TERMINALS 6 AND 7.
3. RATED 75 WATTS FOR PROGRAM SERVICE ONLY. MAXIMUM R.M.S. POWER OUTPUT RATING 75 WATTS ON 1/2 HOUR ON, 1-1/2 HOUR OFF BASIS. (SEE TEXT)

Disc contents

Acoustic
G60T
G80 &G100T
G100T
Allied
Basic 60 Kit
Altec Lansing
1530A
1568A
1569A
1570A
Ampeg
AC12
B-12NB & 15NB
B-12NF & 15NF
B-12N & 15N
B-12X
B-12XT
B-12XT & B-18X
B-12XY
B-15N
B-15S (Rev C)
B-15S (Rev D)
B-18N
B-18V & B-15ND
B-22X
B-25
B-25B
B-25B
B-42X
B-115 & B-410
EJ-12A
ET-1
ET-2B
G-12
G-12
GS-12R
GS-12R
GU-12
G-15
GS-15
GV-15
G-20
GV-22
G-110
J-12D
J-12R
J-12R
J-12 JET
J-12B JET
M-12
M-12A
M-15
M-18
PRE-AMP
PRE-AMP DUETTE & DOLPHIN II
R-12 REVERB ROCKET
R-12A
R-12RB
R-12R ROCKET
R-15R
S-48
SB-12
SB-12
SBT PREAMP
SR4
SST & SBT
SUPER ECHO TWIN
SVT (REV A)
SVT POWER AMP (REV D)
SVT POWER AMP (REV F)
SVT PREAMP
SVT PREAMP
SVT PREAMP (REV B)
SVT PREAMP (REV C)
SVT MODEL 6146B
SVT V9
SVT-2 PWR PREAMP (REV 7)
SVT PRO TUB BD
SVT-2 PRO POWER AMP
SVT PRO POWER AMP
SVT-2 PRO GRAPHIC EQ
SVT-2 PRO PREAMP
V4/VT22 (REV G)
V2/VT40/V4/VT22 DISTORTION
V2 (REV B)
V2/V4/VT22/VT40 (REV D)
V2/VT40/V4/VT22
V3
V4 PREAMP & POWER AMP
V4/VT22
VT22 & V4 (REV A)
V9 PREAMP
Bogen
CHB-20A
E30
BTE
40W POWER AMP
VARICOMP-1
Carvin
FET 100
PREAMP SX
X-30 & X-60
X-AMP
Danelectro
REVERB BOX 9100
Dukane
1A385
Dynaco
PA5-3X
SCA-35
Dynacord
BASS KING T
LE 20
LE 120
LE 120 POWER SUPPLY
G 2000
Engl
480
CONTROL CIRCUIT BOARD
JIVE
PCB LAYOUT
POWER SUPPLY
PREAMP I
PREAMP II
Epiphone
EA-4T, 4TL & 6T
EA-5RVT
EA-7P
EA-8P
EA-10 DELUXE
EA-12RVT
EA-15RVT
EA-16RVT
EA-22RVT
EA-25 CENTURY
EA-26RVT
EA-28RVT
EA-30 TRIUMPH
EA-32RVT
EA-33RVT
EA-35
EA-35T
EA-50
EA-50T
EA-65
EA-70
EA-71
EA-72
EA-300RVT
EA-500T
EA-500T
EA-500T
Fender
Bandmaster
5E7 Layout
6G7 Layout
6G7 Schematic
AC568 Layout
Bantam Bass
CFA7003 Layout
CFA7003 Schematic
Bassman
Old Schematic
10 Schematic
10 (75w) Schematic
20 Schematic
70 Schematic
100 Schematic
135 Schematic
59 (Rev A)
59 (Rev E)
TV Front Layout
TV Front Schematic
5E6 Layout
5E6 Schematic
5E6-A Layout
5E6-A Schematic
5F6 Layout
5F6 Schematic
5F6-A Layout
5F6-A Schematic
6G6-A Layout
6G6-A Schematic
6G6-B Layout
6G6-B Schematic
AA270 Schematic
AA864 Schematic
AB165 Schematic
AC568 Schematic
Blues
Deluxe Schematic
De Ville & De Ville 212 Schematic
Junior (Rev A) Schematic
Junior (Rev D) Schematic
Champ
5C1 Layout
5C1 Schematic
5E1 Layout
5E1 Schematic
Vibro Champ
AA764 Layout
AA764 Schematic
Concert
6G12 Layout
6G12 Schematic
Deluxe
TV Front Layout
5C3 Layout
5C3 Schematic
5D3 Layout
5D3 Schematic
5D4 Layout
5D4 Schematic
5E3 Layout
5E3 Schematic
6G3 Layout
6G3 Schematic
Deluxe Reverb
AB763 Layout
AB763 Schematic
Dual Showman Reverb
AA270 Layout
AA270 Schematic
AA769 Layout
AA769 Schematic
Harvard
5F10 Layout
5F10 Schematic
6G10 Schematic
Princeton
TV Front Schematic
5B2 Schematic
5C2 Schematic
5D2 Layout
5D2 Schematic
5E2 Layout
5E2 Schematic
5F2 Layout
5F2 Schematic
5F2-A Layout
5F2-A Schematic
6G2 Layout
AA964 Layout
AA964 Schematic
Princeton Reverb
AA1164 Layout
AA1164 Schematic
Pro-Amp
5C5 Layout
5C5 Schematic
5D5 Schematic
5E5 Layout
5E5 Schematic
6G5 Layout
6G5 Schematic
6G5-A Layout
6G5-A Schematic
AA763 Layout
AA763 Schematic
AB763 Layout
AB763 Schematic
Pro-Reverb
AA165 Layout
AA165 Schematic
AA1069 Layout
AA1069 Schematic
Reverb
6G15 Layout
6G15 Schematic

Showman
6G14-A Layout
6G14-A Schematic
AB763 Layout
AB763 Schematic
Studio Bass Schematic
Super-Amp
5C4 Layout
5C4 Schematic
5D4 Layout
5D4 Schematic
5E4-A Layout
5E4-A Schematic
5F4 Layout
5F4 Schematic
Super Bassman
CFA7002 Layout
CFA7002 Schematic
Super Reverb
AA1069 Layout
AA1069 Schematic
Tremolux
5E9-A Layout
5E9-A Schematic
5G9 Layout
5G9 Schematic
6G9 Layout
6G9 Schematic
6G9-A Layout
6G9-A Schematic
AA763 Layout
AA763 Schematic
AB763 Layout
AB763 Schematic
Twin-Amp
5C8 Layout
5C8 Schematic
5D8 Layout
5D8 Schematic
5E8 Layout
5E8 Schematic
5E8-A Layout
5E8-A Schematic
5F8 Layout
5F8 Schematic
5F8-A Layout
5F8-A Schematic
6G8 Layout
6G8 Schematic
6G8-A Layout
6G8-A Schematic
Twin Reverb
AA270 Layout
AA270 Schematic
AA769 Layout
AA769 Schematic
AB568 Layout
AB763 Schematic
Vibrasonic
5G13 Layout
5G13 Schematic
Vibroverb
AB763 Layout
AB763 Schematic
Vibrolux
5E11 Layout
5E11 Schematic
6G11 Layout
6G11 Schematic
AB763 Layout
AB763 Schematic
Vibrolux Reverb
AA270 Schematic
CBS 100w Twin Rev/Super 75
Furman
Model RV-1
Garnet
B260D & L260D
Banshee
BTO Overdrive Circuit
BTO PA260
BTO Power Amp
G15T Gnome
G15TR
G15TR
G45TR, G90TR, G100TR & LB100TR
G45TR Power Supply
G90TR Power Supply
G100 Preamp
G100 Power Supply
G100PAR Mixer & Reverb
G100S & G250R
GS100R Power Amp
GS100R Preamp & Reverb
G200 Enforcer
G250FTR & G250D
G250TR & G250D Preamps
G250TR G250D Power Supply
G250TR G250D Reverb & Pres
G250TR, G250D, G200S & G200
G250PA
H-Zog Randy Bachman
H-Zog
L90 & B90
LB90L, LB100 & G45B Preamp
LB100FT
LB100TR & G100TR Power Supply
L190D B190D Power Supply
LB190B & LB260D
LB200F & All 190's
LB200F & 190's Power Supply
LB200F & LB400F
LB200F & LB400F
Lil Rock B90L
M100TAR
Mach 5
Mach 5 Preamp & Reverb
Mach 5 Power Amp
PA90
PA90 & PA90R
PA90 Mixer
PA190 & PA260
PM11 Power
PM11 Mixer
Pro Bass 190 Stinger Fuzz and Pre
PS 250 PA
R90
Rebel Delux
Rebel 11 LB100
Reverb Unit
Giannina
A-120
A-201
A-300
Bulldog Baixo
Bulldog Bass
Bulldog Guitar
Duovox 50B
Duovox 50G
Duovox 100B
Duovox 100G
Duovox 120B
Duovox 120G
Duovox 150B
Duovox 150G
Duovox 240B
Duovox 240G
Mixer A-200/A-201
Mixer PM102
Plus Guitar/Organ
Pre Super Trem
Super Trem
Terra Contra Baixo
Terra Guitar
Thor Baixo
Thor Guitar
Thunder Sound III
Thunder Sound IIIA
Trem II
Trem Amp
Trem Amp III
Trem Compact
True Reverb
Valiante
Gibson
Atlas IV
Atlas Medalist
BA-15RV
BR-series
BR-1
BR-3
BR-4
BR-6
BR-6F
BR-9
Clavioline series
Clavioline
Clavioline Keyboard
Clavioline Stereo Amp
EH-series
EH-100
EH-100
EH-125
EH-150 (Older)
EH-150 (Later)
EH-160
EH-185 6J7 Pre
EH-185 65Q7 Pre
EH-195
EH-195 Circuit Changes
Falcon
GA-series
GA-Custom
GA-1RT-1
GA-1RVT
GAV-1
GA-2RVT
GA-3RV Reverb Unit
GA-4RE
Les Paul GA-5 Early
Les paul Jr. GA-5 Later
GA-5
GA-5T
GA-6
GA-6 Lancer
GA-6
GA-8 6BQ5
GA-8 6V6
GA-8T
GA-9
GA-14 Titan
GA-15
GA-15RVT
GA-16T
GA-17RVT Scout
GA-17RVT
GA-18T
GA-19RVT
GA-20
GA-20T
GA-20T
GA-20T
GA-20RVT
GA-25
GA-25RVT
GA-30 65J7 Preamp
GA-30 12AX7 Preamp
GA-30RV
GA-30RVT
GA-35RVT
Les Paul GA-40
GA-40T
GA-45RVT
GA-45RVT Saturn
GA-46
GA-50
GA-50T
GSS-50
GA-55
GA-55RVT
GA-60
GA-70
GA-75 Early
GA-75 Later
GA-75W
GA-77
GA-77RET
GA-77RVT
GA-78
GA-79
GA-79RVT
GA-80 Varitone
GA-83S Stereo VIB (Preamp)
GA-83S Stereo VIB
GA-83S Stereo VIB (Power Amp)
GA-85
GA-86 Ensemble
GA-88S
GA-90
GA-95 Apollo
GA-100 Bass Amp
G-105
GA-200
GA-300RVT
GA-400
Gibsonette Early
Gibsonette Later
Hawk
KEA
KEH
KEH-R
Duo-Medalist "A"
Duo-Medalist
Super Medalist
Mastertone
Mercury I & II
Reverb-12
Skylark/Tremelo
Thor Bass Amp
Titan-series
Titan Preamp
Titan I-II & V
Titan I-II & V Power/Output
Titan Medalist
Medalist Preamp
Medalist Power
Gotham Audio
V54
V54-ST
Gretsch Chet Atkins
GT Electronics
Tube Direct Unit
Studio 22 Preamp
Guild 50J
Hammond
Organ Reverb AO-35 (Early A-100)
Organ AO-39
Power Amp AO-39 Wiring
Organ Reverb AO-44 (A100)
Organ Reverb AO-44 Exp. (A-100)
Harmony
H-204 (1995)
H-430
H440
Hiwatt
DR 103-112 Output
DR 112-203 Preamp
DR 201-203 Output
DR 504 Output
DR Preamp Stage
L100R + L50R
Power Supply 50w
Power Supply 100w
Power Supply 150w
STA 100-200 Preamp
STA 230-250R-100 PRE
SAT 400 Slave
Hohner
Orgaphon Bass
Juergen Simon
Advanced Tube Bass Preamp
Kalamazoo
Bass 50
Kay
K-500
K-505
K-506
K-520
K-550
K-700
K-703
K-703C
K-805
K-820
K-830
Miscellaneous-1
Miscellaneous-2
Laney
A100H
Leslie
Rotating Speaker Cabinet
Rotor 122 Amplifier
Rotor 122 (Power & Wiring)
Rotor 147 (Power & Wiring)
Lifco
amp 1000a
McIntosh
Voltage & Resistance Chart
175
275
MC-30 Power Amp
MC-30 Type A 116b
MC-60 Type A 121
MC-60 Type A 121 Revised
MC-60 Type A 125
MC-225
MC-2100 Power Supply
Maestro
6V6 Tremelo Amp
Echoplex (1966)
Echoplex EP-2
Echoplex EP-3
Echoplex EP-4
GA-78RV
M1-RVT
M-201
M-216RVT
MA-40RVT
Magnatone
180
480
T-32
Marantz
Hi Fi Amp
Marshall
Charts
50w
50w Master Volume (Model 2204)
50w Master Volume (Model 2204 Revised)
100w
100w PA (Model 1968)
200w PA (Model 1966)
Bass Amp (Model 1992)
Bluesbreaker Reissue (Model 1962)
JCM Bass 800 (Model 1992)
JCM 800 Lead (50w & 100w Power)
JCM 800 Lead (Preamp)
JCM 800 Rev (Model 4140/4145)
Major 200w
Reverb (Models 2205/2810/4210 Revised)
Reverb (2210 Head/4211 Combo)
Studio 15 (Model 4001)
Super PA (Model 1963)
Martin
112
Masco
MA-35N & MA-35RCN
Mesa Boogie
Boogie Mark I
Boogie Mark IIA
Boogie Mark IIB
Boogie Mark IIC & Preamp
Boogie Mark III
Power/Output
Boogie Mark III Preamp
Boogie Bass 400
Musicman
2100-65
2100-130
2165-RD & 2100-RD
2165-RP & 2100-RP
2470-130 & 2275-130
2475-65 & 2275-65
BB-3 (75w)
BB-3 (150w)
GB-2 (2275 & 2475 Chassis)
GD-2A
GP-3A
RD-50
Newcombe
Model 10
Orange
120w Graphic Mark II (Early Model)
120w Graphic Mark II
120w Graphic Mark II (Alternate Diagram)
Model 12S
Peavey
Artist Power Amp
Artist VT Pre-amp
Bravo 112
Classic 30
Classic 120
Classic 120
Clasic 120/120 Power Amp
Patch Bd.
Mace Deuce Preamp
Mace Deuce Power/Output
Rockmaster
Standard PA Preamp
Van Halen 120 Preamp (Diagram1)
Van Halen 120 Preamp (Diagram2)
Premier
88
Pulse Techniques
EQP 1R
MEQ5
Randall
RGT-100
RCA Theremin
1929 Theremin
Revox
Stereo Tape Recorder
Rickenbacker
B16/16D
B Series Guitar
Model M8
Roland
Bolt 60
Selmer
Zodiac Twin 30
Zodiac Twin 50
Seymour Duncan
Convertible
Silvertone
2X6L6 Tremolo Amp (Old)
2X6V6 Amp
2X6V6 Amp (Revised)
1396
1474
1482
1484
1485
Sound
505R
Big 25e & 25rt
Studio-20
Sound City
120w Slave
200w LE
Bass 150w
Concord
L-B 120 Mark IV
L-B Plus 50
L-B 100w M3234
L-B 200 Plus
Sovtek
MIG 50 Master Volume
MIG 60 Lead
St to WU (130)
Standel
1968 Bass Amps
Correction for 1968 Amps
Acoustic magnifiers/PA Amp
Artist & Studio Models
Artist Models Suffix "A"
All Imperial Models Suffix "A"
Imperial Models S110
Custom Models Suffix "A"
Custom & Imperial
Custom Reverb
Duette II
PA 7
PB 25
Power Section
PR5B & PR5R Master Control
Super Imperial & Super Artist
Strobotuner
ST-2
Sunn
1200
1200s
1211 Crossover
1225 Crossover
1226 Crossover
1228 Crossover
2000s
2000s
Model A
Model T Super
100s, Sceptre & Sentura II
Sentura I
Sentura 1 & Solaris
Sentura II
Sonic II/1-40/200s/Sorado
Sonic II, 200s, Sorado & Sonic 1-40
100s+ & Spectrum II
Suprem
Piccolo
2 X E34 Amp
Model 1
Supro
Amp 1947
Late '50S 1 X 6V6 Amp
Reverb
S6625
S6651
S6688
S6698 Trem/Reverb
Steel Guitar Amp 1946
Traynor
TS50
YBA-1 Bass Master
YBA-1 Bass Master
YBA-1 (1966 Version)
YBA-1 (Latest Version)
YBA-1 Power Supply
YBA-1 & YBA-4
YBA-1, YBA-4 & Eng
YBA-1A Bass Master Mk II
YBA-1A & Eng Bass Master Mk II
YBA-2 Bass Mate
YBA-3 Custom Special
YBA-3 & Eng Custom Special
YBA-3 Custom Special (Revised)
YBA-3 Power Amp
YBA-3 Custom Special Preamp
YBA-4
YGA-1 1A
YGL-3 Mk III
YGL-3/3A Mk III
YGM-2 Guitar Mate
YGM-3 & 4
YRM-1
YRM/ISC/Eng
YSR-1
Univox
U-1011 Lead Amp
U-1226 Lead Amp
U-155R Guitar Amp
Valco
6400
6650 TR
Super Reverb (Model 510-33)
Supreme (Model 510-1.B)
Vox
AC10
AC15 Mk II (1959)
AC15 Mk III (1960)
AC30 Reverb (1978)
AC30 Silver Jubilee (Vox Ltd)
AC30 (S.S. Red)
AC30 Top Boost Reissue (1994)
AC30 Twin Reverb w/Mod
AC50/2 (G234 Rect)
AC50/4 Mk II (S.S. Red)
AC50/4 MK III
Berkely II Super Reverb V-8
Buckingham V1121
Buckingham V1123*6
Cambridge Reverb V-3
Percussion Key V829
Royal Guardsman V1131
V15 (Vox Ltd)
Viscount V1154
Westminster V118
Wards Airline
GDR-8518A
GDR-8514A/8515A
GDR-8517A
GDR-9012A
GIM-9151A
GIM-9171A
GIM-9111A
GIM-9131A
GVC-9052
GVC-9058A
GVC-9061A
Watkins
1959 Copicat MK II
Copicat Echo Unit
Dominator 35 Mk IV
Joker 30
W14T & W20T
Watkins/WEM
Clubman Mk 8
Dominator 50
Westminster Mk 9
Western Electric
140A Amp
142A Amp
142A Amp (Converted For 25w)
143A Amp
Wurlitzer
B30
P38 Stereo (Model 1)
P38 Stereo (Model 2)
Miscellaneous
2 x 7886 Master Volume Mod
2 x EL 503
BV 9-190
High Fidelity Audio 15w
High Fidelity Audio 50w
Power Supply

Instructor's Guide for

CalcuLadder

Learning Vitamins® for
Advanced Multiplication and Basic Division
by
Edwin C. Myers, Ph.D.

Another Unit in the **Character and Competence® Series** of Educational Materials

Character, Competence, and Learning Vitamins

Character and competence are central goals of Christian child-rearing. *Character* springs from the context of one's creaturehood, moral condition, stewardship under God, and a right relationship to the Lord Jesus Christ. *Competence* includes both knowledge and know-how: knowledge of what is and what might be done, and the skill to make and do. Christian character provides the motivation to attain competence and to employ it in a life's work of eternal value, while competence makes possible the expression of Christian character in beautiful, productive ways.

As part of the Character and Competence Series of educational materials, Learning Vitamins® exercises are designed to help build character and competence in the maturing child through *brief, potent* drills designed for *daily use,* which *promote growth* in skills that are important for wise stewardship in God's world.

Learning Vitamins emphasize *operational skills mastery*, that is, the nuts-and-bolts *doing* of the things students learn about in texts and workbooks. The focus is on *bridging the gap between learning what to do and being able to do it well.* These supplements help sharpen skills until they become self-reinforcing: students see the benefits of their enhanced skills, and start using these skills on their own.

Learning Vitamins also provide a *biblical and moral frame of reference* for the subjects covered, by means of the Bible texts that appear on every exercise. As you draw attention to these texts and explore them with your students, they become powerful character-building tools.

We at The Providence Project aim to make Learning Vitamins the best and most reasonably priced exercises of their kind. Please pass along to us any comments or suggestions which occur to you as you use these materials. We want them to be everything you're hoping for.

-- Edwin C. Myers

A Note on Copying these Materials

All of the CalcuLadder® 3 materials are copyrighted. Only individuals who purchase these materials to be used at home by members of their immediate family may copy them, and then only for such in-home use by immediate family members. To copy these materials for other purposes you must obtain permission from The Providence Project. Masters of the CalcuLadder 3 drills, suitable for xerographic copying, are available for in-home use as just described, or you may simply use the pages of this unit as copy masters for in-home use.

For further details on these and other materials from The Providence Project, please see the Order Form enclosed with this unit, or contact

The Providence Project
14566 NW 110th St. Whitewater, KS 67154
Telephone toll free: 1–888–776–8776

CalcuLadder 3: What It Is

The CalcuLadder® supplements include 96 one-page Levels of timed Learning Vitamins® drills. This carefully planned, progressive series is divided into 6 groups of 16 Levels each. One set of CalcuLadder 3 contains 12 pages each of Levels 33 through 48, for a total of 192 exercise pages. The student repeats a given Level each day until he achieves sufficient accuracy and speed (See the "How to Use It" section of this Guide for further details). Students often advance one Level about every two weeks. Thus, each CalcuLadder 3 set carries the average student through a full school year, and provides extra exercise copies for times of slower-than-usual progress. The following features make the CalcuLadder 3 drills even more useful and effective:

- **Quality paper**--The pages don't "dissolve" under the pressure of a youngster's pencil eraser.
- **Motivational nuances**--New colors and "growing" graphics on successive Levels encourage rapid student progress, and add beauty and variety to these materials.
- **QuicKeys®**--These clever grading keys make scoring the exercise Levels much faster and easier than using conventional keys.
- **Instructor's Guide**--The Guide provides information, guidelines, and suggestions for using the exercises most effectively.
- **Bible texts**--The Bible texts printed on each page place the skills being learned within a biblical, moral, and character-building frame of reference.
- **Achievement Record**--The CalcuLadder Achievement Record provides added student motivation and a convenient means of recording student progress.

CalcuLadder 3: Its Objective

The objective of CalcuLadder 3 is the student's complete mastery of advanced multiplication and basic division. Multiplication topics covered include multiplication tables through 12x12 (with review of tables covered in CalcuLadder 2) and multiplication of multi-digit numbers in various groupings. Basic division includes concepts and simple division with and without remainders.

By "complete mastery" we mean that the student has both rapid, accurate recall of the basic facts involved, plus confident, efficient skill in the procedures for applying the basic facts to solve lengthier problems. For example, when the student sees 847 x 526 or 68 ÷ 7, he should proceed with confidence, accuracy, and efficiency. Students should select and employ appropriate computational procedures without hesitation--pausing in a problem to think, "Oops! Do I move this partial product one column to the left, or two? Do I add the number I carried before or after multiplying?" should become unnecessary as CalcuLadder 3 is mastered.

In all but the simplest multiplication and division problems, one uses addition and subtraction extensively on the way to finding the answer. Thus, competence in addition and subtraction--such as developed in CalcuLadder 1 and CalcuLadder 2--is important in CalcuLadder 3.

The progressive, timed exercises of CalcuLadder 3 promote *instant recall of multiplication and division facts, and help make the procedural aspects of advanced multiplication and basic division "second nature" for your students, thus imparting both a solid foundation for further progress as well as skills useful in their own right.*

CalcuLadder 3: Who Should Use It

CalcuLadder 3 is for students *at any grade level* who have proper grounding in more basic topics and need the benefit of improved skills in advanced multiplication and basic division. Though CalcuLadder 3 often dovetails nicely with third or fourth grade math courses, it is grade-level independent. It may be used with accelerated students as well as with those who require remedial work. It may be used in full-class, tutorial, "summer brush-up," or self-supervised learning situations. It may be used from grade school through high school, as needs require.

About the Author: A science consultant, educator, and musician, Edwin C. Myers holds the M.A. from Dallas Theological Seminary and the Ph.D. from Carnegie-Mellon University. In addition to authoring the *AlphaBetter®*, *CalcuLadder®*, and (with Mrs. Myers) *ReadyWriter®* Learning Vitamins, Dr. Myers' credits include the design of instruments on NASA Voyager space probes, publications in the fields of optics, magnetics, and geophysics, and a short cantata, *The Road to Emmaus*. Dr. and Mrs. Myers are the grateful parents of twelve home-schooled children.

© Edwin C. Myers 1985, 1989

CalcuLadder 3: How to Use It

- **Use it every day.** To achieve the desired results, it's very important that students take their Learning Vitamins every normal school day. Though circumstances will require occasional skipped days, these exercises should be part of your class's regular routine. For maximum benefit, you should administer these materials during a part of the class period or day in which your students are alert and not drowsy.

- **Use it along with your regular curriculum materials.** Texts and workbooks provide important background, examples, and applications. To achieve real competence, students need *both* information and initial practice such as texts provide, *and* performance drills like CalcuLadder 3.

- **Use it sensitively.** Be sensitive to student emotions and classroom dynamics as you use CalcuLadder 3, so as to promote an optimum mix of fun, competition, desire, love, and humility, with a minimum of trauma.

- **Use it in a somewhat formal setting.** Although taking Learning Vitamins should be fun and motivational, students should understand that they are expected to perform at their best. The drills are, if you will, brief tests, and should be approached soberly. Do not tolerate talking or distracting behavior during the exercise time.

- **Use each Level as a stepping stone to higher Levels.** A "time goal" is printed at the bottom of every CalcuLadder 3 Level. To "graduate" from one Level to the next, students try each day to complete their Level correctly in a time equal to or faster than the time goal. It should be possible for many students to move up one Level approximately every 2 weeks. However, if students find it difficult to meet the time goal within 9 to 12 tries, feel free to allow about 10-15 extra seconds per minute of suggested time (e.g., allow 4 minutes and 40 seconds to complete a "4-minute" Level). While students should be motivated to gain as much speed as possible, some flexibility is clearly permissible. Additionally, a "one error allowed" criterion may sometimes be used.

- **Use a procedure something like this.** As a lead-up to each day's Learning Vitamins, "prime" your students to do their best by asking a few questions or discussing a few examples relating to the Levels they'll be doing. This need take only a few minutes. Then say something like, "Now it's time for our Learning Vitamins. Please clear off all extra materials from your desk(s) and take out a pencil." Students should fold their Learning Vitamins books over so that only one page is facing up, and should cover that page with a sheet of paper until you say, "On your mark. Get set. Go!" Give the "Go!" signal about 3 seconds before the commencement of the timing interval, so that students have time to remove the cover sheet from their exercise. When a student finishes, he should lay his pencil down and quietly say "Finished," or "Done." Note the student's time on his paper. You may be able to let some students time themselves.

- **Use the QuicKeys® and the Achievement Record.** Grade your students' papers using the fast and simple QuicKey grading keys. Read and follow the directions on each QuicKey, and write your students' scores on their papers. (Some students may be able to grade their own work.) When your students pass a Level, recognize their accomplishment by calling them to the front of the class as you fill out the appropriate blanks of the Achievement Record at the back of this unit.

- **Use the exercises completely.** If a student passes a Level after only 3 or 4 tries, encourage him to repeat it a few more times to see how fast he can really go. It's well within the capability of older or faster students to better the nominal time goals by up to a minute or more–and beneficial for them as well! It's also fun and beneficial for students occasionally to review a Level which they have previously passed, just to see how easily they can complete it, and to keep their skills sharp. They can even try to set "world's records"!

- **Use the Bible texts.** The Bible texts printed on the Learning Vitamins Levels show that the subjects which these drills deal with are worth learning because the Bible deals with those same subjects. The texts also often enunciate moral guidelines and principles. Sometimes successive Levels round out a single scriptural thought. Draw your students' attention to the verses. Read them aloud. Explain. Ask questions like, "Who said this?", "When did this happen?", "What does this mean?". Consider using some of the texts as memory verses. These verses should help build our character as well as that of our students.

CalcuLadder 3: The 16 Levels

Levels 33 – 48 of the CalcuLadder® supplements comprise CalcuLadder 3. These Levels review multiplication tables covered in CalcuLadder 2 and cover remaining tables through 12x12. Then they interweave more advanced multiplication with basic division concepts and problems.

Stress to your students the importance of *not writing overly-large or scrawly answers* if they want to achieve maximum speed and accuracy. **Note:** It's a good idea to work through each Level yourself (using a separate sheet of paper, and maybe timing yourself!) to gain greater familiarity with the exercises. Here are further specifics of the CalcuLadder 3 Levels:

- **Level 33 -- Multiplication**   Time Goal: 4 minutes   80 Answers

Level 33 reviews the multiplication tables from 0x2, 1x2, 2x2, . . ., 12x2 through 0x8, 1x8, 2x8, . . ., 12x8, that is, most of the multiplication tables covered previously in CalcuLadder 2. It's important for students to master these facts as a foundation for further progress.

- **Level 34 -- Multiplication**   Time Goal: 4 minutes   76 Answers

Level 34 covers 0x9, 1x9, 2x9, . . ., 12x9 and 0x10, 1x10, 2x10,. . ., 12x10. Students should complete Level 34 by first filling-in the columns of "boxes" on the left-hand side of the exercise page and then filling-in the rows of printed problems. The columns of boxes aid students' understanding of the multiplication process, and provide added practice. Students may use the filled-in boxes to help solve the other problems, but such use diminishes as the pertinent facts are mastered.

- **Levels 35,36 -- Multiplication**   Time Goal: 4 minutes   72 Answers each Level

Level 35 covers 0x11, 1x11, . . ., 12x11 and reviews the "9's" and "10's." Level 36 covers 0x12, 1x12, . . ., 12x12 and reviews the "11's" plus some of the "7's" and "8's." Guidelines for these Levels are like those for Level 34, except that there is only one column of boxes to fill-in.

- **Level 37 -- Multiplication**   Time Goal: 4 minutes   40 Answers

Level 37 introduces problems of the form AB x N and ABC x N which do not involve carrying. The goal here is to provide practice in the *procedure* for solving such problems: Solve ABC x N by taking C x N, then B x N, then A x N. **Note:** You may want to require your students to put commas in answers which are greater than 1,000 or 10,000 in this and later Levels.

- **Levels 38,39 -- Multiplication**   Time Goal: 4 minutes   21, 26 Answers, respectively

These Levels build upon Level 37 by including AB x N-type problems that involve "carrying" (also called "regrouping" or "renaming"). Level 39 requires more speed than does Level 38.

- **Level 40 -- Multiply-Divide**   Time Goal: 4 minutes   84 Answers

Level 40 introduces the concept of division as the "inverse" of multiplication. Orient your students (by discussion and worked examples) to the kind of problems in Level 40 before they take it for the first time. Level 40 should help your students get a solid grasp on what division *is*.

- **Levels 41,43 -- Multiplication**   Time Goal: 4 minutes   15, 10 Answers, respectively

Level 41 has problems of the form ABC x N and ABCD x N, most of which involve carrying. Level 43 has problems of the form AB x MN, in which *partial products* are used for the first time.

- **Levels 42,44 -- Division**   Time Goal: 4 minutes   60, 70 Answers, respectively

Levels 42 and 44 consist of basic division problems without remainders. Students need only write down the answers. These Levels promote mastery of the basic facts of division--the "flipside" of the multiplication tables.

- **Levels 45,47 – Multiplication**   Time Goal: 5 minutes   8, 9 Answers, respectively

Level 45 contains problems of the form AB x MN, ABC x MN, and ABC x LMN, which provide further work with partial products. Level 47 allow practice in helpful "shortcuts." In 534 x 111, for example, the partial products will simply be copies of 534 (1's in the multiplier). In 819 x 545, two of the partial products will be identical (repeated digits in the multiplier). In 438 x 306, only two properly positioned partial products need be written down (embedded zeros in the multiplier). "Trailing zeros" help simplify 7100 x 300 (do 71 x 3 and then "tack on" four zeros), and so on.

- **Levels 46,48 – Division**   Time Goal: 5 minutes   24, 30 Answers, respectively

Level 46 introduces division with remainders. Remainders are labelled with an "r" as shown in the Level's example. Note also that the answer, "2 r2," is written starting above the "4" of the "14." The proper placement of answers helps avoid mistakes later in more difficult division problems, so it's good for students to pay attention to this. Level 48 is like Level 46, but more speed is required.

© Edwin C. Myers 1985, 1989   CalcuLadder 3 *Instructor's Guide*

My name is _____

Today is _____

1	6	11	4	9	2	7	12	5	10
x3	x2	x5	x8	x3	x7	x4	x6	x2	x5

3	8	6	11	4	9	2	7	12	5
x8	x3	x7	x4	x6	x2	x5	x8	x3	x7

10	3	8	6	11	4	9	2	7	12
x4	x6	x2	x5	x8	x3	x7	x4	x6	x2

5	10	7	3	8	6	11	4	9	2
x5	x8	x0	x3	x7	x4	x6	x2	x5	x8

7	12	5	10	3	8	6	11	4	9
x3	x7	x4	x6	x2	x5	x8	x3	x7	x4

2	7	12	5	10	3	8	6	11	0
x6	x2	x5	x8	x3	x7	x4	x6	x2	x5

4	9	2	7	12	5	10	3	8	6
x5	x8	x3	x7	x4	x6	x2	x5	x8	x3

11	4	9	2	7	12	5	10	3	8
x7	x4	x6	x2	x5	x8	x3	x7	x4	x6

The Lord your God has multiplied you, and behold, you are this day as the stars of heaven for multitude. Deut. 1:10

© Edwin C. Myers 1985, 1990   CalcuLadder®   Level 33: Multiplication   4 minutes

My name is _____

Today is _____

1	6	11	4	9	2	7	12	5	10
x3	x2	x5	x8	x3	x7	x4	x6	x2	x5

3	8	6	11	4	9	2	7	12	5
x8	x3	x7	x4	x6	x2	x5	x8	x3	x7

10	3	8	6	11	4	9	2	7	12
x4	x6	x2	x5	x8	x3	x7	x4	x6	x2

5	10	7	3	8	6	11	4	9	2
x5	x8	x0	x3	x7	x4	x6	x2	x5	x8

7	12	5	10	3	8	6	11	4	9
x3	x7	x4	x6	x2	x5	x8	x3	x7	x4

2	7	12	5	10	3	8	6	11	0
x6	x2	x5	x8	x3	x7	x4	x6	x2	x5

4	9	2	7	12	5	10	3	8	6
x5	x8	x3	x7	x4	x6	x2	x5	x8	x3

11	4	9	2	7	12	5	10	3	8
x7	x4	x6	x2	x5	x8	x3	x7	x4	x6

The Lord your God has multiplied you, and behold, you are this day as the stars of heaven for multitude. Deut. 1:10

© Edwin C. Myers 1985, 1990   **CalcuLadder®**   Level 33: Multiplication   4 minutes

My name is _____

Today is _____

1	6	11	4	9	2	7	12	5	10
x3	x2	x5	x8	x3	x7	x4	x6	x2	x5

3	8	6	11	4	9	2	7	12	5
x8	x3	x7	x4	x6	x2	x5	x8	x3	x7

10	3	8	6	11	4	9	2	7	12
x4	x6	x2	x5	x8	x3	x7	x4	x6	x2

5	10	7	3	8	6	11	4	9	2
x5	x8	x0	x3	x7	x4	x6	x2	x5	x8

7	12	5	10	3	8	6	11	4	9
x3	x7	x4	x6	x2	x5	x8	x3	x7	x4

2	7	12	5	10	3	8	6	11	0
x6	x2	x5	x8	x3	x7	x4	x6	x2	x5

4	9	2	7	12	5	10	3	8	6
x5	x8	x3	x7	x4	x6	x2	x5	x8	x3

11	4	9	2	7	12	5	10	3	8
x7	x4	x6	x2	x5	x8	x3	x7	x4	x6

The Lord your God has multiplied you,
and behold, you are this day as the stars of heaven for multitude. Deut. 1:10

© Edwin C. Myers 1985, 1990   **CalcuLadder®**   Level 33: Multiplication   4 minutes

My name is _____

Today is _____

1	6	11	4	9	2	7	12	5	10
x3	x2	x5	x8	x3	x7	x4	x6	x2	x5

3	8	6	11	4	9	2	7	12	5
x8	x3	x7	x4	x6	x2	x5	x8	x3	x7

10	3	8	6	11	4	9	2	7	12
x4	x6	x2	x5	x8	x3	x7	x4	x6	x2

5	10	7	3	8	6	11	4	9	2
x5	x8	x0	x3	x7	x4	x6	x2	x5	x8

7	12	5	10	3	8	6	11	4	9
x3	x7	x4	x6	x2	x5	x8	x3	x7	x4

2	7	12	5	10	3	8	6	11	0
x6	x2	x5	x8	x3	x7	x4	x6	x2	x5

4	9	2	7	12	5	10	3	8	6
x5	x8	x3	x7	x4	x6	x2	x5	x8	x3

11	4	9	2	7	12	5	10	3	8
x7	x4	x6	x2	x5	x8	x3	x7	x4	x6

The Lord your God has multiplied you,
and behold, you are this day as the stars of heaven for multitude.   Deut. 1:10

© Edwin C. Myers 1985,1990   CalcuLadder®   Level 33: Multiplication   4 minutes

My name is _____

Today is _____

1	6	11	4	9	2	7	12	5	10
x3	x2	x5	x8	x3	x7	x4	x6	x2	x5

3	8	6	11	4	9	2	7	12	5
x8	x3	x7	x4	x6	x2	x5	x8	x3	x7

10	3	8	6	11	4	9	2	7	12
x4	x6	x2	x5	x8	x3	x7	x4	x6	x2

5	10	7	3	8	6	11	4	9	2
x5	x8	x0	x3	x7	x4	x6	x2	x5	x8

7	12	5	10	3	8	6	11	4	9
x3	x7	x4	x6	x2	x5	x8	x3	x7	x4

2	7	12	5	10	3	8	6	11	0
x6	x2	x5	x8	x3	x7	x4	x6	x2	x5

4	9	2	7	12	5	10	3	8	6
x5	x8	x3	x7	x4	x6	x2	x5	x8	x3

11	4	9	2	7	12	5	10	3	8
x7	x4	x6	x2	x5	x8	x3	x7	x4	x6

The Lord your God has multiplied you,
and behold, you are this day as the stars of heaven for multitude.   Deut. 1:10

© Edwin C. Myers 1985, 1990   **CalcuLadder**®   Level 33: Multiplication   4 minutes

My name is _____

Today is _____

1	6	11	4	9	2	7	12	5	10
x3	x2	x5	x8	x3	x7	x4	x6	x2	x5

3	8	6	11	4	9	2	7	12	5
x8	x3	x7	x4	x6	x2	x5	x8	x3	x7

10	3	8	6	11	4	9	2	7	12
x4	x6	x2	x5	x8	x3	x7	x4	x6	x2

5	10	7	3	8	6	11	4	9	2
x5	x8	x0	x3	x7	x4	x6	x2	x5	x8

7	12	5	10	3	8	6	11	4	9
x3	x7	x4	x6	x2	x5	x8	x3	x7	x4

2	7	12	5	10	3	8	6	11	0
x6	x2	x5	x8	x3	x7	x4	x6	x2	x5

4	9	2	7	12	5	10	3	8	6
x5	x8	x3	x7	x4	x6	x2	x5	x8	x3

11	4	9	2	7	12	5	10	3	8
x7	x4	x6	x2	x5	x8	x3	x7	x4	x6

The Lord your God has multiplied you,
and behold, you are this day as the stars of heaven for multitude.   Deut. 1:10

© Edwin C. Myers 1985, 1990   CalcuLadder®   Level 33: Multiplication   4 minutes

My name is _____

Today is _____

1	6	11	4	9	2	7	12	5	10
x3	x2	x5	x8	x3	x7	x4	x6	x2	x5

3	8	6	11	4	9	2	7	12	5
x8	x3	x7	x4	x6	x2	x5	x8	x3	x7

10	3	8	6	11	4	9	2	7	12
x4	x6	x2	x5	x8	x3	x7	x4	x6	x2

5	10	7	3	8	6	11	4	9	2
x5	x8	x0	x3	x7	x4	x6	x2	x5	x8

7	12	5	10	3	8	6	11	4	9
x3	x7	x4	x6	x2	x5	x8	x3	x7	x4

2	7	12	5	10	3	8	6	11	0
x6	x2	x5	x8	x3	x7	x4	x6	x2	x5

4	9	2	7	12	5	10	3	8	6
x5	x8	x3	x7	x4	x6	x2	x5	x8	x3

11	4	9	2	7	12	5	10	3	8
x7	x4	x6	x2	x5	x8	x3	x7	x4	x6

The Lord your God has multiplied you, and behold, you are this day as the stars of heaven for multitude.   Deut. 1:10

© Edwin C. Myers 1985, 1990   **CalcuLadder**®   Level 33:   Multiplication   4 minutes

My name is _____

Today is _____

1	6	11	4	9	2	7	12	5	10
x3	x2	x5	x8	x3	x7	x4	x6	x2	x5

3	8	6	11	4	9	2	7	12	5
x8	x3	x7	x4	x6	x2	x5	x8	x3	x7

10	3	8	6	11	4	9	2	7	12
x4	x6	x2	x5	x8	x3	x7	x4	x6	x2

5	10	7	3	8	6	11	4	9	2
x5	x8	x0	x3	x7	x4	x6	x2	x5	x8

7	12	5	10	3	8	6	11	4	9
x3	x7	x4	x6	x2	x5	x8	x3	x7	x4

2	7	12	5	10	3	8	6	11	0
x6	x2	x5	x8	x3	x7	x4	x6	x2	x5

4	9	2	7	12	5	10	3	8	6
x5	x8	x3	x7	x4	x6	x2	x5	x8	x3

11	4	9	2	7	12	5	10	3	8
x7	x4	x6	x2	x5	x8	x3	x7	x4	x6

The Lord your God has multiplied you, and behold, you are this day as the stars of heaven for multitude. Deut. 1:10

© Edwin C. Myers 1985, 1990   **CalcuLadder®**   Level 33: Multiplication   4 minutes

My name is _____
Today is _____

1	6	11	4	9	2	7	12	5	10
x3	x2	x5	x8	x3	x7	x4	x6	x2	x5

3	8	6	11	4	9	2	7	12	5
x8	x3	x7	x4	x6	x2	x5	x8	x3	x7

10	3	8	6	11	4	9	2	7	12
x4	x6	x2	x5	x8	x3	x7	x4	x6	x2

5	10	7	3	8	6	11	4	9	2
x5	x8	x0	x3	x7	x4	x6	x2	x5	x8

7	12	5	10	3	8	6	11	4	9
x3	x7	x4	x6	x2	x5	x8	x3	x7	x4

2	7	12	5	10	3	8	6	11	0
x6	x2	x5	x8	x3	x7	x4	x6	x2	x5

4	9	2	7	12	5	10	3	8	6
x5	x8	x3	x7	x4	x6	x2	x5	x8	x3

11	4	9	2	7	12	5	10	3	8
x7	x4	x6	x2	x5	x8	x3	x7	x4	x6

The Lord your God has multiplied you,
and behold, you are this day as the stars of heaven for multitude.   Deut. 1:10

© Edwin C. Myers 1985,1990   **CalcuLadder**®   Level 33: Multiplication   4 minutes

My name is _____

Today is _____

1	6	11	4	9	2	7	12	5	10
x3	x2	x5	x8	x3	x7	x4	x6	x2	x5

3	8	6	11	4	9	2	7	12	5
x8	x3	x7	x4	x6	x2	x5	x8	x3	x7

10	3	8	6	11	4	9	2	7	12
x4	x6	x2	x5	x8	x3	x7	x4	x6	x2

5	10	7	3	8	6	11	4	9	2
x5	x8	x0	x3	x7	x4	x6	x2	x5	x8

7	12	5	10	3	8	6	11	4	9
x3	x7	x4	x6	x2	x5	x8	x3	x7	x4

2	7	12	5	10	3	8	6	11	0
x6	x2	x5	x8	x3	x7	x4	x6	x2	x5

4	9	2	7	12	5	10	3	8	6
x5	x8	x3	x7	x4	x6	x2	x5	x8	x3

11	4	9	2	7	12	5	10	3	8
x7	x4	x6	x2	x5	x8	x3	x7	x4	x6

The Lord your God has multiplied you,
and behold, you are this day as the stars of heaven for multitude. Deut. 1:10

© Edwin C. Myers 1985,1990   **CalcuLadder®**   Level 33: Multiplication   4 minutes

My name is _____

Today is _____

1	6	11	4	9	2	7	12	5	10
x3	x2	x5	x8	x3	x7	x4	x6	x2	x5

3	8	6	11	4	9	2	7	12	5
x8	x3	x7	x4	x6	x2	x5	x8	x3	x7

10	3	8	6	11	4	9	2	7	12
x4	x6	x2	x5	x8	x3	x7	x4	x6	x2

5	10	7	3	8	6	11	4	9	2
x5	x8	x0	x3	x7	x4	x6	x2	x5	x8

7	12	5	10	3	8	6	11	4	9
x3	x7	x4	x6	x2	x5	x8	x3	x7	x4

2	7	12	5	10	3	8	6	11	0
x6	x2	x5	x8	x3	x7	x4	x6	x2	x5

4	9	2	7	12	5	10	3	8	6
x5	x8	x3	x7	x4	x6	x2	x5	x8	x3

11	4	9	2	7	12	5	10	3	8
x7	x4	x6	x2	x5	x8	x3	x7	x4	x6

The Lord your God has multiplied you, and behold, you are this day as the stars of heaven for multitude.   Deut. 1:10

© Edwin C. Myers 1985, 1990   CalcuLadder®   Level 33: Multiplication   4 minutes

My name is _____

Today is _____

1	6	11	4	9	2	7	12	5	10
x3	x2	x5	x8	x3	x7	x4	x6	x2	x5

3	8	6	11	4	9	2	7	12	5
x8	x3	x7	x4	x6	x2	x5	x8	x3	x7

10	3	8	6	11	4	9	2	7	12
x4	x6	x2	x5	x8	x3	x7	x4	x6	x2

5	10	7	3	8	6	11	4	9	2
x5	x8	x0	x3	x7	x4	x6	x2	x5	x8

7	12	5	10	3	8	6	11	4	9
x3	x7	x4	x6	x2	x5	x8	x3	x7	x4

2	7	12	5	10	3	8	6	11	0
x6	x2	x5	x8	x3	x7	x4	x6	x2	x5

4	9	2	7	12	5	10	3	8	6
x5	x8	x3	x7	x4	x6	x2	x5	x8	x3

11	4	9	2	7	12	5	10	3	8
x7	x4	x6	x2	x5	x8	x3	x7	x4	x6

The Lord your God has multiplied you,
and behold, you are this day as the stars of heaven for multitude. Deut. 1:10

© Edwin C. Myers 1985, 1990   **CalcuLadder®**   Level 33: Multiplication   4 minutes

My name is _____
Today is _____

0×9 = **0**	0×10 = **0**	2	9	4	3	2	9
+9	+10	x9	x9	x9	x10	x9	x9

1×9=☐   1×10=☐
   +9   +10

2×9=☐   2×10=☐   4   8   11   6   1   5
   +9   +10   x9   x10 x 9   x9   x9   x10

3×9=☐   3×10=☐
   +9   +10   11   6   1   10   8   3
4×9=☐   4×10=☐   x 9 x9   x9   x10   x9   x9
   +9   +10

5×9=☐   5×10=☐
   +9   +10   10   7   8   3   10   1
6×9=☐   6×10=☐   x 9 x10 x9   x9   x 9   x10
   +9   +10

7×9=☐   7×10=☐
   +9   +10   5   12   7   11   5   12
8×9=☐   8×10=☐   x9   x 9 x9   x10   x9   x 9
   +9   +10

9×9=☐   9×10=☐
   +9   +10   7   4   2   4   9   6
10×9=☐   10×10=☐   x9   x10 x9   x9   x9   x10
   +9   +10

11×9=☐   11×10=☐
   +9   +10   1   6   11   0   3   10
12×9=☐   12×10=☐   x9   x9   x 9   x10   x9   x 9

  8   9   5   2   12   9   0   12   7   10
  x9   x0   x9   x10 x 9 x10 x9   x10 x9   x 0

And in all matters of wisdom and understanding . . .
he found them ten times better than all the magicians and astrologers . . .   Dan. 1:20

© Edwin C. Myers 1985,1990   **CalcuLadder®**   Level 34: Multiplication   4 minutes

My name is _____

Today is _____

```
0x9 = 0      0x10 = 0     2    9    4    3    2    9
      +9           +10   x9   x9   x9  x10   x9   x9
1x9=[  ]     1x10=[  ]
      +9           +10    4    8   11    6    1    5
2x9=[  ]     2x10=[  ]   x9  x10  x9   x9   x9  x10
      +9           +10
3x9=[  ]     3x10=[  ]
      +9           +10   11    6    1   10    8    3
4x9=[  ]     4x10=[  ]   x9   x9   x9  x10   x9   x9
      +9           +10
5x9=[  ]     5x10=[  ]
      +9           +10   10    7    8    3   10    1
6x9=[  ]     6x10=[  ]   x9  x10   x9   x9   x9  x10
      +9           +10
7x9=[  ]     7x10=[  ]
      +9           +10    5   12    7   11    5   12
8x9=[  ]     8x10=[  ]   x9   x9   x9  x10   x9   x9
      +9           +10
9x9=[  ]     9x10=[  ]
      +9           +10    7    4    2    4    9    6
10x9=[  ]    10x10=[  ]  x9  x10   x9   x9   x9  x10
      +9           +10
11x9=[  ]    11x10=[  ]
      +9           +10    1    6   11    0    3   10
12x9=[  ]    12x10=[  ]  x9   x9   x9  x10   x9   x9


        8    9    5    2   12    9    0   12    7   10
       x9   x0   x9  x10  x9   x10   x9  x10   x9   x0
```

And in all matters of wisdom and understanding . . .
he found them ten times better than all the magicians and astrologers . . .   Dan. 1:20

© Edwin C. Myers 1985,1990   CalcuLadder®   Level 34:   Multiplication   4 minutes

My name is _____
Today is _____

0×9 = **0**	0×10 = **0**	2	9	4	3	2	9
+9	+10	×9	×9	×9	×10	×9	×9
1×9=	1×10=						
+9	+10	4	8	11	6	1	5
2×9=	2×10=	×9	×10	×9	×9	×9	×10
+9	+10						
3×9=	3×10=						
+9	+10	11	6	1	10	8	3
4×9=	4×10=	×9	×9	×9	×10	×9	×9
+9	+10						
5×9=	5×10=						
+9	+10	10	7	8	3	10	1
6×9=	6×10=	×9	×10	×9	×9	×9	×10
+9	+10						
7×9=	7×10=						
+9	+10	5	12	7	11	5	12
8×9=	8×10=	×9	×9	×9	×10	×9	×9
+9	+10						
9×9=	9×10=						
+9	+10	7	4	2	4	9	6
10×9=	10×10=	×9	×10	×9	×9	×9	×10
+9	+10						
11×9=	11×10=						
+9	+10	1	6	11	0	3	10
12×9=	12×10=	×9	×9	×9	×10	×9	×9

8	9	5	2	12	9	0	12	7	10
×9	×0	×9	×10	×9	×10	×9	×10	×9	×0

And in all matters of wisdom and understanding . . .
  he found them ten times better than all the magicians and astrologers . . .   Dan. 1:20

© Edwin C. Myers 1985, 1990   **CalcuLadder**®   Level 34: Multiplication   4 minutes

My name is _____

Today is _____

0×9 = **0**
+9
1×9=☐
+9
2×9=☐
+9
3×9=☐
+9
4×9=☐
+9
5×9=☐
+9
6×9=☐
+9
7×9=☐
+9
8×9=☐
+9
9×9=☐
+9
10×9=☐
+9
11×9=☐
+9
12×9=☐

0×10 = **0**
+10
1×10=☐
+10
2×10=☐
+10
3×10=☐
+10
4×10=☐
+10
5×10=☐
+10
6×10=☐
+10
7×10=☐
+10
8×10=☐
+10
9×10=☐
+10
10×10=☐
+10
11×10=☐
+10
12×10=☐

2	9	4	3	2	9
x9	x9	x9	x10	x9	x9

4	8	11	6	1	5
x9	x10	x9	x9	x9	x10

11	6	1	10	8	3
x9	x9	x9	x10	x9	x9

10	7	8	3	10	1
x9	x10	x9	x9	x9	x10

5	12	7	11	5	12
x9	x9	x9	x10	x9	x9

7	4	2	4	9	6
x9	x10	x9	x9	x9	x10

1	6	11	0	3	10
x9	x9	x9	x10	x9	x9

8	9	5	2	12	9	0	12	7	10
x9	x0	x9	x10	x9	x10	x9	x10	x9	x0

And in all matters of wisdom and understanding . . .
he found them ten times better than all the magicians and astrologers . . . Dan. 1:20

© Edwin C. Myers 1985, 1990 CalcuLadder® Level 34: Multiplication 4 minutes

My name is _____

Today is _____

0x9 = **0**	0x10 = **0**	2 x9	9 x9	4 x9	3 x10	2 x9	9 x9
+9	+10						
1x9= ☐	1x10= ☐						
+9	+10	4 x9	8 x10	11 x 9	6 x9	1 x9	5 x10
2x9= ☐	2x10= ☐						
+9	+10						
3x9= ☐	3x10= ☐						
+9	+10	11 x 9	6 x9	1 x9	10 x10	8 x9	3 x9
4x9= ☐	4x10= ☐						
+9	+10						
5x9= ☐	5x10= ☐						
+9	+10	10 x 9	7 x10	8 x9	3 x9	10 x 9	1 x10
6x9= ☐	6x10= ☐						
+9	+10						
7x9= ☐	7x10= ☐						
+9	+10	5 x9	12 x 9	7 x9	11 x10	5 x9	12 x 9
8x9= ☐	8x10= ☐						
+9	+10						
9x9= ☐	9x10= ☐						
+9	+10	7 x9	4 x10	2 x9	4 x9	9 x9	6 x10
10x9= ☐	10x10= ☐						
+9	+10						
11x9= ☐	11x10= ☐						
+9	+10	1 x9	6 x9	11 x 9	0 x10	3 x9	10 x 9
12x9= ☐	12x10= ☐						

8 x9	9 x0	5 x9	2 x10	12 x 9	9 x10	0 x9	12 x10	7 x9	10 x 0

And in all matters of wisdom and understanding . . .
he found them ten times better than all the magicians and astrologers . . .   Dan. 1:20

© Edwin C. Myers 1985, 1990   **CalcuLadder®**   Level 34: Multiplication   4 minutes

My name is _____
Today is _____

0×9 = **0**	0×10 = **0**	2	9	4	3	2	9
+9	+10	×9	×9	×9	×10	×9	×9
1×9=☐	1×10=☐						
+9	+10	4	8	11	6	1	5
2×9=☐	2×10=☐	×9	×10	×9	×9	×9	×10
+9	+10						
3×9=☐	3×10=☐						
+9	+10	11	6	1	10	8	3
4×9=☐	4×10=☐	×9	×9	×9	×10	×9	×9
+9	+10						
5×9=☐	5×10=☐						
+9	+10	10	7	8	3	10	1
6×9=☐	6×10=☐	×9	×10	×9	×9	×9	×10
+9	+10						
7×9=☐	7×10=☐						
+9	+10	5	12	7	11	5	12
8×9=☐	8×10=☐	×9	×9	×9	×10	×9	×9
+9	+10						
9×9=☐	9×10=☐						
+9	+10	7	4	2	4	9	6
10×9=☐	10×10=☐	×9	×10	×9	×9	×9	×10
+9	+10						
11×9=☐	11×10=☐						
+9	+10	1	6	11	0	3	10
12×9=☐	12×10=☐	×9	×9	×9	×10	×9	×9

8	9	5	2	12	9	0	12	7	10
×9	×0	×9	×10	×9	×10	×9	×10	×9	×0

And in all matters of wisdom and understanding . . .
he found them ten times better than all the magicians and astrologers . . .   Dan. 1:20

© Edwin C. Myers 1985, 1990   CalcuLadder®   Level 34: Multiplication   4 minutes

My name is _____
Today is _____

0×9 = **0**	0×10 = **0**	2	9	4	3	2	9
+9	+10	×9	×9	×9	×10	×9	×9
1×9 = ☐	1×10 = ☐						
+9	+10	4	8	11	6	1	5
2×9 = ☐	2×10 = ☐	×9	×10	×9	×9	×9	×10
+9	+10						
3×9 = ☐	3×10 = ☐						
+9	+10	11	6	1	10	8	3
4×9 = ☐	4×10 = ☐	×9	×9	×9	×10	×9	×9
+9	+10						
5×9 = ☐	5×10 = ☐						
+9	+10	10	7	8	3	10	1
6×9 = ☐	6×10 = ☐	×9	×10	×9	×9	×9	×10
+9	+10						
7×9 = ☐	7×10 = ☐						
+9	+10	5	12	7	11	5	12
8×9 = ☐	8×10 = ☐	×9	×9	×9	×10	×9	×9
+9	+10						
9×9 = ☐	9×10 = ☐						
+9	+10	7	4	2	4	9	6
10×9 = ☐	10×10 = ☐	×9	×10	×9	×9	×9	×10
+9	+10						
11×9 = ☐	11×10 = ☐						
+9	+10	1	6	11	0	3	10
12×9 = ☐	12×10 = ☐	×9	×9	×9	×10	×9	×9

8	9	5	2	12	9	0	12	7	10
×9	×0	×9	×10	×9	×10	×9	×10	×9	×0

And in all matters of wisdom and understanding . . .
he found them ten times better than all the magicians and astrologers . . . Dan. 1:20

© Edwin C. Myers 1985, 1990   **CalcuLadder**®   Level 34: Multiplication   4 minutes

My name is _____

Today is _____

0×9 = 0	0×10 = 0	2	9	4	3	2	9
+9	+10	x9	x9	x9	x10	x9	x9
1×9= ☐	1×10= ☐						
+9	+10	4	8	11	6	1	5
2×9= ☐	2×10= ☐	x9	x10	x 9	x9	x9	x10
+9	+10						
3×9= ☐	3×10= ☐						
+9	+10	11	6	1	10	8	3
4×9= ☐	4×10= ☐	x 9	x9	x9	x10	x9	x9
+9	+10						
5×9= ☐	5×10= ☐						
+9	+10	10	7	8	3	10	1
6×9= ☐	6×10= ☐	x 9	x10	x9	x9	x 9	x10
+9	+10						
7×9= ☐	7×10= ☐						
+9	+10	5	12	7	11	5	12
8×9= ☐	8×10= ☐	x9	x 9	x9	x10	x9	x 9
+9	+10						
9×9= ☐	9×10= ☐						
+9	+10	7	4	2	4	9	6
10×9= ☐	10×10= ☐	x9	x10	x9	x9	x9	x10
+9	+10						
11×9= ☐	11×10= ☐						
+9	+10	1	6	11	0	3	10
12×9= ☐	12×10= ☐	x9	x9	x 9	x10	x9	x 9

8	9	5	2	12	9	0	12	7	10
x9	x0	x9	x10	x 9	x10	x9	x10	x9	x 0

And in all matters of wisdom and understanding . . .
he found them ten times better than all the magicians and astrologers . . .   Dan. 1:20

© Edwin C. Myers 1985,1990   CalcuLadder®   Level 34: Multiplication   4 minutes

My name is _____
Today is _____

0x9 = 0
+9
1x9=
+9
2x9=
+9
3x9=
+9
4x9=
+9
5x9=
+9
6x9=
+9
7x9=
+9
8x9=
+9
9x9=
+9
10x9=
+9
11x9=
+9
12x9=

0x10 = 0
+10
1x10=
+10
2x10=
+10
3x10=
+10
4x10=
+10
5x10=
+10
6x10=
+10
7x10=
+10
8x10=
+10
9x10=
+10
10x10=
+10
11x10=
+10
12x10=

2	9	4	3	2	9
x9	x9	x9	x10	x9	x9

4	8	11	6	1	5
x9	x10	x9	x9	x9	x10

11	6	1	10	8	3
x9	x9	x9	x10	x9	x9

10	7	8	3	10	1
x9	x10	x9	x9	x9	x10

5	12	7	11	5	12
x9	x9	x9	x10	x9	x9

7	4	2	4	9	6
x9	x10	x9	x9	x9	x10

1	6	11	0	3	10
x9	x9	x9	x10	x9	x9

8	9	5	2	12	9	0	12	7	10
x9	x0	x9	x10	x9	x10	x9	x10	x9	x0

And in all matters of wisdom and understanding . . .
he found them ten times better than all the magicians and astrologers . . .   Dan. 1:20

© Edwin C. Myers 1985,1990   CalcuLadder®   Level 34: Multiplication   4 minutes

My name is _____

Today is _____

0x9 = **0**
+9
1x9=☐
+9
2x9=☐
+9
3x9=☐
+9
4x9=☐
+9
5x9=☐
+9
6x9=☐
+9
7x9=☐
+9
8x9=☐
+9
9x9=☐
+9
10x9=☐
+9
11x9=☐
+9
12x9=☐

0x10 = **0**
+10
1x10=☐
+10
2x10=☐
+10
3x10=☐
+10
4x10=☐
+10
5x10=☐
+10
6x10=☐
+10
7x10=☐
+10
8x10=☐
+10
9x10=☐
+10
10x10=☐
+10
11x10=☐
+10
12x10=☐

2 x9	9 x9	4 x9	3 x10	2 x9	9 x9
4 x9	8 x10	11 x9	6 x9	1 x9	5 x10
11 x9	6 x9	1 x9	10 x10	8 x9	3 x9
10 x9	7 x10	8 x9	3 x9	10 x9	1 x10
5 x9	12 x9	7 x9	11 x10	5 x9	12 x9
7 x9	4 x10	2 x9	4 x9	9 x9	6 x10
1 x9	6 x9	11 x9	0 x10	3 x9	10 x9

8 x9	9 x0	5 x9	2 x10	12 x9	9 x10	0 x9	12 x10	7 x9	10 x0

And in all matters of wisdom and understanding . . .
  he found them ten times better than all the magicians and astrologers . . . Dan. 1:20

© Edwin C. Myers 1985,1990   CalcuLadder®   Level 34: Multiplication   4 minutes

My name is _____

Today is _____

0×9 = **0**	0×10 = **0**	2 x9	9 x9	4 x9	3 x10	2 x9	9 x9
+9	+10						
1×9= ☐	1×10= ☐						
+9	+10	4 x9	8 x10	11 x 9	6 x9	1 x9	5 x10
2×9= ☐	2×10= ☐						
+9	+10						
3×9= ☐	3×10= ☐						
+9	+10	11 x 9	6 x9	1 x9	10 x10	8 x9	3 x9
4×9= ☐	4×10= ☐						
+9	+10						
5×9= ☐	5×10= ☐						
+9	+10	10 x 9	7 x10	8 x9	3 x9	10 x 9	1 x10
6×9= ☐	6×10= ☐						
+9	+10						
7×9= ☐	7×10= ☐						
+9	+10	5 x9	12 x 9	7 x9	11 x10	5 x9	12 x 9
8×9= ☐	8×10= ☐						
+9	+10						
9×9= ☐	9×10= ☐						
+9	+10	7 x9	4 x10	2 x9	4 x9	9 x9	6 x10
10×9= ☐	10×10= ☐						
+9	+10						
11×9= ☐	11×10= ☐						
+9	+10	1 x9	6 x9	11 x 9	0 x10	3 x9	10 x 9
12×9= ☐	12×10= ☐						

| 8
x9 | 9
x0 | 5
x9 | 2
x10 | 12
x 9 | 9
x10 | 0
x9 | 12
x10 | 7
x9 | 10
x 0 |

And in all matters of wisdom and understanding . . .
he found them ten times better than all the magicians and astrologers . . . Dan. 1:20

© Edwin C. Myers 1985,1990   CalcuLadder®   Level 34: Multiplication   4 minutes

My name is _____
Today is _____

0×9 = **0**	0×10 = **0**	2 ×9	9 ×9	4 ×9	3 ×10	2 ×9	9 ×9
+9	+10						
1×9=☐	1×10=☐						
+9	+10	4 ×9	8 ×10	11 ×9	6 ×9	1 ×9	5 ×10
2×9=☐	2×10=☐						
+9	+10						
3×9=☐	3×10=☐	11 ×9	6 ×9	1 ×9	10 ×10	8 ×9	3 ×9
+9	+10						
4×9=☐	4×10=☐						
+9	+10	10 ×9	7 ×10	8 ×9	3 ×9	10 ×9	1 ×10
5×9=☐	5×10=☐						
+9	+10						
6×9=☐	6×10=☐	5 ×9	12 ×9	7 ×9	11 ×10	5 ×9	12 ×9
+9	+10						
7×9=☐	7×10=☐						
+9	+10	7 ×9	4 ×10	2 ×9	4 ×9	9 ×9	6 ×10
8×9=☐	8×10=☐						
+9	+10						
9×9=☐	9×10=☐	1 ×9	6 ×9	11 ×9	0 ×10	3 ×9	10 ×9
+9	+10						
10×9=☐	10×10=☐						
+9	+10						
11×9=☐	11×10=☐						
+9	+10						
12×9=☐	12×10=☐						

8 ×9   9 ×0   5 ×9   2 ×10   12 ×9   9 ×10   0 ×9   12 ×10   7 ×9   10 ×0

And in all matters of wisdom and understanding . . .
he found them ten times better than all the magicians and astrologers . . .   Dan. 1:20

© Edwin C. Myers 1985, 1990   **CalcuLadder®**   Level 34: Multiplication   4 minutes

My name is _____

Today is _____

0 × 11 = **0**

1 × 11 = ▢ +11

2 × 11 = ▢ +11

3 × 11 = ▢ +11

4 × 11 = ▢ +11

5 × 11 = ▢ +11

6 × 11 = ▢ +11

7 × 11 = ▢ +11

8 × 11 = ▢ +11

9 × 11 = ▢ +11

10 × 11 = ▢ +11

11 × 11 = ▢ +11

12 × 11 = ▢

5	12	7	5	12	7	2	9
×11	×11	×11	×11	×11	×11	×11	×11

4	2	9	4	11	6	1	11
×11	×11	×11	×11	×11	×11	×11	×11

6	1	8	3	10	8	3	10
×11	×11	×11	×11	×11	×11	×11	×11

11	2	6	5	12	10	12	11
×0	×9	×10	×11	×10	×9	×11	×9

3	0	7	7	4	2	4
×10	×3	×11	×9	×10	×11	×9

9	9	8	3	4	6	5
×9	×11	×10	×9	×11	×1	×10

11	11	6	7	6	8	12
×10	×11	×9	×10	×11	×9	×9

1	9	8	10	3	5	10
×11	×10	×11	×10	×11	×9	×11

Their sorrows shall be multiplied that hasten after another god . . . Psa. 16:4

© Edwin C. Myers 1985, 1990   CalcuLadder®   Level 35: Multiplication   4 minutes

My name is _____

Today is _____

0 x 11 = 0

1 x 11 = [ ]
+11

2 x 11 = [ ]
+11

3 x 11 = [ ]
+11

4 x 11 = [ ]
+11

5 x 11 = [ ]
+11

6 x 11 = [ ]
+11

7 x 11 = [ ]
+11

8 x 11 = [ ]
+11

9 x 11 = [ ]
+11

10 x 11 = [ ]
+11

11 x 11 = [ ]
+11

12 x 11 = [ ]

| 5 | 12 | 7 | 5 | 12 | 7 | 2 | 9 |
| x11 | x11 | x11 | x11 | x11 | x11 | x11 | x11 |

| 4 | 2 | 9 | 4 | 11 | 6 | 1 | 11 |
| x11 | x11 | x11 | x11 | x11 | x11 | x11 | x11 |

| 6 | 1 | 8 | 3 | 10 | 8 | 3 | 10 |
| x11 | x11 | x11 | x11 | x11 | x11 | x11 | x11 |

| 11 | 2 | 6 | 5 | 12 | 10 | 12 | 11 |
| x 0 | x9 | x10 | x11 | x10 | x 9 | x11 | x 9 |

| 3 | 0 | 7 | 7 | 4 | 2 | 4 |
| x10 | x3 | x11 | x9 | x10 | x11 | x9 |

| 9 | 9 | 8 | 3 | 4 | 6 | 5 |
| x9 | x11 | x10 | x9 | x11 | x1 | x10 |

| 11 | 11 | 6 | 7 | 6 | 8 | 12 |
| x10 | x11 | x9 | x10 | x11 | x9 | x 9 |

| 1 | 9 | 8 | 10 | 3 | 5 | 10 |
| x11 | x10 | x11 | x10 | x11 | x9 | x11 |

Their sorrows shall be multiplied that hasten after another god . . . Psa. 16:4

© Edwin C. Myers 1985, 1990   CalcuLadder®   Level 35: Multiplication   4 minutes

My name is _____

Today is _____

```
            5     12     7      5     12     7      2      9
           x11    x11   x11    x11   x11    x11    x11    x11
0x11 = 0

       +11
1x11= [   ]  4      2     9      4     11     6      1     11
             x11    x11   x11    x11   x11    x11    x11    x11
       +11
2x11= [   ]
       +11
3x11= [   ]  6      1     8      3     10     8      3     10
             x11    x11   x11    x11   x11    x11    x11    x11
       +11
4x11= [   ]
       +11
5x11= [   ]  11     2     6      5     12     10     12     11
             x 0    x9    x10    x11   x10    x 9    x11    x 9
       +11
6x11= [   ]
       +11
7x11= [   ]  3      0     7      7      4     2      4
             x10    x3    x11    x9    x10    x11    x9
       +11
8x11= [   ]
       +11
9x11= [   ]  9      9     8      3      4     6      5
             x9     x11   x10    x9    x11    x1     x10
       +11
10x11=[   ]
       +11
11x11=[   ]  11     11    6      7      6     8      12
             x10    x11   x9     x10   x11    x9     x9
       +11
12x11=[   ]
             1      9     8      10     3     5      10
             x11    x10   x11    x10   x11    x9     x11
```

Their sorrows shall be multiplied that hasten after another god... Psa. 16:4

© Edwin C. Myers 1985,1990 **CalcuLadder**® Level 35: Multiplication 4 minutes

My name is _____

Today is _____

0 x 11 = 0

1 x 11 = ☐
+11

2 x 11 = ☐
+11

3 x 11 = ☐
+11

4 x 11 = ☐
+11

5 x 11 = ☐
+11

6 x 11 = ☐
+11

7 x 11 = ☐
+11

8 x 11 = ☐
+11

9 x 11 = ☐
+11

10 x 11 = ☐
+11

11 x 11 = ☐
+11

12 x 11 = ☐

5	12	7	5	12	7	2	9
x11	x11	x11	x11	x11	x11	x11	x11

4	2	9	4	11	6	1	11
x11	x11	x11	x11	x11	x11	x11	x11

6	1	8	3	10	8	3	10
x11	x11	x11	x11	x11	x11	x11	x11

11	2	6	5	12	10	12	11
x 0	x9	x10	x11	x10	x 9	x11	x 9

3	0	7	7	4	2	4
x10	x3	x11	x9	x10	x11	x9

9	9	8	3	4	6	5
x9	x11	x10	x9	x11	x1	x10

11	11	6	7	6	8	12
x10	x11	x9	x10	x11	x9	x9

1	9	8	10	3	5	10
x11	x10	x11	x10	x11	x9	x11

Their sorrows shall be multiplied that hasten after another god . . . Psa. 16:4

© Edwin C. Myers 1985, 1990   CalcuLadder®   Level 35: Multiplication   4 minutes

My name is _____
Today is _____

0x11 = 0

1x11 = [+11]
2x11 = [+11]
3x11 = [+11]
4x11 = [+11]
5x11 = [+11]
6x11 = [+11]
7x11 = [+11]
8x11 = [+11]
9x11 = [+11]
10x11 = [+11]
11x11 = [+11]
12x11 = [ ]

| 5 | 12 | 7 | 5 | 12 | 7 | 2 | 9 |
|x11|x11|x11|x11|x11|x11|x11|x11|

| 4 | 2 | 9 | 4 | 11 | 6 | 1 | 11 |
|x11|x11|x11|x11|x11|x11|x11|x11|

| 6 | 1 | 8 | 3 | 10 | 8 | 3 | 10 |
|x11|x11|x11|x11|x11|x11|x11|x11|

| 11 | 2 | 6 | 5 | 12 | 10 | 12 | 11 |
|x0 |x9 |x10|x11|x10|x9 |x11|x9 |

| 3 | 0 | 7 | 7 | 4 | 2 | 4 |
|x10|x3 |x11|x9 |x10|x11|x9 |

| 9 | 9 | 8 | 3 | 4 | 6 | 5 |
|x9 |x11|x10|x9 |x11|x1 |x10|

| 11 | 11 | 6 | 7 | 6 | 8 | 12 |
|x10|x11|x9 |x10|x11|x9 |x9 |

| 1 | 9 | 8 | 10 | 3 | 5 | 10 |
|x11|x10|x11|x10|x11|x9 |x11|

Their sorrows shall be multiplied that hasten after another god . . . Psa. 16:4

© Edwin C. Myers 1985, 1990   CalcuLadder®   Level 35: Multiplication   4 minutes

My name is _____

Today is _____

		5 x11	12 x11	7 x11	5 x11	12 x11	7 x11	2 x11	9 x11
0x11 =	0								
	+11								
1x11 =	☐	4 x11	2 x11	9 x11	4 x11	11 x11	6 x11	1 x11	11 x11
	+11								
2x11 =	☐								
	+11								
3x11 =	☐	6 x11	1 x11	8 x11	3 x11	10 x11	8 x11	3 x11	10 x11
	+11								
4x11 =	☐								
	+11								
5x11 =	☐	11 x 0	2 x9	6 x10	5 x11	12 x10	10 x 9	12 x11	11 x 9
	+11								
6x11 =	☐								
	+11								
7x11 =	☐	3 x10	0 x3	7 x11	7 x9	4 x10	2 x11		4 x9
	+11								
8x11 =	☐								
	+11								
9x11 =	☐	9 x9	9 x11	8 x10	3 x9	4 x11	6 x1		5 x10
	+11								
10x11 =	☐								
	+11								
11x11 =	☐	11 x10	11 x11	6 x9	7 x10	6 x11	8 x9		12 x 9
	+11								
12x11 =	☐								
		1 x11	9 x10	8 x11	10 x10	3 x11	5 x9		10 x11

Their sorrows shall be multiplied that hasten after another god . . . Psa. 16:4

© Edwin C. Myers 1985,1990   CalcuLadder®   Level 35: Multiplication   4 minutes

My name is _____

Today is _____

```
                 5     12      7      5     12      7      2      9
               x11    x11    x11    x11    x11    x11    x11    x11
0x11 = 0
       +11
        ┌───┐
1x11 =  │   │  4      2      9      4     11      6      1     11
        └───┘x11    x11    x11    x11    x11    x11    x11    x11
       +11
        ┌───┐
2x11 =  │   │
        └───┘
       +11
        ┌───┐ 6      1      8      3     10      8      3     10
3x11 =  │   │x11    x11    x11    x11    x11    x11    x11    x11
        └───┘
       +11
        ┌───┐
4x11 =  │   │
        └───┘
       +11
        ┌───┐11      2      6      5     12     10     12     11
5x11 =  │   │ x0     x9    x10    x11    x10     x9    x11     x9
        └───┘
       +11
        ┌───┐
6x11 =  │   │
        └───┘
       +11
        ┌───┐ 3      0      7      7      4      2      4
7x11 =  │   │x10     x3    x11     x9    x10    x11     x9
        └───┘
       +11
        ┌───┐
8x11 =  │   │
        └───┘
       +11
        ┌───┐ 9      9      8      3      4      6      5
9x11 =  │   │ x9    x11    x10     x9    x11     x1    x10
        └───┘
       +11
        ┌───┐
10x11=  │   │
        └───┘
       +11
        ┌───┐11     11      6      7      6      8     12
11x11=  │   │x10    x11     x9    x10    x11     x9     x9
        └───┘
       +11
        ┌───┐
12x11=  │   │
        └───┘
                 1      9      8     10      3      5     10
               x11    x10    x11    x10    x11     x9    x11
```

Their sorrows shall be multiplied that hasten after another god . . . Psa. 16:4

© Edwin C. Myers 1985, 1990   CalcuLadder®   Level 35: Multiplication   4 minutes

My name is _____

Today is _____

		5 x11	12 x11	7 x11	5 x11	12 x11	7 x11	2 x11	9 x11

0 x 11 = 0

+11

1 x 11 = ☐   4 x11   2 x11   9 x11   4 x11   11 x11   6 x11   1 x11   11 x11

+11

2 x 11 = ☐

+11

3 x 11 = ☐   6 x11   1 x11   8 x11   3 x11   10 x11   8 x11   3 x11   10 x11

+11

4 x 11 = ☐

+11

5 x 11 = ☐   11 x0   2 x9   6 x10   5 x11   12 x10   10 x9   12 x11   11 x9

+11

6 x 11 = ☐

+11

7 x 11 = ☐   3 x10   0 x3   7 x11   7 x9   4 x10   2 x11   4 x9

+11

8 x 11 = ☐

+11

9 x 11 = ☐   9 x9   9 x11   8 x10   3 x9   4 x11   6 x1   5 x10

+11

10 x 11 = ☐

+11

11 x 11 = ☐   11 x10   11 x11   6 x9   7 x10   6 x11   8 x9   12 x9

+11

12 x 11 = ☐

1 x11   9 x10   8 x11   10 x10   3 x11   5 x9   10 x11

Their sorrows shall be multiplied that hasten after another god . . . Psa. 16:4

© Edwin C. Myers 1985, 1990   CalcuLadder®   Level 35: Multiplication   4 minutes

My name is _____

Today is _____

		5 x11	12 x11	7 x11	5 x11	12 x11	7 x11	2 x11	9 x11
0×11 =	0								
1×11=	+11	4 x11	2 x11	9 x11	4 x11	11 x11	6 x11	1 x11	11 x11
2×11=	+11								
3×11=	+11	6 x11	1 x11	8 x11	3 x11	10 x11	8 x11	3 x11	10 x11
4×11=	+11								
5×11=	+11	11 x0	2 x9	6 x10	5 x11	12 x10	10 x9	12 x11	11 x9
6×11=	+11								
7×11=	+11	3 x10	0 x3	7 x11	7 x9	4 x10	2 x11		4 x9
8×11=	+11								
9×11=	+11	9 x9	9 x11	8 x10	3 x9	4 x11	6 x1	5 x10	
10×11=	+11								
11×11=	+11	11 x10	11 x11	6 x9	7 x10	6 x11	8 x9	12 x9	
12×11=									
		1 x11	9 x10	8 x11	10 x10	3 x11	5 x9	10 x11	

Their sorrows shall be multiplied that hasten after another god . . . Psa. 16:4

© Edwin C. Myers 1985, 1990 **CalcuLadder®** Level 35: Multiplication 4 minutes

My name is _____

Today is _____

0 × 11 = 0

1 × 11 = ☐
+11

2 × 11 = ☐
+11

3 × 11 = ☐
+11

4 × 11 = ☐
+11

5 × 11 = ☐
+11

6 × 11 = ☐
+11

7 × 11 = ☐
+11

8 × 11 = ☐
+11

9 × 11 = ☐
+11

10 × 11 = ☐
+11

11 × 11 = ☐
+11

12 × 11 = ☐

5	12	7	5	12	7	2	9
×11	×11	×11	×11	×11	×11	×11	×11

4	2	9	4	11	6	1	11
×11	×11	×11	×11	×11	×11	×11	×11

6	1	8	3	10	8	3	10
×11	×11	×11	×11	×11	×11	×11	×11

11	2	6	5	12	10	12	11
×0	×9	×10	×11	×10	×9	×11	×9

3	0	7	7	4	2	4
×10	×3	×11	×9	×10	×11	×9

9	9	8	3	4	6	5
×9	×11	×10	×9	×11	×1	×10

11	11	6	7	6	8	12
×10	×11	×9	×10	×11	×9	×9

1	9	8	10	3	5	10
×11	×10	×11	×10	×11	×9	×11

Their sorrows shall be multiplied that hasten after another god... Psa. 16:4

© Edwin C. Myers 1985, 1990   **CalcuLadder**®   Level 35: Multiplication   4 minutes

My name is _____

Today is _____

	5	12	7	5	12	7	2	9
0×11 = 0	x11	x11	x11	x11	x11	x11	x11	x11

1×11 = [ ] +11

	4	2	9	4	11	6	1	11
2×11 = [ ] +11	x11	x11	x11	x11	x11	x11	x11	x11

3×11 = [ ] +11

	6	1	8	3	10	8	3	10
4×11 = [ ] +11	x11	x11	x11	x11	x11	x11	x11	x11

5×11 = [ ] +11

	11	2	6	5	12	10	12	11
6×11 = [ ] +11	x0	x9	x10	x11	x10	x9	x11	x9

7×11 = [ ] +11

	3	0	7	7	4	2	4	
8×11 = [ ] +11	x10	x3	x11	x9	x10	x11	x9	

9×11 = [ ] +11

	9	9	8	3	4	6	5	
10×11 = [ ] +11	x9	x11	x10	x9	x11	x1	x10	

11×11 = [ ] +11

	11	11	6	7	6	8	12	
12×11 = [ ]	x10	x11	x9	x10	x11	x9	x9	

1	9	8	10	3	5	10
x11	x10	x11	x10	x11	x9	x11

Their sorrows shall be multiplied that hasten after another god . . . Psa. 16:4

© Edwin C. Myers 1985, 1990   CalcuLadder®   Level 35: Multiplication   4 minutes

My name is _____

Today is _____

5	12	7	5	12	7	2	9
x11	x11	x11	x11	x11	x11	x11	x11

0 x 11 = 0

+11

1 x 11 = ☐

+11

4	2	9	4	11	6	1	11
x11	x11	x11	x11	x11	x11	x11	x11

2 x 11 = ☐

+11

3 x 11 = ☐

+11

6	1	8	3	10	8	3	10
x11	x11	x11	x11	x11	x11	x11	x11

4 x 11 = ☐

+11

5 x 11 = ☐

+11

11	2	6	5	12	10	12	11
x0	x9	x10	x11	x10	x9	x11	x9

6 x 11 = ☐

+11

7 x 11 = ☐

+11

3	0	7	7	4	2	4
x10	x3	x11	x9	x10	x11	x9

8 x 11 = ☐

+11

9 x 11 = ☐

+11

9	9	8	3	4	6	5
x9	x11	x10	x9	x11	x1	x10

10 x 11 = ☐

+11

11 x 11 = ☐

+11

11	11	6	7	6	8	12
x10	x11	x9	x10	x11	x9	x9

12 x 11 = ☐

1	9	8	10	3	5	10
x11	x10	x11	x10	x11	x9	x11

Their sorrows shall be multiplied that hasten after another god . . . Psa. 16:4

© Edwin C. Myers 1985, 1990   **CalcuLadder**®   Level 35: Multiplication   4 minutes

My name is _____

Today is _____

		4 x12	9 x12	2 x12	4 x12	9 x12	2 x12	7 x12	12 x12
0x12 = 0									
	+12								
1x12=	☐	5 x12	7 x12	12 x12	5 x12	10 x12	3 x12	8 x12	10 x12
	+12								
2x12=	☐								
	+12								
3x12=	☐	3 x12	8 x12	1 x12	6 x12	11 x12	1 x12	6 x12	11 x12
	+12								
4x12=	☐								
	+12								
5x12=	☐	8 x7	5 x11	3 x8	4 x12	7 x11	2 x11	9 x12	12 x11
	+12								
6x12=	☐								
	+12								
7x12=	☐	6 x7	12 x0	2 x12	8 x8	9 x11	7 x12	4 x7	
	+12								
8x12=	☐								
	+12								
9x12=	☐	4 x11	12 x12	6 x11	11 x11	5 x12	6 x8	9 x7	
	+12								
10x12=	☐								
	+12								
11x12=	☐	10 x12	8 x11	3 x11	3 x12	5 x7	0 x9	8 x12	
	+12								
12x12=	☐								
		9 x8	10 x11	6 x12	7 x7	11 x12	3 x7	4 x8	

This was the dedication offering for the altar . . .
twelve silver dishes, twelve silver bowls, twelve gold pans . . . Num. 7:84

© Edwin C. Myers 1985, 1990   CalcuLadder®   Level 36: Multiplication   4 minutes

My name is _____

Today is _____

		4 x12	9 x12	2 x12	4 x12	9 x12	2 x12	7 x12	12 x12

0 x 12 = 0

+12

1 x 12 = ☐

+12

2 x 12 = ☐

5 x12   7 x12   12 x12   5 x12   10 x12   3 x12   8 x12   10 x12

+12

3 x 12 = ☐

+12

4 x 12 = ☐

3 x12   8 x12   1 x12   6 x12   11 x12   1 x12   6 x12   11 x12

+12

5 x 12 = ☐

+12

6 x 12 = ☐

8 x7   5 x11   3 x8   4 x12   7 x11   2 x11   9 x12   12 x11

+12

7 x 12 = ☐

+12

8 x 12 = ☐

6 x7   12 x0   2 x12   8 x8   9 x11   7 x12   4 x7

+12

9 x 12 = ☐

+12

10 x 12 = ☐

4 x11   12 x12   6 x11   11 x11   5 x12   6 x8   9 x7

+12

11 x 12 = ☐

+12

12 x 12 = ☐

10 x12   8 x11   3 x11   3 x12   5 x7   0 x9   8 x12

9 x8   10 x11   6 x12   7 x7   11 x12   3 x7   4 x8

This was the dedication offering for the altar . . .
twelve silver dishes, twelve silver bowls, twelve gold pans . . . Num. 7:84

© Edwin C. Myers 1985, 1990   CalcuLadder®   Level 36: Multiplication   4 minutes

My name is _____

Today is _____

		4 x12	9 x12	2 x12	4 x12	9 x12	2 x12	7 x12	12 x12
0x12 = 0	+12								
1x12=	+12	5 x12	7 x12	12 x12	5 x12	10 x12	3 x12	8 x12	10 x12
2x12=	+12								
3x12=	+12	3 x12	8 x12	1 x12	6 x12	11 x12	1 x12	6 x12	11 x12
4x12=	+12								
5x12=	+12	8 x7	5 x11	3 x8	4 x12	7 x11	2 x11	9 x12	12 x11
6x12=	+12								
7x12=	+12	6 x7	12 x0	2 x12	8 x8	9 x11	7 x12	4 x7	
8x12=	+12								
9x12=	+12	4 x11	12 x12	6 x11	11 x11	5 x12	6 x8	9 x7	
10x12=	+12								
11x12=	+12	10 x12	8 x11	3 x11	3 x12	5 x7	0 x9	8 x12	
12x12=									
		9 x8	10 x11	6 x12	7 x7	11 x12	3 x7	4 x8	

This was the dedication offering for the altar . . .
twelve silver dishes, twelve silver bowls, twelve gold pans . . . Num. 7:84

© Edwin C. Myers 1985, 1990   CalcuLadder®   Level 36: Multiplication   4 minutes

My name is _____

Today is _____

		4	9	2	4	9	2	7	12
0×12 = 0		×12	×12	×12	×12	×12	×12	×12	×12

+12

1×12=☐

	5	7	12	5	10	3	8	10
+12	×12	×12	×12	×12	×12	×12	×12	×12

2×12=☐

+12

3×12=☐

	3	8	1	6	11	1	6	11
+12	×12	×12	×12	×12	×12	×12	×12	×12

4×12=☐

+12

5×12=☐

	8	5	3	4	7	2	9	12
+12	×7	×11	×8	×12	×11	×11	×12	×11

6×12=☐

+12

7×12=☐

	6	12	2	8	9	7	4
+12	×7	×0	×12	×8	×11	×12	×7

8×12=☐

+12

9×12=☐

	4	12	6	11	5	6	9
+12	×11	×12	×11	×11	×12	×8	×7

10×12=☐

+12

11×12=☐

	10	8	3	3	5	0	8
+12	×12	×11	×11	×12	×7	×9	×12

12×12=☐

	9	10	6	7	11	3	4
	×8	×11	×12	×7	×12	×7	×8

This was the dedication offering for the altar . . .
twelve silver dishes, twelve silver bowls, twelve gold pans . . . Num. 7:84

© Edwin C. Myers 1985,1990   **CalcuLadder**®   Level 36: Multiplication   4 minutes

My name is _____

Today is _____

		4 x12	9 x12	2 x12	4 x12	9 x12	2 x12	7 x12	12 x12
0x12 = 0									
	+12								
1x12=	☐	5 x12	7 x12	12 x12	5 x12	10 x12	3 x12	8 x12	10 x12
	+12								
2x12=	☐								
	+12								
3x12=	☐	3 x12	8 x12	1 x12	6 x12	11 x12	1 x12	6 x12	11 x12
	+12								
4x12=	☐								
	+12								
5x12=	☐	8 x7	5 x11	3 x8	4 x12	7 x11	2 x11	9 x12	12 x11
	+12								
6x12=	☐								
	+12								
7x12=	☐	6 x7	12 x 0	2 x12	8 x8	9 x11	7 x12		4 x7
	+12								
8x12=	☐								
	+12								
9x12=	☐	4 x11	12 x12	6 x11	11 x11	5 x12	6 x8		9 x7
	+12								
10x12=	☐								
	+12								
11x12=	☐	10 x12	8 x11	3 x11	3 x12	5 x7	0 x9		8 x12
	+12								
12x12=	☐								
		9 x8	10 x11	6 x12	7 x7	11 x12	3 x7		4 x8

This was the dedication offering for the altar . . .
twelve silver dishes, twelve silver bowls, twelve gold pans . . . Num. 7:84

© Edwin C. Myers 1985,1990   **CalcuLadder®**   Level 36: Multiplication   4 minutes

My name is _____

Today is _____

		4	9	2	4	9	2	7	12
0×12 = 0		x12	x12	x12	x12	x12	x12	x12	x12

1×12= [+12]

	5	7	12	5	10	3	8	10
	x12	x12	x12	x12	x12	x12	x12	x12

2×12= [+12]

3×12= [+12]

	3	8	1	6	11	1	6	11
	x12	x12	x12	x12	x12	x12	x12	x12

4×12= [+12]

5×12= [+12]

	8	5	3	4	7	2	9	12
	x7	x11	x8	x12	x11	x11	x12	x11

6×12= [+12]

7×12= [+12]

	6	12	2	8	9	7	4
	x7	x0	x12	x8	x11	x12	x7

8×12= [+12]

9×12= [+12]

	4	12	6	11	5	6	9
	x11	x12	x11	x11	x12	x8	x7

10×12= [+12]

11×12= [+12]

	10	8	3	3	5	0	8
	x12	x11	x11	x12	x7	x9	x12

12×12= [ ]

9	10	6	7	11	3	4
x8	x11	x12	x7	x12	x7	x8

This was the dedication offering for the altar . . .
twelve silver dishes, twelve silver bowls, twelve gold pans . . . Num. 7:84

© Edwin C. Myers 1985,1990   CalcuLadder®   Level 36: Multiplication   4 minutes

My name is _____

Today is _____

	4 x12	9 x12	2 x12	4 x12	9 x12	2 x12	7 x12	12 x12

0x12 = 0

1x12= ☐ +12

2x12= ☐ +12

	5 x12	7 x12	12 x12	5 x12	10 x12	3 x12	8 x12	10 x12

3x12= ☐ +12

	3 x12	8 x12	1 x12	6 x12	11 x12	1 x12	6 x12	11 x12

4x12= ☐ +12

5x12= ☐ +12

	8 x7	5 x11	3 x8	4 x12	7 x11	2 x11	9 x12	12 x11

6x12= ☐ +12

7x12= ☐ +12

	6 x7	12 x0	2 x12	8 x8	9 x11	7 x12	4 x7

8x12= ☐ +12

9x12= ☐ +12

	4 x11	12 x12	6 x11	11 x11	5 x12	6 x8	9 x7

10x12= ☐ +12

11x12= ☐ +12

	10 x12	8 x11	3 x11	3 x12	5 x7	0 x9	8 x12

12x12= ☐

	9 x8	10 x11	6 x12	7 x7	11 x12	3 x7	4 x8

This was the dedication offering for the altar . . .
twelve silver dishes, twelve silver bowls, twelve gold pans . . . Num. 7:84

© Edwin C. Myers 1985,1990   CalcuLadder®   Level 36: Multiplication   4 minutes

My name is _____

Today is _____

$0 \times 12 = 0$

4	9	2	4	9	2	7	12
×12	×12	×12	×12	×12	×12	×12	×12

$1 \times 12 =$ ☐ +12

5	7	12	5	10	3	8	10
×12	×12	×12	×12	×12	×12	×12	×12

$2 \times 12 =$ ☐ +12

$3 \times 12 =$ ☐ +12

3	8	1	6	11	1	6	11
×12	×12	×12	×12	×12	×12	×12	×12

$4 \times 12 =$ ☐ +12

$5 \times 12 =$ ☐ +12

8	5	3	4	7	2	9	12
×7	×11	×8	×12	×11	×11	×12	×11

$6 \times 12 =$ ☐ +12

$7 \times 12 =$ ☐ +12

6	12	2	8	9	7	4
×7	×0	×12	×8	×11	×12	×7

$8 \times 12 =$ ☐ +12

$9 \times 12 =$ ☐ +12

4	12	6	11	5	6	9
×11	×12	×11	×11	×12	×8	×7

$10 \times 12 =$ ☐ +12

$11 \times 12 =$ ☐ +12

10	8	3	3	5	0	8
×12	×11	×11	×12	×7	×9	×12

$12 \times 12 =$ ☐

9	10	6	7	11	3	4
×8	×11	×12	×7	×12	×7	×8

This was the dedication offering for the altar . . .
twelve silver dishes, twelve silver bowls, twelve gold pans . . . Num. 7:84

© Edwin C. Myers 1985, 1990   CalcuLadder®   Level 36: Multiplication   4 minutes

My name is _____

Today is _____

$0 \times 12 = 0$

$1 \times 12 =$ ☐ +12

$2 \times 12 =$ ☐ +12

$3 \times 12 =$ ☐ +12

$4 \times 12 =$ ☐ +12

$5 \times 12 =$ ☐ +12

$6 \times 12 =$ ☐ +12

$7 \times 12 =$ ☐ +12

$8 \times 12 =$ ☐ +12

$9 \times 12 =$ ☐ +12

$10 \times 12 =$ ☐ +12

$11 \times 12 =$ ☐ +12

$12 \times 12 =$ ☐

4 ×12	9 ×12	2 ×12	4 ×12	9 ×12	2 ×12	7 ×12	12 ×12
5 ×12	7 ×12	12 ×12	5 ×12	10 ×12	3 ×12	8 ×12	10 ×12
3 ×12	8 ×12	1 ×12	6 ×12	11 ×12	1 ×12	6 ×12	11 ×12
8 ×7	5 ×11	3 ×8	4 ×12	7 ×11	2 ×11	9 ×12	12 ×11
6 ×7	12 ×0	2 ×12	8 ×8	9 ×11	7 ×12	4 ×7	
4 ×11	12 ×12	6 ×11	11 ×11	5 ×12	6 ×8	9 ×7	
10 ×12	8 ×11	3 ×11	3 ×12	5 ×7	0 ×9	8 ×12	
9 ×8	10 ×11	6 ×12	7 ×7	11 ×12	3 ×7	4 ×8	

This was the dedication offering for the altar . . .
twelve silver dishes, twelve silver bowls, twelve gold pans . . . Num. 7:84

© Edwin C. Myers 1985, 1990   CalcuLadder®   Level 36: Multiplication   4 minutes

My name is _____

Today is _____

$0 \times 12 = 0$

4	9	2	4	9	2	7	12
×12	×12	×12	×12	×12	×12	×12	×12

$1 \times 12 =$ ☐   +12

5	7	12	5	10	3	8	10
×12	×12	×12	×12	×12	×12	×12	×12

$2 \times 12 =$ ☐   +12

$3 \times 12 =$ ☐   +12

3	8	1	6	11	1	6	11
×12	×12	×12	×12	×12	×12	×12	×12

$4 \times 12 =$ ☐   +12

$5 \times 12 =$ ☐   +12

8	5	3	4	7	2	9	12
×7	×11	×8	×12	×11	×11	×12	×11

$6 \times 12 =$ ☐   +12

$7 \times 12 =$ ☐   +12

6	12	2	8	9	7	4
×7	×0	×12	×8	×11	×12	×7

$8 \times 12 =$ ☐   +12

$9 \times 12 =$ ☐   +12

4	12	6	11	5	6	9
×11	×12	×11	×11	×12	×8	×7

$10 \times 12 =$ ☐   +12

$11 \times 12 =$ ☐   +12

10	8	3	3	5	0	8
×12	×11	×11	×12	×7	×9	×12

$12 \times 12 =$ ☐

9	10	6	7	11	3	4
×8	×11	×12	×7	×12	×7	×8

This was the dedication offering for the altar . . .
twelve silver dishes, twelve silver bowls, twelve gold pans . . .   Num. 7:84

© Edwin C. Myers 1985,1990   CalcuLadder®   Level 36: Multiplication   4 minutes

My name is _____

Today is _____

	4	9	2	4	9	2	7	12
	x12	x12	x12	x12	x12	x12	x12	x12

0 x 12 = 0

1 x 12 = [ ] +12

5	7	12	5	10	3	8	10
x12	x12	x12	x12	x12	x12	x12	x12

2 x 12 = [ ] +12

3 x 12 = [ ] +12

3	8	1	6	11	1	6	11
x12	x12	x12	x12	x12	x12	x12	x12

4 x 12 = [ ] +12

5 x 12 = [ ] +12

8	5	3	4	7	2	9	12
x7	x11	x8	x12	x11	x11	x12	x11

6 x 12 = [ ] +12

7 x 12 = [ ] +12

6	12	2	8	9	7	4
x7	x0	x12	x8	x11	x12	x7

8 x 12 = [ ] +12

9 x 12 = [ ] +12

4	12	6	11	5	6	9
x11	x12	x11	x11	x12	x8	x7

10 x 12 = [ ] +12

11 x 12 = [ ] +12

10	8	3	3	5	0	8
x12	x11	x11	x12	x7	x9	x12

12 x 12 = [ ]

9	10	6	7	11	3	4
x8	x11	x12	x7	x12	x7	x8

This was the dedication offering for the altar . . .
twelve silver dishes, twelve silver bowls, twelve gold pans . . . Num. 7:84

© Edwin C. Myers 1985, 1990   CalcuLadder®   Level 36: Multiplication   4 minutes

My name is _____

Today is _____

	4	9	2	4	9	2	7	12
0x12 = 0	x12	x12	x12	x12	x12	x12	x12	x12

1x12 = [ ] +12

	5	7	12	5	10	3	8	10
	x12	x12	x12	x12	x12	x12	x12	x12

2x12 = [ ] +12

3x12 = [ ]

	3	8	1	6	11	1	6	11
	x12	x12	x12	x12	x12	x12	x12	x12

+12

4x12 = [ ] +12

5x12 = [ ]

	8	5	3	4	7	2	9	12
	x7	x11	x8	x12	x11	x11	x12	x11

+12

6x12 = [ ] +12

7x12 = [ ]

	6	12	2	8	9	7	4
	x7	x0	x12	x8	x11	x12	x7

+12

8x12 = [ ] +12

9x12 = [ ]

	4	12	6	11	5	6	9
	x11	x12	x11	x11	x12	x8	x7

+12

10x12 = [ ] +12

11x12 = [ ]

	10	8	3	3	5	0	8
	x12	x11	x11	x12	x7	x9	x12

+12

12x12 = [ ]

	9	10	6	7	11	3	4
	x8	x11	x12	x7	x12	x7	x8

This was the dedication offering for the altar...
twelve silver dishes, twelve silver bowls, twelve gold pans... Num. 7:84

© Edwin C. Myers 1985, 1990   **CalcuLadder®**   Level 36: Multiplication   4 minutes

My name is _____

Today is _____

| 21 | 52 | 93 | 44 | 82 |
| x9 | x4 | x3 | x2 | x3 |

| 33 | 74 | 22 | 63 | 14 |
| x2 | x2 | x4 | x3 | x2 |

| 24 | 71 | 32 | 83 | 42 |
| x2 | x8 | x4 | x3 | x4 |

| 94 | 50 | 64 | 92 | 53 |
| x2 | x9 | x2 | x4 | x3 |

| 91 | 742 | 81 | 30 | 73 |
| x8 | x2 | x7 | x8 | x3 |

| 622 | 61 | 813 | 54 | 92 |
| x3 | x5 | x2 | x2 | x3 |

| 43 | 84 | 234 | 70 | 23 |
| x3 | x2 | x2 | x7 | x3 |

| 62 | 42 | 83 | 34 | 72 |
| x4 | x3 | x2 | x1 | x4 |

For by me thy days shall be multiplied, and the years of thy life shall be increased.

Prov. 9:11

© Edwin C. Myers 1985, 1990   **CalcuLadder**®   Level 37: Multiplication   4 minutes

My name is _____

Today is _____

21	52	93	44	82
x 9	x 4	x 3	x 2	x 3

33	74	22	63	14
x 2	x 2	x 4	x 3	x 2

24	71	32	83	42
x 2	x 8	x 4	x 3	x 4

94	50	64	92	53
x 2	x 9	x 2	x 4	x 3

91	742	81	30	73
x 8	x 2	x 7	x 8	x 3

622	61	813	54	92
x 3	x 5	x 2	x 2	x 3

43	84	234	70	23
x 3	x 2	x 2	x 7	x 3

62	42	83	34	72
x 4	x 3	x 2	x 1	x 4

For by me thy days shall be multiplied, and the years of thy life shall be increased.

Prov. 9:11

© Edwin C. Myers 1985, 1990   CalcuLadder®   Level 37: Multiplication   4 minutes

My name is _____

Today is _____

21	52	93	44	82
x 9	x 4	x 3	x 2	x 3

33	74	22	63	14
x 2	x 2	x 4	x 3	x 2

24	71	32	83	42
x 2	x 8	x 4	x 3	x 4

94	50	64	92	53
x 2	x 9	x 2	x 4	x 3

91	742	81	30	73
x 8	x 2	x 7	x 8	x 3

622	61	813	54	92
x 3	x 5	x 2	x 2	x 3

43	84	234	70	23
x 3	x 2	x 2	x 7	x 3

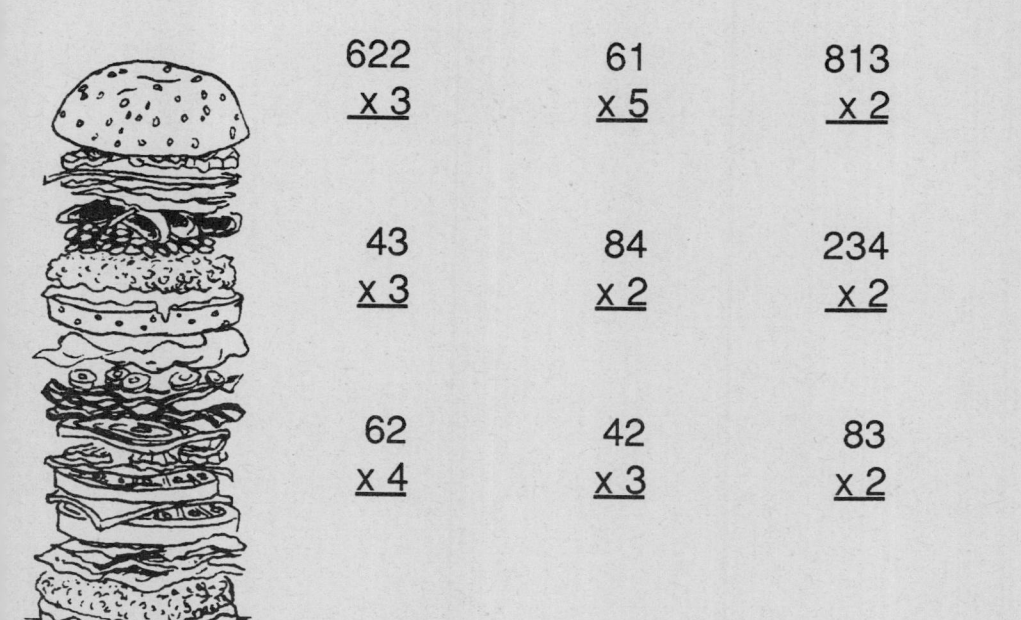

62	42	83	34	72
x 4	x 3	x 2	x 1	x 4

For by me thy days shall be multiplied, and the years of thy life shall be increased.

Prov. 9:11

© Edwin C. Myers 1985, 1990   CalcuLadder®   Level 37: Multiplication   4 minutes

My name is _____

Today is _____

21	52	93	44	82
x 9	x 4	x 3	x 2	x 3

33	74	22	63	14
x 2	x 2	x 4	x 3	x 2

24	71	32	83	42
x 2	x 8	x 4	x 3	x 4

94	50	64	92	53
x 2	x 9	x 2	x 4	x 3

91	742	81	30	73
x 8	x 2	x 7	x 8	x 3

622	61	813	54	92
x 3	x 5	x 2	x 2	x 3

43	84	234	70	23
x 3	x 2	x 2	x 7	x 3

62	42	83	34	72
x 4	x 3	x 2	x 1	x 4

For by me thy days shall be multiplied, and the years of thy life shall be increased.

Prov. 9:11

© Edwin C. Myers 1985,1990   CalcuLadder®   Level 37: Multiplication   4 minutes

My name is _____

Today is _____

| 21 | 52 | 93 | 44 | 82 |
| x9 | x4 | x3 | x2 | x3 |

| 33 | 74 | 22 | 63 | 14 |
| x2 | x2 | x4 | x3 | x2 |

| 24 | 71 | 32 | 83 | 42 |
| x2 | x8 | x4 | x3 | x4 |

| 94 | 50 | 64 | 92 | 53 |
| x2 | x9 | x2 | x4 | x3 |

| 91 | 742 | 81 | 30 | 73 |
| x8 | x2 | x7 | x8 | x3 |

| 622 | 61 | 813 | 54 | 92 |
| x3 | x5 | x2 | x2 | x3 |

| 43 | 84 | 234 | 70 | 23 |
| x3 | x2 | x2 | x7 | x3 |

| 62 | 42 | 83 | 34 | 72 |
| x4 | x3 | x2 | x1 | x4 |

For by me thy days shall be multiplied, and the years of thy life shall be increased.

Prov. 9:11

© Edwin C. Myers 1985, 1990   CalcuLadder®   Level 37: Multiplication   4 minutes

My name is _____

Today is _____

| 21 | 52 | 93 | 44 | 82 |
| x9 | x4 | x3 | x2 | x3 |

| 33 | 74 | 22 | 63 | 14 |
| x2 | x2 | x4 | x3 | x2 |

| 24 | 71 | 32 | 83 | 42 |
| x2 | x8 | x4 | x3 | x4 |

| 94 | 50 | 64 | 92 | 53 |
| x2 | x9 | x2 | x4 | x3 |

| 91 | 742 | 81 | 30 | 73 |
| x8 | x2 | x7 | x8 | x3 |

| 622 | 61 | 813 | 54 | 92 |
| x3 | x5 | x2 | x2 | x3 |

| 43 | 84 | 234 | 70 | 23 |
| x3 | x2 | x2 | x7 | x3 |

| 62 | 42 | 83 | 34 | 72 |
| x4 | x3 | x2 | x1 | x4 |

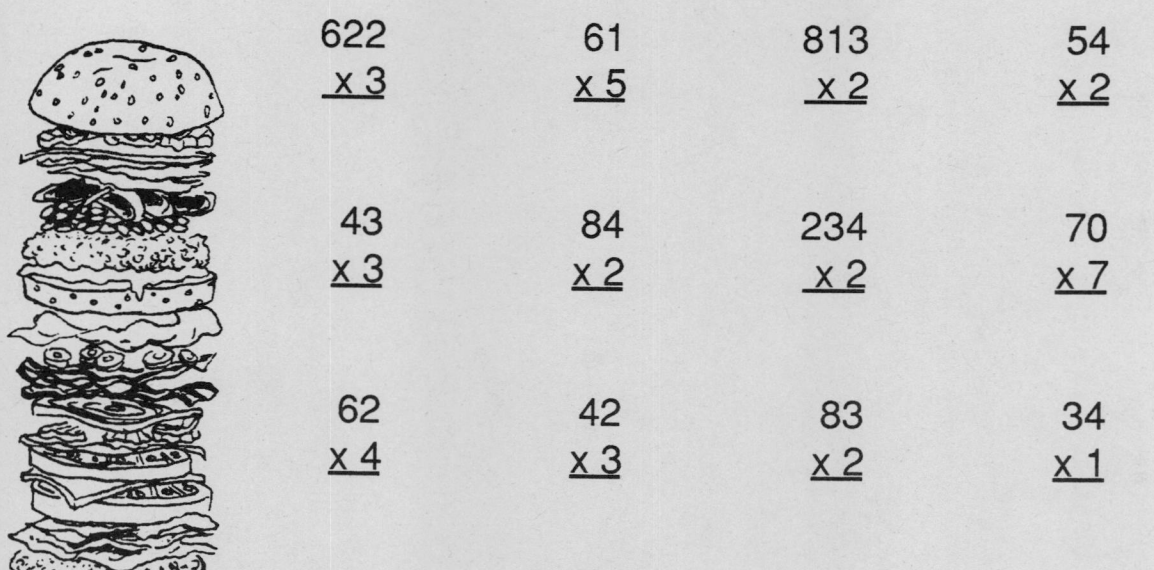

For by me thy days shall be multiplied, and the years of thy life shall be increased.

Prov. 9:11

© Edwin C. Myers 1985, 1990   **CalcuLadder**®   Level 37: Multiplication   4 minutes

My name is _____

Today is _____

| 21 | 52 | 93 | 44 | 82 |
| x9 | x4 | x3 | x2 | x3 |

| 33 | 74 | 22 | 63 | 14 |
| x2 | x2 | x4 | x3 | x2 |

| 24 | 71 | 32 | 83 | 42 |
| x2 | x8 | x4 | x3 | x4 |

| 94 | 50 | 64 | 92 | 53 |
| x2 | x9 | x2 | x4 | x3 |

| 91 | 742 | 81 | 30 | 73 |
| x8 | x2 | x7 | x8 | x3 |

| 622 | 61 | 813 | 54 | 92 |
| x3 | x5 | x2 | x2 | x3 |

| 43 | 84 | 234 | 70 | 23 |
| x3 | x2 | x2 | x7 | x3 |

| 62 | 42 | 83 | 34 | 72 |
| x4 | x3 | x2 | x1 | x4 |

For by me thy days shall be multiplied, and the years of thy life shall be increased.

Prov. 9:11

© Edwin C. Myers 1985,1990   CalcuLadder®   Level 37: Multiplication   4 minutes

My name is _____

Today is _____

| 21 | 52 | 93 | 44 | 82 |
| x9 | x4 | x3 | x2 | x3 |

| 33 | 74 | 22 | 63 | 14 |
| x2 | x2 | x4 | x3 | x2 |

| 24 | 71 | 32 | 83 | 42 |
| x2 | x8 | x4 | x3 | x4 |

| 94 | 50 | 64 | 92 | 53 |
| x2 | x9 | x2 | x4 | x3 |

| 91 | 742 | 81 | 30 | 73 |
| x8 | x2 | x7 | x8 | x3 |

| 622 | 61 | 813 | 54 | 92 |
| x3 | x5 | x2 | x2 | x3 |

| 43 | 84 | 234 | 70 | 23 |
| x3 | x2 | x2 | x7 | x3 |

| 62 | 42 | 83 | 34 | 72 |
| x4 | x3 | x2 | x1 | x4 |

For by me thy days shall be multiplied, and the years of thy life shall be increased.

Prov. 9:11

© Edwin C. Myers 1985, 1990   **CalcuLadder**®   Level 37: Multiplication   4 minutes

My name is _____

Today is _____

21	52	93	44	82
x 9	x 4	x 3	x 2	x 3

33	74	22	63	14
x 2	x 2	x 4	x 3	x 2

24	71	32	83	42
x 2	x 8	x 4	x 3	x 4

94	50	64	92	53
x 2	x 9	x 2	x 4	x 3

91	742	81	30	73
x 8	x 2	x 7	x 8	x 3

622	61	813	54	92
x 3	x 5	x 2	x 2	x 3

43	84	234	70	23
x 3	x 2	x 2	x 7	x 3

62	42	83	34	72
x 4	x 3	x 2	x 1	x 4

For by me thy days shall be multiplied, and the years of thy life shall be increased.

Prov. 9:11

© Edwin C. Myers 1985, 1990   CalcuLadder®   Level 37: Multiplication   4 minutes

My name is _____

Today is _____

| 21 | 52 | 93 | 44 | 82 |
| x9 | x4 | x3 | x2 | x3 |

| 33 | 74 | 22 | 63 | 14 |
| x2 | x2 | x4 | x3 | x2 |

| 24 | 71 | 32 | 83 | 42 |
| x2 | x8 | x4 | x3 | x4 |

| 94 | 50 | 64 | 92 | 53 |
| x2 | x9 | x2 | x4 | x3 |

| 91 | 742 | 81 | 30 | 73 |
| x8 | x2 | x7 | x8 | x3 |

| 622 | 61 | 813 | 54 | 92 |
| x3 | x5 | x2 | x2 | x3 |

| 43 | 84 | 234 | 70 | 23 |
| x3 | x2 | x2 | x7 | x3 |

| 62 | 42 | 83 | 34 | 72 |
| x4 | x3 | x2 | x1 | x4 |

For by me thy days shall be multiplied, and the years of thy life shall be increased.

Prov. 9:11

© Edwin C. Myers 1985, 1990   **CalcuLadder**®   Level 37: Multiplication   4 minutes

My name is _____

Today is _____

| 21 | 52 | 93 | 44 | 82 |
| x9 | x4 | x3 | x2 | x3 |

| 33 | 74 | 22 | 63 | 14 |
| x2 | x2 | x4 | x3 | x2 |

| 24 | 71 | 32 | 83 | 42 |
| x2 | x8 | x4 | x3 | x4 |

| 94 | 50 | 64 | 92 | 53 |
| x2 | x9 | x2 | x4 | x3 |

| 91 | 742 | 81 | 30 | 73 |
| x8 | x2 | x7 | x8 | x3 |

| 622 | 61 | 813 | 54 | 92 |
| x3 | x5 | x2 | x2 | x3 |

| 43 | 84 | 234 | 70 | 23 |
| x3 | x2 | x2 | x7 | x3 |

| 62 | 42 | 83 | 34 | 72 |
| x4 | x3 | x2 | x1 | x4 |

For by me thy days shall be multiplied, and the years of thy life shall be increased.

Prov. 9:11

© Edwin C. Myers 1985, 1990   **CalcuLadder**®   Level 37: Multiplication   4 minutes

My name is _____

Today is _____

21	52	93	44	82
x 9	x 4	x 3	x 2	x 3

33	74	22	63	14
x 2	x 2	x 4	x 3	x 2

24	71	32	83	42
x 2	x 8	x 4	x 3	x 4

94	50	64	92	53
x 2	x 9	x 2	x 4	x 3

91	742	81	30	73
x 8	x 2	x 7	x 8	x 3

622	61	813	54	92
x 3	x 5	x 2	x 2	x 3

43	84	234	70	23
x 3	x 2	x 2	x 7	x 3

62	42	83	34	72
x 4	x 3	x 2	x 1	x 4

For by me thy days shall be multiplied, and the years of thy life shall be increased.

Prov. 9:11

© Edwin C. Myers 1985, 1990   CalcuLadder®   Level 37: Multiplication   4 minutes

My name is _____

Today is _____

19	28	36
x 3	x 2	x 3

53	62	70
x 6	x 9	x 8

97	15	24
x 5	x 6	x 4

42	59	68
x 7	x 4	x 2

86	93	12
x 4	x 5	x 9

30	47	55
x 8	x 7	x 5

74	82	43
x 9	x 8	x 6

. . . I say not unto thee, until seven times, but until seventy times seven.   Matt. 18:22

© Edwin C. Myers 1985, 1990   **CalcuLadder**®   Level 38: Multiplication   4 minutes

My name is _____
Today is _____

19	28	36
x 3	x 2	x 3

53	62	70
x 6	x 9	x 8

97	15	24
x 5	x 6	x 4

42	59	68
x 7	x 4	x 2

86	93	12
x 4	x 5	x 9

30	47	55
x 8	x 7	x 5

74	82	43
x 9	x 8	x 6

. . . I say not unto thee, until seven times, but until seventy times seven.   Matt. 18:22

© Edwin C. Myers 1985, 1990   CalcuLadder®   Level 38: Multiplication   4 minutes

My name is _____

Today is _____

19	28	36
x 3	x 2	x 3

53	62	70
x 6	x 9	x 8

97	15	24
x 5	x 6	x 4

42	59	68
x 7	x 4	x 2

86	93	12
x 4	x 5	x 9

30	47	55
x 8	x 7	x 5

74	82	43
x 9	x 8	x 6

. . . I say not unto thee, until seven times, but until seventy times seven.   Matt. 18:22

© Edwin C. Myers 1985, 1990   **CalcuLadder**®   Level 38: Multiplication   4 minutes

My name is _____
Today is _____

19	28	36
x 3	x 2	x 3

53	62	70
x 6	x 9	x 8

97	15	24
x 5	x 6	x 4

42	59	68
x 7	x 4	x 2

86	93	12
x 4	x 5	x 9

30	47	55
x 8	x 7	x 5

74	82	43
x 9	x 8	x 6

. . . I say not unto thee, until seven times, but until seventy times seven.   Matt. 18:22

© Edwin C. Myers 1985, 1990   CalcuLadder®   Level 38: Multiplication   4 minutes

My name is _____

Today is _____

19	28	36
x 3	x 2	x 3

53	62	70
x 6	x 9	x 8

97	15	24
x 5	x 6	x 4

42	59	68
x 7	x 4	x 2

86	93	12
x 4	x 5	x 9

30	47	55
x 8	x 7	x 5

74	82	43
x 9	x 8	x 6

. . . I say not unto thee, until seven times, but until seventy times seven.   Matt. 18:22

© Edwin C. Myers 1985, 1990   CalcuLadder®   Level 38: Multiplication   4 minutes

My name is _____

Today is _____

19	28	36
x 3	x 2	x 3

53	62	70
x 6	x 9	x 8

97	15	24
x 5	x 6	x 4

42	59	68
x 7	x 4	x 2

86	93	12
x 4	x 5	x 9

30	47	55
x 8	x 7	x 5

74	82	43
x 9	x 8	x 6

. . . I say not unto thee, until seven times, but until seventy times seven.   Matt. 18:22

© Edwin C. Myers 1985, 1990   CalcuLadder®   Level 38: Multiplication   4 minutes

My name is _____

Today is _____

19	28	36
x 3	x 2	x 3

53	62	70
x 6	x 9	x 8

97	15	24
x 5	x 6	x 4

42	59	68
x 7	x 4	x 2

86	93	12
x 4	x 5	x 9

30	47	55
x 8	x 7	x 5

74	82	43
x 9	x 8	x 6

. . . I say not unto thee, until seven times, but until seventy times seven.   Matt. 18:22

© Edwin C. Myers 1985, 1990   CalcuLadder®   Level 38: Multiplication   4 minutes

My name is _____

Today is _____

19	28	36
x 3	x 2	x 3

53	62	70
x 6	x 9	x 8

97	15	24
x 5	x 6	x 4

42	59	68
x 7	x 4	x 2

86	93	12
x 4	x 5	x 9

30	47	55
x 8	x 7	x 5

74	82	43
x 9	x 8	x 6

. . . I say not unto thee, until seven times, but until seventy times seven.   Matt. 18:22

© Edwin C. Myers 1985, 1990   CalcuLadder®   Level 38: Multiplication   4 minutes

My name is _____

Today is _____

19	28	36
x 3	x 2	x 3

53	62	70
x 6	x 9	x 8

97	15	24
x 5	x 6	x 4

42	59	68
x 7	x 4	x 2

86	93	12
x 4	x 5	x 9

30	47	55
x 8	x 7	x 5

74	82	43
x 9	x 8	x 6

. . . I say not unto thee, until seven times, but until seventy times seven.   Matt. 18:22

© Edwin C. Myers 1985, 1990   CalcuLadder®   Level 38: Multiplication   4 minutes

My name is _____

Today is _____

19	28	36
x 3	x 2	x 3

53	62	70
x 6	x 9	x 8

97	15	24
x 5	x 6	x 4

42	59	68
x 7	x 4	x 2

86	93	12
x 4	x 5	x 9

30	47	55
x 8	x 7	x 5

74	82	43
x 9	x 8	x 6

. . . I say not unto thee, until seven times, but until seventy times seven.   Matt. 18:22

© Edwin C. Myers 1985, 1990   CalcuLadder®   Level 38: Multiplication   4 minutes

My name is _____

Today is _____

19	28	36
x 3	x 2	x 3

53	62	70
x 6	x 9	x 8

97	15	24
x 5	x 6	x 4

42	59	68
x 7	x 4	x 2

86	93	12
x 4	x 5	x 9

30	47	55
x 8	x 7	x 5

74	82	43
x 9	x 8	x 6

. . . I say not unto thee, until seven times, but until seventy times seven.   Matt. 18:22

© Edwin C. Myers 1985, 1990   CalcuLadder®   Level 38: Multiplication   4 minutes

My name is _____

Today is _____

19	28	36
x3	x2	x3

53	62	70
x6	x9	x8

97	15	24
x5	x6	x4

42	59	68
x7	x4	x2

86	93	12
x4	x5	x9

30	47	55
x8	x7	x5

74	82	43
x9	x8	x6

. . . I say not unto thee, until seven times, but until seventy times seven.   Matt. 18:22

© Edwin C. Myers 1985, 1990   CalcuLadder®   Level 38: Multiplication   4 minutes

My name is _____

Today is _____

99	52	60	34
x 4	x 6	x 2	x 7

87	95	14	76
x 8	x 5	x 4	x 6

32	83	58	22
x 9	x 4	x 5	x 9

76	49	92	66
x 3	x 9	x 7	x 5

20	37	45	16
x 5	x 8	x 8	x 4

64	89	72
x 7	x 6	x 7

54	18	26
x 3	x 2	x 6

But other seed fell into good ground,
 and brought forth fruit, some a hundredfold, some sixtyfold, some thirtyfold.   Matt. 13:8

© Edwin C. Myers 1985, 1990   CalcuLadder®   Level 39: Multiplication   4 minutes

My name is _____

Today is _____

99	52	60	34
x 4	x 6	x 2	x 7

87	95	14	76
x 8	x 5	x 4	x 6

32	83	58	22
x 9	x 4	x 5	x 9

76	49	92	66
x 3	x 9	x 7	x 5

20	37	45	16
x 5	x 8	x 8	x 4

	64	89	72
	x 7	x 6	x 7

	54	18	26
	x 3	x 2	x 6

But other seed fell into good ground,
  and brought forth fruit, some a hundredfold, some sixtyfold, some thirtyfold.   Matt. 13:8

© Edwin C. Myers 1985, 1990   **CalcuLadder**®   Level 39: Multiplication   4 minutes

My name is _____

Today is _____

99	52	60	34
x 4	x 6	x 2	x 7

87	95	14	76
x 8	x 5	x 4	x 6

32	83	58	22
x 9	x 4	x 5	x 9

76	49	92	66
x 3	x 9	x 7	x 5

20	37	45	16
x 5	x 8	x 8	x 4

	64	89	72
	x 7	x 6	x 7

	54	18	26
	x 3	x 2	x 6

But other seed fell into good ground, and brought forth fruit, some a hundredfold, some sixtyfold, some thirtyfold.   Matt. 13:8

© Edwin C. Myers 1985, 1990   CalcuLadder®   Level 39: Multiplication   4 minutes

My name is _____

Today is _____

99	52	60	34
x 4	x 6	x 2	x 7

87	95	14	76
x 8	x 5	x 4	x 6

32	83	58	22
x 9	x 4	x 5	x 9

76	49	92	66
x 3	x 9	x 7	x 5

20	37	45	16
x 5	x 8	x 8	x 4

64	89	72
x 7	x 6	x 7

54	18	26
x 3	x 2	x 6

But other seed fell into good ground,
and brought forth fruit, some a hundredfold, some sixtyfold, some thirtyfold.   Matt. 13:8

© Edwin C. Myers 1985, 1990   **CalcuLadder**®   Level 39: Multiplication   4 minutes

My name is _____

Today is _____

99	52	60	34
x4	x6	x2	x7

87	95	14	76
x8	x5	x4	x6

32	83	58	22
x9	x4	x5	x9

76	49	92	66
x3	x9	x7	x5

20	37	45	16
x5	x8	x8	x4

	64	89	72
	x7	x6	x7

	54	18	26
	x3	x2	x6

But other seed fell into good ground,
and brought forth fruit, some a hundredfold, some sixtyfold, some thirtyfold. Matt. 13:8

© Edwin C. Myers 1985, 1990   CalcuLadder®   Level 39: Multiplication   4 minutes

My name is _____

Today is _____

99	52	60	34
x 4	x 6	x 2	x 7

87	95	14	76
x 8	x 5	x 4	x 6

32	83	58	22
x 9	x 4	x 5	x 9

76	49	92	66
x 3	x 9	x 7	x 5

20	37	45	16
x 5	x 8	x 8	x 4

64	89	72
x 7	x 6	x 7

54	18	26
x 3	x 2	x 6

But other seed fell into good ground,
  and brought forth fruit, some a hundredfold, some sixtyfold, some thirtyfold.   Matt. 13:8

© Edwin C. Myers 1985, 1990   **CalcuLadder**®   Level 39: Multiplication   4 minutes

My name is _____

Today is _____

99	52	60	34
x 4	x 6	x 2	x 7

87	95	14	76
x 8	x 5	x 4	x 6

32	83	58	22
x 9	x 4	x 5	x 9

76	49	92	66
x 3	x 9	x 7	x 5

20	37	45	16
x 5	x 8	x 8	x 4

64	89	72
x 7	x 6	x 7

54	18	26
x 3	x 2	x 6

But other seed fell into good ground, and brought forth fruit, some a hundredfold, some sixtyfold, some thirtyfold. Matt. 13:8

© Edwin C. Myers 1985, 1990   CalcuLadder®   Level 39: Multiplication   4 minutes

My name is _____

Today is _____

99	52	60	34
x 4	x 6	x 2	x 7

87	95	14	76
x 8	x 5	x 4	x 6

32	83	58	22
x 9	x 4	x 5	x 9

76	49	92	66
x 3	x 9	x 7	x 5

20	37	45	16
x 5	x 8	x 8	x 4

64	89	72
x 7	x 6	x 7

54	18	26
x 3	x 2	x 6

But other seed fell into good ground,
and brought forth fruit, some a hundredfold, some sixtyfold, some thirtyfold.   Matt. 13:8

© Edwin C. Myers 1985, 1990   CalcuLadder®   Level 39: Multiplication   4 minutes

My name is _____

Today is _____

99	52	60	34
x 4	x 6	x 2	x 7

87	95	14	76
x 8	x 5	x 4	x 6

32	83	58	22
x 9	x 4	x 5	x 9

76	49	92	66
x 3	x 9	x 7	x 5

20	37	45	16
x 5	x 8	x 8	x 4

64	89	72
x 7	x 6	x 7

54	18	26
x 3	x 2	x 6

But other seed fell into good ground,
and brought forth fruit, some a hundredfold, some sixtyfold, some thirtyfold. Matt. 13:8

© Edwin C. Myers 1985, 1990   CalcuLadder®   Level 39: Multiplication   4 minutes

My name is _____

Today is _____

99	52	60	34
x 4	x 6	x 2	x 7

87	95	14	76
x 8	x 5	x 4	x 6

32	83	58	22
x 9	x 4	x 5	x 9

76	49	92	66
x 3	x 9	x 7	x 5

20	37	45	16
x 5	x 8	x 8	x 4

64		89	72
x 7		x 6	x 7

54		18	26
x 3		x 2	x 6

But other seed fell into good ground,
and brought forth fruit, some a hundredfold, some sixtyfold, some thirtyfold. Matt. 13:8

© Edwin C. Myers 1985, 1990   **CalcuLadder**®   Level 39: Multiplication   4 minutes

My name is _____

Today is _____

99	52	60	34
x 4	x 6	x 2	x 7

87	95	14	76
x 8	x 5	x 4	x 6

32	83	58	22
x 9	x 4	x 5	x 9

76	49	92	66
x 3	x 9	x 7	x 5

20	37	45	16
x 5	x 8	x 8	x 4

64		89	72
x 7		x 6	x 7

54		18	26
x 3		x 2	x 6

But other seed fell into good ground,
and brought forth fruit, some a hundredfold, some sixtyfold, some thirtyfold. Matt. 13:8

© Edwin C. Myers 1985, 1990   CalcuLadder®   Level 39: Multiplication   4 minutes

My name is _____

Today is _____

99	52	60	34
x4	x6	x2	x7

87	95	14	76
x8	x5	x4	x6

32	83	58	22
x9	x4	x5	x9

76	49	92	66
x3	x9	x7	x5

20	37	45	16
x5	x8	x8	x4

64	89	72
x7	x6	x7

54	18	26
x3	x2	x6

But other seed fell into good ground,
 and brought forth fruit, some a hundredfold, some sixtyfold, some thirtyfold.   Matt. 13:8

© Edwin C. Myers 1985,1990   **CalcuLadder®**   Level 39: Multiplication   4 minutes

My name is _____
Today is _____

Example: __9__ x 8 = 72, so 72 ÷ 8 = __9__

___ x 2 = 16, so 16 ÷ 2 = ___
___ x 8 = 24, so 24 ÷ 8 = ___
___ x 3 = 27, so 27 ÷ 3 = ___
___ x 10 = 0, so 0 ÷ 10 = ___
___ x 11 = 44, so 44 ÷ 11 = ___
___ x 9 = 18, so 18 ÷ 9 = ___
___ x 5 = 5, so 5 ÷ 5 = ___
___ x 7 = 63, so 63 ÷ 7 = ___
___ x 3 = 21, so 21 ÷ 3 = ___
___ x 1 = 4, so 4 ÷ 1 = ___
___ x 11 = 22, so 22 ÷ 11 = ___
___ x 9 = 9, so 9 ÷ 9 = ___
___ x 5 = 45, so 45 ÷ 5 = ___
___ x 7 = 49, so 49 ÷ 7 = ___
___ x 3 = 12, so 12 ÷ 3 = ___
___ x 1 = 11, so 11 ÷ 1 = ___
___ x 11 = 66, so 66 ÷ 11 = ___
___ x 9 = 45, so 45 ÷ 9 = ___
___ x 5 = 15, so 15 ÷ 5 = ___
___ x 7 = 70, so 70 ÷ 7 = ___
___ x 3 = 24, so 24 ÷ 3 = ___

___ x 5 = 10, so 10 ÷ 5 = ___
___ x 10 = 50, so 50 ÷ 10 = ___
___ x 12 = 72, so 72 ÷ 12 = ___
___ x 6 = 42, so 42 ÷ 6 = ___
___ x 4 = 32, so 32 ÷ 4 = ___
___ x 2 = 12, so 12 ÷ 2 = ___
___ x 12 = 60, so 60 ÷ 12 = ___
___ x 10 = 30, so 30 ÷ 10 = ___
___ x 8 = 80, so 80 ÷ 8 = ___
___ x 6 = 48, so 48 ÷ 6 = ___
___ x 4 = 24, so 24 ÷ 4 = ___
___ x 2 = 10, so 10 ÷ 2 = ___
___ x 12 = 36, so 36 ÷ 12 = ___
___ x 7 = 35, so 35 ÷ 7 = ___
___ x 8 = 64, so 64 ÷ 8 = ___
___ x 6 = 12, so 12 ÷ 6 = ___
___ x 4 = 0, so 0 ÷ 4 = ___
___ x 2 = 18, so 18 ÷ 2 = ___
___ x 12 = 84, so 84 ÷ 12 = ___
___ x 10 = 40, so 40 ÷ 10 = ___
___ x 8 = 16, so 16 ÷ 8 = ___

He divides the sea with his power... Job 26:12

© Edwin C. Myers 1985,1990   CalcuLadder®   Level 40: Multiply-Divide   4 minutes

My name is _____

Today is _____

Example: <u>9</u> x 8 = 72, so 72 ÷ 8 = <u>9</u>

___ x 2 = 16, so 16 ÷ 2 = ___
___ x 8 = 24, so 24 ÷ 8 = ___
___ x 3 = 27, so 27 ÷ 3 = ___
___ x 10 = 0, so 0 ÷ 10 = ___
___ x 11 = 44, so 44 ÷ 11 = ___
___ x 9 = 18, so 18 ÷ 9 = ___
___ x 5 = 5, so 5 ÷ 5 = ___
___ x 7 = 63, so 63 ÷ 7 = ___
___ x 3 = 21, so 21 ÷ 3 = ___
___ x 1 = 4, so 4 ÷ 1 = ___
___ x 11 = 22, so 22 ÷ 11 = ___
___ x 9 = 9, so 9 ÷ 9 = ___
___ x 5 = 45, so 45 ÷ 5 = ___
___ x 7 = 49, so 49 ÷ 7 = ___
___ x 3 = 12, so 12 ÷ 3 = ___
___ x 1 = 11, so 11 ÷ 1 = ___
___ x 11 = 66, so 66 ÷ 11 = ___
___ x 9 = 45, so 45 ÷ 9 = ___
___ x 5 = 15, so 15 ÷ 5 = ___
___ x 7 = 70, so 70 ÷ 7 = ___
___ x 3 = 24, so 24 ÷ 3 = ___

___ x 5 = 10, so 10 ÷ 5 = ___
___ x 10 = 50, so 50 ÷ 10 = ___
___ x 12 = 72, so 72 ÷ 12 = ___
___ x 6 = 42, so 42 ÷ 6 = ___
___ x 4 = 32, so 32 ÷ 4 = ___
___ x 2 = 12, so 12 ÷ 2 = ___
___ x 12 = 60, so 60 ÷ 12 = ___
___ x 10 = 30, so 30 ÷ 10 = ___
___ x 8 = 80, so 80 ÷ 8 = ___
___ x 6 = 48, so 48 ÷ 6 = ___
___ x 4 = 24, so 24 ÷ 4 = ___
___ x 2 = 10, so 10 ÷ 2 = ___
___ x 12 = 36, so 36 ÷ 12 = ___
___ x 7 = 35, so 35 ÷ 7 = ___
___ x 8 = 64, so 64 ÷ 8 = ___
___ x 6 = 12, so 12 ÷ 6 = ___
___ x 4 = 0, so 0 ÷ 4 = ___
___ x 2 = 18, so 18 ÷ 2 = ___
___ x 12 = 84, so 84 ÷ 12 = ___
___ x 10 = 40, so 40 ÷ 10 = ___
___ x 8 = 16, so 16 ÷ 8 = ___

He divides the sea with his power... Job 26:12

© Edwin C. Myers 1985,1990   **CalcuLadder**®   Level 40: Multiply-Divide   4 minutes

My name is _____

Today is _____

Example: __9__ x 8 = 72, so 72 ÷ 8 = __9__

___ x 2 = 16, so 16 ÷ 2 = ___   ___ x 5 = 10, so 10 ÷ 5 = ___
___ x 8 = 24, so 24 ÷ 8 = ___   ___ x 10 = 50, so 50 ÷ 10 = ___
___ x 3 = 27, so 27 ÷ 3 = ___   ___ x 12 = 72, so 72 ÷ 12 = ___
___ x 10 = 0, so 0 ÷ 10 = ___   ___ x 6 = 42, so 42 ÷ 6 = ___
___ x 11 = 44, so 44 ÷ 11 = ___   ___ x 4 = 32, so 32 ÷ 4 = ___
___ x 9 = 18, so 18 ÷ 9 = ___   ___ x 2 = 12, so 12 ÷ 2 = ___
___ x 5 = 5, so 5 ÷ 5 = ___   ___ x 12 = 60, so 60 ÷ 12 = ___
___ x 7 = 63, so 63 ÷ 7 = ___   ___ x 10 = 30, so 30 ÷ 10 = ___
___ x 3 = 21, so 21 ÷ 3 = ___   ___ x 8 = 80, so 80 ÷ 8 = ___
___ x 1 = 4, so 4 ÷ 1 = ___   ___ x 6 = 48, so 48 ÷ 6 = ___
___ x 11 = 22, so 22 ÷ 11 = ___   ___ x 4 = 24, so 24 ÷ 4 = ___
___ x 9 = 9, so 9 ÷ 9 = ___   ___ x 2 = 10, so 10 ÷ 2 = ___
___ x 5 = 45, so 45 ÷ 5 = ___   ___ x 12 = 36, so 36 ÷ 12 = ___
___ x 7 = 49, so 49 ÷ 7 = ___   ___ x 7 = 35, so 35 ÷ 7 = ___
___ x 3 = 12, so 12 ÷ 3 = ___   ___ x 8 = 64, so 64 ÷ 8 = ___
___ x 1 = 11, so 11 ÷ 1 = ___   ___ x 6 = 12, so 12 ÷ 6 = ___
___ x 11 = 66, so 66 ÷ 11 = ___   ___ x 4 = 0, so 0 ÷ 4 = ___
___ x 9 = 45, so 45 ÷ 9 = ___   ___ x 2 = 18, so 18 ÷ 2 = ___
___ x 5 = 15, so 15 ÷ 5 = ___   ___ x 12 = 84, so 84 ÷ 12 = ___
___ x 7 = 70, so 70 ÷ 7 = ___   ___ x 10 = 40, so 40 ÷ 10 = ___
___ x 3 = 24, so 24 ÷ 3 = ___   ___ x 8 = 16, so 16 ÷ 8 = ___

He divides the sea with his power. . .   Job 26:12

© Edwin C. Myers 1985, 1990   CalcuLadder®   Level 40: Multiply-Divide   4 minutes

My name is _____

Today is _____

Example: __9__ x 8 = 72, so 72 ÷ 8 = __9__

___ x 2 = 16, so 16 ÷ 2 = ___
___ x 8 = 24, so 24 ÷ 8 = ___
___ x 3 = 27, so 27 ÷ 3 = ___
___ x 10 = 0, so 0 ÷ 10 = ___
___ x 11 = 44, so 44 ÷ 11 = ___
___ x 9 = 18, so 18 ÷ 9 = ___
___ x 5 = 5, so 5 ÷ 5 = ___
___ x 7 = 63, so 63 ÷ 7 = ___
___ x 3 = 21, so 21 ÷ 3 = ___
___ x 1 = 4, so 4 ÷ 1 = ___
___ x 11 = 22, so 22 ÷ 11 = ___
___ x 9 = 9, so 9 ÷ 9 = ___
___ x 5 = 45, so 45 ÷ 5 = ___
___ x 7 = 49, so 49 ÷ 7 = ___
___ x 3 = 12, so 12 ÷ 3 = ___
___ x 1 = 11, so 11 ÷ 1 = ___
___ x 11 = 66, so 66 ÷ 11 = ___
___ x 9 = 45, so 45 ÷ 9 = ___
___ x 5 = 15, so 15 ÷ 5 = ___
___ x 7 = 70, so 70 ÷ 7 = ___
___ x 3 = 24, so 24 ÷ 3 = ___

___ x 5 = 10, so 10 ÷ 5 = ___
___ x 10 = 50, so 50 ÷ 10 = ___
___ x 12 = 72, so 72 ÷ 12 = ___
___ x 6 = 42, so 42 ÷ 6 = ___
___ x 4 = 32, so 32 ÷ 4 = ___
___ x 2 = 12, so 12 ÷ 2 = ___
___ x 12 = 60, so 60 ÷ 12 = ___
___ x 10 = 30, so 30 ÷ 10 = ___
___ x 8 = 80, so 80 ÷ 8 = ___
___ x 6 = 48, so 48 ÷ 6 = ___
___ x 4 = 24, so 24 ÷ 4 = ___
___ x 2 = 10, so 10 ÷ 2 = ___
___ x 12 = 36, so 36 ÷ 12 = ___
___ x 7 = 35, so 35 ÷ 7 = ___
___ x 8 = 64, so 64 ÷ 8 = ___
___ x 6 = 12, so 12 ÷ 6 = ___
___ x 4 = 0, so 0 ÷ 4 = ___
___ x 2 = 18, so 18 ÷ 2 = ___
___ x 12 = 84, so 84 ÷ 12 = ___
___ x 10 = 40, so 40 ÷ 10 = ___
___ x 8 = 16, so 16 ÷ 8 = ___

He divides the sea with his power. . .   Job 26:12

© Edwin C. Myers 1985, 1990   CalcuLadder®   Level 40: Multiply-Divide   4 minutes

My name is _____

Today is _____

Example: __9__ x 8 = 72, so 72 ÷ 8 = __9__

___ x 2 = 16, so 16 ÷ 2 = ___
___ x 8 = 24, so 24 ÷ 8 = ___
___ x 3 = 27, so 27 ÷ 3 = ___
___ x 10 = 0, so 0 ÷ 10 = ___
___ x 11 = 44, so 44 ÷ 11 = ___
___ x 9 = 18, so 18 ÷ 9 = ___
___ x 5 = 5, so 5 ÷ 5 = ___
___ x 7 = 63, so 63 ÷ 7 = ___
___ x 3 = 21, so 21 ÷ 3 = ___
___ x 1 = 4, so 4 ÷ 1 = ___
___ x 11 = 22, so 22 ÷ 11 = ___
___ x 9 = 9, so 9 ÷ 9 = ___
___ x 5 = 45, so 45 ÷ 5 = ___
___ x 7 = 49, so 49 ÷ 7 = ___
___ x 3 = 12, so 12 ÷ 3 = ___
___ x 1 = 11, so 11 ÷ 1 = ___
___ x 11 = 66, so 66 ÷ 11 = ___
___ x 9 = 45, so 45 ÷ 9 = ___
___ x 5 = 15, so 15 ÷ 5 = ___
___ x 7 = 70, so 70 ÷ 7 = ___
___ x 3 = 24, so 24 ÷ 3 = ___

___ x 5 = 10, so 10 ÷ 5 = ___
___ x 10 = 50, so 50 ÷ 10 = ___
___ x 12 = 72, so 72 ÷ 12 = ___
___ x 6 = 42, so 42 ÷ 6 = ___
___ x 4 = 32, so 32 ÷ 4 = ___
___ x 2 = 12, so 12 ÷ 2 = ___
___ x 12 = 60, so 60 ÷ 12 = ___
___ x 10 = 30, so 30 ÷ 10 = ___
___ x 8 = 80, so 80 ÷ 8 = ___
___ x 6 = 48, so 48 ÷ 6 = ___
___ x 4 = 24, so 24 ÷ 4 = ___
___ x 2 = 10, so 10 ÷ 2 = ___
___ x 12 = 36, so 36 ÷ 12 = ___
___ x 7 = 35, so 35 ÷ 7 = ___
___ x 8 = 64, so 64 ÷ 8 = ___
___ x 6 = 12, so 12 ÷ 6 = ___
___ x 4 = 0, so 0 ÷ 4 = ___
___ x 2 = 18, so 18 ÷ 2 = ___
___ x 12 = 84, so 84 ÷ 12 = ___
___ x 10 = 40, so 40 ÷ 10 = ___
___ x 8 = 16, so 16 ÷ 8 = ___

He divides the sea with his power. . .   Job 26:12

© Edwin C. Myers 1985,1990   CalcuLadder®   Level 40: Multiply-Divide   4 minutes

My name is _____

Today is _____

Example: _9_ x 8 = 72, so 72 ÷ 8 = _9_

___ x 2 = 16, so 16 ÷ 2 = ___
___ x 8 = 24, so 24 ÷ 8 = ___
___ x 3 = 27, so 27 ÷ 3 = ___
___ x 10 = 0, so 0 ÷ 10 = ___
___ x 11 = 44, so 44 ÷ 11 = ___
___ x 9 = 18, so 18 ÷ 9 = ___
___ x 5 = 5, so 5 ÷ 5 = ___
___ x 7 = 63, so 63 ÷ 7 = ___
___ x 3 = 21, so 21 ÷ 3 = ___
___ x 1 = 4, so 4 ÷ 1 = ___
___ x 11 = 22, so 22 ÷ 11 = ___
___ x 9 = 9, so 9 ÷ 9 = ___
___ x 5 = 45, so 45 ÷ 5 = ___
___ x 7 = 49, so 49 ÷ 7 = ___
___ x 3 = 12, so 12 ÷ 3 = ___
___ x 1 = 11, so 11 ÷ 1 = ___
___ x 11 = 66, so 66 ÷ 11 = ___
___ x 9 = 45, so 45 ÷ 9 = ___
___ x 5 = 15, so 15 ÷ 5 = ___
___ x 7 = 70, so 70 ÷ 7 = ___
___ x 3 = 24, so 24 ÷ 3 = ___

___ x 5 = 10, so 10 ÷ 5 = ___
___ x 10 = 50, so 50 ÷ 10 = ___
___ x 12 = 72, so 72 ÷ 12 = ___
___ x 6 = 42, so 42 ÷ 6 = ___
___ x 4 = 32, so 32 ÷ 4 = ___
___ x 2 = 12, so 12 ÷ 2 = ___
___ x 12 = 60, so 60 ÷ 12 = ___
___ x 10 = 30, so 30 ÷ 10 = ___
___ x 8 = 80, so 80 ÷ 8 = ___
___ x 6 = 48, so 48 ÷ 6 = ___
___ x 4 = 24, so 24 ÷ 4 = ___
___ x 2 = 10, so 10 ÷ 2 = ___
___ x 12 = 36, so 36 ÷ 12 = ___
___ x 7 = 35, so 35 ÷ 7 = ___
___ x 8 = 64, so 64 ÷ 8 = ___
___ x 6 = 12, so 12 ÷ 6 = ___
___ x 4 = 0, so 0 ÷ 4 = ___
___ x 2 = 18, so 18 ÷ 2 = ___
___ x 12 = 84, so 84 ÷ 12 = ___
___ x 10 = 40, so 40 ÷ 10 = ___
___ x 8 = 16, so 16 ÷ 8 = ___

He divides the sea with his power. . .   Job 26:12

© Edwin C. Myers 1985,1990   **CalcuLadder®**   Level 40: Multiply-Divide   4 minutes

My name is _____
Today is _____

Example: __9__ x 8 = 72, so 72 ÷ 8 = __9__

___ x 2 = 16, so 16 ÷ 2 = ___　　　___ x 5 = 10, so 10 ÷ 5 = ___
___ x 8 = 24, so 24 ÷ 8 = ___　　　___ x 10 = 50, so 50 ÷ 10 = ___
___ x 3 = 27, so 27 ÷ 3 = ___　　　___ x 12 = 72, so 72 ÷ 12 = ___
___ x 10 = 0, so 0 ÷ 10 = ___　　　___ x 6 = 42, so 42 ÷ 6 = ___
___ x 11 = 44, so 44 ÷ 11 = ___　　___ x 4 = 32, so 32 ÷ 4 = ___
___ x 9 = 18, so 18 ÷ 9 = ___　　　___ x 2 = 12, so 12 ÷ 2 = ___
___ x 5 = 5, so 5 ÷ 5 = ___　　　　___ x 12 = 60, so 60 ÷ 12 = ___
___ x 7 = 63, so 63 ÷ 7 = ___　　　___ x 10 = 30, so 30 ÷ 10 = ___
___ x 3 = 21, so 21 ÷ 3 = ___　　　___ x 8 = 80, so 80 ÷ 8 = ___
___ x 1 = 4, so 4 ÷ 1 = ___　　　　___ x 6 = 48, so 48 ÷ 6 = ___
___ x 11 = 22, so 22 ÷ 11 = ___　　___ x 4 = 24, so 24 ÷ 4 = ___
___ x 9 = 9, so 9 ÷ 9 = ___　　　　___ x 2 = 10, so 10 ÷ 2 = ___
___ x 5 = 45, so 45 ÷ 5 = ___　　　___ x 12 = 36, so 36 ÷ 12 = ___
___ x 7 = 49, so 49 ÷ 7 = ___　　　___ x 7 = 35, so 35 ÷ 7 = ___
___ x 3 = 12, so 12 ÷ 3 = ___　　　___ x 8 = 64, so 64 ÷ 8 = ___
___ x 1 = 11, so 11 ÷ 1 = ___　　　___ x 6 = 12, so 12 ÷ 6 = ___
___ x 11 = 66, so 66 ÷ 11 = ___　　___ x 4 = 0, so 0 ÷ 4 = ___
___ x 9 = 45, so 45 ÷ 9 = ___　　　___ x 2 = 18, so 18 ÷ 2 = ___
___ x 5 = 15, so 15 ÷ 5 = ___　　　___ x 12 = 84, so 84 ÷ 12 = ___
___ x 7 = 70, so 70 ÷ 7 = ___　　　___ x 10 = 40, so 40 ÷ 10 = ___
___ x 3 = 24, so 24 ÷ 3 = ___　　　___ x 8 = 16, so 16 ÷ 8 = ___

He divides the sea with his power. . .　　Job 26:12

© Edwin C. Myers 1985, 1990　　　CalcuLadder®　　Level 40: Multiply-Divide　　4 minutes

My name is _____

Today is _____

Example: __9__ x 8 = 72, so 72 ÷ 8 = __9__

___ x 2 = 16, so 16 ÷ 2 = ___
___ x 8 = 24, so 24 ÷ 8 = ___
___ x 3 = 27, so 27 ÷ 3 = ___
___ x 10 = 0, so 0 ÷ 10 = ___
___ x 11 = 44, so 44 ÷ 11 = ___
___ x 9 = 18, so 18 ÷ 9 = ___
___ x 5 = 5, so 5 ÷ 5 = ___
___ x 7 = 63, so 63 ÷ 7 = ___
___ x 3 = 21, so 21 ÷ 3 = ___
___ x 1 = 4, so 4 ÷ 1 = ___
___ x 11 = 22, so 22 ÷ 11 = ___
___ x 9 = 9, so 9 ÷ 9 = ___
___ x 5 = 45, so 45 ÷ 5 = ___
___ x 7 = 49, so 49 ÷ 7 = ___
___ x 3 = 12, so 12 ÷ 3 = ___
___ x 1 = 11, so 11 ÷ 1 = ___
___ x 11 = 66, so 66 ÷ 11 = ___
___ x 9 = 45, so 45 ÷ 9 = ___
___ x 5 = 15, so 15 ÷ 5 = ___
___ x 7 = 70, so 70 ÷ 7 = ___
___ x 3 = 24, so 24 ÷ 3 = ___

___ x 5 = 10, so 10 ÷ 5 = ___
___ x 10 = 50, so 50 ÷ 10 = ___
___ x 12 = 72, so 72 ÷ 12 = ___
___ x 6 = 42, so 42 ÷ 6 = ___
___ x 4 = 32, so 32 ÷ 4 = ___
___ x 2 = 12, so 12 ÷ 2 = ___
___ x 12 = 60, so 60 ÷ 12 = ___
___ x 10 = 30, so 30 ÷ 10 = ___
___ x 8 = 80, so 80 ÷ 8 = ___
___ x 6 = 48, so 48 ÷ 6 = ___
___ x 4 = 24, so 24 ÷ 4 = ___
___ x 2 = 10, so 10 ÷ 2 = ___
___ x 12 = 36, so 36 ÷ 12 = ___
___ x 7 = 35, so 35 ÷ 7 = ___
___ x 8 = 64, so 64 ÷ 8 = ___
___ x 6 = 12, so 12 ÷ 6 = ___
___ x 4 = 0, so 0 ÷ 4 = ___
___ x 2 = 18, so 18 ÷ 2 = ___
___ x 12 = 84, so 84 ÷ 12 = ___
___ x 10 = 40, so 40 ÷ 10 = ___
___ x 8 = 16, so 16 ÷ 8 = ___

He divides the sea with his power... Job 26:12

© Edwin C. Myers 1985, 1990   CalcuLadder®   Level 40: Multiply-Divide   4 minutes

My name is _____

Today is _____

Example: __9__ x 8 = 72, so 72 ÷ 8 = __9__

___ x 2 = 16, so 16 ÷ 2 = ___
___ x 8 = 24, so 24 ÷ 8 = ___
___ x 3 = 27, so 27 ÷ 3 = ___
___ x 10 = 0, so 0 ÷ 10 = ___
___ x 11 = 44, so 44 ÷ 11 = ___
___ x 9 = 18, so 18 ÷ 9 = ___
___ x 5 = 5, so 5 ÷ 5 = ___
___ x 7 = 63, so 63 ÷ 7 = ___
___ x 3 = 21, so 21 ÷ 3 = ___
___ x 1 = 4, so 4 ÷ 1 = ___
___ x 11 = 22, so 22 ÷ 11 = ___
___ x 9 = 9, so 9 ÷ 9 = ___
___ x 5 = 45, so 45 ÷ 5 = ___
___ x 7 = 49, so 49 ÷ 7 = ___
___ x 3 = 12, so 12 ÷ 3 = ___
___ x 1 = 11, so 11 ÷ 1 = ___
___ x 11 = 66, so 66 ÷ 11 = ___
___ x 9 = 45, so 45 ÷ 9 = ___
___ x 5 = 15, so 15 ÷ 5 = ___
___ x 7 = 70, so 70 ÷ 7 = ___
___ x 3 = 24, so 24 ÷ 3 = ___

___ x 5 = 10, so 10 ÷ 5 = ___
___ x 10 = 50, so 50 ÷ 10 = ___
___ x 12 = 72, so 72 ÷ 12 = ___
___ x 6 = 42, so 42 ÷ 6 = ___
___ x 4 = 32, so 32 ÷ 4 = ___
___ x 2 = 12, so 12 ÷ 2 = ___
___ x 12 = 60, so 60 ÷ 12 = ___
___ x 10 = 30, so 30 ÷ 10 = ___
___ x 8 = 80, so 80 ÷ 8 = ___
___ x 6 = 48, so 48 ÷ 6 = ___
___ x 4 = 24, so 24 ÷ 4 = ___
___ x 2 = 10, so 10 ÷ 2 = ___
___ x 12 = 36, so 36 ÷ 12 = ___
___ x 7 = 35, so 35 ÷ 7 = ___
___ x 8 = 64, so 64 ÷ 8 = ___
___ x 6 = 12, so 12 ÷ 6 = ___
___ x 4 = 0, so 0 ÷ 4 = ___
___ x 2 = 18, so 18 ÷ 2 = ___
___ x 12 = 84, so 84 ÷ 12 = ___
___ x 10 = 40, so 40 ÷ 10 = ___
___ x 8 = 16, so 16 ÷ 8 = ___

He divides the sea with his power. . . Job 26:12

© Edwin C. Myers 1985,1990   CalcuLadder®   Level 40: Multiply-Divide   4 minutes

My name is _____

Today is _____

Example: __9__ x 8 = 72, so 72 ÷ 8 = __9__

___ x 2 = 16, so 16 ÷ 2 = ___   ___ x 5 = 10, so 10 ÷ 5 = ___
___ x 8 = 24, so 24 ÷ 8 = ___   ___ x 10 = 50, so 50 ÷ 10 = ___
___ x 3 = 27, so 27 ÷ 3 = ___   ___ x 12 = 72, so 72 ÷ 12 = ___
___ x 10 = 0, so 0 ÷ 10 = ___   ___ x 6 = 42, so 42 ÷ 6 = ___
___ x 11 = 44, so 44 ÷ 11 = ___   ___ x 4 = 32, so 32 ÷ 4 = ___
___ x 9 = 18, so 18 ÷ 9 = ___   ___ x 2 = 12, so 12 ÷ 2 = ___
___ x 5 = 5, so 5 ÷ 5 = ___   ___ x 12 = 60, so 60 ÷ 12 = ___
___ x 7 = 63, so 63 ÷ 7 = ___   ___ x 10 = 30, so 30 ÷ 10 = ___
___ x 3 = 21, so 21 ÷ 3 = ___   ___ x 8 = 80, so 80 ÷ 8 = ___
___ x 1 = 4, so 4 ÷ 1 = ___   ___ x 6 = 48, so 48 ÷ 6 = ___
___ x 11 = 22, so 22 ÷ 11 = ___   ___ x 4 = 24, so 24 ÷ 4 = ___
___ x 9 = 9, so 9 ÷ 9 = ___   ___ x 2 = 10, so 10 ÷ 2 = ___
___ x 5 = 45, so 45 ÷ 5 = ___   ___ x 12 = 36, so 36 ÷ 12 = ___
___ x 7 = 49, so 49 ÷ 7 = ___   ___ x 7 = 35, so 35 ÷ 7 = ___
___ x 3 = 12, so 12 ÷ 3 = ___   ___ x 8 = 64, so 64 ÷ 8 = ___
___ x 1 = 11, so 11 ÷ 1 = ___   ___ x 6 = 12, so 12 ÷ 6 = ___
___ x 11 = 66, so 66 ÷ 11 = ___   ___ x 4 = 0, so 0 ÷ 4 = ___
___ x 9 = 45, so 45 ÷ 9 = ___   ___ x 2 = 18, so 18 ÷ 2 = ___
___ x 5 = 15, so 15 ÷ 5 = ___   ___ x 12 = 84, so 84 ÷ 12 = ___
___ x 7 = 70, so 70 ÷ 7 = ___   ___ x 10 = 40, so 40 ÷ 10 = ___
___ x 3 = 24, so 24 ÷ 3 = ___   ___ x 8 = 16, so 16 ÷ 8 = ___

He divides the sea with his power... Job 26:12

© Edwin C. Myers 1985, 1990   **CalcuLadder**®   Level 40: Multiply-Divide   4 minutes

My name is _____

Today is _____

Example: __9__ x 8 = 72, so 72 ÷ 8 = __9__

___ x 2 = 16, so 16 ÷ 2 = ___
___ x 8 = 24, so 24 ÷ 8 = ___
___ x 3 = 27, so 27 ÷ 3 = ___
___ x 10 = 0, so 0 ÷ 10 = ___
___ x 11 = 44, so 44 ÷ 11 = ___
___ x 9 = 18, so 18 ÷ 9 = ___
___ x 5 = 5, so 5 ÷ 5 = ___
___ x 7 = 63, so 63 ÷ 7 = ___
___ x 3 = 21, so 21 ÷ 3 = ___
___ x 1 = 4, so 4 ÷ 1 = ___
___ x 11 = 22, so 22 ÷ 11 = ___
___ x 9 = 9, so 9 ÷ 9 = ___
___ x 5 = 45, so 45 ÷ 5 = ___
___ x 7 = 49, so 49 ÷ 7 = ___
___ x 3 = 12, so 12 ÷ 3 = ___
___ x 1 = 11, so 11 ÷ 1 = ___
___ x 11 = 66, so 66 ÷ 11 = ___
___ x 9 = 45, so 45 ÷ 9 = ___
___ x 5 = 15, so 15 ÷ 5 = ___
___ x 7 = 70, so 70 ÷ 7 = ___
___ x 3 = 24, so 24 ÷ 3 = ___

___ x 5 = 10, so 10 ÷ 5 = ___
___ x 10 = 50, so 50 ÷ 10 = ___
___ x 12 = 72, so 72 ÷ 12 = ___
___ x 6 = 42, so 42 ÷ 6 = ___
___ x 4 = 32, so 32 ÷ 4 = ___
___ x 2 = 12, so 12 ÷ 2 = ___
___ x 12 = 60, so 60 ÷ 12 = ___
___ x 10 = 30, so 30 ÷ 10 = ___
___ x 8 = 80, so 80 ÷ 8 = ___
___ x 6 = 48, so 48 ÷ 6 = ___
___ x 4 = 24, so 24 ÷ 4 = ___
___ x 2 = 10, so 10 ÷ 2 = ___
___ x 12 = 36, so 36 ÷ 12 = ___
___ x 7 = 35, so 35 ÷ 7 = ___
___ x 8 = 64, so 64 ÷ 8 = ___
___ x 6 = 12, so 12 ÷ 6 = ___
___ x 4 = 0, so 0 ÷ 4 = ___
___ x 2 = 18, so 18 ÷ 2 = ___
___ x 12 = 84, so 84 ÷ 12 = ___
___ x 10 = 40, so 40 ÷ 10 = ___
___ x 8 = 16, so 16 ÷ 8 = ___

He divides the sea with his power... Job 26:12

© Edwin C. Myers 1985, 1990   CalcuLadder®   Level 40: Multiply-Divide   4 minutes

My name is _____

Today is _____

Example: __9__ x 8 = 72, so 72 ÷ 8 = __9__

___ x 2 = 16, so 16 ÷ 2 = ___
___ x 8 = 24, so 24 ÷ 8 = ___
___ x 3 = 27, so 27 ÷ 3 = ___
___ x 10 = 0, so 0 ÷ 10 = ___
___ x 11 = 44, so 44 ÷ 11 = ___
___ x 9 = 18, so 18 ÷ 9 = ___
___ x 5 = 5, so 5 ÷ 5 = ___
___ x 7 = 63, so 63 ÷ 7 = ___
___ x 3 = 21, so 21 ÷ 3 = ___
___ x 1 = 4, so 4 ÷ 1 = ___
___ x 11 = 22, so 22 ÷ 11 = ___
___ x 9 = 9, so 9 ÷ 9 = ___
___ x 5 = 45, so 45 ÷ 5 = ___
___ x 7 = 49, so 49 ÷ 7 = ___
___ x 3 = 12, so 12 ÷ 3 = ___
___ x 1 = 11, so 11 ÷ 1 = ___
___ x 11 = 66, so 66 ÷ 11 = ___
___ x 9 = 45, so 45 ÷ 9 = ___
___ x 5 = 15, so 15 ÷ 5 = ___
___ x 7 = 70, so 70 ÷ 7 = ___
___ x 3 = 24, so 24 ÷ 3 = ___

___ x 5 = 10, so 10 ÷ 5 = ___
___ x 10 = 50, so 50 ÷ 10 = ___
___ x 12 = 72, so 72 ÷ 12 = ___
___ x 6 = 42, so 42 ÷ 6 = ___
___ x 4 = 32, so 32 ÷ 4 = ___
___ x 2 = 12, so 12 ÷ 2 = ___
___ x 12 = 60, so 60 ÷ 12 = ___
___ x 10 = 30, so 30 ÷ 10 = ___
___ x 8 = 80, so 80 ÷ 8 = ___
___ x 6 = 48, so 48 ÷ 6 = ___
___ x 4 = 24, so 24 ÷ 4 = ___
___ x 2 = 10, so 10 ÷ 2 = ___
___ x 12 = 36, so 36 ÷ 12 = ___
___ x 7 = 35, so 35 ÷ 7 = ___
___ x 8 = 64, so 64 ÷ 8 = ___
___ x 6 = 12, so 12 ÷ 6 = ___
___ x 4 = 0, so 0 ÷ 4 = ___
___ x 2 = 18, so 18 ÷ 2 = ___
___ x 12 = 84, so 84 ÷ 12 = ___
___ x 10 = 40, so 40 ÷ 10 = ___
___ x 8 = 16, so 16 ÷ 8 = ___

He divides the sea with his power... Job 26:12

© Edwin C. Myers 1985, 1990   CalcuLadder®   Level 40: Multiply-Divide   4 minutes

My name is _____

Today is _____

```
  293         432         205
×   3       ×   4       ×   6


  672         318         436
×   2       ×   5       ×   7


  158         375         182
×   3       ×   4       ×   9


 1564         762        3246
×   2       ×   6       ×   3


 2738        4507        8514
×   4       ×   5       ×   6
```

A faithful man shall abound with blessings . . . Prov. 28:20

© Edwin C. Myers 1985, 1990   **CalcuLadder®**   Level 41: Multiplication   4 minutes

My name is _____

Today is _____

293	432	205
x 3	x 4	x 6

672	318	436
x 2	x 5	x 7

158	375	182
x 3	x 4	x 9

1564	762	3246
x 2	x 6	x 3

2738	4507	8514
x 4	x 5	x 6

A faithful man shall abound with blessings . . . Prov. 28:20

© Edwin C. Myers 1985, 1990   **CalcuLadder**®   Level 41: Multiplication   4 minutes

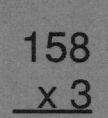

My name is _____

Today is _____

293	432	205
x 3	x 4	x 6

672	318	436
x 2	x 5	x 7

158	375	182
x 3	x 4	x 9

1564	762	3246
x 2	x 6	x 3

2738	4507	8514
x 4	x 5	x 6

A faithful man shall abound with blessings . . .   Prov. 28:20

© Edwin C. Myers 1985, 1990   **CalcuLadder**®   Level 41: Multiplication   4 minutes

My name is _____

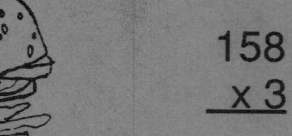

Today is _____

293	432	205
x 3	x 4	x 6

672	318	436
x 2	x 5	x 7

158	375	182
x 3	x 4	x 9

1564	762	3246
x 2	x 6	x 3

2738	4507	8514
x 4	x 5	x 6

A faithful man shall abound with blessings . . . Prov. 28:20

© Edwin C. Myers 1985, 1990   **CalcuLadder®**   Level 41: Multiplication   4 minutes

My name is _____

Today is _____

293	432	205
x 3	x 4	x 6

672	318	436
x 2	x 5	x 7

158	375	182
x 3	x 4	x 9

1564	762	3246
x 2	x 6	x 3

2738	4507	8514
x 4	x 5	x 6

A faithful man shall abound with blessings . . .   Prov. 28:20

© Edwin C. Myers 1985, 1990   **CalcuLadder**®   Level 41: Multiplication   4 minutes

My name is _____

Today is _____

293	432	205
x 3	x 4	x 6

672	318	436
x 2	x 5	x 7

158	375	182
x 3	x 4	x 9

1564	762	3246
x 2	x 6	x 3

2738	4507	8514
x 4	x 5	x 6

A faithful man shall abound with blessings . . . Prov. 28:20

© Edwin C. Myers 1985, 1990 **CalcuLadder**® Level 41: Multiplication 4 minutes

My name is _____

Today is _____

293	432	205
x 3	x 4	x 6

672	318	436
x 2	x 5	x 7

158	375	182
x 3	x 4	x 9

1564	762	3246
x 2	x 6	x 3

2738	4507	8514
x 4	x 5	x 6

A faithful man shall abound with blessings . . . Prov. 28:20

© Edwin C. Myers 1985, 1990   **CalcuLadder®**   Level 41: Multiplication   4 minutes

My name is _____

Today is _____

293 × 3	432 × 4	205 × 6
672 × 2	318 × 5	436 × 7
158 × 3	375 × 4	182 × 9
1564 × 2	762 × 6	3246 × 3
2738 × 4	4507 × 5	8514 × 6

A faithful man shall abound with blessings . . . Prov. 28:20

© Edwin C. Myers 1985, 1990   CalcuLadder®   Level 41: Multiplication   4 minutes

My name is _____

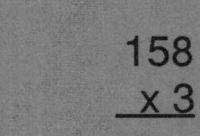

Today is _____

```
  293        432        205
 x  3       x  4       x  6


  672        318        436
 x  2       x  5       x  7


  158        375        182
 x  3       x  4       x  9


 1564        762       3246
 x  2       x  6       x  3


 2738       4507       8514
 x  4       x  5       x  6
```

A faithful man shall abound with blessings . . .   Prov. 28:20

© Edwin C. Myers 1985, 1990   CalcuLadder®   Level 41: Multiplication   4 minutes

My name is _____

Today is _____

293 x 3	432 x 4	205 x 6
672 x 2	318 x 5	436 x 7
158 x 3	375 x 4	182 x 9
1564 x 2	762 x 6	3246 x 3
2738 x 4	4507 x 5	8514 x 6

A faithful man shall abound with blessings . . .   Prov. 28:20

© Edwin C. Myers 1985, 1990   **CalcuLadder**®   Level 41: Multiplication   4 minutes

My name is _____

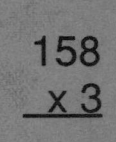

Today is _____

293	432	205
x 3	x 4	x 6

672	318	436
x 2	x 5	x 7

158	375	182
x 3	x 4	x 9

1564	762	3246
x 2	x 6	x 3

2738	4507	8514
x 4	x 5	x 6

A faithful man shall abound with blessings . . . Prov. 28:20

© Edwin C. Myers 1985, 1990   CalcuLadder®   Level 41: Multiplication   4 minutes

My name is _____

Today is _____

293	432	205
x 3	x 4	x 6

672	318	436
x 2	x 5	x 7

158	375	182
x 3	x 4	x 9

1564	762	3246
x 2	x 6	x 3

2738	4507	8514
x 4	x 5	x 6

A faithful man shall abound with blessings . . .   Prov. 28:20

© Edwin C. Myers 1985, 1990   **CalcuLadder**®   Level 41: Multiplication   4 minutes

My name is _____
Today is _____

3)3	2)8	6)12	7)35	9)27	4)28
6)48	5)5	10)60	5)40	8)16	3)9
9)45	1)6	4)8	7)63	10)40	2)4
9)9	11)33	3)21	7)7	12)48	8)80
1)3	5)15	2)12	9)18	4)32	8)48
6)60	3)27	7)14	2)16	4)24	9)36
6)18	8)32	7)28	8)72	3)24	5)35
2)6	12)24	9)54	6)30	8)40	3)18
10)50	7)42	2)10	9)63	5)30	4)28
6)42	7)49	8)24	3)12	5)25	1)8

And God saw the light, that it was good, and God divided the light from the darkness.
Gen. 1:4

© Edwin C. Myers 1985, 1990   **CalcuLadder®**   Level 42: Division   4 minutes

My name is _____

Today is _____

3)3̄	2)8̄	6)1̄2̄	7)3̄5̄	9)2̄7̄	4)2̄8̄
6)4̄8̄	5)5̄	10)6̄0̄	5)4̄0̄	8)1̄6̄	3)9̄
9)4̄5̄	1)6̄	4)8̄	7)6̄3̄	10)4̄0̄	2)4̄
9)9̄	11)3̄3̄	3)2̄1̄	7)7̄	12)4̄8̄	8)8̄0̄
1)3̄	5)1̄5̄	2)1̄2̄	9)1̄8̄	4)3̄2̄	8)4̄8̄
6)6̄0̄	3)2̄7̄	7)1̄4̄	2)1̄6̄	4)2̄4̄	9)3̄6̄
6)1̄8̄	8)3̄2̄	7)2̄8̄	8)7̄2̄	3)2̄4̄	5)3̄5̄
2)6̄	12)2̄4̄	9)5̄4̄	6)3̄0̄	8)4̄0̄	3)1̄8̄
10)5̄0̄	7)4̄2̄	2)1̄0̄	9)6̄3̄	5)3̄0̄	4)2̄8̄
6)4̄2̄	7)4̄9̄	8)2̄4̄	3)1̄2̄	5)2̄5̄	1)8̄

And God saw the light, that it was good, and God divided the light from the darkness.

Gen. 1:4

© Edwin C. Myers 1985, 1990 **CalcuLadder**® Level 42: Division 4 minutes

My name is _____

Today is _____

3)3	2)8	6)12	7)35	9)27	4)28
6)48	5)5	10)60	5)40	8)16	3)9
9)45	1)6	4)8	7)63	10)40	2)4
9)9	11)33	3)21	7)7	12)48	8)80
1)3	5)15	2)12	9)18	4)32	8)48
6)60	3)27	7)14	2)16	4)24	9)36
6)18	8)32	7)28	8)72	3)24	5)35
2)6	12)24	9)54	6)30	8)40	3)18
10)50	7)42	2)10	9)63	5)30	4)28
6)42	7)49	8)24	3)12	5)25	1)8

And God saw the light, that it was good, and God divided the light from the darkness.

Gen. 1:4

© Edwin C. Myers 1985,1990   CalcuLadder®   Level 42: Division   4 minutes

My name is _____

Today is _____

$3\overline{)3}$ $2\overline{)8}$ $6\overline{)12}$ $7\overline{)35}$ $9\overline{)27}$ $4\overline{)28}$

$6\overline{)48}$ $5\overline{)5}$ $10\overline{)60}$ $5\overline{)40}$ $8\overline{)16}$ $3\overline{)9}$

$9\overline{)45}$ $1\overline{)6}$ $4\overline{)8}$ $7\overline{)63}$ $10\overline{)40}$ $2\overline{)4}$

$9\overline{)9}$ $11\overline{)33}$ $3\overline{)21}$ $7\overline{)7}$ $12\overline{)48}$ $8\overline{)80}$

$1\overline{)3}$ $5\overline{)15}$ $2\overline{)12}$ $9\overline{)18}$ $4\overline{)32}$ $8\overline{)48}$

$6\overline{)60}$ $3\overline{)27}$ $7\overline{)14}$ $2\overline{)16}$ $4\overline{)24}$ $9\overline{)36}$

$6\overline{)18}$ $8\overline{)32}$ $7\overline{)28}$ $8\overline{)72}$ $3\overline{)24}$ $5\overline{)35}$

$2\overline{)6}$ $12\overline{)24}$ $9\overline{)54}$ $6\overline{)30}$ $8\overline{)40}$ $3\overline{)18}$

$10\overline{)50}$ $7\overline{)42}$ $2\overline{)10}$ $9\overline{)63}$ $5\overline{)30}$ $4\overline{)28}$

$6\overline{)42}$ $7\overline{)49}$ $8\overline{)24}$ $3\overline{)12}$ $5\overline{)25}$ $1\overline{)8}$

And God saw the light, that it was good, and God divided the light from the darkness.

Gen. 1:4

© Edwin C. Myers 1985, 1990   **CalcuLadder**®   Level 42: Division   4 minutes

My name is _____

Today is _____

3)3̄	2)8̄	6)1̄2̄	7)3̄5̄	9)2̄7̄	4)2̄8̄
6)4̄8̄	5)5̄	10)6̄0̄	5)4̄0̄	8)1̄6̄	3)9̄
9)4̄5̄	1)6̄	4)8̄	7)6̄3̄	10)4̄0̄	2)4̄
9)9̄	11)3̄3̄	3)2̄1̄	7)7̄	12)4̄8̄	8)8̄0̄
1)3̄	5)1̄5̄	2)1̄2̄	9)1̄8̄	4)3̄2̄	8)4̄8̄
6)6̄0̄	3)2̄7̄	7)1̄4̄	2)1̄6̄	4)2̄4̄	9)3̄6̄
6)1̄8̄	8)3̄2̄	7)2̄8̄	8)7̄2̄	3)2̄4̄	5)3̄5̄
2)6̄	12)2̄4̄	9)5̄4̄	6)3̄0̄	8)4̄0̄	3)1̄8̄
10)5̄0̄	7)4̄2̄	2)1̄0̄	9)6̄3̄	5)3̄0̄	4)2̄8̄
6)4̄2̄	7)4̄9̄	8)2̄4̄	3)1̄2̄	5)2̄5̄	1)8̄

And God saw the light, that it was good, and God divided the light from the darkness.

Gen. 1:4

© Edwin C. Myers 1985, 1990   **CalcuLadder**®   Level 42: Division   4 minutes

My name is _____

Today is _____

$3\overline{)3}$ $2\overline{)8}$ $6\overline{)12}$ $7\overline{)35}$ $9\overline{)27}$ $4\overline{)28}$

$6\overline{)48}$ $5\overline{)5}$ $10\overline{)60}$ $5\overline{)40}$ $8\overline{)16}$ $3\overline{)9}$

$9\overline{)45}$ $1\overline{)6}$ $4\overline{)8}$ $7\overline{)63}$ $10\overline{)40}$ $2\overline{)4}$

$9\overline{)9}$ $11\overline{)33}$ $3\overline{)21}$ $7\overline{)7}$ $12\overline{)48}$ $8\overline{)80}$

$1\overline{)3}$ $5\overline{)15}$ $2\overline{)12}$ $9\overline{)18}$ $4\overline{)32}$ $8\overline{)48}$

$6\overline{)60}$ $3\overline{)27}$ $7\overline{)14}$ $2\overline{)16}$ $4\overline{)24}$ $9\overline{)36}$

$6\overline{)18}$ $8\overline{)32}$ $7\overline{)28}$ $8\overline{)72}$ $3\overline{)24}$ $5\overline{)35}$

$2\overline{)6}$ $12\overline{)24}$ $9\overline{)54}$ $6\overline{)30}$ $8\overline{)40}$ $3\overline{)18}$

$10\overline{)50}$ $7\overline{)42}$ $2\overline{)10}$ $9\overline{)63}$ $5\overline{)30}$ $4\overline{)28}$

$6\overline{)42}$ $7\overline{)49}$ $8\overline{)24}$ $3\overline{)12}$ $5\overline{)25}$ $1\overline{)8}$

And God saw the light, that it was good, and God divided the light from the darkness.

Gen. 1:4

© Edwin C. Myers 1985, 1990   **CalcuLadder**®   Level 42: Division   4 minutes

My name is _____

Today is _____

3)3̄	2)8̄	6)1̄2̄	7)3̄5̄	9)2̄7̄	4)2̄8̄
6)4̄8̄	5)5̄	10)6̄0̄	5)4̄0̄	8)1̄6̄	3)9̄
9)4̄5̄	1)6̄	4)8̄	7)6̄3̄	10)4̄0̄	2)4̄
9)9̄	11)3̄3̄	3)2̄1̄	7)7̄	12)4̄8̄	8)8̄0̄
1)3̄	5)1̄5̄	2)1̄2̄	9)1̄8̄	4)3̄2̄	8)4̄8̄
6)6̄0̄	3)2̄7̄	7)1̄4̄	2)1̄6̄	4)2̄4̄	9)3̄6̄
6)1̄8̄	8)3̄2̄	7)2̄8̄	8)7̄2̄	3)2̄4̄	5)3̄5̄
2)6̄	12)2̄4̄	9)5̄4̄	6)3̄0̄	8)4̄0̄	3)1̄8̄
10)5̄0̄	7)4̄2̄	2)1̄0̄	9)6̄3̄	5)3̄0̄	4)2̄8̄
6)4̄2̄	7)4̄9̄	8)2̄4̄	3)1̄2̄	5)2̄5̄	1)8̄

And God saw the light, that it was good, and God divided the light from the darkness.

Gen. 1:4

© Edwin C. Myers 1985,1990   CalcuLadder®   Level 42: Division   4 minutes

My name is _____

Today is _____

$3\overline{)3}$ $2\overline{)8}$ $6\overline{)12}$ $7\overline{)35}$ $9\overline{)27}$ $4\overline{)28}$

$6\overline{)48}$ $5\overline{)5}$ $10\overline{)60}$ $5\overline{)40}$ $8\overline{)16}$ $3\overline{)9}$

$9\overline{)45}$ $1\overline{)6}$ $4\overline{)8}$ $7\overline{)63}$ $10\overline{)40}$ $2\overline{)4}$

$9\overline{)9}$ $11\overline{)33}$ $3\overline{)21}$ $7\overline{)7}$ $12\overline{)48}$ $8\overline{)80}$

$1\overline{)3}$ $5\overline{)15}$ $2\overline{)12}$ $9\overline{)18}$ $4\overline{)32}$ $8\overline{)48}$

$6\overline{)60}$ $3\overline{)27}$ $7\overline{)14}$ $2\overline{)16}$ $4\overline{)24}$ $9\overline{)36}$

$6\overline{)18}$ $8\overline{)32}$ $7\overline{)28}$ $8\overline{)72}$ $3\overline{)24}$ $5\overline{)35}$

$2\overline{)6}$ $12\overline{)24}$ $9\overline{)54}$ $6\overline{)30}$ $8\overline{)40}$ $3\overline{)18}$

$10\overline{)50}$ $7\overline{)42}$ $2\overline{)10}$ $9\overline{)63}$ $5\overline{)30}$ $4\overline{)28}$

$6\overline{)42}$ $7\overline{)49}$ $8\overline{)24}$ $3\overline{)12}$ $5\overline{)25}$ $1\overline{)8}$

And God saw the light, that it was good, and God divided the light from the darkness.

Gen. 1:4

© Edwin C. Myers 1985, 1990   **CalcuLadder**®   Level 42: Division   4 minutes

My name is _____

Today is _____

$3\overline{)3}$ $2\overline{)8}$ $6\overline{)12}$ $7\overline{)35}$ $9\overline{)27}$ $4\overline{)28}$

$6\overline{)48}$ $5\overline{)5}$ $10\overline{)60}$ $5\overline{)40}$ $8\overline{)16}$ $3\overline{)9}$

$9\overline{)45}$ $1\overline{)6}$ $4\overline{)8}$ $7\overline{)63}$ $10\overline{)40}$ $2\overline{)4}$

$9\overline{)9}$ $11\overline{)33}$ $3\overline{)21}$ $7\overline{)7}$ $12\overline{)48}$ $8\overline{)80}$

$1\overline{)3}$ $5\overline{)15}$ $2\overline{)12}$ $9\overline{)18}$ $4\overline{)32}$ $8\overline{)48}$

$6\overline{)60}$ $3\overline{)27}$ $7\overline{)14}$ $2\overline{)16}$ $4\overline{)24}$ $9\overline{)36}$

$6\overline{)18}$ $8\overline{)32}$ $7\overline{)28}$ $8\overline{)72}$ $3\overline{)24}$ $5\overline{)35}$

$2\overline{)6}$ $12\overline{)24}$ $9\overline{)54}$ $6\overline{)30}$ $8\overline{)40}$ $3\overline{)18}$

$10\overline{)50}$ $7\overline{)42}$ $2\overline{)10}$ $9\overline{)63}$ $5\overline{)30}$ $4\overline{)28}$

$6\overline{)42}$ $7\overline{)49}$ $8\overline{)24}$ $3\overline{)12}$ $5\overline{)25}$ $1\overline{)8}$

And God saw the light, that it was good, and God divided the light from the darkness.

Gen. 1:4

© Edwin C. Myers 1985, 1990   **CalcuLadder**®   Level 42: Division   4 minutes

My name is _____

Today is _____

3)3̄	2)8̄	6)1̄2̄	7)3̄5̄	9)2̄7̄	4)2̄8̄
6)4̄8̄	5)5̄	10)6̄0̄	5)4̄0̄	8)1̄6̄	3)9̄
9)4̄5̄	1)6̄	4)8̄	7)6̄3̄	10)4̄0̄	2)4̄
9)9̄	11)3̄3̄	3)2̄1̄	7)7̄	12)4̄8̄	8)8̄0̄
1)3̄	5)1̄5̄	2)1̄2̄	9)1̄8̄	4)3̄2̄	8)4̄8̄
6)6̄0̄	3)2̄7̄	7)1̄4̄	2)1̄6̄	4)2̄4̄	9)3̄6̄
6)1̄8̄	8)3̄2̄	7)2̄8̄	8)7̄2̄	3)2̄4̄	5)3̄5̄
2)6̄	12)2̄4̄	9)5̄4̄	6)3̄0̄	8)4̄0̄	3)1̄8̄
10)5̄0̄	7)4̄2̄	2)1̄0̄	9)6̄3̄	5)3̄0̄	4)2̄8̄
6)4̄2̄	7)4̄9̄	8)2̄4̄	3)1̄2̄	5)2̄5̄	1)8̄

And God saw the light, that it was good, and God divided the light from the darkness.
Gen. 1:4

© Edwin C. Myers 1985, 1990   **CalcuLadder®**   Level 42: Division   4 minutes

My name is _____

Today is _____

$3\overline{)3}$ $2\overline{)8}$ $6\overline{)12}$ $7\overline{)35}$ $9\overline{)27}$ $4\overline{)28}$

$6\overline{)48}$ $5\overline{)5}$ $10\overline{)60}$ $5\overline{)40}$ $8\overline{)16}$ $3\overline{)9}$

$9\overline{)45}$ $1\overline{)6}$ $4\overline{)8}$ $7\overline{)63}$ $10\overline{)40}$ $2\overline{)4}$

$9\overline{)9}$ $11\overline{)33}$ $3\overline{)21}$ $7\overline{)7}$ $12\overline{)48}$ $8\overline{)80}$

$1\overline{)3}$ $5\overline{)15}$ $2\overline{)12}$ $9\overline{)18}$ $4\overline{)32}$ $8\overline{)48}$

$6\overline{)60}$ $3\overline{)27}$ $7\overline{)14}$ $2\overline{)16}$ $4\overline{)24}$ $9\overline{)36}$

$6\overline{)18}$ $8\overline{)32}$ $7\overline{)28}$ $8\overline{)72}$ $3\overline{)24}$ $5\overline{)35}$

$2\overline{)6}$ $12\overline{)24}$ $9\overline{)54}$ $6\overline{)30}$ $8\overline{)40}$ $3\overline{)18}$

$10\overline{)50}$ $7\overline{)42}$ $2\overline{)10}$ $9\overline{)63}$ $5\overline{)30}$ $4\overline{)28}$

$6\overline{)42}$ $7\overline{)49}$ $8\overline{)24}$ $3\overline{)12}$ $5\overline{)25}$ $1\overline{)8}$

And God saw the light, that it was good, and God divided the light from the darkness.

Gen. 1:4

© Edwin C. Myers 1985, 1990   CalcuLadder®   Level 42: Division   4 minutes

My name is _____

Today is _____

$3\overline{)3}$	$2\overline{)8}$	$6\overline{)12}$	$7\overline{)35}$	$9\overline{)27}$	$4\overline{)28}$
$6\overline{)48}$	$5\overline{)5}$	$10\overline{)60}$	$5\overline{)40}$	$8\overline{)16}$	$3\overline{)9}$
$9\overline{)45}$	$1\overline{)6}$	$4\overline{)8}$	$7\overline{)63}$	$10\overline{)40}$	$2\overline{)4}$
$9\overline{)9}$	$11\overline{)33}$	$3\overline{)21}$	$7\overline{)7}$	$12\overline{)48}$	$8\overline{)80}$
$1\overline{)3}$	$5\overline{)15}$	$2\overline{)12}$	$9\overline{)18}$	$4\overline{)32}$	$8\overline{)48}$
$6\overline{)60}$	$3\overline{)27}$	$7\overline{)14}$	$2\overline{)16}$	$4\overline{)24}$	$9\overline{)36}$
$6\overline{)18}$	$8\overline{)32}$	$7\overline{)28}$	$8\overline{)72}$	$3\overline{)24}$	$5\overline{)35}$
$2\overline{)6}$	$12\overline{)24}$	$9\overline{)54}$	$6\overline{)30}$	$8\overline{)40}$	$3\overline{)18}$
$10\overline{)50}$	$7\overline{)42}$	$2\overline{)10}$	$9\overline{)63}$	$5\overline{)30}$	$4\overline{)28}$
$6\overline{)42}$	$7\overline{)49}$	$8\overline{)24}$	$3\overline{)12}$	$5\overline{)25}$	$1\overline{)8}$

And God saw the light, that it was good, and God divided the light from the darkness.

Gen. 1:4

© Edwin C. Myers 1985, 1990   **CalcuLadder**®   Level 42: Division   4 minutes

My name is _____

Today is _____

$$\begin{array}{r}32\\ \times 23\\ \hline\end{array} \qquad \begin{array}{r}51\\ \times 14\\ \hline\end{array} \qquad \begin{array}{r}36\\ \times 71\\ \hline\end{array}$$

$$\begin{array}{r}24\\ \times 25\\ \hline\end{array} \qquad \begin{array}{r}72\\ \times 64\\ \hline\end{array} \qquad \begin{array}{r}83\\ \times 45\\ \hline\end{array}$$

$$\begin{array}{r}93\\ \times 34\\ \hline\end{array} \qquad \begin{array}{r}12\\ \times 72\\ \hline\end{array}$$

$$\begin{array}{r}67\\ \times 56\\ \hline\end{array} \qquad \begin{array}{r}44\\ \times 28\\ \hline\end{array}$$

. . . a great multitude, which no man could number, of all nations, and kindreds, and people, and tongues, stood before the throne, and before the Lamb . . . Rev. 7:9

© Edwin C. Myers 1985, 1990 CalcuLadder® Level 43: Multiplication 4 minutes

My name is _____

Today is _____

```
   32              51              36
  x23             x14             x71


   24              72              83
  x25             x64             x45


        93              12
       x34             x72


        67              44
       x56             x28
```

. . . a great multitude, which no man could number, of all nations, and kindreds, and people, and tongues, stood before the throne, and before the Lamb . . .   Rev. 7:9

© Edwin C. Myers 1985, 1990   **CalcuLadder®**   Level 43: Multiplication   4 minutes

My name is _____

Today is _____

```
  32        51        36
 x23       x14       x71


  24        72        83
 x25       x64       x45


  93        12
 x34       x72


  67        44
 x56       x28
```

. . . a great multitude, which no man could number, of all nations, and kindreds, and people, and tongues, stood before the throne, and before the Lamb . . . Rev. 7:9

© Edwin C. Myers 1985, 1990 CalcuLadder® Level 43: Multiplication 4 minutes

My name is _____

Today is _____

```
    32              51              36
   x23             x14             x71
```

```
    24              72              83
   x25             x64             x45
```

```
    93              12
   x34             x72
```

```
    67              44
   x56             x28
```

. . . a great multitude, which no man could number, of all nations, and kindreds, and people, and tongues, stood before the throne, and before the Lamb . . . Rev. 7:9

© Edwin C. Myers 1985, 1990   CalcuLadder®   Level 43: Multiplication   4 minutes

My name is _____

Today is _____

32	51	36
x23	x14	x71

24	72	83
x25	x64	x45

93		12
x34		x72

67		44
x56		x28

. . . a great multitude, which no man could number, of all nations, and kindreds, and people, and tongues, stood before the throne, and before the Lamb . . . Rev. 7:9

© Edwin C. Myers 1985, 1990 CalcuLadder® Level 43: Multiplication 4 minutes

My name is _____

Today is _____

32	51	36
x23	x14	x71

24	72	83
x25	x64	x45

93		12
x34		x72

67		44
x56		x28

. . . a great multitude, which no man could number, of all nations, and kindreds, and people, and tongues, stood before the throne, and before the Lamb . . .   Rev. 7:9

© Edwin C. Myers 1985, 1990   CalcuLadder®   Level 43: Multiplication   4 minutes

My name is _____

Today is _____

32	51	36
x23	x14	x71

24	72	83
x25	x64	x45

93		12
x34		x72

67		44
x56		x28

. . . a great multitude, which no man could number, of all nations, and kindreds, and people, and tongues, stood before the throne, and before the Lamb . . .   Rev. 7:9

© Edwin C. Myers 1985, 1990   CalcuLadder®   Level 43: Multiplication   4 minutes

My name is _____

Today is _____

```
  32        51        36
 x23       x14       x71
```

```
  24        72        83
 x25       x64       x45
```

```
  93                  12
 x34                 x72
```

```
  67                  44
 x56                 x28
```

. . . a great multitude, which no man could number, of all nations, and kindreds, and people, and tongues, stood before the throne, and before the Lamb . . .   Rev. 7:9

© Edwin C. Myers 1985, 1990   CalcuLadder®   Level 43: Multiplication   4 minutes

My name is _____

Today is _____

    32   51   36
x23   x14   x71

    24   72   83
x25   x64   x45

    93   12
x34   x72

    67   44
x56   x28

. . . a great multitude, which no man could number, of all nations, and kindreds, and people, and tongues, stood before the throne, and before the Lamb . . .   Rev. 7:9

© Edwin C. Myers 1985, 1990   CalcuLadder®   Level 43: Multiplication   4 minutes

My name is _____

Today is _____

```
   32          51          36
  x23         x14         x71
```

```
   24          72          83
  x25         x64         x45
```

```
   93                      12
  x34                     x72
```

```
   67                      44
  x56                     x28
```

. . . a great multitude, which no man could number, of all nations, and kindreds, and people, and tongues, stood before the throne, and before the Lamb . . .   Rev. 7:9

© Edwin C. Myers 1985, 1990   **CalcuLadder**®   Level 43: Multiplication   4 minutes

My name is _____

Today is _____

```
   32              51              36
  x23             x14             x71


   24              72              83
  x25             x64             x45


                   93              12
                  x34             x72


                   67              44
                  x56             x28
```

... a great multitude, which no man could number, of all nations, and kindreds, and people, and tongues, stood before the throne, and before the Lamb... Rev. 7:9

© Edwin C. Myers 1985, 1990   CalcuLadder®   Level 43: Multiplication   4 minutes

My name is _____

Today is _____

32	51	36
x23	x14	x71

24	72	83
x25	x64	x45

93		12
x34		x72

67		44
x56		x28

. . . a great multitude, which no man could number, of all nations, and kindreds, and people, and tongues, stood before the throne, and before the Lamb . . .   Rev. 7:9

© Edwin C. Myers 1985, 1990   CalcuLadder®   Level 43: Multiplication   4 minutes

My name is _____

Today is _____

$4\overline{)12}$	$9\overline{)90}$	$2\overline{)18}$	$5\overline{)10}$	$12\overline{)36}$	$7\overline{)56}$	$10\overline{)70}$
$3\overline{)6}$	$8\overline{)56}$	$1\overline{)7}$	$6\overline{)54}$	$11\overline{)22}$	$4\overline{)20}$	$9\overline{)72}$
$2\overline{)14}$	$5\overline{)25}$	$12\overline{)60}$	$7\overline{)70}$	$10\overline{)90}$	$3\overline{)12}$	$8\overline{)24}$
$7\overline{)49}$	$6\overline{)42}$	$4\overline{)28}$	$9\overline{)54}$	$2\overline{)10}$	$5\overline{)30}$	$7\overline{)42}$
$10\overline{)50}$	$3\overline{)18}$	$8\overline{)40}$	$1\overline{)8}$	$6\overline{)30}$	$11\overline{)44}$	$4\overline{)36}$
$9\overline{)36}$	$2\overline{)6}$	$5\overline{)35}$	$12\overline{)84}$	$7\overline{)28}$	$8\overline{)72}$	$3\overline{)24}$
$8\overline{)32}$	$10\overline{)20}$	$6\overline{)18}$	$4\overline{)24}$	$9\overline{)18}$	$2\overline{)16}$	$5\overline{)50}$
$7\overline{)14}$	$3\overline{)27}$	$10\overline{)30}$	$8\overline{)48}$	$1\overline{)4}$	$6\overline{)60}$	$11\overline{)88}$
$4\overline{)32}$	$9\overline{)81}$	$2\overline{)12}$	$5\overline{)15}$	$12\overline{)72}$	$7\overline{)21}$	$6\overline{)36}$
$3\overline{)15}$	$8\overline{)64}$	$5\overline{)45}$	$6\overline{)24}$	$11\overline{)66}$	$4\overline{)16}$	$9\overline{)63}$

. . . and there Joshua divided the land to the children of Israel. . . Josh. 18:10

© Edwin C. Myers 1985, 1990   CalcuLadder®   Level 44: Division   4 minutes

My name is _____

Today is _____

4)12	9)90	2)18	5)10	12)36	7)56	10)70
3)6	8)56	1)7	6)54	11)22	4)20	9)72
2)14	5)25	12)60	7)70	10)90	3)12	8)24
7)49	6)42	4)28	9)54	2)10	5)30	7)42
10)50	3)18	8)40	1)8	6)30	11)44	4)36
9)36	2)6	5)35	12)84	7)28	8)72	3)24
8)32	10)20	6)18	4)24	9)18	2)16	5)50
7)14	3)27	10)30	8)48	1)4	6)60	11)88
4)32	9)81	2)12	5)15	12)72	7)21	6)36
3)15	8)64	5)45	6)24	11)66	4)16	9)63

. . . and there Joshua divided the land to the children of Israel. . .   Josh. 18:10

© Edwin C. Myers 1985, 1990   **CalcuLadder®**   Level 44:   Division   4 minutes

My name is _____

Today is _____

4)12	9)90	2)18	5)10	12)36	7)56	10)70
3)6	8)56	1)7	6)54	11)22	4)20	9)72
2)14	5)25	12)60	7)70	10)90	3)12	8)24
7)49	6)42	4)28	9)54	2)10	5)30	7)42
10)50	3)18	8)40	1)8	6)30	11)44	4)36
9)36	2)6	5)35	12)84	7)28	8)72	3)24
8)32	10)20	6)18	4)24	9)18	2)16	5)50
7)14	3)27	10)30	8)48	1)4	6)60	11)88
4)32	9)81	2)12	5)15	12)72	7)21	6)36
3)15	8)64	5)45	6)24	11)66	4)16	9)63

. . . and there Joshua divided the land to the children of Israel. . . Josh. 18:10

© Edwin C. Myers 1985, 1990   CalcuLadder®   Level 44: Division   4 minutes

My name is _____

Today is _____

4)12	9)90	2)18	5)10	12)36	7)56	10)70
3)6	8)56	1)7	6)54	11)22	4)20	9)72
2)14	5)25	12)60	7)70	10)90	3)12	8)24
7)49	6)42	4)28	9)54	2)10	5)30	7)42
10)50	3)18	8)40	1)8	6)30	11)44	4)36
9)36	2)6	5)35	12)84	7)28	8)72	3)24
8)32	10)20	6)18	4)24	9)18	2)16	5)50
7)14	3)27	10)30	8)48	1)4	6)60	11)88
4)32	9)81	2)12	5)15	12)72	7)21	6)36
3)15	8)64	5)45	6)24	11)66	4)16	9)63

. . . and there Joshua divided the land to the children of Israel. . . Josh. 18:10

© Edwin C. Myers 1985, 1990   CalcuLadder®   Level 44: Division   4 minutes

My name is _____

Today is _____

$4\overline{)12}$ $9\overline{)90}$ $2\overline{)18}$ $5\overline{)10}$ $12\overline{)36}$ $7\overline{)56}$ $10\overline{)70}$

$3\overline{)6}$ $8\overline{)56}$ $1\overline{)7}$ $6\overline{)54}$ $11\overline{)22}$ $4\overline{)20}$ $9\overline{)72}$

$2\overline{)14}$ $5\overline{)25}$ $12\overline{)60}$ $7\overline{)70}$ $10\overline{)90}$ $3\overline{)12}$ $8\overline{)24}$

$7\overline{)49}$ $6\overline{)42}$ $4\overline{)28}$ $9\overline{)54}$ $2\overline{)10}$ $5\overline{)30}$ $7\overline{)42}$

$10\overline{)50}$ $3\overline{)18}$ $8\overline{)40}$ $1\overline{)8}$ $6\overline{)30}$ $11\overline{)44}$ $4\overline{)36}$

$9\overline{)36}$ $2\overline{)6}$ $5\overline{)35}$ $12\overline{)84}$ $7\overline{)28}$ $8\overline{)72}$ $3\overline{)24}$

$8\overline{)32}$ $10\overline{)20}$ $6\overline{)18}$ $4\overline{)24}$ $9\overline{)18}$ $2\overline{)16}$ $5\overline{)50}$

$7\overline{)14}$ $3\overline{)27}$ $10\overline{)30}$ $8\overline{)48}$ $1\overline{)4}$ $6\overline{)60}$ $11\overline{)88}$

$4\overline{)32}$ $9\overline{)81}$ $2\overline{)12}$ $5\overline{)15}$ $12\overline{)72}$ $7\overline{)21}$ $6\overline{)36}$

$3\overline{)15}$ $8\overline{)64}$ $5\overline{)45}$ $6\overline{)24}$ $11\overline{)66}$ $4\overline{)16}$ $9\overline{)63}$

. . . and there Joshua divided the land to the children of Israel. . .   Josh. 18:10

© Edwin C. Myers 1985,1990   **CalcuLadder®**   Level 44: Division   4 minutes

My name is _____

Today is _____

4)12	9)90	2)18	5)10	12)36	7)56	10)70
3)6	8)56	1)7	6)54	11)22	4)20	9)72
2)14	5)25	12)60	7)70	10)90	3)12	8)24
7)49	6)42	4)28	9)54	2)10	5)30	7)42
10)50	3)18	8)40	1)8	6)30	11)44	4)36
9)36	2)6	5)35	12)84	7)28	8)72	3)24
8)32	10)20	6)18	4)24	9)18	2)16	5)50
7)14	3)27	10)30	8)48	1)4	6)60	11)88
4)32	9)81	2)12	5)15	12)72	7)21	6)36
3)15	8)64	5)45	6)24	11)66	4)16	9)63

. . . and there Joshua divided the land to the children of Israel. . . Josh. 18:10

© Edwin C. Myers 1985, 1990   **CalcuLadder®**   Level 44: Division   4 minutes

My name is _____

Today is _____

4)12	9)90	2)18	5)10	12)36	7)56	10)70
3)6	8)56	1)7	6)54	11)22	4)20	9)72
2)14	5)25	12)60	7)70	10)90	3)12	8)24
7)49	6)42	4)28	9)54	2)10	5)30	7)42
10)50	3)18	8)40	1)8	6)30	11)44	4)36
9)36	2)6	5)35	12)84	7)28	8)72	3)24
8)32	10)20	6)18	4)24	9)18	2)16	5)50
7)14	3)27	10)30	8)48	1)4	6)60	11)88
4)32	9)81	2)12	5)15	12)72	7)21	6)36
3)15	8)64	5)45	6)24	11)66	4)16	9)63

. . . and there Joshua divided the land to the children of Israel. . .   Josh. 18:10

© Edwin C. Myers 1985,1990   CalcuLadder®   Level 44: Division   4 minutes

My name is _____

Today is _____

4)12	9)90	2)18	5)10	12)36	7)56	10)70
3)6	8)56	1)7	6)54	11)22	4)20	9)72
2)14	5)25	12)60	7)70	10)90	3)12	8)24
7)49	6)42	4)28	9)54	2)10	5)30	7)42
10)50	3)18	8)40	1)8	6)30	11)44	4)36
9)36	2)6	5)35	12)84	7)28	8)72	3)24
8)32	10)20	6)18	4)24	9)18	2)16	5)50
7)14	3)27	10)30	8)48	1)4	6)60	11)88
4)32	9)81	2)12	5)15	12)72	7)21	6)36
3)15	8)64	5)45	6)24	11)66	4)16	9)63

. . . and there Joshua divided the land to the children of Israel. . . Josh. 18:10

© Edwin C. Myers 1985, 1990   **CalcuLadder**®   Level 44: Division   4 minutes

My name is _____

Today is _____

4)12   9)90   2)18   5)10   12)36   7)56   10)70

3)6   8)56   1)7   6)54   11)22   4)20   9)72

2)14   5)25   12)60 7)70   10)90   3)12   8)24

7)49   6)42   4)28   9)54   2)10   5)30   7)42

10)50 3)18   8)40   1)8   6)30   11)44 4)36

9)36   2)6   5)35   12)84 7)28   8)72   3)24

8)32   10)20 6)18   4)24   9)18   2)16   5)50

7)14   3)27   10)30 8)48   1)4   6)60   11)88

4)32   9)81   2)12   5)15   12)72   7)21   6)36

3)15   8)64   5)45   6)24   11)66   4)16   9)63

. . . and there Joshua divided the land to the children of Israel. . . Josh. 18:10

© Edwin C. Myers 1985, 1990   CalcuLadder®   Level 44: Division   4 minutes

My name is _____

Today is _____

4)12	9)90	2)18	5)10	12)36	7)56	10)70
3)6	8)56	1)7	6)54	11)22	4)20	9)72
2)14	5)25	12)60	7)70	10)90	3)12	8)24
7)49	6)42	4)28	9)54	2)10	5)30	7)42
10)50	3)18	8)40	1)8	6)30	11)44	4)36
9)36	2)6	5)35	12)84	7)28	8)72	3)24
8)32	10)20	6)18	4)24	9)18	2)16	5)50
7)14	3)27	10)30	8)48	1)4	6)60	11)88
4)32	9)81	2)12	5)15	12)72	7)21	6)36
3)15	8)64	5)45	6)24	11)66	4)16	9)63

. . . and there Joshua divided the land to the children of Israel. . . Josh. 18:10

© Edwin C. Myers 1985,1990   **CalcuLadder**®   Level 44: Division   4 minutes

My name is _____

Today is _____

4)12	9)90	2)18	5)10	12)36	7)56	10)70
3)6	8)56	1)7	6)54	11)22	4)20	9)72
2)14	5)25	12)60	7)70	10)90	3)12	8)24
7)49	6)42	4)28	9)54	2)10	5)30	7)42
10)50	3)18	8)40	1)8	6)30	11)44	4)36
9)36	2)6	5)35	12)84	7)28	8)72	3)24
8)32	10)20	6)18	4)24	9)18	2)16	5)50
7)14	3)27	10)30	8)48	1)4	6)60	11)88
4)32	9)81	2)12	5)15	12)72	7)21	6)36
3)15	8)64	5)45	6)24	11)66	4)16	9)63

. . . and there Joshua divided the land to the children of Israel. . .   Josh. 18:10

© Edwin C. Myers 1985, 1990   **CalcuLadder®**   Level 44: Division   4 minutes

My name is _____

Today is _____

4)12	9)90	2)18	5)10	12)36	7)56	10)70
3)6	8)56	1)7	6)54	11)22	4)20	9)72
2)14	5)25	12)60	7)70	10)90	3)12	8)24
7)49	6)42	4)28	9)54	2)10	5)30	7)42
10)50	3)18	8)40	1)8	6)30	11)44	4)36
9)36	2)6	5)35	12)84	7)28	8)72	3)24
8)32	10)20	6)18	4)24	9)18	2)16	5)50
7)14	3)27	10)30	8)48	1)4	6)60	11)88
4)32	9)81	2)12	5)15	12)72	7)21	6)36
3)15	8)64	5)45	6)24	11)66	4)16	9)63

. . . and there Joshua divided the land to the children of Israel. . . Josh. 18:10

© Edwin C. Myers 1985, 1990   **CalcuLadder**®   Level 44: Division   4 minutes

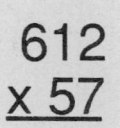

My name is _____

Today is _____

```
   34          52          96
  x27         x61         x43
```

```
  612          87         392
  x 57        x58         x 36
```

```
  407                     325
 x248                    x673
```

The Lord God...make you a thousand times so many as you are, and bless you... Deut. 1:11

© Edwin C. Myers 1985,1990   CalcuLadder®   Level 45: Multiplication   5 minutes

My name is _____

Today is _____

```
   34          52          96
  x27         x61         x43


  612          87         392
  x 57        x58         x 36


  407                     325
 x248                    x673
```

The Lord God...make you a thousand times so many as you are, and bless you... Deut. 1:11

© Edwin C. Myers 1985, 1990   **CalcuLadder**®   Level 45: Multiplication   5 minutes

My name is _____

Today is _____

```
   34            52            96
  x27           x61           x43
```

```
  612            87           392
  x 57          x58           x 36
```

```
  407                         325
 x248                        x673
```

The Lord God...make you a thousand times so many as you are, and bless you... Deut. 1:11

© Edwin C. Myers 1985, 1990   **CalcuLadder**®   Level 45: Multiplication   5 minutes

My name is _____

Today is _____

```
   34          52          96
  x27         x61         x43
```

```
  612          87         392
  x 57        x58         x 36
```

```
   407                    325
  x248                   x673
```

The Lord God...make you a thousand times so many as you are, and bless you... Deut. 1:11

© Edwin C. Myers 1985, 1990   CalcuLadder®   Level 45: Multiplication   5 minutes

My name is _____

Today is _____

34	52	96
x27	x61	x43

612	87	392
x 57	x58	x 36

407		325
x248		x673

The Lord God...make you a thousand times so many as you are, and bless you... Deut. 1:11

© Edwin C. Myers 1985, 1990   **CalcuLadder®**   Level 45: Multiplication   5 minutes

My name is _____

Today is _____

34 x27	52 x61	96 x43
612 x 57	87 x58	392 x 36
407 x248		325 x673

The Lord God...make you a thousand times so many as you are, and bless you... Deut. 1:11

© Edwin C. Myers 1985, 1990   CalcuLadder®   Level 45: Multiplication   5 minutes

My name is _____

Today is _____

34 x27	52 x61	96 x43
612 x 57	87 x58	392 x 36
407 x248	325 x673	

The Lord God...make you a thousand times so many as you are, and bless you... Deut. 1:11

© Edwin C. Myers 1985,1990   **CalcuLadder**®   Level 45: Multiplication   5 minutes

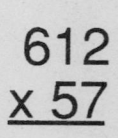

My name is _____

Today is _____

$$\begin{array}{r}34\\ \times 27\\ \hline\end{array} \qquad \begin{array}{r}52\\ \times 61\\ \hline\end{array} \qquad \begin{array}{r}96\\ \times 43\\ \hline\end{array}$$

$$\begin{array}{r}612\\ \times\ 57\\ \hline\end{array} \qquad \begin{array}{r}87\\ \times 58\\ \hline\end{array} \qquad \begin{array}{r}392\\ \times\ 36\\ \hline\end{array}$$

$$\begin{array}{r}407\\ \times 248\\ \hline\end{array} \qquad \begin{array}{r}325\\ \times 673\\ \hline\end{array}$$

The Lord God...make you a thousand times so many as you are, and bless you... Deut. 1:11

© Edwin C. Myers 1985, 1990   **CalcuLadder®**   Level 45: Multiplication   5 minutes

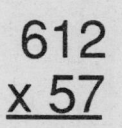

My name is _____

Today is _____

```
   34          52          96
  x27         x61         x43


  612          87         392
  x 57        x58         x 36


  407                     325
 x248                    x673
```

The Lord God...make you a thousand times so many as you are, and bless you... Deut. 1:11

© Edwin C. Myers 1985, 1990   CalcuLadder®   Level 45: Multiplication   5 minutes

My name is _____

Today is _____

```
    34         52         96
   x27        x61        x43


   612         87        392
   x 57       x58        x 36


   407                   325
  x248                  x673
```

The Lord God...make you a thousand times so many as you are, and bless you... Deut. 1:11

© Edwin C. Myers 1985,1990   **CalcuLadder®**   Level 45: Multiplication   5 minutes

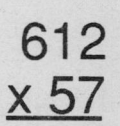

My name is _____

Today is _____

$$\begin{array}{r}34\\ \times 27\\ \hline\end{array}\qquad\begin{array}{r}52\\ \times 61\\ \hline\end{array}\qquad\begin{array}{r}96\\ \times 43\\ \hline\end{array}$$

$$\begin{array}{r}612\\ \times\ 57\\ \hline\end{array}\qquad\begin{array}{r}87\\ \times 58\\ \hline\end{array}\qquad\begin{array}{r}392\\ \times\ 36\\ \hline\end{array}$$

$$\begin{array}{r}407\\ \times 248\\ \hline\end{array}\qquad\begin{array}{r}325\\ \times 673\\ \hline\end{array}$$

The Lord God...make you a thousand times so many as you are, and bless you... Deut. 1:11

© Edwin C. Myers 1985, 1990   **CalcuLadder®**   Level 45: Multiplication   5 minutes

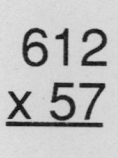

My name is _____

Today is _____

```
   34         52         96
  x27        x61        x43


  612         87        392
  x 57       x58        x 36


  407                   325
 x248                  x673
```

The Lord God...make you a thousand times so many as you are, and bless you... Deut. 1:11

© Edwin C. Myers 1985, 1990   **CalcuLadder®**   Level 45: Multiplication   5 minutes

My name is _____

Today is _____

$$6)\overline{14} \quad \begin{array}{r} 2\ r2 \\ 12 \\ \ 2 \end{array}$$

2)$\overline{9}$ 4)$\overline{26}$ 5)$\overline{8}$ 6)$\overline{34}$

3)$\overline{10}$ 7)$\overline{11}$ 9)$\overline{29}$ 5)$\overline{44}$

3)$\overline{4}$ 10)$\overline{66}$ 7)$\overline{59}$ 8)$\overline{19}$

4)$\overline{35}$ 9)$\overline{42}$ 8)$\overline{75}$ 6)$\overline{49}$

2)$\overline{5}$ 3)$\overline{23}$ 12)$\overline{49}$ 7)$\overline{31}$

10)$\overline{44}$ 9)$\overline{10}$ 4)$\overline{11}$ 5)$\overline{12}$

. . . and the two fishes divided he among them all.   Mark 6:41

© Edwin C. Myers 1985, 1990   CalcuLadder®   Level 46: Division   5 minutes

My name is _____

Today is _____

$$\begin{array}{r}2\ r2\\6\overline{)14}\\12\\\hline 2\end{array}$$ $2\overline{)9}$ $4\overline{)26}$ $5\overline{)8}$ $6\overline{)34}$

$3\overline{)10}$ $7\overline{)11}$ $9\overline{)29}$ $5\overline{)44}$

$3\overline{)4}$ $10\overline{)66}$ $7\overline{)59}$ $8\overline{)19}$

$4\overline{)35}$ $9\overline{)42}$ $8\overline{)75}$ $6\overline{)49}$

$2\overline{)5}$ $3\overline{)23}$ $12\overline{)49}$ $7\overline{)31}$

$10\overline{)44}$ $9\overline{)10}$ $4\overline{)11}$ $5\overline{)12}$

. . . and the two fishes divided he among them all.   Mark 6:41

© Edwin C. Myers 1985, 1990   **CalcuLadder**®   Level 46: Division   5 minutes

My name is _____

Today is _____

$$6)\overline{14}^{2\,r2}$$
$$\underline{12}$$
$$2$$

2)9   4)26   5)8   6)34

3)10   7)11   9)29   5)44

3)4   10)66   7)59   8)19

4)35   9)42   8)75   6)49

2)5   3)23   12)49   7)31

10)44   9)10   4)11   5)12

. . . and the two fishes divided he among them all.   Mark 6:41

© Edwin C. Myers 1985, 1990   CalcuLadder®   Level 46: Division   5 minutes

My name is _____

Today is _____

$$\begin{array}{r}2\ r2\\6\overline{)14}\\12\\\hline 2\end{array}$$

2)9   4)26   5)8   6)34

3)10   7)11   9)29   5)44

3)4   10)66 7)59   8)19

4)35   9)42   8)75   6)49

2)5   3)23   12)49 7)31

10)44 9)10   4)11   5)12

. . . and the two fishes divided he among them all.   Mark 6:41

© Edwin C. Myers 1985, 1990   CalcuLadder®   Level 46: Division   5 minutes

My name is _____

Today is _____

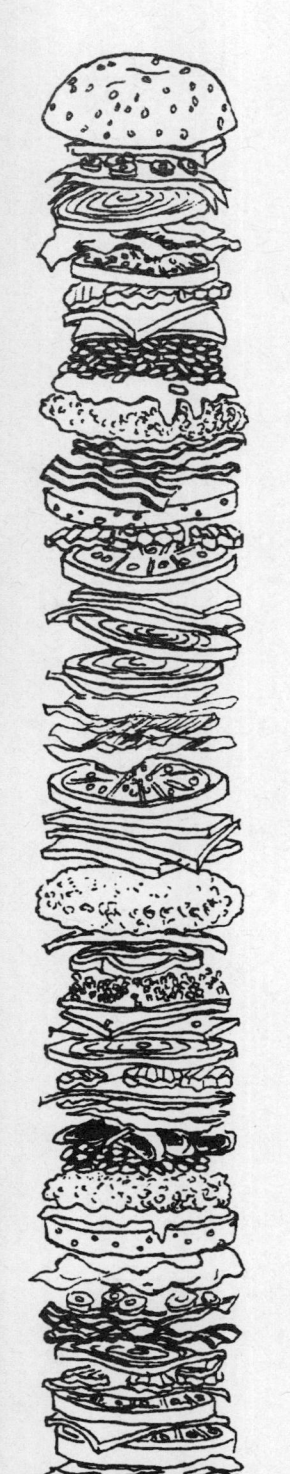

$$\begin{array}{r}2\ r2\\6\overline{)14}\\\underline{12}\\2\end{array}$$ $2\overline{)9}$ $4\overline{)26}$ $5\overline{)8}$ $6\overline{)34}$

$3\overline{)10}$ $7\overline{)11}$ $9\overline{)29}$ $5\overline{)44}$

$3\overline{)4}$ $10\overline{)66}$ $7\overline{)59}$ $8\overline{)19}$

$4\overline{)35}$ $9\overline{)42}$ $8\overline{)75}$ $6\overline{)49}$

$2\overline{)5}$ $3\overline{)23}$ $12\overline{)49}$ $7\overline{)31}$

$10\overline{)44}$ $9\overline{)10}$ $4\overline{)11}$ $5\overline{)12}$

. . . and the two fishes divided he among them all.   Mark 6:41

© Edwin C. Myers 1985, 1990   CalcuLadder®   Level 46: Division   5 minutes

My name is _____

Today is _____

$$\begin{array}{r}2\ r2\\6\overline{)14}\\12\\\hline 2\end{array}$$

$2\overline{)9}$ $4\overline{)26}$ $5\overline{)8}$ $6\overline{)34}$

$3\overline{)10}$ $7\overline{)11}$ $9\overline{)29}$ $5\overline{)44}$

$3\overline{)4}$ $10\overline{)66}$ $7\overline{)59}$ $8\overline{)19}$

$4\overline{)35}$ $9\overline{)42}$ $8\overline{)75}$ $6\overline{)49}$

$2\overline{)5}$ $3\overline{)23}$ $12\overline{)49}$ $7\overline{)31}$

$10\overline{)44}$ $9\overline{)10}$ $4\overline{)11}$ $5\overline{)12}$

. . . and the two fishes divided he among them all.   Mark 6:41

© Edwin C. Myers 1985, 1990   **CalcuLadder**®   Level 46: Division   5 minutes

My name is _____

Today is _____

$$\begin{array}{r}2\ r2\\6\overline{)14}\\12\\\hline 2\end{array}$$ $2\overline{)9}$ $4\overline{)26}$ $5\overline{)8}$ $6\overline{)34}$

$3\overline{)10}$ $7\overline{)11}$ $9\overline{)29}$ $5\overline{)44}$

$3\overline{)4}$ $10\overline{)66}$ $7\overline{)59}$ $8\overline{)19}$

$4\overline{)35}$ $9\overline{)42}$ $8\overline{)75}$ $6\overline{)49}$

$2\overline{)5}$ $3\overline{)23}$ $12\overline{)49}$ $7\overline{)31}$

$10\overline{)44}$ $9\overline{)10}$ $4\overline{)11}$ $5\overline{)12}$

. . . and the two fishes divided he among them all.   Mark 6:41

© Edwin C. Myers 1985, 1990   CalcuLadder®   Level 46: Division   5 minutes

My name is _____

Today is _____

$$\begin{array}{r} 2\ r2 \\ 6\overline{)14} \\ \underline{12} \\ 2 \end{array}$$

$2\overline{)9}$ $4\overline{)26}$ $5\overline{)8}$ $6\overline{)34}$

$3\overline{)10}$ $7\overline{)11}$ $9\overline{)29}$ $5\overline{)44}$

$3\overline{)4}$ $10\overline{)66}$ $7\overline{)59}$ $8\overline{)19}$

$4\overline{)35}$ $9\overline{)42}$ $8\overline{)75}$ $6\overline{)49}$

$2\overline{)5}$ $3\overline{)23}$ $12\overline{)49}$ $7\overline{)31}$

$10\overline{)44}$ $9\overline{)10}$ $4\overline{)11}$ $5\overline{)12}$

. . . and the two fishes divided he among them all.   Mark 6:41

© Edwin C. Myers 1985, 1990   CalcuLadder®   Level 46: Division   5 minutes

My name is _____

Today is _____

$$6\overline{)14} \quad 2\,r2$$
$$\underline{12}$$
$$2$$

$2\overline{)9}$ $4\overline{)26}$ $5\overline{)8}$ $6\overline{)34}$

$3\overline{)10}$ $7\overline{)11}$ $9\overline{)29}$ $5\overline{)44}$

$3\overline{)4}$ $10\overline{)66}$ $7\overline{)59}$ $8\overline{)19}$

$4\overline{)35}$ $9\overline{)42}$ $8\overline{)75}$ $6\overline{)49}$

$2\overline{)5}$ $3\overline{)23}$ $12\overline{)49}$ $7\overline{)31}$

$10\overline{)44}$ $9\overline{)10}$ $4\overline{)11}$ $5\overline{)12}$

. . . and the two fishes divided he among them all.   Mark 6:41

© Edwin C. Myers 1985, 1990   CalcuLadder®   Level 46: Division   5 minutes

My name is _____

Today is _____

$$\begin{array}{r}2\ r2\\6\overline{)14}\\12\\\hline 2\end{array}$$

2)9   4)26   5)8   6)34

3)10   7)11   9)29   5)44

3)4   10)66   7)59   8)19

4)35   9)42   8)75   6)49

2)5   3)23   12)49   7)31

10)44   9)10   4)11   5)12

. . . and the two fishes divided he among them all.   Mark 6:41

© Edwin C. Myers 1985, 1990   CalcuLadder®   Level 46: Division   5 minutes

My name is _____

Today is _____

$$\begin{array}{r} 2\ r2 \\ 6\overline{)14} \\ \underline{12} \\ 2 \end{array}$$ $2\overline{)9}$ $4\overline{)26}$ $5\overline{)8}$ $6\overline{)34}$

$3\overline{)10}$ $7\overline{)11}$ $9\overline{)29}$ $5\overline{)44}$

$3\overline{)4}$ $10\overline{)66}$ $7\overline{)59}$ $8\overline{)19}$

$4\overline{)35}$ $9\overline{)42}$ $8\overline{)75}$ $6\overline{)49}$

$2\overline{)5}$ $3\overline{)23}$ $12\overline{)49}$ $7\overline{)31}$

$10\overline{)44}$ $9\overline{)10}$ $4\overline{)11}$ $5\overline{)12}$

. . . and the two fishes divided he among them all.   Mark 6:41

© Edwin C. Myers 1985, 1990   CalcuLadder®   Level 46: Division   5 minutes

My name is _____

Today is _____

$$\begin{array}{r}2\ r2\\6\overline{)14}\\12\\\hline 2\end{array}$$ $2\overline{)9}$ $4\overline{)26}$ $5\overline{)8}$ $6\overline{)34}$

$3\overline{)10}$ $7\overline{)11}$ $9\overline{)29}$ $5\overline{)44}$

$3\overline{)4}$ $10\overline{)66}$ $7\overline{)59}$ $8\overline{)19}$

$4\overline{)35}$ $9\overline{)42}$ $8\overline{)75}$ $6\overline{)49}$

$2\overline{)5}$ $3\overline{)23}$ $12\overline{)49}$ $7\overline{)31}$

$10\overline{)44}$ $9\overline{)10}$ $4\overline{)11}$ $5\overline{)12}$

. . . and the two fishes divided he among them all.   Mark 6:41

© Edwin C. Myers 1985, 1990   CalcuLadder®   Level 46: Division   5 minutes

My name is _____

Today is _____

534	276	819
x111	x331	x545

625	438	579
x713	x306	x401

623	7100	356
x 10	x 300	x 2500

Now he who supplies seed to the sower and bread for food, will supply and multiply your seed for sowing and increase the harvest of your righteousness. 2 Cor. 9:10

© Edwin C. Myers 1985, 1990   **CalcuLadder**®   Level 47: Multiplication   5 minutes

My name is _____

Today is _____

534	276	819
x111	x331	x545

625	438	579
x713	x306	x401

623	7100	356
x 10	x 300	x 2500

Now he who supplies seed to the sower and bread for food, will supply and multiply your seed for sowing and increase the harvest of your righteousness. 2 Cor. 9:10

© Edwin C. Myers 1985, 1990   CalcuLadder®   Level 47: Multiplication   5 minutes

My name is _____

Today is _____

```
   534          276          819
  x111         x331         x545
```

```
   625          438          579
  x713         x306         x401
```

```
   623         7100          356
  x  10        x 300        x 2500
```

Now he who supplies seed to the sower and bread for food, will supply and multiply your seed for sowing and increase the harvest of your righteousness. 2 Cor. 9:10

© Edwin C. Myers 1985,1990   **CalcuLadder®**   Level 47: Multiplication   5 minutes

My name is _____

Today is _____

534	276	819
x111	x331	x545

625	438	579
x713	x306	x401

623	7100	356
x 10	x 300	x 2500

Now he who supplies seed to the sower and bread for food, will supply and multiply your seed for sowing and increase the harvest of your righteousness. 2 Cor. 9:10

© Edwin C. Myers 1985, 1990   **CalcuLadder**®   Level 47: Multiplication   5 minutes

My name is _____

Today is _____

```
  534          276          819
x111         x331         x545
```

```
  625          438          579
x713         x306         x401
```

```
  623         7100          356
x  10        x 300        x 2500
```

Now he who supplies seed to the sower and bread for food, will supply and multiply your seed for sowing and increase the harvest of your righteousness. 2 Cor. 9:10

© Edwin C. Myers 1985, 1990   **CalcuLadder®**   Level 47: Multiplication   5 minutes

My name is _____

Today is _____

```
   534          276          819
  x111         x331         x545
```

```
   625          438          579
  x713         x306         x401
```

```
   623         7100          356
  x 10        x 300       x 2500
```

Now he who supplies seed to the sower and bread for food, will supply and multiply your seed for sowing and increase the harvest of your righteousness. 2 Cor. 9:10

© Edwin C. Myers 1985,1990   **CalcuLadder®**   Level 47: Multiplication   5 minutes

My name is _____

Today is _____

534	276	819
x111	x331	x545

625	438	579
x713	x306	x401

623	7100	356
x 10	x 300	x 2500

Now he who supplies seed to the sower and bread for food, will supply and multiply your seed for sowing and increase the harvest of your righteousness. 2 Cor. 9:10

© Edwin C. Myers 1985, 1990   CalcuLadder®   Level 47: Multiplication   5 minutes

My name is _____

Today is _____

534	276	819
x111	x331	x545

625	438	579
x713	x306	x401

623	7100	356
x 10	x 300	x 2500

Now he who supplies seed to the sower and bread for food, will supply and multiply your seed for sowing and increase the harvest of your righteousness. 2 Cor. 9:10

© Edwin C. Myers 1985, 1990   **CalcuLadder**®   Level 47: Multiplication   5 minutes

My name is _____

Today is _____

```
   534          276          819
  x111         x331         x545
  ————         ————         ————


   625          438          579
  x713         x306         x401
  ————         ————         ————


   623         7100          356
  x 10        x 300        x 2500
  ————        —————        ——————
```

Now he who supplies seed to the sower and bread for food, will supply and multiply your seed for sowing and increase the harvest of your righteousness.   2 Cor. 9:10

© Edwin C. Myers 1985, 1990   **CalcuLadder®**   Level 47: Multiplication   5 minutes

My name is _____

Today is _____

```
   534          276          819
  x111         x331         x545
```

```
   625          438          579
  x713         x306         x401
```

```
   623         7100          356
  x 10        x 300        x 2500
```

Now he who supplies seed to the sower and bread for food, will supply and multiply your seed for sowing and increase the harvest of your righteousness. 2 Cor. 9:10

© Edwin C. Myers 1985, 1990   CalcuLadder®   Level 47: Multiplication   5 minutes

My name is _____

Today is _____

```
  534        276        819
x111       x331       x545
```

```
  625        438        579
x713       x306       x401
```

```
  623       7100        356
x  10      x 300      x 2500
```

Now he who supplies seed to the sower and bread for food, will supply and
multiply your seed for sowing and increase the harvest of your righteousness. 2 Cor. 9:10

© Edwin C. Myers 1985, 1990   **CalcuLadder**®   Level 47: Multiplication   5 minutes

My name is _____

Today is _____

```
  534          276          819
x111         x331         x545
```

```
  625          438          579
x713         x306         x401
```

```
  623         7100          356
x  10        x 300        x 2500
```

Now he who supplies seed to the sower and bread for food, will supply and multiply your seed for sowing and increase the harvest of your righteousness. 2 Cor. 9:10

© Edwin C. Myers 1985,1990   CalcuLadder®   Level 47: Multiplication   5 minutes

My name is _____
Today is _____

$4\overline{)13}$ $5\overline{)13}$ $9\overline{)68}$ $6\overline{)41}$

$7\overline{)53}$ $3\overline{)16}$ $2\overline{)19}$ $11\overline{)35}$

$8\overline{)27}$ $10\overline{)98}$ $4\overline{)37}$ $9\overline{)84}$

$2\overline{)3}$ $5\overline{)29}$ $12\overline{)43}$ $7\overline{)40}$

$6\overline{)58}$ $10\overline{)63}$ $3\overline{)26}$ $8\overline{)62}$

$11\overline{)47}$ $4\overline{)22}$ $9\overline{)39}$ $5\overline{)32}$ $6\overline{)15}$

$7\overline{)31}$ $6\overline{)23}$ $3\overline{)11}$ $8\overline{)46}$ $12\overline{)93}$

Study to show yourself approved to God, a workman who does not need to be ashamed, rightly dividing the word of truth. 2 Tim. 2:15

© Edwin C. Myers 1985, 1990   CalcuLadder®   Level 48: Division   5 minutes

My name is _____

Today is _____

$4\overline{)13}$ $5\overline{)13}$ $9\overline{)68}$ $6\overline{)41}$

$7\overline{)53}$ $3\overline{)16}$ $2\overline{)19}$ $11\overline{)35}$

$8\overline{)27}$ $10\overline{)98}$ $4\overline{)37}$ $9\overline{)84}$

$2\overline{)3}$ $5\overline{)29}$ $12\overline{)43}$ $7\overline{)40}$

$6\overline{)58}$ $10\overline{)63}$ $3\overline{)26}$ $8\overline{)62}$

$11\overline{)47}$ $4\overline{)22}$ $9\overline{)39}$ $5\overline{)32}$ $6\overline{)15}$

$7\overline{)31}$ $6\overline{)23}$ $3\overline{)11}$ $8\overline{)46}$ $12\overline{)93}$

Study to show yourself approved to God, a workman
who does not need to be ashamed, rightly dividing the word of truth.   2 Tim. 2:15

© Edwin C. Myers 1985, 1990   CalcuLadder®   Level 48: Division   5 minutes

My name is _____

Today is _____

$4\overline{)13}$ $5\overline{)13}$ $9\overline{)68}$ $6\overline{)41}$

$7\overline{)53}$ $3\overline{)16}$ $2\overline{)19}$ $11\overline{)35}$

$8\overline{)27}$ $10\overline{)98}$ $4\overline{)37}$ $9\overline{)84}$

$2\overline{)3}$ $5\overline{)29}$ $12\overline{)43}$ $7\overline{)40}$

$6\overline{)58}$ $10\overline{)63}$ $3\overline{)26}$ $8\overline{)62}$

$11\overline{)47}$ $4\overline{)22}$ $9\overline{)39}$ $5\overline{)32}$ $6\overline{)15}$

$7\overline{)31}$ $6\overline{)23}$ $3\overline{)11}$ $8\overline{)46}$ $12\overline{)93}$

Study to show yourself approved to God, a workman who does not need to be ashamed, rightly dividing the word of truth. 2 Tim. 2:15

© Edwin C. Myers 1985, 1990   **CalcuLadder**®   Level 48: Division   5 minutes

My name is _____

Today is _____

$4\overline{)13}$ $5\overline{)13}$ $9\overline{)68}$ $6\overline{)41}$

$7\overline{)53}$ $3\overline{)16}$ $2\overline{)19}$ $11\overline{)35}$

$8\overline{)27}$ $10\overline{)98}$ $4\overline{)37}$ $9\overline{)84}$

$2\overline{)3}$ $5\overline{)29}$ $12\overline{)43}$ $7\overline{)40}$

$6\overline{)58}$ $10\overline{)63}$ $3\overline{)26}$ $8\overline{)62}$

$11\overline{)47}$ $4\overline{)22}$ $9\overline{)39}$ $5\overline{)32}$ $6\overline{)15}$

$7\overline{)31}$ $6\overline{)23}$ $3\overline{)11}$ $8\overline{)46}$ $12\overline{)93}$

Study to show yourself approved to God, a workman
who does not need to be ashamed, rightly dividing the word of truth.   2 Tim. 2:15

© Edwin C. Myers 1985, 1990   CalcuLadder®   Level 48: Division   5 minutes

My name is _____

Today is _____

4)13	5)13	9)68	6)41
7)53	3)16	2)19	11)35
8)27	10)98	4)37	9)84
2)3	5)29	12)43	7)40
6)58	10)63	3)26	8)62
11)47	4)22	9)39	5)32	6)15
7)31	6)23	3)11	8)46	12)93

Study to show yourself approved to God, a workman
who does not need to be ashamed, rightly dividing the word of truth.   2 Tim. 2:15

© Edwin C. Myers 1985,1990   **CalcuLadder**®   Level 48: Division   5 minutes

My name is _____

Today is _____

4)13̄	5)13̄	9)68̄	6)41̄
7)53̄	3)16̄	2)19̄	11)35̄
8)27̄	10)98̄	4)37̄	9)84̄
2)3̄	5)29̄	12)43̄	7)40̄
6)58̄	10)63̄	3)26̄	8)62̄
11)47̄	4)22̄	9)39̄	5)32̄	6)15̄
7)31̄	6)23̄	3)11̄	8)46̄	12)93̄

Study to show yourself approved to God, a workman
who does not need to be ashamed, rightly dividing the word of truth. 2 Tim. 2:15

© Edwin C. Myers 1985, 1990   CalcuLadder®   Level 48: Division   5 minutes

My name is _____

Today is _____

4)13	5)13	9)68	6)41
7)53	3)16	2)19	11)35
8)27	10)98	4)37	9)84
2)3	5)29	12)43	7)40
6)58	10)63	3)26	8)62
11)47	4)22	9)39	5)32	6)15
7)31	6)23	3)11	8)46	12)93

Study to show yourself approved to God, a workman who does not need to be ashamed, rightly dividing the word of truth. 2 Tim. 2:15

© Edwin C. Myers 1985, 1990   CalcuLadder®   Level 48: Division   5 minutes

My name is _____

Today is _____

4)13	5)13	9)68	6)41
7)53	3)16	2)19	11)35
8)27	10)98	4)37	9)84
2)3	5)29	12)43	7)40
6)58	10)63	3)26	8)62
11)47	4)22	9)39	5)32	6)15
7)31	6)23	3)11	8)46	12)93

Study to show yourself approved to God, a workman
who does not need to be ashamed, rightly dividing the word of truth.   2 Tim. 2:15

© Edwin C. Myers 1985, 1990   CalcuLadder®   Level 48: Division   5 minutes

My name is _____

Today is _____

4)13   5)13   9)68   6)41

7)53   3)16   2)19   11)35

8)27   10)98   4)37   9)84

2)3   5)29   12)43   7)40

6)58   10)63   3)26   8)62

11)47   4)22   9)39   5)32   6)15

7)31   6)23   3)11   8)46   12)93

Study to show yourself approved to God, a workman who does not need to be ashamed, rightly dividing the word of truth.   2 Tim. 2:15

© Edwin C. Myers 1985,1990   CalcuLadder®   Level 48: Division   5 minutes

My name is _____

Today is _____

4)13   5)13   9)68   6)41

7)53   3)16   2)19   11)35

8)27   10)98   4)37   9)84

2)3   5)29   12)43   7)40

6)58   10)63   3)26   8)62

11)47   4)22   9)39   5)32   6)15

7)31   6)23   3)11   8)46   12)93

Study to show yourself approved to God, a workman
who does not need to be ashamed, rightly dividing the word of truth.   2 Tim. 2:15

© Edwin C. Myers 1985, 1990   CalcuLadder®   Level 48: Division   5 minutes

My name is _____

Today is _____

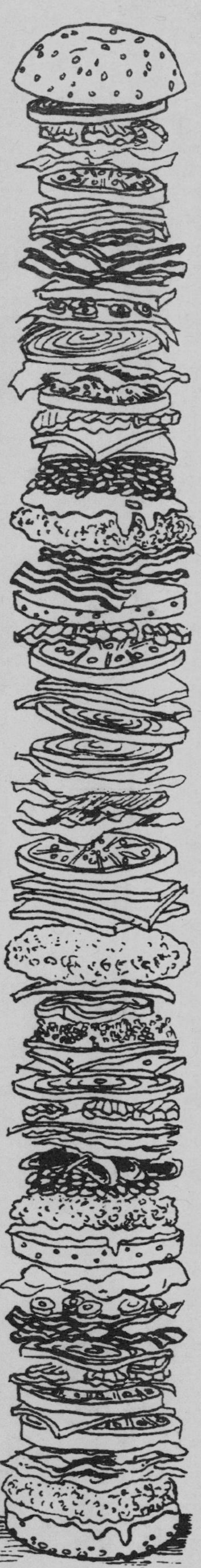

4)13	5)13	9)68	6)41
7)53	3)16	2)19	11)35
8)27	10)98	4)37	9)84
2)3	5)29	12)43	7)40
6)58	10)63	3)26	8)62
11)47	4)22	9)39	5)32	6)15
7)31	6)23	3)11	8)46	12)93

Study to show yourself approved to God, a workman
who does not need to be ashamed, rightly dividing the word of truth.   2 Tim. 2:15

© Edwin C. Myers 1985, 1990   CalcuLadder®   Level 48: Division   5 minutes

My name is _____

Today is _____

4)13   5)13   9)68   6)41

7)53   3)16   2)19   11)35

8)27   10)98   4)37   9)84

2)3   5)29   12)43   7)40

6)58   10)63   3)26   8)62

11)47   4)22   9)39   5)32   6)15

7)31   6)23   3)11   8)46   12)93

Study to show yourself approved to God, a workman
who does not need to be ashamed, rightly dividing the word of truth.   2 Tim. 2:15

© Edwin C. Myers 1985, 1990   CalcuLadder®   Level 48: Division   5 minutes

CalcuLadder Achievement Record

*This Record certifies that,
with God's enablement and provision,*

_____
student's name

*mastered the indicated CalcuLadder™ Levels
on the dates shown.*

Level _____ on _____ attested by _____.
                          Date   Instructor

Level _____ on _____ attested by _____.
Level _____ on _____ attested by _____.
Level _____ on _____ attested by _____.
Level _____ on _____ attested by _____.
Level _____ on _____ attested by _____.
Level _____ on _____ attested by _____.
Level _____ on _____ attested by _____.
Level _____ on _____ attested by _____.
Level _____ on _____ attested by _____.
Level _____ on _____ attested by _____.
Level _____ on _____ attested by _____.
Level _____ on _____ attested by _____.
Level _____ on _____ attested by _____.
Level _____ on _____ attested by _____.
Level _____ on _____ attested by _____.

"Well done, thou good and faithful servant. Thou hast been faithful over a few things; I will make thee ruler over many things. Enter thou into the joy of thy Lord." —Matt. 25:21